MW00453175

Rockets and People

Volume III: *Hot Days of the Cold War*

ISBN 978-0-16-081733-5

ISBN 978-0-16-081733-5

Rockets and People

Volume III: *Hot Days of the Cold War*

Boris Chertok

Asif Siddiqi, Series Editor

The NASA History Series

National Aeronautics and Space Administration
NASA History Division
Office of External Relations
Washington, DC
May 2009
NASA SP-2009-4110

Library of Congress Cataloging-in-Publication Data

Chertok, B. E. (Boris Evseevich), 1912–
    [Rakety i lyudi. English]
    Rockets and People: Hot Days of the Cold War (Volume III) / by
Boris E. Chertok ;
    [edited by] Asif A. Siddiqi.
    p. cm. — (NASA History Series) (NASA SP-2009-4110)
  Includes bibliographical references and index.
    1. Chertok, B. E. (Boris Evseevich), 1912– 2. Astronautics—
Soviet Union—Biography. 3. Aerospace engineers—Soviet Union—
Biography. 4. Astronautics—Soviet Union—History.
I. Siddiqi, Asif A., 1966– II. Title. III. Series. IV. SP-2009-4110.
TL789.85.C48C4813 2009
629.1'092—dc22

                             2009020825

*I dedicate this book
to the cherished memory
of my wife and friend,
Yekaterina Semyonova Golubkina.*

Contents

Series Introduction

In an extraordinary century, Academician Boris Yevseyevich Chertok has lived an extraordinary life. He has witnessed and participated in many important technological milestones of the 20th century, and in these volumes, he recollects them with clarity, humanity, and humility. Chertok began his career as an electrician in 1930 at an aviation factory near Moscow. Thirty years later, he was one of the senior designers in charge of the Soviet Union's crowning achievement as a space power: the launch of Yuriy Gagarin, the world's first space voyager. Chertok's 60-year-long career, punctuated by the extraordinary accomplishments of both Sputnik and Gagarin, and continuing to the many successes and failures of the Soviet space program, constitutes the core of his memoirs, *Rockets and People*. In these four volumes, Academician Chertok not only describes and remembers, but also elicits and extracts profound insights from an epic story about a society's quest to explore the cosmos.

Academician Chertok's memoirs, forged from experience in the Cold War, provide a compelling perspective into a past that is indispensable to understanding the present relationship between the American and Russian space programs. From the end of the World War II to the present day, the missile and space efforts of the United States and the Soviet Union (and now, Russia) have been inextricably linked. As such, although Chertok's work focuses exclusively on Soviet programs to explore space, it also prompts us to reconsider the entire history of spaceflight, both Russian and American.

Chertok's narrative underlines how, from the beginning of the Cold War, the rocketry projects of the two nations evolved in independent but parallel paths. Chertok's first-hand recollections of the extraordinary Soviet efforts to collect, catalog, and reproduce German rocket technology after the World War II provide a parallel view to what historian John Gimbel has called the Western "exploitation and plunder" of German technology after the war.[1] Chertok

---

1. John Gimbel, *Science, Technology, and Reparations: Exploitation and Plunder in Postwar Germany* (Stanford: Stanford University Press, 1990).

describes how the Soviet design team under the famous Chief Designer Sergey Pavlovich Korolev quickly outgrew German missile technology. By the late 1950s, his team produced the majestic R-7, the world's first intercontinental ballistic missile. Using this rocket, the Soviet Union launched the first Sputnik satellite on 4 October 1957 from a launch site in remote central Asia.

The early Soviet accomplishments in space exploration, particularly the launch of Sputnik in 1957 and the remarkable flight of Yuriy Gagarin in 1961, were benchmarks of the Cold War. Spurred by the Soviet successes, the United States formed a governmental agency, the National Aeronautics and Space Administration (NASA), to conduct civilian space exploration. As a result of Gagarin's triumphant flight, in 1961, the Kennedy Administration charged NASA to achieve the goal of "landing a man on the Moon and returning him safely to the Earth before the end of the decade."[2] Such an achievement would demonstrate American supremacy in the arena of spaceflight at a time when both American and Soviet politicians believed that victory in space would be tantamount to preeminence on the global stage. The space programs of both countries grew in leaps and bounds in the 1960s, but the Americans crossed the finish line first when Apollo astronauts Neil A. Armstrong and Edwin E. "Buzz" Aldrin, Jr. disembarked on the Moon's surface in July 1969.

Shadowing Apollo's success was an absent question: What happened to the Soviets who had succeeded so brilliantly with Sputnik and Gagarin? Unknown to most, the Soviets tried and failed to reach the Moon in a secret program that came to naught. As a result of that disastrous failure, the Soviet Union pursued a gradual and consistent space station program in the 1970s and 1980s that eventually led to the Mir space station. The Americans developed a reusable space transportation system known as the Space Shuttle. Despite their seemingly separate paths, the space programs of the two powers remained dependent on each other for rationale and direction. When the Soviet Union disintegrated in 1991, cooperation replaced competition as the two countries embarked on a joint program to establish the first permanent human habitation in space through the International Space Station (ISS).

Academician Chertok's reminiscences are particularly important because he played key roles in almost every major milestone of the Soviet missile and space programs, from the beginning of World War II to the dissolution of the Soviet Union in 1991. During the war, he served on the team that developed

---

2. U.S. Congress, *Senate Committee on Aeronautical and Space Sciences, Documents on International Aspects of the Exploration and Uses of Outer Space, 1954–1962, 88th Cong., 1st sess., S. Doc. 18* (Washington, DC: GPO, 1963), pp. 202–204.

the Soviet Union's first rocket-powered airplane, the BI. In the immediate aftermath of the war, Chertok, then in his early 30s, played a key role in studying and collecting captured German rocket technology. In the latter days of the Stalinist era, he worked to develop long-range missiles as deputy chief engineer of the main research institute, the NII-88 (pronounced "nee-88") near Moscow. In 1956, Korolev's famous OKB-1 design bureau spun off from the institute and assumed a leading position in the emerging Soviet space program. As a deputy chief designer at OKB-1, Chertok continued with his contributions to the most important Soviet space projects of the day: Vostok, Voskhod, Soyuz, the world's first space station Salyut, the Energiya superbooster, and the Buran space shuttle.

Chertok's emergence from the secret world of the Soviet military-industrial complex, into his current status as the most recognized living legacy of the Soviet space program, coincided with the dismantling of the Soviet Union as a political entity. Throughout most of his career, Chertok's name remained a state secret. When he occasionally wrote for the public, he used the pseudonym "Boris Yevseyev."[3] Like others writing on the Soviet space program during the Cold War, Chertok was not allowed to reveal any institutional or technical details in his writings. What the state censors permitted for publication said little; one could read a book several hundred pages long comprised of nothing beyond tedious and long personal anecdotes between anonymous participants extolling the virtues of the Communist Party. The formerly immutable limits on free expression in the Soviet Union irrevocably expanded only after Mikhail Gorbachev's rise to power in 1985 and the introduction of *glasnost'* (openness).

Chertok's name first appeared in print in the newspaper *Izvestiya* in an article commemorating the 30th anniversary of the launch of Sputnik in 1987. In a wide-ranging interview on the creation of Sputnik, Chertok spoke with the utmost respect for his former boss, the late Korolev. He also eloquently balanced love for his country with criticisms of the widespread inertia and inefficiency that characterized late-period Soviet society.[4] His first written works in the *glasnost'* period, published in early 1988 in the Air Force journal *Aviatsiya i kosmonavtika* (Aviation and Cosmonautics), underlined Korolev's central role in the foundation and growth of the Soviet space program.[5] By

---

3. See for example, his article "Chelovek or avtomat?" (Human or Automation?) in the book M. Vasilyev, ed., *Shagi k zvezdam* (Footsteps to the Stars) (Moscow: Molodaya gvardiya, 1972), pp. 281–287.

4. B. Konovalov, "Ryvok k zvezdam" (Dash to the Stars), *Izvestiya*, October 1, 1987, p. 3.

5. B. Chertok, "Lider" (Leader), *Aviatsiya i kosmonavtika* no. 1 (1988): pp. 30–31 and no. 2 (1988): pp. 40–41.

this time, it was as if all the patched up straps that held together a stagnant empire were falling apart one by one; even as Russia was in the midst of one of its most historic transformations, the floodgates of free expression were transforming the country's own history. People like Chertok were now free to speak about their experiences with candor. Readers could now learn about episodes such as Korolev's brutal incarceration in the late 1930s, the dramatic story behind the fatal space mission of Soyuz-1 in 1967, and details of the failed and abandoned Moon project in the 1960s.[6] Chertok himself shed light on a missing piece of history in a series of five articles published in *Izvestiya* in early 1992 on the German contribution to the foundation of the Soviet missile program after World War II.[7]

Using these works as a starting point, Academician Chertok began working on his memoirs. Originally, he had only intended to write about his experiences from the postwar years in one volume, maybe two. Readers responded so positively to the first volume, *Rakety i liudi* (Rockets and People) published in 1994, that Chertok continued to write, eventually producing four substantial volumes, published in 1996, 1997, and 1999, covering the entire history of the Soviet missile and space programs.[8]

My initial interest in the memoirs was purely historical: I was fascinated by the wealth of technical arcana in the books, specifically projects and concepts that had remained hidden throughout much of the Cold War. Those interested in dates, statistics, and the "nuts and bolts" of history will

---

6. For early references to Korolev's imprisonment, see Ye. Manucharova, "Kharakter glavnogo konstruktora" (The Character of the Chief Designer), *Izvestiya*, January 11, 1987, p. 3. For early revelations on Soyuz-1 and the Moon program, see L. N. Kamanin, "Zvezdy Komarova" (Komarov's Star), *Poisk* no. 5 (June 1989): pp. 4–5 and L. N. Kamanin, "S zemli na lunu i obratno" (From the Earth to the Moon and Back), *Poisk* no. 12 (July 1989): pp. 7–8.

7. *Izvestiya* correspondent Boris Konovalov prepared these publications, which had the general title "U Sovetskikh raketnykh triumfov bylo nemetskoye nachalo" (Soviets Rocket Triumphs Had German Origins). See *Izvestiya*, March 4, 1992, p. 5; March 5, 1992, p. 5; March 6, 1992, p. 5; March 7, 1992, p. 5; and March 9, 1992, p. 3. Konovalov also published a sixth article on the German contribution to American rocketry. See "U amerikanskikh raketnykh triumfov takzhe bylo nemetskoye nachalo" (American Rocket Triumphs Also Had German Origins), *Izvestiya*, March 10, 1992, p. 7. Konovalov later synthesized the five original articles into a longer work that included the reminiscences of other participants in the German mission such as Vladimir Barmin and Vasiliy Mishin. See Boris Konovalov, *Tayna Sovetskogo raketnogo oruzhiya* (Secrets of Soviet Rocket Armaments) (Moscow: ZEVS, 1992).

8. *Rakety i lyudi* (Rockets and People) (Moscow: Mashinostroyeniye, 1994); *Rakety i lyudi: Fili Podlipki Tyuratam* (Rockets and People: Fili Podlipki Tyuratam) (Moscow: Mashinostroyeniye, 1996); *Rakety i lyudi: goryachiye dni kholodnoy voyny* (Rockets and People: Hot Days of the Cold War) (Moscow: Mashinostroyeniye, 1997); *Rakety i lyudi: lunnaya gonka* (Rockets and People: The Moon Race) (Moscow: Mashinostroyeniye, 1999). All four volumes were subsequently translated and published in Germany.

find much that is useful in these pages. As I continued to read, however, I became engrossed by the overall rhythm of Academician Chertok's narrative, which gave voice and humanity to a story ostensibly about mathematics and technology. In his writings, I found a richness that had been nearly absent in most of the disembodied, clinical, and often speculative writing by Westerners studying the Soviet space program. Because of Chertok's story-telling skills, his memoir is a much needed corrective to the outdated Western view of Soviet space achievements as a mishmash of propaganda, self-delusion, and Cold War rhetoric. In Chertok's story, we meet real people with real dreams who achieved extraordinary successes under very difficult conditions.

Chertok's reminiscences are remarkably sharp and descriptive. In being self-reflective, Chertok avoids the kind of solipsistic ruminations that often characterize memoirs. He is both proud of his country's accomplishments and willing to admit failings with honesty. For example, Chertok juxtaposes accounts of the famous aviation exploits of Soviet pilots in the 1930s, especially those to the Arctic, with the much darker costs of the Great Terror in the late 1930s when Stalin's vicious purges decimated the Soviet aviation industry.

Chertok's descriptive powers are particularly evident in describing the chaotic nature of the Soviet mission to recover and collect rocketry equipment in Germany after World War II. Interspersed with his contemporary diary entries, his language conveys the combination of joy, confusion, and often anti-climax that the end of the war presaged for Soviet representatives in Germany. In one breath, Chertok and his team are looking for hidden caches of German matériel in an underground mine, while in another they are face to face with the deadly consequences of a soldier who had raped a young German woman (Volume I, Chapter 21).[9] There are many such seemingly incongruous anecdotes during Chertok's time in Germany, from the experience of visiting the Nazi slave labor camp at Dora soon after liberation in 1945, to the deportation of hundreds of German scientists to the USSR in 1946. Chertok's massive work is of great consequence for another reason—he cogently provides context. Since the breakup of the Soviet Union in 1991, many participants have openly written about their experiences, but few have successfully placed Soviet space achievements in the broader context of the history of Soviet science, the history of the Soviet military-industrial complex,

---

9. For the problem of rape in occupied Germany after the war, see Norman M. Naimark, *The Russians in Germany: A History of the Soviet Zone of Occupation, 1945–1949* (Cambridge, MA: The Belknap Press of Harvard University Press, 1995), pp. 69–140.

or indeed Soviet history in general.[10] The volumes of memoirs compiled by the Russian State Archive of Scientific-Technical Documentation in the early 1990s under the series, *Dorogi v kosmos* (Roads to Space), provided an undeniably rich and in-depth view of the origins of the Soviet space program, but they were, for the most part, personal narratives, i.e., fish-eye views of the world around them.[11] Chertok's memoirs are a rare exception in that they strive to locate the Soviet missile and space program in the fabric of broader social, political, industrial, and scientific developments in the former Soviet Union.

This combination—Chertok's participation in the most important Soviet space achievements, his capacity to lucidly communicate them to the reader, and his skill in providing a broader social context—make this work, in my opinion, one of the most important memoirs written by a veteran of the Soviet space program. The series will also be an important contribution to the history of Soviet science and technology.[12]

In reading Academician Chertok's recollections, we should not lose sight of the fact that these chapters, although full of history, have their particular perspective. In conveying to us the complex vista of the Soviet space program, he has given us one man's memories of a huge undertaking. Other participants of these very same events will remember things differently.

---

10. For the two most important histories of the Soviet military-industrial complex, see N. S. Simonov, *Voyenno-promyshlennyy kompleks SSSR v 1920-1950-ye gody: tempy ekonomicheskogo rosta, struktura, organizatsiya proizvodstva i upravleniye* (The Military-Industrial Complex of the USSR in the 1920s to 1950s: Rate of Economic Growth, Structure, Organization of Production and Control) (Moscow: ROSSPEN, 1996); and I. V. Bystrova, *Voyenno-promyshlennyy kompleks sssr v gody kholodnoy voyny (vtoraya polovina 40-kh – nachalo 60-kh godov)* [The Military-Industrial Complex of the USSR in the Years of the Cold War (The Late 1940s to the Early 1960s)] (Moscow: IRI RAN, 2000). For a history in English that builds on these seminal works and complements them with original research, see John Barber and Mark Harrison, eds., *The Soviet Defence-Industry Complex from Stalin to Khrushchev* (Houndmills, UK: Macmillan Press, 2000).

11. Yu. A. Mozzhorin et al., eds., *Dorogi v kosmos: Vospominaniya veteranov raketno-kosmicheskoy tekhniki i kosmonavtiki, tom I i II* (Roads to Space: Recollections of Veterans of Rocket-Space Technology and Cosmonautics: Volumes I and II) (Moscow: MAI, 1992) and Yu. A. Mozzhorin et al., eds., *Nachalo kosmicheskoy ery: vospominaniya veteranov raketno-kosmicheskoy tekhniki i kosmonavtiki: vypusk vtoroy* (The Beginning of the Space Era: Recollections of Veterans of Rocket-Space Technology and Cosmonautics: Second Issue) (Moscow: RNITsKD, 1994). For a poorly translated and edited English version of the series, see John Rhea, ed., *Roads to Space: An Oral History of the Soviet Space Program* (New York: Aviation Week Group, 1995).

12. For key works on the history of Soviet science and technology, see Kendall E. Bailes, *Technology and Society under Lenin and Stalin: Origins of the Soviet Technical Intelligentsia, 1917–1941* (Princeton, NJ: Princeton University Press, 1978); Loren R. Graham, *Science in Russia and the Soviet Union: A Short History* (Cambridge: Cambridge University Press, 1993); and Nikolai Krementsov, *Stalinist Science* (Princeton, NJ: Princeton University Press, 1997).

Soviet space history, like any discipline of history, exists as a continuous process of revision and restatement. Few historians in the 21st century would claim to be completely objective.[13] Memoirists would make even less of a claim to the "truth." In his introduction, Chertok acknowledges this, saying, "I . . . must warn the reader that in no way do I have pretensions to the laurels of a scholarly historian. Correspondingly, my books are not examples of strict historical research. In any memoirs, narrative and thought are inevitably subjective." Chertok ably illustrates, however, that avoiding the pursuit of scholarly history does not necessarily lessen the relevance of his story, especially because it represents the opinion of an influential member of the postwar scientific and technical intelligentsia in the Soviet Union.

Some, for example, might not share Chertok's strong belief in the power of scientists and engineers to solve social problems, a view that influenced many who sought to transform the Soviet Union with modern science after the Russian Revolution in 1917. Historians of Soviet science such as Loren Graham have argued that narrowly technocratic views of social development cost the Soviet Union dearly.[14] Technological hubris was, of course, not unique to the Soviet scientific community, but absent democratic processes of accountability, many huge Soviet government projects—such as the construction of the Great Dnepr Dam and the great Siberian railway in the 1970s and 1980s—ended up as costly failures with many adverse social and environmental repercussions. Whether one agrees or disagrees with Chertok's views, they are important to understand because they represent the ideas of a generation who passionately believed in the power of science to eliminate the ills of society. As such, his memoirs add an important dimension to understanding the *mentalité* of the Soviets' drive to become a modern, industrialized state in the 20th century.

Chertok's memoirs are part of the second generation of publications on Soviet space history, one that eclipsed the (heavily censored) first generation published during the Communist era. Memoirs constituted a large part of the second generation. In the 1990s, when it was finally possible to write candidly about Soviet space history, a wave of personal recollections flooded the market. Not only Boris Chertok, but also such luminaries as Vasiliy Mishin,

---

13. For the American historical discipline's relationship to the changing standards of objectivity, see Peter Novick, *That Noble Dream: The 'Objectivity' Question and the American Historical Profession* (Cambridge, UK: Cambridge University Press, 1988).

14. For technological hubris, see for example, Loren Graham, *The Ghost of the Executed Engineer: Technology and the Fall of the Soviet Union* (Cambridge, MA: Harvard University Press, 1993).

Kerim Kerimov, Boris Gubanov, Yuriy Mozzhorin, Konstantin Feoktistov, Vyacheslav Filin, and others finally published their reminiscences.[15] Official organizational histories and journalistic accounts complemented these memoirs, written by individuals with access to secret archival documents. Yaroslav Golovanov's magisterial *Korolev: Fakty i Mify* (Korolev: Facts and Myths), as well as key institutional works from the Energiya corporation and the Russian Military Space Forces, added richly to the canon.[16] The diaries of Air Force General Nikolay Kamanin from the 1960s to the early 1970s, published in four volumes in the late 1990s, also gave scholars a candid look at the vicissitudes of the Soviet human spaceflight program.[17]

The flood of works in Russian allowed Westerners to publish the first works in English. Memoirs—for example, from Sergey Khrushchev and Roald Sagdeev—appeared in their English translations. James Harford published his 1997 biography of Sergey Korolev based upon extensive interviews with veterans of the Soviet space program.[18] My own book, *Challenge to Apollo: The Soviet Union and the Space Race, 1945–1974*, was an early attempt

---

15. V. M. Filin, *Vospominaniya o lunnom korablye* (Recollections on the Lunar Ship) (Moscow: Kultura, 1992); Kerim Kerimov, *Dorogi v kosmos (zapiski predsedatelya Gosudarstvennoy komissii)* [Roads to Space (Notes of the Chairman of the State Commission)] (Baku: Azerbaijan, 1995); V. M. Filin, *Put k 'Energii'* (Path to Energiya) (Moscow: 'GRAAL',' 1996); V. P. Mishin, *Ot sozdaniya ballisticheskikh raket k raketno-kosmicheskomu mashinostroyeniy*u (From the Creation of the Ballistic Rocket to Rocket-Space Machine Building) (Moscow: 'Inform-Znaniye,' 1998); B. I. Gubanov, *Triumf i tragediya 'energii': razmyshleniya glavnogo konstruktora* (The Triumph and Tragedy of Energiya: The Reflections of a Chief Designer) (Nizhniy novgorod: NIER, four volumes in 1998–2000); Konstantin Feoktistov, *Trayektoriya zhizni: mezhdu vchera i zavtra* (Life's Trajectory: Between Yesterday and Tomorrow) (Moscow: Vagrius, 2000); N. A. Anifimov, ed., *Tak eto bylo—Memuary Yu. A. Mozzhorin: Mozzhorin v vospominaniyakh sovremennikov* (How it Was—Memoirs of Yu. A. Mozzhorin: Mozzhorin in the Recollections of his Contemporaries) (Moscow: ZAO 'Mezhdunarodnaya programma obrazovaniya, 2000).

16. Yaroslav Golovanov, *Korolev: fakty i mify* (Korolev: Facts and Myths) (Moscow: Nauka, 1994); Yu. P. Semenov, ed., *Raketno-Kosmicheskaya Korporatsiya "Energiya" imeni S. P. Koroleva* (Energiya Rocket-Space Corporation Named After S. P. Korolev) (Korolev: RKK Energiya, 1996); V. V. Favorskiy and I. V. Meshcheryakov, eds., *Voyenno-kosmicheskiye sily (voyenno-istoricheskiy trud): kniga I* [Military-Space Forces (A Military-Historical Work): Book I] (Moscow: VKS, 1997). Subsequent volumes were published in 1998 and 2001.

17. The first published volume was N. P. Kamanin, *Skrytiy kosmos: kniga pervaya, 1960–1963 gg.* (Hidden Space: Book One, 1960–1963) (Moscow: Infortekst IF, 1995). Subsequent volumes covering 1964–1966, 1967–1968, and 1969–1978 were published in 1997, 1999, and 2001 respectively.

18. Sergei N. Khrushchev, *Nikita Khrushchev and the Creation of a Superpower* (University Park, PA: The Pennsylvania State University Press, 2000); Roald Z. Sagdeev, *The Making of a Soviet Scientist: My Adventures in Nuclear Fusion and Space From Stalin to Star Wars* (New York: John Wiley & Sons, 1993); James Harford, *Korolev: How One Man Masterminded the Soviet Drive to Beat America to the Moon* (New York: John Wiley & Sons, 1997).

to synthesize the wealth of information and narrate a complete history of the early Soviet human spaceflight program.[19] Steven Zaloga provided an indispensable counterpoint to these space histories in *The Kremlin's Nuclear Sword: The Rise and Fall of Russia's Strategic Nuclear Forces, 1945–2000*, which reconstructed the story of the Soviet efforts to develop strategic weapons.[20]

With any new field of history that is bursting with information based primarily on recollection and interviews, there are naturally many contradictions and inconsistencies. For example, even on such a seemingly trivial issue as the name of the earliest institute in Soviet-occupied Germany, "Institute Rabe," there is no firm agreement on the reason it was given this title. Chertok's recollections contradict the recollection of another Soviet veteran, Georgiy Dyadin.[21] In another case, many veterans have claimed that artillery general Lev Gaydukov's meeting with Stalin in 1945 was a key turning point in the early Soviet missile program; Stalin apparently entrusted Gaydukov with the responsibility to choose an industrial sector to assign the development of long-range rockets (Volume I, Chapter 22). Lists of visitors to Stalin's office during that period—declassified only very recently—do not, however, show that Gaydukov ever met with Stalin in 1945.[22] Similarly, many Russian sources note that the "Second Main Directorate" of the USSR Council of Ministers managed Soviet missile development in the early 1950s, when in fact, this body actually supervised uranium procurement for the A-bomb project.[23] In many cases, memoirs provide different and contradictory information on the very same event (different dates, designations, locations, people involved, etc.).

19. Asif A. Siddiqi, *Challenge to Apollo: The Soviet Union and the Space Race, 1945–1974* (Washington, D.C.: NASA SP-2000–4408, 2000). The book was republished as a two-volume work as *Sputnik and the Soviet Space Challenge* (Gainesville, FL: University Press of Florida, 2003) and *The Soviet Space Race with Apollo* (Gainesville, FL: University Press of Florida, 2003).

20. Steven J. Zaloga, *The Kremlin's Nuclear Sword: The Rise and Fall of Russia's Strategic Nuclear Forces, 1945–2000* (Washington, DC: Smithsonian Institution Press, 2002).

21. G. V. Dyadin, D. N. Filippovykh, and V. I. Ivkin, *Pamyatnyye starty* (Memorable Launches) (Moscow: TsIPK, 2001), p. 69.

22. A. V. Korotkov, A. D. Chernev, and A. A. Chernobayev, "Alfavitnyi ukazatel posetitelei kremlevskogo kabineta I. V. Stalina" ("Alphabetical List of Visitors to the Kremlin Office of I. V. Stalin"), *Istoricheskii arkhiv* no. 4 (1998): p. 50.

23. Vladislav Zubok and Constantine Pleshakov, *Inside the Kremlin's Cold War: From Stalin to Khrushchev* (Cambridge, MA: Harvard University Press), p. 172; Golovanov, *Korolev*, p. 454. For the correct citation on the Second Main Directorate, established on December 27, 1949, see Simonov, *Voyenno-promyshlennyy kompleks sssr*, pp. 225–226.

Academician Chertok's wonderful memoirs point to a solution to these discrepancies: a "third generation" of Soviet space history, one that builds on the rich trove of the first and second generations, but is primarily based on *documentary* evidence. During the Soviet era, historians could not write history based on documents since they could not obtain access to state and design bureau archives. As the Soviet Union began to fall apart, historians such as Georgiy Vetrov began to take the first steps in document-based history. Vetrov, a former engineer at Korolev's design bureau, eventually compiled and published two extraordinary collections of primary documents relating to Korolev's legacy.[24] Now that all the state archives in Moscow—such as the State Archive of the Russian Federation (GARF), the Russian State Archive of the Economy (RGAE), and the Archive of the Russian Academy of Sciences (ARAN)—are open to researchers, more results of this "third generation" are beginning to appear. German historians such as Matthias Uhl and Cristoph Mick and those in the United States such as myself have been fortunate to work in Russian archives.[25] I would also note the enormous contributions of the Russian monthly journal *Novosti kosmonavtiki* (News of Cosmonautics) as well as the Belgian historian Bart Hendrickx in advancing the state of Soviet space history. The new work has opened opportunities for future research. For example, we no longer have to guess about the government's decision to approve development of the Soyuz spacecraft, we can see the original decree issued on 4 December 1963.[26] Similarly, instead of speculating about the famous decree of 3 August 1964 that committed the Soviet Union to compete

---

24. M. V. Keldysh, ed., *Tvorcheskoye naslediye Akademika Sergeya Pavlovicha Koroleva: izbrannyye trudy i dokumenty* (The Creative Legacy of Sergey Pavlovich Korolev: Selected Works and Documents) (Moscow: Nauka, 1980); G. S. Vetrov and B. V. Raushenbakh, eds., *S. P. Korolev i ego delo: svet i teni v istorii kosmonavtiki: izbrannyye trudy i dokumenty* (S. P. Korolev and His Cause: Shadow and Light in the History of Cosmonautics) (Moscow: Nauka, 1998). For two other published collections of primary documents, see V. S. Avduyevskiy and T. M. Eneyev, eds. *M. V. Keldysh: izbrannyye trudy: raketnaya tekhnika i kosmonavtika* (M. V. Keldysh: Selected Works: Rocket Technology and Cosmonautics) (Moscow: Nauka, 1988); B. V. Raushenbakh, ed., *Materialy po istorii kosmicheskogo korablya 'vostok': k 30-letiyu pervogo poleta cheloveka v kosmicheskoye prostranstvo* (Materials on the History of the 'Vostok' Space Ship: On the 30th Anniversary of the First Flight of a Human in Space) (Moscow: Nauka, 1991).

25. Matthias Uhl, *Stalins V-2: Der Technolgietransfer der deutschen Fernlen-kwaffentechnik in die UdSSR und der Aufbau der sowjetischen Raketenindustrie 1945 bis 1959* (Bonn, Germany: Bernard & Graefe-Verlag, 2001); Christoph Mick, *Forschen für Stalin: Deutsche Fachleute in der sowjetischen Rüstungsindustrie 1945–1958* (Munich: R. Oldenbourg, 2000); Asif A. Siddiqi, "The Rockets' Red Glare: Spaceflight and the Russian Imagination, 1857–1957," Ph.D. dissertation, Carnegie Mellon University, 2004.

26. "O sozdaniia kompleksa 'Soyuz'" (On the Creation of the Soyuz Complex), December 4, 1963, RGAE, f. 298, op. 1, d. 3495, ll. 167–292.

with the American Apollo program, we can study the actual government document issued on that date.[27] Academician Chertok deserves much credit for opening the doors for future historians, since his memoirs have guided many to look even deeper.

The distribution of material spanning the four volumes of Chertok's memoirs is roughly chronological. In the first English volume, Chertok describes his childhood, his formative years as an engineer at the aviation Plant No. 22 in Fili, his experiences during World War II, and the mission to Germany in 1945–46 to study captured German missile technology.

In the second volume, he continues the story with his return to the Soviet Union, the reproduction of a Soviet version of the German V-2 and the development of a domestic Soviet rocket industry at the famed NII-88 institute in the Moscow suburb of Podlipki (now called Korolev). He describes the development of the world's first intercontinental ballistic missile, the R-7; the launch of Sputnik; and the first generation probes sent to the Moon, Mars, and Venus.

In the third volume, he begins with the historic flight of Yuriy Gagarin, the first human in space. He discusses several different aspects of the burgeoning Soviet missile and space programs of the early 1960s, including the development of early ICBMs, reconnaissance satellites, the Cuban missile crisis, the first Soviet communications satellite Molniya-1, the early spectacular missions of the Vostok and Voskhod programs, the dramatic Luna program to land a probe on the Moon, and Sergey Korolev's last days. He then continues into chapters about the early development of the Soyuz spacecraft, with an in-depth discussion of the tragic mission of Vladimir Komarov.

The fourth and final volume is largely devoted to the Soviet project to send cosmonauts to the Moon in the 1960s, covering all aspects of the development of the giant N-1 rocket. The last portion of this volume covers the origins of the Salyut and Mir space station programs, ending with a fascinating description of the massive Energiya-Buran project, developed as a countermeasure to the American Space Shuttle.

It was my great fortune to meet with Academician Chertok in the summer of 2003. During the meeting, Chertok, a sprightly 91 years old, spoke passionately and emphatically about his life's work and remained justifiably proud of the achievements of the Russian space program. As I left

---

27. "Tsentralnyy komitet KPSS i Sovet ministrov SSSR, postanovleniye" (Central Committee KPSS and SSSR Council of Ministers Decree), 3 August 1964, RGAE, f. 29, op. 1, d. 3441, ll. 299–300. For an English-language summary, see Asif A. Siddiqi, "A Secret Uncovered: The Soviet Decision to Land Cosmonauts on the Moon," *Spaceflight* 46 (2004): pp. 205–213.

the meeting, I was reminded of something that Chertok had said in one of his first public interviews in 1987. In describing the contradictions of Sergey Korolev's personality, Chertok had noted: "This realist, this calculating, [and] farsighted individual was, in his soul, an incorrigible romantic."[28] Such a description would also be an apt encapsulation of the contradictions of the entire Soviet drive to explore space, one which was characterized by equal amounts of hard-headed realism and romantic idealism. Academician Boris Yevseyevich Chertok has communicated that idea very capably in his memoirs, and it is my hope that we have managed to do justice to his own vision by bringing that story to an English-speaking audience.

ASIF A. SIDDIQI
Series Editor
October 2004

---

28. Konovalov, "Ryvok k zvezdam."

Introduction to Volume III

This, the third volume of Boris Chertok's four-volume memoirs, continues the narrative arc which he began in the first volume. If the first volume covered his apprenticeship as an engineer and the second, the birth of the Soviet postwar missile program, in the third volume, we finally have what might be called the full bloom of the Soviet space program. Here, Chertok describes his impressions of the apex of Soviet achievements in space exploration, from the halcyon days of the launch of Yuri Gagarin into orbit in 1961 to the first piloted Soyuz mission in 1967.

Chertok devotes a significant portion of the volume to the early years of Soviet human spaceflight. These include a chapter on the Vostok and Voskhod programs, which left an indelible mark on early years of the "space race," a lengthy meditation on the origins and early missions of the Soyuz program, and a gripping account of one of the most tragic episodes of the Soviet space program: the flight and death of cosmonaut Vladimir Komarov during the very first piloted Soyuz flight in 1967. Additional chapters cover robotic programs such as the Molniya communications satellite system, the Zenit spy satellite program, and the Luna series of probes that culminated in the world's first survivable landing of a probe on the surface of the Moon. Chertok also devotes several chapters to the development of early generations of Soviet intercontinental ballistic missiles (ICBMs) and missile defense systems; his narrative here skillfully combines technical, political, personal, and strategic concerns, highlighting how these considerations were often difficult to separate into neat categories. In particular, we learn about the Soviet drive to develop a workable solid propellant ICBM and the subsequent arguments over the development of second general ICBMs in the late 1960s, a fight so acrimonious that contemporaries called it "the little civil war."

Chertok's chapter on the Cuban Missile Crisis provides a radically unique perspective on the crisis, from the point of view of those who would have been responsible for unleashing nuclear Armageddon in 1962 had Kennedy and Khrushchev not been able to agree on a stalemate. Two further chapters cover the untimely deaths of the most important luminaries of the era: Sergey

Korolev and Yuriy Gagarin. Each of these chapters is a tour de force, as Chertok uses a vast array of published accounts to enrich his own personal recollections of the episodes. Finally, historians of Soviet science will find much of interest in the concluding chapter focused on the relationship between the space program and the Soviet Academy of Sciences. This chapter represents one of the most insightful descriptions of the formation of a Soviet "aerospace" elite during the post-World War II era.

During the period covered by Chertok, from 1961 to 1967, the Soviet Union achieved an unprecedented series of firsts; Russians still typically associate this era with a "golden age" of Soviet space exploration. Much as the Apollo missions indelibly convey a nostalgic sense of the possibilities of American space exploration, the visages of young "hero" cosmonauts from the early 1960s at parades in Red Square continue to exemplify the immense political and cultural cache of space exploration during the Cold War.

The central figure in Chertok's tale is Sergey Pavlovich Korolev, the "chief designer" of the leading missile and spacecraft design organization, who many consider the most important architect of the Soviet push for space; he is still eulogized in saintly terms in the post-Soviet landscape. Westerners who have written about the history of the Soviet space program typically fixate on Korolev to the exclusion of other actors. There are compelling reasons to do so: Korolev was an extraordinarily charismatic figure whose biography encompassed equal parts tragedy and redemption. His biographer, Yaroslav Golovanov, astutely noted that, "Korolev was a most exact reflection of an epoch. . . . He knew all its triumphs and drained the cup of its bitterness to the dregs. Korolev's biography is the concretization of the history of our land in one man . . ."[1] Chertok's description of Korolev, particularly his last days, gives Westerners an unprecedented perspective into the life of one of the most important scientific managers in the 20th century. Although Korolev is square and central in Chertok's narrative, the author offers a much more nuanced perspective of the Soviet space program, one that includes a panoply of other characters, from top Communist Party officials who managed the projects, to junior engineers who produced many of the technical innovations. One marvels at his memory—Chertok is able to remember a vast assortment of names of people present at important managerial meetings. Much of this detail is derived from notes made in his contemporaneous diaries from the 1960s and 1970s (the originals of which

---

1. Yaroslav Golovanov, *Sergei Korolev: The Apprenticeship of a Space Pioneer* (Moscow: Mir Publishers, 1975), 293.

have since been donated for storage to the archives of the National Air and Space Museum in Washington, DC). The chapters in Volume III also highlight his ability to bring to life previously unknown or lesser known individuals in the history of the Soviet space program. For example, in Chapter 15 on the development of the first Soviet communications satellite, Molniya-1, we find touching profiles of brilliant engineers such as 27-year old Vyacheslav Dudnikov, the principal personage behind the design of the satellite, and Murad Kaplanov, the descendent of a royal family of Kumyks, an ethnic minority in the Soviet republic of Dagestan, who designed Molniya's payload. Other more powerful luminaries in the Soviet space program, such as the gifted but irascible Vasiliy Mishin who succeeded Sergey Korolev in 1966, are humanized in a manner that contrasts starkly with the wooden depictions of Soviet space personalities so common in Western narratives.

Chertok does not shy away from his obviously high evaluation of scientists and engineers. Like many of his generation, i.e., those that came of age in the 1930s and went on to leading industrial and government positions after World War II, his faith in the power of science and technology to solve the world's problems remains undiminished. In this technocratic view of the ideal human society, Chertok sees a prominent and positive role for scientists and engineers in the functioning of an advanced society. The problems with science and technology are not with those who produce them but rather those, especially politicians and bureaucrats, who use them. It's not surprising that Chertok's account of Minister of Defense Rodion Malinovskiy's visit to Baykonur is scathing; he recalls how the minister had little interest in learning anything about the technology at the launch site, waving away a colonel's report by saying "I don't need you to tell me what's what. You already take me for a complete fool. Instead, why don't you tell me where the latrine is around here." (Chapter 12, p. 353).

The richness of Chertok's writing should not obscure the fact that this is a memoir written by a historical participant, not a tome authored by a professional historian. In other words, the opinions presented here are by definition subjective and thus prone to the same kinds of limitations inherent in any recollection, especially one made over four decades after the events. Partly to correct his own fallibilities, Chertok does an excellent job of using supporting evidence to buttress his impressions. For example, he makes liberal use of recently published material in the Russian press, such as primary documents published in various books or ground-breaking articles by Russian journalists, which have uncovered previously unknown aspects of the Soviet space program. Similarly, since the publication of the first edition in the 1990s, a number of direct participants of space-related events have offered Chertok their own impressions, which he has generously reproduced at various points in the narrative. A recent landmark collection of original government documents on the early history of the Soviet space program appeared too late for Chertok to use in these memoirs, but future historians will find it useful to juxtapose Chertok's accounts with the evidence from these primary documents.[2]

I conclude with a few final words on the implementation of the project. Working on this series continues to be an extraordinary honor and pleasure. I owe a debt of gratitude to many for their hard work in bringing these stories to the English-speaking world. As before, I must thank historian Steve Garber, who supervised the entire project at the NASA History Division. He also provided insightful comments at every stage of the editorial process. Similarly, thanks are due to Jesco von Puttkamer for his continuing support in facilitating communications between the two parties in Russia and the United States. Without his enthusiasm, sponsorship, and support, this project would not have been possible. Many others at NASA Headquarters contributed to publication of these memoirs, including NASA Chief Historian Steven J. Dick, Nadine J. Andreassen, William P. Barry, and others.

As series editor, my work was not to translate, a job that was very capably done by a team at award-winning TechTrans International, Inc., based in Houston, Texas. Their team included: Cynthia Reiser (translator), Laurel Nolen (editor), Alexandra Tussing (postediting), and Lev Genson (documents control), and Daryl Gandy (translation lead).

Thanks also are due to the staff of the Communications Support Services Center (CSSC) at NASA Headquarters. Editors Andrew Jarvis and Stacie Dapoz carefully copyedited and proofread the volume. As the designer, Ann Marie Wildman capably laid it out. Printing specialist Tun Hla expertly handled this final crucial stage. Gail Carter-Kane and Cindy Miller professionally managed this project. Kudos also go to Michael Crnkovic and Tom Powers, who provided useful production oversight.

I would also like to thank Peter Gorin, Oleg Gurko, Christian Lardier, Timofei Varfolomeyev, and Dave Woods for kindly providing some of the photographs for use in Volume III. Unless otherwise noted, all images are from the collection of Chertok. I would also like to thank Anoo Siddiqi for her unwavering support for my work on this project.

---

2. Yu. M. Baturin, ed., *Sovetskaya kosmicheskaya initsiativa i gosudarstvennykh dokumentakh, 1946-1964 gg.* [*Soviet Space Initiatives in State Documents, 1946-1964*] (Moscow: RTSoft, 2008).

As the series editor, my job was first and foremost to ensure that the English language version was as faithful to Chertok's original Russian version as possible. At the same time, I also had to account for the stylistic considerations of English-language readers who may be put off by literal translations. The process involved communicating directly with Chertok in many cases and, with his permission, taking liberties to restructure paragraphs and chapters to convey his original spirit. I also made sure that technical terms and descriptions of rocket and spacecraft design satisfied the demands of both Chertok and the English-speaking audience. Finally, I provided many explanatory footnotes to elucidate points that may not be evident to readers unversed in the intricacies of Russian history. Readers should be aware that all of the footnotes are mine unless cited as "author's note," in which case they were provided by Chertok.

Asif A. Siddiqi
Series Editor
April 2009

A Few Notes about Transliteration and Translation

THE RUSSIAN LANGUAGE IS WRITTEN using the Cyrillic alphabet, which consists of 33 letters. While some of the sounds that these letters symbolize have equivalents in the English language, many have no equivalent, and two of the letters have no sound of their own, but instead "soften" or "harden" the preceding letter. Because of the lack of direct correlation, a number of systems for transliterating Russian (i.e., rendering words using the Latin alphabet), have been devised, all of them different.

Russian Alphabet	Pronunciation	US Board on Geographic Names	Library of Congress
А, а	a̅	a	a
Б, б	b	b	b
В, в	v	v	v
Г, г	g	g	g
Д, д	d	d	d
Е, е	ye	ye* / e	e
Ё, ё	yō	yë* / ë	ë
Ж, ж	zh	zh	zh
З, з	z	z	z
И, и	ē	i	i
Й, й	shortened ē	y	ĭ
К, к	k	k	k
Л, л	l	l	l
М, м	m	m	m
Н, н	n	n	n
О, о	o	o	o
П, п	p	p	p
Р, р	r	r	r
С, с	s	s	s
Т, т	t	t	t
У, у	ū	u	u
Ф, ф	f	f	f
Х, х	kh	kh	kh
Ц, ц	ts	ts	ts
Ч, ч	ch	ch	ch
Ш, ш	sh	sh	sh
Щ, щ	shch	shch	shch
Ъ	(hard sign)	"	"
Ы	gutteral ē	y	y
Ь	(soft sign)	'	'
Э, э	ĕ	e	ĭ
Ю, ю	yū	yu	iu
Я, я	ya̅	ya	ia

*Initially and after vowels

For this series, Editor Asif Siddiqi selected a modification of the U.S. Board on Geographic Names system, also known as the University of Chicago system, as he felt it better suited for a memoir such as Chertok's, where the intricacies of the Russion language are less important than accessibility to the reader. The modifications are as follows:

- the Russian letters "ь" and "ъ" are not transliterated, in order to make reading easier;
- Russian letter "ё" is denoted by the English "e" (or "ye" initially and after vowels)—hence, the transliteration "Korolev," though it is pronounced "Korolyov".

The reader may find some familiar names to be rendered in an unfamiliar way. This occurs when a name has become known under its phonetic spelling, such as "Yuri" versus the transliterated "Yuriy," or under a different transliteration system, such as "Baikonur" (LoC) versus "Baykonur" (USBGN).

In translating *Rakety i lyudi,* we on the TTI team strove to find the balance between faithfulness to the original text and clear, idiomatic English. For issues of technical nomenclature, we consulted with Asif Siddiqi to determine the standards for this series. The cultural references, linguistic nuances, and "old sayings" Chertok uses in his memoirs required a different approach from the technical passages. They cannot be translated literally: the favorite saying of Flight Mechanic Nikolay Godovikov (Vol. 1, Chapter 7) would mean nothing to an English speaker if given as, "There was a ball, there is no ball," but makes perfect sense when translated as "Now you see it, now you don't." The jargon used by aircraft engineers and rocket engine developers in the 1930s and 1940s posed yet another challenge. At times, we had to do linguistic detective work to come up with a translation that conveyed both the idea and the "flavor" of the original. Puns and plays on words are explained in footnotes. *Rakety i lyudi* has been a very interesting project, and we have enjoyed the challenge of bringing Chertok's voice to the English-speaking world.

TTI TRANSLATION TEAM
Houston, TX
October 2004

List of Abbreviations

ABM	Anti-Ballistic Missile
ABMA	Army Ballistic Missile Agency
ALS	automatic lunar station
AMS	automatic interplanetary station
APO	emergency object destruction
APR	automatic missile destruction
ASO	automatic attitude control system
ASP	automatic preparation system
ASU	sanitation unit
ATC	air traffic control
AVD	emergency engine shutdown
AVDU	emergency engine unit shutdown
BON	Special Purpose Brigade
BRK	lateral radio correction
BUS	rendezvous control unit
BVDPO	docking and orientation engines control assembly
CEP	circular error probability
CIA	Central Intelligence Agency
CPSU	Communist Party of the Soviet Union
DKD	backup correction engine
DO	attitude control engines
DoD	Department of Defense
DPO	approach and attitude control engine
DRK	long-range radio complex
DROB	discrete tank depletion control system
DRS	long-range radio communications
DUS	angular rate sensor
EPAS	Experimental Apollo-Soyuz Flight
FAB	high-explosive bombs
FEP	photovoltaic converters
GAZ	Gorky Automobile Factory
GDL	Gas Dynamics Laboratory

GIPKh	State Institute of Applied Chemistry
GIRD	Group for the Study of Reactive Motion
GKOT	State Committee for Defense Technology
GOELRO	State Commission for the Electrification of Russia
GOGU	main operations control group
Gosplan	State Planning Commission
Gossnab	State Committee for Material and Technical Supply
GRU	Main Intelligence Directorate
GSKB	State Union Design Bureau
GSO	control system readiness
GTO	Ready for Labor and Defense
GTsP	State Central Firing Range
GUKOS	Main Directorate of Space Assets
GURVO	Main Directorate of Rocket Armaments
ICBM	intercontinental ballistic missile
IGY	International Geophysical Year
IIYeT	Institute for the History of Natural Sciences and Technology
IKV	infrared vertical
IO	ionic orientation
IP	Tracking Station
IPM	Institute of Applied Mathematics
IS	Satellite Killer
JPL	Jet Propulsion Laboratory
KB	Design Bureau
KB Khimmash	Design Bureau of Chemical Machine Building
KD	correcting engine
KDU	correcting engine unit
KGB	Committee of State Security
KIK	Command and Measurement Complex
KIS	monitoring and test facility
KKP	space monitoring system
KOSS	Space Communications System
KP	command post
KRL	command radio-link
KTDU	correcting engine braking unit
KUNG	standard all-purpose truck trailer
KV	short-wave
KVO	circular error probability
KVTs	Coordination Computation Center

LII	Flight-Research Institute
LK	Lunar (Landing) Craft
LKI	Flight-Development Test
LOK	Lunar Orbital Craft
LPI	Leningrad Polytechnic Institute
MAD	Mutually Assured Destruction
MAI	Moscow Aviation Institute
MAP	Ministry of the Aviation Industry
MEI	Moscow Power Institute
MIK	Assembly and Testing Building
MIRV	Multiple Independently targetable Reentry Vehicle
MNTS-KI	Interdepartmental Scientific-Technical Council for Space Research
MOM	Ministry of General Machine Building
MPSS	Ministry of the Communications Equipment Industry
MRP	Ministry of the Radio Engineering Industry
MSM	Ministry of Medium Machine Building
MVTU	Moscow Higher Technical School
NASA	National Aeronautics and Space Administration
NATO	North Atlantic Treaty Organization
NIEI PDS	Scientific-Research and Experimental Institute of the Airborne Assault Service
NII	Scientific-Research Institute
NIIAP	Scientific-Research Institute of Automatics and Instrument Building
NIIERAT	Scientific-Research Institute for the Operation and Repair of Aviation Technology
NIIP	Scientific-Research Test Site
NIP	Ground Tracking Station
NIR	scientific-research project
NIIRP	Scientific-Research Institute of Radio Instrument Building
NPO	Scientific-Production Association
NTS	Scientific and Technical Council
NZ	emergency supply
OKB	Experimental Design Bureau
OPM	Department of Applied Mathematics
OSP	primary parachute system
OTP	Technical-Operational Management
PGU	First Main Directorate

PKO	anti-space defense
PO	Production Association
PVO	Air Defense (Troops)
PVU	onboard sequencer
RGCh IN	separating payloads with individual targeting of the warhead
RKK Energiya	Energiya Rocket-Space Corporation
RKO	missile-space defense
RKS	apparent velocity control
RKT	radio trajectory monitoring (system)
RNII	Reactive Scientific-Research Institute
ROKS	aircraft coordinate radio locator
RVSN	Strategic Rocket Forces
SA	descent module
SALT	Strategic Arms Limitation Talks
SAN	astronavigation system
SAS	Emergency Rescue System
SB-1	Special Bureau-1
SBN	launch vehicle safety system
SDI	Strategic Defense Initiative
SDUK	remote control and monitoring system
SECAM	Sequential Color with Memory
SEP	power supply system
ShPU	Shaft Launch Unit
SKB	Special Design Bureau
SKD	approach and correction engine
SKDU	approach and correction engine unit
SOB	tank emptying system
Sovnarkhoz	Council of the National Economy
SRPN	missile attack early warning system
SUBK	on-board complex control system
SUS	descent control system
TDU	braking engine unit
TGU	Third Main Directorate
TNA	turbopump assembly
TP	engineering facility
TsAGI	Central Aerohydrodynamics Institute
TsDSA	Central House of the Soviet Army
TsIK	Central Executive Committee
TsKB	Central Design Bureau

TsKBEM	Central Design Bureau of Experimental Machine Building
TsNIIMash	Central Scientific-Research Institute of Machine Building
TsNIIS	Central Scientific-Research Institute of Communications
TsNPO	Central Scientific-Production Association
TsPK	Cosmonaut Training Center
TsSKB	Central Specialized Design Bureau
TsUKOS	Central Directorate of Space Assets
TsUP	Flight Control Center
TsVM	electronic computer
TZ	Technical Assignment
VAK	Supreme Certification Commission
VDNKh	Exhibition of Achievements of the National Economy
VEI	All-Union Electrical Institute
VGU	Second Main Directorate
VKP(b)	All-Union Communist Party (Bolshevik)
VN	all normal
VNIIEF	All-Russian Scientific-Research Institute of Experimental Physics
VNIIEM	All-Union Scientific-Research Institute of Electromechanics
VNIIIT	All-Union Scientific-Research Institute of Current Sources
VNIIT	All-Union Scientific-Research Institute of Television
VVIA	N. Ye. Zhukovskiy Air Force Engineering Academy
VPK	Military-Industrial Commission
ZATO	closed administrative territorial formation
ZEM	Experimental Machine Building Factory
ZhBK	onboard service lines conduit
ZKI	factory checkout tests
ZIKh	M. V. Khrunichev Factory
ZiS	Stalin Factory
ZSP	backup parachute system
ZU	memory unit

Chapter 1
The Cold War

It would be appropriate to begin the third book of my memoirs by describing the preparations for and execution of the first piloted spaceflight. However, I need to interrupt the chronology with a brief political and philosophical digression to provide a better understanding of the general situation that existed in the early 1960s.

During World War II, two fundamentally new types of strategic weapons were developed: the long-range ballistic guided missile in Germany and the atomic bomb in the United States. The history of organizing the development and production of these weapons (which are completely different technologies) is instructive because there is much in common in the organizational process itself. Both Germany and the U.S. had powerful industrial corporations, world-famous companies capable of producing the most sophisticated technology of that time for all types of weaponry. Nevertheless, not one of those companies was entrusted to develop Germany's missiles or America's atomic bomb.

Given the need for strict secrecy, the operational scales and time constraints required for producing these new weapons were such that both nations were compelled to create powerful state-run organizations that were responsible for the entire gamut of operations spanning conceptual vetting, research, development, production, and application.

Despite the very different power structures of Hitler's totalitarian Germany and the democratic U.S., the leaders of both governments were forced to contend with the fact that conducting operations on such a scale required goal-oriented, systemic management independent of corporate interests. Industrial conglomerates, institutions of higher learning, and the engineering services of each country's army and navy were tapped for the projects. The cooperation, number of participants, and funding of these projects was unprecedented in the history of technology. Had it not been for the degree of secrecy, each of these problems could have been openly declared a national goal.

Through intelligence channels, the Soviet leadership obtained an idea of how the Manhattan Project was organized in the U.S. There are quite a number of publications on this subject.[1] On the other hand, we studied the organization of missile technology operations in Germany firsthand.

I have already written about how we—my comrades from the aviation and defense industries, our like-minded comrades from the command of the Guards' Mortar Units, and I—convinced the leaders of the aviation and defense industry that, if we wanted to master the new missile technology, then the immense scale of the operations required a fundamentally new organization. Subsuming it within just one People's Commissariat (or ministry) was unacceptable.

In 1946, Korolev joined this project with his innate initiative, drive, and energy. The team of enthusiasts that had rallied around him received effective support from the young generals and officers of the Guards' Mortar Units rather than from the Ministry of Defense. Up until the last days of the war these units had been directly subordinate to the Headquarters of the Supreme Commander-in-Chief.[2]

Our industry still had positive prewar experience from replicating foreign technology. Now, instead of scrupulously replicating automobiles and aircraft, the time had come for us to change our way of thinking in order to use that experience for organizing scientific and production projects of national importance. The subsequent fate of the Soviet Union depended on the results of the goal-oriented joint collaboration of civilian and military scientists and industry under the Cold War conditions that had set in.

The year 1996 marked the 50th anniversary of the official founding of many rocket technology organizations. In view of this, various commemorative gatherings reflected on the half of a century of work in this field. We had the opportunity to recall that not only in the fields of science and technology, but also within the scope and the methods used to organize national defense projects, we had created our own methods and schools, in many respects surpassing our Cold War enemy.

The anniversary year 1996 served as an occasion to analyze politically the role of nuclear-tipped missiles in the Cold War. Before the reforms of the late 1980s, it had been difficult for historians and authors of memoirs to

---

1. The classic English language work on the history of the Soviet atomic bomb project is David Holloway's *Stalin and the Bomb: The Soviet Union and Atomic Energy, 1939–1956* (New Haven: Yale University Press, 1994).

2. The Guards Mortar Units of the Supreme Command Headquarters were responsible for operating the famed *Katyusha* solid propellant rockets during World War II.

objectively assess this period due to top secret classifications. In June 1996, the leadership of the Energiya Rocket-Space Corporation (RKK Energiya) delegated me to participate in a session of the scientific-technical council of the Ministry of Defense's Fourth Central Scientific-Research Institute (Fourth Central NII), which was celebrating its 50th anniversary, and to give a congratulatory speech befitting this occasion on behalf of the institute's close neighbor, i.e., Energiya.[3] The organization being celebrated was the military institute that for many years we had simply called NII-4. It was located in Bolshevo, a mere 25-minute stroll from our doorstep. The institute had been created as a result of the historic USSR Council of Ministers decree dated 13 May 1946, and the order of the minister of the USSR Armed Forces dated 24 May 1946.

It wasn't just location that brought together NII-4 in Bolshevo with NII-88 in Podlipki, which the very same decree had created. The two also had missiles in common. After I was appointed NII-88 deputy chief engineer and control systems department chief in 1946, I often dealt with NII-4 specialists due to the nature of my work. I was acquainted with the first NII-4 directors, Generals Aleksey Nesterenko and Andrey Sokolov, and their deputies from our work in Germany.

NII-4 was staffed with military specialists from the Guards' Mortar Units and drew in military scientists from other organizations. For example, Mikhail Tikhonravov, Nikolay Chernyshev, and Ivan Gvay, and several other officer-scientists from NII-1 (formerly NII-3 and before that, RNII) transferred to NII-4. This entitled me to mention, in my welcoming remarks, that the first postwar directors of not only NII-4, but also the State Central Firing Range in Kapustin Yar and the new departments of the Main Artillery Directorate had come to large missile technology from the simple and small solid-propellant projectiles of the Katyusha.

When "hot" World War II ended, the Cold War began for all of us. The heroic participants in the battles of the Great Patriotic War—generals, officers, and scientists who had labored at the home front—switched to creative work that was no less heroic to produce the new missile weaponry.[4] During those years, NII-4 became the Ministry of Defense's largest scientific organization. Today's Fourth Central NII can rightfully claim to hold a place of honor in maintaining strategic parity, the balance of nuclear-tipped missiles between the two superpowers.

---

3. NII—*Nauchno-issledovatelskiy institut.*
4. Russian speakers use the term "Great Patriotic War" to refer to World War II.

NII-4 not only conducted work of a purely practical nature for specific missile systems and their combat readiness and operation, but it was our sole organization that attempted to use a scientific approach to develop a doctrine of nuclear missile stalemate with the goal of maintaining a balance of forces between the two camps.

We interacted closely with military scientists such as Yakov Shor, Georgiy Narimanov, Gennadiy Melnikov, Ivan Meshcheryakov, Pavel Elyasberg, Pavel Agadzhanov, Grigoriy Levin, Nikolay Fadeyev, Vladimir Yastrebov, and Mikhail Kislik. We worked with these scientists while discussing projects at scientific-technical and academic councils, preparing missile flight test programs, coordinating operational requirements, writing reports, and conducting many other operations related to the construction of missile complexes. In my case, we mainly discussed control problems, trajectory measurements, and telemetry monitoring. At the anniversary scientific-technical council session, only Yuriy Mozzhorin (formerly Senior Lieutenant, then Lieutenant-General, Professor, Doctor of Technical Sciences, Hero of Socialist Labor, laureate of many prizes, and recipient of many orders) was my equal in recalling the hot days of the Cold War and of "the battles where [we] fought together."[5] Institute specialists had participated firsthand in the testing of German A4 missiles in 1947 and in the subsequent testing of the R-1 and R-2 missiles. Together with our ballistics experts, they devised the mathematical methods for modeling missile flight and the first firing tables.

My speech, in which I referred to all council participants regardless of military rank or civilian title, as soldiers of the Cold War, was met with applause. The institute chief's speech, and all the speeches thereafter, in one way or another had to do with the problems of creating the nuclear-missile shield and, in due course, top secret Cold War operations.[6] Many of those operations were perhaps more vital for the future of humankind than the great battles of World War II. After the meeting, we adjourned to another hall where we raised toasts "to the soldiers of the nuclear-missile shield." Joining the toasts, I noted that rather than a shield, we had hammered out a sword.

For people far removed from missile and nuclear technology, the term "nuclear-missile shield" is associated with a solid line of fortifications along the nation's borders packed with missiles armed with nuclear warheads. In the minds of the uninformed public, these missiles were there to protect us

---

5. Reference to Pushkin's poem "The Song of the Wise Oleg" (*Pesn o veshchem Olege*), and the line "They are remembering former days and the battles, where they fought together."

6. The NII-4 director in 1996 was V. Z. Dvorkin.

from possible U.S. or NATO missile or aviation attacks. There is an element of truth to that; the anti-aircraft missiles designed to strike aircraft and the anti-ballistic missiles designed to combat ballistic missiles can rightfully be called a "shield." They were actually designed for defense, and not for attack. However, it is by no means necessary to use nuclear warheads for such a missile shield; effective weapons were invented to destroy the aircraft and missiles of a potential enemy. At one time these weapons included a fantastic particle-beam weapon. Missile systems equipped with non-nuclear warheads have been used in recent years in localized wars (Desert Storm, Afghanistan, Chechnya).

The term "nuclear-missile" should refer to a "sword" rather than a "shield." If the missile is carrying a nuclear warhead, it ceases to be a simple missile. In military and political terminology, such a missile falls into the category of "offensive strategic arms" or "strategic nuclear assets." A strategic weapon is intended not to protect, but to destroy vitally important targets and people in enemy territory. "Strategic" targets include strategic missile launch site areas and key political and economic centers.

CIA

It seems that the "shield" concept, both in its literal sense and in its figurative and allegorical sense, has been wrongly attributed to offensive arms, which perform the role of a "sword" rather than a "shield." Nevertheless, in our literature, the Strategic Rocket Forces (RVSN) are considered to

This famous image of the "Gagarin Launch Pad" at Tyura–Tam was taken by the CIA's U–2 spy plane in the summer of 1957 at a time of great tension between the two superpowers. The photograph provided valuable information to American intelligence services on the status of the Soviet ICBM program.

be the guardians of the "nuclear-missile shield."[7] In this case we are dealing with a shift of the "shield" concept. Strategic missile complexes have such destructive force that in and of themselves their "peaceful" presence on our planet acts as a "shield"—a guarantee against attack. In this sense, strategic nuclear arms during peacetime are the means for deterring the outbreak of a "hot war."

THE HISTORY OF THE SECOND HALF OF THE 20TH CENTURY IS DEFINED BY THE COLD WAR, a military and ideological confrontation between the two superpowers. The nuclear and missile arms race threatened humankind with total annihilation. Paradoxically, this race maintained peace for more than 50 years.

One can gain some insight into the different objectives of the "shield" and the "sword" by delving into what are now unclassified figures. By the end of the Cold War in 1991, the U.S. and the nations in the Commonwealth of Independent States (CIS) had more than 50,000 nuclear warheads on various vehicles.[8] If you take 0.5 megatons as the average yield of a single warhead, then the total nuclear potential of the strategic offensive forces was 25,000 megatons. Since we had declared that we had achieved parity, one must assume that the nuclear potential was divided about 50-50 between the USSR and NATO. In the event of the simultaneous use of the strategic potential of all their nuclear assets, the USSR, U.S., and NATO were capable of detonating a total of at least 20,000 megatons (let us say that they simply would not have time to use the other 5,000 megatons). One megaton is equivalent to 50 bombs of the sort dropped on Hiroshima, which killed approximately 100,000 people. All told, both superpowers could detonate the equivalent of 1,000,000 such bombs. That means they could destroy a million towns with a total population of 100 billion people. And there are roughly only about 6 billion people on this planet. The two superpowers and their Cold War allies were capable of annihilating the human race more than 20 times! Even if I have erred in my calculations by a factor of 10, the USSR, U.S., and NATO had still amassed twice as many nuclear assets as were needed to completely wipe out all of humanity. If there is an error in the calculations, it might be that people have proven to be considerably more resilient than the professional strategists had reckoned. I am intentionally not touching on the ecological consequences of a broad-scale nuclear war. This is the subject of

---

7. RVSN—*Raketnyye voyska strategicheskogo naznacheniya* (literally, Rocket Forces of Strategic Designation but more commonly Strategic Rocket Forces).

8. "Diskussiya o yadernom oruzhii," *Vestnik Rossiyskoy Akademii Nauk* no. 5 (1992). ["Discussion of Nuclear Weaponry," *Bulletin of the Russian Academy of Sciences* no. 5 (1992)].

special studies, which show that the survivors of nuclear blasts soon die from the ensuing ecological disasters.

During the Vietnam War, the Americans used heavy B-52 bombers to drop more bombs (in terms of weight) on North Vietnam than were dropped by all the warring nations during World War II. Nevertheless, North Vietnam won the war, using among other things, our anti-aircraft missile systems to destroy the most powerful bombers of that day. The wars in Korea, Vietnam, the Middle East, and Afghanistan were local "hot" wars that occurred during the 50-year Cold War.

The majority of professional historians searching for the origins of the Cold War cite political factors associated with fundamental ideological differences. Here, they fail to appreciate the influence of military-technical progress in the field of nuclear and missile technology. I agree with those historians who hold that the Cold War had already begun during World War II.

Atomic bombs and long-range ballistic missiles were produced as the main weapon for World War III. However, they proved to be an effective means for waging the Cold War and deterring it from becoming a "hot war."

The root causes of the confrontation between the two superpowers at the beginning of the Cold War included the struggle to get the upper hand in atomic bomb production and the search to acquire the German missile legacy. Work to produce an atomic bomb began in the U.S. in 1939. Everyone who has written a history of the creation of the atomic bomb has felt compelled to mention Albert Einstein's letter to President Franklin D. Roosevelt. Hungarian émigré physicist Leo Szilárd composed the draft of the letter and Dr. Alexander Sachs, vice president of one of the leading industrial corporations, an economist, and a dynamic Russian immigrant who enjoyed the president's confidence, handed it to Roosevelt.[9] Sachs' first visit to Roosevelt was unsuccessful. However, Roosevelt proposed that Sachs join him for breakfast the following day. During this second meeting Roosevelt summoned his military aide, General Edwin "Pa" Watson. Handing him the letter that Sachs had brought with him, Roosevelt said: "Pa, this requires action."[10]

And as simply as that hand-off, on 12 October 1939, a decision of enormous historic magnitude was made. At first, progress was just fair-to-middling. The bureaucratic machine in the U.S. was in no big hurry to expend great resources on projects whose expediency even famous scientists doubted.

---

9. Sachs, born in Lithuania, was a vice president at Lehman Corporation from 1936 to 1943.

10. V. L. Malkov, *Mankhettenskiy proyekt* [*The Manhattan Project*] (Moscow: Nauka,1995). The classic English-language work on the history of the Manhattan Project is Richard Rhodes' *The Making of the Atomic Bomb* (New York: Simon & Schuster, 1986).

The pace for implementing all the measures picked up after Nazi Germany attacked the Soviet Union. The scientists convinced Roosevelt and Churchill that the Germans might produce the bomb first. The history of the development of 20th century military technology abounds with examples of new hardware developed at the initiative of scientists rather than generals and marshals.[11]

In March 1942, the President was informed that all calculations and experiments showed the feasibility of the primary objective—the production of bombs in 1944. The race had begun for their practical realization. America's first powerful scientific-industrial organization was created, operating under strict secrecy and directly subordinate to the President. Even the U.S. Congress was denied the right to control expenditures on the Manhattan Project. Roosevelt's inner circle in the U.S. and British Prime Minister Churchill considered it essential to set up a principle that later became a postulate: "An atomic weapons monopoly is the trump card for maintaining a peaceful equilibrium given the new balance of powers that has taken shape on battlefields with the victory of the anti-fascist coalition."

By 1944, without even leaving the laboratory, the atomic bomb had become a crucial factor in world diplomacy. The German V-1 cruise missile and V-2 ballistic missile were used for the first time that same year. However, their mass use did not have a substantial impact on the course of World War II. These new weapons in no way hindered the Soviet Army's offensive operations. The time had not yet come for the integration of the missile and the atomic bomb.

As early as the spring of 1944, U.S. military analysts had predicted that the Soviet Union would end the war possessing enormous military strength that not a single European power or even a coalition of European states would be capable of resisting.

Doubts as to Stalin's postwar policy in Europe kept Roosevelt from even hinting at projects for the rapid development of "super bombs." Consequently, when the "Big Three" met at the Yalta Conference at Livadia Palace in February 1945, Roosevelt told Stalin nothing in this regard. Nevertheless, during the last months of the war, Roosevelt's position was unambiguously defined: he felt that postwar cooperation with the USSR was necessary and possible. These same attitudes prevailed in U.S. scientific and engineering circles.

We in the Soviet Union still vividly recalled the terrible times of the prewar repressions when accusations of collaboration with "Western imperialists"

---

11. Although it is true that a number of major military technologies were championed by the scientific and engineering communities, almost all such technologies (radar, nuclear power, the jet engine, long-range ballistic missiles, the proximity fuse, etc.) required a coalition of interests involving scientists, engineers, the military, industry, and political commitment to create working systems.

were most widespread.[12] For that reason, the Soviet people tried not to display their hopes for postwar friendship with their wartime allies and did not dare exhibit any eagerness to develop scientific-technical contacts with the Americans or British.

On 12 April 1945, at 3:45 p.m., Roosevelt was gone. With little hesitation, new U.S. President Harry S. Truman made the decision to switch from a policy of collaboration to one of confrontation with the Soviet Union. Possession of the atomic bomb secret inspired Truman. He used the atomic weapon as the best trump card of American diplomacy during the first years of the Cold War. The dropping of atomic bombs on Hiroshima and Nagasaki was not so much the last act of World War II as it was the first big operation of the Cold War with Russia.[13]

We did not behave as belligerently as Truman did. In 1945, under the scientific leadership of Igor Kurchatov, the organizational leadership of Boris Vannikov, and the supervision of Lavrentiy Beriya, work on our own nuclear arms resumed on a broad scale. State Committee No. 1, later reorganized into the First Main Directorate under the USSR Council of Ministers, was created to manage these projects.[14]

During that same year of 1945, in secret from one another, wartime allies America and the USSR began a race to exploit German missile expertise. I wrote about this in detail in volume one of Rockets and People.[15] After Hiroshima and Nagasaki, Soviet engineers and missile specialists, and presumably our American counterparts and the Germans who worked with them, began to ponder what an atomic bomb really is and whether it might be possible to integrate it with a missile. But the level of secrecy surrounding the bomb was so high that it wasn't until 1953 that we were permitted to communicate with our atomic specialists, while the U.S. missile specialists had to wait another year.

In 1945, without any inter-Allied agreements, we seized material and intellectual spoils, that is, the remains of German achievements in the field of missile and nuclear technology. There were no political accords governing

---

12. Chertok is referring to Stalin's Great Purges in 1937–39 when one of the common charges leveled against the hundreds of thousands arrested was collaboration with agents of Western nations.

13. G. M. Korniyenko, *Kholodnaya voyna: svidetelstvo yeye uchastnika* [*The Cold War: A Participant's Testimony*] (Moscow: Mezhdunarodnyye otnosheniya, 1994). An updated version of this book was published in 2001.

14. This first "Special Committee" (for the atomic bomb) was established as a result of a government decree issued on 20 August 1945. The First Main Directorate that Chertok mentions was a subordinate body of the Special Committee.

15. See Chapters 15 to 26 of Boris Chertok, *Rockets and People, Vol. I*, ed. Asif A. Siddiqi (Washington, DC: NASA SP-2005-4110, 2005).

this spoils system. Each party rushed to seize them first, literally snatching materials right out of the grasp of its anti-Nazi ally. And, having previously been our common enemy, the Germans soon became comrades-in-arms in the development of missiles for both superpowers drawn into the Cold War.

Contemporary historians believe that Truman took the initiative in the Cold War. Stalin did not seek conciliation and compromise with Truman as he did in his relations with Roosevelt and Churchill during World War II. Stalin also used the foreign policy side of the Cold War to harden domestic policy and restore a regime of senseless repressions.

In 1995, a British television company started working on a TV series on the history of the Cold War. The series authors contacted me, as a participant in this war, and asked me to comment on the rocket and space aspects of its origins. I spent about two hours giving the British my interpretation of the origin of the missile portion of the Cold War. At best, 5 minutes of the interview will make it on screen.[16] The creators of the series had already met with scientists, politicians, and U.S. diplomats. They were not particularly surprised by my conviction that the initiative for the onset of the Cold War belonged to the U.S. "Even former U.S. Secretary of Defense Robert S. McNamara, whom we filmed, had the same opinion," said the young British TV crew.

"This is going to be a 20-part chronological narrative documentary series about three generations of Russians, Americans, and Germans who lived through the Cold War. It will be a film that will allow you to relive and catch a glimpse of the years to which the greater part of your life was devoted," they said in parting.

In March 1947, Truman unveiled the doctrine that bears his name, proclaiming virtually the entire globe to be the sphere of "U.S. national interests."[17] Its paramount and priority objective was the struggle to contain Soviet communism. New U.S. Secretary of Defense James V. Forrestal displayed extraordinary energy in building up the military and intensifying the confrontation with the USSR. Between 1946 and 1949, operational plans were prepared one after the other, with feverish haste, for a pre-emptive nuclear war against the Soviet Union.

---

16. Chertok is referring to a 24-part series, "The Cold War," sponsored by CNN that involved a British team led by documentary maker Jeremy Isaacs. The series was unveiled in 1998, after Chertok had written these words.

17. The Truman Doctrine was originally formulated as a policy to provide aid to the Greek and Turkish governments, who were seen as being under threat from local communists (in Greece) and Soviet pressure (in Turkey).

The U.S. leadership, however, refrained from starting a nuclear war. And it wasn't because the U.S. didn't already have thousands of atomic bombs. According to American intelligence data, the Soviet Union was not ready for nuclear war. But according to the findings of American military analysts, U.S. infantry was not capable of countering its Soviet counterpart. In the opinion of American analysts, the Soviet Army was capable of moving victoriously through Europe in two weeks and capturing all the American bases. The Americans would be forced to flee and leave their European allies at the mercy of the Soviet Army. Soviet cities might be destroyed, but European cities would be occupied by the Soviet Army.

In 1949, the Americans had decisive superiority in long-range bombers and had already geared up for the series production of atomic bombs. For these types of strategic arms, the correlation of forces favored the U.S. This enabled the U.S. leadership to implement policy from a position of power. At the same time, the Americans understood that an attempt to destroy communism on Soviet territory with an atomic bomb attack would run the risk of establishing communism over all of Europe and the Middle East.

It wasn't until 29 August 1949 that the Soviet Union conducted its first test of an atomic bomb. This was not yet a bomb, but rather a device that confirmed we had mastered the principle. Academician Yuliy Khariton, the "father" of the Soviet atomic bomb, recalled: "Kurchatov once told me that at a meeting with Stalin before the detonation of the first bomb, the leader announced: 'The atomic bomb must be made at any cost.' And when its detonation took place and the awards were handed out, Stalin remarked: 'If we had been delayed 12 to 18 months with the atomic bomb, then we probably would have "tested" it on ourselves'."[18] Stalin and our "infantry" generals understood that the Soviet Army infantry forces in Europe were a temporary deterring factor. We needed to catch up with and outdistance the Americans in nuclear arms technology and, considering how we lagged behind in aviation technology, hurry to develop new types of armaments—first and foremost, missiles.

The historic decree signed by Stalin that launched the era of our missile technology was issued on 13 May 1946. That is why 1996 was a special anniversary for the rocket-space industry, which had played a decisive role in

---

18. *Izvestiya*, 1992. No. 265. Yuliy Borisovich Khariton (1904–96) was the chief designer of the first Soviet atomic bomb although Igor Vasilyevich Kurchatov (1903–60) is typically considered the scientific head of the project.

the balance of powers.[19] By analogy with the previously issued atomic decree [signed on 20 August 1945], this missile decree called for the creation of a "special committee" for the coordination of all projects. At first, this special committee was called [Special] Committee No. 2, then the Second Main Directorate (VGU), then once again Committee No. 2.[20]

The sum total of the actions stipulated by the 13 May 1946 decree, in terms of scale, organizational interactions, and the general scope of the projects, was unprecedented in our history—even for the atomic technology of that time. This occurred in a nation that should have been healing from the horrible wounds inflicted by war and thinking about how to provide minimal living conditions for the millions of people in the ravaged cities and villages.

After the devastating war, we could not in our wildest dreams have imagined the possibility of enjoying the achievements of civilization that were firmly entrenched in the American way of life. But perhaps this was to our advantage over the Americans. We didn't think about having a normal roof over our heads, while for them fundamental comfort was as essential as air.

IN LATE 1949, AFTER LISTENING TO [MINISTER OF ARMAMENTS] DMITRIY USTINOV AND [ARTILLERY FORCES COMMANDER-IN-CHIEF MARSHAL] NIKOLAY VORONOV REPORT on the results of the R-1 missile firing range tests, the work on the R-2 missile, and scientific research on the advanced R-3 missile, Stalin understood that we were still far from being capable of threatening America by dropping an atomic bomb on its territory. That is when the task of creating a "shield" to defend against a U.S. nuclear attack was formulated. To begin, we needed at least to secure Moscow.

Back then, a U.S. nuclear attack on Moscow would only have been possible using aircraft. If the Americans would destroy our cities, our armies would move into Europe and Turkey to destroy American air force bases, but at the same time, Moscow would have to be completely inaccessible to enemy aircraft. The capital of the USSR would have to be protected by a missile shield that could not be penetrated by aircraft.

---

19. In Russian vernacular, the linked descriptor "rocket-space" is comparable to the Western usage of the term "space program," although the former constitutes both the development of space vehicles and also ballistic missiles and launch vehicles.

20. VGU—*Vtoroye glavnoye upravleniye*. Recently declassified documents now indicate that VGU had no connection with the development of missiles. The VGU was established in December 1949 to manage uranium procurement for the A-bomb project. For a detailed description of the organizational aspects of both the nuclear and missile programs, see the second volume in this series: Boris Chertok, *Rockets and People, Vol. II: Creating a Rocket Industry*, ed. Asif A. Siddiqi (Washington, DC: NASA SP-2006-4110, 2006).

It is now difficult to determine who actually came up with the idea of creating an absolutely impenetrable air defense system. According to the memoirs of the creators of this system, it was Stalin who first conceived it as a directive.[21] In 1950, he summoned Pavel Kuksenko, the distinguished radio specialist and chief designer at Ministry of Armaments' SB-1 (later KB-1), and assigned him the following task: create an air defense system for Moscow that not a single aircraft can penetrate.[22] A special Council of Ministers decree was issued in August 1950 to solve this problem. The Third Main Directorate (TGU) was created under the USSR Council of Ministers to facilitate all the projects for Moscow's missile defense under the aegis of Lavrentiy Beriya.[23]

Thus, during the first five years of the Cold War, three government agencies were created in the USSR—three special "main directorates"—that solved the nation's three main defense problems. Ministers, scientific-research institutes (NIIs), design bureaus (KBs), and factories were assigned to each of these directorates when necessary.[24] Achieving the goals set for each of these three main

Minatom

This is a model of the first Soviet atomic bomb, known as RDS-1, which was exploded on 29 August 1949. It had a yield of 22 kilotons.

21. K. S. Alperovich, *Rakety vokrug Moskvy* [*Missiles around Moscow*] (Moscow: Voyenizdat, 1995).

22. SB-1—*Spetsialnoye byuro 1* (Special Bureau 1); KB-1—*Konstruktorskoye byuro 1* (Design Bureau 1).

23. TGU—*Tretye glavnoye upravleniye*. For a detailed description of the work of the TGU, see Chertok, *Rockets and People, Vol. 2*, Chapter 22.

24. KB—*Konstruktorskoye byuro.*

directorates required more than just solving enormously difficult scientific and technological problems. The problem of organizing and managing projects of such a scale required not only competent scientific directors and talented chief designers, but also leaders at a high governmental level. At the dawn of the Cold War, these "marshals" were Boris Vannikov for the atomic field, Dmitriy Ustinov for missiles, and Vasiliy Ryabikov for air defense.

The development of large defense systems during the Cold War, using fundamental research achievements and the most sophisticated technology, serves as an instructive example of highly competent technocratic management.

In the years to come, the three main directorates served as the basis for the creation of the governmental Commission on Military-Industrial Affairs (VPK) under the USSR Council of Ministers.[25] The VPK integrated the entire military-industrial complex. In 1965, three main ministries subordinate to the VPK inherited the profiles of the three former main directorates: the Ministry of Medium Machine Building (MSM) for all nuclear technology; the Ministry of General Machine Building (MOM) for missile technology, including space technology; and the Ministry of the Radio Engineering Industry (MRP) for anti-aircraft and anti-ballistic missile technology.[26]

---

25. The full name of this body was the Commission on Military-Industrial Issues of the USSR Council of Ministers, more commonly known as VPK—*Voyenno-promyshlennaya komissiya* (Military-Industrial Commission). Officially formed in December 1957, the VPK was the top management body for the entire Soviet defense industry.

26. MSM—*Ministerstvo srednego mashinostroyeniya*; MOM—*Ministerstvo obshchego mashinostroyeniya*; MRP—*Ministerstvo radiotekhnicheskoy promyshlennosti*.

Chapter 2
Preparation for Piloted Flights

According to official historiography, the era of piloted spaceflight began with the flight of Yuriy Gagarin on 12 April 1961. For those of us in the field of spacecraft development and production, the actual date of reference is 15 May 1960.

Ever since 15 May 1942—Bakhchivandzhi's first flight in the BI-1 rocket-propelled airplane—15 May has been associated with some sort of rocket-space event: 15 May 1957—the first R-7; 15 May 1958—*Sputnik-3*; 15 May 1960—the first *Korabl-Sputnik*.[1]

And subsequently there were more outstanding events associated with that date. If I were to study astrology, most likely I would find an explanation for this phenomenon. In the meantime, let's just chalk up such coincidences to chance.

Thirty years after Gagarin's triumph, there were certain journalists who made an easy living on the sensational exposure of the secrets of Soviet cosmonautics. Without any references to source documents, they wrote about the tragic death of cosmonauts who had flown into space even before Gagarin. One of those cosmonauts supposedly remained forever in space; another perished when the launch vehicle exploded on the pad; a third "broadcast" the beating of his heart to the Earth, some ham radio operators heard groaning and crying. To lend their stories credibility, the reporters even provided the last names of the deceased cosmonauts. All of these revelations were unadulterated lies!

This didn't just happen in our country. In the U.S., three years after the astronauts landed on the Moon, a book was published claiming that there had been no flight to the Moon. Supposedly it had all been television hocus-pocus—a performance staged at special secret studios. NASA rushed

---

1. *Korabl-Sputnik*, which literally means "satellite-ship," was the generic name given to the Vostok precursor flights launched in 1960 and 1961.

to refute the claim. A scandal ensued, contributing to the enormous success of the slanderous publication. The author and publisher made a killing by spreading bald-faced lies.

I hereby testify and once again solemnly declare that, before Gagarin, not a single human being flew into space from the territory of the Soviet Union! Before the flight of Gagarin, on 3 March 1961, Valentin Bondarenko, a cadet from the first detachment of cosmonauts, died as a result of a fire in a pressure chamber during a training session. This was the fault of ground personnel and had nothing to do with space technology. Over the course of the entire space age, cosmonauts Komarov, Dobrovolskiy, Volkov, and Patsayev perished as a result of accidents in Soviet spacecraft.[2] In the U.S. during that same period, 17 astronauts perished: three burned to death during an Apollo ground test; seven were lost in the *Challenger* explosion during liftoff; and seven more died on the *Columbia* as it was returning from space.

Cosmonauts and astronauts are still mortals after their return to Earth. Accidents and illnesses can take the life of any of them. Yuriy Gagarin died in an airplane crash.[3] Pavel Belyayev miraculously returned to Earth from space and five years later died in a hospital.[4] Physician Boris Yegorov was a member of the three-man *Voskhod* crew that was the first to fly without spacesuits, that is, without the requisite safety net. He died many years later of a heart attack. Cosmonauts German Titov, Nikolay Rukavishnikov, and Gennadiy Strekalov went through extreme emergency situations, but they were saved. They died not in space, but on the ground from ordinary cardiovascular or oncological diseases.[5]

Accidents did happen involving spacecraft before the flight of Gagarin, but these spacecraft were either automated or carried dogs. The ideological conditions at that time forced us to describe only our successes and conceal our failures. This policy of unnecessary secrecy did more harm than good. Because of the need to preserve state secrets, the creators of space technology received due assessment of their scientific and creative work, usually, only in their obituaries.

---

2. Cosmonaut Vladimir Komarov was killed during the *Soyuz-1* mission in 1967. See Chapter 20 in this volume. Cosmonauts Georgiy Dobrovolskiy, Vladislav Volkov, and Viktor Patsayev were killed during the *Soyuz-11* mission in 1971. Both accidents occurred during the reentry portion of the respective flights.

3. See Chapter 25 in this volume.

4. Yuriy Gagarin was killed in an air accident in March 1968. Pavel Belyayev, who commanded the *Voskhod-2* mission in 1965, died as a result of complications from pneumonia after an operation for pancreatitis in January 1970.

5. Cosmonauts Yegorov, Titov, Rukavishnikov, and Strekalov passed away in 1994, 2000, 2002, and 2004, respectively.

Incidentally, for reasons having nothing to do with secrecy, the world does not know the names of many of the true architects of the U.S. lunar program, with the exception of Wernher von Braun. He directed the development of the launch vehicle. Who were the chief designers of the lunar spacecraft, the engines, and control system? The names of the corporations and scientific centers, their addresses, and the names of their presidents were widely publicized, but the true architects are known only to a tight circle of specialists. Our chief designers at least had monuments erected in their honor posthumously. While still living, they were given the gold medals of Heroes of Socialist Labor, laureate medals, decorations, and awarded academic degrees, while the "poor" Americans received dollars, comfortable homes, impressive automobiles, and other hallmarks of civilized prosperity.[6]

This digression in my writing was prompted by television broadcasts in which millions of viewers have been shown documentary footage of launches and interviews with cosmonauts and presidents—but very rarely with the creators of rocket and space technology! The creation of rocket and space technology fell to the lot of the engineers—those who carried the primary burden of creation and responsibility—to perform invisible feats over the course of many years.

Here I shall allow myself a small digression into the history of the Vostoks. The preliminary investigations into the feasibility of producing an orbital satellite "with a man on board" date back to August 1958. Mikhail Tikhonravov and Konstantin Feoktistov were the individuals calling the shots for this project. In late 1958, work got under way to develop the control system, life support, and other systems.

Feoktistov acted as chief conceptual designer. Through all the programs involving the design of piloted vehicles, from the Vostok series to Soyuz, he showed himself to have the sharpest mind of all the designers that I have ever worked with. It was amazing to observe how patiently Korolev endured Feoktistov's stubbornness—and his occasional and excessive devotion to principle bordering on fanaticism. Some of my comrades sometimes complained of Feoktistov's dictatorial, even despotic, style when he discussed design matters. This pertained only to designs and not to interpersonal relations. In the latter, Feoktistov might serve as a model of cultured decency. His fanaticism can also be explained by the fact that he himself dreamed of

---

6. The "Hero of Socialist Labor," first awarded in 1939, was the highest national honor awarded to civilians during the Soviet era.

flying into space. He got this opportunity thanks to Korolev's persistence, but not until three years after Gagarin's flight.[7]

In April 1959, a secret "draft plan of the Vostok spacecraft" was produced, while the first ballistic calculations with orbital descent options came out in May.[8] It was possible to move forward with the proposal for human flight only with the support of the military, since one way or another, every R-7 rocket needed for the new programs depended on them. We had already extended the patience of the Ministry of Defense, using its firing range, its contingent of military specialists, and troop units for launches to the Moon, Mars, and Venus. The R-7 rocket, in its updated R-7A (8K74) version and with a third stage—the Block Ye—added, was already capable in 1959 of inserting a payload with a mass of up to five metric tons into near-Earth orbit.[9] This was sufficient to begin experimental piloted launches.

It wasn't the first time, nor would it be the last, that the Americans rendered indirect support to our new program. On the CIA's initiative, they began the development of spy satellites. The film from the Discoverer satellites was returned to Earth in special capsules. I have to admit that the Americans had gotten ahead of us with these satellites. In 1959, we did not yet have the technology to return payloads from orbit.[10]

The question of returning from orbit was the primary problem both for a piloted flight and for a flight carrying materials for photography and all sorts of other reconnaissance activities. A top secret governmental decree on the Vostok program was issued on 22 May 1959, with the aim of combining these interests. This decree placed OKB-1 in charge of the experimental development of the primary systems and design of an automatic spy satellite. The development of a satellite for reconnaissance and navigation was declared an urgent defense mission. With the help of Mstislav Keldysh and Konstantin Rudnev, Korolev succeeded in inserting seven additional

---

7. See Chapter 9 in this volume.

8. The development of a "draft plan" (*eskiznyy proyekt*) was a major part of the Soviet R&D process, involving the detailed layout of an entire system and its subsystems, projection of its performance characteristics, and a plan for its certification process. Usually, a draft plan would span several substantial volumes of paperwork.

9. The Soviets typically assigned consecutive letters from the Cyrillic alphabet to each separate stage of a rocket. A "block" represented a stage. In the case of the R-7, Blocks A, B, V, G, and D denoted the core and the four stra9ons. Block Ye was an added upper stage used originally for the early robotic lunar missions in 1958–60 and then upgraded for use in the Vostok human spaceflight program.

10. The first successful recovery of a Discoverer reentry capsule was in August 1960 during the *Discoverer 14* mission. Discoverer was the cover name for the top secret CORONA reconnaissance satellite program run by the CIA and the National Reconnaissance Office (NRO).

words into this decree: " . . . and also a sputnik for human flight."[11] Such a tactical merger of two seemingly completely different missions within a single decree subsequently also resulted in the technical unification of the main design elements of the piloted Vostoks and the Zenit spacecraft—the first photoreconnaissance spacecraft.

The office of the State Committee for Defense Technology and VPK drew up the decree with input from Korolev and the other chief designers. Nikita Khrushchev reviewed and signed it without any red tape. Khrushchev understood even better than the very authors of the proposal what the flight of a Soviet cosmonaut into space meant in terms of national prestige and proof of the superiority of the socialist system.

The experience gained working on the first complex spacecraft of the lunar and Mars-Venus programs suggested that we needed a much more rigorous approach to the problem of reliability, which affected the design of the vehicle.

As we went through the process of selecting a configuration for returning to Earth, we considered several options for the form and structure of the descent module. At one of those stormy sessions when all of their time had been eaten up in disputes on that subject, under pressure from Tikhonravov and Feoktistov, Korolev backed the ballistic landing configuration and a spherical descent module. This type of descent module with a reliable thermal shield was the simplest both for the aerodynamics specialists and for the designers. All the hardware that didn't need to be returned was attached to the descent module in two detachable compartments—an instrument compartment and an assembly compartment—which separated before entry into the atmosphere. Unlike an aircraft, we had the opportunity to test out the reliability of a piloted vehicle without the pilot!

The objective of carrying out the first piloted flights into space had already been clearly stipulated in the government decree issued on 10 December 1959. A draft plan for the automatic 1K variant of the piloted spaceship was developed over a three-month period. Korolev approved it on 26 April 1960. This made it possible to develop a launch program for the first experimental vehicles and to legitimize it with yet another decree on 4 June 1960, which called for the performance of flight tests from May through December 1960.

The launch program for the unmanned versions had not yet been completed, but there was a great deal of hype surrounding a piloted flight,

---

11. At the time, among his many duties, Mstislav Vsevolodovich Keldysh (1911–78) was chairman of the Special Commission of the USSR Academy of Sciences on Artificial Satellites of the Earth. Konstantin Nikolayevich Rudnev (1911–80) was chairman of the State Committee for Defense Technology, i.e., the "ministry" overseeing the missile and space programs.

stemming from reports of U.S. preparations for a piloted launch on an Atlas launch vehicle. Flight tests of this combat missile had begun on 11 June 1957—almost simultaneously with our *Semyorka*. However, it did not reach its design range until the eleventh launch on 28 August 1958.[12] After a series of modifications, the rocket was capable of inserting a payload weighing up to 1,300 kilograms into orbit. This enabled the Americans to design the piloted Mercury capsule and to schedule a piloted flight for 1961.

Relinquishing precedence to the Americans in launching a human being into space after all of our space triumphs was unthinkable. On 11 October 1960, Khrushchev signed a decree stipulating the production of a piloted Vostok spacecraft as a critical mission.

In early 1960, a special "3KA Policy" was issued, "3KA" being the factory drawing designation for the piloted Vostok. For the first time the policy spelled out in the form of a directive the procedure for manufacturing and conducting factory tests for all the systems to be used in the piloted flights. The assemblies, instruments, and systems constituting the Vostok would have to be labeled and have a "Suitable for 3KA" entry in their logbooks. It was forbidden to deliver any articles that would be integrated into the 3KA unless they had undergone a complete cycle of factory tests. The military representatives were required to conduct the strictest quality and reliability control. The chief designers and various organization heads were personally responsible for the quality of the articles labeled "Suitable for 3KA." They did not have the right to delegate their signature to any of their deputies. The "3KA Policy" played a great disciplinary role in our industry.

For the time being, the draft plan assigned just one objective for 3KA spacecraft: to support a multi-hour piloted space flight in Earth orbit and to safely return that man to the ground. No scientific, applied science, or military assignments were stipulated for the future cosmonaut. Just fly and stay alive. And then we'll see!

After the initial days of "canine euphoria," during a routine meeting with [Deputy Chief Designer] Konstantin Bushuyev and me regarding preparation of the next vehicles, Korolev said that Vladimir Yazdovskiy and the other

---

12. There were six primary versions of the Atlas missile. The first version, an Atlas A with a range of about 950 kilometers, was launched for the first time on 11 June 1957. The first full-range flight was performed by an Atlas B on 29 November 1958 when the missile flew about 9,600 kilometers downrange. Atlas D was the first operational ICBM version.

physicians were very troubled by the dog Belka's in-flight behavior.[13] During orbital passes four to six, the dog struggled and tried to break free of the safety harness. Thorough analysis determined that she had vomited. The health professionals believed this was an effect of weightlessness.

"They're pressuring me," said Korolev, "to stipulate, when we have the flight program approved in the Central Committee, that the first piloted flight last no longer than one orbit."

THE CREATION OF THE KORABL-SPUTNIK TEST BEDS FOR PILOTED FLIGHTS PROVED TO BE A NEW AND COMPLEX TASK NOT ONLY TECHNICALLY, but also in terms of organization. During the first decade of the missile age at least 90 percent of the scientific-technical problems fell within the sphere of the first Council of Chief Designers. A sort of closed "caste" of rocket scientists, engineers, factories, ballistics centers, and test firing ranges formed around this Council. As we embarked on the development of the piloted flight program, we understood the need to substantially broaden the scope of our cooperation.

The program for the production of vehicles for piloted spaceflight was developed at OKB-1. Many people participated in the development of the program, but its sole creator—its real creator—was Korolev. To implement the plan that had been conceived for the experimental development of the spacecraft and all the related support systems, the first governmental decree was painstakingly prepared and issued in May 1959. The primary administrators were listed in the decree. The administrators did not always accept our proposals enthusiastically. Sometimes a great deal of effort had to be applied for the necessary chief designer to be included in a governmental decree.

The space launches of 1957 to 1959 and the first Moon shots showed that the new tasks far exceeded the scope of work, interests, and capabilities of the first historic group of six chief designers.[14] Annoyed by something at one of our encounters at the "cottages," Vladimir Barmin told Voskresenskiy and me: "There used to be a Council of six chiefs. Now there's just one—

---

13. The "canine euphoria" is a reference to a series of Vostok precursor flights in 1960 and 1961 involving the launch of dogs into orbit. Two dogs, Belka and Strelka, were launched into orbit and successfully recovered in August 1960 after a spaceflight in a 1KA spacecraft.

14. Chertok is referring to the "Big Six" chief designers who began their cooperation designing long-range missiles in the late 1940s: Sergey Pavlovich Korolev (overall missile design), Valentin Petrovich Glushko (rocket engines), Nikolay Alekseyevich Pilyugin (guidance and control), Mikhail Sergeyevich Ryazanskiy (guidance and control), Vladimir Pavlovich Barmin (launch complexes), and Viktor Ivanovich Kuznetsov (gyroscope systems).

Korolev." But he was wrong. It was not just Korolev's whim that the number of Chiefs grew.

Young radio engineers were the first to make inroads into the monopoly of the "Big Six." Competitors working on radio engineering problems who did not recognize the NII-885 monopoly capitalized on the abundance of new assignments and boldly elbowed their way into Mikhail Ryazanskiy's sphere of activity. Ryazanskiy's own staff was not large enough to handle the rapidly expanding front of operations.

After taking over management of the Moscow Power Institute (MEI) radio staff from Academician Vladimir Kotelnikov, Aleksey Bogomolov set up an OKB that began to develop and produce new telemetry and trajectory monitoring systems.[15] A special noise-resistant command radio-link (KRL) was required to transmit the myriad of radio control commands to the *Third Artificial Satellite* (*Sputnik-3*) and then to all Korabl-Sputniks.[16] Chief designer Armen Mnatsakanyan worked on this problem at NII-648.

Space television got its own radio-link and became an essential accessory for every spacecraft. For NII-380 Director Igor Aleksandrovich Rosselevich and two chief developers, Igor Valik and Petr Bratslavets, reporting on and being responsible for each television session were made mandatory.

NII-695 and its chief designer, Yuriy Bykov, received the rights to develop a special radio intercom line, whose ground stations have been known to the entire world since the days of Gagarin's flight by the call sign *Zarya* (Dawn). Bykov developed the short-wave *Signal* operational telemetry system and direction-finding system to search for the descent module. Each of the radio systems was placed in ground stations as well as Command and Measurement Complex stations.[17]

After the end of the radio systems monopoly, the next to fall was Glushko's monopoly on engine development. Aleksey Isayev was the first to intrude, having developed the engine for the R-11 missile. At that time, back in 1954, Glushko took it in stride. But now, Isayev's pupil Mikhail Melnikov was butting into his R-7 propulsion system. Melnikov had developed its vernier engines. Then Semyon Kosberg, the developer

---

15. MEI—*Moskovskiy energeticheskiy institut.*

16. KRL—*Komandnaya radioliniya.* In official documentation, the Soviets use generic titles to refer to the first three satellites: the *First Artificial Satellite of the Earth*, the *Second Artificial Satellite of the Earth*, and the *Third Artificial Satellite of the Earth*. In the West, these three are more commonly known as *Sputnik-1, Sputnik-2,* and *Sputnik-3.*

17. The official Soviet designation for the ground tracking network was KIK—*Komandno-izmeritelnyy kompleks* (Command and Measurement Complex).

of the third stage propulsion system, came into the picture, and once again Melnikov grappled with the challenge of creating the engine for the fourth stage of that same R-7 (the four-stage version of the R-7 was known as article 8K78).

No sooner had Korolev's OKB-1 begun designing the piloted vehicle than the problem of a braking rocket engine occurred. The braking engine's thrust would have to be fired in the opposite direction of the spacecraft's orbital velocity vector. The spacecraft would descend from orbit and enter the atmosphere only after the engine executed a braking burn. The atmosphere's braking effect would cancel the remaining energy that the launch vehicle had imparted to the spacecraft upon insertion into space. Two more chief designers were needed to develop a system for returning the piloted vehicle to the ground—one for the braking engine and one for the landing parachute system.

New chief designers and new organizations were rapidly getting involved in our bustling activity, seduced by the attractive prospects. If there was no volunteer spirit, Korolev exercised his diplomatic skills; coercion from above was used only in extreme cases. Back then, Korolev did not have the title of "general designer" that became fashionable later on.[18] He was a chief among the other chiefs. His authority did not depend on his position or his title.

As the era of piloted spaceflight began, without destroying the old Council of six chiefs, Korolev was obliged to bring in at least another 15 individuals with full voting status, including Keldysh, chief designers of new systems, leaders from the Command and Measurement Complex, ballistics centers, the Institute of Aviation Medicine, and the Air Force Command. Korolev managed, without offending anyone, to take over the technical management of the entire complex, dividing the responsibility among his deputies, each of whom cooperated with their "subcontracting" chief designers. Korolev's designers Bushuyev, Tikhonravov, and Feoktistov, who had supervised the conceptual design of all spacecraft, were now responsible for more than just themselves. They drew up requirements for the Braking Engine Unit (TDU), life support, landing, medical monitoring, and even food systems in space.[19] The chief designers of other organizations were directly responsible for developing each of these systems.

The large, complex system, now called the rocket-space complex, needed control in its space segment. At the very beginning of the space age, the chief designer of the first missile control systems, Nikolay Pilyugin, showed no desire

---

18. Unlike the title "chief designer," the title "general designer" was originally used only in the aviation industry. From the late 1960s on, after Korolev's death, the latter term was adopted in most other sectors of the defense industry.

19. TDU—*Tormoznaya dvigatelnaya ustanovka.*

to be in charge of developing this space segment. There was no time to search for a new chief designer for all the spacecraft control systems, and no one was found on the "outside" to work on this area. Pilyugin's refusal to develop spacecraft control systems did not dismay Korolev. He understood that during the initial phase, he needed to be right on top of the whole complex of spaceflight control problems. Korolev made Vasiliy Mishin and I his first deputies and tasked us with coordinating the work on the whole complex of control problems for the new spacecraft, including all the radio engineering problems.[20]

The development of systems by his own OKB-1 staff and the coordination of the other chief designers' projects were closely intertwined. It was difficult to determine what required more attention. The merger with TsNII-58 and the transfer of Boris Rauschenbach's team from NII-1 enhanced the potential capabilities of OKB-1 so much that we didn't flinch and weren't frightened by new extremely complex assignments.[21]

In the newly established departments, the young engineers took on the responsibility for creating new systems with a boldness and enthusiasm that one could attribute to the fact that they did not yet grasp the degree of risk. Incidentally, this is inherent in all pioneers and trailblazers. If each of them had known what lay ahead, many discoveries might never have been made.

From 1959 to 1961, an organizational structure evolved around the projects for the development of spacecraft on-board flight control systems. This structure would endure for many years to come. Moreover, as the new space-related organizations of chief designers Vladimir Chelomey, Georgiy Babakin, and then Dmitriy Kozlov and Mikhail Reshetnev were established, they adopted our principle: a spacecraft's control system is developed by the head design bureau (KB) in cooperation with specialized NIIs and KBs that are headed by chief designers who are members of a common Council of Chiefs.[22] Thus, for spacecraft, I was responsible for the entire control systems complex. Within our OKB-1, responsibility was allocated without particular disputes or conflicts. Rauschenbach was in charge of developing the attitude control and guided descent systems. Igor Yurasov was responsible for the landing control system, spacecraft stabilization during the TDU burn phase, and for

---

20. In Soviet institutional culture, it was common to have a "first deputy" who was basically "first among all the deputies."

21. For the mergers of TsNII-58 and Rauschenbach's team from NII-1 with OKB-1, see *Rockets and People, Vol. II*, Chapters 26 and 27, respectively.

22. Vladimir Nikolayevich Chelomey (1914–84), Georgiy Nikolayevich Babakin (1914–71), Dmitriy Ilyich Kozlov (1919–), and Mikhail Fedorovich Reshetnev (1924–96) were leading Soviet spacecraft designers each heading their own major design bureaus.

Several senior engineers from OKB-1 at the November festivities (i.e., celebrations for the anniversary of the Russian Revolution) in Kaliningrad. From left to right are Boris Chertok, V. A. Kalashnikov, Ye. A. Tumovskiy, and I. F. Alyshevskiy.

From the author's archives.

developing the on-board electrical equipment. Viktor Kalashnikov was left with the design departments, instrument testing, all the electromechanical issues, launch vehicle control surface actuators, and thermal control system pumps. I kept Anatoliy Shustov's radio engineering department, Mikhail Krayushkin's antenna laboratory, and Petr Kupriyanchik's ground testing equipment department immediately subordinate to myself.

The chief designers with whom we "control" specialists happened to work in those days fell provisionally into three categories. Those working on components of systems developed at OKB-1 were the first group. This group included Vladimir Khrustalev (optical sensors for the attitude control system), Viktor Kuznetsov and Yevgeniy Antipov (gyros), Andronik Iosifyan (all types of power converters), Nikolay Lidorenko (solar arrays and chemical current sources), and Ruben Chachikyan (pressurization elements within the landing system).

The chief designers for the space radio systems comprised the second group: Yevgeniy Boguslavskiy, Aleksey Bogomolov, Yuriy Bykov, Armen Mnatsakanyan, and Igor Rosselevich. They joked that they served three masters: Korolev, to whom they answered for everything; Chertok, who put their equipment on board, made the antennas, and provided electric power; and the military, who would commission and maintain the ground stations throughout the enormous Soviet territory.

The third group of chief designers included everyone who needed control commands, electric power, and telemetry for their systems. They had their own key people at OKB-1, but they could not have managed without our control specialists.

There is an old saying: "Even the Sun has spots." If I were asked to recall the "spots" of the chief designers with whom I embarked on the era of piloted spaceflight, I would be in a very difficult situation. My memory, notes, and documents have preserved a lot of interesting material. But among all of this, there is not a single fact that casts a shadow on the competency or professional and human qualities of the chief designers of the space cohort of those first years. There were errors of ignorance. Experience came with each new launch. Together we studied at our cosmic universities. At times these lessons were very costly and grim. Sometimes within our narrow, trusted inner circle we would grumble that the "iron curtain" prevented us from associating with specialists abroad. But we had faith in our own strengths.

In this context, I shall recall two instances from the history of how Chief Designers Aleksey Isayev and Yuriy Bykov came to join piloted cosmonautics. While discussing the problem of returning from space, an argument heated up over the selection of the TDU. Korolev proposed that we immediately order a solid-propellant engine and sought Yuriy Pobedonostsev's advice in this regard. The latter had switched to the solid-propellant field some time ago.[23] Mishin, Bushuyev, and Melnikov objected. They contended that it was much easier to control the burn magnitude of a liquid-propellant engine than that of a solid-propellant engine. The ballistics experts calculated that a solid-propellant TDU would cause a deviation of 400 to 500 kilometers from the designated landing site. For a liquid-propellant engine, possible errors were reduced by a factor of 10.

They finally won Korolev over. He instructed Mishin and Melnikov to immediately meet with Isayev and persuade him to develop a special liquid-propellant engine for the TDU. Isayev flatly refused. Korolev did not back down. Aware of my longstanding friendship with Isayev, he asked me to stop by Isayev's KB the next morning and do whatever it might take to bring him to a meeting. Isayev met me in an aggressively defensive position. He was really up to his ears with engine projects for Semyon Lavochkin's anti-aircraft missile system.[24] He had fought hard to master the manufacturing process for large-scale series production. Hundreds of reliable engines were needed. Petr Grushin, new chief designer of missiles, had joined their ranks. He set up shop in Khimki at the site of the former Factory No. 293—our alma mater. Grushin urgently needed new engines for anti-ballistic missiles![25]

---

23. Pobedonostsev originally served as chief engineer of NII-88, where Korolev developed his early missiles after the war. At the time of Vostok development, Pobedonostsev was a senior engineer at NII-125, which specialized in research on solid propellants for rockets.

24. At the time, Isayev's design bureau (OKB-2 of the State Committee of Defense Technology) was involved in design and series production of engines for a number of different versions of Lavochkin's missiles for the S-25 air defense system.

25. Grushin's design bureau (OKB-2 of the State Committee of Aviation Technology) developed the missiles for System A, the Soviet Union's first experimental anti-ballistic missile system.

"Beriya is long gone," said Isayev, "but his business lives on. The Third Main Directorate is twisting my arm. Everyday I'm disrupting some schedule or another. The missiles for submarines are also waiting for engines. They've got bandits at the series production plant, not process engineers. There is absolutely nothing keeping them in line. And you and Korolev want to slip another noose around my neck. When one of Lavochkin or Grushin's engines blows up, no one knows about that and it's something that can be fixed. But if a man fails to return to Earth because of me? All I can do is blow my brains out![26] No, Korolev isn't going to talk me into this. It's better that we go to the shop. I'll show you a new idea. We want to make a submersible engine. It will be in the fuel tank of a submarine missile."[27]

Aleksey was beaming with new enthusiasm, and I went with him to the shop. But I fulfilled my assignment and brought Isayev to Korolev. In the reception room he asked me to wait, assuring me that Korolev would kick him out of his office 5 minutes later. But rather than 5 minutes, 45 minutes passed. Isayev emerged from Korolev's office, red-faced and perplexed. When he saw my quizzical look, he threw up his arms and said simply:

"What an artiste! A great artiste!"

I was tired from waiting 45 minutes in the reception room and replied: "You're both great artistes. What kind of arrangement did you come up with?"

Taking his time, Isayev lit up a Belomor. "I made a deal with Korolev that you have to do all the electrical engineering."

As a result of this historical meeting of the "two artistes," Isayev's engines bring all cosmonauts back to the ground. Isayev's team obtained a monopoly on spacecraft engines.

And now I'll turn to Bykov. After the dogs Belka and Strelka were successfully launched into space [in August 1960], everyone was in an excellent mood. We returned to Moscow from the firing range without the top brass. Each of us used the 8-hour flight on board the Il-14 as he saw fit. Those who were tired slept. Voskresenskiy organized a card game. I tore Yuriy Bykov away from his reading and proposed that we discuss problems of radio communications with the spacecraft that would be launched in the near future carrying a human being.

---

26. As Chertok described in *Rockets and People, Vol. I* (p. 108), "blow my brains out" was one of Isayev's favorite and most famous exclamations.

27. Besides engines for air defense missiles, Isayev was also involved in producing engines for early generations of several Soviet submarine-launched ballistic missiles developed by the design bureau (SKB-385) of Viktor Makeyev.

Bykov and I had known one another since before the war. Both of us had graduated from MEI, but had studied in different departments and became acquainted after we had already become engineers. Bykov graduated from the radio physics department and was a radio engineering enthusiast. My old grade school friend, Sergey Losyakov, introduced us in 1940.[28]

In 1944, Bykov, Losyakov, and I developed devices to suppress the radio interference produced by the ignition systems of new types of fighter aircraft. In 1945, Veniamin Smirnov, Nikolay Chistyakov, and I were sent on temporary assignment to Germany, where I worked with them in Greater Berlin. After many reorganizations, fate scattered Professor Gerts Levin's former radio team. Chistyakov became a professor at a communications institute. I found Smirnov in Leningrad. In 1964 he was chief engineer at a new radio organization, which we wrangled into developing one of the versions of the rendezvous control radio system. Later, Professor Smirnov was a department head at the Leningrad Electrical Engineering Institute. Professor Viktor Milshteyn, author of classic works on electrical metrology, was cut down in his prime by terminal cancer. Subsequent reorganizations tossed Losyakov and Bykov over to NII-695 on Bolshaya Kalitnikovskaya Street in Moscow near the famous Ptichiy Market. Here Bykov worked as chief designer of aircraft radios. With his former youthful enthusiasm, he leapt at the offer to develop a radio communications system for a man in space. After talks with institute Director Leonid Gusev, work "came to a boil" not only in the laboratories, but also in governmental offices. An ensuing decree named Bykov as chief designer of communications and direction-finding systems for our piloted space launch program.

Discussing new ideas with Bykov on the airplane made it easier for us to endure the severe turbulence. He told me about his ideas for the experimental development and testing of a radio intercom link. Losyakov, who managed the department of radio receivers at NII-695, proposed testing communications reliability using the relay method. To do this he developed an on-board receiver that was supposed to receive transmissions from conventional broadcast radios and then relay them through the future cosmonaut's standard on-board radio telephone transmitter. At that time I doubted the advisability of that idea, from the standpoint that the radio ranges of broadcast stations were not designed to penetrate into space. But Bykov convinced me with a simple argument: the experiment was cheap—what will be, will be.

This experiment was placed on one of the unmanned Korabl-Sputniks. Voice reception on the ground after relay was unintelligible. Music was

---

28. See Chertok, *Rockets and People, Vol. I*, pp. 202, 213.

distorted by noise and loss of reception to the point that popular songs were completely unrecognizable. This experiment was probably the reason why Italian ham radio operators reported in 1960 that they had picked up transmissions of rambling speech, groans, and wailing from space.

Initially Korolev had sized up Bykov, assessing him as a future partner on direct telephone communications with cosmonauts. Bykov's pronounced formality and his outward and inward refinement put Korolev on his guard. Would he flinch at a decisive and difficult moment when a cosmonaut's life and the nation's prestige might be at stake? Soon Korolev not only left behind his suspicions, but other chief designers involved in piloted launches envied his obvious affection for, and complete faith in, Bykov.

Having returned briefly to Podlipki and Moscow from the firing range after two failed launches of the Ye-3 [in April 1960], we focused our main efforts on the 1KP spacecraft.[29] This was the first prototype of the future single-seater spacecraft for piloted flights. The 1KP program of flights was at risk of cancellation due to our failures to meet all conceivable deadlines for developing equipment for the October Mars launches.[30] Korolev understood everything, but he did not want to hear any arguments or excuses. Once he was free of the lunar project, he retreated entirely into the risky race to put a man into space. Who would be the first man in space, a Russian or an American? We understood very well that it was unacceptable to yield priority to the Americans three years after the launch of the first *Sputnik*.

Sometimes it seemed that Sergey Pavlovich Korolev (S.P.) knew even better than I what to do in my subordinate departments and with my development work. He knew how to obtain and use information so that we always felt we were under his watchful eye. I tried to report my own troubles and *bobiks* to Korolev in advance with a touch of optimism; I did not want him to end up in a bad mood from information reaching him by unknown means, resulting in premature "lightning strikes" against the guilty parties.[31] But the optimistic reports that Rauschenbach and I delivered regarding the attitude control system for the 1KP spacecraft did not calm Korolev.

---

29. Chertok is referring to two attempts to launch the Ye-3 lunar probe in April 1960. See Chertok, *Rockets and People, Vol. II*, pp. 558–562.

30. See Chertok's description of the two Mars launches in October 1960 in Chertok, *Rockets and People, Vol. II*, Chapter 31.

31. Beginning with the first and only ground-firing test in 1947 from Kapustin Yar, testers often used the words *bob* or *bobik* to describe a technical glitch requiring hours to identify and eliminate.

The main group of 1KP developers and testers flew out to the firing range on 28 April. The spacecraft itself was shipped to the firing range on an An-12 airplane. It managed to get there before the assemblers who were supposed to receive it at the airfield and accompany it to the engineering facility (TP).[32] As soon as they arrived, hundreds of engineers and military testers began to roll out test equipment and cable networks; inspect and check the readiness of endless lists of launch site and engineering facility systems; stations of the Command and Measurement Complex; communications stations; and also inspect and check hotels, dining halls, and automobile transport. Those with the most gumption began preparing for operations by filling out requisitions for alcohol. This preliminary work during the very first hours after arriving at the firing range showed how much we had forgotten in the commotion before setting off on our mission. Lead Designer Oleg Ivanovskiy, who had just arrived, sent one radiogram after another to the factory demanding that they immediately make up for a shortage.[33]

The mindset of many managers changed during the eight-hour flight from Moscow to the firing range. Before the flight, conscious of their personal responsibilities, they each tried to prepare everything necessary for the work at the firing range; and once they discovered a shortage of documentation, equipment, or instrumentation during the very first hours there, they were outraged: "Where are their heads at? Slackers! Send a radiogram immediately!" Nevertheless, the schedule of operations mapped out in days, hours, and even in minutes, was drawn up by the lead designer on the assumption that they had everything they needed and there would be no *bobiks*.

S.P. demanded that I not fly out to the firing range until the bugs in the attitude control and descent systems were worked out. The spacecraft had been delivered to the firing range without it. Rauschenbach's team called the motion control system *Chayka* (Seagull). Subsequently that name became firmly rooted in the vernacular. To this day the motion control system of piloted spacecraft is called *Chayka*. The modern *Chayka* bears as little resemblance to that first one as the latest model of the *Moskvich* automobile does to the first *Moskvich* model 401.[34]

---

32. The phrase *Tekhnicheskaya Pozitsiya* (TP) literally means "Technical Position," but in the context of the missile industry more typically can be translated as "engineering facility."

33. "Lead designers" were different from "chief designers," in that the former were more junior in the design bureau hierarchy. Whereas chief and deputy chief designers were deeply involved in the R&D stages of a program, lead designers were typically responsible for the production phase.

34. *Moskvich* automobiles were produced in the Soviet Union in the postwar era. Early models were basically copies of the German Opel Kadett car.

Everyone who could had already left for the firing range, while I, receiving Korolev's daily missives of extreme dissatisfaction, stayed on at shop No. 39 in Podlipki with the new "Wunderkinder" and my battle-tested experienced electricians to debug the first *Chayka*.

For its time, the first *Chayka* for the 1KP spacecraft was a fundamentally new system made up of complex components. A high degree of reliability was required in the attitude control process during the execution of the braking maneuver to guarantee the return of the descent module to Earth—and not simply to Earth, but to Soviet territory. For reliability, the *Chayka* contained two independent control loops: primary and backup. The primary loop was supposed to maintain three-axis stabilization with the help of an infrared vertical sensor (IKV) and gyroscopic orbit.[35] Vladimir Khrustalev and Boris Medvedev developed the IKV at the Geofizika Special Design Bureau. This instrument differentiated the boundary between Earth over its entire circumference and space. After processing the signals issued from the IKV, the control system was supposed to orient the spacecraft so that one axis was pointing toward Earth's center. In order that it not spin randomly about this axis, the gyroscopic orbit would orient it in the direction of the velocity vector. The gyroscopic orbit was an invention proposed by Yevgeniy Tokar, who was at that time a young engineer and would later become a professor. After a great deal of squabbling, Chief Designer Viktor Kuznetsov accepted it for design development and production. Viktor was quite averse to using the inventions of outsiders. But in this case he deigned to give in, since there were no other proposals.

Three gyroscopic angular rate sensors (DUS) were supposed to dampen the Korabl-Sputnik's vibratory movements.[36] They were developed at the Aviapribor Factory KB. The chief designer there was Yevgeniy Antipov, the same Antipov who back in 1934 had tried to persuade me not to invent the electronic bomb release for TB-3 airplanes. At that time Antipov was in the midst of developing an electromechanical bomb release, and the troubles of a 22-year-old inventor from Factory No. 22 simply got in his way.[37] Now I—the former inventor—was not asking but demanding, an order backed up by government resolutions, that he develop especially reliable DUSs for spacecraft.

The Korabl-Sputnik also had a backup attitude control system. Proposed by Boris Rauschenbach and Viktor Legostayev, it was comparatively simple. It contained an optical sensor for solar orientation and the same vibration

---

35. IKV—*Infrakrasnaya vertikal.*
36. DUS—*Datchik uglovykh skorostey.*
37. See Chertok, *Rockets and People, Vol. II*, pp. 90–92.

damping DUSs. Both systems had control relay units that issued commands to the pneumatic valves of the attitude control thrusters. All of this great variety of instrumentation was assembled together for the first time and connected to each other, the power supply system, command radio link, telemetry, and test panels in the assembly shop. Such systems, no matter how ingenious their developers, never work the first time they are switched on. It's a good sign when smoke doesn't pour out of them from short circuits.

OKB-1 Factory Director Roman Turkov, who had visited the assembly shop three times a day and had not had the opportunity to meddle first-hand in the debugging process, chuckled at me: "You and your 'Wunderkinder' are going to give Korolev a heart attack if you don't wind up in the hospital first."

But I couldn't complain about the "Wunderkinder." The reality of the situation had a more powerful effect. Brought to the point of exasperation by Korolev's ire and the abundance of *bobiks*, I ordered that the entire

From the author's archives.

G. M. Markov, chief of the spacecraft final assembly shop at OKB-1's factory during the early 1960s.

Chayka be dismantled, packed up, and loaded onto an airplane: "We're going to finish the system at the firing range. At least we can announce that we have already arrived for flight tests."

Contrary to the old saying "answer for one trouble, answer for seven," that was hardly the case here. The primary control block and gyroscopes were shipped from Podlipki to the airport on a truck without an escort. To start, the driver, not knowing what kind of precious cargo he was hauling, really tore it up when he drove over a railroad crossing. The system's impact performance was further "tested" when he firmed up his resolve behind the wheel by consuming an undetermined amount of alcohol that he had snatched from the assembly shop, and in a state of "moderate inebriation," smashed the truck into a tree.

Such occurrences have a considerably greater impact on a manager's nervous system than a rocket failure with all the enormous cost that entails. The latter is considered a natural phenomenon in the experimental development of a complex system. An ordinary traffic accident caused by our sloppiness was viewed as extraordinary, verging on criminal negligence. It wasn't just those who were directly culpable who were punished, but also the many chiefs throughout the entire production chain of command, all the way down to the engineers who failed to monitor the transport of their priceless instruments until they reached their destination.

Despite the extra baggage we all carried after they threw the book at us, imposing every conceivable penalty, we finally boarded an Il-14 airplane with the *Chayka*. The first thunderstorm of May, however, did not care for Korolev's threatening orders. All flights from Vnukovo airport were cancelled. For us this meant yet another sleepless night. The flight was finally cleared for takeoff to Uralsk on the morning of 3 May. Buses, trucks, and cars were already waiting for us impatiently at *Lastochka*— the Tyura-Tam airfield. Precisely at midnight on 5 May, according to lead designer Ivanovskiy's schedule, the *Chayka* began its 1KP vehicle-level component tests.

It is not until you are here at the Assembly and Testing Building (MIK) at Site No. 2's engineering facility, where finally the whole kit and caboodle has been assembled, that you understand what a vast variety of ideas, systems, instruments and assemblies we have crammed into our new 4,600-kilogram Korabl-Sputnik.[38]

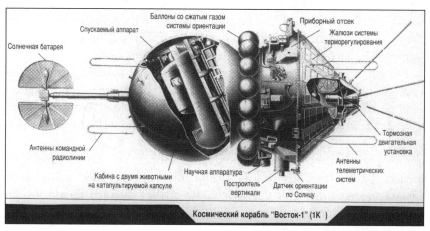

From the author's archives.

This drawing depicts the 1K vehicle launched into orbit in May 1960, publicly announced as *Korabl Sputnik*. This was the precursor spacecraft to the piloted 3KA Vostok that took Yuriy Gagarin into orbit less than a year later. The legend in Russian reads clockwise from top left: solar panels; descent module; spheres with compressed gases for the attitude control system; Instrument Compartment; system of blinds for the thermal control system; the Braking Engine Unit (TDU); antennae for the telemetry system; solar orientation sensor; vertical; scientific equipment; cabin with two animals installed on an ejectable capsule; and antennae of the command radio-link.

---

38. MIK—*Montazhno-ispytatelnyy korpus*—was the Soviet equivalent of the Vehicle Assembly Building.

Vostok descent modules in shop No. 44 of the Experimental Machine Building in Kaliningrad (now Korolev).

From the author's archives.

How were we going to manage to test out all of this? A brigade of developers with their own diagrams, instructions, test panels, and desires to replace the already installed on-board instruments with more reliable ones was plugging away on each system. No one had time for tests. Everybody needed the wiring women to resolve faulty connections or lengthen short cables.

During just seven days of continuous assembly and testing operations, the 1KP spacecraft was brought into a state suitable for the simultaneous activation of all systems as per the flight program. On 9 May—Victory Day—we wanted to celebrate by conducting integrated tests and analyzing the telemetry tapes.[39]

Actually we didn't begin until 12 May. We were delayed by dozens of unforeseeable but necessary checks and rechecks of the pyrotechnic cartridges; the passage of commands over the radio control link; repeated activation of *Chayka* modes; turning of the solar arrays as they oriented themselves on electric bulb simulators; and many of the things that you find out only during the first tests of new systems.

The first unmanned 1KP spacecraft was considered as the "simplest" because it did not have a life support system and thermal protection. But from the perspective of control specialists and electrical engineers, it was more complicated than subsequent ones.[40] For the first time, swiveling solar arrays were installed on it. We assumed that this would enable it to be tested

---

39. The Soviets (and now Russians) celebrate the end of the World War II in Europe on 9 May with "Victory Day." Americans and Western Europeans mark the same occasion on 8 May with "V-E Day." The actual German surrender was on 7 May 1945.

40. The "P" in the "1KP" stood for *prostyy* (simple).

in a long-duration flight. Subsequent spacecraft obtained their electric power from silver-zinc batteries. When they lost their charge it was tantamount to the death of the spacecraft.

Instead of the scheduled 4 hours, the entire day on 13 May was devoted to the final assembly and integration of the descent module and the instrument compartment. After that, for end-to-end tests of the *Chayka,* a crane hoisted the multi-ton Korabl-Sputnik on a flexible hanger, and the testers began to manually spin it about three axes. To everyone's delight, the thrusters "snorted," confirming that they hadn't gotten their command address wires crossed when the last resoldering operations were done on board. Instead of the scheduled 9 hours, we spent 20 hours on the final integration with the launch vehicle. The strenuous time schedule we had set for ourselves did not take into account alignment problems in the mating hardware and cables that were tattered due to neglect.

Finally, instead of 12 May, we drove out to the launch site on the night of 14 May. In the bunker and at the launch pad, everyone was amazed by the great variety, the different sizes, and complete lack of interface among the test panels that each system had "contrived." Understanding that this was not the proper time, I tried to persuade everyone who could still absorb anything after a week of sleeplessness, that things could not continue like this in the future; we had to think about standardization.

At the launch site, the first thing they checked was the erector boom with the "funicular" for the future cosmonaut. Back then, this supplemental structure, to which even television viewers have grown accustomed after all these years, seemed absolutely fantastic.

At 11 p.m., Chairman of the State Commission Chief Marshal of the Artillery Mitrofan Nedelin opened the traditional readiness reports proceedings. Everything was going smoothly until Korolev abruptly announced that he would require that all chiefs observe safety regulations and evacuate to a distance of 5 kilometers or stay in the bunker. At this point, the security service announced an evacuation plan for all "non-essentials" and shelter for those who might be needed in the event of launch mishaps in trenches specially dug for them.

"Hey, Boris, look at the lengths they've gone to because of your banged-up knee," said Voskresenskiy rather loudly.[41]

Interpreting this joke as a subversion of safety measures, Korolev addressed Voskresenskiy using the formal "you": "Comrade Voskresenskiy, as my deputy

---

41. Chertok injured his knee while running from a lunar launch failure in April 1960. See Chertok, *Rockets and People, Vol. II,* pp. 561–562.

for testing, you personally should watch out for the safety of your people. I insist that in the event of a violation of regulations the State Commission bar the guilty party from returning to work. If the chief designers want to be at Tracking Station (IP) No. 1, then let them ask the Chief Marshal's permission."[42]

Leonid, who was sitting next to me, gave me a sharp elbow in the ribs and whispered: "This spectacle is especially for Valentin."

I later learned that before their departure from Moscow, Korolev and Valentin Glushko had had a very serious clash over engines for the R-9 intercontinental ballistic missile (ICBM). Glushko had approached Lev Grishin, deputy chairman of the State Committee for Defense Technology, requesting that he be released from Korolev's dictates when selecting engine layouts. Keeping his cool, Nedelin requested that each perform his own duty.

At 5 a.m., the eastern sky gradually turned from a dark violet hue to a rosy pink. When the air is clear on the Kazakh steppe, the colors of May sunrises and sunsets are uniquely soft and brilliant at the same time.

The security patrol at IP-1 relentlessly drove everyone into the trenches that had been dug deep enough to stand up in; and because they were fortified with timbers, it would be impossible to observe the launch from them. At T-minus 5 minutes, I managed to duck under the canvas cover over the truck bed holding the *Kama* radio system and secretly move to an "exposed position." The launch was proceeding normally. This time I had a firm conviction that the rocket would perform properly. The Sun, which was just creeping over the horizon, spotlighted the crosspiece separating the first stage strap-on boosters. Three hundred seconds into the launch, a telemetry operator, who had leaned out of the truck, flashed me a "thumbs up" sign! But after 460 seconds, according to their report, the signal grew weak and the recording became garbled.

With our heads hanging down, fully convinced that there had been an explosion or fire in the Block Ye (the third stage), we plodded over to our cars and drove to the cabin dignified with the name *Ekspeditsiya* (Expedition). Here, in a cramped little room, was the only machine for high-frequency communications with Moscow and the NII-4 Coordination Computation Center (KVTs), which received information from all the tracking stations.[43] About 20 people were already packed into our tight quarters adorned with grimy wallpaper.

---

42. IP—*Izmeritelnyy punkt.*
43. KVTs—*Koordinatsionno-vychislitelnyy tsentr.*

Reports came from NII-4 that Yeniseysk, Sary-Shagan, and Ulan-Ude had reliably identified the normal shutdown of the third stage from the integrator. All the radios on the *Korabl-Sputnik* were functioning, and consequently, the antennas had deployed and the solar arrays were rotating. To be absolutely sure, we moved to the room called the "cinema hall." There, equipment had been installed for direct reception from the on-board *Signal* transmitter operating in the short-wave range. Well-wishers packed the hall to overflowing. Those who hadn't been able to squeeze into the room crowded around the open windows. Yuriy Bykov, master of the *Signal*, was urging his operator to stop twiddling with the radio receiver's tuning knobs.

The loudspeaker emitted first—quiet, then ever-more distinct telegraphic dispatches from space. Everyone cheered! The euphoria was greater than for the launch of the first *Sputnik* in October 1957. As was our tradition, Voskresenskiy, Kuznetsov, and I headed off to our cottage to celebrate such an historic event. After sleepless nights, cognac has a stronger effect than usual. Leonid lay down on his bed, and after mumbling that he wasn't needed to write the press release, he fell asleep quick as a wink.

I returned to the cabin. Data about the orbit was already arriving there from the KVTs. According to preliminary calculations the *Korabl-Sputnik* would survive there a long time, and there was no need to hurry with the vehicle's descent. The orbital parameters were: apogee—369 kilometers; perigee—312 kilometers; inclination—65 degrees; orbital period—91.2 minutes.

Korolev, Keldysh, Ishlinskiy, and Grishin composed the press release.[44] The Chief Marshal listened to their bickering, received reports from Moscow, and it seems, was more worried than anybody. He scarcely got so agitated at the frontline command posts. Finally, a historic decision was taken to call the 1KP a "spaceship" (*kosmicheskiy korabl*).

"And why not," said Korolev, "There is the seaship, the rivership, the airship, and now there will be spaceship!"

When the text of the press release had been typed up and transmitted to Moscow, Grishin, who had dozed off, woke up: "Comrades, do you understand what we have written? The word 'spaceship'—this is revolutionary! The hair on the back of my neck is standing on end!"

Nedelin was holding a communication session with Moscow and was in a hurry to get on the air. Back in Moscow, it was only 7 a.m., and they were in no rush to broadcast the press release. They cautioned him saying that "it might be at the end of the latest news." Instead of this it was the usual weather report.

---

44. Aleksandr Yulevich Ishlinskiy (1913–2003) was at the time a leading scientist at NII-944. Later, from 1964 to 1990, he headed the Academy of Sciences' Institute of Mechanics.

Finally, the stirring Moscow radio jingle known throughout the world came over the air waves and then the voice of Yuriy Levitan, who had been rushed to the studio: *"This is Moscow speaking! All the radio stations of the Soviet Union are operating . . . "* Levitan read with such pathos that we felt as if we were *just* learning from him about the "preparation for the flight of a human being into space."

Nedelin very much liked the update that the KVTs had inserted into the press release based on the calculation results: *"At 0738 hours Moscow time the Soviet* Korabl-Sputnik *passed over Paris . . . At 1036 hours Moscow time the spacecraft will pass over New York."* Nedelin enthusiastically explained to us: "That is why they delayed going on the air! Nikita Sergeyevich [Khrushchev] is in Paris now, and it was necessary to wake him up and let him know! What a black eye for them! Over New York—that's another black eye!" Once again, we rendered support to Khrushchev's politics from space.

The State Commission and chiefs decided to fly back to Moscow in order to be at the information reception and processing center—for both space and political information. They needed to capitalize on the euphoria of the success to accelerate the preparation of other spacecraft and to decide about a piloted flight. First they decided that the reentry [of the *Korabl-Sputnik*] would take place on 18 or 19 May. Our first descent module—the sphere—had no thermal protection. Therefore, upon entry into the atmosphere it would burn up anyway. But the attitude control process before braking, the TDU operation, and entry into the atmosphere must be tested.

In those days, two groups were formed to conduct operations: Group M in Moscow headed by Korolev, who took on the role of general leader, and Group T at the firing range, to which I was assigned to lead. Group T comprised representatives from each system that was to be tested in space. The advantage of our Group T was the capability for direct analysis of telemetry information, which we received when the *Korabl-Sputnik* passed through the coverage zone of IP-1. The Group T telemetry specialists analyzed the tapes after each communications session together with the systems' developers. Next, we all gathered together, compiled our general findings and sent them to Group M. There, information was coming in from tracking stations all over the Soviet Union, but only information, and not the tapes themselves. The Group T tracking station was the only one with the capability to perform an expert analysis of the tapes—there were no systems specialists at the other tracking stations. These circumstances led to a conflict between groups T and M. After the attitude control system test sessions, we questioned whether the infrared vertical sensor designed to orient the spacecraft about a local vertical was operating properly. From session to session, the rotation rate of the infrared-radiation-sensitive sensor scanning the horizon diminished. Finally we realized that the sensor had stopped. Evidently, the electric motor had failed or a malfunction had occurred. At the same time, all other parameters indicated the primary attitude control system was functioning normally.

The backup (solar) attitude control system had shown no contra-indications during use. We deliberated over the high-frequency communications line and argued with Legostayev, Bashkin, and Khrustalev, who insisted on using the primary system for attitude control before descent. At a Group T meeting, Branets, a young engineer from Raushchenbach's department, and Medvedev, Khrustalev's deputy, categorically opposed the use of the primary system. Convinced of the great risk of relying on the primary system for attitude control, I spent a long time on the line with Korolev persuading him to opt for descent using the backup system. He called together almost every last member of the technical review team, deliberated with Keldysh, and despite my objections, the State Commission decided to execute descent using the primary system. All the instructions were issued to the command radio link stations of the eastern tracking stations, and commands were transmitted on board for an orientation session using the primary system and for a TDU burn at the designated time. We did not yet have a reliable attitude control system readiness (GSO) criterion.[45] Today, modern technology enables us, with the aid of an on-board computer, to perform system diagnostics and authorize the firing of an orbital correction engine or descent only when the GSO flag is present. Intervention from the ground is required only if a GSO does not appear at the proper time. But we were still far from the computer age. In terms of navigational support, our first spacecraft was as different from today's spacecraft as Christopher Columbus' caravel was from a nuclear submarine.

The commands arrived on board, but the TDU fired with the braking pulse aimed in an inexplicably random direction. The subsequent TASS report admitted in this regard that:

> . . . *the braking engine fired while the* Korabl-Sputnik *was being stabilized during the operation of the propulsion system. However, as a result of a malfunction that had occurred by that time in one of the instruments of the spacecraft's attitude control system, the braking pulse deviated from the calculated direction. Consequently, instead of reducing the spacecraft's speed, its speed was increased somewhat, and the* Korabl-Sputnik *moved into a new elliptical orbit that lies in its previous plane, but with a significantly greater apogee. The pressurized cabin separated from the* Korabl-Sputnik *and, meanwhile, normal operation of the cabin stabilization*

---

45. GSO—*Gotovnost sistemy orientatsii.*

system was recorded. As a result of the first launch, a series of vital scientific and technical problems have been solved. The spacecraft's systems operated normally and supported conditions required for future piloted flight . . . The results of the work performed make it possible to move on to subsequent phases of tests.

THE PRESS RELEASE WAS DELIVERED IN A CALM TONE. BUT IN PRACTICE, we realized (during the very first launch) the real danger of an error that might cause a future cosmonaut to stay in orbit for many years. Instead of a braking pulse, we had an accelerating one, which inserted the spacecraft into an elliptical orbit with a 307-kilometer perigee and a 690-kilometer apogee. It was on this orbit that, according to the logic loaded into the automatic controls, the compartments separated. Unlike Gagarin's *Vostok*, which stayed up for an hour-and-a-half, the first *Korabl-Sputnik* stayed up for 28 months and five days! It wasn't until 15 October 1965 that the first *Korabl-Sputnik* launched on 15 May 1960 began to brake and burned up in the dense atmospheric layers!

After returning to Moscow, I spent a long time sorting out relations with my colleagues in Group M. Because of their obstinacy, the spacecraft had been dumped into an orbit where, according to the prognosis, it would survive from three to six years.

"Imagine what would happen if a human being ended up in a situation like that," I dramatized the situation to make them repent. "The whole world would follow his sufferings. He would die from lack of oxygen sooner than from hunger. Then we'd identify the systems' failures as the power reserves were depleted. The *Signal* system would fade away, then telemetry. And this before the eyes of the entire world!"

They agreed with me, but they could not convincingly explain their reasons for making the erroneous decision. The error upset Legostayev and Bashkin more than the others. The two of them and Feoktistov had persuaded Korolev to decline Group T's recommendation for descent using the backup system.

Many years later, when piloted space flight had ceased to be a novelty, the Americans made a science-fiction film entitled "Prisoners of the Universe." The failure of their TDU prevents two Americans from returning in their spacecraft to Earth. A Soviet spacecraft very similar to a Vostok flies out to render assistance, but a defective control system prevents it from docking with the Americans' spacecraft. After using up a lot of fuel, the Soviet rescue craft is forced to return to Earth. To prolong the life of his younger comrade, the older astronaut ejects himself into space. Several days later, when just hours of life support reserves remain, a secret U.S. military reusable winged spacecraft is launched for the first time and saves the astronaut. This film came out several years before the flight of the Space Shuttle and acted as a vivid advertisement campaign for NASA to increase its budgetary allocations for this program.[46]

THE GOVERNMENTAL DECREE DATED 4 JUNE 1960 AND ENTITLED "PLAN FOR THE EXPLORATION OF SPACE FOR 1960 AND THE FIRST HALF OF 1961," had called for the manufacture of two 1KP spacecraft without life support system and thermal shielding. However, Korolev convinced the State Commission that the experience we had already obtained was sufficient to move on to launches of unmanned 1K spacecraft—without the letter "P," that is, *with* a life support system and thermal protection that would guarantee the return of a descent module to Earth. The decree called for testing three 1K spacecraft before August 1960 for the experimental development of the spacecraft systems and intelligence-gathering equipment.

The next Korabl-Sputnik's descent module was protected with standard thermal insulation. For the first time in history it was supposed to return to Earth from space carrying live dogs, Lisichka (Little Fox) and Chayka (Seagull).[47]

Korolev was very fond of the affectionate, ginger-colored dog Lisichka. The medical technicians were preparing her for a fit check in the ejectable capsule of the descent module at the MIK. We were sorting out the latest glitch in the interface between the electrical circuits of the ejectable "dog" container and the descent module. Lisichka remained completely calm despite our bickering and the general roughness of testing. Korolev came over to us. I was about to give him an update, but he brushed me aside and, without asking the medical technicians, he scooped up Lisichka into his arms. She trustingly nuzzled up to him. S.P. gently stroked the dog and, not caring that others were present around him, he said: "I so want you to come back." The expression on Korolev's face was unusually sad. He held her for a few more seconds then handed her to someone in a white coat and, without looking back, plodded slowly into the bustling MIK hall.

---

46. Chertok's description refers to the movie *Marooned* (1969) which itself was based on Martin Caidin's famous novel of the same name which was published in 1964. Caidin's novel was published in Russian as Martin Keydin [Caidin], *V plenu orbity* [*Trapped in Orbit*] (Moscow: Mir, 1967) with a foreword by second cosmonaut German Titov.

47. *Author's note*: In 1957, the British Royal Society for the Prevention of Cruelty to Animals sent a letter of protest to Khrushchev concerning the death of Layka in space on the second Sputnik. Layka, whose death was preplanned, was the first victim of the space exploration—she was knowingly sent to her death, but on the other hand, she was immortalized in the history of cosmonautics.

During our years of working together, Korolev and I had been in difficult real-life situations many times. Depending on the circumstances, I had experienced various, sometimes contradictory, feelings toward him. This episode on a hot July day in 1960 is etched on my memory. Korolev was petting Lisichka, and for the first time I was struck with such a feeling of pity for him that I got choked up.

And perhaps this was a presentiment. On 28 July at 1231 hours, the 8K72 launch vehicle lifted off. Korabl-Sputnik 1K vehicle no. 1 with Lisichka and Chayka on board was equipped and prepared considerably better than the preceding 1KP. This time we had painstakingly worked through all possible situations to avoid making a mistake when selecting the attitude control system and issuing the command to descend from orbit. However, Lisichka and Chayka were not destined for space. The spacecraft crashed not far from the launch site when the first stage of the launch vehicle failed. This occurred after 38 seconds of flight.

The accident investigation commission concluded that the disintegration of the combustion chamber of the Block D strap-on due to high-frequency vibrations was most probably what caused the crash of the launch vehicle and spacecraft. Glushko was unable to give a clear explanation as to why these vibrations suddenly occurred. They attributed it to deviations in the manufacturing process at the Kuybyshev factory "where Comrade Chechenya is director."[48]

This time, watching the launch vehicle explode and burn, I did not jeopardize my still-healing knee again, and I had jumped down into the trench beforehand. The launch vehicle broke up 38 seconds into the flight and the core and straporns scattered over the steppe, causing no harm. Someone in the trench muttered: "We shouldn't have put the little red pooch on that rocket." This crash showed how urgent it was to develop a descent module rescue system for use immediately after liftoff. The death of Lisichka and Chayka would act as a stimulus for the development of such a system. Soon thereafter the Emergency Rescue System (SAS) was developed.[49] TASS issued no reports on the results of the failed launch on 28 July 1960.

IN AUGUST WE WERE STILL SUFFERING IN THE TYURA-TAM HEAT. The temperature in the shade reached 40°C [104°F] when we began to prepare the third spacecraft, 1K vehicle no. 2. This was a full-fledged Korabl-Sputnik

---

48. This is a reference to Leonid Stepanovich Chechenya who was director of Factory No. 24 in Kuybyshev (Samara).

49. SAS—*Sistema avariynogo spaseniya.*

lavishly furnished with equipment for science experiments. The biomedical experiments were designed to study the vital activities of animals under spaceflight conditions, the effect of space radiation on plant organisms, and the performance of a waste recovery, food supply, water supply, and sanitation systems. For these purposes, there were two white rats and many black and white mice in a pressurized cabin.

However, the main feature was the presence of two passengers—Belka and Strelka. These were such active and vivacious little dogs that no one doubted their successful return. We created very comfortable conditions for the dogs. They were placed in the pressurized cabin of an ejectable container equipped with life support systems.

Knowing Khrushchev's predilection, seeds from various sorts of corn were loaded into the descent module in order to later verify the effect of weightlessness on crop yield.[50] At the insistence of Keldysh and Korolev, the desire of many scientists to join the newly opened field of agricultural research was accommodated to the maximum extent. Facilities for microbiological, cytological, genetic, and immunological experiments took up a considerable portion of the descent module's space.

Our old colleagues from the study of cosmic rays competed with biologists and health professionals in the battle for a place in the descent module. For the first time, our conceptual designers did not protest the heavy blocks of plates coated with a thick layer of nuclear photo emulsion. The head of this research, Professor Sergey Vernov, managed to put a unit capable of instant development on board the spacecraft. The stand-alone device that commanded these operations was developed with our participation. And the specialists on solar ultraviolet and X-ray radiation didn't pass up the chance to load their instruments on board.

Recalling this scientific jumble 45 years later, I cannot say that any outstanding scientific discoveries were made during the flight of that *Korabl-Sputnik*. However, the persistence of the two "Ks"—Keldysh and Korolev—with regard to conducting fundamental research, would inspire respect even in our time.

---

50. This is a reference to Khrushchev's "Virgin Lands" program of the late 1950s in which large expanses of unplowed land within the Soviet Union were used for massive seeding to produce wheat.

43

THE OLD CADRE OF MISSILE TESTERS HAD RETREATED INTO THE SHADOWS DURING SPACECRAFT PREPARATION THIS TIME. The engineers of the aviation industry and physicians of aviation medicine were now the heroes of the hot days and sweltering nights. Medical Service Colonel Vladimir Yazdovskiy was in charge of the biomedical team. He had been collaborating with us since the days when we were conducting high-altitude launches of dogs on R-1Ye missiles at Kapustin Yar. He was on a first-name basis with Korolev, Voskresenskiy, all the chiefs, and myself, and he was considered a rocket industry insider.

Cosmonaut Yuriy Gagarin during training on the Vostok ejection seat.

From the author's archives.

At the time, four aviation industry organizations got involved with space technology.[51] Chief Designer Semyon Alekseyev at Factory No. 918 developed ejection seats, spacesuits for future cosmonauts, a breathing oxygen mixture supply system, a pressurized cabin for animals, and a sanitation unit (ASU).[52] Grigoriy Voronin, the chief designer at an aviation industry component factory developed a thermostatic control system and recovery systems for all types of metabolic wastes. Chief Designer Fedor Tkachev, who headed the Scientific-Research Institute of the Airborne Assault Service (NIIPDS), developed parachute systems for the descent module and the ejection seat for a piloted flight, and the ejecting pressurized cabin for flights carrying animals.[53] The descent module parachutes opened at an altitude of around 8,000 meters. When it descended to 5,000 meters the hatch cover was jettisoned and the animal container was ejected from the descent module. In the future, this

---

51. Most of the organizations involved in the space program so far (up to 1960) came from the armaments industry and not the aviation industry.

52. ASU—*Assenizatsionnoye ustroystvo*.

53. NIIPDS—*Nauchno-issledovatelskiy institut parashyutno-desantnoy sluzhby*.

was to be an ejection seat with a cosmonaut in a spacesuit descending in his own parachute. A system of barometric altitude sensors issued signals to open the parachutes. These were sophisticated redundant units developed by an aviation factory where the chief designer was Ruben Chachikyan.

The parachute complex with our automatic electric controls was the most worrisome system. Together with NIIPDS Director Tkachev, his deputy Nikolay Lobanov, and their fearless paratroopers, we developed the logic and circuitry to jettison the hatches, and the sequence for issuing commands to the parachute system and for seat ejection. By that time, combat aviation had accumulated a rich assortment of all kinds of headaches. We had to take all of that into account in our circuits and have the automatics department take over all the paratrooper's duties. Engineer Valentin Gradusov was responsible for the detailed development of the automatic controls triggering the jettisoning of the parachute hatches and issuing the commands to open the parachutes. Studying the landing automatics test results, Korolev said to me, "Your guy Gradusov should understand that he's pulling the ring instead of the pilot. Keep in mind, that if the parachutes don't deploy . . . " he trailed off for rhetorical effect. The landing system on Korabl-Sputniks (and then on Vostoks) was more complicated than contemporary systems. The descent module and the

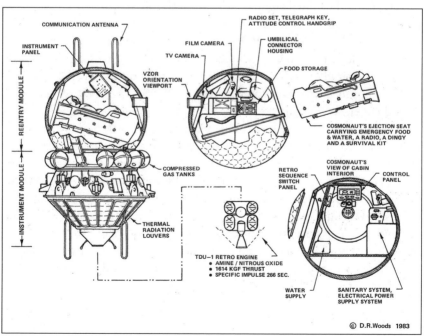

Dave Woods, 1983

Details of the world's first piloted spacecraft, the Vostok 3KA, designed to carry a single cosmonaut into orbit.

cosmonaut ejected from it needed to be recovered separately; there needed to be two landing systems. The Voskhod, Soyuz, and Apollo spacecraft had only one system; the Vostok ejection process was soon abandoned.

For a long time, the use of a two-step landing system on Vostok spacecraft whereby the cosmonaut did not land in the descent module, but was ejected from it and descended in his own parachute, was considered secret. For some reason, there was the "fear" that the International Astronautical Federation would not accept a world record if the cosmonaut did not land in the descent module. The ejection process was also a compulsory measure because the descent module's speed of impact with the ground exceeded the permissible speed guaranteeing the cosmonaut's safe landing. A special "soft landing" system needed to be developed. It was in fact developed, but it was not used until 1964 on the *Voskhod* spacecraft.

THE STAFF OF NII-380 (FOR TELEVISION RESEARCH), hardened at the firing range and now working with Yazdovskiy's physicians, combined two specially fabricated *Seliger* system television cameras so that through the container hatches one could broadcast frontal images of Belka, while the other would broadcast profile images of Strelka.

Observing the heated arguments between the physicians and the television brethren, I couldn't pass up the opportunity to toss some snide remarks at both sides: "So far Bratslavets and Valik only have experience broadcasting images of the far side of the Moon. If you set up the camera so that it gets a shot of Belka's tail instead of her face, it'll be a success right off the bat." "You do not understand the whole majesty of the moment!" said Yazdovskiy. After having a good laugh, we continued to make adjustments, test, and argue. The dogs, tongues hanging out, were languishing from the heat worse than we were.

Preparation of all the spacecraft's systems took 12 days at the engineering facility. On 16 August, the next triumphant rollout to the launch pad took place with a goal of launching the following day. Unexpectedly, the main oxygen valve on the launch vehicle was rejected and the launch had to be delayed until they could bring in a new one on a special flight from Kuybyshev.[54] This upset the medics more than anyone. They were convinced that the dogs would go crazy from the unfamiliar circumstances on the launch pad before they made it to space. Mother Nature heard their entreaties. She took pity; it grew cooler.

---

54. R-7-based launch vehicles were manufactured at Factory No. 1 (later Progress Factory) in Kuybyshev.

19 August was a blindingly clear day. The valve was replaced, everything was rechecked for the fifth time, and at 1144 hours 7 seconds, the launch vehicle carrying spacecraft 1K vehicle No. 2 lifted off. The State Commission, chief designers, and their "equivalents" crammed into the snug little room of operational Group T at Site No. 2. Everyone displayed such suppressed excitement and composure, even if drenched with perspiration, until first Yeniseysk and then Kamchatka confirmed that "separation was normal, the spacecraft entered Earth orbit."

That night we crowded into the space television receiving station. Bratslavets was overcome with emotion. While the spacecraft passed over the firing range, we had a perfect shot of both dogs barking. At that time, the U.S. reflector satellite *Echo-1* was passing over the Tyura-Tam area, easily visible on that clear night. The sphere, which was inflated to a diameter of 30 meters, did a good job of reflecting not just the Sun's light, but also radio signals. The Americans had launched it on 12 August, and it was inserted into a circular orbit at an altitude of 1,500 kilometers. The dogs' barking coinciding with the passage of the American satellite triggered a rousing reaction:

"Our dogs are barking at the American Echo. It would be great if they would pee at the same time!"

Yazdovskiy was happy: "If the dogs are woofing rather than whining that means they're coming back."

After the next orbit, the ground tracking stations (NIPs) sent their information—A-Okay on board.[55] The latest press release for Levitan was composed with a great deal of hype and edited several times. Chief Designer Aleksey Bogomolov insisted that the TASS report had to contain a paragraph devoted to television. This got through and caused a sensation in the press and in television broadcasts about the success of "Soviet space television technology."

In the dining hall during breakfast, State Committee of Defense Technology Deputy Chairman Lev Grishin was amusing everyone with his original anecdotes. He never told the same joke twice. "Today we built up to the preliminary 'dog' stage," he said, "and tomorrow we'll progress to the main 'human' stage." We took this occasion to drink down "fifteen drops" each.[56] At the height of our celebratory breakfast, Korolev was called to the telephone. Nikolay Golunskiy was on the telephone. When Korolev returned to the table, his facial expression had changed. He announced, "The telemetry experts say that the IKV has failed again just like it did on the 1KP spacecraft."

---

55. NIP—*Nazemnyy izmeritelnyy punkt.*
56. "Fifteen drops" denotes a small celebratory drink.

"Look, Sergey Pavlovich," I proposed, "It's a convenient opportunity to test the backup attitude control system."

Korolev looked daggers at me and started commanding: "Quick, send a radiogram. And tell Ryazanskiy and Bushuyev there at NII-4 to stop looking at pictures and prepare a descent program with the ballistics experts."

My Group T joined up with the State Commission. We held a meeting and argued all afternoon and evening until 2 a.m. At 9:30 a.m. in the morning, the meetings and consultations with Moscow resumed. Finally the decision was made to execute a landing using the backup emergency system.

Ground Tracking Station 4 (NIP-4) at Yeniseysk gives the descent cycle start command. This means that the *Granit* (Granite) system, the sequencer developed by Isaak Sosnovik and manufactured at the Plastik Factory in Moscow is starting up. NIP-6 at Kamchatka confirms that the descent cycle is in progress—*Granit* is sending control time tags. This means the commands are getting through. The TDU will start up somewhere over Africa at 1032 hours. But will the system hold the solar orientation until the engine starts up? We will only see the playback of information recorded over Africa when the instrument compartment comes over us after separation. And maybe, if everything is OK, we will hear *Signal*, the shortwave radio telemetry system that Chief Designer Yuriy Bykov developed.

At 10:50 a.m., we hear the *Signal* beep. That means the TDU has fired. If the descent module is flying toward the atmosphere and not out to space, as happened with the 1KP spacecraft, then at 1057 hours, *Signal* should fall silent: the module will enter the atmosphere and the antennas will burn up in the hot stream of enveloping plasma.

The tension was almost unbearable when, at 1057 hours, IP-1 and Moscow confirm that *Signal* can still be heard. Ten seconds later, *Signal* has weakened and is drowned out by noise. Everyone is cheering. Now we await reports about signal reception from the *Peleng* transmitter.[57] This will confirm the deployment of the descent module parachutes—the antennas are embedded in the parachute straps.

At 1104 hours, a jubilant shout comes over the telephone: "I hear P-3!" That is probably how a lookout would have shouted, "Land ho!" from the mast of an ocean-going caravel.

---

57. The *Peleng* system was a short-wave beacon for direction-finding system employing triangulation to locate the landing site.

Moscow confirms: "We've heard P-1, P-2, P-3. We have ejection, too!"

Surveillance services of the Air Defense Troops and the KGB mobilized to track the signals report: "Landing on 18th orbit in Orsk-Kustanay-Amangeldy triangle—just 10 kilometers deviation from designated point."

Korolev is keyed up—he demands that instructions be given over all communications lines to protect the dogs and everything that has returned from space. Finally, information arrives via unknown channels reporting: "The descent module and container were found near a state-owned farm (*sovkhoz*).[58] Police squads are being deployed to the landing site."

GRISHIN SUMMARIZED THE SITUATION SAYING, "Such scandalous affairs can't come to an end without the police. First of all, they'll make off with the parachute silk." Korolev and Nedelin picked out a team of "eager beavers," physicians, and factory assemblers, and decided to fly out immediately to the landing site. Everyone else was allowed and advised to fly back to Moscow immediately.

Celebrating these joyful events with Voskresenskiy in our cottage, we argued about the behavior of State Commission chairman, Nedelin.

"Korolev—that's understandable. For him it's very important to be at the landing site in person and see the dogs before they're retrained," I speculated. "But the Chief Marshal of the Artillery—what's all this to him?"

"He's hot-tempered and easily carried away," answered Voskresenskiy. "It's his spirit that's moving him, not his sense of duty."

We both concluded that we had had good luck in our work.

"Nedelin is not a martinet; he's our ally, even in 'canine affairs.'"

After polishing off what was left of the three-star Armenian cognac with all the friends who had descended on us, we piled into our cars and drove to the airfield.

In *Rockets and People, Volume II*, I wrote that two days after returning from the firing range, Korolev, Bushuyev, Ostashev, and I went to SKB-567.[59] After raising hell with them for the delay in developing the radio complex for the 1M (Mars spacecraft), Korolev changed the subject to Belka and Strelka.

Instead of Podlipki, we drove to a mansion on Frunze Street. This was Nedelin's residence. Nedelin received us in a spacious hall with a vaulted ceiling decorated with all kinds of amorous ornamentation. Keldysh and someone from the Central Committee drove up. Everyone took a seat in old fashioned,

---

58. A *Sovkhoz*—*Sovetskoye khozyaystvo* (Soviet Household)—was a state farm during the Soviet era.

59. SKB—*Spetsialnoye konstruktorskoye byuro* (Special Design Bureau). For the return to Earth and the visit to SKB-567, see Chertok, *Rockets and People, Vol. II*, pp. 571–572.

high-backed chairs around an enormous table. As host and chairman of the State Commission, Nedelin opened the meeting with a discussion of what information to release to the public from such a successful flight of a Korabl-Sputnik and how best to release it. At this moment, they singled out a group to prepare a press release and started to argue about whom to give the right for the first publication.

At the height of the debate, someone announced that at 12 noon, Physicians Oleg Gazenko and Lyudmila Radkevich from the Institute of Aviation Medicine had brought the precious dogs to TASS on Tverskoy Boulevard for a press conference. In addition to our law-abiding correspondents, an American from the Associated Press and a Frenchman from the French newspaper *Le Matin* had managed to get into the press conference. A ruckus ensued: who granted them permission, and when and why? Now photographs of the dogs would appear in New York and Paris without our censorship before they appeared in *Pravda*.

Nedelin's adjutants started calling the Ministry of Foreign Affairs, TASS, and other places demanding on behalf of the State Commission that they halt the independent action. But they were too late. While we were arguing, TASS had broadcast the press conference over the radio. A decision needed to be made at this moment: to show Belka and Strelka on TV that evening. At 9:30 p.m. on our TV screens at home we could watch Gazenko dressed in the civilian suit of a medical service colonel and Lyudmila Radkevich, exhausted from the canine commotion but dressed up and happy.

It was evening, and at Nedelin's home the arguing continued: should an obelisk be erected on the site of the successful landing or some other monument? Keldysh smiled ironically and recalled that Academician Ivan Pavlov erected a monument to an anonymous dog in Koltushi. Were our dogs less deserving? We decided that such a delicate subject should be handed over for review to Ivan Serbin in the Central Committee Defense Department.[60]

The most time was spent discussing what to show and write in open publications concerning the rocket and technical layout of the spacecraft itself, and also whether the descent module could be displayed at the USSR

---

60. Ivan Dmitriyevich Serbin (1910–81), one of the most powerful Communist Party *apparatchiks* in charge of the Soviet space program, was officially the chief of the Central Committee's Department of the Defense Industry at the time.

Exhibition of Achievements of the National Economy (VDNKh).[61] The majority were in favor of a detailed description and putting it on exhibition.

Nedelin recapped: "I'm flying to the south tomorrow and I'll consult with Nikita Sergeyevich [Khrushchev] on vacation there. You know how bold he can be in deciding issues when we are wavering. Take Cuba, for example. After the successful 8K74 [R-7A ICBM] launches into the Pacific, he said point blank: 'If need be, they'll also reach Cuba'."

After the triumphant flight of Belka and Strelka—the world's first dogs to successfully return to Earth after completing multiple orbits—there were plans for two more canine flights on the 1K spacecraft. Preparations were under way at a fever pitch. On 10 September 1960, without waiting for the results of subsequent launches, D. F. Ustinov, R. Ya. Malinovskiy, K. N. Rudnev, M. V. Keldysh, and S. P. Korolev sent a memorandum to the Central Committee, that is, to Khrushchev personally, in which they proposed accelerating work to carry out a piloted flight.[62]

The memorandum proposed that they perform one or two launches with dogs in October–November 1960; two more launches using 3KA spacecraft in November–December, also with dogs on board; and then in December 1960, they should carry out the first piloted flight. Such overly optimistic dates were proposed under the pretext that the "Americans have announced they intend to launch a man into space in 1960."

Bushuyev complained to Voskresenskiy and me: "I dared to ask Keldysh how we knew that the Americans were launching a man into space in 1960. With that sly smile of his, Keldysh told me that we have no reliable information, but we proposed those dates to be on the safe side."

As I recall, this memorandum, which directly affected all the chief designers, was the first ever signed by Korolev without preliminary discussion in the Council of Chiefs. This gave Barmin the occasion once more to grumble: "It used to be a Council of six chiefs; now there is only one chief left on the Council." Barmin usually did his grumbling in the company of Voskresenskiy, Bushuyev, and I, assuming that we would inform Korolev of the other chiefs' frame of mind. However, the general atmosphere of the great race was such that we loyal Korolevian deputies had no desire to get involved in the internal disputes of the "Big Six."

---

61. The VDNKh—*Vystavka dostizheniy narodnogo khozyaystva*—traced its origins back to the All-Union Agricultural Exhibition (VSKhV), which opened in 1939 in Moscow. In 1954, the original complex was expanded to 80 pavilions spread over nearly 600 acres to highlight all Soviet economic achievements. One of the most notable pavilions at the VDNKh was the Kosmos Pavilion that showcased models and replicas of various Soviet spacecraft.

62. An edited version of the memorandum that Chertok mentions was first declassified and published in 1991. See V. Belyanov et al., "Zvezdnyy reys Yuriya Gagarina" ["The Starry Race of Yuriy Gagarin"], *Izvestiya TsK KPSS* no. 5 (1991): 101–129.

The memorandum reached Khrushchev, and on 11 October 1960, the CPSU Central Committee and Council of Ministers issued a decree calling for the "preparation and launch of a Vostok spacecraft (3KA) with a man on board in December 1960, and to consider this a crucial mission."[63]

At the factory, they began fabricating the 3KA series spacecraft under Turkov's unremitting scrutiny. The rigid demands for reliability resulted in the most painstaking examination of the smallest deviation from the documentation. In and of itself, the 3KA policy was by no means a safety formality, but proved to be a very effective means for increasing reliability. However, its execution clashed with the deadlines set by the government decree calling for a piloted flight in December. Korolev respected Turkov and, as I recall, tried to avoid conflict with him. He artfully displayed his indignation at the failure to meet the deadlines for the manufacture of the first 3KA spacecraft, pouncing on the shop chiefs or designers *subordinate* to Turkov who, through negligence, had allowed errors in the technical documentation to go undetected, requiring that changes be made during production.

I spent late September through early October at the firing range up to my ears in problems involving the conquest of Mars. Two attempts to insert Automatic Interplanetary Stations (AMSs) on a trajectory to Mars on 10 and 14 October were unsuccessful through the fault of the 8K78 rocket, later called Molniya, which had required more attention than I was able to give to it.[64] We were forced to divide our enthusiasm, our work time (limited to 24 hours in a day), and all our stamina among four problems being solved simultaneously: the conquests of Venus and Mars, launching a man into space, and producing the new R-9 ICBM. Four decades later, I am still amazed at how we managed to accomplish, maybe not everything, but the main thing—launching a man.

The disaster of 24 October 1960, which I recounted in detail in the second volume of this series, was one of the reasons why the next series 1K spacecraft launches were delayed.

The third *Korabl-Sputnik* (1K vehicle no. 5) carrying the dogs Pchelka and Mushka wasn't launched until 1 December 1960. In addition to the dogs, the medical professionals and biologists crammed various other small animals into the spacecraft. The one-day flight proceeded normally and the team that worked around the clock analyzing the telemetry tapes reported that there were no significant glitches. However, after the TDU fired up, it turned out

---

63. CPSU—Communist Party of the Soviet Union.

64. AMS—*Avtomaticheskaya mezhplanetnaya stantsiya*. For Chertok's account of these launches, see *Rockets and People, Vol. II*, Chapter 31.

that the pulse produced was insufficient to land on our territory. According to the map that Svyatoslav Lavrov's ballistics experts quickly drew, the full descent trajectory extended into eastern China.

"What are we going to do?" asked new State Commission Chairman Konstantin Rudnev, addressing the crowd packed according to tradition into the MIK control room at Site No. 2. Everyone had been caught quite off-guard.

"We need to warn the Ministry of Foreign Affairs immediately," suggested someone. Then it dawned on me: "We don't need to do anything," I said. "The Emergency Object Destruction (APO) system will blow up the spacecraft.[65] We didn't issue the commands to shut it down. Small fragments might reach the ground, but there won't be any diplomatic complications."

"And what if a man were on board?" asked Rudnev.

"We had the APO system installed only on Zenits and other *automated* spacecraft."[66]

While this discussion was going on, a report came in from a far eastern tracking station—Ussuriysk. There they had picked up a transmission from the spacecraft that had suddenly broken off before entry into the atmosphere. Usually, loss of communications due to ionization began at altitudes of 100 to 90 kilometers. The test of the APO had confirmed system reliability the first time.

"We congratulate Chertok and all the saboteurs," declared Anatoliy Kirillov.

I withdrew to call the design bureau and congratulate the chief developers of the electrical automatic controls, Aleksandr Melikov and Aleksandr Pronin, on their success. The system understood that the spacecraft was landing on foreign territory and destroyed itself.

On 22 December 1960, launch of the last of the 1K series Korabl-Sputniks (vehicle no. 6) took place. Its passengers were dogs Kometa and Shutka. The gas generator of the third stage propulsion system failed 425 seconds into the flight. An emergency shutdown of the third stage engine occurred and the spacecraft failed to go into orbit. Instead it executed a suborbital flight with a maximum altitude of 214 kilometers.

According to the telemetry data, all the landing system automatics functioned normally. Korolev insisted on a meeting of the State Commission to discuss the need for an emergency search and transport of the descent module. A group headed by Arvid Pallo was equipped to carry out searches

---

65. APO—*Avariynyy podryv obyekta.*

66. The Zenit, first launched (unsuccessfully) in 1961, was the robotic photo-reconnaissance variant of the Vostok spacecraft.

and recovery. Before they departed, the entire group was instructed on how to avoid tripping the APO charge.

"Pallo has already been 'blown up' once in 1942 during tests of the BI airplane engine with the pilot Bakhchivandzhi.[67] A second blowup is unlikely," I reassured Korolev, who above all feared an unpredictable triggering of the APO.

We had both known Arvid Pallo for a long time as a very experienced tester. I was not worried for him. Two days passed before the descent module was spotted from the Li-2 search plane 60 kilometers from the town of Tur in the Krasnoyarsk Territory. It was day four before the expedition led by Pallo managed to reach the descent module. The rescuers were pleasantly surprised. Despite the -40°C [-40°F] temperature, the dogs were alive. The ejection seat system holding the dog container had failed to actuate. This saved their lives. Inside the descent module, the temperature at the end of four sub-zero days was compatible with life for the dogs. Kometa and Shutka were not honored with any ticker-tape parades, press conferences, or even TASS reports. Their miraculous rescue was the result of one of those "not-in-your-wildest-dreams" technical failures. It would have been an excellent subject for the media. However, the classified nature of the event prevented journalists from seizing that opportunity.

---

67. Pallo had been involved during World War II in tests of the BI rocket-plane. See Chertok, *Rockets and People, Vol. I*, Chapter 11.

Chapter 3
The First Piloted Spaceflight: "We're Off"

Many compare the nationwide jubilation on 12 April 1961 with that of Victory Day, 9 May 1945, in terms of its magnitude. To me, such a comparison based on superficial resemblance doesn't seem right. Victory Day was inevitable, long-awaited, and preplanned by history itself as a holiday for the entire nation "with tears in our eyes." The official announcement of final victory—Germany's signing of the act of unconditional surrender—served as the signal to openly express elation and grief. The massive celebration was historically appropriate.

Preparation for the piloted space flight was secret, as were all of our space programs. The announcement of Major Gagarin's flight into space, which was completely unknown, came as a total surprise to Earth's inhabitants and caused jubilation all over the world. Smiling Muscovites poured into the streets and filled Red Square carrying homemade signs: "Everyone into space!" The whole nation celebrated!

The flight of the first human being into space, and the success of Soviet science and technology, served to unite spiritually all social strata. The "Khrushchev Thaw" had slipped into decline, and ideological pressure of the Cold War was already felt.[1] Gagarin's flight once again infused us with hope for a bright future. It was still difficult to realize what exactly human spaceflight would give to the Motherland and to humanity, but each citizen of the Soviet Union felt that he or she personally had a part in this great enterprise: It wasn't an American or a European, but our man from Smolensk who, through the labor of our scientists and the efforts of our entire nation, had accomplished this feat.

---

1. The "Khrushchev Thaw" describes the brief period in the late 1950s and early 1960s when there was significant loosening of social and cultural restrictions in the USSR. Many ideas previously censored in Soviet public life were allowed in print, radio, television, and the arts.

I was not in Moscow on 9 May 1945, or on the day of Gagarin's triumphant reception. There is a considerable amount of publications, photographs, and film footage of what took place on those days. Years later, serious books have been published about the preparation of the spacecraft, its flight, and about Gagarin himself. Among the authors of the most objective works of the literary memoir genre, I feel compelled to single out Mark Gallay, a distinguished test pilot and instructor/trainer of the first group of cosmonauts: Oleg Ivanovskiy, the lead designer of the Vostok spacecraft; Nikolay Kamanin, the assistant to the Air Force Commander-in-Chief responsible for cosmonaut training and in charge of piloted cosmonautics in the Air Force; Yuriy Gagarin himself (recorded and edited by S. Borzenko and N. Denisov); German Titov, Gagarin's comrade and the world's second cosmonaut; and Yaroslav Golovanov, an engineer who became a professional journalist and a writer with very close ties to the rocket-space community.[2]

Golovanov's literary work *Korolev* was a clever synthesis of historical facts, portrayals of the individuals involved in the events of that time, and of the general atmosphere of the "day of cosmonautics."[3] This book, in addition to all its other merits, has the advantage over the preceding publications in that it came out at a time when it was permissible to print actual names. The rigid and, for the most part, wrong-headed censorship of the former years kept the authors on a short leash.

In his remarkable book *First Stages*, Oleg Ivanovskiy renamed himself A. Ivanov.[4] Konstantin Davydovich (Bushuyev) became Konstantin Davydov; Aleksey Fedorovich (Bogomolov) became Vasiliy Fedorovich; Nikolay Alekseyevich (Pilyugin) became Nikolay Aleksandrovich; Vilnitskiy became

---

2. Mark Gallay's memoirs were published as *Cherez nevidimyye baryery* [*Through the Invisible Barrier*] (Moscow, 1962). The book was updated and republished in 1965 and 1969. Nikolay Kamanin's diaries were published in four volumes as *Skrytyy kosmos* [*Hidden Space*] (Moscow: Infortekst IF, 1995–2001). Yuriy Gagarin's recollections about his life were compiled and published as *Doroga v kosmos: zapiski letchika-kosmonavta sssr* [*Road to Space: Notes of a Pilot-Cosmonaut*], eds. N. Denisov and S. Borzenko (Moscow: Pravda, 1961). The book was updated and republished in 1963, 1969, and 1978. An English edition was also published in 1962. German Titov's recollections were published as *Semsot tysyach kilometrov v kosmose* [*Seven Hundred Thousand Kilometers Through Space*] (Moscow, 1961). An updated version as well as an English-language version were published in 1963.

3. Yaroslav Golovanov, *Korolev: fakty i mify* [*Korolev: Facts and Myths*] (Moscow: Nauka, 1994).

4. Oleg Ivanovskiy's books were published under the pseudonym "A. Ivanov." The first edition was published as *Pervyye stupeni* [*First Stages*] (Moscow: Molodaya gvardiya, 1970). Updated editions were published in 1975 and 1982. The latter edition had the different title *Vpervyye: zapiski vedushchego konstruktora* [*The First Time: Notes of a Leading Designer*] (Moscow: Moskovskiy rabochiy, 1982). Ivanovskiy wrote newer editions (with different titles) under his own name in 1988 and 2005.

Valchitskiy; I was called Boris Yevseyev, and so on. The uninitiated would have to read publications of the first years of the space age with the aid of a special "Who's Who" handbook. Now it is possible to write and recount events without obvious misrepresentation. However, time has taken away the unique flavor of those days and nights.

Rereading many memoirs about the first piloted spaceflights, I feel something akin to envy toward the authors. In my memoirs, I am not capable of placing emphasis on Gagarin's story and on his flight on 12 April *per se*. If only I were writing when the tracks were still fresh! Now 44 years removed, the pictures of those historic days have become blurred by other events floating over them. It's like what happens when you inadvertently load your camera with a used roll of film. When you develop it, various scenes overlap one another.

In March and April 1961, I was in Tyura-Tam. During and after preparation for Gagarin's flight, other events that were directly related to our projects were also taking place at the firing range. I describe some of these temporally parallel events in Chapter 5.

The practices in effect at the firing range during the days of preparation to launch a man into space differed little from those used previously when Korabl-Sputniks lifted off for space, before they were called Vostoks. The stress and insomnia during preparation of the first rockets, even the first *Sputnik*, had been greater. Now there was noticeably more fundamental order. Something intangible hung in the atmosphere at Tyura-Tam. A person arriving at the firing range after a long break might have noticed that the "old-timers" had developed a sense of self-esteem.

In those days, many new people were arriving. After landing in this atmosphere for the first time, they quickly adapted to the way of life at the firing range. Rather than annoying them, the packed hotels and dining halls reconciled people and brought them together. We firing range old-timers didn't even notice that the steppe had burst into bloom with the stubby tulips earlier than usual. But all those arriving from Moscow noticed it right away. Everyone sensed the imminence of the historic event. But no one showed heightened emotions or enthusiastically uttered words of triumph. Perhaps as they greeted one another when they met, people smiled more often and broadly than usual.

CANDIDATES FOR THE FIRST SPACEFLIGHTS WERE SELECTED in accordance with special CPSU Central Committee and USSR Council of Ministers decrees issued on 5 January and 22 May 1959. Based on these top secret decrees, promising pilots were selected from various aviation units for spaceflight training. Beginning in October 1959, they arrived in groups at the Central

Military Scientific-Research Aviation Hospital to undergo extensive medical examinations. Hospital Director Dr. Nikolay Nikolayevich Gurovskiy was in charge of this new sphere of medical activity. Later, we availed ourselves of Gurovskiy's services in rendering medical care not only to cosmonauts, but also to the developers of space technology. Gurovskiy proved to be not only an experienced medical professional, but also a very good person. A medical commission selected the cosmonauts. The commission had at its disposal what were for those times the most state-of-the-art equipment and procedures. After performing tests to see that the candidates were in perfect health, a multitude of tests were performed during the process of various training sessions in rotating chairs and swings, in a centrifuge, and a 10-day stay in an isolation chamber. The commission tested memory, mental agility, resourcefulness in stressful situations, power of observation, hypoxia tolerance, and much more. In addition, the commission conducted physical training and parachute jumps. Out of 250 pilots sent before the commission, only 20 were pronounced fit. The Air Force Command made the final decision about placement in the first cosmonaut detachment.

The year 1960 should be considered the beginning of the Soviet Air Force's experience with rocket-space technology. Until then, all cosmonautics problems in the Ministry of Defense system had been the monopoly of the Strategic Rocket Forces. The future Cosmonauts Training Center—military unit 26266—was established by order of the Air Force Commander-in-Chief in March 1960. Out of the 20 who were selected, only 12 candidates became cosmonauts. They were accepted as students at the Cosmonauts Training Center. Colonel Yevgeniy Anatolyevich Karpov was named chief of

the future Cosmonauts Training Center. I should note that in my subsequent close acquaintanceship with him, he proved to be a highly erudite and exceptionally kind person who was poorly suited to the hard-boiled political, ideological, and disciplinary requirements of the military.

Yevgeniy Karpov served as the first chief of the Cosmonaut Training Center, from 1960 to 1963. Sitting on the right is Olga Tikhonravova, the wife of the famous rocket and space pioneer Mikhail Tikhonravov.

From the author's archives.

Considering the situation, the Air Force Commander-in-Chief managed to get the Minister of Defense to authorize a new position—assistant to the Air Force Commander-in-Chief for space matters. This individual would be the director of cosmonaut training. Lieutenant General Nikolay Petrovich Kamanin was appointed to this new post. The name Kamanin was well known to anyone who remembered the *Chelyuskin* epic of 1935. Military pilot Kamanin received the Hero of the Soviet Union Gold Star for his rescue of the passengers stranded on an ice floe after their ship *Chelyuskin* was crushed by ice.[5]

From the author's archives.

Korolev and his (second) wife, Nina Ivanovna Koroleva, with the original 1960 group of the cosmonaut corps in May 1961 at the Sochi resort. There are a vast number of iconic pictures from this famous trip, many of them of this particular setting. Sitting in front from left to right: cosmonauts Pavel Popovich, Viktor Gorbatko, Yevgeniy Khrunov, and Yuri Gagarin, Chief Designer Sergey Korolev, his wife Nina Koroleva with Popovich's daughter Natasha, Cosmonaut Training Center Director Yevgeniy Karpov, parachute trainer Nikolay Nikitin, and physician Yevgeniy Fedorov. Standing in the second row from left to right: cosmonauts Aleksey Leonov, Andrian Nikolayev, Mars Rafikov, Dmitriy Zaykin, Boris Volynov, German Titov, Grigoriy Nelyubov, Valeriy Bykovskiy, and Georgiy Shonin. In the back from left to right: cosmonauts Valentin Filatyev (obscured), Ivan Anikeyev, and Pavel Belyayev. Four cosmonauts were missing from the photograph: Anatoliy Kartashov and Valentin Varlamov had both been dropped from training because of injuries; Valentin Bondarenko died in a training accident a few months before; and Vladimir Komarov was indisposed.

---

5. The goal of the *Chelyuskin* expedition was to travel from the Barents Sea through the Arctic Ocean to the Sea of Japan during a single navigational season. The expedition crew was stranded in a heavy ice floe in the Chukchi Sea on 13 February 1934, and was later saved in a daring rescue that was widely publicized in the Soviet Union.

Only 12 very young fighter pilots finally learned why they had been selected and where they were going.[6] Their academic classes and parachute training began in March 1960. Anatoliy Karpov selected the six most promising candidates, in his view, for top-priority training: Captains Pavel Popovich and Andrian Nikolayev; Senior Lieutenants Yuriy Gagarin, German Titov, Valentin Varlamov, and Anatoliy Kartashev. The six were given priority during training sessions and access to the first Vostok simulator. The remaining students followed a less intensive training program. During the training process, Kartashev and Varlamov sustained injuries and dropped out of the first group of six. In their place Grigoriy Nelyubov and Valeriy Bykovskiy joined the team selected by medics.

In April 1960, Korolev summoned Tikhonravov, Bushuyev, Feoktistov, myself, and some others, and introduced us to the newly-appointed assistant to the Air Force Commander-in-Chief for space matters, Air Force Lieutenant General Kamanin. Also present during this meeting were military physician Vladimir Yazdovskiy, organizer of all the previous "dog" flights, who had joined our rocket-space community, and the cosmonauts' commanding officer, Yevgeniy Karpov. Both of these colonel physicians devoted a great deal of effort not only to medical monitoring during physical training for spaceflights, but also to the postflight examination of cosmonauts.

While discussing the cosmonaut selection procedures, turning to Yazdovskiy I said: "Not one of your little mutts would pass this." Yazdovskiy was about to answer, but Korolev abruptly cut him off: "Enough of your stupid jokes! We are going to entrust you, comrade Chertok, with setting up a cosmonaut training program in radio communications and systems control. Coordinate it with Nikolay Petrovich!" The fact that he had addressed me using the formal "you" and called me "comrade Chertok" to boot was indicative of the extreme degree to which my flippant behavior had irritated Korolev. Rauschenbach was busy setting up programs for the yet-to-be-developed manual attitude control. For radio communications, I persuaded Bykov, the master of *Zarya*, to put together a program. Bykov, in turn, assigned this project to his deputy, Meshcheryakov.

---

6. Candidates I. N. Anikeyev, V. F. Bykovskiy, Yu. A. Gagarin, V. V. Gorbatko, V. M. Komarov, A. A. Leonov, G. G. Nelyubov, A. G. Nikolayev, P. R. Popovich, G. S. Shonin, G. S. Titov, and B. V. Volynov joined the detachment by order of the Air Force Commander-in-Chief on 7 March 1960. The remaining eight joined later: Ye. V. Khrunov on 9 March, V. I. Filatyev and D. A. Zaykin on 25 March, P. I. Belyayev, V. V. Bondarenko, V. S. Varlamov, and M. Z. Rafikov on 28 April, and A. Ya. Kartashov on 17 June.

Sergey Darevskiy took on all the responsibilities for instruction in the use of the pilot's console. He was chief designer of the Special Design Bureau of the Flight-Research Institute (SKB LII) where this console was being developed and where the first space simulator was produced.[7] Kamanin was not offended by what Korolev had perceived as my flippant attitude. He invited Bushuyev and me to meet the future cosmonauts. They were being housed temporarily in one of the buildings of the M. V. Frunze Central Airfield almost directly opposite the Dinamo metro (subway) station. I had spent two days with Bolkhovitinov in that building in August 1937, when there was still hope of restoring radio contact with Levanevskiy's crew.[8]

I must confess that after seeing these potential cosmonauts for the first time, I was disappointed. I remember them as lieutenants who, because of their youth, looked alike and did not appear very serious. Since the war years, I'd had a completely different mental image of the concept of "fighter pilot." If someone had told us that several years later, one after the other, these boys would become Heroes of the Soviet Union, and some even generals, I would have answered that that would only be possible during wartime. It turned out that World War III was not at all necessary for this. There was an ongoing battle at the front line of the Cold War's scientific-technical front. Rather than soldiers, it was scientists, engineers, the "generals" of industry, and workers who determined the battles' outcome. And warriors of another sort came on the scene—cosmonauts!

One way or another, we were dealing with fighter pilots. Fighter planes had been designed with pilots in mind. We were faced with developing a vehicle that would have a "man on board" rather than a pilot flying! Would he need a manual control system?

There are several versions of the history of the advent of the manual control system. In the first version, Korolev demanded the system as a concession to the pilots. Korolev had not forgotten that once upon a time he had controlled an aircraft.[9] The second version is more credible: When Rauschenbach, Legostayev, Bashkin, and Feoktistov were discussing the automatic attitude control and braking engine firing system, the following thought "automatically" occurred to them: "What would it cost to add on a manual system to these automatics?" It turned out that it was not a complicated matter to produce a manual system on the basis of the obligatory automatic

---

7. SKB LII—*Spetsialnoye konstruktorskoye byuro Letno-issledovatelskogo instituta.*

8. See Chertok, *Rockets and People, Vol. I,* Chapter 7.

9. During his youth, Korolev was an expert glider pilot and received a number of major accolades for both his design sense and his piloting skill.

one. The manual control system was developed to provide the capability to back up a failed automatic system (even though the system had redundancy) for return to the ground. We designed and deliberated over everything that had to do with manual control technology with great interest, although we considered that it was there "just in case."

For the first flight, out of concern for the cosmonaut's reasoning power, someone proposed introducing a cipher lock. Only after punching in "125" would it be possible to power on the manual control system. For the first flight, this code was conveyed to the cosmonaut in a sealed envelope. We believed that if he was able to get the envelope out of the instruction folder, open it, read the code, and punch the code in, then he was in his right mind and could be trusted to perform manual control.

After the flight, Oleg Ivanovskiy and Mark Gallay confessed that they had secretly informed Gagarin of the code "125" before he got into the spacecraft, violating the State Commission decision. In addition to the automatic and manual systems, a "ballistic" system was also provided to guarantee return. If the braking engine were to fail, an orbit had been selected that was so low that the speed of the spacecraft would gradually decrease due to drag in the upper atmosphere; and in no more than five to seven days, the spacecraft would burrow through and end up on the ground. True, it would land in an unpredictable area—according to probability theory, in the ocean!

The automatic and manual control systems were successes! They were extremely simple and reliable. It is amazing that now no one would propose such simple and reliable systems. Any expert nowadays would say that if it doesn't have a computer, it's simply a joke!

Objectively speaking, among all the events that prefaced the success of the world's first piloted spaceflight, I would rank in first place the decision to affiliate Vasiliy Grabin's team (NII-58) with OKB-1, as well as the transfer of Boris Rauschenbach's team there.[10] All in all, from 1959 to 1960, we gained intellectual potential for an integrated, goal-oriented activity that no other organization in our country possessed at that time. And not only in our country! When foreign scientists got the opportunity to familiarize themselves with the Vostok control principles, they admired their simplicity and reliability compared to the first American Mercury piloted spacecraft.

To this day, Rauschenbach, Legostayev, Tokar, Skotnikov, and Bashkin have every right to be proud of their automatic solar attitude control system and backup manual attitude control system, as do their co-workers and our

---

10. See Chertok, *Rockets and People, Vol. II*, Chapters 26 and 27.

subcontractors who developed the optical sensors and automatic and visual attitude control instruments at the Geofizika organization on Stromynka Street. Activation of the systems via command radio link from Earth was backed up by possible control from the "pilot's" console. For these purposes, the electricians, including Karpov and Shevelev (recent graduates of the Taganrog Radio Engineering Institute who were known for their height and the equally large number of ideas they contributed) developed relay-controlled automatic logics interfaced with the manual control loop, sequencer, and command radio link. A team from NII-648, headed at that time by Armen Mnatsakanyan, adapted the command radio link for the Vostoks.

The Americans did not manage to develop such a reliable automatic system and relied to a much greater degree on the human being. It wasn't until 1965 that the two-man American Gemini spacecraft surpassed the Vostoks in terms of basic parameters. After that, we needed another three years to once again pull ahead with the Soyuz vehicles. True, this race cost Vladimir Komarov his life. But that did not happen through the fault of the control systems developers.

Radio systems connected the first piloted spacecraft with Earth. The *Zarya* 1-meter band radio voice link, the shortwave *Signal* link developed by Chief Designer Yuriy Bykov's scientific-research institute, the *Rubin* (Ruby) super high-frequency link for trajectory monitoring, and the new *Tral-P* (Trawl-P) telemetry link were developed with inconceivable deadlines at OKB MEI in cooperation with the Leningrad All-Union Scientific-Research Institute of Television. This television system had very modest specifications: lines per frame—100; frame transmission frequency—10 Hz; number of tonal gradations—8. But this was the first space television!

Of all the Vostok systems, the landing system was extremely complex. They did not risk having the cosmonaut descend in the module out of fear of the g-load during impact with the ground, and so they came up with a two-stage system. The descent module and the cosmonaut landed separately! After entry into the atmosphere, at an altitude of seven kilometers, the hatch blew off and the cosmonaut was catapulted through it in his seat. The cosmonaut was in free fall until an altitude of four kilometers, awaiting the opening of his parachute. Finally, his main parachute deployed, and his seat detached and went into free fall. The descent module with its own parachute landed nearby without the cosmonaut. Thus, there were two landing systems, and the space-suited cosmonaut was supposed to make contact with the ground according to all the rules of parachute jumping. Optimizing the circuits involved in blowing out the hatch, ejecting the cosmonaut, and deploying the parachutes caused the electricians more trouble than all the other systems. Here, there were no manual systems to save the cosmonaut's life in the event of a random failure.

The first pair of Air Force cosmonauts selected, Gagarin and Titov, spent a lot of time at the firing range with the assistance of Semyon Alekseyev, Chief Designer of the spacesuit and seat, working on the individual fit of the complex harness system. Fedor Tkachev, the chief designer of the parachutes, thought his system was very simple. For me, it was much more difficult to figure out dozens of cords, straps, and locks than it was to figure out electrical circuits. But I had to sort it out because the parachute system would not deploy without electrical commands.

If you take today's standards of reliability for launch vehicles as a point of reference, then by April 1961, we had no grounds for optimism. Even for commercial launches of unmanned automatic spacecraft, specifically communications satellites, according to international standards in the 1980s, it was required that the launch vehicles have at least eight successful launches in a row. Of the five Korabl-Sputniks launched in 1960 for the experimental development of the systems, four got off the ground. Of those four, three made it into orbit, and two landed. Of the two that returned, only one landed normally![11] Before a man was launched, it would be absolutely necessary to have another two or three successful unmanned launches.

During the two Venus launches on 4 and 12 February 1961, the first three stages behaved normally. On 9 March, Korabl-Sputnik 3KA vehicle no. 1 carrying a dummy and a dog named Chernushka was launched according to the program proposed for the piloted flight. After completing one orbit, the vehicle landed in the prescribed area, 260 kilometers from Kuybyshev.[12] Vladimir Yazdovskiy showed Chernushka to the local collective farmers. The single orbit of the 3KA simulated the single orbit that we hoped would be an additional guarantee that a cosmonaut would return to Earth alive.

The postflight inspection of the descent module showed that after the TDU shut down, the sealed connector of the cable tower connecting the descent module and the instrument compartment burning up in the atmosphere failed to blow off. Both parts entered the atmosphere mechanically separated along the main structural ring, but connected by a thick cable. Final separation took place only after the cable tower burned up in the atmosphere. In the frenzy of preparation for the next launch, no one attached any significance to the fact that the compartments separated not because of the engine shutdown command, but due to heat sensor signals that were not issued until the descent module was heating up in the atmosphere.

---

11. The five launches were on 15 May (the first *Korabl-Sputnik*, not designed for recovery), 28 July (failed to reach orbit), 19 August (*Second Korabl-Sputnik*, successfully returned), 1 December (*Third Korabl-Sputnik*, failed to return safely), and 22 December (failed to reach orbit).

12. In the Soviet press, the vehicle was publicly referred to as the *Fourth Korabl-Sputnik*.

On 25 March 1961, another 3KA spacecraft was launched. It had exactly the same radio equipment that Yuriy Bykov had developed for the flight version of the piloted spacecraft. The launch was a success. The cosmonauts present at the firing range could see for themselves the reliability of the radio communications during orbital insertion and in-orbit flight while the spacecraft was in our tracking stations' coverage zone. The landing took place successfully forty-five kilometers southeast of Votkinsk.[13] "Ivan Ivanovich," the dummy, and Zvezdochka (Little Star), the dog, came back to Earth. Gagarin had suggested her name on the eve of the launch. Dogs proved to be "man's best friend" in the sphere of cosmonautics, too. Ordinary little mongrel dogs blazed the trail for humankind's entry into space.

The same failure as had occurred on the preceding launch—the mistimed separation of the cable tower—was also recorded on the descent module of 3KA vehicle no. 2. It overshot the estimated landing point by 660 kilometers. Neither I nor any of the individuals still living who were involved in those historical launches can recall why such serious glitches in the last two unmanned launches were not the subject of a report before the State Commission or even a discussion with Korolev.

On the morning of 29 March 1961, the State Commission under the chairmanship of Konstantin Rudnev listened to Korolev's recommendation for the launch of a Vostok spacecraft with a man on board. Korolev did not inform the State Commission of the cable tower separation failures.[14]

That same evening, after hearing Korolev's recommendation to launch a piloted Vostok spacecraft, a meeting of the Military-Industrial Commission (VPK) took place at the Kremlin.[15] VPK Chairman Ustinov conducted the meeting. He sensed the historic significance of the impending decision and perhaps that is why he asked each chief designer to express his opinion. After receiving assurances about the readiness of each system and the support of the State Committee chairmen, Ustinov drew up the resolution: "To accept the recommendation of the chief designers . . . "[16] Thus, Ustinov should be

---

13. In the Soviet press, this vehicle was publicly known as the *Fifth Korabl-Sputnik*.

14. "State Commissions" were temporary and ad hoc bodies established to conduct testing and eventually certify all new weapons systems in the Soviet defense industry. Their membership was usually composed of leading industrial managers, chief designers (or deputy chief designers), and military service officers. For important space projects, the chairman of a State Commission could be (as was the case with Vostok) an individual with a ministerial rank, such as Rudnev.

15. VPK—*Voyenno-promyshelennaya komissiya*.

16. This document that Chertok cites was declassified in 1991 and published in V. Belyanov et al., "Zvezdnyy reys Yuriya Gagarina" ["The Starry Race of Yuriy Gagarin"], *Izvestiya TsK KPSS* no. 5 (1991): 101–129.

considered the first high-ranking government leader to give the "green light" to launch a man into space.

After the unanimous decision for a piloted launch, the Military-Industrial Commission edited and signed a report to the Central Committee announcing readiness to carry out the world's first flight of a human being into space. Three proposed TASS reports were attached to the report for preliminary approval:

- Successful flight (to be announced right after insertion on orbit)
- Successful landing (to be announced right after landing)
- Emergency landing in the ocean or on foreign territory with a request for those states to assist in the cosmonaut's rescue.

The State Commission was told to remove the emergency destruction system from the piloted spacecraft. The launch was to take place when preparation was completed during the time frame of 10 to 20 April. On 3 April, Khrushchev held a meeting of the Central Committee Presidium.[17] Based on Ustinov's report, the Central Committee Presidium accepted the resolution authorizing the launch of a man into space.

The State Commission, the VPK, and the Central Committee made this historic decision, despite the fact that the ground experimental testing program had not yet been completed for the landing and ejection systems. These tests were conducted at LII. They included ejection from an Il-28 airplane as well as from the descent module at ground level, and dropping the descent module from a height of five meters. All the tests were completed successfully.

The entire day of 4 April at OKB-1 was taken up with organizational headaches concerning matching up airplanes with people, documentation, and cargo for delivery to the firing range. The Central Committee Presidium decision had certainly come as no surprise. Nevertheless, each of the chief designers or their deputies found excuses for staying a day or two longer in Moscow to urgently prepare one more thing, finish up something, and insert spare parts and tools into the flight manifest. Besides spare parts and materials, it became necessary to bring along specialists whose names had not appeared on any lists, without whom "launch would be impossible." I responded to such requests as directed: "Go talk to Sergey Pavlovich personally." In the majority of cases, knowing Korolev's attitude toward this kind of forgetfulness and disorganization, these individuals didn't risk going to him with their requests.

On 4 April, I was supposed to depart from Vnukovo with Mishin and Keldysh. Keldysh was delayed the whole day and our departure was held

---

17. The Soviet Central Committee's Politburo was known as the Presidium from 1952 to 1966.

up until 11:00 p.m. Vnukovo was hit by an unusual April snowstorm. The airplane was covered with such a thick layer of wet snow that they spent an entire hour dousing it with hot water. We took off at midnight and immediately dozed off until we reached Aktyubinsk.

The Sun was shining at *Lastochka* (Swallow), the name given to the firing range airfield. In honor of Keldysh's arrival, firing range chief General Aleksandr Zakharov showed up in person with his entourage to meet us.

Korolev felt it necessary to arrive at the firing range several hours before the cosmonauts. Veteran test pilot Mark Gallay and all the chief designers flew in with him. On 5 April, the entire team of cosmonauts arrived at the firing range accompanied by physicians, motion-picture cameramen, and reporters, with Kamanin in charge. On 6 April, Konstantin Rudnev, the chairman of the State Committee for Defense Technology (GKOT), arrived.[18] He had been appointed State Commission chairman after the death of Nedelin. That same day, Korolev, Keldysh, and Kamanin approved the cosmonaut's mission of a single-orbit flight. They indicated the flight objective and cosmonaut's actions during its normal course and also "in special cases." They decided to announce the mission at a session of the State Commission.

On 8 April, at the session of the State Commission, history's first human spaceflight mission was approved: "Execute a single-orbit flight around Earth at an altitude of 180 to 230 kilometers lasting 1 hour 30 minutes with a landing in a predetermined area. The flight objective is to verify a human being's ability to stay in a specially equipped spacecraft, test the spacecraft equipment in flight, test the spacecraft's communications with the ground, and confirm the reliability of the spacecraft's and cosmonaut's landing systems."

After the open portion of the commission meeting, the "inner circle" remained and confirmed Kamanin's proposal to accept Gagarin for the flight and have Titov as backup. Now it seems laughable, but back in 1961, the State Commission in all seriousness made the decision "not to allow the disclosure of secret information about the firing range and launch vehicle" when publicizing the results of the flight and recording it as a world record. Thus, in 1961 the world did not find out from where Gagarin was launched and what rocket carried him into space.

It seemed that everything, including the weather, inspired confidence in success. But we managed to create difficulties for ourselves in order to heroically overcome them. Three days before Gagarin's launch, on 9 April, we decided to conduct the first launch of the new R-9 ICBM with the unclassified designation "article 8K75." This event cut into preparation for Gagarin's

---

18. GKOT—*Gosudarstvenniy komitet po oboronnoy tekhnike.*

flight and spoiled the subsequent celebration for many of us. The R-9 launch was set for 0500 hours on 9 April. It actually took place at 1215 hours. The missile stood filled with oxygen for 7 extra hours while they looked for errors in the circuit of the on-ground automatics that controlled fuelling. After long and arduous attempts to achieve readiness, the missile lifted off with a speed that we were not accustomed to seeing.

Despite the report that the second stage had shut down prematurely, the first launch of the new ICBM was celebrated at the launch site with a dress parade of all the military and civilian participants. Strategic Rocket Forces Commander-in-Chief Marshal Kirill Moskalenko gave a speech before the assembled ranks and congratulated everyone for their great success. He was followed by Rudnev, while Korolev expressed his gratitude to the testers. He was the only one who mentioned that not everything had gone smoothly. The missile missed the target and we had a lot of work yet to do on it.

Right there on the launch pad, after the dress parade, Korolev informed Moskalenko and Rudnev that he would assign Mishin and me to immediately begin investigating what had caused all the problems that had occurred during preparation for the R-9 launch. Then, calling the two of us aside and smiling like the cat that swallowed the canary, he announced that the following day, 10 April, we should have a "friendly meeting" with him on the bank of the Syr Darya River at building zero.[19] "And bring Leonid [Voskresenskiy] with you," he added.

Rudnev had proposed the meeting on the bank of the Syr Darya. He had persuaded Moskalenko to hold an informal meeting with the future cosmonauts for a chosen few to give them an opportunity to let their hair down "without any agenda." They had even planned for us to go boating!

For this gathering, we took advantage of the open-air veranda built on the bank of the river on the premises of the "marshals' building zero" at Site No. 10. The veranda was designed to protect the highest-ranking military leadership against the baking Sun when they were resting and strolling. For our informal conversation, the veranda, later historically referred to as "Gagarin's gazebo," was set up with tables laid out with modest *zakuski* and various non-alcoholic beverages.[20] A really select group of about 25 individuals were assembled, including six future cosmonauts.

---

19. Building zero was a hotel complex for VIPs on the bank of the Syr Darya River by Site No. 10.

20. *Zakuski* are Russian *hors d'oeuvres*.

From the author's archives.

Officials standing on the terrace of building "zero" at Tyura-Tam (later the Baykonur Cosmodrome) on the eve of Yuriy Gagarin's launch in April 1961. From left to right in the front are cosmonaut Valeriy Bykovskiy, Marshal Kirill Moskalenko (commander-in-chief of the Strategic Rocket Forces), cosmonaut Yuriy Gagarin, Chief Designer Sergey Korolev, and cosmonaut Pavel Popovich.

Senior Lieutenants Gagarin and Titov were seated next to Soviet Field Marshal Moskalenko, State Commission Chairman (and Minister) Rudnev, Chief Designer Korolev, and Chief Theoretician of Cosmonautics Keldysh. I liked that neither of the cosmonauts was particularly timid. Evidently all the previous training procedures had already seasoned them. "Prohibition" did not promote an atmosphere of conviviality around the table. Nevertheless, all the conversations interspersed with mineral water and fruit juice toasts were quite genial compared with the formal reports in the VPK and State Commissions.

Korolev spoke very simply, without pathos: "We have here six cosmonauts, each ready to complete a flight. It has been decided that Gagarin will fly first. Others will fly after him . . . We wish you success, Yuriy Alekseyevich!"

I first listened attentively and evaluated Gagarin when he spoke to the assembled elite of the rocket-space community about the task he had been assigned. He didn't use a lot of fancy words. He was simple, clear, and quite charming. "Yes, you have made the right choice," I thought, recalling the conversations and the long, drawn-out procedures for selecting candidates for the first flight.

Before that meeting, we had had behind-the-scenes debates: Gagarin or Titov? I recall that Ryazanskiy preferred Titov. Voskresenskiy said that Gagarin had a hidden daring streak that we didn't see. Rauschenbach, who

69

had tested the cosmonauts, liked both equally. Feoktistov tried very hard, but could not hide his desire to be in their place. Before our meeting on the bank of the Syr Darya, it seemed to me that both candidates were too young for their imminent worldwide fame.

"But you know," said Isayev, "I remember Bakhchi. We're somehow to blame for what happened to him. I wouldn't have any doubts about such a person. The risk of flying on a Vostok is, if anything, greater than it was on the BI. But my emotions have become so dulled that it's a lot easier for me to live through this here than it was then in the Urals." Of all those gathered at the firing range at that time, only Isayev and I remembered Bakhchivandzhi, whose death on 27 March 1943 had been a terrible blow for us. But back then, the war was going on!

Rudnev, Moskalenko, Kamanin, and Karpov also delivered calm speeches without excessive references to their great responsibility to the Party and the people. Besides Gagarin, Titov and Nelyubov thanked us for our confidence. Korolev's words, "others will fly after him . . . " referred to the candidates seated there. They proved to be prophetic, but not completely. Of the candidates that were present that day on the bank of the Syr Darya, everyone flew except for Nelyubov.[21]

Yes, they were right about Gagarin. It's too bad that on that sunny April day, due to strict security rules, there was only one "classified" movie camera belonging to *Mosnauchfilm* cameraman Volodya Suvorov.[22]

Thirty-four years later I was once again at "Gagarin's gazebo." This time I was with Korolev's daughter and grandsons. There was quite enough photo and video equipment. But the equipment couldn't resurrect the images of those who had been there on 10 April 1961. It was sad to see the now shallow Syr Darya. "Even a canoe would get stuck here," I mused. In 1995, I could not even respond to the simple question: "Besides myself and four cosmonauts (Titov, Popovich, Nikolayev, and Bykovskiy), of those 25 persons who were at the gazebo before the historic flight, who is still alive?" I am updating my memoirs in 2005. Of the four cosmonauts I mentioned in 1995, only two remain alive—Popovich and Bykovskiy.[23]

---

21. Nelyubov died from an apparent suicide on 18 February 1966.

22. *Mosnauchfilm* was the official governmental authority responsible for producing scientific films and documentaries. Suvorov later published a memoir about his experiences. See Vladimir Suvorov and Alexander Sabelnikov, *The First Manned Spaceflight: Russia's Quest for Space* (Commack, NY: Nova Science Publishers, 1997).

23. Titov and Nikolayev passed away on 20 September 2000 and 3 July 2004, respectively.

In those days Kamanin was keeping secret diaries that were published after his death. On 5 April he made the following entry: "So, who will it be—Gagarin or Titov? It is difficult to decide who to send to certain death . . . "[24]

We developers of the launch vehicle and Vostok spacecraft were a lot more optimistic. Moreover, almost all of my friends I talked to during those pre-flight days felt inwardly confident of success. On the evening of 10 April 1961, in a solemn atmosphere in cramped quarters under the blinding lights for the movie cameras, a meeting of the State Commission took place. Many people gathered for this meeting. Everyone spoke clearly, briefly, and solemnly for the film and audio recordings. All the decisions had already been made in a closed session. But even this single news film about the State Commission meeting was not declassified and cleared for screening at open sessions until 10 years later.

On 11 April, all the prescribed launch vehicle and spacecraft tests were conducted at the launch site. Before singing off in the logbook for the operation performed, almost everyone responsible for a system recited the words: "Knock on wood—no glitches!" And indeed, by the morning of 12 April, everything was ready and signed with no glitches.

At T-minus 4 hours, the fueling process began. At T-minus 2 hours, the bus carrying the cosmonauts drove up to the launch pad. The number of people who accompanied Gagarin and hugged him before he took his seat in the elevator was considerably more than had been stipulated somewhere in the specified schedule. Fortunately some, though not many, authentic frames of the newsreel have survived. To a large extent, this was to the credit of the cameramen of the *Mosnauchfilm* studio and, in particular, the aforementioned indefatigable Volodya Suvorov. Now on the anniversary dates at grand gatherings, they show the frames of the send-off and Gagarin being seated in the elevator. Voskresenskiy and Ivanovskiy accompany him to the elevator. Ivanovskiy rides up with Gagarin on the elevator and then helps him get situated in the descent module.

I went down into the bunker and looked around to make sure that those inside were not running into glitches either. In the bunker were the totally focused military operators of the launch vehicle control panels—Pilyugin

---

24. The complete text that Chertok cites from Kamanin's diary reads: "So, who will it be—Gagarin or Titov? I still have few days to make a final decision on this question. It is difficult to decide who to send to certain death and so difficult to decide who among the two or three worthy to make world famous and forever preserve his name in the history of humanity." See N. P. Kamanin, *Skrytyy kosmos: kniga pervaya, 1960–1963 gg.* [*Hidden Space: Book 1, 1960–1963*] (Moscow: Infortekst, 1995), p. 45.

huddling in a corner of the control panel room with his consultants; Moskalenko and Rudnev in the guest room; and Yuriy Karpov, who was monitoring the control panels of the "object."

Yuriy Bykov began the *Zarya* test from the bunker. Ten minutes later, communications with *Kedr* (Cedar)—that's the call sign they gave to Gagarin—had been set up. Before Korolev descended into the bunker, Kamanin and Popovich had communicated with Gagarin from the "guestroom."

I climbed out of the bunker and informed Korolev, who was on the launch pad, that there were no glitches in my systems. He gave me the all-clear to go to IP-1. Right after I arrived at the tracking station, I found out that after Gagarin had been strapped into the spacecraft, an emergency situation had occurred after all: on the control panel in the bunker the display light confirming closure of the spacecraft entry hatch had failed to light up. Ivanovskiy and assembler Morozov quickly reopened and closed the hatch and checked the limit contact. Spanning the years, for lack of any other glitches that required the launch control team's heroism, this episode gradually took on a life of its own; it accumulated dramatic details and entered the oral and written recollections of the launch of the first man into space.

The conversations between *Zarya* and *Kedr* were relayed to IP-1. The last reports from the bunker: "Ignition," "Preliminary," "Main," and "Lift off!" connected all of us to the departing rocket. Gagarin's jaunty exclamation "We're off!" was drowned out by the surging roar of the engines. We raced to Site No. 2. There they were already conversing with Gagarin. The audio was excellent. Bykov was beaming. For the first time, his *Zarya* was talking from space with the voice of a live human being.

"Visibility is excellent! Out the window I see Earth, clouds . . . I see rivers . . . It's beautiful!"

The most agonizing thing that day was waiting for the report of a successful landing. But it was all over very quickly. Everyone who had gathered at the firing range scurried back to their hotels, grabbed their suitcases, piled into

V. I. Morozov, the fitter who helped close the hatch of the *Vostok* spacecraft for Yuriy Gagarin's historic launch.

From the author's archives.

cars, and rushed to the airfield. Before taking a seat in the car waiting for him, and with a very guilty smile, Korolev turned to Mishin, Khomyakov, and me while standing outside of his cabin:

"We can't all leave. You've got to prepare the *Devyatka*."[25]

Avoiding eye contact, Korolev shook hands, and hurried to the refuge of the car. All those who had been selected to attend a reception with Gagarin in Saratov hurried after him to squeeze into cars. We said farewell without letting on that we were jealous of everyone departing for the reception with Gagarin and for jubilant Moscow.

The day after Gagarin's launch, those of us who remained at the firing range through "Korolev's ill will," as Kalashnikov put it, joined in the jubilation of the entire nation, from time to time switching on radios. I consoled my friends with the fact that we were also the "world's first" to have the opportunity to study telemetry recordings of the in-flight behavior of the historic launch vehicle and spacecraft systems. After examining the tapes, we realized that all three launch vehicle stages had operated glitch-free, except for the range radio-control system and the integrating accelerometers that issue the command to shut down the engine of Block A (core stage).

Mikhail Borisenko's radio operators explained that the DC-AC current converter had failed. But Pilyugin's beloved electrolytic integrators in the core block had also distinguished themselves. An error of 0.25 meters per second caused the apogee altitude to increase by 40 kilometers over the design value.[26] If Isayev's TDU had not kicked in, the *Vostok* would have stayed in orbit for 15 to 20 days rather than the estimated 5 to 7. The telemetry specialists headed by Nikolay Golunskiy, who had mastered the processing of information about the behavior of spacecraft systems, used the frenzy of activity to their advantage and also departed.

Gagarin was alive. It had all ended happily in a long multi-link chain of probabilities; and those of us who had stayed behind at the firing range wasted no time doing an in-depth study of the *Vostok* systems' behavior during its return to Earth. I shall cite the main specifications of Gagarin's *Vostok* spacecraft for posterity.

Launch mass	4,725 kilograms
• Descent module	2,460 kilograms
• Instrument aggregate compartment (with TDU)	2,265 kilograms

---

25. *Devyatka* (Niner) was the nickname for the R-9 ICBM.
26. Probably this value should be 25 meters per second instead of 0.25 meters per second.

Dimensions:

- Body length   4.3 meters
- Body diameter (maximum)   2.5 meters
- Descent module diameter   2.3 meters
- Free volume of descent module   1.6 cubic meters

Mass distribution:

- Structure   20 percent
- Heat shield   17 percent
- On-board systems   21.5 percent
- Power supply system   12.5 percent
- TDU (with propellant)   8.5 percent
- Landing system   3.2 percent
- Seat and cosmonaut   7.1 percent
- Gases   1 percent

WE LEARNED THE DETAILS ABOUT THE CELEBRATIONS IN MOSCOW, the rally at Red Square, the reception at the Kremlin, and the enthusiastic responses from around the world from the news reports of Levitan and the BBC! Our hard feelings toward Korolev intensified even more after we found out from the duty officer in Podlipki (during a conversation via high-frequency communications) that the government service had sent an invitation from the Kremlin to Mishin and me at our homes "to attend an evening reception with your spouses."[27]

IN MOSCOW, ON THE EVENING OF 14 APRIL, THERE WAS A GRAND FIREWORKS DISPLAY. Meanwhile, at Tyura-Tam, according to schedule, we launched the latest 8K74 (R-7A) from Site No. 31. Mishin, Ostashev, and I drove out to Site No. 31 figuring that this launch would also be a pretty good salute. The military squad executed the launch completely. Our salute took place, but the next day studying the *Tral* telemetry tapes, we realized that at around 250 seconds into the flight, the pressure had fallen in the gas generator and then in the combustion chambers. For a while, the missile was carried by the vernier engines, and then it fell into a spin. The Emergency Engine Shutdown (AVD) system was triggered: the engines cut out and the payload separated.[28] And this was an armed missile! True, it didn't have a nuclear warhead.

---

27. Despite these invitations, Chertok and Mishin would be unable to attend since they were stationed at Tyura-Tam.

28. AVD—*Avariynoye vyklyucheniye dvigatelya.*

It grieved us when we heard a radio report that on 15 April, a press conference had been held at the House of Scientists in honor of the flight of the first man into space. After Gagarin, Academicians Vasiliy Parin, Yevgeniy Fedorov, Norair Sisakyan, and Aleksandr Nesmeyanov spoke at the press conference. And there was no mention, not a word about the other academicians—the real heroes of this event. Their very presence in the hall even proved to be a nuisance. In the "heat of the moment," had someone failed to think it through? No, this harmful extra layer of security had been thought through and deliberately carried out. The official report made public on 16 April somewhat placated the masses of anonymous heroes:

The CPSU Central Committee and USSR Council of Ministers has deemed it necessary to award orders and medals to the scientists, workers, engineers, and technicians who participated in the production of the Vostok *orbital spacecraft and who supported the world's first successful flight of a Soviet man into space. The appropriate ministries and branches have been instructed to submit the names of those individuals who were involved in the production of the* Vostok *and its flight support so that they might be awarded.*

Academy of Sciences President Aleksandr Nesmeyanov had done nothing against space research, but he had somehow failed to suit Khrushchev and Suslov.[29] On 19 May, a general assembly of the Academy of Sciences accepted his "request" to resign from his high post and Mstislav Keldysh was elected president. On 19 May, the new president opened a general assembly of the Academy of Sciences dedicated entirely to human spaceflight. Everything that he said in his opening address was unconventional, and for the uninitiated, very new. However, when listing all the achievements, only two names were mentioned—Gagarin and Tsiolkovskiy. Academician Anatoliy Blagonravov gave the main report at the assembly. He dwelled on the technical problems of spaceflight—a section that had been prepared at our own OKB-1 and carefully edited by Korolev before it was handed over to the speaker.

It was history's desire that during the launch of the first *Sputnik*, Blagonravov was on a science-related trip to the United States. During Gagarin's flight, Blagonravov was in Italy. At the conclusion of his report, he said: "I witnessed first-hand what elation and admiration the news of the historic flight of our

---

29. Mikhail Andreyevich Suslov (1902–82) was a Politburo member and "chief ideologue" of the Soviet Communist Party during the Brezhnev era.

first cosmonaut caused among the broad masses of the Italian people." You couldn't blame Blagonravov for trying to grab a bit of borrowed glory. He had resisted and did not want to give this report, believing that it dishonored Korolev and the other members of the Council of Chiefs, and the academicians as well. But the Presidium of the Academy, on instructions from the Central Committee, obliged him to do so. The majority of the assembly participants understood fully that if the esteemed academician had been in Italy on the day of Gagarin's

From the author's archives.

Korolev and Gagarin after his 1961 mission.

flight, then he bore no responsibility for this particular success of our science and technology.

On 26 March 1962, the second USSR cosmonaut after Gagarin, Major German Titov, sent a message to the CPSU Central Committee in which he wrote:

> *The flight of Major Gagarin was an outstanding feat of the Soviet people, a victory for leading-edge Soviet science and technology in the conquest of space. This historic achievement was a contribution to the treasury of worldwide science and technology. 12 April should be celebrated as special day for space research and conquest.*
>
> *It would be appropriate:*
>
> *To observe the date of 12 April annually as "Cosmonautics Day."*
>
> *To propose to the United Nations on behalf of the government of the Soviet Union that 12 April be established as International Cosmonautics Day.*
>
> *People all over the world have shown tremendous interest in the flights of Soviet cosmonauts. The latter, in response to official invitations, have visited 26 foreign countries in Europe, Asia, Africa, and America. I presume that this proposal will be supported by many nations, especially since the International Aviation Federation has confirmed the flights of the Soviet cosmonauts as world records.*
>
> *USSR pilot-cosmonaut—G. Titov*
>
> *26 March 1962.*[30]

---

30. This letter reproduced here by Chertok was first published in the journal *Novosti kosmonavtiki* [*News of Cosmonautics*] no. 4 (2005): 1.

Titov's initiative went into effect. A decree of the Presidium of the USSR Supreme Soviet dated 9 April 1962 established 12 April as Cosmonautics Day in commemoration of Gagarin's flight. That day, in the hall of the Kremlin Palace of Congresses, a function was held celebrating the first anniversary of the flight. Gagarin gave a speech. Not a single chief designer was in the Presidium! Not a single one of the actual participants in the production of the rocket and spacecraft!

After his flight, Gagarin spent almost an entire year traveling all over the world. The burden of fame—physical, emotional, and spiritual—was as much as a human being could endure. But Yuriy withstood it all honorably. Throughout all of Gagarin's visits to dozens of countries and hundreds of cities, there was never an occasion to reproach the first human being to look down on Earth from space. During the first postflight year, Gagarin visited Czechoslavakia, Finland, England, Iceland, Brazil, Canada, Hungary, France, India, Afghanistan, Cuba, and Ceylon. And that's not counting the receptions he attended and trips he made around his own country. For the glory of his Motherland, Gagarin literally worked on Earth to the point of exhaustion.

When he finally returned to work at the Cosmonaut Training Center (TsPK) after Titov's one-day flight, Gagarin became actively involved in training his comrades for flights.[31] He began studying the Soyuz design and training for flights on the future spacecraft. In 1963, Gagarin was deservedly appointed TsPK deputy chief for cosmonaut training. Before each piloted launch, Gagarin flew to the firing range to see his comrades off on their flights. He conducted communications via *Zarya* after the spacecraft entered orbit. He kept his call sign, *Kedr*, on the ground too.

From the author's archives.

Yuriy Gagarin with Nina Ivanovna Koroleva and Sergey Pavlovich Korolev in May 1961 at the Sochi resort.

The Central Committee and Council of Ministers directive calling for giving awards to those who participated in preparing and executing the first piloted spaceflight was fulfilled two months after Gagarin's flight. Considering the multi-step selection process conducted "in the ministries and branches" of hundreds of individuals deserving of awards, no one was offended by the amount

---

31. TsPK—*Tsentr podgotovki kosmonavtov.*

of time it took. But in the final statement about the decree dated 17 June 1961, the names of the decorated individuals were not mentioned. Among the individuals recognized, I was awarded the title Hero of Socialist Labor.

Another picture from the Sochi trip: Yuriy Gagarin (left), his wife, Valentina, Nina Koroleva, and Sergey Korolev.

From the author's archives.

However, until the end of their days, Korolev and all the other chief designers could not compete with Gagarin in terms of the number of different awards. He received the highest governmental awards in almost every country that he visited. According to our unwritten laws of the Cold War, no scientist involved in rocket-space technology, no matter how great his merits, was supposed to be known abroad and had no claim to fame in his own country. Academician Petr Kapitsa advocated awarding the Nobel Prize to the scientist who set up the experiment for the creation of the first *Sputnik*. The scientist who realized mankind's dream of flying into space was all the more deserving of the Nobel Prize. But Kapitsa's appeal remained "the voice of him that crieth in the wilderness."[32]

Gagarin's flight provoked a great a shock in the U.S. with the report of the first *Sputnik* on 4 October 1957. Astronaut Alan B. Shepard, Jr. completed a suborbital flight in a Mercury capsule on 5 May 1961, after eight test launches of the Redstone rocket on ballistic trajectories.[33] The first American to complete a real spaceflight was Astronaut John H. Glenn, Jr. He completed a flight with three orbits in an updated Mercury capsule launched on an Atlas D launch vehicle on 20 February 1962.

---

32. Petr Leonidovich Kapitsa (1894–1984), one of the Soviet Union's most famous physicists, received the Nobel Prize for physics in 1978 for his inventions and discoveries in the area of low-temperature physics.

33. Prior to Shepard's mission, there were three test launches (MR-1, MR-2, and MR-BD) of the Redstone in 1960 and 1961.

President John F. Kennedy criticized his predecessor, Dwight D. Eisenhower, for failing to appreciate the space achievements, and in particular, piloted flights. On 25 May 1961, Kennedy gave his famous address to Congress, announcing to the people that the U.S. would send a man to the Moon before the end of the decade. This was a long-term program, during the fulfillment of which the United States would have to win the battle for superiority in space. Yuriy Gagarin's flight was the strongest stimulus for the development of the American piloted programs, which were crowned by the lunar landing expeditions. I contend that if Gagarin's flight on 12 April 1961 had ended in failure, U.S. Astronaut Neil A. Armstrong would not have set foot on the Moon on 20 July 1969.

I AM REVISING THIS BOOK FOR ENGLISH-SPEAKING READERS IN APRIL 2005, 44 years after Gagarin's flight. From this length of time, it is easier to assess the enormous importance of this achievement in the history of civilization.

At the beginning of the Cold War, which historians place in 1946, the USSR and U.S. were drawn into a grueling nuclear arms race. Rather than diminish it, the flight of the first *Sputnik* stimulated the competition to produce strategic missiles. By the early 1960s, nuclear parity did not yet exist between the USSR and the West. The nuclear missile shield became the basis of the Soviet Union's security later on.

Gagarin's flight served to stimulate the beginning of competition on another plane in a field that objectively led to the weakening of the positions of Cold War apologists. The historic paradox of cosmonautics was that the achievements of missile technology stimulated confrontation between the two superpowers, while the successes of the piloted space programs based on these achievements promoted rapprochement, cooperation, and a desire to exchange ideas and experience. The flights of our cosmonauts and American astronauts diverted a great deal of resources from weapons technology and did not contribute to meeting military challenges. Each new piloted flight around our shared planet objectively served as a call to unite and to reduce confrontation.

Paying tribute to the heroism of Gagarin, of all subsequent cosmonauts, and the American astronauts, and to the self-sacrificing work of scientists and all the developers of rocket-space technology, one must not forget the role of the two leaders of the opposing Cold War superpowers—Nikita S. Khrushchev and John F. Kennedy. Both of them displayed extraordinary courage and initiative, using their authority to realize the piloted space programs. For people throughout the world, the flights of Vostoks, Voskhods, Soyuzes, Salyuts, the *Mir* space station, Mercuries, Geminis, Apollos, and Space Shuttles relieved the fear of potential nuclear destruction. These flights objectively resulted in the merger of scientific ideas in cosmonautics—humankind's grand new field. But before that was achieved, cosmonautics endured a grueling Cold War.

IN 1995, WHILE WORKING ON MY MEMOIRS, I approached Yuriy Karpov, who had developed the Vostok's main electrical system, and asked him to "tie up some loose ends" and write an article for open publication. On 12 April 1961, he had been in the bunker. He heard the reports and then participated in the postflight analysis. To study the history of the first flight, Karpov recruited a group of colleagues comprising V. G. Berkut, V. V. Kalinin, V. F. Kirillov, V. G. Kirsanov, V. V. Kukovnikov, V. I. Staroverov, G. N. Formin, V. K. Shelev, and A. A. Shchukin. On 12 April 2001, the newspaper *Krasnaya Zvezda* (Red Star) published Karpov's article "There Was No Emergency Situation during Descent." Soon, an article by Igor Lisov appeared in the journal *Novosti Kosmonavtiki* (News of Cosmonautics) with the demand "We Need the Whole Truth."[34] By that time Yuriy Karpov had died. German Formin, who had worked with Karpov for many years, agreed to continue to search for the truth. I discussed various documents and publications with him on a regular basis. The result of almost a year of historical research, G. Formin's article "The Truth about Yuriy Gagarin's Return" was published in *Novosti Kosmonavtiki* in 2002.[35]

Forty-one years after the flight, it was a specialist rather than a journalist who penned an open publication summarizing the documented information and recollections of those involved in the events, which provide a credible explanation of the causes for the off-nominal behavior of the *Vostok* spacecraft during Yuriy Gagarin's return. I dare say that if Gagarin were alive, he would find the content of G. Formin's publication interesting and new. I have no answer for myself or for the readers as to why, after Gagarin's flight, Korolev did not put forth this matter for discussion before the review team or the State Commission to analyze the causes of the systems' off-nominal behavior. After all, it would have been an occasion for boasting: "Look—despite the failures in the orbital return systems, the safety margin was enough that everything ended well!" So what really happened?

The launch took place at 0906 hours 59.7 seconds. In accordance with the logic that we had loaded into the on-board automatic control system, the nominal descent cycle of Gagarin's one-orbit flight was to begin automatically, triggered by a contact when the spacecraft separated from the launch vehicle.

---

34. I. Lisov, "Polet Gagarina: nuzhna vsya pravda!" ["Gagarin's Flight: We Need the Whole Truth!"], *Novosti kosmonavtiki* [*News of Cosmonautics*] no. 6 (2001): 10.

35. G. Fomin, "Pravda o vozvrashchenii Yuriya Gagarina" ["The Truth on the Return of Yuriy Gagarin"], *Novosti kosmonavtiki* [*News of Cosmonautics*] no. 4 (2002): 2–4.

When this occurred, the *Granit* sequencer activated.[36] The spacecraft separated from the launch vehicle at 0918 hours 7 seconds. According to the timeline loaded into the sequencer, 39 minutes into the flight the automatic attitude control system (ASO) would switch on—this was the first active descent preparation command.[37] Sixty-four minutes into the flight, power was switched on to the gyros and angular rate sensors. After 70 minutes, an ASO test was activated. During a period of 1 minute, this test identified the presence of the "ASO readiness" flag authorizing the firing of the TDU. Seventy-one minutes into the flight, having received authorization, the TDU began to fire and the following occurred: the gyroscopes of the spacecraft stabilization system were uncaged when the TDU fired, and the integrator that measured apparent velocity—and was supposed to issue a command to shut down the TDU—was uncaged. When a rate of 136 meters per second was reached, a circuit switched on, permitting the output of a command to separate the compartments from the temperature sensors.

When the TDU is activated, the spherical pressurization tank begins to pressurize the fuel and oxidizer tanks through the bladders contained in them. These bladders force the fluid toward the intake ports under zero-gravity conditions. The fuel and oxidizer, forced by the bladders out of the tanks, flow into the pump inlets of the turbopump assembly, which is started up by a special powder charge. The fuel and oxidizer flow under pressure simultaneously into the combustion chamber and gas generator where the components self-ignite. The gas generator spins the turbine with its hot gas and the pumps feed propellant into the combustion chamber to build the engine up to a stable operating mode.

After engine build-up, acceleration develops, forcing the fuel and oxidizer in their tanks toward the intake ports. Tank pressurization is maintained as gas is fed directly to the surface of the propellant components in the tanks. To this end, pyrovalves go off, feeding pressure directly into the tanks, bypassing the bladders, which are no longer needed. At the sequencer 71-minute marker (TDU firing), the check valve that enabled the preliminary purging of the combustion chamber was supposed to close. However, the valve failed to close or didn't close completely. Part of the fuel flowing from the turbopump assembly (TNA) to the combustion chamber flowed off-nominally through this unclosed check valve into the bladder in the fuel tank.[38] Having entered

---

36. The literal abbreviation for timer in Russian parlance is PVU—*Programmno-vremennoye ustroystvo* (Programmed-Timing Device).

37. ASO—*Avtomaticheskaya sistema orientatsii*.

38. TNA—*Turbonasosnyy agregat*.

the bladder, the fuel could not subsequently be fed into the combustion chamber. The off-design loss of fuel caused the engine to stop operating 1 second before the integrator generated the primary command for nominal engine shutdown.

A ship in the Gulf of Guinea received the telemetry. The telemetry recordings were meticulously processed two months later. The off-nominal spontaneous engine stoppage (instead of nominal shutdown) led to a loss of burn intensity of approximately 4 meters per second (132 meters per second instead of 136 meters per second).

The Navy men identified the execution of the braking burn and reported it immediately. Presumably, no one attached significance right away to the fact that there had been no report that the integrator had issued the primary command. The TDU had operated for the designated time, and that was great!

Because the braking burn did not reach full intensity, there was no nominal execution of the primary command. As a result, the *Granit* sequencer did not launch cycle No. 6, which generates subsequent commands. The spacecraft compartments were supposed to separate after 10 seconds. Gagarin waited for it, but after 10 seconds separation had not occurred.

After the TDU shut down, inexplicable perturbation occurred, causing the spacecraft to spin about all three axes at rates up to 30 rotations per second. Gagarin reported this to the State Commission and later wrote the following in a report:

As soon as the TDU shut down, there was a sudden jerk. The spacecraft began to spin about its axes very rapidly. The rate of rotation was around 30 degrees per second, at least. I was waiting for separation. Separation didn't happen. I knew that per design separation was supposed to occur 10 to 12 seconds after TDU shutdown . . .

Separation occurred at 10 hours 35 minutes rather than 10 hours 25 minutes, as I had expected, i.e., approximately 10 minutes after TDU shutdown.

What was the source of the disturbing torque that caused the spacecraft to spin so rapidly? Today the explanation is simple, but it came only after studying the "Report Based on the Analysis of the Operation of TDU SP5-4 on Object 3KA No. 3," which had been stored all these years in the secret archives of the A.M. Isayev Khimmash Design Bureau.[39] The engine shut down because the fuel supply to the combustion chamber ended prematurely.

---

39. Isayev's design bureau, which was known as OKB-2 at the time, developed the reentry engine for the Vostok.

There was no nominal electrical command for engine shutdown. Therefore, all of the propellant component feed lines remained open. Pressurizing gas and oxidizer under around 60 atmospheres of pressure continued to pour into the combustion chamber and vernier nozzles, generating powerful perturbation about all the axes. Ten minutes after the engine shut down, the spacecraft reached the upper boundary of the atmosphere and its surface began to heat up. This actuated the heat sensors, which also gave a command to separate the descent module from the instrument aggregate compartment. After that, everything proceeded normally.

Due to the weak braking burn, Gagarin overshot the landing. At 1048 hours, the surveillance radar at the Engels airfield control center identified the target in a southwesterly direction at an altitude of 8 kilometers and a range of 33 kilometers. Radar tracked the target to the ground.

After a successful ejection, Gagarin landed not far from the village of Smelovka in the Engels region of Saratov Oblast at 1100 hours. The first people to meet Gagarin on the ground were Anna Akimovna Takhtarova and her granddaughter Rita. They were followed by machine operators who came running from the field camp of the Shevchenko collective farm in the village of Uzmorye. Major Akhmed Gasiyev arrived on a ZiL-151 tractor. He delivered the cosmonaut to the missile battalion in the village of Podgorye. After Gagarin had reported to the Air Defense Troops battalion commander, he was taken at his request to the spacecraft from which he had ejected.

The report of his landing had already spread through all the services. The search group's Mi-4 helicopter found Gagarin and took him to Engels and then to the general command residence in Kuybyshev. Korolev and the entire State Commission, along with their elated escorts and authorized correspondents, flew to Kuybyshev to meet with Gagarin and debrief him.

On the evening of 13 April, Korolev and the State Commission departed for Moscow. Gagarin rested, and on 14 April he flew out separately for a red carpet welcome. An honorary escort of fighter planes accompanied his airplane during the approach to the capital. Gagarin's march from the airplane down the red carpet with one of his shoes untied was the subject of one of the few openly published photos and movie and television footage of the historic event.

Chapter 4
The Cuban Missile Crisis . . . and Mars

The launches of Soviet cosmonauts and then U.S. astronauts into space could have become an important vehicle for the rapprochement of the two superpowers. The magnitude of these events for all of mankind was so great that there was every reason to end the harsh conditions of the Cold War and make the transition toward cooperation in space.

The Cold War demanded a cancellation grueling arms race—no time for a breather, no cessation of hostilities. Whatever unflattering statements were written and said about Khrushchev in our time, he was one of the few leaders of that time who understood and strove to de-escalate this punishing race. At one time, it seemed that he had found mutual understanding with President Eisenhower. To do this, Khrushchev had taken full advantage of our lunar successes. America's real leaders didn't like this and, as everyone knows, the thaw that had begun froze again in May 1960.[1]

Gagarin's flight had provided one more opportunity for rapprochement under John F. Kennedy, the new young president of the United States. Our nation's top political leaders tried to launch a campaign to ease the tension, relying on humanity's genuine, whole-hearted admiration of this event.

All of America's radio stations interrupted regularly-scheduled programs to broadcast news of the "amazing new triumph of the Russians in their competition

From the author's archives.

Yuriy Gagarin in Cuba as a guest of Fidel Castro.

---

1. This is a reference to the shootdown over Soviet territory on 1 May 1960 of the American U-2 spy plane piloted by Francis Gary Powers.

with the U.S. in space." But this didn't last long. Three days later, alarming reports from Cuba began to fill the front pages of newspapers. American newspapers shouted about the communist threat of Fidel Castro's regime. The front pages of our newspapers printed Gagarin's pre-flight declaration: "I dedicate this flight to the communist peoples" and such headlines as: "New crimes of imperialism. Armed intervention against Cuba under way" and "We're with you, Cuba!" Three days later the headlines read: "Counterrevolution crushed" and "A lesson to the warmongers."[2]

Many of my comrades and I were contemporaries of and active participants in two simultaneous efforts in human history: the peaceful space effort, designed in advance for a happy ending, and the nuclear missile effort with an unpredictable finale.

During the Cuban Missile Crisis, for the first time since World War II, the world was just 24 hours away from the possible beginning of World War III. Missile weaponry played the starring role in this international crisis and generated the crisis itself. Therefore, professional historians refer to it not simply as the Cuban or Caribbean crisis, but as a missile crisis.[3] I am trying to put the events in chronological order, taking into consideration that many people have forgotten the individual facts; younger generations don't have the time to take interest in primary source history; and much of what actually took place was not declassified until the late 1980s.

On 8 November 1960, John Fitzgerald Kennedy won the presidential election; and on 20 January 1961, he received the keys to the White House. Based on his first speeches and statements to Congress, there was hope that Kennedy would become the new Roosevelt. We genuinely wanted a relaxation of the tensions that had occurred after the U-2 incident and dreamed of business contacts with U.S. space scientists. After Kennedy's famous address to Congress on 25 May about preparing for a mission to the Moon, even Korolev once let it slip: "It would be nice to fly across the ocean and have a look at how they're planning to do this." Keldysh, who had become president of the Academy of Sciences, also let it be known that he was continuing his efforts to set up scientific contacts between the real creators of piloted spacecraft.[4]

---

2. This is a reference to the disastrous Bay of Pigs invasion of 15–19 April 1961, when U.S.-supported Cuban exiles attempted to invade Cuba and overthrow the regime of Fidel Castro. The failure of the invasion significantly worsened American-Cuban relations and embarassed the Kennedy Administration.

3. What Westerners call the Cuban Missile Crisis, Russians have always known as "The Caribbean Crisis." In this text, for ease of reading, we use the former rather than the latter.

4. Keldysh became president of the USSR Academy of Sciences in May 1961.

These hopes were very quickly dashed. In a speech delivered to the Senate in June 1961, Kennedy assigned this mission: "Make the United States' capability to deliver a retaliatory nuclear strike invulnerable. To do this, upgrade the missile fleet and strengthen military bases."[5]

U.S. military men, differing on the details, openly declared doctrines for drastically stepping up the nuclear arms race. Admirals demanded that the construction of Polaris missile-carrying nuclear submarines be speeded up. The Air Force command was interested in producing hundreds of Minuteman ICBMs to deliver a "preventive strike" from underground launch sites.

To catch up, two factories—Factory No.1 in Kuybyshev and Factory No. 586 in Dnepropetrovsk—worked three shifts to produce nuclear warhead-carrying intercontinental missiles—the R-7A, R-9, and R-16. By late 1962, many missiles had already been manufactured. But, as before, there were just four R-7A launch pads. The R-9 and R-16 had not yet been optimized for launches from silos. The first dozens of these missiles were immediately put on duty at ground positions that weren't protected against nuclear strike.

In the U.S., the Army, and Air Force were developing ballistic missiles separately. The first Redstone short-range combat missile was developed in the U.S. under the supervision of Wernher von Braun at the Redstone Arsenal. Almost all of the German rocket specialists brought out of Germany in 1945 were concentrated there. The medium-range Jupiter rocket was also developed there in 1956 with the direct participation of the Germans.

The Jupiter rocket went into service in the summer of 1958. The Germans remained loyal to liquid oxygen, but they replaced the fuel—ethyl alcohol—with kerosene. The Jupiter had an autonomous inertial guidance and control system and carried a thermonuclear warhead with a one-megaton yield. Squadrons of Jupiter strategic missiles were deployed in Turkey and Italy. Their maximum firing range of 3,180 kilometers enabled them to be aimed at targets all over Ukraine, the Caucasus, and southern and central Russia.[6]

---

5. The passage Chertok cites here appears to be a somewhat misreported passage from Kennedy's Special Message to Congress on the Defense Budget from 28 March 1961 in which he called for the largest and most rapid defense buildup in U.S. peacetime history. Kennedy argued that: "A retaliatory capacity based on adequate numbers of [weapons from hidden, moving, or invulnerable bases] would deter any aggressor from launching or even threatening an attack—an attack he knew could not find or destroy enough of our force to prevent his own destruction."

6. The effective range of the Jupiter intermediate range ballistic missile (IRBM) was about 2,415 kilometers.

Work was going less successfully at the Douglas Aircraft Corporation, from which the Air Force had ordered a medium-range missile comparable to the Jupiter rocket. This missile was given the name "Thor." Despite a series of unsuccessful launches, by all measures the Thor missile had caught up to the Jupiter rocket by late 1958, and it was put into service in the U.S. Air Force. The Thor missile was also an oxygen-kerosene missile with an autonomous guidance and control system. The missile's nosecone contained a thermonuclear warhead with a 1.5-megaton yield.

By 1962, the U.S. Army and Air Force stationed more than 100 Jupiter and Thor missiles in Turkey and Europe capable of delivering a nuclear strike against targets in the Soviet Union with a total yield of up to 125 megatons. The U.S. had 10 times more missiles on submarines than we did. I did not find precise data in official publications on the actual ratio of nuclear warheads that the two sides were capable of raining down on each other in 1962. The following figures are cited in memoirs: the U.S. had 5,000 units of nuclear weaponry supported by delivery systems (to strike the territory of the USSR) against the USSR's 300 units.[7] The U.S. arsenal included nuclear warheads that could be delivered by intercontinental missiles, heavy bombers, nuclear submarines, and intermediate-range missiles based on the territories of America's NATO allies.

Unlike Nedelin, the military leaders of that time almost never consulted on strategic military matters of the nuclear missile era with Korolev and the other chiefs. Once, after returning from some meeting at the Ministry of Defense, S.P. confided in Bushuyev and me (we had been waiting for him in his office): "They just won't accept that their time has passed. There are young, on-the-ball officers there. But they stand in their way and don't listen to them." Our military neighbors from NII-4 who fell into the "young" category also complained that no one "at the top" wanted to listen to them.

One of these "young" people was Nikolay Smirnitskiy. I had sat in an armored vehicle with Captain Smirnitskiy in 1947 before the first firing rig test and during the first launches at Kapustin Yar. In 1967, Smirnitskiy was appointed chief of the Main Directorate of Missile Armaments (GURVO) and simultaneously advanced to the post of deputy commander-in-chief of the Strategic Rocket Forces, but in 1975, having become a lieutenant general,

---

7. L. N. Nezhinskiy, ed., *Sovetskaya vneshnyaya politika v gody 'kholodnoy voyny' (1945–1985): novoye prochteniye* [*Soviet Foreign Policy during the Cold War Years (1945–1985): A New Reading*] (Moscow: Mezhdunarodnyye otnosheniya, 1995).

he was sent into retirement.[8] In 1962, he had served in GURVO at the rank of colonel. Our paths rarely crossed. Once he very anxiously told me: "You have no idea how difficult this is. The ministers and commanders-in-chief have different views. Each of the chiefs—Korolev, Yangel, Chelomey—has his own point of view. We're in a terrible hurry, and we still don't have a unified concept. Everyone wants to go to Khrushchev personally and convince him that they're right. But, after all, developing a missile war strategy isn't his job. Everything is very complicated. Back then, in the armored vehicle, it was a lot more clear-cut for us."

Later I heard some details about the "missile crisis" from Smirnitskiy. He was actively involved in preparing the deployment of the missile sites in Cuba, and he knew what had been hidden from our people for many years despite open publications abroad.

To continue with the chronicle of events, on 17 April 1961, Cuban counterrevolutionary squads and mercenaries, with support from the U.S. Navy and Air Force, invaded Cuba at La Playa Girón (the Bay of Pigs). The Cuban army, using the experience gained from our advisers and using our weapons, completely defeated the invading troops. The Vienna Summit between Khrushchev and Kennedy took place on 3 and 4 June.[9] Despite two days of negotiations, this meeting was not a turning point in Soviet-American relations.

Central Intelligence Agency (CIA) Director Allen W. Dulles, Secretary of Defense Robert S. MacNamara, and Vice President Lyndon B. Johnson, hard-nosed proponents of the Cold War, pressured Kennedy and demanded a harsher policy toward the Soviet Union. While paying us a visit, Ustinov, who was VPK chairman at that time, said that after the summit Khrushchev had said of Kennedy, "That pretty boy is not to be envied." Events in Europe also poured oil into the fire of the Cold War, in particular the building of the Berlin Wall in August 1961.

At the Pentagon, the "Cuba Project" was developed and then approved on 20 February 1962. It specified October 1962 as the deadline for Castro's overthrow.[10] The U.S. Congress passed a resolution granting the president the right to undertake military action against Cuba if the action was required "to protect U.S. interests."

---

8. GURVO—*Glavnoye upravleniye raketnogo vooruzheniya*. GURVO was responsible for acquisition and certification of missiles for the Strategic Rocket Forces.

9. This was the first meeting between the two superpower leaders.

10. The "Cuba Project" (or "Operation Mongoose") was initiated by Kennedy in November 1961. The practical dimensions were approved in February 1962.

In April, June, and July 1962, we inserted three Zenit spy satellites into space.[11] The results obtained confirmed the exceptional value of this type of intelligence-gathering tool. After studying the processed film returned from space at the General Staff's Main Intelligence Directorate (GRU) center, I was quite gratified knowing that I had been involved in developing such an effective surveillance and intelligence gathering system.[12] True, the GRU officers complained about the quality of domestically-produced photographic film: "If we had American Kodak film, we'd have seen much more in there."

There, for the first time, I heard someone "under the strictest secrecy" allude to preparation for the shipment of R-12 missiles to Cuba. The GRU received the assignment to verify whether the ground launch positions of these missiles could be detected and identified. They determined that if aerial photoreconnaissance photographs were obtained, then our missiles would certainly be detected. Although possible, it would be more difficult to do this from a satellite.

At Castro's request, in response to the threat of American invasion, Khrushchev consented to establish groups of Soviet armed forces in Cuba. But the Americans were not frightened by conventional forces. To counter a real military threat, Khrushchev made an exceptionally bold decision: bring nuclear missiles right to the U.S. border. Neither the minister of defense, the elderly Marshal Rodion Malinovskiy, nor the members of the Central Committee Presidium would have resolved to take such a risky step. The final decision was made after General Staff Chief Marshal Sergey Biryuzov made a reconnaissance inspection trip to Cuba heading up a group of military missile specialists led by Smirnitskiy. The 43rd Division of the Strategic Rocket Forces was supposed to be the main group of combat personnel of the Soviet armed forces in Cuba. The division comprised three regiments armed with R-12 missiles (24 launchers), and two regiments armed with R-14 missiles (16 launchers).

For the first time, only missiles from Yangel's Dnepropetrovsk KB were involved in the balance between war and peace. Did Yangel know about this? He knew and was proud of the fact that not only his still "green" R-16, but also medium-range missiles designated for Europe were capable of intimidating the Americans. Afterwards, military missile men reproached those of us who worked on Korolev's projects, saying: "See, Yangel is working for us, while Korolev is working for TASS."

---

11. There were three attempts to launch Zenit-2 spy satellites into orbit: 26 April (became *Kosmos-4*), 1 June (failed to reach orbit), and 28 July (became *Kosmos-7*).

12. The GRU—*Glavnoye razvedyvatelnogo upravleniye*—was the USSR's military intelligence-gathering organization.

If a missile division launched all of its missiles first (a second launch wouldn't be possible), the division would be capable of destroying at least 40 vital strategic military targets over almost the entire U.S. territory (except Alaska). The total nuclear potential of an entire division in the first and only launch, if each missile has a launch site and reaches its target, could reach—depending on the type of warhead—up to 70 megatons.

Two anti-aircraft missile divisions and a fighter aviation regiment provided cover to protect the missiles from air strikes. Four motorized rifle regiments were supposed to guard our missiles against a U.S. ground attack in the event that the U.S. invaded the island. In addition, boats armed with salvo-firing missile systems (updated *Katyushas*) and Il-28 bombers were designated to combat amphibious assault landings.

Preparing the equipment and the contingent of troops and transferring it all to Cuba by ship under the pretense of transporting fuel and other peaceful cargo was an enormous operation. Loading, camouflaging, and stowing the missiles, nuclear warheads, and airplanes on ships and then on shore was a particularly dicey operation. One must take into consideration that all the armaments required sophisticated support services, fuel supply, communications, and a secluded area to house all the personnel. The first Soviet troop combat units headed by Army General I. A. Pliyev arrived in Cuba in July and early August. The thoroughly camouflaged deployment of the nuclear warheads began in mid-August.

Operation Anadyr, to prepare and ship the nuclear missile expedition to "Freedom Island" (what we called Cuba at the time), was conducted so secretly that none of us who associated with the rocket-space elite suspected for what purpose the missiles developed by our friends in Dnepropetrovsk were being prepared. In September, the R-12 missiles and Il-28 airplanes began arriving in Cuba.

Now declassified archives have shown that by 20 October, according to the Pentagon's plan, U.S. strategic forces were supposed to be fully prepared for nuclear war. This information provided to our top political leadership apparently served as the reason for the publication of a TASS statement on 11 September:

The Soviet government considers it its duty to display vigilance under the circumstances that have developed and to instruct the USSR Ministry of Defense and the Soviet Army Command to take all measures to bring our armed forced into the highest state of combat readiness.[13]

---

13. *Pravda*, 12 September 1962.

At the same time, the Soviet government called on the government of the United States to halt its unbridled anti-Cuban propaganda and to restore diplomatic relations with Fidel Castro. Only after the TASS report did we begin to fear that the military-diplomatic games might somehow interfere with the upcoming Mars launches in October.

The first R-12 missile was prepared in Cuba for fueling and to be mated with the nuclear warhead on 4 October 1962. Before 10 October, another 10 missiles were ready for installation on launch tables; by 20 October, there were 20 missiles. Some sources mention that the missiles installed in Cuba were fully prepared for launch. What do historians of the Cuban Missile Crisis mean by that? I heard from Smirnitskiy that nuclear warheads hadn't been mated with a single missile. But if that was the case, then not a single missile was actually ready for launch. I have no intention of trying to clear up that contradiction. That is a matter for military historians— let them hunt for the truth in declassified archives or interview still-living witnesses. In naming the readiness dates, I have only used publications that remain undisputed.

It's no surprise that for more than 30 years, we did not know exactly what happened in Cuba. There's something else that's amazing. The remarkab.. U.S. intelligence services did not figure out in time that nuclear missiles had been delivered to Cuba.

Despite active air defense, American U-2 airplanes relentlessly penetrated Cuban air space and intensely photographed the enemy island. On 14 October, a U-2 returned to base after a routine flight. When they processed the photos, the U.S. intelligence officers were shocked. Soviet missiles were discovered! President Kennedy was informed on 16 October, after thorough examination of the photos. He hardly expected to find Soviet missiles armed with nuclear warheads right under his nose! Khrushchev had outsmarted him.

To a certain extent, history had repeated itself. The Germans in Peenemünde could not conceal the A4 missiles during their launch preparation and they worked in the open. Nevertheless, British photoreconnaissance detected them with great difficulty, and not until a year after their launches had begun.

Not a single Cuban was allowed to operate our nuclear missiles. The launch of just one such missile from Cuban territory would not amount to a Cuban attack on the United States, but a Soviet attack. Cuba had been converted into a Soviet nuclear missile base! The deployment of intermediate-range missiles on Cuban soil compensated for the dearth of intercontinental missile launch pads on Soviet territory. American historians have attested that, according to the final data of aerial reconnaissance, 24 missile launch sites were detected in Cuba.

The U.S. Joint Chiefs of Staff proposed that a series of massive air raids on Cuba be prepared and executed immediately. Kennedy found the inner strength to withstand this pressure and reject this proposal. Had such a plan been executed, World War III would have begun the next day. Some historians believe that by confronting the military and more aggressive members of his cabinet, Kennedy signed his own death sentence.

At this time, in addition to the R-12 missiles already delivered to Cuba, our transport ships began moving toward Cuba with camouflaged R-14 missiles on their decks. A squadron of submarines was ordered to escort these missile transports.

On 15 October, after a brief stay at the OKB, I flew to the firing range as part of a large team whose primary task was to launch the 2MV automatic interplanetary station (AMS) toward Mars.

On 17 October the fifth Zenit-2 spy satellite was successfully launched.[14] After a tragic beginning, which I shall describe later, we launched Zenits one after the other.[15] The photographs had ceased to be exotic, and there was a continuous rush job under way at GRU headquarters to interpret them. NII-10's new command radio link (KRL) was being tested for the first time on this fifth Zenit. We called it "Petelin's KRL"—using the surname of the institute's director. At one time Kalmykov had been in charge there.[16] Chief Designer Armen Mnatsakanyan's KRL was given over primarily for piloted vehicles because the military decided that Petelin's link had a superior message-bearing capacity and was more noise resistant.

On 17 October, our intelligence services reported the latest failure of a Minuteman ICBM during launch from a silo at Cape Canaveral.[17] On 18 October, a report came in that the Americans had conducted their

---

14. This spacecraft was disguised under the open designation *Kosmos-10*.
15. See Chapter 12 for the development of the Zenit.
16. Mikhail Pavlovich Petelin (1906–1990) was directior of NII-10 at the time. Valeriy Dmitriyevich Kalmykov (1908–1974) had been director in the late 1940s before moving on to become minister of the radio engineering industry.
17. A Minuteman ICBM was launched on 17 October 1962 on an R&D flight from Cape Canaveral but failed to reach its target.

30th atmospheric nuclear test over Johnston Island in the Pacific Ocean.[18] On 19 October, apparently to intimidate us, the Americans launched an Atlas ICBM carrying a nuclear warhead and detonated it in space at an altitude of around 150 kilometers.[19]

On 21 October, a State Commission headed by Leonid Smirnov departed for the firing range. Keldysh, Ishlinskiy, Ryazanskiy, and as Kirillov put it, "other missing Martians," flew in. Unaware of the degree of danger in the events taking place across the ocean, we were preparing to launch the four-stage 8K78 rocket with a mission objective of hitting Mars or, if worst came to worst, flying past the planet at close range.

All the "interplanetary elite" had gathered at the firing range except for Korolev. He had arranged with Smirov for me to be entrusted with technical management of spacecraft preparation, while Voskresenskiy would be in charge of rocket preparation and launch. Keldysh felt compelled to personally monitor preparations and pay particular attention to the condition of the equipment that would perform research in near-Mars space.

Drawing from the experience that Arkadiy Ostashev and I had gained the year before, we set up two 12-hour work shifts at the engineering facility. Three launch vehicles and three spacecraft were prepared concurrently with slight schedule lags. The planetary windows forced us to schedule the launch of all three rockets during a time frame from 24 October through 4 November. If there was a delay, the launches would become pointless and would have to be postponed until the following year. Together with Colonel Anatoliy Kirillov, we developed a very rigid schedule and the State Commission approved it. Right after the first launch on 24 October, if three stages operated normally, the next rocket would be transported to the launch site 24 hours later. We would repeat the same cycle for the next rocket.

Kirillov was chief of the first, or "Korolevian," directorate at the firing range and was in charge of all the military personnel who worked with us at the engineering facility and launch site. On launch days, he personally directed launch preparation and during the last minutes, standing in the bunker at the

---

18. The U.S. conducted a 1.59 megaton nuclear test on 18 October 1962, known as "Chama," as part of Operation Dominic, a series of 105 above-ground nuclear tests carried out in 1962 and 1963.

19. Chertok is probably referring to the Checkmate test which was part of Operation Fishbowl during which a solid propellant XM-33 Strypi rocket was launched on 20 October 1962 from the Johnston Atoll in the Pacific Ocean. The missile carried a 60 kiloton warhead, which was exploded at about 147 kilometers altitude. The Soviets also carried out three high altitude nuclear tests during this exact period (on 22 October 1962, 28 October 1962, and 1 November 1962) under their Operation K.

periscope, he gave the commands "Key for lift off!" and "Launch!" After the last three years of hectic work at the firing range, Kirillov and I had developed a friendship. Unlike the rest of us who had lost contact with newspapers and the latest news broadcasts, by dint of his duty position, Kirillov was obliged not only to keep up with the most important political events, but also to receive instructive information in that regard from the firing range political department. But when we were making the decision to roll out the rocket for the first Mars shot that year, no one—not the firing range command nor even the State Commission chairman—knew that the 20th R-12 missile was about to be put into combat readiness in Cuba.

On the morning of 21 October, the first Mars rocket was rolled out without particular fanfare and installed on the launch site. The rocket's round-the-clock preparation began.

On 22 October, President Kennedy delivered a television address to the American people informing them of "offensive missile sites" that had been detected in Cuba. He called for "a strict quarantine on all offensive military equipment under shipment to Cuba."[20] We, the first missile specialists in the USSR, had just found out about the new deployment of the missiles developed by our Dnepropetrovsk colleagues from Kennedy's speech. The Americans concentrated 180 warships near Cuba and brought the U.S. Navy, Air Force, and Marine Corps into combat readiness status. Troops in Western Europe were also were also put into complete combat readiness status. Sorties of strategic bombers carrying nuclear bombs were flown around the clock.

On 23 October, President Kennedy signed an executive order calling for a naval blockade of Cuba. This order threatened our ships carrying R-14 missiles under submarine escort with an encounter with the U.S. Navy.

Khrushchev sent Castro a message characterizing the U.S. actions as an unprecedented intervention into the internal affairs of Cuba and an act amounting to provocation against the Soviet Union. The Soviet government issued an announcement about the "unprecedented aggressive actions of the United States, which was ready to push the world to the brink of military catastrophe. If the aggressors unleash war, then the Soviet Union will deliver a powerful retaliatory strike." The Soviet government demanded a session of the U.N. Security Council. That same day, U.S. intelligence informed Kennedy that there were 24 R-12 (SS-4 in American parlance) missile sites and 20 Il-28 bombers in Cuba.

---

20. The complete quote from Kennedy's speech is: "Within the past week, unmistakable evidence has established the fact that a series of offensive missile sites is now in preparation on that imprisoned island." He added, "To halt this offensive buildup, a strict quarantine on all offensive military equipment under shipment to Cuba is being initiated."

A rare picture of the 8K78 three-stage rocket, which launched a series of Soviet lunar and deep space probes in the early 1960s. This particular image shows the launch of a 2MV-type space probe.

T. Varfolomeyev

On the morning of 24 October, a "78" (the 8K78 launch vehicle) carrying spacecraft 2MV-4 vehicle no. 3 lifted off. All three stages functioned properly. The fourth stage—Block L—failed to start up, and the Mars spacecraft became a useless Earth satellite. We didn't inform anyone about the launch preparation for our Mars shot. Under those tense conditions, U.S. air defense systems could have perceived it as a combat launch. Fortunately, radar technology, and perhaps preliminary intelligence as well, had already enabled them to distinguish space launches from combat ones.[21]

On 25 October, the next 8K78 carrying spacecraft 2MV-4 vehicle no. 4 was rolled out to the launch pad with a view to launching no later than 29 October. This Mars spacecraft was on the launch pad at Site No. 1 in the climactic hours of the missile crisis. The world was standing on the brink of thermonuclear war, and we were calmly preparing to launch a rocket to Mars in the hope of satisfying mankind's age-old curiosity: "Is there life on Mars?"

On 25 October, the ships carrying the R-14 missiles and warheads received the order from Moscow to turn back without haste. They did this, despite the fact that Pliyev had reported to Moscow that the U.S. was preparing an air strike against our installations on Cuba on the night of 26 to 27 October. It was decided that all Air Defense Troops assets would be used in the event of an air strike. As if confirming the effectiveness of our Air Defense Troop missile systems, that day they shot down an American U-2 spy plane over Cuba. Nevertheless, President Kennedy did not give in to the military's demand for an immediate air strike.

Early in the morning on 27 October, after a sleepless night, convinced that we had mastered all the latest headaches at the Assembly and Testing Building

---

21. This launch (as well as further ones in the Mars series of 1962) were quickly detected by U.S. intelligence, and their existence revealed to the open press soon after.

(MIK), I headed home to rest. Awakened by a strange uneasiness, I quickly grabbed something to eat at the empty "Petit bourgeois" (what we called the dining hall for managerial staff) and set out on foot to the MIK.

At the gatehouse, there was usually a lone soldier on duty who would give my pass a cursory glance. Now suddenly I saw a group of soldiers wielding sub-machine guns, and they thoroughly scrutinized my pass. Finally they admitted me to the facility grounds and there, to my surprise, I again saw sub-machine-gun-wielding soldiers who had climbed up the fire escape to the roof of the MIK. Other groups of soldiers in full combat gear, even wearing gas masks, were running about the periphery of the secure area. When I stopped in at the MIK, I immediately saw that the "duty" R-7A combat missile, which had always been covered and standing up against the wall, which we had always ignored, was uncovered. Soldiers and officers were scurrying around it, but not a soul was near our third Mars rocket.

Our testers surrounded me, pelting me with baffled questions and complaints. A couple of hours ago all the servicemen had been ordered to halt operations on the Mars launch vehicle and to immediately prepare to roll the duty combat vehicle package out to the launch site. While I was mulling over what to do, Kirillov showed up in the assembly hall. Instead of his usual genial smile, he greeted me with the somber, melancholy expression that one might expect at a funeral. Without releasing my hand that I'd extended for our handshake, he quietly said: "Boris Yevseyevich, I have something of urgent importance I must tell you."

Kirillov and I had been on a first-name basis for some time and when he addressed me using the formal "you," it immediately discouraged me from lodging a complaint with him about the suspension of testing operations in the MIK.

We went into his office on the second floor. Here, visibly upset, Kirillov told me: "Last night I was summoned to headquarters to see the chief of the [Tyura-Tam] firing range. The chiefs of the directorates and commanders of the troop units were gathered there. We were told that the firing range must be brought into a state of battle readiness immediately. Due to the events in Cuba, air attacks, bombardment, and even U.S. airborne assaults are possible. All Air Defense Troops assets have already been put into combat readiness. Flights of our transport airplanes are forbidden. All facilities and launch sites have been put under heightened security. Highway transport is drastically restricted. But most important—I received the order to open an envelope that has been stored in a special safe and to act in accordance with its contents. According to the order, I must immediately prepare the duty combat missile at the engineering facility and mate the warhead located in a special depot, roll the missile out to the launch site, position it, test it, fuel it, aim it, and wait for a special launch command. All of this has already been executed at Site No. 31.[22] I have also given all the necessary commands here at Site No. 2. Therefore, the crews have been removed from the Mars shot and shifted over to preparation of the combat missile. The nosecone and warhead will be delivered here in 2 hours. Then everyone not involved with mating the warhead and missile will be moved away."

"Moved away where?" I couldn't keep myself from asking. "The warhead is 3 megatons! You can't move us 100 kilometers away!"

"I don't know anything about the megatons, 3 or 10—that's not my problem, rules are rules. When working with warheads, unauthorized individuals must not be in the vicinity. And here's the worst thing. We're removing the Mars rocket from the launch site, we're freeing up the space. I have already reported all of this to the State Commission chairman and asked to give instructions to inform all services that readiness for launch on 29 October has been cancelled. The chairman did not agree and said that this command could be given tomorrow. He was trying to telephone Moscow, but all the communications lines with Moscow are being strictly monitored and, except for orders and instructions to rocket troop headquarters and readiness reports, no conversations are allowed." Having rendered me speechless with this information, Kirillov told me that Keldysh and Voskresenskiy were in the "Marshal's" cottage and had requested that he pass on the word that they wanted to see me.

"Anatoliy Semyonovich [Kirillov]," I implored, "can't we hold off removing the vehicle from the launch pad? Before you know it, the launch against Washington or New York will be cancelled, so why disrupt the Mars launch?" "We can always argue that removing such a complex rocket requires many hours. There is still the hope that in that time we'll manage to put a call through to Moscow, to Korolev, Ustinov, or Khrushchev himself, and persuade them not to disrupt our work."

Kirillov's face broke into a broad smile.

"I didn't expect you to be so naïve. In the first place, for failing to execute an order I will be put before a military tribunal, and secondly, I repeat—it's impossible to get a call through to Moscow, much less to Korolev, Ustinov, and even Khrushchev."

"Yes, sir! You're in charge! But, Anatoliy Semyonovich! Just between you and me—do you have the courage to give the 'Launch!' command, knowing full well that this means not just the death of hundreds of thousands from

---

22. Site No. 31 was the location of one of the two R-7A pads at Tyura-Tam. The other was at Site No. 1.

that specific thermonuclear warhead, but perhaps the beginning of the end for everyone? You commanded a battery at the front, and when you shouted 'Fire!' that was quite another matter."

"There's no need to torment me. I am a soldier now; I carry out an order just as I did at the front. A missile officer just like me, not a Kirillov, but some Jones or other, is standing at a periscope and waiting for the order to give the 'Launch!' command against Moscow or our firing range. Therefore, I advise you to hurry over to the cottage. You can take my car for 5 minutes."

The "Marshal's" cottage is now called the "Gagarin" cottage. Gagarin, Titov, and subsequently all the other cosmonauts in the first group of five, spent the last night before launch in that cottage. Before Gagarin, the cottage was intended as a place for Marshal Nedelin to rest. After his death, Keldysh and Ishlinskiy sometimes stayed there, and in their absence, State Commission chairmen found refuge there.

When I entered, Voskresenskiy, Ishlinskiy, Keldysh, and Finogeyev were seated around a table in this now historic cottage playing cards. In an adjacent room, Bogomolov was trying to glean the latest news from a radio receiver. Lena, the housekeeper of all the cottages, was in the kitchen drying wine glasses.

"Boris," called out Voskresenskiy, breaking away from his card game. He realized that I already knew everything. "Since you despise cards, we humbly beseech that you get the watermelon out of the refrigerator and help Lena prepare the *zakuski*.

I took the enormous watermelon and two bottles of cognac out of the fridge. When everything was ready, we heard a report that U.N. Secretary General U Thant had sent personal messages to Khrushchev and Kennedy. Once again, Voskresenskiy took the initiative and proposed the first toast: "To the health of U Thant, and may God grant that this not be our last drink!"

This time we all drank down our toast in silence and very solemnly, realizing how close we now were to a situation in which this cognac and this watermelon could be our last. Keldysh confirmed that there was no possibility of talking with Moscow. It was pointless to argue with the firing range brass. He added, "I have a feeling that everything will work out. I don't believe that Nikita Sergeyevich would cave in to that boy's provocation."

I retorted that that boy had promised the whole world to land an American on the Moon, and we still hadn't responded to that challenge. An argument started up in which everyone agreed that the president had promised, but engineers just like us had to deliver on that promise, and this would hardly be achieved easily and quickly.

At one time, in strict confidence, someone had divulged to me how to call Moscow using all sorts of switchboards to bypass all the secure lines of communications. Even under normal circumstances it was "not kosher" to use this line from the firing range. In a situation like today, it had all the more reason to be shut down. But, anything's possible.

I left this good company and went to Korolev's nearby cottage, sat down at the telephone, and began to experiment. I no longer recall how many passwords I had to mention before I heard the Moscow switchboard operator to whom I dictated Korolev's phone number. While I waited, I felt streams of cold sweat running down my back. If his secretary picks up, what do I tell her? And how do I explain myself at all under these circumstances? Do you really think no one is monitoring this line?

It rang and rang. Oh, please, don't let me lose this connection. Hurrah!

"Antonina Alekseyevna! This is Chertok. It's urgent. Put me through to Sergey Pavlovich!"

And suddenly a calm voice, as if we had just been conversing: "Boris! I know all about it. Don't do anything stupid. We're working and eliminating the glitches. Give my regards. Understood?"

"Yes, sir!"

Then I heard the beep-beep-beep of the hang-up signal.

My conversation with Moscow had lasted less than 1 minute. But it had been so stressful that I had to stop by my room, change out of my sweat-drenched undershirt, and have a cold sponge-down.

On 27 October, the president's brother, Robert F. Kennedy, visited the Soviet Embassy and proposed that the Soviet missiles be dismantled in exchange for assurance that there would be no invasion of Cuba. He requested that Khrushchev be informed immediately: "The President hopes that the head of the Soviet government will understand him correctly."[23] Khrushchev "understood correctly." In such a critical situation, the nerves of the military in one of the two countries could snap, regardless of the will of the heads of state. Because there was no time, Khrushchev made the unusual decision to broadcast the Soviet government's message in open text over the radio.

---

23. The President's brother, Attorney General Robert F. Kennedy, visited the Soviet Embassy to speak to Soviet ambassador Anatoliy Dobrynin on the night of 26 October, putting on the table future discussion of American Jupiter missiles stationed in Turkey. The next evening, on 27 October, Robert F. Kennedy met again with Dobrynin in the Justice Department. This meeting was to reiterate the contents of a letter from President Kennedy to Nikita Khrushchev sent earlier in the day in which the President called for removing Soviet missiles from Cuba under U.N. monitoring in exchange for lifting the blockade around Cuba and a guaranteee that the U.S. would not invade the island.

It was already dark when I returned to the Marshal's cottage. On the road, a Gazik came to an abrupt halt.[24] Kirillov jumped out of it, saw me, swept me up in a hug, and practically screamed: "All clear!" We burst into the cottage and demanded that they pour "not our last drink," but alas! The bottles were empty. While everyone excitedly discussed the historic significance of the "All clear" command, Lena brought out a bottle of "three star" cognac from some secret stash. Once again the Mars rockets were waiting for us at the launch site and in the MIK.

THE CRISIS WAS OVER; BUT ON 28 OCTOBER 1962, A NUCLEAR EXPLOSION DID INDEED TAKE PLACE. At Site No. 51 at a *Desna* launch complex, preparations were underway to launch the two last R-9 missiles in the second phase of flight development tests. There was a reason for these tests: Operation K-4—a nuclear test in space at an altitude of more than 150 kilometers—was scheduled for that same day. R-12 missiles were being used for a series of nuclear detonation experiments in space. They were considered the most reliable for these purposes. There was a reason these particular missiles had been taken to Cuba.

According to the concept of the experiment's authors, two of our R-9 missiles were supposed to fly through space as close as possible to the epicenter of the blast 1 second after detonation. Synchronizing the launches of the R-12 from Kapustin Yar and the R-9 from Tyuratam proved to be no simple task for the communications specialists and command staff of the two firing ranges.

At 0936 hours, under clear, calm weather conditions, the *Devyatka* (R-9) finally lifted off. It rose just 20 meters off the ground; its flame suddenly pulled to the side; the missile foundered and fell to the launch pad. An explosion followed, and immediately two fires broke out. Two vertical columns of smoke already stretched high into the sky when suddenly a second sun exploded in the north. For an instant, it was "brighter than a thousand suns." In the pale autumn sky, at tremendous speed, expanding rings shot out from the point of the nuclear explosion, as if a pebble had been tossed into water. Complete silence ensued after these two very different explosions. Soldiers emerged from an adjacent chemical laboratory and, as if nothing had happened, they began to smack a ball about over the volleyball net. The fire at Site No. 51 was localized within 30 minutes. According to Kirillov's report, there were no "fatalities or injuries."

---

24. A Gazik was an automobile produced by GAZ—*Gorkovskiy avtomobilnyy zavod*—the Gorkovskiy Automobile Factory.

The second *Devyatka* was launched simultaneously from the *Desna* ground launch site. It also "returned to its base," destroying the launch facilities. That is how "high frequency" failed us during the K-4 operation.

"As you can see," added Kirillov, "Cuba's not the only place where missile sites are being dismantled, while nuclear explosions are occurring far from America."

THE MISSILE CRISIS WAS OVER. MARS LAUNCHES CONTINUED. The next launch on 1 November 1962 went down in the history of world cosmonautics as *Mars-1* after all. However, not a single historic reference links this launch to the "god of war's" attempts in those days to unleash World War III.

The "all clear" signal on the issue of nuclear war enabled us to continue launch preparations for AMS 2MV-4 vehicle no. 4. Its primary mission was to fly over Mars's surface to take photographs and then transmit the images to Earth via a radio link. At the same time, we hoped to study the properties of near-Mars space, identify the existence of a magnetic field, and determine the field's intensity and the position of the magnetic poles relative to the planet's rotational axis. This time, we managed to execute the launch on the optimal date of the launch window, and it wasn't necessary to remove science equipment to reduce the station mass.

The launch was a success. It was the ninth launch of the four-stage *Semyorka* 8K78, known today as Molniya (Lightning). In the TASS report, the AMS was called *Mars-1*. Weighing 893.5 kilograms in its complete configuration, it flew to Mars on its computed optimal-energy trajectory. At the postlaunch State Commission meeting, the ballistics experts said it would take seven months and approximately three days for the station to reach Mars. Someone cracked under the solemnity of the moment and remarked: "One more combat missile readiness alert like the Cuban one and there won't be anyone to investigate whether there's life on Mars." Having arrived right before the launch, Korolev was in an excellent mood and responded to the situation accordingly: "Instead of cracking silly jokes, you'd better get ready for the next launch. This was a fly-by, but it would be a lot better to send a descent module to Mars!"

On 4 November, we carried out the third and last Mars shot of 1962. Spacecraft 2MV-3 vehicle no. 1 was supposed to fly to Mars and deploy a descent module. However, luck was no longer smiling on us. The fourth stage failed once again, and the AMS remained with it in Earth orbit. We soon forgot about the missile crisis.

Items, including a miniature globe, which were launched on a probe to Venus (unsuccessfully) in 1964.

From the author's archives.

Few had been aware of the actual threat of a potential nuclear missile war at that time. In any event, one did not see the usual lines for salt, matches, and kerosene that form during the threat of war. Life continued with its usual day-to-day joys, woes, and cares. When the world really was on the verge of a nuclear catastrophe, only a very small number of people in the USSR and the United States realized it. Khrushchev and Kennedy exercised restraint and did not give in to their emotions. Moreover, the military leaders of both sides did not display any independent initiative nor did they deviate at all from the orders of their respective heads of state.

Very likely, Khrushchev wasn't just guided by the pursuit of peace "at any cost." He knew that the U.S. nuclear arsenal was many times greater than ours. The Cubans did not know this and viewed Moscow's order to call off missile preparation and dismantle the launch sites as a betrayal of Cuba's interests. President Kennedy had no doubt as to the United States' nuclear supremacy. The possibility of a single nuclear warhead striking New York kept him from starting a nuclear war. Indeed, this could have been the warhead on the R-7A missile that they didn't roll out of the MIK to the pad at Site No. 1.

Chapter 5
Strategic Missile Selection

In 1960 and 1961, Chief Designer Korolev's OKB-1 team and the dozens of subcontractors working in partnership with them were involved simultaneously in two interrelated fields of endeavor. The same people worked in both fields. The first—combat missiles—was given priority in time and production. The second—space—was inconceivable without the first.

Working in the missile field, together with the nuclear scientists, we created weapons systems with monstrous destructive force. Our mission was to deliver, through space and through atmospheric plasma that was heated to incandescence by velocity, a very precisely constructed device—a thermonuclear warhead—to any point on the globe. The device was called a payload. We developed professional, trusting relationships with the creators of this cargo. We had profound respect for one another. This helped us solve the difficult problems of transporting such an extremely delicate physical object as a nuclear warhead on a ballistic missile with its mechanical, acoustical, and thermal loads. New telemetry systems helped us investigate deviations from the prescribed behavioral norms of the automatic devices. At the same time, caught up in the complexity of solving engineering problems, we never gave a thought to what would actually happen "there" and "then" if our "payload" actually reached its destination.

In the same week, and sometimes on the same day, according to our own plans or government decrees, we were transformed into romantics who actually created space technology rather than just dreaming of it. In this field, we could give free rein not only to engineering thought, but also dream about fantastic projects involving manned flights to the Moon, Venus, and Mars. Our Chief Designer, Korolev, was unequaled in his ability to be in these two fields virtually simultaneously. He could be demanding to the point of cruelty for the sake of producing a combat missile within a timeframe that was barely workable, while being a genial dreamer when he found time for semi-fantastic designs of interplanetary automatic and piloted spacecraft.

Our first R-strategic missile, the R-5M (or SS-3), went into service in 1957.[1] The second, Mikhail Yangel's R-12 (or SS-4), went into service in 1959. The R-5M missile had a range of 1,200 kilometers and the R-12—2,000 kilometers. Both were single-stage, but differed fundamentally in terms of propellant. The R-5M used oxygen-kerosene propellant, while the R-12 used nitrogen tetroxide and unsymmetrical dimethyl hydrazine (UDMH). Yangel thought that using liquid oxygen for combat missiles was unacceptable and said: "Korolev's attraction to liquid oxygen will lead our missile technology into a dead end."

Even when Yangel and Korolev had worked together at NII-88 from 1950 to 1954, they had failed to develop a relationship. At one time I was Yangel's deputy and he, in turn, was Korolev's deputy. Then Yangel, as NII-88 director, was Korolev's boss, and I became Korolev's deputy. I witnessed firsthand the incompatibility of these two intelligent and creatively obsessed individuals. With the active support of the Ministry of Defense, Yangel's arrival at OKB-586 in Dnepropetrovsk [in 1954] marked the beginning of the development of a series of strategic ballistic missiles using high-boiling propellant components. The R-12 was the first of these. Later he would create the R-14 (or SS-5) missile with a range up to 4,500 kilometers using the high-boiling components nitrogen tetroxide and toxic dimethyl hydrazine. Intermediate-range missiles were mastered while being produced in large batches. This high-technology production was a large burden for our industries. The army had the opportunity to train and organize the rocket forces. The Kapustin Yar test firing range switched to a routine of round-the-clock testing operations.

I categorize the first R-7 intercontinental missile or its modification, the R-7A (or SS-6), which went into service soon thereafter, as the "zero" generation of intercontinental missiles. Korolev's R-9 and R-9A and Yangel's R-16 and R-16U should be considered the actual effective beginning of the creation of the "nuclear missile shield." I was directly involved in developing and putting the R-9 missile line into service. Below I shall delve into the history of R-9 missile development, which coincided with the beginning of the era of practical manned cosmonautics and interplanetary flights.

In early 1961, we had only four actual launch sites for R-7A intercontinental missiles—two in Tyura-Tam and two in Plesetsk. If, after 24 hours of preparation, all four missiles had reached the United States, a total of 12 megatons would have rained down there.

---

1. Because actual Soviet designations were unknown in the West at the time, Western analysts used their own system of identifying Soviet missiles. The R-5M was known as the SS-3 by the U.S. Department of Defense (DOD) and as the Shyster by NATO. Service records of the Strategic Rocket Forces show that the R-5M was officially declared operational on 21 June 1956.

To what extent did Korolev have to pursue the combat field after his brilliant triumphs in space? Why had we created difficulties for ourselves on the pathway to space that had opened up before us, when the burden of building a nuclear missile "sword" could have been entrusted to others? If we had stopped developing combat missiles, then design and production capacity would have been freed up to expand space programs. If Korolev had made his peace with the fact that Mikhail Yangel, Vladimir Chelomey, and Viktor Makeyev were sufficient for the development of combat missiles, neither Khrushchev, much less Ustinov, who had been appointed deputy chairman of the USSR Council of Ministers and VPK chairman in December 1957, would have considered forcing us to develop a new generation of intercontinental missiles.

However, having created the first R-7 intercontinental missile and its modification, the R-7A, we could not abandon the risky race to be able to deliver nuclear warheads to any point on the globe. Back then, none of us had given much thought to what would happen in the target area if we were to drop a real warhead with a yield of 1.5 to 3 megatons. It was assumed that that would never happen.

Our team had more than enough individuals who supported working on combat missiles. If we gave up the combat field, we risked losing the support of the Ministry of Defense that we so badly needed and falling out of favor with Khrushchev himself. I, too, was considered a member of the unofficial party of missile "hawks" headed by Mishin and Okhapkin. The actual process of creating combat missiles captivated us a great deal more than the ultimate objective. We faced, without enthusiasm, the natural process of losing our monopoly on the production of intercontinental strategic missiles. We were jealous of the work that our subcontractors were doing with other chief designers.

In 1954, having become the head of OKB-586, Yangel immediately organized work on designing missiles using high-boiling propellant components. In August 1955, Yangel managed to have a government decree issued on the production of the R-12 missile with a design range of 2,000 kilometers, which surpassed our R-5M by 800 kilometers. The R-12 missile also had a separating nose cone. However, instead of an atomic warhead, this one had a thermonuclear warhead with a 1-megaton yield. Very influential military officers, including our old comrades-in-arms Mrykin and Smirnitskiy, believed that its primary advantage over the R-5M was not just its range and the yield of its warhead, but also its use of high-boiling propellant components (nitric acid oxidizer and kerosene-type hydrocarbon fuel). For the first time in a missile of this range, great losses of oxygen to evaporation during standby alert had been eliminated.

Moreover, Pilyugin talked Yangel into using a completely autonomous guidance and control system for the R-12. As a result, there was no need for the inconvenient location of radio control stations far from the launch site. But to be honest, the R-12 did not worry us at OKB-1 very much.

On 2 February 1956, the historic launch of our R-5M carrying a live atomic warhead took place. The missile, manufactured in Dnepropetrovsk as per Korolev's documentation, was put into service. Korolev, Glushko, Pilyugin, Mishin, Barmin, Ryazanskiy, and Kuznetsov all received the title of Hero of Socialist Labor.[2]

But when Yangel left us, he did not go to Dnepropetrovsk to perfect Korolev's liquid-oxygen missiles. His own R-12 missile was developed there within a very short period of time. Its flight testing began on 22 June 1957 in Kapustin Yar. It was confirmed that the missile's range exceeded 2,000 kilometers.

The R-12 missile was launched from a ground-based launcher. It was installed on the launcher unfueled and already mated with the nuclear warhead. The total launch preparation time exceeded 3 hours. The strictly autonomous guidance and control system ensured a circular error probability within 2.3 kilometers. Immediately after being put into service in March 1959, this missile was put into large-scale series production at the factory and became the primary armament of the Strategic Rocket Forces, which were created in December 1959.

But even earlier, in December 1956, Yangel had managed, with the direct support of Ustinov, to get the Council of Ministers to issue a decree on the creation of the new R-16 intercontinental missile with flight development tests (LKI) to begin in July 1961.[3] The first R-7 intercontinental missile hadn't even flown *once*, and Khrushchev had already given his consent for the development of another missile! Despite the fact that our *Semyorka* had been given the green light and we had no grounds to complain about a lack of attention from the top brass, the decision on the R-16 served as a serious warning to us.

Common sense suggested that it was advisable to develop an intermediate range missile capable of striking any U.S. base in Europe and Asia from Soviet territory before developing the R-16. Yangel took that into consideration

---

2. The seven received the Hero of Socialist Labor award on 20 April 1956. Besides these seven men, a number of other individuals associated with the Soviet nuclear and missile industry also received the same honor at this time, including A. M. Isayev, M. V. Keldysh, Yu. B. Khariton, A. D. Sakharov, and Ya. B. Zeldovich.

3. To describe standard flight-testing of missiles, Russians use the term "LKI"—literally, *Letno-konstruktorskiye ispytaniya* (Flight-Design Testing).

and accelerated the development of the R-14 missile, which had a range of 4,000 kilometers. In July 1958, a decree was issued calling for the development of this missile, stipulating that flight development tests would begin in 1960—a year before the R-16.

The first R-7 pre-launch preparation experience was accumulated over the course of two years of testing in its combat version and launch vehicle modifications. From the moment it was installed in the launching system, it took 8 to 10 hours under optimal conditions to achieve missile launch readiness. This time could be reduced only if the missile stood on combat alert in its fueled state. But after being fueled with liquid oxygen, it could only stand on alert for a few dozen hours. The enormous losses of oxygen to evaporation during transport, during storage, and after fueling were a fundamental drawback of liquid oxygen missiles. Cold War strategy made more and more rigorous demands on the combat readiness cycle. It was now a matter of minutes rather than hours!

In late 1958, we received detailed information about the U.S. design of the Titan I missile developed by the Glenn L. Martin Company.[4] This was a two-stage missile design using liquid oxygen and kerosene propellant. With a launch mass of 98.5 metric tons, the Titan was designed to carry nuclear warheads with yields ranging from 4 to 7 megatons. The Titan I missile was based in protected silo launch installations, and after fueling it had a launch readiness of 15 minutes. For the time being, not one of our missiles could achieve that. In 1958, Yangel's R-16 proposal was the only realistic response to the American challenge. At that time, underground silo shelters for *Semyorka* launches were out of the question.

An alternative to the R-16 might be a liquid-oxygen missile that was fundamentally new in terms of layout, duty mode, and time required for fueling and achieving launch readiness. Mishin came up with the first proposals for the development of a new intercontinental missile that was given the secret name R-9. He was infatuated with civilian spaceflight to a lesser degree than Korolev and kept a close watch for information about U.S. developments in combat missiles. He was one of the first among us to understand that the R-7 missile was not for war. It was unrealistic to match up the very complicated, exposed, and quite vulnerable R-7 launch sites against the dozens of American launch installations for the longer-range and heavier Titans.

In September 1958, Khrushchev arranged for a review of missile technology at Kapustin Yar. He was shown launches of the R-5M already

---

4. The Glenn L. Martin Company became the Martin Company in 1957.

in service and its competitor, the R-12. Korolev reported to Khrushchev the proposals about the promising R-9 intercontinental missile for the first time. The development of the R-9 required the recruitment of a large number of subcontracting organizations and enormous expenditures, all of which could not be done without issuing a special decree.

It took a lot of effort for Korolev to convince the Council of Chiefs to sign the R-9 proposals so that they could be submitted to the government. Barmin and Pilyugin expressed doubt as to the feasibility of the stated launch mass, which was only 100 metric tons, while the R-16 was 30 metric tons heavier. At first, Glushko did not agree to the liquid-oxygen engine with a ground thrust of 140 metric tons.[5] Consequently, Korolev came up with a proposal to task Nikolay Kuznetsov—the chief designer of aviation turbojet engines at OKB-276 in Kuybyshev—with developing an alternate version of the engine.[6] Mishin was particularly hard-nosed in his debates with Glushko. He argued that if they were to develop a liquid-oxygen engine using the so-called closed cycle configuration, then its thrust could be increased by another 10 to 15 metric tons.[7] This would make it possible to increase either the yield of the nuclear warhead or the flight range.

After a coaxing process that was long and trying, the Ministry of Defense, the State Committee on Defense Technology headed by Rudnev, and Ustinov agreed to our proposal for the development of the new R-9 intercontinental missile using non-toxic propellant components: oxygen and kerosene. Under pressure, *Gosplan* and the Ministry of Finance signed off on the draft decree.[8] This decree, which committed us to develop the R-9 and begin flight tests in 1961, was signed by Khrushchev on 13 May 1959—a historic day for rocket technology. This was the 13th anniversary of the decree dated 13 May 1946, signed by Stalin. However, to speed up the production of R-14 and R-16 missiles, this same decree called for OKB-586 to be released from developing

---

5. Glushko originally proposed a version of the R-9 called the R-9B, which used storable propellants.

6. The variant of the R-9 using Kuznetsov's engines was known as the R-9M.

7. In the West, closed cycle rocket engines are typically known as "staged combustion cycle" rocket engines. The basic premise of such engines is that part of the propellant is burned in a pre-burner. The resulting hot gas is then used to spin the engines pumps and turbines, and then injects the exhausted gas and the rest of the propellant into the combustion chamber. The advantage is that all of the engine cycles' gases and heat go through the combustion chamber, i.e., there are no pumping losses at all, thus increasing the overall efficiency of the engine. Closed cycle rocket engines also allow engines to have very high chamber pressures.

8. *Gosplan—Gosudarstvennaya planovaya komissiya* (State Planning Committee)—founded in 1921 by the Council of People's Commissars—was the top governmental body responsible for economic planning, and formulated the Soviet Five-Year Plans.

missiles for the Navy (all of these projects were transferred to SKB-385 in Miass) and for all work to be halted on S. P. Korolev's naval missile projects.[9]

On the heels of this gift, which dramatically reduced the Dnepropetrovsk staff's workload, came a decree awarding orders and medals to a large number of the Dnepropetrovsk missile men. Yangel, Smirnov, and Budnik received the gold medals of Heroes of Socialist Labor.[10]

JUST BEFORE THE NEW YEAR ON 30 DECEMBER 1959, KOROLEV CONVENED AN EXPANDED COUNCIL OF CHIEFS TO DISCUSS CURRENT ISSUES AND FUTURE PLANS. The Council included Glushko, Barmin, Ryazanskiy, and Pilyugin. The main issue was the status of the new R-9 project. As he opened the meeting, S.P. felt he needed to inform the Council that our new R-9 missile had been a topic for discussion at the December Plenum session of the Central Committee [of the Communist Party]. S.P. added: "Evidently, they are starting to become disillusioned with Yangel's R-16 and there will be serious pressure on the R-9." Under Mishin's influence, Korolev said that if Valentin Glushko was busy developing engines for Yangel, then Nikolay Kuznetsov should be assigned to develop the engine for the R-9. Kuznetsov had an excellent production facility and experience producing aircraft engines.

S.P. cast an expressive glance toward Glushko. But, as usual, the latter sat there with a deadpan, inscrutable expression, as if this matter didn't interest him in the least. Unlike us, Glushko knew that such statements were just wishful thinking on Korolev's part. Korolev's threat about transferring the development of the R-9 engine to Nikolay Kuznetsov was the beginning of a rift between Korolev and Glushko, two luminary pioneers of our rocket technology.

There was one issue under discussion that elicited quite a bit of interest from Pilyugin and Glushko—the latter of whom perked up. This was my proposal to introduce what we referred to as a "central drive" on the R-9. Instead of the jet vanes used prior to the R-7, and instead of the steering chambers that first appeared on the R-7, we proposed deflecting the main rocket engines to control the missile. The first stage engine had four chambers with a single common turbopump assembly. Turning each of the four chambers required very high moment, which our conventional electric control-surface actuators could not provide. Therefore, we proposed the idea of a "central drive." Kerosene was used as the working fluid for the hydraulic cylinders controlling the deflection of

---

9. A government decree on 3 December 1958 had ordered Yangel's OKB-586 to begin work on the R-21, a submarine-launched ballistic missile. The May 1959 decree transferred this project to Makeyev's SKB-385.

10. The three men were awarded the Hero of Socialist Labor medal on 10 July 1959.

each chamber. It was tapped from the main delivery line feeding all four chambers downstream from the turbopump assembly. The pressure of the kerosene downstream from the pump was 140 atmospheres. This made it possible to make the actuating cylinders relatively small. Each of the four chambers deflected more or less than six degrees relative to the neutral position. The moment generated by these deflections was quite sufficient to control the missile.

The actuating cylinders together with control relays and feedback potentiometers were configured as a single assembly, which was contained in the main propulsion system. The young specialist Viktor Shutenko developed the new steering assembly under the supervision of Kalashnikov and Vilnitskiy. Shutenko began working in early 1959 and, enthralled with his crucial assignment, brought the new idea to fruition.

To my amazement, Glushko spoke glowingly of the new idea in the Council. He announced that the central drive steering assemblies had already been configured into the new engine, and that the kerosene uptake system had been cleared with him but that new control-surface actuators would have to be delivered to OKB-456 for joint tests on firing rigs. For my colleagues and me, the decision to introduce the central drive was a creative triumph that started a new trend in control technology. The central drive concept that Kalashnikov, Vilnitskiy, and I proposed gained broad acceptance.

Viktor Kalashnikov, organizer of the steering drive project, passed away long ago and brilliant Designer Lev Vilnitskiy has retired. But the central drive concept lives on and continues to evolve. Twenty years later, Vadim Kudryavtsev, the new leader of this direction of work, the gray-haired Viktor Shutenko, who had become the chief of a large department, and dozens of new engineers brought digital control drives to the highest degree of sophistication and reliability. They supported the first successful flights of Energiya, the world's most powerful rocket, and regulated the modes of its engines, the most powerful liquid-oxygen rocket engines in the world. Alas, Vadim Kudryavtsev, the man most devoted to digital control drives, is no longer alive. But if rocket technology in Russia continues to live, then so will the memory of the steering drive pioneers. The decision made about the central drive in the Council of Chiefs right before the New Year of 1960 had historical significance for our technology. A whole school grew up around this seminal technical idea. Currently the central drive is used on many other types of domestic and foreign rockets. Despite the antagonism that was brewing between Glushko and Mishin, they both supported me in this matter.

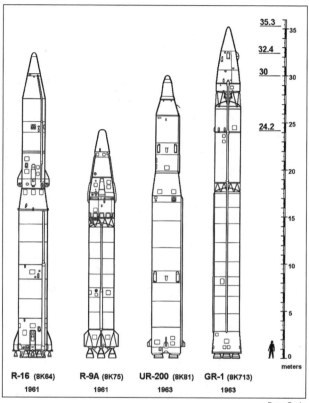

The development of the "first" generation of Soviet ICBMs shown here involved a diversity of producers. The R-16, R-9A, and the UR-200 were developed by the Yangel, Korolev, and Chelomey design bureaus, respectively. The GR-1 on the right was a proposed "global missile," which was cancelled in 1965 before any development flights. The older R-7 is considered the "zero" generation of Soviet ICBMs.

35.3	35
32.4	
30	30
24.2	25
	20
	15
	10
	5
	0
	meters

R-16 (8K64) 1961   **R-9A** (8K75) 1961   **UR-200** (8K81) 1963   **GR-1** (8K713) 1963

Peter Gorin

After making the decision about the central drive for the R-9, Korolev proposed discussing schedules for designing a new super-long-range missile that he referred to as a "global" missile. The concept involved adding a third stage to the R-9. The flight range would be unlimited. The third stage was even capable of entering Earth orbit. The guidance and control system of the last stage and its nuclear "payload" required the use of astronavigation. According to Korolev, Khrushchev had reacted enthusiastically to the proposal. However, Korolev's decision to put Kuznetsov's engines on the first stage instead of Glushko's prevented the 8K713—as the global missile was called—from reaching the flight-testing stage.[11] Mishin insisted on Kuznetsov's engines. But at that time those engines simply weren't there yet. Moreover,

---

11. The design designation of this "global missile" was the GR-1.

soon thereafter international restrictions were imposed that prohibited nuclear warheads from being inserted into space.[12]

As for astronavigation for ballistic missiles, decades later, our colleagues from Geofizika implemented that idea on Makeyev's missiles. The astronavigation system was installed on the gyro-stabilized platform of a submarine-launched missile. After exiting the water and flying beyond the range of possible cloud cover, the missile determined its location and turned toward the target using the astronavigation system operating according to the program loaded into its on-board computer. The astronavigation system developments that we had begun together with Geofizika also found the widest application for combat missiles and spacecraft. This was facilitated by progress in the field of microelectronics that one couldn't have dreamed of in the 1960s.

Several years later, just two of the 8K713 "global" missiles were manufactured and "put into service" strictly for military parades on Red Square. To intimidate military attaches and the entire diplomatic corps, this missile would float around Red Square as the grand finale of military parades to rousing applause from the review stands, but the engines on the "parade" missiles were mockups.[13]

The engineering work on the R-9 design, which Mishin took pains to speed up, showed that a payload with a mass of 1.7 metric tons could accommodate a warhead with a yield of 1.65 megatons. A range of 11,000 to 14,000 kilometers was achievable with a missile that had a launch mass of 90 to 100 metric tons rather than the 150 metric tons of the U.S. Titan II or the 148 metric tons of Yangel's R-16.

Mishin assigned Yakov Kolyako to supervise the design work on the R-9. Kolyako had participated in the Battle of Moscow in 1941 and been severely wounded. He stood out among the designers as remarkably calm and self-controlled in the stressful situations that cropped up in the process of working with Mishin. More than once I had the occasion to witness sharp exchanges when Mishin announced to the designers: "I'll teach you how to work!" Nevertheless, he once said to me: "That Kolyako understands me."

---

12. These international restrictions did not, however, limit testing of orbital bombardment systems such as the 8K713. Yangel later developed the R-36-O "global missile" which was tested from 1965 to 1971.

13. NATO assigned the designation "Scrag" to these missiles, which were paraded through Red Square usually during the ceremonies to mark May Day (in May 1965 and May 1966) or the October Revolution (in November 1965 and November 1966). Western intelligence analysts did not know the program had already been cancelled in 1965.

Sergey Okhapkin, who had worked with Tupolev in his youth, said that Andrey Nikolayevich [Tupolev] knew how to assess an aircraft's flight performance "by sight." Tupolev used to say, "For an airplane to fly well, it must be beautiful." Our *Devyatka* also turned out to be beautiful. At least that's what we, her creators, thought. To a great extent, she owed her beauty to Kolyako's design talent. Until the end of his life, Yakov Kolyako was my neighbor. Every year on 9 May he got together with his old comrades-in-arms. Having miraculously survived the grueling battles for the defense of Moscow in 1941, Kolyako died crossing Korolev Street in Moscow in 1993. He was killed by a "new Russian" in a hit-and-run accident.

At the very beginning of the conceptual design process, we realized that the easy life that we had allowed ourselves when distributing the mass on the *Semyorka* was out of the question here. We needed fundamentally new ideas. As far as I remember, Mishin was the first to express the revolutionary idea of using super-cooled liquid oxygen. If, instead of minus 183°C [-297.4°F], which was close to the boiling point of oxygen, we were to lower its temperature to minus 200°C [-328°F], or even better—to minus 210°C [-346°F], then first of all, it would take up less space, and second, it would drastically reduce losses to evaporation. If we succeeded in maintaining this temperature, then a rapid fueling process could be performed. The oxygen entering the warm tank would not boil furiously as was the case on all our rockets from the R-1 through the R-7.

Obtaining, transporting, and storing super-cooled liquid oxygen proved to be so serious a problem that it ceased to be a strictly missile-related issue. Beginning with Mishin's backing and later with Korolev's involvement in solving these problems, it became an effort of national and economic significance.

Academician Petr Leonidovich Kapitsa was called in as an advisor on the oxygen problem. Academician Sergey Arkadiyevich Vekshinskiy was a consultant on problems of maintaining a high vacuum in large volumes for thermal insulation.[14] I used to meet with Vekshinskiy back during the war when Roman Popov and I were designing the aircraft coordinate radio locator (ROKS).[15] At that time, Vekshinskiy was one of the leading Soviet electronic tube specialists; now he was director of a large vacuum technology NII, having worked in the fields of radar and atomic science. I arranged to introduce Korolev and Mishin to Vekshinskiy. This meeting was held specifically for the R-9. When I reminded Vekshinskiy about our work during the war, he said wistfully: "For some reason, it was easier to work on an empty stomach back then."

---

14. Academician Sergey Arkadyevich Vekshinskiy (1896–1974) was a noted Soviet scientist in the area of electrical vacuum technology. From 1947 to 1974, he served as the director of the Scientific-Research Institute of Vacuum Technology.

15. ROKS—*Radioopredelitel koordinat samoleta.*

From the author's archives.

Firing tests at NII–229 Sergiyev Posad (now Peresvet). This was the premiere rocket engine testing facility of the Soviet missile and space program.

In August 1960, firing tests began on the R-16 missile in Zagorsk.[16] Glushko's unsymmetrical dimethyl hydrazine and nitrogen tetroxide propellant engines operated soundly. At the same time, "high frequency" began to shake and destroy the new liquid-oxygen propellant engines for the R-9 on the rigs at OKB-456.

Glushko's supporters said that the trouble that hounded the initial period of experimental development on the liquid-oxygen engines for the R-9 stemmed from the fundamental impossibility of producing a powerful liquid-oxygen engine with a stable operating mode at this time. Even Isayev, who had no desire to openly engage in the arguments, in a private conversation told me roughly the following: "It's not like Glushko doesn't want it. For the time being he simply can't and doesn't know how to make the process stable using oxygen in such large chambers. I don't know either. And in my opinion, for the time being no one knows the true causes for the occurrence of high frequency."

After receiving information to the effect that the Americans use liquid oxygen on the Titan I, Korolev said—both in the Council of Chiefs and in negotiations over the direct line with the Kremlin—that this confirmed that

---

16. Zagorsk (now Sergiyev Posad) was the location of the Scientific-Research Institute No. 229 (NII-229), the leading Soviet institution which conducted static testing of major liquid propellant rocket engines before certification for flight.

we were on the right track developing the R-9. He believed that we had been right in selecting the R-9A with a liquid-oxygen engine rather than the R-9B with the high-boiling component engine that Glushko had insisted upon.

However, in late 1961, we received information that the same Martin Company had produced a Titan II missile designed to strike crucial strategic targets. The Titan II autonomous guidance and control system had a targeting accuracy of 1.5 kilometers at a range of 16,000 kilometers! Depending on the range, the nose cone was equipped with a warhead with a yield ranging from 10 to 15 megatons. Titan II missiles were placed in individual launch silos in a fueled state and could be launched one minute after receiving the command. The Americans had rejected liquid oxygen and were using high-boiling components. At the same time, information came in that the Titan I was being taken out of service because its readiness time could not be reduced due to the use of liquid oxygen. Now it was Glushko's turn to gloat. The conflict over the selection of engines for the R-9, which had begun in 1958, subsequently exacerbated tensions in both personal and work relations, damaging both them and the common cause.

Meanwhile, the construction of the engineering facility and launch site for the R-16 (or SS-7) was completed. These were the so-called "forties sites."[17] Warmed up by Nedelin, Yangel sought to fulfill his obligations to begin the R-16 flight tests ahead of schedule before 1960 was over. But haste makes waste. On 24 October 1960, during preparation for the first R-16 missile launch, the worst disaster in the history of our rocket technology took place. I gave a detailed account of this in volume two of *Rockets and People*.[18] Despite the blow to morale and the loss of life, the OKB-586 staff found the strength to successfully complete flight tests of the predecessor R-14 missile in December 1960.

R-16 launch Site No. 41, which had been destroyed by the explosion and fire, was restored in three months. The control system was modified and "cured" of its fatal flaws under the leadership of the new Chief Designer, Vladimir Sergeyev. Before the disaster he had worked as department chief with Pilyugin. After Boris Konoplev died in the explosion, Sergeyev accepted the Central Committee's offer to take over Konoplev's position at NII-692 in Kharkov. Pilyugin and the specialists from neighboring factories in Kharkov rendered a great deal of assistance to Sergeyev and the team that he joined. This enabled them to quickly recover from the shock and resume flight tests three months later.

---

17. In other words, these were areas at the launch range with site numbers in the forties.
18. See Chertok, *Rockets and People, Vol. II*, Chapter 32.

In January 1961, Viktor Kuznetsov and I went to the "Yangelian" forties sites. Lieutenant General Andrey Illarionovich Sokolov, the new chairman of the State Commission for the continuation of R-16 testing, had invited us. Since the time of our joint work in Peenemünde in 1945, I had had the occasion to get together with him many times. Many of my acquaintances and I thought that he was a tough and demanding general, but also a thinking man who was capable of listening and showed no signs of overbearing willfulness. At that time Sokolov was chief of NII-4—the head institute of the Rocket Forces. After becoming chairman of the State Commission in place of deceased Chief Artillery Marshal Nedelin, he wanted to consult with "outside" specialists before the new phase of tests on such a dangerous missile.

Sokolov took us to the engineering facility where tests were under way on the R-16 missile that had been modified after the disaster, and had arrived from Dnepropetrovsk. He said that despite the round-the-clock work schedule, he had established defined work shifts and personal accountability when glitches occurred. All testing processes were rigidly and thoroughly regulated by documents and the signatures of military officers and industrial representatives. "There was no need for me to force the strictest discipline on anyone during testing," said Sokolov. "Since the October disaster, people have been showing their personal commitment to order and attention to the smallest details."

Several hours spent at the R-16 engineering facility convinced me that at least the organization of all the testing operations here was now better than at our "first" directorate. I should explain that the primary structural military subdivisions at the firing range were the "testing directorates," each of which managed the projects of one of the chief designers. The manager of Korolev's "first" directorate was Anatoliy Kirillov while Aleksandr Kurushin, who had transferred from Kapustin Yar, had been appointed the new manager of Yangel's "second" directorate.

On 2 February 1961, a modified R-16 was launched for the first time since the disaster. Andronik Iosifyan, who was involved in the launch, told us that he had been more concerned about Yangel's well-being than the failure of the launch. For the most part, the launch was successful. One could forgive the large range error this time, considering that electromagnetic compatibility issues had not been entirely worked out. Subsequently, the flight development tests proceeded so well that Strategic Rocket Forces (RVSN) Commander-in-Chief Moskalenko, with the backing of Sokolov and Yangel, proposed that the R-16 be put into service in April "battle-equipped," albeit for the time being, in above ground positions.

Flight-testing of the ground based R-16 missile was not completed until February 1962. By that same time, construction of silos had been completed at the Tyura-Tam firing range and flight tests of the silo-based R-16 missiles began.[19]

I FEEL THAT I SHOULD EXPLAIN TO THE READER, WHO MIGHT NOT BE VERY FAMILIAR WITH ROCKET TECHNOLOGY, that one of the operational features of modern long-range missiles is that they are significantly more dependent on the ground than winged airborne vehicles. All that airplanes need for take off and landing is a horizontal area—an airfield. At the dawn of aviation, airplanes didn't require concrete runways at all. A meadow with an expanse of a hundred meters was quite suitable. Long-range ballistic missiles of all generations require complex ground-based launch equipment. Unlike an airplane designer, a missile designer—from the first days of the conceptual design to the beginning of flight operation or the placement of the missiles on combat alert—cannot defend his design or conduct flight tests without working jointly with the designer of the ground-based systems. The missile and the ground-based system absolutely must be developed jointly, as a unified system. The same is true for naval missiles. In this case, a submarine performs the role of the ground.

The missile "ground" problem was complicated not only by the advent of heavy intercontinental missiles, but also by the heightening tensions of the Cold War. Each of the opposing sides feared being the first one subjected to a nuclear missile attack. Moreover, they believed that the missile-equipped belligerent side would deliver a strike not only against the country's most important vital centers, but above all, it would try to destroy all the reconnoitered enemy missile launch systems, forestalling the possibility of a retaliatory strike. All nuclear strategy theoreticians envision a retaliatory strike. The concept of a retaliatory strike makes two contradictory demands on the designers of combat missile systems.

First, for a retaliatory strike, missiles must be launched as soon as it is known that the enemy's missiles have already been launched. This situation is fraught with the most acute time pressure. Before the advent of missiles, the side subjected to attack had days, or at the very least, hours to prepare a retaliatory strike. With modern intelligence-gathering means, at the very

---

19. The R-16 silo systems were called *Sheksna-V* and consisted of three silo launchers arranged in a straight line about 60 meters apart. Flight tests of the silo-based R-16 variant began in January 1962 although the first actual silo launch was in July 1962.

least it should take several hours to detect just the preparation for attack using conventional arms. Putting long-range ballistic missiles into service changes the strategy radically. If the attacker has tens or hundreds of missiles on combat alert for months and even years, it is impossible to guess when they will be launched.

The first strike side's missiles need 30 minutes to reach their target; missile defense systems use up the majority of that time detecting, identifying, and transmitting credible information. Considering the time needed to make a decision about a retaliatory strike and transmit the order to the Strategic Rocket Forces command, the missiles themselves have just a few minutes for preparation and departure from their launch positions. The first requirement boils down to having the missiles of the attacking side strike already empty launch positions.

Second, in the event that the first requirement is not fulfilled and the enemy missiles reach the targets before the side under attack launches its own missiles, the launch positions must not become disabled during the nuclear explosions in the immediate vicinity. The launch positions must be protected against the shock wave, high temperature, electromagnetic and radiation exposure, and all other effects of nuclear explosions.

Each missile must have its own "reinforced concrete pillbox." As early as 1944, the Germans attempted to design launchers to launch A4 missiles from bomb shelters. The Americans were far ahead of us in developing the idea of sheltering missiles in vertical silos, which functioned as both nuclear bomb shelters and launch positions. Titan II missiles, and after them hundreds of solid-propellant Minuteman missiles, went into service beginning in 1960 and were put on stand-by duty in silo launchers.[20]

Only after studying intelligence about the U.S. missile silos did the Soviet leadership belatedly make the decision in 1960 to build silo launchers for the R-12, R-14, R-16, R-9, and subsequent modifications. Building the silo launchers required the development of new underground preparation and launch systems. Top secret facilities named after rivers were built at the firing ranges: *Dvina, Chusovaya, Sheksna,* and *Desna.* Each river had its own missile assigned to it. Due to the requirements of automatic launching and the adaptation of the missiles to the conditions of long-term storage and fail-safe liftoff from the silos, problems cropped up one after the other. The total amount of work and, consequently, the expenditures for building

---

20. The precise term that Russians use for silo launchers is ShPU—*Shakhtnaya puskovaya ustanovka* (Shaft Launch Unit).

the silo launchers far exceeded the corresponding amounts for exposed, above-ground launch positions. The first R-16s in the above-ground launch version were put on stand-by duty in April 1961. It took another two years of hard work to put the silo-launched version of this same missile into service.

Our R-9 would first have to be taught to launch from an exposed above-ground launch pad and then to "hide" in a silo. To speed up this process the decision was made to build a temporary launch position in the immediate vicinity of the first *Semyorka* launch pad. The temporary R-9 launch position was assigned the number 51. It was located in a depression just 300 meters from the hill where the structures of Site No. 1 loomed. This proximity made it possible to use the Assembly and Testing Building (MIK) at Site No. 2 and the fueling, ground, electric power equipment, existing communications lines, and other conveniences of Site No. 1 for preparation. It was also very convenient that the most highly qualified specialists, including the chief designers, could devote the proper attention to the new missile without having to commute anywhere.

In March 1961, the R-9 was installed on the launch table for the first time for a fit check and we had the opportunity to admire it. The austere and perfect lines of the still enigmatic *Devyatka* contrasted sharply with those of the *Semyorka*, which had known all the burdens of firing range life, fettered by the multi-story service tower and umbilical towers. The R-9 really was designed much better than its older sister in terms of launch mass. Its range was equal to or even greater than that of the R-7A and its nosecone carried a warhead with a yield of 1.65 megatons. I recall that the *Semyorka* carried 3.5 megatons. But was there really such a big difference between reducing a city to ashes by hitting it with 80 Hiroshima bombs or hitting it with 175?

The beauty and austerity of the *Devyatka's* lines came at a price. A relentless battle was waged against extra kilograms of dry mass. We struggled for kilometers of range using a stringent weight policy and by perfecting the performance characteristics of all systems. Despite his fear of spontaneous "high frequency" vibrations, Glushko increased the pressure in the chambers higher than that of the *Semyorka* and designed a very compact RD-111 engine for the *Devyatka* that was almost the same size as the *Semyorka's* RD-107. It developed 140 metric tons of thrust on the ground (as opposed to 82 metric tons of thrust for the RD-107 engine) and the pressure in the chamber reached 80 atmospheres (60 atmospheres for the RD-107). Increasing the pressure was also one of the possible reasons for the occurrence of "high frequency." The RD-111 had four combustion chambers with one common turbopump assembly (as did the RD-107).[21] A fundamentally new design feature was the installation of the chambers on an engine frame in bearings, the axes of which were located in the yaw and pitch planes. Complete flight control was achieved by turning the chambers using the hydraulic actuators of the central drive during the first stage trajectory phase. The turbopump assembly fit very compactly over the chambers and was connected to it with flexible hoses. Unlike the *Semyorka's* engines, hydrogen peroxide was not needed to drive the turbopump assembly. The gas to drive the turbine was generated in the gas generator by burning a small amount of propellant. A powder starter performed the initial startup of the turbopump assembly. We developed special electric drives to regulate the engine's thrust and the ratio of the propellant components.

To use all of the propellant, rather than leaving hundreds of kilograms in the form of "guaranteed reserves," we developed the discrete (or in modern vernacular, digital) tank depletion control system (DROB).[22] Konstantin Marks, Pavel Kulish, and Vladimir Voroskalevskiy had every reason to be proud of the capacitance-type sensors in the tanks and the transistor logic. The system proved to be more reliable and simpler than a similar one used on *Semyorka*. In addition to all the other benefits, the revolutionary idea of having the central drive rock the engine chambers made it possible to reduce the size and substantially reduce the mass of the on-board batteries.

Yet another revolutionary proposal was the on-board service lines conduit (ZhBK).[23] The hydraulic and electric service lines necessary to connect the missile with the "ground" until the very last seconds were run through this conduit, which extended along the generatrix from the second stage to the launch table. Normally, a myriad of pipelines and cables would extend over the missile's framework to the ground equipment for communication with the "ground" and then take off with the missile as an unnecessary burden in flight. We had a motto—"Only that which is essential for the flight should take off"—following which we "resettled" hundreds of kilograms of various and sundry service lines from "on board" to the ZhBK conduit. The ZhBK, which itself was of impressive dimensions, was jettisoned from the missile and landed with a thunderous crash on the concrete launch pad seconds before liftoff. An open-work truss connected the second stage with the first. After the stages separated, the aft section of the second stage came off. Thus, in flight, the second stage became 800 kilograms lighter right away.

---

21. For "turbopump assembly," Russian engineers use the abbreviation TNA—*Turbonasosnyy agregat.*
22. DROB—*Diskretnaya sistema regulirovaniya oporozhneniya bakov.*
23. ZhBK—*Zhelob bortovykh kommunikatsiy.*

Semyon Kosberg's second stage engine for the *Devyatka* with 30 metric tons of thrust was destined to go down in cosmonautics history for a long time. After modification, the second stage with this engine took the place of the Soyuz rocket's third stage, having received the name "Block I."[24] Kosberg had developed a reliable oxygen-kerosene engine. The generator gas that had been used in the engine's turbopump assembly was used as the working medium in the steering control nozzles.

The modern-day Soyuz-U three-stage spacecraft launch vehicle is the synthesis of two combat intercontinental missiles: the first and second stages are the R-7A missile, and the third stage is the second stage of the R-9 missile.

AFTER TWO WEEKS OF GROUND SIMULATIONS AND ELIMINATING GLITCHES, THE FIRST LAUNCH OF THE FIRST R-9 MISSILE WAS SCHEDULED FOR 9 APRIL 1961. This coincided with the most stressful days of preparation for Gagarin's flight. During the day, many testers were busy with the *Semyorka* for Gagarin's Vostok, while at night they were preparing the first *Devyatka*. Even new Strategic Rocket Forces Commander-in-Chief Marshal Moskalenko, who had not yet managed to study the fine points of missile technology, posed the question: "Couldn't this launch be postponed?"

Chairman of the State Commission on the Gagarin flight Rudnev also wondered why we had this overlap. But Korolev assured him that after the first manned launch, regardless of the outcome, our hands would be too full for *Devyatka*. In this regard, he was right. Even during this first R-9 launch, Korolev was not present in the new cramped bunker at Site No. 51. He was busy negotiating with Moscow, with Khrushchev personally, for the final decision on the manned launch. Voskresenskiy, Kirillov, Dorofeyev, Ostashev, and Khomyakov, the "lead designer" for the *Devyatka*, were entrusted to conduct the first R-9 launch. Korolev ordered Mishin and me to take part in the launch "as commissars." He had already told me: "You will answer for that central drive of yours with your head. Make sure there's not a drop of oil!"

Korolev was referring to the hydraulic system actuating cylinders that controlled the tilting of the chambers of the first-stage engines. They were filled with light oil before launch to check out the control system. A special ground-based assembly generated the necessary pressure in the hydraulic

---

24. The original engine for the R-9's second stage was the 8D715. This engine was significantly modified before use on the Block I (third stage) of the Soyuz launch vehicle. An uprated version (the 8D715P) was first used on the three-stage 11A57 (Voskhod) launch vehicle from 1963. Another uprated version (11D55) was then used on the three-stage 11A511 (Soyuz) launch vehicle from 1966.

system of the central drive. If the seals in the joints of the oil lines and flexible hoses failed, leaks could form, which supposedly posed a hazard if liquid oxygen came into contact with them.

I terrorized Kalashnikov, Vilnitskiy, and Shutenko. I personally looked through the aft section hatches until I was assured that everything was dry and clean. But who knows what will happen when you fill a rocket with liquid oxygen? The oil lines were cut off from the ground assembly before engine startup. Kerosene from the turbo pump assembly entered the hydraulic system under high pressure and forced oil into the kerosene tank. The oil was not used during the flight, but ingress might occur before liftoff itself.

Preparation for the first missile launch was far behind schedule. Errors that hampered readiness setup had been detected in the automatic ground systems for fueling. We finally reached T-minus 15 minutes, a full 5 hours late. Voskresenskiy, who was standing at the periscope, suddenly announced: "Give all the services a 15-minute break."

He turned to us and said that there was a noticeable oxygen leak coming out of the flange connection by the launch table.

"I'm going out to take a look. Ostashev is coming with me; the rest of you, don't leave the bunker!"

Horrible thoughts creep into your mind in such situations. Korolev just had to remind me about the steering oil. Now there's an oxygen leak after everyone has left the pad. Will oil suddenly begin to leak? Mishin and I watched through the periscope. The two of them ambled over to the launch table, which was shrouded in white vapor. As always, Voskresenskiy was wearing his traditional beret. Mishin couldn't help but comment, "Lenya and that swagger of his." In emergencies Voskresenskiy did not hurry. With perfect posture, never glancing down, he strode with his own unique gait. He did not hurry because, in his duel with one more unexpected defect, he was concentrating and mulling over the impending solution.

Having inspected the steaming joint, Voskresenskiy and Ostashev ambled off and disappeared behind the nearby wall of the launch building. A couple of minutes later Voskresenskiy reappeared in our field of vision, but his beret was missing. Now his stride was brisk and purposeful. He was carrying something in his outstretched hand and, walking up to the launch table, he applied that "something" to the steaming flange. Ostashev joined him and, judging by their gesticulations, both of them were satisfied with the solution they had come up with. After standing briefly by the launch table, they turned and headed for the bunker. When the ambling figures stepped away from the missile, it became clear that the leak had stopped: the swirling white clouds of vapor were gone. After returning to the bunker without his beret,

Voskresenskiy took his place at the periscope and, without any explanation, he once again announced T-minus 15 minutes.

At 1215 hours the missile was enveloped in flame, launch debris scattered, and with a roar, it suddenly lifted off toward the Sun. The first stage completed its 100 seconds of operation. The telemetry operators announced over the public address system: "We have separation; the adapter compartment has been jettisoned."

This report came at 155 seconds into the flight: "Malfunctions, malfunctions! . . . Malfunctions accompanied by loss of stabilization!"

This wasn't bad for the first launch. We had checked out the first stage, its engine, the control system, the central drive, startup of the second-stage engine, hot separation, and the release of the second-stage aft compartment. Next came the usual report that the films would be taken to the MIK for development.

"I'm going to go look for my beret," said Voskresenskiy somewhat vaguely as he headed off toward "ground zero." One of the soldiers who had joined the search found the beret about 20 meters from the launch table, but instead of putting it on, Voskresenskiy carried it in his hand, without even trying to stuff it in his pocket. Responding to my unvoiced question he said: "It needs to be cleaned."

We found out the details of the improvised oxygen line repair job from Ostashev. Having taken shelter from the oxygen vapor behind the near wall, Voskresenskiy had taken off his beret, thrown it on the ground, and . . . peed on it. Ostashev joined in and added his liquid contribution. Then Voskresenskiy quickly carried his wet beret over to the leaky flange and with the virtuosity of an experienced surgeon he applied it precisely to the site of the leak. After several seconds, a solid crust of ice "mended" the missile's oxygen feed line.

Among the specialists who had flown in to the firing range for the manned launch were women who, in Voskresenskiy's opinion, out of considerations of modesty, did not need to know about this "Hussar" feat of his.[25]

That evening, having gathered in cottage number three, we couldn't pass up the chance to have a good time and poke some fun at the handymen. We advised Voskresenskiy that in the future he should provide the launch crew with urine samples to prove that it was explosion-proof. The beret was cleaned and after that it was used for its intended purpose. This method for repairing oxygen lines entered rocket mythology.

---

25. The usage of "hussar" suggests the stereotype of a hard-living individual.

There were also tragic instances associated with oxygen line leaks. During preparation for the launch of a modified *Semyorka* (an 11A511U) carrying a spy satellite from the Plesetsk firing range on 18 March 1980, a fire broke out on the launch pad after the rocket was filled with oxygen. The fire quickly engulfed the fueled rocket. Dozens of people died in the fire. I wrote in detail about this disaster in volume two of my memoirs.[26]

WE DETERMINED WHAT HAD CAUSED THE SECOND-STAGE ENGINE OF THE R-9 MISSILE TO SHUT DOWN DURING ITS FIRST FLIGHT ON 9 APRIL THAT VERY SAME DAY. Khomyakov flew out to the second-stage impact area. Among the wreckage, he found the valve that had shut off the gas supply to the turbopump assembly. The cause of the valve failure was determined definitively. The valves were modified for subsequent launches.

Knowing well the true state of affairs, Khrushchev was bluffing, putting our missile power up against hundreds of U.S. B-52 flying fortresses carrying nuclear weapons and dozens of Titan and Atlas missiles. U.S. intelligence could easily prove America's indisputable nuclear superiority.

We at OKB-1, and our friends and competitors in Dnepropetrovsk, understood quite well that only the R-9A (which had a more powerful warhead and greater payload mass than the R-9) or the R-16, were capable in the next one to two years of radically changing the ratio of strategic forces.

Back then, at the beginning of R-9A flight tests in April 1961, one could not yet say "R-9A *and* R-16." "Or" was implied. It all boiled down to which of the missiles would first show the following attributes: sufficient reliability, at least a 20-minute combat readiness, and a high degree of target striking accuracy. No one had been able to fix the errors in lateral direction and range, or circular error probability (CEP).[27] This was as great as 3 to 5 kilometers for the R-7A. The Americans boasted that the Titan II and the new Minuteman would have a CEP no greater than 1 kilometer by 1963.

Korolev understood that the R-9A was at least six months behind the R-16. The advantages of the supercooled oxygen used to fuel the R-9 missile were still virtually unproven. We needed to hurry. We needed to show that piloted launches did not interfere with the solution of crucial military problems. On 12 April 1961, after exchanging heartfelt congratulations, embraces, and hurried toasts over Gagarin's mission, it wasn't easy for Korolev to announce that Mishin, Dorofeyev, Khomyakov, Kalashnikov, and I would

---

26. See Chertok, *Rockets and People, Vol. II*, Chapter 32.
27. The Russian term for CEP is KVO—*Krugovoye veroyatnoye otkloneniye.*

not be flying to the Vostok landing site or to Moscow for Gagarin's reception. We were obliged to remain at the firing range and prepare the subsequent R-9 launches. We stayed and prepared. The second launch on 21 April from Site No. 51 proceeded successfully. The payload reached Kamchatka. After receiving the report from Kamchatka, Kirillov said: "It seems this missile can fly even without the help of Voskresenskiy's beret!"

Korolev was not present for this launch. He was still held up in Moscow after the festivities and demonstrations. Without losing stride, the schedule for the next manned launch needed to be developed. Directives conferring awards, decrees about benefits for Kaliningrad, housing construction, and a new heavy launch vehicle needed to be pushed through before the euphoria of the Party officials and government bureaucracy wore off.

Korolev arrived at the firing range on 23 April overflowing with impressions from the enthusiastic responses that flooded into Moscow from around the world. But not a single telegram was addressed to Korolev personally, nor did the other chiefs receive telegrams. S.P. felt that he had slighted us somewhat and tried as best he could to tell us in detail about everything that had happened in Moscow.

On 25 April, the third *Devyatka* launch took place, this time with Korolev present. After 3.85 seconds (this precise timing was determined after analyzing the telemetry tapes), one of the four chambers abruptly went "to the stop," then the pressure in it fell; the missile began to sink down and toppled over on the launch pad. The usual fire associated with such situations broke out. The kerosene that burned in the oxygen atmosphere melted not only metal, but also the concrete surface of the launch pad.

We holed up in the bunker until the fire fighters had doused the surrounding area to keep the fire from spreading to the fueling tanks, which still had some oxygen in them. After "we tended then our many wounds, and every man recalled a fallen friend," the first reports were reassuring: "No fatalities, no injuries."[28] However, after 2 hours, it was discovered that an officer was missing. Soon thereafter, his body was found in one of the underground service galleries where, against all instructions, he had taken refuge before the launch. The hot smoke had suffocated him.

An accident investigation took place the next day. Everyone gathered to hear reports on the results of the telemetry tape analysis and processing of the data from the launch system automatic recorders. The first impression was that the actuator for the second chamber had, for some unknown reason,

---

28. The quote about the "many wounds" is from "Borodino" (1837) by Russian romantic writer and poet Mikhail Yurevich Lermontov (1814–41).

pushed the chamber to the limit angle. The gyroscopes reacted, but the system had not been able to cope with the disturbance; and after rising barely 30 meters off the ground, the missile toppled over on the launch pad.

"This is probably something you guys botched," said Korolev confidently, turning to Kalashnikov and me. For the time being we had no proof of our innocence.

The more they studied the behavior of all the guidance and control system parameters, the more S.P. was convinced that his accusations were justified, and we were not able to come up with other convincing hypotheses. This is how things worked with our team: if you wanted to prove that you weren't the culprit in a given accident, then it was up to you to put forth a plausible theory. Not having another such theory, my sole vindication was the fact that, according to the recordings, there was no proof of a failure in the steering drive system. The control system commands were a reaction to some very strong external disturbance, the source of which was as yet unknown.

Korolev demanded that we show him a diagram of the central drive, and we began to explain, not for the first time, what might happen given various combinations of any two failures. During the uproarious argument, S.P. noticed that Mishin was missing.

"Mishin and Engineer Colonel [Vsevolod] Bokov requested assistance with the inspection of the accident site debris," answered Kirillov.[29] "I assigned them two officers and a soldier. Perhaps they'll find something interesting."

"And here they are! Speak of the devil," said Korolev when he saw Mishin and Bokov enter the room. "We've already figured out without you that the problem is in the steering drive."

"The steering drive has nothing to do with it," announced Mishin loudly and, smiling victoriously, he raised a formless piece of mangled steel over his head. Bokov also lifted up other grimy pieces of debris in his bandaged hand and showed them.

"I hurt myself pulling this material evidence out of the heap," he explained.

Turning to me, Mishin announced cheerfully, "Boris, you owe me a bottle of cognac: the steering drive couldn't restrain the chamber, which exploded and fell apart. This is what's left of it. This kind of destruction is typical of 'high frequency'."

---

29. Vsevolod Andreyevich Bokov (1921–) was at the time chief of the department of analysis at the firing range at Tyura-Tam. He later served in senior positions in the Russian military space forces.

Glushko's first deputy, Vladimir Kurbatov, inspected the piece of metal in silence. After a long pause, he said: "Yes, those are pieces of our chamber. But we still have to make sure that it fell apart before impact and not after."

There was such commotion and arguing that we forgot about steering drives. Soon thereafter, an official protocol was drawn up and the cause of the accident was reported to the brass—disintegration of the combustion chamber, probably due to high-frequency pressure vibrations. Urgent measures needed to be taken. One of these was to conduct preliminary firing tests on the factory firing rig. After the factory firing test, the engine was given a preventative cleaning, dried, and sent for installation in missiles. Without waiting for flight development testing to be completed, the Progress Factory in Kuybyshev began series production of the R-9. An accident during the third R-9 launch showed that continuing flight development tests from Site No. 51 posed a threat to the well-developed facilities at Site No. 1 and might disrupt *Semyorka* launches.

The R-9 flight development tests were supposed to be conducted from the *Desna-N* standard combat complex, which had been built according to Barmin's design by early summer. However, it soon became clear that at this complex, which did not have a highly automated design, launch preparation required more than 2 hours. They therefore decided to continue flight development tests from the "temporary" Site No. 51 for the time being.

Korolev supported Mishin's very active stand. The latter had proposed using an unusual version of the *Dolina* (Valley) combat launcher. This design called for the unfueled missile to be on standby duty on the *Dolina* in a horizontal position in a special dug-out shelter. The propellant storage facilities were also located there. The oxygen in the storage facility was supercooled in a tank covered with an insulation blanket. Special compressors were needed to maintain the vacuum in the large tanks. Our industry did not produce them. Korolev managed to get a VPK decision to set up production of these compressors based on a Philips Company model. Of course, the company had no knowledge of this.

OKB-1 and subcontracting organizations designed and built the first *Dolina* launch complexes with virtually no input from Barmin, the chief designer of the ground-based launch systems. The law stating that "any initiative is punishable" turned out to be right. All the hardships of designing, building, and putting the *Dolina* into service fell to the lot of OKB-1. Anatoliy Abramov, Boris Dorofeyev, Vladimir Karashtin, Viktor Ovchinnikov, and many others of our specialists were switched over to this top-priority project. This also placed a large additional burden on the factory.

Dolina's story is very illustrative from the standpoint of the conduct of the missile system's chief designer. This is one example emblematic of the way operations were organized not only in Korolev's design bureau, but also in the bureaus of Yangel, Chelomey, Makeyev, and later of Nadiradze.[30]

The advantage of the original liquid-oxygen storage system developed by OKB-1 had been very conclusively demonstrated. Losses were hundreds of times lower than those sustained with the old fueling equipment. However, not everything shaped up well with the automation of all the operations to transport and install the missile in a vertical position, its subsequent rapid fueling, targeting, and final pre-launch testing.

Development of the automatic preparation system (ASP) for launch was entrusted to Vladimir Mikhailovich Karashtin, a graduate of the Taganrog Radio Engineering Institute.[31] He was sent to us along with Karpov and Shevelev, who had seized the forefront in developing spacecraft automatic control systems beginning with *Sputnik-3*. These young engineers who ended up in the stream of our programs had a vast range in which to apply their creative abilities. Together with the "telephone" specialists from the Krasnaya Zarya Factory in Leningrad, they succeeded in reducing the R-9 launch preparation time to 20 minutes, beginning from the horizontal position. It came as a surprise that it was the gyro run-up time to nominal RPMs (60,000) that would determine the further reduction of the readiness cycle, rather than the fueling process. This process required 15 minutes. How had the Americans managed to reduce their readiness time to 2 to 3 minutes? Soon thereafter we received the information that the gyro rotors on U.S. missiles spin continuously the entire time they are on standby. In this regard, Viktor Kuznetsov said that our industry could not produce precision bearings with a continuous operation service life of one year.

Many years of work had led to the development of gyroscopes using new principles. The rigid requirements for the combat readiness of missiles that stood on duty for years in combat positions in a state of less than 1 minute readiness led to the development of various inertial missile navigation systems, elements of which were successfully applied to other areas of motion control technology.

I should note that documents discussing combat missiles did not indicate that the firing range was a cosmodrome, let alone that it was Baykonur; rather, it was called NIIP-5 of the Ministry of Defense.[32] At this NIIP-5,

---

30. During the Soviet era, almost all strategic long-range missiles were produced in the design bureaus headed by Sergey Pavlovich Korolev (1906–66), Mikhail Klavdiyevich Yangel (1911–71), Vladimir Nikolayevich Chelomey (1914–84), Viktor Petrovich Makeyev (1925–85), and Aleksandr Davydovich Nadiradze (1914–87).

31. ASP—*Avtomaticheskaya sistema podgotovki*.

32. NIIP—*Nauchno-issledovatelskiy ispytatelnyy poligon* (Scientific-Research Test Site).

not only were missiles and launch systems tested, but experimental launch complex areas were created for conducting drills in the control of combat missile forces. After the *Dolina* for the R-9, a complex comprising three silos and one common command post (KP) was built at NIIP-5.[33] One KP for three silos was what distinguished this launch complex area.

A vigorous debate flared up between OKB-1 and the leadership of the Rocket Forces over the construction of silos for the on-duty R-9 missiles. Mishin proposed the idea of having one silo next to a populated area. Korolev ardently supported that idea. The arguments were most practical: it would be very economical because it wouldn't be necessary to build special garrisons with all the community services in remote, hard-to-reach areas. In 1961, Korolev showed me the draft of a letter to Marshal Moskalenko, in which, defending the single-silo version, he wrote:

> . . . *it is worthwhile to consider that a decisive military conflict can be prevented only if the socialist nations have a definite military-strategic predominance over the capitalist nations . . . The expropriation of large tracts of land for construction in remote areas and the significant cargo traffic to them will pique the interest of enemy intelligence services. The autonomous single-silo version next to populated areas could easily be kept secret . . .*

As far as I know, this letter was never sent. The military leadership received Khrushchev's decisive support, and the version that the military proposed for the R-9—one underground command post for three silos with their auxiliary services rather far from populated areas in central Russia—was accepted. The desire to quickly pass through the "design" phase of the first flight development test was very great. Fifteen missiles were spent on this phase during 1961 alone. The last launch from Site No. 51 took place on 3 August 1961, three days before the launch of German Titov on *Vostok-2*. This time the R-9 did not lift off. It simply rose up slightly and 0.3 seconds later, it "sat down" on the launch pad and burned up. Despite the failed R-9 launch, Korolev did not dare leave us at the firing range after German Titov's successful flight. Without finishing up deliberations in the R-9 accident investigation commission, all available chief designers and their deputies departed for the festivities in Moscow.

The second phase of R-9 flight development tests was conducted primarily from silos. From March through November 1962, 14 launches were conducted. Of these, nine were considered successful. The majority of

---

33. KP—*Komandnyy punkt.*

the failed R-9 launches were attributed to propulsion systems and instrument control systems. There was not a single failure due to the central drive.

A third phase of flight development tests was scheduled to finally decide whether the R-9 could be put into service. This phase was referred to as "joint flight development tests," meaning that authorized military details carried out the main work, while industrial representatives served primarily as observers.

During the year from 11 February 1963 through 2 February 1964, 25 missiles were launched. Of these, 17 reached their targets. In all, during three phases of flight development tests, 54 missiles were used in a little less than three years. Despite the less-than-reassuring final reliability numbers, the R-9 missile was given the designation R-9A and put into service on 21 July 1965. The experience gained during the launches and the improvement of series production standards at the Progress Factory had done their job. During so-called "series test firings" during the period from 15 May 1964 through 16 December 1968, 14 out of 16 missiles reached their targets!

Two models of single nuclear warheads were developed for the *Devyatka*: standard and heavy. The standard model had a yield of 1.65 megatons; carrying it, the missile could reach a range up to 14,000 kilometers. The "heavy" warhead had a yield of 2.5 megatons and could be delivered a distance of 12,500 kilometers. When a control radio channel was used, the CEP of the R-9A missile did not exceed 1,600 meters.

IN THE EARLY 1960S, WE THOUGHT OUR ARGUMENTS IN FAVOR OF LIQUID-OXYGEN PROPELLANT MISSILES WERE VERY CONVINCING. But as the military gained operational experience, they were more and more inclined to favor high-boiling component missiles. With all the shortcomings of high-boiling oxidizers, they had indisputable advantages over liquid oxygen—they enabled the missile to stand on duty in a fueled state. Military officers reminded us of our own work on missiles for submarines. It didn't occur to anyone (except the Germans during the World War II) to propose that missiles filled with liquid oxygen spend months on submarine cruises.

An indisputable advantage of the R-9A missile was its mass and overall dimensions. With a launch mass of 80 metric tons, the R-9A was 68 metric tons lighter than the R-16. In terms of the number of silos that were on duty awaiting a possible nuclear war, however, the R-9A lagged considerably behind the silo-launched version of the R-16U, which went into service in July 1963, two years before the R-9.[34] The R-16 missile had a fully autonomous guidance and control system. The warhead CEP was 2,700 meters. The missile could be equipped with a light warhead with a 3-megaton yield and a heavy warhead with a 6-megaton yield.

When arguments flared up about the advantages and shortcomings of missiles operating on high-boiling versus liquid-oxygen components, we defended ourselves pointing out that the readiness time for the R-16 was scarcely less than that of the *Devyatka*. The R-16 could not stand in a fueled state for long—the corrosive components might ruin the fittings. For this reason, the missiles stood on duty with empty tanks.

Although the R-16 became the main missile for the formation of the powerful Strategic Rocket Forces units by 1965, like the R-9A, it was in many respects inferior to the U.S. ICBMs. By the time the R-16 and then the R-9A had been put into service, these missiles were obsolete. Both of the first-generation ICBM systems were taken out of service in the mid-1970s.

Around 30 silos armed with R-9A missiles comprised the arsenal of the Strategic Rocket Forces for almost 15 years. In the "general balance sheet" of our nuclear-missile forces, this is not much. But in the 1960s they couldn't imagine that 20 years later this "general balance sheet" would total more than 1,000 missile silos and more than 10,000 nuclear warheads.

I advise everyone who wants to see for themselves that an engineering product such as a combat missile really can be beautiful to visit Soviet Army Street in Moscow. After being taken out of service, one of the *Devyatkas* was mounted on a launch table at the entrance to the Central Museum of the Armed Forces. The R-9 missile was the last joint creation of all six members of the first Korolevian Council of Chief Designers. Forty-four years after the first R-9 missile launch, on the 50th anniversary of the Baikonur Cosmodrome in 2005, I posed in front of this missile for the shooting of a historical documentary "The Council of Chief Designers."

While creating the first generation of ICBMs, we understood that this nuclear missile shield was not durable. New, more advanced combat missile systems were needed. Would we have time before World War III started? I have not studied the history of the creation of all the generations of U.S. ICBMs, and I am not about to judge how this work of colossal technological proportions was organized. For our nation, the nuclear arms race was a very

---

34. The R-16U was a "unified" or "standardized" version of the R-16, which was produced in two different versions, an above-ground variant and a silo-launched variant. The two versions were declared operational on 15 June 1963 and 15 July 1963 respectively.

heavy economic and technological burden. Hidden from the world, the second and subsequent generations of the nuclear missile shield were created under conditions of acute conflict among the chief designers and our nation's highest ranking military and political leaders. I can explain the internal conflicts, which occurred unbeknownst to our Cold War enemies, by the fact that right from the start there were no doctrinal guidelines for the future of "war and peace." The Ministry of Defense should have developed this doctrine, and it was the duty of the nation's top political leadership to accept it as mandatory law.

The development of the state's strategic doctrine was not part of the plans and duties of Korolev and his deputies. But that is the Russian way. In our free time, we pondered and debated enthusiastically on global strategic problems and the prospects of creating a new world order using strategic missiles.

Vasiliy Mishin came up with uncompromising and radical ideas faster than anyone. Konstantin Bushuyev expressed himself cautiously. Leonid Voskresenskiy attempted to predict the future skeptically, but with wit and cynicism. Sergey Okhapkin usually listened and limited himself to comments like "this, my dear old friends, is none of our concern."[35] If Korolev happened to be present during such highly intellectual conversation, he turned to me encouragingly as if waiting for a synopsis. Referring to the classics of Soviet satirical literature, I told him that we were just deliberating like the " . . . piqué vests of the city of Chernomorsk. As you know, they contended that 'Chamberlain is a brain. Briand is also a brain. Chamberlain and Briand are both brains. And you can't trust them farther than you can throw them.'"[36] Korolev cracked a big smile and switched the conversation to another more pressing subject.

---

35. Although his name is little-known in the West, Sergey Osipovich Okhapkin (1910–80) was one of the leading deputies under Korolev. After Korolev's death, he served as first deputy chief designer ("first among the deputies") of the design bureau, i.e., the same position that Mishin served under Korolev.

36. Here, Chertok is alluding to a scene from *The Golden Calf* (*Zolotoy telenok*) (1931), a humorous novel by popular Soviet satirists Ilya Ilf (Ilya Arnolodovich Faynzilberg) and Yevgeniy Petrov (Yevgeniy Petrovich Katayev). In Chapter 14 of this book, a group of elderly gentlemen (whose common mode of dress seems to be white pique vests and straw boaters), who had been businessmen before the revolution, habitually gather at what used to be a thriving café and is now a Soviet public dining room. There they discuss events on the world political scene. In this instance, their conversation revolves around numerous European leaders in the news at that time, including British Foreign Secretary Austen Chamberlain and French Foreign Minister Aristide Briand, who were awarded the Nobel Peace Prize for their work negotiating the Lucarno Pact of 1925. The original Russian contains puns based on idiomatic expressions, which are lost in English translation.

Actually, I had been keen on the doctrine of Italian General Giulio Douhet since my Komsomol days; and then shaken by the defeats of the first year of World War II, I was tormented by the thought that 1941 could be repeated, but this time with missiles.[37] But I knew the men who played a big role in developing our missile strategy; I had met them back in Germany. Back in 1945, being one of the founders of the Institute RABE, I was quite well acquainted with Lieutenant Colonel Tyulin, Captain Smirnitskiy, and later with Senior Lieutenant Mozzhorin. It wasn't too long before these three men became generals. Soon after his return from Germany, Georgiy Tyulin was appointed deputy chief of scientific work at NII-4. In 1959, a Central Committee decision removed him from the Ministry of Defense system and, retaining all his military privileges, he was appointed director of the head institute for missile technology, NII-88. In 1961, Tyulin was transferred to the high post of first deputy minister in the State Committee for Defense Technology, which inherited the duties of the Ministry of Armaments after the creation of *Sovnarkhozy* in accordance with Khrushchev's reforms.[38] In 1965, Major General Tyulin was appointed to the post of first deputy minister of general machine building. In August 1961, Major General Mozzhorin, Doctor of Technical Sciences, was appointed NII-88 director in place of Tyulin. Mozzhorin continued to work in this post for 29 years!

In Kapustin Yar in 1947, I had sat shoulder to shoulder with Captain Smirnitskiy in our cramped armored vehicle during the first firing rig tests of the German V-2 missiles and then during subsequent launches of the V-2s that we had assembled. Smirnitskiy was never a careerist. Nevertheless, by the early 1960s he held the high post of Chief of the Main Directorate of Missile Armaments (GURVO) and Deputy Commander-in-Chief of the Strategic Rocket Forces (RVSN). After entering the war as enlisted men, Tyulin, Mozzhorin, and Smirnitskiy rose to high generals' ranks during the following 20 years. They were exceptionally cultured and very high-minded generals. I remained close friends with all of them until the end of their days. It so happened that in the early 1960s, these three men took on the responsibility

37. Giulio Douhet (1869–1930), one of the leading air power theorists of the early 20th century, was a proponent of strategic air bombing. The reference to "1941" is an allusion to the shock Nazi invasion of the Soviet Union.

38. *Sovnarkhoz—Soviet narodnogo khozyaystva* (Council of the National Economy)—was a system introduced by Nikita Khrushchev in 1957 as a way to decentralize the Soviet economy. *Sovnarkhozy* (the plural of *Sovnarkhoz*) were introduced all over the country to replace the centralized ministry system. After Khrushchev's overthrow in 1964, Brezhnev and Kosygin reverted back to the ministry system, partly as a repudiation of Khrushchev's policies and partly because the decentralization experiment had been an economic failure.

of developing the national strategic doctrine that would determine the government's policy for the production of missile weaponry. In the course of dealing with each of these three kindred spirits on various matters, I got a charge from hearing their thoughts and concerns. Yuriy Mozzhorin achieved the greatest results in the doctrine's practical development, enabling specific requirements to be generated for the chief designers and industry. He organized the development of this problem on a scientific level at NII-88.

Ultimately, the nation's top leadership should dictate to the chief designers what they must create, rather than let themselves be pushed around. It was obvious that implementing the enticing designs of all the chief designers was simply beyond the nation's power. The two head institutes—the military NII-4 and the civilian NII-88—conducted systemic research and strove to find the optimal version of a strategic doctrine that would provide the Ministry of Defense, the military-industrial complex, and the Supreme Commander-in-Chief with the capability to select strategic missile systems; though they worked separately, their efforts were coordinated and virtually free of conflict. The two institutes conducted their systemic research on the basis of the specifications of already produced domestic and U.S. strategic nuclear arms. Unlike their colleagues at NII-88, the scientists of NII-4 were risking their careers if their work, even if classified "top secret," happened to be at odds with the subjective and opportunistic views of the minister of defense.

The development of various versions of nuclear missiles began with the study of U.S. nuclear warhead delivery systems and U.S. strategic doctrines. Beginning in 1946, the U.S. regularly developed plans for nuclear attack on the Soviet Union. Scientists from the Russian Federal Nuclear Center-All-Russian Scientific-Research Institute of Experimental Physics (VNIIEF)—more commonly known as "Arzamas-16"—published data from American sources about these secret plans for nuclear attack on the USSR.[39] In June 1946, a Pentagon plan under the code name "Pincers" called for the use of 50 airborne nuclear bombs against 20 cities in the USSR. The "Sizzle" plan, adopted in 1948, called for using 133 nuclear bombs against 70 Soviet cities. The "Dropshot" plan increased these numbers to 300 nuclear bombs dropped on 200 Soviet cities. In December 1960, the Pentagon developed and approved its latest plan under the code name "SIOP-62," which called for a nuclear strike against 3,423 targets on the territory of the USSR.[40] The Pentagon's strategic

---

39. VNIIEF—*Vsesoyuznyy nauchno-issledovatelskiy institut eksperimentalnoy fiziki*—was the Soviet Union's (and now Russia's) equivalent to the American Los Alamos National Laboratory.
    40. SIOP—Single Integrated Operational Plan.

doctrine viewed the USSR as the main threat to the U.S. and its allies. The elimination of this threat would be accomplished by a preventive nuclear strike, and certainly not by seeking paths of peaceful cooperation.

After long debates, three possible scenarios were concisely formulated for the defense of our country in the event that the use of force could not be avoided in resolving international conflicts:

- preemptive nuclear missile strike against an aggressor that had prepared for an attack
- strategic missile launch-under-attack before the aggressor's nuclear warheads reach our territory, and
- guaranteed retaliatory launch-under-attack.[41]

Research has shown that, given the number of nuclear delivery assets that both sides possessed by the mid-1960s, the first and second scenarios would lead to the loss of all life in the USSR, U.S., and Europe. Both institutes agreed that the only reasonable defense doctrine was a guaranteed retaliatory launch-under-attack. Hence, it followed that missile systems must possess those characteristics that would not only withstand a massive enemy nuclear strike, but would have the capability, using the remaining missile systems, to deliver a guaranteed retaliatory strike that would cause intolerable losses for the enemy.

The experience gained developing R-16 missiles and mastering their high-volume production helped Yangel's team develop the new powerful R-36 (or SS-9) missile within a short timeframe. Flight development tests began on this missile from a ground launch position in September 1963. Strategic Rocket Forces Deputy Commander-in-chief Lieutenant General Mikhail Grigoryev was chairman of the State Commission for testing. He and I had become well acquainted not during missile testing, but while working on Daniil Khrabrovitskiy's film *Taming the Fire*.[42] We were both recruited as consultants. Mikhail Grigoryevich told me that the first R-36 launch was such a failure that many Commission members were skeptical of this development's prospects.

"But I believed in this missile," said Grigoryev. "I knew Yangel's and Glushko's teams quite well and insisted on continuing the work, but under the condition that a long list of provisions would be implemented. The main task was to start up series production at the same time tests were being

---

41. Western strategists typically use the terms "pre-emption," "launch-on-warning," and "launch-on-attack" to describe these three scenarios.

42. *Ukroshcheniye ogonya* (*Taming the Fire*) (1972) was a fictional movie based loosely on the inside history of the Soviet space program. Daniyl Khrabrovitskiy, the director of the film, consulted with many "secret" designers, including Chertok, while making the film.

conducted. They worked heroically, but the missile was not put into service until four years after the first launch.[43] The advent of the R-36 missile marked the beginning of the second generation of ICBMs."

Models of several of Yangel's early missiles on display at a museum. From left to right are the R-12 IRBM (deployed 1959), the R-14 IRBM (1961), the R-16U ICBM (1963), and the original R-36 ICBM (1967). A later model of the R-36, the R-36P, was the first Soviet ICBM to be equipped with multiple independently targeted reentry vehicles.

Asif Siddiqi

The R-36 missile in various modifications would become one of our most formidable strategic assets. With its single warhead payload, the R-36 was capable of carrying one of two types of thermonuclear warheads: a "light" one with an 18-megaton yield or a "heavy" one with a 25-megaton yield. The inertial guidance and control system, based on a gyro-stabilized platform and on-board computer, ensured a CEP of 1,200 meters. The R-36 missile surpassed the Titan II in all parameters. However, the Americans threw down a new challenge, having replaced the single warhead payload with multiple independently targetable reentry vehicles (MIRV).[44] This breakthrough in control and navigation technology led to the next turn in the missile arms race and to the creation of the third generation.

In 1962, a "third power" joined the competition between the schools of Korolev and Yangel for the production of strategic missiles. This was OKB-52 of the aviation industry, headed by Vladimir Chelomey. For Yangel, his face-off with Korolev went to the back burner. A new serious ideological rival had come

---

43. The original variant of the R-36 ICBM was officially declared operational on 21 July 1967.

44. The Russian abbreviation for MIRVs is RGCh IN—*Razdelyayushchayasya golovnaya chast s individualnym navedeniyem boyegolovok*—Separating Payloads with Individual Targeting of the Warhead.

on the scene. In March 1963, OKB-52 received the assignment directly from Khrushchev to develop an ICBM that surpassed the R-9A and R-16 in all parameters. Chelomey had "swooped" into ballistic missile technology on the wings of cruise missiles.

I first heard about Vladimir Chelomey in the early 1950s. Having arrived at NII-885, as always with a multitude of questions about preparation for R-2 missile launches, I found Ryazanskiy and Pilyugin more distraught and preoccupied than usual. Ryazanskiy was reluctant to talk about the work that they had supposedly messed up while Pilyugin merely noted with sarcasm that "they recommended that Mikhail and I start drying bread crusts."[45]

"The aviation industry has a certain inventor named Chelomey. He first invented the 10 and then the 16 *eksem.*"

"What do you mean '*eksem*'?" I asked, somewhat perplexed. "Have you gotten involved with chemical weapons?"

"No, no. It's a *samolet-snaryad* (an airplane-missile)—a further development of the German V-1.[46] But, with the Germans everything was fundamentally simple, and with us it's turned out considerably more complex. To begin with, Chelomey was reproducing the V-1. Thank God, they didn't go into series production. Some people were smart enough to say that nobody needed it any more. Then he suspended the airplane-missile from a real airplane, modified the engine, and achieved a speed of 800 kilometers per hour instead of the Germans' 600. But the word '*ekses*' comes from what we call '*ikses*.' The British say '*eks*,' not '*iks*.' "[47]

The guys working with us who were assigned to salvage the loused up work over at the Ministry of Aviation Industry (MAP) don't say 10 *ikses*, 14 *ikses*, or 16 *ikses*, but so many *ekses*.[48] And then the XM designation cropped up, and in shop slang it's become "*eksem.*"

"So what are you doing here?," I asked. "I know that Antipov at Factory No. 122 is the one who has been assigned to make the guidance and control system for cruise missiles."

---

45. The comment on "bread crusts" was a contemporary euphemism about the threat of imprisonment based around a morbid joke on the lack of food in prison.

46. From the 1930s to the 1950s, the Soviets often used the term *samolet-snaryad* (literally "airplane-projectile") to describe cruise missiles, especially those that bore a strong resemblance to airplanes.

47. The Russian pronunciation of the Latin letter "X" is *iks* and not *eks*, hence the confusion. Chelomey used a designation system for his cruise missiles that used a number followed by the letter 'X' format, e.g., 10X or 16X. When Soviet engineers would describe these systems, they would have normally said 10-*iks* or 16-*iks* but in this case said 10-*eks* or 16-*eks*.

48. MAP—*Ministerstvo aviatsionnoy promyshlennosti.*

"That's right, we didn't know anything until Chelomey devised a way to increase the accuracy. The accuracy of the autonomous system was no better than that of the German V-1, which could barely hit a target the size of London. So here's the 16X—an airplane-missile with radio control. It is suspended under the belly of a Tu-4 or Tu-2 delivery aircraft. After the airplane-missile is dropped from the delivery aircraft, the aircraft crew is supposed to control it via radio. But the control system has an ingenious design. A television camera that is supposed to detect the target is installed on the airplane-missile. A television signal is sent from the airplane-missile to the delivery aircraft. There, on a screen, the crew identifies the target and issues commands correcting the autopilot of the airplane-missile. At the proper moment, the command is given to dive to strike the target. If everything works, then we hope to attain an error for the 16X of ±4.5 kilometers instead of the ±15 kilometers that the Germans had. And everything would be great if MAP had developers who could come up with a system as ingenious as radio remote control."

"This assignment [to develop control systems for Chelomey's cruise missiles] was 'rammed' into NII-885 a year ago. Somewhere in the system it was suggested: 'They know how to do everything at NII-885. They do radio correction for ballistic missiles; they can shut down engines via radio; they know how to measure velocity via radio; they perform telemetry; they made the radio system for the *Wasserfall*, and so they'll be able to cope with this trivial task'."[49] But when we studied the situation . . . fat chance! It's not enough that we're backing into a garage that itself is on wheels and veering away. On top of that, instead of a driver sitting behind the wheel, you have an operator sitting on a scooter at a console. He's being shot at from the ground and from the air, and meanwhile some officer who urgently needs to report that the target has been hit is cussing him out over the radio."

"We're in over our heads with our own problems, and to tell you the truth, we weren't following this work. You've got enthusiasts who botched up something, missed deadlines, and didn't produce any system. We are guilty of thwarting the decree. It's a good thing that Ryabikov got involved.[50] He understands how complex this assignment is. What's more, it turned

---

49. The *Wasserfall* was a ground-launched air defense missile developed by the Germans during World War II which Soviet engineers tried to reproduce and test at NII-88 in the late 1940s and early 1950s.

50. Vasiliy Mikhaylovich Ryabikov (1907–74) was a top administrator in the Soviet military-industrial complex. Between 1951 and 1955, in a number of different senior defense industry positions, he oversaw the development of Moscow's top secret air defense system known as *Berkut*.

out that Sergey Beriya is conducting similar work at KB-1.[51] He proposed a radio-controlled airborne torpedo for firing on naval ships back when he was working on his graduation thesis. Ryabikov promised to stick up for us, especially since the Air Force wants to back away from this work." Ryazanskiy concluded his story on that note.

The fiasco involving the work on airplane-missiles threatened to create enormous problems not only for the NII-885 leadership, but also for Chelomey himself if it got to the point of an investigation under Stalin. Stalin's death removed the threat of severe punishment, but the work gradually came to an end.[52]

In 1955, Chelomey succeeded in reassembling a team of cruise missile enthusiasts in an organization named OKB-52. Unlike the Air Force, leaders of the Navy showed more attention to airplane-missiles. Chelomey took considerable pains to arm submarines with cruise missiles (as they started calling airplane-missiles).[53] Chelomey had to withstand fierce competition with eminent aviation industry designers Mikoyan, Ilyushin, and Beriyev to win a place for his missiles on submarines.[54] The P-5 (or SS-N-3c) missile developed at OKB-52 possessed substantial advantages. Its primary advantage was its automatic wing deployment system. The missile was compactly stowed with retracted wings in a launch container on a submarine. Two powerful solid-fuel booster engines started up to launch it. Immediately after the missile exited the container, automatic controls deployed the wings, the booster engines were jettisoned, and the flight continued using a turbojet cruise engine at a velocity exceeding the speed of sound. Cruise missiles with automatically deployed wings were an invention that made it possible to move ahead of the Americans in terms of the quality of our submarine weapons. Submarines armed with P-5 missiles went into service in 1959—three years after submarines armed with the Korolev-Makeyev R-11FM ballistic missiles went into service. Chelomey's cruise missiles and Makeyev's ballistic missiles

---

51. Sergey Lavrentyevich Beriya (1924–2000), the son of the dreaded Soviet security services Chief Lavrentiy Pavlovich Beriya (1899–1953), was a chief designer at KB-1, one of the largest and most secret Soviet missile design organizations responsible for the development of the Moscow air defense system in the early 1950s. After his father's death, he assumed the name Sergey Alekseyevich Gerechkori.

52. Chelomey's design bureau at Factory No. 51 was dissolved by an order from Stalin dated 19 February 1953, only a few weeks before Stalin's death.

53. The more modern Russian term for cruise missiles derives from the phrase *krylataya raketa*, which literally means "winged missile."

54. Artem Ivanovich Mikoyan (1905–70), Sergey Vladimirovich Ilyushin (1894–1977), and Georgiy Mikhaylovich Beriyev (1903–79) headed OKB-155, OKB-240, and OKB-49, respectively.

had one common shortcoming: the submarine had to surface to launch the missiles. Submarines armed with cruise missiles had appeared on the scene in the U.S. before we had them. But they could not withstand the competition with the American Polaris underwater-launched ballistic missile system.

In the early 1950s, Chelomey and the NII-885 had not managed to develop a system for remote-controlling a cruise missile from a delivery aircraft. Approximately 10 years later, Chelomey put into service a system in which a submarine replaced the delivery aircraft. After a surface launch, the cruise missile maintained radar contact with the submarine right up until the moment of target acquisition using the missile's radar homing head. The radar image was relayed to the submarine where the operating officer selected the most important target in the naval force. After this, the command was issued from the submarine to lock on to the selected target, and after that the missile was controlled by signals from its own homing system. This combined guidance and control system was developed at NII-49, where the chief designer was Vyacheslav Arefyev.

The development of underwater-launched cruise missiles was Chelomey's undisputed achievement. In 1968, the *Ametist* anti-ship cruise missile system was put into service. It had a range of 80 kilometers. During 1969, using its wealth of experience and production stock, OKB-52 developed *Granit* long-range anti-ship missiles, which went into service in 1983. In 1976, OKB-52, which by then was called the Central Design Bureau of Machine Building, began work on the *Meteorit-M* all-purpose long-range naval cruise missile. As a result of many failures during testing, this work was discontinued. However, despite being loaded down with a combat ballistic *Sotka* (UR-100) missile, Proton launch vehicles, and Almaz orbital stations, Chelomey's team continued to develop new generations of cruise missiles during the 1980s.[55] Among all the "great" chief and general designers, only Chelomey worked on strategic cruise missiles and ballistic missiles simultaneously. In addition, he was general designer of the Almaz orbital station, the space "satellite killer" (*Istrebitel Sputnikov*, IS), and a series of designs for heavy launch vehicles.[56]

THE HISTORY OF STRATEGIC MISSILE WEAPONRY DEVELOPMENTS IS REPLETE WITH TECHNOLOGICAL COMPETITION (in the best sense of this word) between the schools of Korolev, Yangel, and Chelomey. As a rule, the initiative for the development of combat missile systems originated not with military customers, but from the design schools of Korolev, Yangel, Chelomey, and

---

55. *Sotka* is a derivation of the Russian word for "hundred."
56. The IS was a co-orbital anti-satellite (ASAT) system developed and tested in the 1960s.

later, Nadiradze. Each came forward with his own concept, while the Ministry of Defense strategists who had developed various military operational requirements were guided sometimes by instructions from the top brass, sometimes by one chief designer or the other. Now it's easy to analyze the mistakes of the past. "Things are clearer from a distance," the historians say. Back then it was really very difficult to make the right choice.

During the last years of Khrushchev's administration, disputes over the selection of a missile system were smoothed out thanks to his constant attention and personal involvement in the discussion process. Khrushchev simply did not have time before October 1964 to make the final decision for the selection of the optimal version of an intercontinental missile system.

Yangel and Chelomey came forward as the chief designers who had each proposed his own promising system for the third-generation nuclear missile shield. Khrushchev still viewed Korolev as the main "space" force that made it possible to win one political victory after the other without a "hot war." A comparatively small number of Korolev's space rockets and Yangel's intermediate-range missiles had demonstrated to the world the USSR's potential rocket power and the supremacy of the socialist system, and they had afforded the opportunity to announce from a lofty podium that "we can produce missiles as if they were sausages."

Of course, it is understood that this includes the production of many hundreds of missiles installed in reliably protected silos. The expenditures on this "sausage" would exceed expenditures on space projects many times over. Khrushchev understood this and sought to make a choice that also provided an economic advantage.

It was difficult for him to choose between Yangel and Chelomey. Minister of Defense Malinovskiy could not be a reliable advisor in this matter. Ustinov, who had expended a lot of energy establishing the Dnepropetrovsk Factory and creating OKB-586, supported Yangel, of course. Leonid Vasiliyevich Smirnov, former director of the Dnepropetrovsk Factory, who was appointed VPK Chairman in place of Ustinov in March 1963, was also on Yangel's side. For the other ministers and State Committee chairmen, Chelomey was, to a certain extent, a new figure. It was thought that Khrushchev would support Chelomey's proposal because Khrushchev's son Sergey worked at OKB-52 in Reutov, where he supervised the development of the guidance and control system. Rumors about Sergey's influence on his father were, in my opinion, greatly exaggerated. There was a lot of talk to the effect that nepotism helped in those organizational successes that Chelomey achieved. This was primarily a reference to the reorganization of OKB-52, particularly its absorption of V.M. Myasishchev's former organization OKB-23 and the restructuring of the M. V. Khrunichev Factory (ZIKh) for missile production for Chelomey.[57] Actually, the largest design and production complex in the aviation industry located in Fili was not so much a branch of OKB-52, but the coat that they sewed to the single button in Reutov. Sergey Khrushchev criticized me for my "button" analogy. He thought that Reutov was hardly a button sewn to the Fili facility but rather the lining of a high-quality coat.

During the time when Khrushchev was eviscerating aviation projects, a joke classified as "secret" was in circulation:

"Did you hear they shut down the Bolshoy Theater?"

"What happened?"

"It's going to be turned into a 'red corner' for Chelomey's OKB."[58]

The missile elite was convinced that after Khrushchev's fall and Ustinov's appointment as Central Committee secretary for defense matters, Chelomey's star would set.[59] But it turned out that in the offices of the Central Committee, the Council of Ministers, and Ministry of Defense, Chelomey's standing was rather strong. New Minister of Defense Andrey Grechko explicitly supported putting Chelomey's *Sotka* into service as the Rocket Forces' primary armament.

During the return to the ministerial system in 1965, Chelomey's OKB, its branch in Fili, and the M. V. Khrunichev Factory (ZIKh) were transferred from the aviation industry to the Ministry of General Machine Building—

---

57. ZIKh—*Zavod imeni M. V. Khrunicheva* (Factory Named After M. V. Khrunichev). Myasishchev's design bureau was attached to Chelomey's organization in October 1960 as the latter's Branch No. 1. Facilities at the Khrunichev Factory were also prioritized for Chelomey's work with that order.

58. A "red corner" (*krasnyy ugolok*) was a special room or area in a Soviet institution set aside for ideological and propaganda purposes. Khrushchev "eviscerating aviation projects" is an allusion to Khrushchev's negative attitude to the aviation industry at the time. Because of his fascination with missiles, Khrushchev tried to limit funding for aviation projects and diverted major resources to missile development. As a result, many prominent Soviet aviation designers were left without contracts and either had to shut down operations or switch to being subcontractors for missile development. Chelomey's organization, OKB-52, was one of the few in the aviation industry that grew during the Khrushchev era.

59. Ustinov became Secretary of the Central Committee in charge of defense industries and space in 1965. As such, he became the top policy-maker in charge of the Soviet space program.

the new rocket-space ministry.[60] Now the new minister, Sergey Afanasyev, was obliged to look after General Designer Chelomey and the field of endeavor entrusted to him, as much as he did for Korolev and Yangel. Heated interdepartmental disputes became intradepartmental as well.

A decisive factor in the competition between the schools of Korolev and Yangel, which contributed to the rapid advancement of Chelomey's projects, was the high degree of intellectual potential and the design and production culture of the aviation industry teams in Fili. This was evident in the technology for high-volume series production of UR-100 ICBMs.

Chelomey proposed creating a new generation of nuclear-tipped missiles based on the comparatively simple and inexpensive UR-100 universal rockets, rather than Korolev's R-9A and Yangel's R-16.

Enjoying Khrushchev's active support, Chelomey "wedged his way in," having come forward with a proposal to develop two types of missiles: the UR-200 and UR-100. The UR-200 was, according to all parameters, a heavy missile that should have replaced Yangel's R-16 and prevented the development of the up-and-coming R-36. The UR-100 was proposed as a universal rocket. According to Chelomey's colorful posters, it could be used as a ground-based ICBM, as an ICBM to arm submarines and surface vessels, and finally, it could provide cover for the Soviet Union in space, destroying enemy warheads. In other words, it radically solved the anti-ballistic missile defense problem.

High-volume series production of these missiles was mastered by factories of the Ministry of General Machine Building, created in 1965. However, each industry-produced intercontinental missile was not supposed to be warehoused but loaded into a silo launcher. The construction and activation of silo launchers cost as much as the missiles installed in them. Unlike the Pentagon, the USSR Ministry of Defense did not spend money from its budget on the development or modernization of missiles. *Gosplan* and the Ministry of Finance allocated the funds necessary for this directly to the ministries in charge. Therefore, in official publications of the annual governmental budget, the "defense expenditures" column contained the modest figure of 12–14 percent of the total annual budget. This 12–14 percent for the Ministry of Defense was hardly enough for routine expenses: clothing, shoeing, and feeding the army; supporting generals; maintaining military academies, firing ranges, airfields, etc. The actual expenditures to produce all the new types of

---

60. In 1965, after Khrushchev's fall, the new Soviet leadership returned to the old ministry system of government. One of the new ministries created at the time was the Ministry of General Machine Building (MOM) which was put in charge of developing strategic missiles, space launch vehicles, and spacecraft.

armament were not shown in the budget. Therefore, the Ministry of Defense, General Staff, and other military channels did not object to the development of an excessive array of missiles, if resources for this were found within the industrial ministries. Thus, the Ministry of General Machine Building provided funding for the development of the second generation of UR-100 and R-36 ICBMs. This ministry had been allocated all the necessary funds, money earmarked strictly in accordance with a decree of the Central Committee and Council of Ministers. I should mention that the Commission on Military-Industrial Matters under the Council of Ministers could not fund anything itself. However, its job entailed drawing up and presenting to the Central Committee and Council of Ministers draft resolutions outlining whom to support, how, and by what means.

The development of the UR-200 (or SS-10) and UR-100 (or SS-11) missile systems coincided with the R-16 and then the R-9A going into service. The far-fetched proposals concerning the UR-100's universality were soon rejected. The naval and anti-ballistic missile versions were window dressing. However, its advantages as an inexpensive intercontinental missile were alluring. Experienced aviation designers from Myasishchev's old team considered the weak points of Korolev's and Yangel's intercontinental missiles not just based on design data but also based on the first operational experience. The greatest advantage of the UR-100 was the fact that, for the first time in the history of domestic missile building, a missile standing on duty was isolated from the external environment. It was enclosed in a capsule—a special container filled with inert gas.[61] The process of monitoring the technical state of the missiles, pre-launch preparation, and launch had the advantage of being fully automated.

The first UR-100 test launch took place in April 1965, and in autumn 1966, a UR-100 missile system was put into service. It had the following unique features: strengthened defense against the destructive factors of nuclear weaponry, the capability to remain on high alert longer than the R-9A and R-16, new methods for remote-controlled launches, the capability to monitor the condition of 10 missiles and launch equipment from a command post, and, at the same time, the capability for autonomous missile preparation and launching. With a launch mass of 50 metric tons, the missile's flight range was 10,000 kilometers, and the single warhead payload could be delivered with an accuracy of 1,400 meters. The warhead yield was 1 megaton. At that time, this was the lightest of the intercontinental missiles. You have to give Chelomey and his team of UR-100 developers credit; this missile was designed with a view toward modernization.

---

61. Russians call such an arrangement "ampulization."

The "first edition" of the UR-100 and the R-36 comprised the basis of our nuclear missile strength during the second half of the 1960s. Top-secret activities surrounding the subsequent design of the "nuclear shield"—the concept of the third generation of missiles—unfolded from the late 1960s through the early 1970s.

The updated UR-100K, called the RS-10 in the open press and the SS-11 in the U.S. DOD, with the same launch mass, had a maximum range of 12,000 kilometers and a CEP of 900 meters.[62] It was put into service in December 1975. The first test launch of the last update, UR-100NU (also known as RS-18 and SS-19) took place in October 1977, and it went into service in November 1979. The launch mass of this missile was doubled to 103.4 metric tons. Its maximum range was 10,000 kilometers. The missile was capable of carrying six warheads with a yield of 0.75 megatons each. Its CEP did not exceed 350 meters. The missile systems equipped with these *Sotki* to this day comprise the main components of our missile "sword." These updated *Sotki* were Chelomey's contribution to the third generation.

After Yangel's death, Vladimir Utkin was in charge of KB Yuzhnoye.[63] Under the conditions of a Cold War with an external enemy and a "civil war" with skeptics in the Ministry of Defense, the Dnepropetrovsk team, together with guidance and control specialists Pilyugin and Sergeyev, through desperate efforts, restored ideological and then real "parity." In its new version, the R-36 was assigned the designation RS-20V or R-36MU. Instead of a single warhead, it was equipped with ten, each with a yield of 0.5 megatons and a maximum range of up to 11,000 kilometers. Each warhead homed in on its target with an accuracy of 500 meters. The single warhead payload on this missile reached a range of 16,000 kilometers. Its launch mass was 217 metric tons. Essentially, this was a new missile rather than a modification of the [original] R-36 missile taken out of service in the late 1970s. There was a reason that the Americans, who had assigned this missile the DOD designation SS-18, called it "Satan" and during nuclear disarmament negotiations, their main demand was the dismantling of these multiple warhead missiles. The well-known Tsiklon launch vehicle is one of the modifications of this missile adapted for space missions.[64]

---

62. The "RS-10" designation was an artificial designation created by Soviet delegations participating in arms control talks during the 1970s and 1980s. U.S. officials had insisted that the Soviets provide designations for their weapons for enumeration in arms control treaties. The Soviets reluctantly manufactured these fake designations of the type "RS-n" for this express purpose.

63. Yangel's OKB-586 was renamed KB Yuzhnoye in 1966.

64. The Tsiklon launch vehicle was derived from the original R-36 ICBM, not the updated R-36MU.

THE FATE OF MANY THOUSANDS OF PEOPLE AT MISSILE ENTERPRISES, factories, and subcontracting production facilities, as well as people working on the construction of launch sites, and even those serving in the army, depended on the selection of one type of missile system or another. Disputes within technical councils, ministry staffs, and expert commissions took on such a violent nature that this period in the history of our military missile technology (from approximately 1965 through 1975) was called the "little civil war." "Brother against brother, father against son," they said ironically in the ministry halls. When old acquaintances and friends met up, they sorted out their relationships from the standpoint of "whose side are you on?" And if they didn't see eye to eye, a rift formed in their personal relationship.

Even the Academy of Sciences got involved in the "civil war." In 1968, having started to regularly attend meetings of my Department of Mechanics and Control Processes, general assemblies, and sessions of various councils, I was amazed that the separation into "us" and "them" existed in academic circles as well.[65]

The main cause of the "little civil war" was a fundamental disagreement in the building of the third generation of missile systems. At the beginning of the second half of the 1960s, the United States deployed and put on duty more than 1,000 Minuteman solid-fuel light missiles and 54 Titan II heavy missiles using high-boiling components. Our missile industry was producing UR-100 and R-36 missiles at a stepped-up pace in order to achieve numerical superiority. Work on the construction of silos for these missiles and their handover to the Strategic Rocket Forces went according to plan, and all told, hundreds of thousands of people were occupied with this work in industry and the army. By the late 1960s, we not only caught up with the U.S., but we surpassed them in terms of the total number of missiles standing on duty.

However, the United States jumped ahead of us, having put into service a brand new, third-generation Minuteman III missile. They had not one, but three warheads with individual homing devices to hit different targets. The availability of missiles with separating warheads gave the United States substantial advantages. In addition, Minuteman III missiles were installed in specially protected silos and were capable of being launched even after a nuclear explosion in the immediate vicinity. A high degree of accuracy was achieved by using on-board computers.

---

65. Every full and corresponding member of the Academy of Sciences belonged to a particular department denoting their broad specialty. Many missile designers (like Chertok) belonged to the Academy's Department of Mechanics and Control Processes. Chertok became a corresponding member of the Academy in November 1968.

To offset the advantages that the Americans had achieved in the strategic nuclear arms race, at Mozzhorin's initiative, NII-88 came forward with the proposal to increase the degree of protection of already existing UR-100 and R-36 silo systems and immediately develop a new generation of missiles with on-board computers for the individual guiding of each warhead to its own target.

Yangel, VPK Chairman Smirnov, First Deputy Minister of General Machine Building Tyulin, and GURVO Chief Smirnitskiy vigorously supported NII-88's proposal, as did Central Committee Secretary for Defense Matters Ustinov.

This proposal triggered shrill opposition from the leadership of the Ministry of Defense, our Minister Afanasyev, and General Designer Chelomey. They believed that the already well-developed cooperation of smooth-running missile production and acceptance would be ruined. Factory workloads would now be threatened, and people would need to master the technology and operation of a guidance and control system that used computers. The Rocket Forces would have to train new specialists.

The new Minister of Defense, Marshal of the Soviet Union Andrey Antonovich Grechko, supported Chelomey, who had given a hostile reception to the new proposals of NII-88 and Yangel. During a meeting with Mozzhorin at the latest Soyuz flight debriefing, he bitterly recounted that at a meeting of the Scientific and Technical Council (NTS) in Reutov, Chelomey had spoken out along the following lines: "The MIRV concept is just a fad.[66] As far as installing on-board computers is concerned, I don't understand how a calculator installed on board a missile can increase firing accuracy."

Chelomey thought that increasing the degree of protection of existing launchers was unrealistic. To practically demonstrate the feasibility of their proposals, Yangel and NII-88 developed an updated version of the R-36 missile with multiple warheads and a fundamentally high-protection silo launcher with a so-called "mortar launch."[67]

---

66. NTS—*Nauchno-tekhnicheskiy sovet*. All major R&D institutions in the Soviet defense industry (like the Chelomey design bureau) had scientific and technical councils where new technical proposals were evaluated for their prospects.

67. The "mortar launch" technique involves using pressure from steam formed underneath a missile (with its tail unit protected beneath a shroud) to "pop" a missile out of a silo.

Chief Designer Nikolay Pilyugin speaking at an evening devoted to the 20th anniversary of OKB-1 in 1966. In the background is Vasiliy Mishin, just recently appointed chief designer of OKB-1.

Among the chief designers close to me, Nikolay Pilyugin showed the greatest interest in the "debate of the century." During every meeting in his office, he pulled models of printed circuit boards for Yangel's future computer out of his desk drawer and told me about the joint work and cooperation with the organizations under the Ministry of the Radio Engineering Industry.

The face-off between the strategic missile concepts reached its apogee during the second half of 1969. KB Yuzhnoye, headed by Yangel, and NII-88, headed by Mozzhorin, with the support of First Deputy Minister Tyulin, VPK Chairman Smirnov, and the loners that had joined them, proposed three basic measures:

- significantly increasing the degree of protection of already existing UR-100 and R-36 launch systems
- developing a new generation of intercontinental MR UR-100 and R-36M missiles carrying multiple independently targetable warheads using a mortar launch, and installing them in well protected silos, and
- curtailing the production and buildup of the number of RVSN launchers.

TsKB Mashinostroyeniye, headed by Chelomey, and the Ministry of Defense, headed by Marshal Grechko, proposed the following:

- modernizing the light UR-100K missiles to replace the UR-100 in existing silos

- developing a long-term strategic system of light UR-100N missiles with multiple independently targetable warheads and increased protection for their silo launchers
- significantly building up the number of strategic missile groupings, and
- gradually phasing out R-36 heavy missiles, replacing them with the UR-100N.[68]

Various levels of academic councils were unable to find a compromise. Moreover, military officers who disagreed with the position of the minister of defense were threatened with no future legitimate promotions, forced retirement from the army, or even dismissal from this sector of the military industry. The matter ended up in proceedings in the Defense Council, chaired by Central Committee Secretary Leonid Brezhnev.[69] Mozzhorin described this session at Stalin's former dacha in the Crimea in his memoirs published in 2000, two years after his death on 15 May 1998.

The last years of his life, Mozzhorin helped me a great deal in conducting my academic "Korolev Readings."[70] My first two volumes of *Rockets and People* had already been published, and he had promised to supplement the story with his own memoirs. Unfortunately, he didn't live to see their publication. Everything concerning the "debate of the century" described here is described in much greater detail and more eloquently in his book *This is How it Was . . . .*[71]

At the Defense Council session in the Crimea, Brezhnev expressed extreme dissatisfaction with the fact that the matter was being brought before the Council in an atmosphere of bitter discord. As a result of the Defense Council's decision, higher political and governmental leadership gave their approval for the doctrine of guaranteed retaliatory strike—the deterrent doctrine—and the need to seriously increase the degree of protection for existing launch silos. It was proposed that Yangel design promising highly-protected missile systems with multiple independently targetable warheads. It was proposed

---

68. TsKB Mashinostroyeniye—*Tsentralnoye konstruktorskoye byuro mashinostroyeniye* (Central Design Bureau of Machine Building).

69. The Defense Council was the highest decision-making body in the Soviet Communist Party and government on national security issues.

70. The "Korolev Readings" are annual conferences held usually in January and February of each year in Moscow dedicated to space technology. A major portion of the conferences involves papers from veterans of the space program.

71. Yu. A. Mozzhorin, *Tak eto bylo . . .* [*This is How it Was . . .*] (Moscow: ZAO Mezhdunarodnaya programma obrazovaniya, 2000). For a detailed description of the "little civil war," see in particular pp. 144–188. For an English language summary, see Steven J. Zaloga, *The Kremlin's Nuclear Sword: The Rise and Fall of Russia's Strategic Nuclear Forces, 1945–2000* (Washington, DC: Smithsonian Institution Press, 2002), pp. 135–141.

that Chelomey develop his own new, promising UR-100N missile system to be installed in strengthened second-generation UR-100 launching facilities. It was proposed that Yangel develop a ground-based mobile missile system.

In the end, the "little civil war" promoted the attainment of parity with the U.S. and even superiority in the area of nuclear missile weapons. Minister of Defense Grechko suffered an ideological defeat. He did not forgive the military specialists who actively supported the stance of Yangel and Mozzhorin. Lieutenant General Smirnitskiy was forced to leave the army. General Sokolov, the NII-4 director, was removed from his office. Other military scientists also suffered. The relationship between Minister Afanasyev and Tyulin, his first deputy, completely soured.[72]

On 26 August 1976, in the 73rd year of his life, Minister of Defense Marshal of the Soviet Union Grechko passed away suddenly. The Secretary of the Central Committee appointed former minister of armaments, Colonel General Dmitriy Ustinov to be Minister of Defense. He was elected to the Politburo and soon given the title Marshal of the Soviet Union. In due course, Minister of General Machine Building Afanasyev "forgave" General Designer Utkin (who had replaced Yangel in 1971) and other subordinates in the ministry who had opposed him in the "little civil war." But Ustinov did not forgive Afanasyev for his opposition. In 1983, per Politburo decision, two-time Hero of Socialist Labor and Lenin and State Prize laureate [Afanasyev] was "dismissed" from the ministry that he personally had created and headed for 18 years and was appointed minister of heavy and transport machine building.

Per the Defense Council's decision, Yangel's and Chelomey's missile systems went into high-volume series production and into service for the rocket forces. Both systems had Glushko's liquid-propellant engines using the same components: nitrogen tetroxide as the oxidizer and unsymmetrical dimethyl hydrazine as the fuel. Now it seemed that Chief Designer Glushko would hold all the power of our nuclear missile forces. Even the engine on the first stage of the R-9A, although it ran on liquid oxygen, was his!

Other chief designers of systems also had to work for all the "warring" sides and not let on that they preferred one "top designer" more than another. Pilyugin, Kuznetsov, and Sergeyev found themselves in a difficult position. Pilyugin had made inertial guidance and control systems since the first years of the missile era. During 1961, availing himself of the formidable instrument

---

72. Smirnitskiy was fired as chief of GURVO in December 1975, Sokolov lost his job as director of NII-4 in April 1970, and Tyulin was forced into retirement from being first deputy minister of general machine building in 1976.

production capacity in Kharkov, Sergeyev continued Konoplev's work producing guidance and control systems for Yangel's R-16. Pilyugin was very reluctant to work for Chelomey, especially after a conflict in connection with the spontaneous oscillations of the control-surface actuators on the UR-200 missile. Chelomey had designed this missile as an alternative to the R-36.

The first pre-launch tests of the UR-200 missile were in 1963. When the control-surface actuators were activated, they began to shake the chambers even though no control commands had been issued to them. At first this caused slight jerking reactions in the missile, but then the spontaneous vibrations started to develop so intensely that power had to be removed from the control-surface actuators and launch preparation had to be halted. After observing this phenomenon, Chelomey declared that Pilyugin had deliberately picked out these parameters for the control system.

"Pilyugin should be arrested for this monkey business."

Under Stalin's regime, such declarations could have sealed Pilyugin's fate.

"Come on, he said it in the heat of the moment. Take it easy, Nikolay," said Korolev in a conciliatory tone.

This conversation took place around the table in our "deluxe" dining hall at Site No. 2.

"I will not work with him anymore," declared Pilyugin.

"Don't be stupid, Nikolay," I said, taking Korolev's side, "Install a filter. We've already been through this many times."

Pilyugin continued to work; there was nowhere else to go. He produced the system for the UR-100 and the heavy UR-500 launch vehicle, i.e., the *Sotka* and Proton respectively. The UR-200 missile had no prospects and did not become the launch vehicle for anything.[73] Despite being one of the proponents for Yangel's concept, Pilyugin was obliged to carry out the decree of the Central Committee and government—to produce a guidance and control system for Chelomey's missiles.

When the need arose to create a control system for the special stage that would split the R-36 missile warheads among various targets, Yangel persuaded Pilyugin to take on this difficult task. This assignment captivated him. And you have to give credit to the NIIAP team that did an excellent job with it.[74]

---

73. The UR-200 program was cancelled in 1965 after a series of test flights during 1963 and 1964.

74. In 1963, Pilyugin's group separated from NII-885 and became the independent NIIAP—*Nauchno-issledovatelskiy institut avtomatiki i priborostroyeniya* (Scientific-Research Institute of Automatics and Instrument Building).

Chelomey did not create a Council of Chiefs like the old Korolevian one. Actually, he presided over a review team in the form of a consultative agency attached to the general designer, rather than our understanding of a Council of Chiefs. Pilyugin tried to avoid attending these councils, sending his deputies instead.

The second serious conflict between Pilyugin and Chelomey occurred regarding the theory of the pendulous accelerometer that was used to control flight range. Chelomey, who justifiably believed himself to be a specialist in oscillation theory, tried to prove that great range errors during *Sotka* development launches should be attributed to imperfections in the range control system, which used this invention of Pilyugin's. Viktor Kuznetsov, who had lost his gyroscope monopoly, added fuel to the fire. Pilyugin managed to set up a new facility in southwest Moscow for the development and precision production of gyroscope technology.

Other chiefs often found themselves in difficult situations, especially Iosifyan, the chief designer of on-board electrical equipment and Goltsman, the chief designer of extensive ground-based electric power equipment.

PERFORMING COMBAT READINESS INSPECTIONS AND DIAGNOSTICS ON EACH MISSILE IN A SILO AND CONSTANT REPAIRS REQUIRED TRAINING THOUSANDS OF QUALIFIED MILITARY SPECIALISTS. Troop units—missile brigades and divisions—were also divided according to a sort of clannish principle that depended on the chief designer or missile system developer. The doctrine of deterrence that the Defense Council had virtually approved logically led to more rigid requirements for the missile systems under construction and required that they be updated substantially. This work demanded enormous organizational efforts and material expenditures.

Our theoreticians were armed with Robert S. MacNamara's arguments. As Secretary of Defense in John F. Kennedy's Administration, he had organized broad studies of various nuclear war scenarios. Like it or not, MacNamara's concept made our military leaders listen. One of MacNamara's primary postulates was the thesis of reducing U.S. losses by delivering strikes against the Soviet Union's strategic assets. After calculating the possible number of missiles that one side or the other could produce in three to five years, the American strategists realized that it was a situation of "mutually assured destruction (MAD)." MacNamara concluded that the destruction of one-fourth to one-third of the population, and half to two-thirds of their industrial potential was unacceptable damage for the two sides. According to his estimates, this would be achieved by a nuclear strike with a total yield of 400 megatons. If you take the average yield of a nuclear warhead as 1 megaton and consider that, for one reason or another, no more than 50 percent of the warheads would reach their targets on enemy territory, then the number of missiles continuously ready for launch is 800. Taking into consideration the missiles undergoing repair, preventive maintenance, and those damaged during the first strike, from 1,000 to 1,500 silo launchers needed to be in service!

According to MacNamara's strategy, with both sides having such levels of nuclear potential, it was impossible in the event of full-scale nuclear war to reduce damage to an acceptable level. Therefore, the nuclear weapon doctrine of "deterrence" by threatening to inflict damage on superior scales gained recognition. American military theoreticians proposed considering mutually assured destruction as the primary guarantor of peace.

Under this concept, neither side abandoned the enhancement of their strategic arsenals. That is why we had to work not only on the flight range and accuracy of missiles and on the yield of the warheads, but also on protecting launchers, the technology for managing combat readiness, and testing the readiness of each silo. That is why the selection of the primary systems comprising this very strategic potential was so complex and led to the "little civil war."

From the very beginning, the Americans showed greater consistency when making this selection. They concentrated their efforts on the development and successive replacement of generations of one basic type of solid-propellant missiles on land and similarly set about replacing generations of solid-propellant missiles on submarines. For a long time during the post-Khrushchev period, we continued to develop and produce several *parallel* lines of strategic missiles, allowing unjustified redundancy.

By early 1968, the Americans had only two types of silo-based ICBMs on duty: 1,000 solid-propellant Minuteman and no more than 50 liquid-propellant Titan II missiles. As updated Minuteman III missiles went into service, obsolete Titan II missiles were decommissioned and one type of ICBM remained on duty in well-protected silos until the latest modernization. By that same time in the U.S., there were 656 solid-propellant Polaris missiles in service on submarines. Having essentially a single type of land-based missile and a single type of naval missile enabled the Americans to concentrate their industrial efforts on the successive modernization of these missiles and their systematic replacement.

Almost 2,000 missiles (taking into consideration missiles that were fired) were manufactured in the U.S. and deployed in silos for duty on the ground and underwater during the period from 1961 through 1967—in just more than five years! Each on-duty missile needed a silo with complex launching equipment! The launch position areas needed to be equipped with extremely reliable communications systems, tactical control, and security.

We were faced with having to build at least all of that to be able to say with certainty that we had achieved parity in the extremely challenging nuclear missile race with the United States. But through it all, we tried to remain leaders in cosmonautics. To the end of his days, Korolev tried to solve the strategic missile selection problem. He sought the optimal way to resolve the conflicts that arose in connection with the problems of selecting the optimal version of a strategic missile. It needed to be proven in practice that solid-propellant missiles were capable of defusing the conflicts between Korolev and Yangel, and by the same token, between Yangel and Chelomey. Moreover, it needed to be shown that OKB-1, which had opened the space age, was still capable of a new feat in the field of combat missile technology, which neither Yangel nor Chelomey could accomplish. I will describe this fruitful initiative of Korolev in the chapter "Correcting the Great Ones' Mistakes" (Chapter 6).

The early 1970s can be considered the peak period in the race between the Soviet Union and the United States to produce intercontinental missile systems. Thirty years have since passed! The other nuclear super power—the Soviet Union—no longer exists. A shoot-out between the U.S. and Russia using the intercontinental missiles that many thousands of scientists and hundreds of thousands of engineers labored to produce is highly unlikely. New problems have arisen requiring the development of new nuclear missile strategic doctrines. Nuclear weapons and the technologies for their manufacture are no longer the monopoly of the U.S. and Russia. One must remember that by 2010–2015, at least 30 nations will possess intermediate range nuclear missiles. By that time, China (and perhaps India) will become superpowers, surpassing Russia in terms of military-strategic might. However, in terms of its massive reserves of hydrocarbon raw materials—oil and gas—and fresh water, Russia will remain the mightiest superpower.

The problem of providing humanity with energy using thermonuclear fusion will not be solved in the next decade. The goal of local wars or, God forbid, World War III, will not be to destroy an enemy, but to gain access to areas containing natural resources and fresh water. China might need Lake Baikal to provide drinking water for its population of 1.5 billion. To withstand the new threat from the east, the United States and Europe will need to establish control over the oil and gas of Western Siberia. We are not capable of predicting how far the United States will go to preserve its virtual superiority in the epoch of globalization and to protect "U.S. interests" on any point of Earth's surface and in space. We cannot foresee when leaders will come to power in China who, unlike the current, relatively peace-loving leadership, will be capable of practically implementing the idea of Chinese rule over all of Asia as far as the Ural Mountains. Under those conditions, it appears that the strategic significance of high-precision, non-nuclear weaponry together with intermediate- and even short-range tactical nuclear weapons might become a factor in deterring a large war just as ICBMs were in the 20th century.

Chapter 6
Correcting the Great Ones' Mistakes

During the first half of the 1960s, our primacy in space was undeniable, but despite the heroism displayed in the work of the Korolev, Yangel, Chelomey, and Makeyev teams, and of the subcontracting organizations and production facilities affiliated with them, we lagged ever farther behind in the production of ICBMs. Our rocket-space propaganda, which relied not only on domestic enthusiasm, but also on prestige abroad, prematurely created the myth that we enjoyed overwhelming nuclear missile superiority. Only the inner circle of missile specialists had no illusions about the actual balance of intercontinental nuclear missile forces.

Our leading-edge successes in space were hardly due to general economic and technological superiority over the U.S. In this specific sector of scientific and technical progress, we had managed to create an intellectual, conceptual, and organizational advantage, thanks in large part to the initiatives and monopolistic position of Korolev's OKB-1.

Space spectaculars more readily resonated with the nation's top leaders than did the controversial and complex problems of selecting specific strategic weapons systems. During Khrushchev's administration, missiles were given indisputable priority over all other branches of the strategic armed forces. But which missiles?

To be fair, one must acknowledge that the Americans shared the successes they enjoyed in space during the 1960s with German specialists. Then again, one must also acknowledge that with the advent of the era of large-volume series production of ICBMs and submarine-launched ballistic missiles, the Americans followed their own path. Here, throughout the 1960s and during the first half of the 1970s, they had indisputable supremacy. The total number of ICBMs installed in silos in the U.S. had reached 850 by the end of 1965. The total nuclear yield of the warheads was approximately 1,000 to 1,200 megatons. In light of the reliability of the missiles at that time, the missiles of the United States in 1965 could have destroyed the USSR twice alone!

By that time, our potential response was at most 150 intercontinental missiles in all, with a total nuclear yield (even taken together with submarine-launched ballistic missiles) of 250 megatons. The United States' offensive strategic nuclear missile arsenal was at least four times greater than ours! If you were to add to that the nuclear warheads belonging to strategic aviation, then their overwhelming superiority would be five to six times greater.

A study of the rates of the proliferation of intercontinental strategic missiles shows the fallacy of the thesis that the Americans were counting on strategic aviation and therefore had not attached the same value to missiles that we had. That might have been the case before 1957, before we launched our first *Sputnik*. In the U.S., design of the solid-propellant Minuteman I missile was under way in 1957–58. Their production began in 1959, and in 1961, silo launchers with these missiles went into service.[1] By late 1963, there were 450 missiles standing on duty in silos! Beginning in 1963, Minuteman I missiles were gradually replaced with the more advanced Minuteman II, and then with the even more advanced Minuteman III. Submarines were being built and armed at rapid rates. In 1963, there were 160 solid-fuel Polaris missiles in service on U.S. submarines.

Besides their overwhelming numerical superiority, one more advantage the Americans had was accuracy. Minuteman missiles, with a strictly inertial guidance and control system, had a Circular Error Probability (CEP) of 1,000 to 1,500 meters; for Polaris missiles, it was 1,600 meters. Our R-9A, which used a radio correction system in accordance with the approved military operational requirements, had an error three times greater. The Minuteman was twice as accurate as Yangel's R-16, which had a strictly inertial system. The Americans had gained quantitative and qualitative superiority, mainly because they did not have internal disputes over the advantages or shortcomings of low- and high-boiling propellant components for combat missiles. Yangel vehemently insisted that Korolev's enthusiasm for liquid oxygen was driving our missile technology to a dead end.

Now, citing the many years of American experience, one can claim just as vehemently that Korolev was not alone: Yangel himself, and subsequently Chelomey, while they may not have been driving missile technology to a dead end, were certainly following a path that was needlessly complicated. The great chief and general designers—Korolev, Yangel, and Chelomey—all made the same mistake. Korolev was the first to understand and attempt to correct it.

---

1. The Strategic Air Command (SAC) placed the first 10 Minuteman I ICBMs on active duty on 22 October 1962.

The Americans suddenly overtook us where we had considered ourselves the strongest after the war. We were rightfully proud of our *Katyushas*.[2] Our military historians asserted that neither the Germans nor the Allies had managed during, and immediately after, the war to produce such effective reactive solid-propellant projectiles using special nitroglycerin powder propellant. Our projectiles had solid-propellant rocket engines that were a great deal simpler, more reliable, and less expensive than any liquid-propellant engines.

The history of powder-fuel engines usually begins with an account of how the solid-fuel rocket engine was the very first to find practical application in missile technology. I do not want to repeat the textbook narrative from Chinese fireworks to the *Katyusha*. I only will mention that non-explosive combustion was one of the problems of developing the solid-propellant engines for the *Katyusha's* rounds. The development of the manufacturing process for explosive cartridges with a diameter up to 150 to 200 mm allowed the combustion process to become stable for several seconds. Our chemists specializing in explosives had every right to be proud of these powders. But they proved to be quite unsuitable for [guided] missiles having a powered flight segment that lasts tens or hundreds of seconds. In solid-propellant missiles, the charge and the engine are all one piece and cannot cool the nozzle the way the fuel in a liquid-propellant rocket engine does during the combustion process. The intensity of the combustion product's thermal effect on the shell of the rocket engine casing becomes intolerably high when the engine operates for a prolonged period of time. Moreover, during prolonged storage or exposure to operating pressure, the propellant charge developed cracks, the lateral surfaces of the charge ignited, and the temperature became so high that the body burned through. Charges made of stable, smokeless, granular powder containing special solvents proved to be good for missile projectiles, but quite unsuitable for large rockets. Conventional solid-propellant rocket engines also had a low thrust performance index compared with liquid-propellant rocket engines.

Since the times of the classic works of the rocket technology pioneers, it has been considered an inviolable truth that solid propellant—a variety of powders—is used in those cases "when a simple, inexpensive, briefly-operating engine is required."[3] Only liquid propellant should be used for long-range missiles. It continued like this until the early 1950s

---

2. The *Katyushas* were multiple launch units for unguided short-range solid propellant rocket projectiles. The Soviets used them very effectively during World War II against the Nazis.

3. G. E. Langemak and V. P. Glushko, *Rakety, ikh ustroystvo i primeneniye* [*Rockets, Their Design and Application*] (Moscow-Leningrad, 1935).

when the California Institute of Technology's Jet Propulsion Laboratory (JPL) developed a composite solid propellant.[4] This was not a powder at all. The only thing it had in common with powders was the fact that the fuel did not require a secondary oxidizer—this was contained in the propellant itself.

Composite solid propellant, invented in the United States, was far superior to all the types of powders that we used in our rocket artillery in terms of its energy characteristics. Prompted by missile specialists, the powerful American chemical industry appreciated the prospects of the invention and developed the technology for large-scale production.

Composite solid propellant is a mechanical mixture of fine, solid particles of oxidizer and powdered metal or metal hydride evenly distributed in an organic polymer, and the composite contains up to 10 to 12 components. It uses oxygen-rich nitrates and perchlorates and organic nitro compounds as oxidizers. Metals in the form of high-dispersion powders constitute the primary fuel. The cheapest and most readily available fuel is aluminum powder. Even with a streamlined manufacturing process, composite propellants are still significantly more expensive than the best liquid components (in terms of energy parameters).

When the propellant is poured into the missile's body, an internal combustion channel is formed. The engine body is also protected against thermal effect by a layer of propellant. It became possible to produce a solid-propellant rocket engine with an operating time of dozens and hundreds of seconds.

The new munitions technology, the greater level of safety, and composite propellants' capacity for stable combustion made it possible to manufacture larger charges and thereby to create a high coefficient of mass perfection, despite the fact that the thrust performance index of solid-propellant rocket engines—even for the best component formulas—was substantially lower than for contemporary liquid-propellant rocket engines.[5] However, their structural simplicity—no turbopump assembly, complex fittings, or pipelines—and the high density of the solid propellant made it possible to create a rocket with a higher Tsiolkovskiy number (initial-to-final weight ratio). The clashes between Korolev and Yangel as well as the ensuing "civil war"—the rivalry between Yangel's and Chelomey's schools—might have been completely different if our industry had mastered the production of composite solid propellant, say, five years earlier.

---

4. The JPL invented a castable composite propellant charge much earlier, during World War II, which was tested on such short-range missiles as the Private A and Private F.

5. The "coefficient of mass perfection" is denoted by m/v (the ratio of mass to volume).

The NII-4 Institute undertook the first attempt to develop a solid-propellant long-range ballistic missile during 1955–59. General Andrey Sokolov was director of NII-4 at that time, and Sokolov's deputy was Colonel Georgiy Tyulin.

Boris Zhitkov, Doctor of Technical Sciences, supervised the development of the PR-1 solid-fuel missile with a range of 60 to 70 kilometers. This missile was successfully tested at Kapustin Yar in 1959. NII-4 obtained a Council of Ministers special decree for the development of the PR-2 solid-fuel guided missile. With a missile mass of 6.2 metric tons, it was capable of carrying a warhead weighing 900 kilograms to a range of 250 kilometers. This missile was the solid-propellant analog of the R-11 liquid-oxygen missile developed by Korolev's OKB-1. While working on these projects, formulas for high-energy composite solid-propellants, thermal protective coatings, erosion-resistant materials, and swiveling control nozzles were developed. However, industry and even the Ministry of Defense failed to support the initiative of NII-4 scientists.

Korolev understood that in competition with Yangel and Chelomey, the R-9 missile and any of its modifications would lose simply because "high-boiling" component missiles could be stored in a fueled state. They would always have a higher degree of readiness. He needed a trigger—an impetus to begin the selection process, to seek a fundamentally different third path. Not one but three concurrent factors made Korolev the first of our chief designers and missile strategists to rethink, to alter the choice whereby strategic missile weaponry had been oriented exclusively toward liquid-oxygen missiles. For various reasons, historical works on rocket-space technology and studies of Korolev's creative legacy fail to devote proper attention to his work in this area.

The **first** impetus to begin work at OKB-1 on solid-propellant missiles was the abundance of information that had accumulated in early 1958 on the Americans' intent to develop a new type of three-stage intercontinental missile. I no longer remember when we received the first information about Minuteman missiles, but I was in Mishin's office on business of some sort and listened to a conversation about the authenticity of this information. One of the conceptual designers was reporting to him about how the information received matched the conceptions that we had at that time about the capabilities of solid-propellant missiles. The collective opinion was unanimous: it was impossible in our time to develop a missile with a launch mass of just 30 metric tons, a warhead mass of 0.5 metric tons, and a range of 10,000 kilometers.

On that note, things settled down temporarily. But not for long. On his way to the Northern Fleet, Viktor Makeyev dropped in to see us. He spoke with Korolev and Mishin about naval affairs and problems and then he dropped in on me with his guidance specialists. The conversation had to do with our assisting in the development of more powerful control surface actuators. We quickly reached an agreement on this matter. At the end of the meeting, he said that he had passed on information about the U.S. Polaris missile to S.P. If this was not disinformation, then the Americans had the capability to immediately arm their submarines with solid-propellant missiles that were much better suited for naval conditions.

"Imagine. No leaks, no off-gassing, no evaporation anywhere. You can cruise underwater as long as you want, and there are no frightening smells."

Makeyev had already had his fill of trouble with our legacy—the R-11FM fueled with "nitric acid" and then with his own R-13 (or SS-N-4). The latter project, inspired by Aleksey Isayev, was a missile with the "submerged" engine. To reduce the overall length of the missile, Isayev had proposed "submerging" the entire propulsion system in the propellant tank.[6] But fueling, storage, corrosion, and leakage problems still remained.

Makeyev said, "I got the feeling that our S.P. didn't know whether we could believe this information on the Polaris."

The **second** impetus to begin operations on solid-propellant missiles came from Yuriy Pobedonostsev, our old compatriot from GIRD, RNII, and NII-88.[7] An accidental encounter in Sokolniki helped me find out about this.

My old residence on Korolenko Street and the new one in Ostankino were very convenient for short cross-country skiing outings. Just 5 minutes after leaving home, you could be on your skis. If I happened to have a free Sunday, sometimes I would take the boys skiing through Sokolniki to the distant wooded park on Losinyy Island. I recall that once, in the company of two 11-year-old boys, my son and his school friend Igor Shchennikov (who lived in our apartment building), we ventured so far that on the way home we got lost in a snowstorm. I led the boys home completely exhausted and got a well-deserved chewing out from their mothers. Incidentally, this episode also has relevance to solid-propellant missiles. Thirty years later, one of the leading designers of solid-propellant missiles, Igor Shchennikov, still remembers that skiing expedition.

---

6. The engine was known as the S2.713.

7. GIRD—*Gruppa po izucheniyu reaktivnogo dvizheniya* (Group for the Study of Reactive Motion)—was a semi-amateur group from the early 1930s dedicated to rocket experimentation. Its four primary leaders were Fridrikh Tsander, Sergey Korolev, Mikhail Tikhonravov, and Yuriy Pobedonostsev. RNII—*Reaktivnyy nauchno-issledovatelskiy institut* (Reactive Scientific-Research Institute)—was the first state-sponsored rocketry research institute organized in the Soviet Union. It was established in 1933 by combining teams from both GIRD in Moscow and the Gas Dynamics Laboratory (GDL) in Leningrad.

In early March 1958, there was still enough snow cover in Sokolniki for cross-country skiers. During one of my skiing outings, I unexpectedly ran into Pobedonostsev. He was strolling with his wife along a cleared walkway, and I was coming from the opposite direction on a parallel ski trail. We were both very happy to see one another. I took off my skis and we walked together. Ten years before that encounter I had been the deputy to NII-88 chief engineer Yuriy Aleksandrovich Pobedonostsev. It was both easy and interesting to talk with him. His optimism had not failed him during the most difficult situations of the first years of NII-88's existence. I too was touched by his optimism.

Pobedonostsev had left NII-88 to be provost of the Academy of Managerial Staff of the Defense Industry.[8] When OKB-1 was made an independent organization, it was rumored that he might return, but to Korolev rather than to NII-88. However, the old comrades-in-arms decided that to maintain good relations, it was better that one of them not be subordinate to the other.

During his time in Germany, Pobedonostsev was in excellent health. He was the only one who dared to swim in the cold water of the Baltic Sea during the spring and fall of 1945 in Peenemünde. During extended visits to Kapustin Yar, Pobedonostsev did not miss an opportunity to run over the steppe, while we preferred "horizontal testing" back in the compartment of our special train. Now he complained that his heart had started to fail, but he wouldn't give up.

"I lecture students," said Pobedonostsev, "and I work at NII-125 with Boris Petrovich Zhukov. I'm trying to revive old ideas with the help of new solid-propellant technology. I've met with Sergey a couple of times and persuaded him to get involved with us on a solid-propellant model. The only thing that aggravates me is Mishin's intransigence. He doesn't want to hear about our proposals. Sergey promised me that he would select a group that would answer directly to him. If the project moves forward, you and I will get together and discuss guidance problems. They'll be a lot different from the liquid-oxygen systems."

I soon recalled my encounter in Sokolniki after I received a direct reference from Korolev. He asked me over the telephone whether I had ever met Igor Sadovskiy. I confirmed that I not only had met him, but that I knew the handsome, young Igor Sadovskiy very well from NII-88. In 1948, Sadovskiy had tried to lure me to the field of anti-aircraft missile guidance and control.

---

8. The Academy of the Defense Industry (its formal name) was an educational institution established in the postwar years to train individuals for bureaucratic jobs in the Soviet defense industry.

He was working as a conceptual designer on these problems until the field was transferred to the Ministry of the Aviation Industry.

"Yeah," S.P. interrupted me, "he wanted to become Chairman of the Council of Ministers, then Minister of Medium Machine Building, but neither of those plans worked out. He has returned to us and is working with Lavrov in a more modest position, but on an interesting new subject."

I caught on that Sadovskiy was in Korolev's office and was listening to our conversation.

"In a week or two he will tell you everything. Think about which of your capable people you want to hook up with him for consultations. Only for advice for the time being, and then we'll see."

A week later Sadovskiy dropped by to see me and briefed me on the situation. For the past few years he really had worked in the office of the Council of Ministers and then in the Ministry of Medium Machine Building—the atomic ministry. But missile technology once again had a strong pull on him. He realized that bureaucratic work was not for him. He quickly made an arrangement with Korolev and was named deputy to Svyatoslav Lavrov, Chief of the Ballistics Design Department. Sadovskiy drew volunteers and assembled a small "underground" group to prepare proposals for solid-propellant ballistic missiles. The main core consisted of three young specialists, Verbin, Sungurov, and Titov.

"These kids are still green, but very capable," said Sadovskiy. "I have divided three main tasks among them: internal ballistics, external ballistics, and design. My former bureaucratic connections have helped me. I managed to reach an agreement with Boris Petrovich Zhukov, director of NII-125 (this is our main institute for missile and special solid propellants) on our joint—and for the time being theoretical—study. And our old boss Pobedonostsev is in charge of the laboratory at NII-125. They're not just working on paper, they're conducting experiments on the development of larger solid-propellant charges with a new composition." Sadovskiy had told Korolev about his "underground" activity.

Korolev immediately arranged with Zhukov and Pobedonostsev to "emerge from the underground," and they began to develop a design for a medium range solid-propellant missile.

I told Sadovskiy about my encounter with Pobedonostsev.

"Hey, I am that 'special task force' that S.P. promised Pobedonostsev he would organize at OKB-1 for joint work."

At the risk of tiring the reader, I have been dwelling on meetings and conversations that would seem not to be particularly interesting. But now, in the grand scheme of history, they seem rather important to me. In trying to reconstruct the history of this period, I maintain that Korolev, Pobedonostsev, Sadovskiy, and Zhukov—that is specifically the order that seems the most correct to me—were the first active individuals responsible for the resurgence of medium range solid-propellant ballistic missile technology in the Soviet Union.

As for the aforementioned work of NII-4, its work on solid propellant missiles was an example of an initiative of military engineers that found no support from their own ministry and was not picked up by a single one of the "powerful" industry designers. Prior to 1959, all the chief designers were so carried away with the competition to develop liquid-propellant missiles that they simply brushed aside the NII-4 projects, despite the fact that credible information was already available about the U.S. Minuteman and Polaris projects.

After the names of our "solid-propellant pioneers," i.e., Korolev, Pobedonostsev, Sadovskiy, and Zhukov, I would list Nikolay Pilyugin, Yakov Tregub, Vladilen Finogeyev, Aleksandr Nadiradze, and finally, Dmitriy Ustinov as being important to the development of solid propellant missiles in the Soviet Union. Communist Party Central Committee Secretary Ustinov was the first of the great political leaders who appreciated the prospects of this new—and at the same time the oldest—field.

Sometimes actions that seem insignificant at first glance play the role of a trigger in history. And actually, the missile development process snowballed from then on. Together with NII-125, a so-called group, in which 40-year-old Sadovskiy was the most experienced solid-propellant expert, published a three-volume report proving that it was possible to produce a medium-range missile using "ballistite" powder, which was supposed to be produced in the form of large-diameter pressed powder charges.[9] The powder for ballistic missiles was called "ballistite" rather than "ballistic." The theoreticians explained that that was a tribute to artillery traditions.

For Pobedonostsev, it was as if his creative engineering work had come full circle. During 1934–35, the young engineer Pobedonostsev had been involved in the testing of solid-propellant reactive shells and had studied the theory of combustion of various powders. During World War II, through their truly heroic work, scientists, engineers, and workers—"powder specialists"—developed nitroglycerine ballistite powders and a high-volume manufacturing process for producing shells. This made it possible to use solid-propellant reactive shells on a massive scale with a very high degree of effectiveness. After the war, they continued to perfect the manufacturing process of shells for

---

9. "Ballistite" was a type of smokeless powder made from two explosive components, nitrocellulose and nitroglycerin, patented in 1887 by Alfred Nobel (1833–96).

the salvo-fire rocket artillery that had been developed, drawing on the combat experience of the *Katyusha*. These projects were conducted at the head institute of the solid-propellant industry, NII-125. There, Yuriy Pobedonostsev returned once again to the old subject matter of RNII.

At NII-125, Pobedonostsev was one of those who initiated the development of a manufacturing process for shells in the form of a set of charges with a diameter as large as 0.8 to 1 meter and a total length of up to 6 meters.

Unlike their liquid-propellant counterparts, solid-propellant missiles required the solution of another whole series of new problems. First and foremost, ready-made materials needed to be found or new temperature-resistant materials and nozzle cluster designs needed to be developed. Additionally, guidance methods needed to be devised without resorting to control vanes. Unlike liquid-propellant engines, solid-propellant engines could not be regulated and shut down, which made the precise control of flight range difficult. "Once they're ignited, all you can do is wait until they burn out." That's how the young specialists from Sadovskiy's group explained the situation to us at first. There were no testing rigs or experimental facilities of the proper size for developmental tests on solid-propellant engines. There was the danger that the design process would be stifled and it would fizzle out, especially since Mishin and others among our conceptual designers were skeptical and opposed it.

Korolev undertook what was for those days a risky, but "very strong move," as chess players would say. Literally about 10 days after the order was issued for our merger with TsNII-58, he requested that all specialists in the fields of projectiles, powders, and ballistics assemble in the Red Hall behind what had formerly been Vasiliy Grabin's office.[10] I was not at that meeting. Later Sadovskiy enthusiastically told me that the tiny hall was "packed with around 100 of Grabin's people."

Korolev arrived at the meeting with Sadovskiy. He began by speaking about the American Minuteman and Polaris missiles, brandishing a sheet of paper that listed their specifications. Turning to Grabin's specialists, Korolev called on them to join the project to produce Soviet solid-propellant missiles. He emphasized that while we had a clear advantage in liquid-propellant missiles, we not only lagged behind for the time being in terms of solid-propellant missiles but simply had nothing. Korolev introduced Sadovskiy as head of

---

10. This is a reference to the 1959 government decision which attached the former TsNII-58 as a branch of Korolev's main OKB-1 organization. TsNII-58 had been headed by former tank designer V. G. Grabin. See Chertok, *Rockets and People, Vol. II*, Chapter 27.

the work and announced that he would be his, Korolev's, deputy in this new field. Grabin's people, who had feared being left without any work after the merger with OKB-1, suddenly saw a very promising future for their creative activity. They received the proposal enthusiastically.

Over the course of several days, two departments were formed under Sadovskiy's supervision, one for design and one for testing. Together with Sadovskiy and Yurasov, I went to see Pilyugin to persuade him to take on guidance and control problems. It turned out that Korolev had already softened him up. And we didn't so much persuade him as discuss with him immediate and urgent problems. Pilyugin's deputy, Mikhail Khitrik, didn't miss the opportunity to ask a difficult question to trip us up: "The Americans are building Minuteman missiles with a range of 10,000 kilometers and a mass of just 30 metric tons. And all you can manage is 2,000 kilometers with a greater launch mass. What's the deal? Incidentally, both their missiles and your missiles have three stages."

Sadovskiy was already sufficiently prepared to provide a cogent answer to that question, which opponents had posed before.

"The Americans aren't lying. They have succeeded in developing a fundamentally new high-efficiency propellant that chemists call composite solid-propellant. Our industry doesn't yet know how to make charges out of this propellant. Research has just begun on our initiative. Perhaps, in a year or two the formulas and manufacturing process for composite solid-propellant charges will be developed. But for the time being, we will make use of the achievements of NII-125. We will use ballistite powder charges, and rather than fill the missile with powder as the Americans do, we will insert charges into the body of a missile that already has a ready-made nozzle. In other words, the missile already is the engine."

Right there in Pilyugin's office we reached an agreement on range control methods: the forward head of each engine of each stage would have reverse thrust nozzles that would open on a command sent from the control system using a detonating fuse. Thus, with a single command, we could immediately neutralize the thrust and ensure accuracy in range, which would not be impaired by the after-effect pulse that liquid-propellant rocket engines always have. As for steering, we would design special exterior-mounted solid-propellant steering control engines and swivel them using powerful control surface actuators. For this, we would humble ourselves before Nikolay Lidorenko so that he would come up with special high-current batteries.[11] The people in Pilyugin's department began working, buoyed by their own enthusiasm, before any orders were issued.

In November 1959, Korolev's clout and the irksome information from abroad came into play at the highest level. A governmental decree came out calling for the development of a missile using ballistite powder charges with a range of 2,500 kilometers and a warhead mass of 800 kilograms. The missile was called RT-1. This was the governmental decree calling for the production in the Soviet Union of a solid-propellant long-range ballistic missile, the

RKK Energiya

The RT-1, developed by the Korolev design bureau in the early 1960s, was the first Soviet strategic missile to be powered by solid propellants.

chief designer of which was Korolev. As soon as the decree was issued, the missile was designated 8K95.

At OKB-1 in early 1960, despite Mishin's resentment, around 500 people were already working on this missile. A mockup of the missile was put on display in shop No. 39 during Brezhnev's visit to OKB-1, and Sadovskiy was awarded the honor of giving a report to the Central Committee Secretary.[12] At Mishin's initiative, a mockup of a competing design—Isayev's "potbellied" liquid-propellant engine—was put on display there. Korolev made concessions in this case, wishing to demonstrate his objectivity. This did not thrill Isayev. The "potbellied" engine went no further than the exhibition. The RT-1 was designed within an unusually brief period of time and was put into production in a cooperative arrangement that was new for the missile industry.

---

11. Nikolay Stepanovich Lidorenko (1916–) was the chief designer of power sources for Soviet spacecraft. For several decades he worked at the Scientific-Research Institute of Current Sources (NII IT), now known as NPO Kvant.

12. At the time, Brezhnev was the Central Committee Secretary in charge of the defense industries, which included the ballistic missile program.

OKB–MEI Chief Designer Aleksey
Bogomolov, and OKB–1 Deputy Chief
Designers Boris Chertok and Yevgeniy
Shabarov.

From the author's archives.

For the first time, chemists,
powder specialists, and specialists
in the textile manufacturing process
of the fiberglass housings—and not
heavy equipment manufacturers—
determined the missile's manufacturing process. NII-885 manufactured
all instrumentation for the guidance
and control system while OKB MEI
provided the *Tral* telemetry system.
My departments designed the control surface actuators and the automatic missile destruction (APR)
system.[13] The year 1961 was taken
up with production and experimental development on test rigs. In the spring
of 1962, Korolev appointed Yevgeniy Shabarov to be the head of flight-testing
of our first solid-propellant missile at the State Central Firing Range (GTsP)
in Kapustin Yar. Permanent GTsP Chief General Vasiliy Voznyuk agreed to
be chairman of the State Commission.

This was the first time that three-stage solid-propellant missiles had
been sent to Kapustin Yar for flight development tests (LKI). With a
launch mass of 35.5 metric tons, the missile was designed for a range of
2,500 kilometers. Each of the missile's three stages consisted of four
solid-propellant engines connected mechanically and operationally. The
diameter of the powder charges of each first-stage engine was 800 millimeters,
while for the second and third stages the diameter was 700 millimeters. The
control devices of the first and third stages were swiveling engines, and for the
second stage the control devices were aerodynamic control surfaces.

The testers of liquid-propellant missiles at firing ranges consider fueling to
be the most hazardous and unpleasant process. "Fueling" the RT-1 delighted
the testers. Prefabricated powder charges arrived from NII-125. According to
the instructions, they needed to be thoroughly wiped off with medical-grade

---

13. APR—*Avtomaticheskiy podryv rakety.*

alcohol before loading them into the body of each missile block. Naturally, the aroma of alcohol put the testers in a better mood than the acrid vapors of nitric acid and the annoying smell of kerosene.

On 28 April 1962, the first launch of the RT-1—the first Soviet solid-propellant medium range missile—took place. The first and subsequent two launches ended in failure, with the command of the APR system that we had developed. It blew up the detonating fuses, which opened up the engines and "zeroed" the thrust. This revealed the need to modify the charges and control system.

Flight development tests resumed in March 1963. In all, nine missiles were tested in flight. The last launch took place in June 1963. The warhead reached the target with a deviation to the right of 2.7 kilometers and with an overshot of 12.4 kilometers. In terms of accuracy, the results were disappointing.

The VPK and the RVSN command needed to decide whether to continue work to perfect the RT-1. There were already two medium-range missiles in service, Yangel's R-12 and R-14. There were no zealous supporters of an initiative to put one more medium-range missile into service. By that time Korolev and Sadovskiy had obtained a Council of Ministers decree to set up broad-scale operations on composite solid propellant. The State Institute of Applied Chemistry (GIPKh) was named as the head organization for the development of composite solid propellants.[14] Vladimir Stepanovich Shpak was its director and chief designer.

The search for formulas and the development of a technology for the industrial production of composite solid propellants spread out "from the southern mountains to the northern seas."[15] Institutes, design bureaus, and factories were at work in Biysk, Perm, Moscow, Leningrad, Votkinsk, and in Krasnozavodsk on the outskirts of Moscow. New chief designers came forward for the first- and second-stage blocks. Each one dreamed of being the first to pluck a feather from the Firebird's tail![16] Relations with NII-125 became strained, as the institute felt that it risked losing its leading role and insisted on continuing its work on large ballistite powder charges.

Starting work on a new subject, Korolev took a broad approach to problem solving, which often irritated high-ranking officials. He abhorred the principle of "let's get started and then we'll figure things out" that very

---

14. GIPKh—*Gosudarstvennyy institut prikladnoy khimii.*

15. The common idiom "from the southern mountains to the northern seas" is analogous to the English "from sea to shining sea." The former is a quote from "March about the Motherland" by Vasiliy Ivanovich Lebedev-Kumach (1898–1949), a famous Soviet poet, lyricist, and song-writer.

16. This is a reference to the Russian folk tale about the Firebird (*zhar-ptitsa*) whose beautiful tail feathers had magical powers.

authoritative figures sometimes followed. From the very beginning of working on a new problem, Korolev sought to engage as many new organizations and competent specialists as possible, and he encouraged developing several alternative scenarios for the sake of achieving a single goal. This broad-based approach to a problem often resulted in other previously unplanned tasks being solved "along the way" to the ultimate goal.

The decree for the development of the RT-2 intercontinental solid-propellant missile is an example of this type of broad-based problem. In the process of solving the ultimate task, two others were solved: from the three stages of the intercontinental missile, a medium-range and a "shorter" range missile were formed. The decree dated 4 January 1961, which came out before testing of the RT-1 (or 8K95) missile was completed, was in preparation for a long time. Korolev patiently conducted elaborate and tedious negotiations with people who were new to him and with the heads of agencies that were not always loyal. The decree approved and accepted for implementation three interconnected solid-propellant engine designs that made it possible to create the following three mutually complementary missile systems:

1. The silo- and ground-based RT-2 ICBM system comprising a three-stage missile operating on composite solid propellant with a range of at least 10,000 kilometers and an inertial guidance and control system. The RT-2 missile system was initially intended to have an integrated payload carrying the same warhead as had been developed for the R-9 and R-16, with a yield of 1.65 megatons. The decree named Korolev as the missile system's chief designer. The RT-2 missile was assigned the designation 8K98.

From the author's archives.

2. A medium-range missile system—up to 5,000 kilometers—ground-based and using the 8K98 first and third stages. This missile was assigned the designation 8K97. Mikhail Tsirulnikov, Chief Designer at the Perm Machine Building Design Bureau was named chief designer of this medium-range system. He had developed the engines for the first and third stages of the 8K98.

Petr Aleksandrovich Tyurin (1917–2000) served as chief designer of TsKB-7 (later KB Arsenal) from 1953 to 1981 during which time the design bureau's profile significantly expanded into solid propellant ballistic missiles and reconnaissance spacecraft.

3. The RT-15 track-mounted mobile missile system, which could also be launched from a silo, had a range of up to 2,500 kilometers. This mobile-launch missile was assigned the designation 8K96. The 8K98 second- and third-stage engines were used for it. TsKB-7 was selected as the head organization for the development of the mobile system.[17] The chief designer was Petr Tyurin. TsKB-7 (soon thereafter renamed KB Arsenal) had a great deal of experience developing artillery systems for the Navy by the time it began the missile construction operations.

Korolev was the chairman of the Council of Chief Designers for all three missile systems.

Over the course of the design operations, it became clear that it was senseless to develop the 8K97 missile because the 8K98 provided a range of 5,000 kilometers when the guidance and control system was readjusted.

According to the draft plan, the RT-2 had a launch mass of 46.1 metric tons and a maximum range of 10,500 kilometers. Even at OKB-1, there were enough skeptics who doubted whether it was possible to build a missile with that range but with half the weight of the R-9. There was also the option of installing a more powerful warhead on the RT-2. In this case, the range was reduced to 4,500 kilometers.

Despite the successful results of the institute's research, the industrial production of composite solid propellant having the necessary degree of effectiveness was lagging. Obtaining propellant with a high specific thrust, capable of retaining its elastic properties for many years, proved to be the most difficult problem. Chelomey, who had asserted that cracks would most certainly form in charges during long-term storage, making them unsuitable for use, was one of the influential and vocal opponents of solid-propellant missiles. It was not possible to detect the presence of cracks before launching, according to him. Therefore, we were supposedly risking "dropping" a missile carrying nuclear warheads on our own territory. The arguments were chilling.

However, the surging activity that Korolev and Sadovskiy were conducting in the development of composite solid-propellant charges yielded its own fruit. There was a sharp increase in the number of visitors, previously unknown to us, who stopped by Korolev's office to discuss these new problems. Korolev said that soon, we would need to find time and get involved with the 8K98 in earnest.

Design operations for the entire missile system were being conducted simultaneously. Sadovskiy became more and more of a coordinator and supervisor rather than a designer and developer. Real authority transferred

---

17. TsKB—*Tsentralnoye konstruktorskoye byuro* (Central Design Bureau).

to those developing the structure and specific systems. But the combat missile system did not have a bona fide manager for some time. The special solid-propellant departments headed by Donskoy and Smerdov—specialists in the field of artillery systems from Grabin's former NII—were simply not sufficient. Korolev placed additional responsibility for design operations on Sergey Okhapkin, who with his inherent exuberant efficiency recruited designers for the job from his own departments, Severov's materials engineers, and factory process engineers.

The difficulty of project design lay not so much in the specific development of technical documentation for production, as it did in the mushrooming volume of things that needed to be coordinated between design bureaus located in different cities. TsKB-7 Chief Designer Tyurin was tremendously helpful to Okhapkin in the design development of the third stage. The second and third stages were manufactured at the Perm Machine Building Factory.[18] After prolonged research, the best composite solid propellant turned out to be the butyl rubber proposed by the powder specialists in Biysk.[19] Silo launchers and command posts were designed in Leningrad. Integrated electrical testing of the entire missile took place at our Experimental Machine Building Factory (ZEM), specifically in shop No. 39's monitoring and test facility (KIS).[20]

Pilyugin started up work on the guidance and control system at his own new facility in southwest Moscow. Having received the assignment to develop a fully-automated launch preparation system with a readiness time of no more than 3 minutes, he decided to also grab up a subject that was optional for his organization, the remote control and monitoring system (SDUK).[21] This system was supposed to encompass monitoring, diagnostics, and the issuance of commands for all silos and connect the command posts of all the scattered regions with Strategic Rocket Forces headquarters. Different ideas resulted in professional conflicts, in the sense that each developer had proof of the reliability of his own proposal and of the unsuitability of the structure or elemental base of his competitor's system.

Korolev would not be witness to the fruits of this work. His life was just 20 days short of letting him see a soft landing on the Moon, 45 days short

---

18. This factory was originally known as Factory No. 172.

19. Butyl rubber is a synthetic rubber first developed by Standard Oil (later Exxon) company in the late 1930s.

20. ZEM—*Zavod eksperimentalnogo mashinostroyeniya*—was the official name of OKB-1's pilot production plant, co-located with the design bureau in Kaliningrad. KIS—*Kontrolnaya ispytatelnaya stantsiya*.

21. SDUK—*Sistema distantsionnogo upravleniya i kontrolya*.

of letting him learn that the Soviet Union's pennant had reached Venus, and 10 months short of letting him see the first launch of his brainchild, the intercontinental solid-propellant missile.[22] Firing range tests of the RT-2 missile system began at the State Central Firing Range at Kapustin Yar in November 1966. The first launch was a success.

After Korolev's death, the stress that accompanied work on the entire 8K98 missile system eased up for a time. The VPK, MOM, and the Strategic Rocket Forces command were so swamped with fulfilling production plans, construction of hundreds of new silo launchers, and putting the missile systems of Yangel and Chelomey on combat duty that the missed deadlines for the beginning of flight development tests on the 8K98 did not trouble them much. Mishin thought that Korolev's traditions should be preserved and that an independent branch should be created for series production. That is how Korolev dealt with all the missiles: the R-1, R-2, R-5, and R-5M were transferred to Dnepropetrovsk; the R-7 and R-9 to Kuybyshev; the R-11 to Krasnoyarsk; and the R-11FM and all naval projects to Miass.[23] For the RT-2, it was also proposed that a branch and a series production OKB be created in Gorkiy. Korolev did not live to see the implementation of this idea. However, when Sadovskiy received the offer to take on the leadership of the branch in Gorkiy and to eventually become chief designer, he was hardly thrilled about it. No one was tempted by the move to Gorkiy, which involved setting up at a new site.

Sadovskiy did not conceal his fears that Mishin would not support the solid-propellant project; and taking advantage of its lack of popularity at the still-shaky new MOM, he would not defend its right to life with the same passion that Korolev had. During the development of the new composite solid-propellant engines, relations between Sadovskiy and Zhukov had become strained. Zhukov began to search for new allies and soon found support in the Ministry of the Defense Industry. Work was being conducted there on medium-range missiles under the leadership of the talented scientist and inventor Aleksandr Nadiradze.

There was now a real danger that the RT-2 would not fly at all for lack of a bona fide manager. However, the project had already come too far. Dozens

---

22. Korolev died on 14 January 1966. The *Luna-9* spacecraft performed the first survivable landing on the Moon on 3 February 1966. The *Venera-3* probe accomplished the first impact on the surface of Venus on 1 March 1966.

23. OKB-586 (M. K. Yangel) was located at Dnepropetrovsk; OKB-1 Branch No. 3 (D. I. Kozlov) was located at Kuybyshev; OKB-10 (M. F. Reshetnev) was located at Kransoyarsk; and SKB-385 (V. P. Makeyev) was located at Miass.

of scientific-research organizations and factories were at work. All of them had plans, schedules, obligations, and reports to be made to higher-ranking agencies. In this critical situation, OKB-1's new Deputy Chief Designer for Testing, Yakov Isayevich Tregub, showed himself to be a real fighter.

Readers of volume one of my book *Rockets and People* will recall that Captain Tregub was sent off on temporary assignment at the disposal of General Aleksandr Tveretskiy, Commander of the Special Purpose Brigade (BON) back in Germany.[24] In 1947, in Kapustin Yar, General Voznyuk appointed Major Tregub to be the chief of the first launch-control team. The staff of the first electrical firing department—Voskresenskiy, Pilyugin, Chertok, and Smirnitskiy—was directly subordinate to Tregub during the launches conducted in the fall of 1947.

With the beginning of broad-scale operations to produce Air Defense Troops missile systems, they also began to set up the first firing range for the testing of Air Defense Troops missile systems, not far from our first missile range, the State Central Firing Range. Tregub was transferred to that firing range, and up until 1964, his entire life was wrapped up in the development and testing of anti-aircraft and anti-ballistic missile systems. Moving up the ladder of the military-engineering hierarchy, this World War II veteran, participant in the launches of the first long-range ballistic missiles, this tester of radar air defense missile systems, occupied a managerial post in the head NII for air defense systems of the Ministry of Defense in the early 1960s.[25] In the call of duty, he became familiar with our rocket-space projects and found the weak points in the systemic structures. However, the quiet work at the NII was clearly not Major General Tregub's cup of tea.

In 1964, after his conversations with Korolev and Mishin, they reached a full mutual understanding. Korolev appealed personally to the commander-in-chief of the Air Defense Troops with the request that Tregub return to the lap of missile technology. Marshal Pavel Batitskiy concurred.[26] Thus, Major General Yakov Isayevich Tregub moved to OKB-1 as deputy for testing during the last year of Korolev's life, and then under Mishin until 1973.

After familiarizing himself with the state of affairs with the RT-2, Tregub found that an ample number of competent specialists were working on problems involving engines, propellants, and materials; Sadovskiy took the leading role in this endeavor. With the same energy and enthusiasm that he

---

24. BON—*Brigada osobogo naznacheniya*. For Tregub, see Chertok, *Rockets and People, Vol. I*, pp. 355–356.

25. This was NII-2 of the Ministry of Defense.

26. Marshal Pavel Fedorovich Batitskiy (1910–84) served as commander-in-chief of the Soviet Air Defense Troops (PVO) from 1966 to 1978.

had displayed since his days at Kapustin Yar, Tregub took full responsibility for the combat missile system, including the construction of silos, setting up launch site areas, and automatic remote control and monitoring systems. He talked me into giving him supervisory authority over routine matters concerning the missile system's flight control, radio tracking, targeting, and power supply systems.

The ministry (i.e., MOM) bureaucracy, which was overloaded with a multitude of decrees calling for the construction of silos, needed to set up a special "silo directorate." Deputy Minister Grigoriy Rafailovich Udarov was assigned personal responsibility for the production of silo launchers for all the missile systems. He was in charge of dozens of design and construction organizations.

Udarov belonged to the generation of Young Communist League (*Komsomol*) members from the 1920s who were virtually wiped out without a trace during the repressions of 1937–38.[27] He was surprised that he was spared. The oldest leader in MOM, he organized silo construction operations in a style reminiscent of the old *Komsomol* "shock force" construction projects. Udarov tasked TsKB-34 in Leningrad, headed by (A.M.) Shakhov, to design the silo launcher for the 8K98.

Udarov supported Tregub's enterprising work, which encompassed an entire set of problems leading up to the moment when the first three silo launchers were handed over in turnkey condition for flight development tests with their own control posts (KP); the launch position area comprised 10 silos with a single common KP. The 8K98 needed no storage facilities for the propellant components—the missiles were delivered, already fueled, to be placed on long-term duty or for the next launch. There was no threat to the silos of the hazardous off-gassing of oxygen or toxic fumes from high-boiling propellants.

In 1965, Korolev still had the opportunity to be the arbiter in the conflict over the selection of the developer for a remote control and monitoring system. Konstantin Marks was the first to respond to the needs of the 8K98 in the automatic monitoring and launch control system. All sorts of contrived systems and tank depletion automatics had already flown; creative energy needed to be directed into a new field.

In collaboration with the automation design bureau in Zaporozhye, Marks proposed his own version of automatic preparation and launch for the 8K98 solely using relay technology principles. It is possible that his version

---

27. The *Komsomol—kommunisticheskiy soyuz molodezhi* (Communist Union of Youth, or more generally, the Young Communist League) was the youth wing of the Communist Party of the Soviet Union. Formed in 1918 soon after the Russian Revolution, the *Komsomol* served as an ideological training ground for young Soviet citizens.

might have been accepted, but at that time we received a visit from our friend from Germany and Kapustin Yar, Colonel Grigoriy Ioffe. At Kapustin Yar, Captain Ioffe had been famous not only as the leading military specialist on the electrical testing of missiles, but also as a fanatic angler who enjoyed quiet fishing expeditions on the Akhtuba River.

I was only able to join Ioffe on fishing expeditions three times during temporary assignments to Kapustin Yar. I marveled at his skill, how he concentrated his attention on the bobber, enduring the attacks of swarms of mosquitoes without moving a muscle. Unlike me, his patience was rewarded with a superb catch of sturgeon that would later make a magnificent fish soup.

After finding out about of our troubles, Colonel Ioffe perked up and told us that he was serving as senior military representative in Taras Sokolov's OKB, which had been set up at the Leningrad Polytechnical Institute. Sokolov had already developed the *Signal* system—a remote control and monitoring system using non-contact ferrite elements.[28]

"In Moscow you punch in a code, press a few buttons, and depending on what you want, the missiles fly out of the silos one at a time or as a salvo in the launch position area that you selected. The system is being developed at the request of the Ministry of Defense for the launch control of Chelomey's *Sotka* [UR-100] series."

Right then and there, we struck a deal with our old friend and sent our highly qualified and objective specialists Petr Kupriyanchik and Vyacheslav Khorunov to Leningrad to see Sokolov. Upon their return, they came out in favor of developing a system based on Sokolov's ideas and proposed that we abandon the operations with Marks' Zaporozhye design bureau. The differing points of view were reported to Korolev and he ordered that a special commission be set up to select a subcontractor. The commission, headed by Tregub and with fierce resistance from Marks, came out in favor of the Leningrad version. Korolev decided to finance both versions and make the final selection based on the results of comparative tests.

The first models of the systems were presented for comparative testing after Korolev had passed away. The testers confirmed the superiority of the Leningrad version, especially since the internal control automatics, diagnostics, and readiness check of each missile interfaced with the communications and

---

28. Taras Nikolayevich Sokolov (1911–79) served as chief designer of OKB of the Leningrad Polytechnical Institute (LPI) (later known as NPO Impuls) and led the development of the *Signal* automated strategic command-and-control system, which was deployed in 1967 and 1968.

remote control system that Sokolov had already developed. Marks disagreed with the commission's decisions and announced to Mishin that if the tactical control system that he and the Zaporozhye design bureau had proposed was not accepted for implementation, then he would not work. Mishin flew into a rage and retorted, "If you don't want to work, then hand in your resignation!"

Marks immediately wrote a letter of resignation, to which Mishin straight away added his own concurrence, accepting Marks' voluntary withdrawal from his position. A while later, when the dust had settled, Marks completed the paperwork for a transfer to the State Union Design Bureau of Special Machine Building (GSKB Spetsmash) to work with Vladimir Barmin.[29] He didn't calm down over there either and independently developed his own version of the SDUK. Marks reported the advantages of his system to the ministry. But it was too late. Industry and the assembly organizations were busy installing Sokolov's systems in hundreds of different silos. However, the "little civil war" over SDUKs did not end there.

In 1967, Central Committee Secretary Ustinov called Pilyugin in and proposed that he become the chief designer of a flight guidance system for mobile-launch solid-propellant missiles being developed by the as-yet obscure Chief Designer Nadiradze who worked in the Ministry of the Defense Industry. Pilyugin replied that he was terribly loaded down with work; his own Minister Afanasyev didn't help him much; and if he were to agree to develop a system for an outside ministry, then it would be impossible for him to turn to MOM, his own ministry. Ustinov reassured him that everything would be worked out. But things didn't go that smoothly. For a long time, Afanasyev reminded Pilyugin that he had gone around his own ministry to take on the project.

When Nadiradze arrived to approve the design specifications for Pilyugin, the latter stipulated a condition: "I will take this only in combination with the strategic control system." Using Pilyugin's own ideas, his deputy, Nikolay Tishchenko, was developing a third version of the SDUK, which they would soon try to "impose" onto the updated version of the 8K98. This time, Tregub and the rest of us had to defend the system that Sokolov had developed in Leningrad.

LET'S RETURN TO THE HISTORY OF THE RT-2. The 49th anniversary of the Great October Socialist Revolution was approaching in late 1966. Everyone was preparing gifts to be presented on behalf of their workplaces.[30] It was our duty to begin flight tests on the RT-2 before the anniversary.

---

29. GSKB Spetsmash—*Gosudarstvennoye soyuznoye konstruktorskoye byuro*.

30. Giving "gifts" to the Party and government during major holidays was a customary practice during the Soviet era. Typically, organizations prepared major events or produced notable successes to coincide with the holidays.

The first launches of previously developed intercontinental missiles usually proceeded in keeping with the notorious "first-pancake-is-always-lumpy" principle. The first 8K98 launch on 4 November 1966 from a silo launcher at the Plesetsk firing range was a success. True, the payload overshot the CEP boundaries declared for the Minuteman missiles. But no one attached particular significance to that. The chairman of the State Commission for the tests that had begun was Major General Anatoliy Vasilyev, former deputy of Voznyuk, who had headed the A. F. Mozhayskiy Military Engineering Academy.[31] Sadovskiy was the technical head and effectively had the rights of a chief designer. Tregub was in charge of preparation and launch. Okhapkin came for the first solid-propellant missile launch. He announced that Mishin had supposedly ordered him to come. Actually, having the genuine instincts and common sense of an experienced designer, Okhapkin quickly grasped the potential of solid-propellant missiles.

The subsequent two launches in December 1966 were unsuccessful. After two failures, some were inclined to curtail the launches and shift over to the long process of testing and verifying the engines and control system on the ground. Tregub categorically objected. He reasoned that the ground-based experimental facilities were very primitive and each launch gave us invaluable experience. Missile production was running smoothly now, and from the standpoint of shortening the testing and verification cycle, it was economically and politically advantageous to perform modifications on the missiles during flight development testing. In this case, he was right and State Commission Chairman Vasilyev supported him.

Between 4 November 1966 and 3 October 1968, a total of 25 launches were conducted in the process of flight development testing. Of these, 16 were successful. Seven missiles were selected and delivered to silos for long-term duty and subsequent verification by firing.

During 1968, 8K98 missiles were put into service and began to be installed in silos on Russian territory. A missile division comprising six regiments was armed with RT-2 missiles standing on alert. In terms of firing power, a single missile division surpassed many times over that of all the divisions of the warring nations of World War II combined!

---

31. Lt.-Gen.-Engineer Anatoliy Alekseyevich Vasilyev (1921–73) served under Voznyuk from 1952 to 1955. He was later the chief of the Main Directorate of Reactive Armaments (GURVO) from 1964 to 1967 before serving as the head of the A. F. Mozhayskiy Military Engineering Academy from 1969 to 1972.

The first Soviet solid-propellant strategic intercontinental missile appeared seven years after the first American one. By that time, there were 1,000 Minuteman I and Minuteman II missiles standing on alert in Montana, South Dakota, North Dakota, Missouri, and Wyoming, each armed with nuclear warheads with a TNT equivalent yield of from 0.5 to 1 megaton. Those very same 400 megatons that MacNamara had spoken of were covered and then some by Minuteman missiles alone. By that time, "to be on the safe side," 41 submarines were carrying more than 650 Polaris solid-propellant missiles with a total TNT equivalent of 400 megatons.

Those involved in the RT-2 flight tests decided to perform a very impressive experiment: try out the SDUK in a salvo firing configuration. Twenty seconds after "pushing the button" in Moscow, three missiles simultaneously launched out of three silos at the Plesetsk firing range and headed for their targets. There was also a new missile division ready now to "conduct fire" this way.

Similar drills rather demonstrably confirmed the possibility of beginning World War III by pressing several buttons in succession rather than with an attack by several dozen armored divisions and a sortie of 1,000 airplanes. To prevent such an event, both sides developed similar deterrence doctrines based on the inevitability of a retaliatory strike, which could also be initiated by pressing a button. Predictions made back then that the provocative push of a button might cause a global catastrophe remain far from absurd to this day. And it's not even necessary that thousands of missiles be launched at once. The MX first-strike ICBM, developed to replace the Minuteman III, carries 10 nuclear warheads with a total yield of 6 megatons, which exceeds the power of all the ordnance detonated during World War II. A single silo is all it takes!

STARTING IN OCTOBER 1968, MISSILES WERE ROUTINELY FIRED FROM THE PLESETSK FIRING RANGE TO TEST THE THEORY OF SOLID-PROPELLANT CRACKING THAT CHELOMEY HAD PREDICTED. Missiles with various storage periods were selected for this. The cracking theory was not proven to be true. From January 1970 through late 1972, the first production runs of the 8K98 were replaced with the updated RT-2P (8K98P). TsKB-7 performed a considerable amount of the work to update the RT-2 independently. The updated missiles had a launch mass of 51 metric tons. The new inertial control system developed at Pilyugin's NIIAP had a precision gyro-stabilized platform with floated accelerometers and a computer, which ensured a CEP no greater than 1,500 meters. A completely new payload was developed for RT-2P missiles. Arzamas-16 Chief Designer Samvel Kocheryants developed a more compact nuclear warhead. In addition, the missile was equipped with decoy targets to thwart anti-ballistic missile systems. The updated missile went into service in 1972.

During 1973, documents and the technical rights of chief designer began to be transferred to KB Arsenal in Leningrad. TsKBEM and then NPO Energiya retained the duty of field supervision over missiles in service in the launch position areas.[32] Initially, the guaranteed service life of an RT-2 missile was determined to be seven years. During the process of firing missiles that had been standing on duty, their reliability was demonstrated after 15 years of storage! In all, during developmental testing and routine firings through the year 1994, 100 missiles were fired at intermediate and full ranges! Beginning in the 1970s, the RT-2P missile proved itself as one of the most reliable.

By the 1980s, the nation's missile arsenal was supersaturated, and the production of the RT-2P gradually folded up. It wasn't until 1995 that 60 RT-2P missiles ended their combat duty. For 25 years, these missile systems had served honorably. Together with other liquid- and solid-propellant missiles, they had provided strategic parity as a guarantee of peace.

However, for the history of our domestic missile building industry, the most important thing was not the number of silos occupied by RT-2 missiles, but the fact that the RT-2 had paved the way for other types of solid-propellant missiles. Having seen for himself advantages of solid-propellant missiles, Ustinov, who was still Central Committee Secretary, did everything possible to develop a new organization—the Institute of Thermal Technology— whose general designer was Aleksandr Davidovich Nadiradze. Nadiradze, who went on to become an academician and was honored twice as a Hero of Socialist Labor, was not about to compete with Yangel and Chelomey, who had sheltered their missiles in silos.[33] He placed his missiles on wheels, thus giving rise to the mobile launch systems, i.e., "land-based submarines." It was assumed that during a "contingency period" the self-propelled missile launchers would leave their hangars and roll out in directions that were unknown and unexpected for a potential enemy before launching. This, rather than a fortified silo, would save them from a possible nuclear strike.

---

32. TsKBEM—*Tsentralnoye konstruktorskoye byuro eksperimentalnogo mashinostroyeniya* (Central Design Bureau of Experimental Machine Building) was the name given to Korolev's OKB-1 in 1966, soon after Korolev's death. In 1974, TsKBEM united with a number of other production and design organizations to become NPO Energiya—*Nauchno-proizvodstvennoye obyedineniye 'Energiya'* (Energiya Scientific-Production Association).

33. The organizational roots of Nadiradze's institute, NII-1, actually date back to 1945 with the formation of the State Central Design Bureau No. 1 (GTsKB-1). In 1947, it was named NII-1 and in 1966, renamed the Moscow Institute of Thermal Technology (MITT), a name which it still holds at the present. Nadiradze was awarded the Hero of Socialist Labor in 1976 and 1982. He became an academician in 1988.

Mobile missile systems introduced a certain degree of rapprochement into the protracted "civil war" between the missile schools of Yangel and Chelomey. For them, the increasing rigor of the survivability requirements had resulted in a relentless competition to build protected silo launchers for a guaranteed retaliatory strike. The military chiefs of NII-4 demonstrated the futility of further large-scale operations to increase the protection of silo launchers and insisted on the need to deploy more survivable mobile ground-based and rail-based systems. This concept was approved after Ustinov became the Minister of Defense in 1976.

The control system was one of the most difficult problems for a mobile launch. Ustinov offered Pilyugin's deputy, Finogeyev, the post of deputy minister of the defense industry to coordinate all control operations for this new type of missile system. After obtaining Pilyugin's agreement, Vladilen Finogeyev occupied such a seat of responsibility and promoted the creation of the *Temp* (Pace), *Pioner* (Pioneer), *Start* (Launch), and *Topol* (Poplar) mobile-launch solid-propellant missiles, which NATO later assigned the designations SS-20, SS-24, SS-25, etc.[34]

Now the Moscow Institute of Thermal Technology is seeking buyers for its decommissioned four- and five-stage launch vehicles, which are capable of inserting small spacecraft into orbit.[35] The RT-2 did not live to see the day of strategic missile rummage sales.

KB Yuzhnoye was not able to hold its ground in the face of "solid-propellant" ideas. They began developing a solid-propellant missile at this traditionally "liquid-propellant" design bureau while Yangel was still alive. The new General Designer Vladimir Utkin implemented the idea in the form of the RT-23 (RS-22) stationary and mobile railroad system. The RT-23, a three-stage missile carrying 10 independently targeted warheads, was an analog of the United States' MX missile.

Petr Tyurin, who had gained a wealth of experience in the field of ground-based solid-propellant missiles, entered into competition with Makeyev and developed the first domestically-produced solid-propellant submarine-launched ballistic missile at KB Arsenal. Tyurin's design, which

---

34. The "SS"-type designations were given by the U.S. DOD while NATO assigned names such as Scaleboard, Sinner, etc. The following denote the equivalent names from the Soviet side, DOD, and NATO: *Temp* (SS-12, Scaleboard-A), *Temp-S* (SS-12M, Scaleboard-B), *Temp-2S* (SS-16, Sinner), *Pioner* (SS-20, Saber), *Topol* (SS-25, Sickle), *Topol-M* (SS-27, no NATO assignment).

35. Using technology from the *Pioner* and *Topol* missiles, this institute has developed a series of orbital launch vehicles for under the name *Start*. The first *Start-1* was launched in 1993.

was assigned the designation D-11, saw service on a single solitary submarine, which was decommissioned in 1991.[36]

Despite his real successes in the development of domestic solid-propellant missiles, Makeyev, with the most active support of Isayev, stubbornly worked to perfect liquid-propellant missiles for submarines. The teams of Makeyev and Isayev and all the subcontracting firms that cooperated with them attained truly remarkable results in this field. The most recently developed RSM-54 (or SS-N-23 Skiff) liquid-propellant submarine-launched ballistic missile (SLBM) is unequaled in the world in terms of its energy-mass perfection and other specifications.

Nevertheless, although it was quite late in coming, Makeyev's team fully developed the RSM-52 (or SS-N-20 Sturgeon) solid-propellant submarine-launched ballistic missile. Missiles of this type went into service in 1980. They were placed on third generation *Tayfun* (Typhoon) ballistic missile submarines. *Tayfun* submarines were the largest nuclear submarines in naval history. Each submarine was armed with 20 RSM-52 missiles.

The end of the Cold War in no way signified the termination of operations to improve intercontinental strategic missile weaponry. Almost 10 years after us, the Americans arrived at a concept for mobile strategic launch positions similar to ours and developed the Midgetman ICBM. The mobile Midgetman missiles were supposed to become the United States' only ground-based strategic missiles designed to survive under the conditions of a nuclear missile attack. The plan was that, after receiving notification of a missile attack, missile transporting-launching vehicles would emerge from shelters onto a system of roads and would disperse a safe distance from the base during the period of time required for the enemy missiles to reach U.S. territory. The dispersal could take place so rapidly that the only way to destroy the missile system might be a nuclear barrage, which would make an attack on these missiles highly unlikely. During peacetime, missile transporting-launching vehicles must be randomly shifted from one basing site to another so that a potential enemy would not know their location.[37]

Thanks to the continuing improvement of missile systems, both missile powers—the United States and Russia—possess mobile ground-based and naval strategic nuclear systems that can survive after the buttons are pushed on both sides. Humankind will be given the opportunity to start all over.

---

36. During the Soviet era, the Soviet Navy would use a "D-number"-type designation to denote a SLBM system. The D-11 included the R-31 SLBM. In the West, the system was known as SS-N-17 (by the DOD) and Snipe (by NATO).

37. *Aviation Week and Space Technology* (Vol. 126, No. 20, 1987, p. 47). The Midgetman was tested a few times in the late 1980s and early 1990s, but the program was cancelled in 1992.

Chapter 7
After Gagarin, Others Will Fly

"Every reminiscence is colored by today's being what it is, and therefore by a deceptive point of view . . . ."[1] I understood the validity of these words spoken by Einstein in his "creative autobiography" as soon as I began to work on my own memoirs. In an effort to free oneself of "coloring," one feels compelled to publish documents or diaries. Documents from the historical period that I am writing about are now available; putting them into some sort of order would be labor-intensive, however, and their publication would require annotations that would inevitably be colored by what is happening now. This is particularly relevant to what I can write about both *Vostok-2* and Cosmonaut No. 2, German Titov.

The launch of *Vostok-2* was scheduled for 6 August 1961. Right before that launch, on 3 August, an attempt was made to launch the latest R-9 missile from launch Site No. 51, just as before Gagarin's launch. The missile exploded, partially destroying its launch pad facilities. This was a heavy blow to the R-9 program, but did not put a crimp in the optimistic wit of the testers participating in both events: "If Korolev didn't let Mishin and Chertok go to Moscow after Gagarin's flight because of the R-9 failure, they are certainly not going to see Moscow now!" Others protested that the R-9 explosion was ostensibly a good sign before a piloted flight and that the same people should not be punished twice, especially since they "have received prestigious government awards."

Despite the nervousness that followed the R-9 failure, *Vostok-2* preparations proceeded in a calm and orderly fashion. Flight duration was the problem that caused a heated argument between Korolev and the Air Force brass. After consulting with the medical professionals, Korolev insisted that the flight last at least 24 hours. Relying on the authority of those same medical professionals, Kamanin, the cosmonauts' chief mentor, exercised caution and proposed no more than three orbits.

---

1. Albert Einstein, "Avtobiograficheskiy zametki" ["Autobiographical Notes"] in *Sobraniye nauchnykh trudov, t. 4* [*Collected Scientific Works, Vol. 4*] (Moscow: Nauka, 1967), p. 259.

Bushuyev and Feoktistov developed a list of modifications for the spacecraft based on the experience of the previous flight, allowing it to remain in orbit for at least a week. We performed a series of modifications that increased the reliability and operational ease of radio communications. Together with Rosselevich, we installed a more "respectable" television system. Frankly, we were all somewhat embarrassed about the television broadcasts of Gagarin's flight. Bogomolov's telemetry system was supplemented by the *Signal* system that Bykov had proposed. This short-wave system served as the spacecraft's direction finder and provided backup transmission of the most vital medical parameters.

The developer of each piece of hardware accessed by the cosmonaut during flight strove to include its checkout procedures in the program. Cosmonaut No. 2 was not a "guinea pig." He was actually given quite a substantial workload.

There had been complete unanimity as to the candidate for the 24-hour flight. All were in favor of Titov. Titov was supposed to test manual control of the spacecraft twice; perform visual observations through the windows and note what he had seen; conduct VHF communications sessions during each pass over the USSR and shortwave sessions two times per hour; perform calisthenics; eat lunch and dinner; use the toilet facilities; and finally, sleep!

Sleep in space! If anything, that was one of the most important experiments. If a person could sleep in space in a spacesuit in weightlessness without a comforter and pillows, that meant he could live and work! This is why Korolev had fought with Kamanin, arguing in favor of a 24-hour flight. For three orbits, one could forgo sleep and all other physiological needs, including a tasty dinner, until returning to Earth.

In 1961, Gagarin was allowed to travel only if accompanied by Kamanin. During the preparation for Titov's launch, neither Gagarin nor Kamanin was present at the firing range because they were in Canada. Thus, during that time, there was no one to argue with Korolev about the flight duration.

The general mood at the firing range, compared with that of the Gagarin launch, was considerably calmer and more matter-of-fact. Each

From the author's archives.

A recent view of the launch pad at Site No. 1 at Baykonur Cosmodrome from where both *Sputnik* **and Gagarin were launched.**

From the author's archives.

Another recent view of Site No. 1 showing the overhanging "apron" at the launch pad.

of us felt inwardly certain that it would succeed. Sharing thoughts and considerations that we had not voiced at official meetings, in private we all agreed that the spacecraft's hardware had been soundly developed and tested. No one had any doubts about Titov either. Boris Rauschenbach, who had conducted test exercises with him on manual attitude control, Sergey Darevskiy, who had checked out his skills on the pilot's console, and Chief Medic Vladimir Yazdovskiy were all sure that Titov would not let them down.

At the firing range, when they "sounded muster" before a crucial launch, we spontaneously formed groups "according to interest." In such groups you could share your innermost thoughts about our missiles' reliabilities; talk about the dispositions of Korolev or Glushko; tell jokes; speculate about future prospects; and perform harmless pranks.

I joined the group formed by Ryzanskiy, Voskresenskiy, Pilyugin, Kuznetsov, and Bogomolov. Our group was firmly convinced that, for the time being, the weakest spot was the launch vehicle. After Gagarin's launch, Pilyugin acknowledged that Rauschenbach and I had produced a fundamentally simple and reliable guidance and control system. He was no longer jealous, but he sulked if Vityunya (as we called Viktor Kuznetsov) suggested that we had managed without his (Pilyugin's) system on the spacecraft.

One evening after the R-9 failure, we were sharing a bottle of cognac among the six of us and trying to pick up *Voice of America* on the radio in the hopes of hearing firsthand what the Americans thought about Gagarin's visit to Canada. Barmin dropped in on us. He remarked unequivocally that since the beginning of the human spaceflight era, the old Council of Chiefs had fallen apart. Only one "chief designer" remained along with his "chief theoretician of cosmonautics." Now that the "chief theoretician" had also become the president of the Academy of Sciences, the role of the old Council would be insignificant. Our conversations inevitably got caught up in a discussion of the nation's general political situation.

On 30 July, a draft of the new CPSU program had been published for discussion. The program was submitted for approval at the 22nd Communist Party Congress, which was supposed to convene in October.[2] The program stated that, "The Party proudly proclaims: this generation of Soviet people shall live under communism!" We read the program and discussed it in fits and starts. Much of what it contained coincided with our aspirations, but despite our orthodoxy, its concluding notions evoked skeptical smiles.

So here and now in the heart of our turf, I couldn't restrain myself and told Barmin that the Council of Chiefs must, according to the program, also survive "until we have communism," especially since the Council itself was made up of communists. I once again recounted a story well known to the majority of our technocratic community about how Lavrentiy Beriya had neutralized discord between two chief designers. In 1952, Beriya was supposed to review and approve the latest timetable concerning the construction of the famous air defense ring around Moscow. An aide reported to him that the two chief designers had not initialed the timetable. They simply could not agree on how to divide the responsibilities and work between one another. The aide requested that Lavrentiy Pavlovich hear them out.

"Tell them," said Beriya, "that if two communists cannot come to terms with each other, then one of them is an enemy. I don't have time to sort out which one of them is actually the enemy. Give them another 24 hours to reach an agreement."

The aide left, and 5 minutes later he returned to the office and placed before Beriya the timetable initialed by both chiefs. Beriya's time had long since passed, and now there were many who wanted to exacerbate antagonisms.

I had started to argue that there were now issues that none of the old group of six chief designers, except for Korolev, showed any interest in. This included cosmonaut selection and training, life support systems, problems of controlling a piloted flight and its return to Earth, and a myriad of unforeseeable troubles. Barmin was not about to argue, but he hinted that if there wasn't going to be solidarity, then new forces would weaken the influence of or even break up the Council. The future proved Barmin's prediction to be true. But I'll deal with that later.

---

2. The Congress of the Communist Party of the Soviet Union was an irregularly timed gathering of Communist Party delegates from across the country. On paper, the Congress was the ruling body of the Communist Party, but effectively the Central Committee (and smaller bodies such as its Secretariat and Politburo) had de facto power. After Stalin's death, the Party Congress was typically held approximately every five years.

From the author's archives.

Sergey Korolev sees off German Titov on his *Vostok-2* flight in 1961. On the lower right is launch pad Chief Designer Vladimir Barmin.

THE DUTY OFFICERS WOKE US UP AT 3 A.M. Everyone who was supposed to be at the launch area assembled at 4 a.m. On 6 August at 5 a.m., the State Commission gave the green light to fuel the vehicle and proceed with the launch.

All the routine preflight procedures were observed. Yevgeniy Anatolyevich Karpov—physician and first director of the Cosmonaut Training Center (TsPK)—woke up Titov and his backup, Nikolayev. At T-minus 2 hours, dressed in their spacesuits, they were brought out to the rocket at the launch pad. Titov read his prepared speech saying that his spaceflight was dedicated to the 22nd Party Congress. He thanked the creators of the beautiful *Vostok-2* spacecraft and once again thanked the Central Committee of Lenin's dear party and the Soviet government for their confidence and assured them that he would fulfill this honorable and important mission. The public address system carried Titov's voice over the hushed launch site.

On that hot morning of 6 August at 0900 hours Moscow Time, the second Soviet citizen, German Titov, lifted off into space. Although the text of the TASS report had been prepared in advance, the State Commission did not give permission to broadcast it until it was certain that the vehicle had entered an orbit close to the design orbit and that the cosmonaut himself was in shipshape. Keldysh and Ishlinskiy edited the text of the communiqué. Korolev asked that publication be delayed until they had received reports from the ballistics centers. At 0920 hours they learned from the ground tracking station (NIP) reports and from Titov that he was in orbit and everything was A-OK on board.

191

Moscow demanded reports about the launch immediately. The Central Committee, TASS, and radio services were a lot more jittery than we were here at the firing range. Finally, they placed the approved text in front of Yuriy Levitan at the studio. And then after the familiar stirring tones of the station jingle we heard: "Attention! This is Moscow speaking . . . All radio stations of the Soviet Union are operating . . . ." After the TASS report about the launch, Levitan read Titov's declaration. Someone recalled that the launch date coincided with the 16th anniversary of the atomic bombing of Hiroshima.

"Our propaganda will take every possible advantage of that," noted Keldysh.

At the control post (KP), chiefs were being assigned their shifts. Korolev wanted to be awakened for the manual control sessions and insisted that Chertok, Rauschenbach, Feoktistov, and Bykov be there too. This demand made a shambles of the strict shift arrangement. All the lead specialists actually went without sleep for more than 24 hours, hoping to catch up on their sleep en route to or back in Moscow.

At 1530 hours, Titov reported the following: "I'm withstanding weightlessness very well." He was putting up a brave front. Later he confessed that he was slightly nauseated and had motion sickness. Particularly unpleasant sensations occurred when he moved his head abruptly. He tried to turn his head slowly or hold it still. But his assignments to shoot film footage and observe Earth through the windows required movement. Titov learned to ward off dizziness by adopting a calm, deliberate posture. Back on the ground, reporting to the State Commission, he described in detail his actions and well-being. But meanwhile, we could only conjecture and argue.

Studying the telemetry, our health professionals at the control post also suspected something, but Yazdovskiy and Karpov reassured them: "Once he takes a nap, it'll pass."

I was at the control post when Titov announced, with a certain degree of reproach directed at *Zarya-1*: "You can do whatever you want. I'm going to sleep!" And he really did try to doze off.

After the seventh orbit, the spacecraft left the coverage zone of our NIPs and did not reappear until the 12th pass in the Kamchatka reception zone. The "dead" orbits took up almost 8 hours. During that time, all hope was placed on "unreliable" communication via shortwave and monitoring using the *Signal* system.

I returned to the control post with Voskresenskiy at 2 a.m. on what was now 7 August. A slight panic had begun there. Titov did not respond to *Zarya-1* calls. Bykov called his subordinates at all (control) posts. Korolev could not bear the uncertainty and suspense. The lack of reports from Kamchatka and Ussuriysk about communications with the cosmonaut had to be somebody's fault.

"Your soldiers and officers have fallen asleep," he lashed out at the duty officer for Military Unit 32103.[3]

Everything turned out to be simple. It wasn't the soldiers and officers who fell asleep; it was the cosmonaut. It became clear that not only can one sleep in space, but also one can oversleep! Right then and there, I was assigned to develop design specifications for the clock-making industry to produce a space alarm clock.

The next nerve-racking experience at the control post began during the wait for a report about the descent cycle startup, and finally, a report from the landing area. Everything ended well, but during the two-parachute descent, cosmonaut number two landed just 10 meters or so from a railroad on which a train was traveling at that time. The search and rescue service did not report those matters to General Aleksandr Kutasin; but when this bit of information came to light, it was proposed that a Ministry of Railways representative be included on the State Commission to coordinate the railroad schedules with the launch programs.

From the author's archives.

German Titov reports to Nikita Khrushchev on the successful completion of his *Vostok-2* flight.

Vostok-2 landed near the village of Krasnyy Kut in Saratov *oblast*.[4] Just a week later, after the descent module was delivered to us at the factory, we discovered that there had been a repeat of Gagarin's off-nominal situation. After the braking engine unit (TDU) shut down, pyro cartridges were supposed to jettison the feed-through plate and cable bundles between the descent module and the instrument compartment. The compartments themselves were to separate only after the jettisoning procedure. In this case, the feed-through plate was not jettisoned and the compartments entered the

---

3. Military Unit no. 32103 operated the Soviet ground communications segment, i.e., the Command and Measurement Complex (KIK).

4. The *oblast* is a type of administrative and geographical division in Soviet and eastern European countries, analogous to "region" or "province."

atmosphere mechanically separated, but connected by the thick bundle of cables. In the dense atmospheric layers, the cable burned up, and then everything proceeded normally. This was a repeat of a defect in the fourth and fifth spacecraft.[5] Without waiting for instructions from Korolev, Voskresenskiy and I organized a small group, which we tasked with figuring out and doing away with this "hazardous mess." An investigation showed that rather than a circuitry error, the improper running of the cable had caused the problem. At the "separate compartments" command, a special pyro knife cut the cable for transmitting the plate jettison command a fraction of 1 second before this command was issued. An error had been made in running the cable.

Local transportation quickly delivered Titov to the nearest regional office of the Communist Party. He communicated with Moscow and reported to "dear Nikita Sergeyevich Khrushchev" that the mission had been accomplished. The next day, the newspapers printed a nice photograph of Titov reporting to Khrushchev from the office of the secretary of the regional office (*raykom*) of the Communist Party about his successful return to Earth.

Gagarin found out about the flight of *Vostok-2* after its second orbit when he was in Canada. He and Kamanin were visiting Farmer Cyrus Eaton, who they jokingly called "Khrushchev's best friend."[6] Gagarin and Kamanin departed from Halifax on an Il-18 after the sixth orbit. Nevertheless, they managed to reach the regional committee *dacha* on the bank of the Volga River shortly after Titov arrived there. Thus, the first two cosmonauts met at the same *dacha* of the Saratov *oblast* Party committee where Gagarin had rested after his return from space.

On 7 August, in a special edition, *Pravda* published a suite of messages: the TASS report about the completion of the flight; and the addresses of the Central Committee, Presidium of the Supreme Soviet of the USSR, and the government to the Communist Party, to the people of the Soviet Union, to the peoples and governments of all nations, and to all progressive peoples.[7] As I later learned, this address contained the following words dictated personally by Khrushchev:

---

5. This is a reference to the *Fourth* and *Fifth Korabl Sputniks* launched in early 1961.

6. Cyrus Stephen Eaton (1883–1979) was a succesful philanthropist who gained fame in the early Cold War years as an outspoken critic of U.S. foreign policy, particularly toward Communist countries. He was a founder of the original Pugwash group, which established direct contacts between Soviet and American intellectuals with the goal of reducing military tensions between the two nations.

7. Phrases such as "progressive peoples" were commonly used in official Soviet language as euphemisms for socialist-leaning populations across the globe.

Everything in the name of mankind! Everything for the good of mankind! This is our highest goal. The space flights of the Soviet people signify the steadfast will, the steadfast aspirations of all Soviet people for lasting peace throughout the world. We are placing our achievements in space research at the service of peace, scientific progress, and the good of all the peoples of our planet.

At the same time the address of these same three supreme authorities to all scientists, designers, engineers, technicians, workers, and all the staffs and organizations that participated in the successful conduct of the latest spaceflight was published. They congratulated us warmly and triumphantly: "Hail to Soviet scientists, designers, engineers, technicians, and workers—conquerors of space!"

This was followed by words that the few who have retained their faith in the old ideals would hesitate to cite today, even at anti-government meetings: "Hail to our people—who have created, conquered, and who, under the leadership of the Communist Party, are blazing the trail to the bright future of all of mankind—to communism!"

Why do I remember all of this? My contemporaries and I were people who sincerely believed in the ideals and ultimate objectives proclaimed in these pronouncements. We were hardly naïve fanatics and were not trying to close our eyes to reality with its array of contradictions. It is very difficult to convey to the reader the external and internal circumstances that determined our spirit, our collectivism, and ideological commitment. Here I will venture to say that my contemporaries, the very ones to whom those salutary words were addressed, were neither hypocrites nor sanctimonious individuals. Their interests in life did not conflict with the calls to find the road to communist paradise. If a triumph in space brought us closer to those goals, it meant that these resources, labors, and expenditures were justified by such a great goal.

Although Mishin and I were allowed to leave the firing range and fly back to Moscow, Korolev did not take us to the meeting with Titov near Saratov. He had his own reason for this. Both of us were his deputies and were not directly responsible for cosmonaut training. Korolev relied more on Bushuyev in these matters.[8] We needed fewer distractions from our immediate tasks, the creation of reliable technology.

---

8. In the early 1960s, Korolev had about a dozen deputy chief designers and a single "first deputy chief designer," each of them responsible for a particular profile within the design bureau. The first deputy was V. P. Mishin, while the deputy chief designers included A. P. Abramov, K. D. Bushuyev, B. Ye. Chertok, D. I. Kozlov, S. S. Kryukov, P. I. Meleshin, M. V. Melnikov, S. O. Okhapkin, Ye. V. Shabarov, M. K. Tikhonravov, P. V. Tsybin, and L. A. Voskresenskiy. Among them, Bushuyev was responsible for all piloted space projects.

However, those who didn't get to go to the *dacha* harbored no hard feelings. We got home a day before Moscow's triumphant reception of Titov. Katya, my wife, was openly overjoyed. Now the courier-delivered invitation to come to the Great Kremlin Palace with my spouse would be put to use. What's more, I received a pass to the viewing stands on Red Square where the rally began at 1500 hours on 9 August.

From the author's archives.

German Titov, Nikita Khrushchev, and Yuriy Gagarin on the reviewing stand atop the Lenin Mausoleum at Red Square in 1961.

A photo made the rounds in all the newspapers and magazines those days, one of Yuriy Gagarin and German Titov on the viewing stand by Lenin's mausoleum. Just behind them stood a beaming Khrushchev with his arms around their shoulders like a proud father with his sons. All three were smiling and happy.

We formed a large group on Red Square, especially since all the chiefs, Korolev's deputies, and other chiefs' deputies had been invited along with their spouses. Many of our wives had known each other since their days in Bleicherode (in Germany). Others had entered our closed society later. All in all, our group formed a tight-knit column that set off after the rally for the Great Kremlin Palace. When we came in through the grand entrance, our group decided that there was no use lingering on the steps leading to Georgiyevskiy Hall—the guests who had already experienced Gagarin's reception climbed the steps in a solid stream.

"Don't rush into the hall," the Ryazanskiys managed to tip us off. They were more experienced when it came to Kremlin receptions.

From the author's archives.

Chertok and his wife, Yekaterina (Katya) Golubkina, by a fire on a camping trip.

"Let's wait a minute," Katya said, pulling me aside. "This is a once-in-a-lifetime opportunity."

We stopped on the very top steps. On both sides of the stairway, girls in snow-white dresses stood like statues, and military academy cadets were stock-still in preparation for a trumpet salute. While we milled about trying to find the best spot to observe the now-empty stairway, the cadets raised their golden trumpets in fanfare announcing the appearance of the great hero. At the bottom of the stairway Khrushchev appeared, having beckoned with a broad gesture for Titov, Gagarin, their wives, and many relatives to follow. From who knows where, the members of the Central Committee Presidium suddenly appeared. They climbed the stairs and entered the hall en masse, while out of view a choir and orchestra played Glinka's "The Patriotic Song."[9]

"Chertoks, step lively now," said Ryazanskiy, sweeping us along with his wife Lesha into the already thronging and noisy hall. We jostled through the crowd until the Klyucharevs snatched us away from the main stream and settled us in at a table where all our friends were crowded shoulder to shoulder, already getting ready for the first toast. One glance at the table was all it took to assure oneself that we would not be wanting. The tables were laid out more than lavishly.

---

9. Mikhail Ivanovich Glinka (1804–57) was an influential Russian composer of classical music who created a distinctly Russian sensibility with his compositions. "The Patriotic Song," one of his lesser works, was written as a piece for piano, without lyrics. In 1990–2000, this piece, with new lyrics, was made the Russian national anthem.

Yuriy Gagarin, his wife, Valentina Gagarina, and Nikolay Kamanin as guests of Indian Prime Minister Jawaharlal Nehru.

> *"One drink and then you're through," said Nehru.*
> *"Drink all they pour! Finito!" said Tito.*
> *"Drink 'til you've drunk enough," said Khrushchev.*[10]

These lines composed by an anonymous "folk" poet were repeated often at our convivial table.

Our table glittered with recently received Heroes of Socialist Labor and Lenin Prize gold medals and decorations. These were from Gagarin's flight. There was every reason to be in a good mood. Khrushchev's speech only further validated our feelings: "Today we are in a special mood and we believe that we have every reason for this . . . Our success in the exploration of space is remarkable. We take pride in it. I propose a toast to German Titov and his wife, to their parents, and to Yuriy Gagarin and his wife."

When Titov was brought to the microphone, there was such a clinking of glasses at our table that no one exactly heard his toast. We were draining glasses primarily for our own local toasts. In the hall, a sort of "Brownian

---

10. Jawaharlal Nehru (1889–1964) was the first prime minister of India while Josip Broz Tito (1892–1980) was the postwar leader of Yugoslavia. Both were men of giant stature in the newly post-colonial world where the Soviet Union engaged in sustained efforts during the Cold War to gain the allegience of millions.

motion" began. Guests moved from table to table meeting with old and new acquaintances, celebrating each meeting with a toast.

And nevertheless, relative quiet descended on the hall when Brezhnev read the decree awarding Titov the title "Hero of the Soviet Union" and presented him with a Gold Star medal, an Order of Lenin, and a USSR Pilot-Cosmonaut badge. German returned to the microphone and proposed a toast to the beloved Party, to its Leninist Central Committee, and to Nikita Sergeyevich Khrushchev.

Comparing notes about this reception once we had sobered up, and recalling particularly outstanding "feats," we unanimously decided that we had breathed some life into the Great Kremlin Palace. We particularly enjoyed the choral performance in the Georgiyevskiy Hall under the direction of Klyucharev of his favorite anthem:

Bravely we will go into battle
For the power of the Soviets
And as one we shall die
Fighting for it

In vain, I tried to egg on the chorus to perform its signature number "Over Hill and Dale." Even our "contingent," completely intoxicated, determined that I completely lacked the requisite ear for music for the chambers of the Kremlin.

From the author's archives.

"We have witnessed the triumph of Klyucharev and the total failure of Chertok." That was how Golunskiy described the results of the "choral competition" at the Kremlin.

After exhausting the supply of first-class wines and hors d'oeuvres and enjoying the opportunity to socialize with members of the government in Georgiyevskiy Hall, we crossed over in a boisterous crowd to the Faceted Palace. Here we found Rauschenbach carrying on

A picture of German Titov autographed and dedicated to Boris Chertok soon after his landing.

conversations with high-ranking clerics from three religions. Voskresenskiy proposed assembling his compatriots and quoting Ostap Bender in unison: "There is no God" and "Religion is the opium of the people."[11] Our mindful wives quashed this attempt to disturb the peace.

After taking a seat opposite the rabbi and mufti, I proposed a toast to Cosmonaut Titov, who during 24 hours in space did not find God. "God is within each of us," said one of the spiritual leaders. On that note, around the table everyone began to polish off the last drops of the famous Khvanchkara wine.[12]

It was past 8 p.m. when courteous young people began to hint to us that it was time to go. We had already guessed that the party was over. We exited onto Red Square in a boisterous, jolly crowd and boarded busses that took us home. Our sons weren't at home. They had both joyfully celebrated Titov's flight at their own parties. Titov's 24-hour flight and postflight celebration have remained a brilliant highlight in my memory. And not only in mine. I would say that after the victory celebration in 1945, the space victories of 1961 have remained the brightest spot in Soviet society.

One day later, a closed meeting was held at our factory. All the chief designers, chief theoreticians, unknown scientists, and engineers had the opportunity to openly stand at the podium before a large crowd of the *true* creators of missiles and spacecraft.

A press conference was held on 11 August. The assembly hall of Moscow State University on the Lenin Hills could not accommodate all those—numbering more than 1,000—wishing to attend. Traditionally, Keldysh delivered the introductory remarks. The president of the USSR Academy of Sciences spoke openly here. None of the uninitiated journalists were supposed to suspect that he was the very "theoretician of cosmonautics" referred to by correspondents who had been let in on the big secret. Keldysh said nothing about the chief designer, or about other scientists. On behalf of the Academy, he awarded Titov a Tsiolkovskiy Gold Medal.[13] Titov's speech was a well-prepared report about the flight results.

---

11. In Ilya Ilf and Yevgeniy Petrov's satirical novel *Dvenadtsat stulyev* (*The Twelve Chairs*, 1928), the main character Ostap Bender "unmasks" an incognito priest by shouting these phrases outside his hotel room.

12. Khvanchkara is a wine, available since the early 20th century, made from grapes cultivated in the Khvanchkara vineyards in now-independent Georgia.

13. In 1954, the USSR Academy of Sciences established the Tsiolkovskiy Gold Medal to be awarded once every three years in honor of a notable contribution to cosmonautics. The first medals were awarded in 1958 (in secret) to Korolev, Glushko, and Pilyugin.

During his postflight report before the State Commission, Titov frankly admitted that he had experienced bouts of dizziness and motion sickness that began after the third orbit. At the press conference, however, he was only allowed to mention loss of appetite. Yazdovskiy discussed this in greater detail. The only report devoted to technology was Academician Vladimir Kotelnikov's story about Earth-to-*Vostok-2* radio communications. The official portion of the conference ended with a speech by Academician Leonid Sedov.

It wasn't until 8 September that *Pravda* posted a detailed description of the spacecraft's layout, its communications and life support systems, and all the flight stages. The article was written at OKB-1 and carefully edited by Korolev. Its grand conclusion was that a human being can live and work in space.

Asif Siddiqi

Vladimir Ivanovich Yazdovskiy (1913–1999) was the leading space biomedicine specialist in the Soviet Union. For many years, he worked at the Soviet Air Force's Institute of Aviation and Space Medicine.

Beginning with *Vostok* (-1), two possible methods were mentioned when describing the landing system: in the descent module itself or by ejection of the cosmonaut's seat, which had its own parachute. Supposedly the cosmonaut selected the landing method. To this day, no one can understand why we needed this ruse back then. From the very beginning, mandatory ejection logics had been loaded into the automatic controls. The landing speed of the descent module at contact reached 10 meters per second. This was hazardous and threatened serious injury to the cosmonaut.

I have gone into so much detail with *Vostok-2* and German Titov's flight not because I was more involved with this event than other piloted flights but for two other reasons. First, despite the many subsequent successful and impressive piloted flights, I feel that German Titov was the first to prove that it was possible to work in space. Gagarin's single-orbit flight was a historic breakthrough. But he did not dispel doubts and convince skeptics. Titov's 24-hour flight blazed the trail into space for mankind, not just for fighter pilots. This flight convinced us engineers that a well-trained cosmonaut could be trusted a great deal more than we had assumed, due to our love for smart automatic controls. For the creators of piloted programs, this flight had considerably greater significance psychologically than many subsequent ones. It was more significant in the history of cosmonautics than Lindbergh's 1927 transatlantic flight and the Chkalov expedition's 1937 transpolar flight to the U.S. had been for the history of aviation.

The second reason is that, albeit somewhat belatedly, I felt it necessary to once again remind the reader of Khrushchev's role in the history of our piloted cosmonautics. Each historian in the field of cosmonautics feels compelled to mention President Kennedy's initiative in the lunar program. I agree with American historians that, had it not been for his boldness and initiative, Americans would not have been destined to land on the Moon in 1969. On the other hand, Khrushchev's role in the history of our cosmonautics is either ignored altogether or very much downplayed in publications and works that came out after the "October Revolution of 1964."[14]

The voluminous Academy of Sciences publication, *The Mastery of Cosmic Space in the USSR*, issued in 1971, cites official TASS reports and central press materials during the period 1957–67.[15] Khrushchev's name was, however, removed from all the cosmonauts' speeches, addresses, and reports beginning with Gagarin on *Vostok* up to Komarov on *Soyuz-1*.

In 1977, the Military Publishing House (*Voyenizdat*) published German Titov's documentary narrative *My Blue Planet*. In 1981, the same publishing house released Yuriy Gagarin's *The Road to Space*.[16] Khrushchev is not mentioned in either work. I do not blame either Gagarin or Titov. If the authors had tried to quote even a tiny portion of the kind words and gratitude that they had expressed to the former general secretary of the Party before, during, and especially after their flights, then these mainly truthful books would not have been published.

And one more thing. In all the TASS reports, publications, and coverage from 1961, there is no mention of the geographic location Baykonur. No mention is ever made of the launch site. True, the term "cosmodrome" started to find its way into print. It took us a long time to get used to the word "cosmodrome." Even now in conversation and when we run into each other, we usually say: "Remember when we were at the firing range . . . ." The words "firing range" and "Tyura-Tam" are what stuck in veterans' memories.

---

14. The "October Revolution of 1964" is a reference to the ouster of Nikita Khrushchev by powerful Communist Party and government leaders led by Leonid Brezhnev.

15. *Osvoyeniye kosmicheskogo prostranstva v sssr* [*The Mastery of Cosmic Space in the USSR*] (*Moscow: Nauka, 1971*).

16. G. S. Titov, *Golubaya moya planeta* [*My Blue Planet*] (Moscow: Voyenizdat, 1973). Revised editions were published in 1977 and 1982. Yu. A. Gagarin, *Doroga v kosmos: zapiski letchika-kosmonavta sssr* [*The Road to Space: Notes of a Pilot-Cosmonaut of the USSR*] (Moscow: Pravda, 1961). Updated editions of the latter were published in 1963, 1969, 1971, 1978, and 1981.

In late December 1961, at the initiative of Kamanin and Rudenko, the Air Force Commander-in-Chief Konstanin Vershinin appealed to the Central Committee—specifically the Central Committee rather than the VPK—with a proposal to select a new corps of cosmonauts, including five women. The Presidium of the Central Committee endorsed this initiative and the Council of Ministers issued a resolution afterward.

Korolev did not support this initiative. It bypassed him. Once, in his inner circle, he said that many generals envy our success; they want to have their piece of the cosmic pie, but the first time something goes wrong they'll run off or they'll stomp on us, saying "we warned you!"

However, by no means did all generals attempt to join in with the successes in human spaceflight. Minister of Defense Rodion Malinovskiy, his Deputy Andrey Grechko, and Chief of the General Staff Matvey Zakharov made it very clear that they wanted nothing to do with Vostoks. Korolev and Keldysh were supposed to maneuver—develop a closer relationship with the Air Force—but at the same time show the top military leadership, specifically the "ground pounder" marshals (as Colonel Tsybin loved to refer jokingly to the top brass), that cosmonautics was essential to defense.

"Who is going to convince Malinovskiy that now it's time for broads to fly into space?" Tsybin posed this clearly loaded question to Korolev.

"Let your friends Kamanin and Rudenko handle that. It's time for us to deal with selection and training."

The production of all technology was in one way or another under the authority of Korolev and the Council of Chiefs, while from the very beginning, cosmonaut selection and training passed them by. This irritated Korolev. And now they were going to make us launch women!

Second cosmonaut German Titov (right) appears with NASA astronaut John Glenn and President John F. Kennedy at the White House in 1962. Titov was in Washington, D.C. to give his account of the *Vostok-2* spaceflight to the Committee on Space Research (COSPAR). The 25-year-old Titov was the youngest person ever to go into space—a record that still stands to this day.

Asif Siddiqi

203

In February 1962, training began for a joint flight of two spacecraft. Ustinov had supported this proposal of Korolev's very actively. The first U.S. successes had served as the stimulus for this flight. John Glenn had finally completed a three-orbit flight in a Mercury spacecraft on 20 February, after seven failed launch attempts.[17]

At the next meeting on the MV [Mars-Venera] project, I heard from Keldysh that Ustinov had rushed to report the joint flight proposal to Khrushchev and he had requested that it be expedited in every way possible so that we might rub the Americans' noses in it and show the whole world that they have fallen hopelessly behind us. The Air Force selected Andriyan Grigoryevich Nikolayev and Pavel Romanovich Popovich as the primary candidates to be trained for the dual flight. I received instructions from Korolev to ensure that in-flight communications between the two spacecraft were reliable and that the "ground" could hear their conversations. Not stopping at that, with me present, Korolev called Bykov and explained how important this was.

We began joint engineering work with radio specialists, antenna specialists, and with the services of Military Unit 32103. Above all, this required the precise work of the ground services at the ground tracking stations (NIPs) and on the Ministry of Communications telephone channels that were transferred to the Ministry of Defense during the spaceflight.

For the majority of us in Korolev's inner circle, the arguments that had flared up concerning the duration of the upcoming flight of Nikolayev and Popovich seemed frivolous. Korolev and Keldysh insisted on a three-day flight. The Air Force, represented by Kamanin, vehemently insisted on a 24-hour scenario; the flight would be prolonged by another day only if the cosmonauts felt well.

In early July, Barmin assured us that, after routine maintenance, Site No. 1 would be ready for operation by 1 August. The launch of Nikolayev and Popovich could be scheduled for August. In connection with this, Korolev called a meeting at which Bushuyev, Feoktistov, Rauschenbach, and I were present. Kamanin, Karpov, and Yazdovskiy from the Air Force were there. S.P. told us about his meeting with Khrushchev.

"He is in favor of the three-day scenario."

When Kamanin started to object, S.P. quickly flew off the handle and, raising his voice, began to upbraid him for the showboating that the Air Force had been doing while trotting off Gagarin and Titov abroad.

---

17. Strictly speaking, there were no "failed launch attempts," merely several launch postponements. The launch was successively planned for 16 January, 23 January, 27 January, and 1 February. Glenn was launched on *Mercury-Atlas 6* on 20 February 1962 on a three-orbit mission lasting 4 hours 55 minutes.

"We are going to prepare our own people for the Soyuz," said Korolev.

In my opinion, S.P. said a lot of unnecessary things. It wouldn't be wise to spoil our relationship with the Air Force. However, one could understand Korolev's irritation. The entire world adored Gagarin and Titov. General Kamanin accompanied them on their travels. Other military men in flight uniform and medical officers were shown in public, but not a word was said about the chief designer! But this was not the fault of the Air Force. This was the policy of the Central Committee. Even the Committee for State Security (KGB) was obliged to strictly enforce the Party directive calling for the complete anonymity of the creators of space technology.[18]

Supporting Korolev, I affirmed that in terms of electric power reserves, we would be able to ensure a flight duration of up to seven days, and in an energy-conserving mode, up to 10 days. Bushuyev confirmed that there was also a double margin in terms of life support. Rauschenbach supported Korolev, reasoning that we already had experience with a 24-hour flight; there were rocket propellant reserves left over for attitude control; and we could guarantee a three-day flight and even have manual control drills.

On 16 July, Leonid Smirnov convened a session of the VPK to discuss the launch of the two Vostoks. At Keldysh's suggestion, they first listened to Sergey Vernov's report about the radiation that had occurred in space in connection with the American Argus nuclear test program in space.[19] Vernov verified that the radiation situation would be normal in five to 10 days. Disputes about flight duration flared up with such intensity that during the break Smirnov congregated his "inner circle." The dispute ended with Korolev being instructed "to review, coordinate, and report once again." From my perspective, the dispute between the Air Force and OKB-1 was a complete waste of time. Bushuyev and I suggested to S.P. that he come to grips with the fact that, depending on how the cosmonauts felt, we might land the cosmonauts after 24 hours or even the second orbit. But he tore into us: "This is a matter of principle, not form. We should dictate our own conditions, but you are 'yes' men."

Korolev set up a meeting with the cosmonauts and, without hesitation, signed off on a flight mission lasting up to three days. At Korolev's insistence we got

---

18. KGB—*Komitet gosudarstvennoy bezopasnosti.*

19. Chertok is probably referring to Starfish Prime, a high-altitude nuclear explosion (1.4 Megaton) carried out on 9 July 1962 in a joint project of the Defense Atomic Support Agency and the Atomic Energy Commission. Project Argus was carried out in August–September 1958. Sergey Nikolayevich Vernov (1910–82) was one of the leading space physics specialists in the Soviet Union. He served as director of the Scientific-Research Institute of Nuclear Physics of the Moscow State University from 1960 to 1982.

ready, and on 28 July we launched a Zenit-2 satellite so that we could once again be sure of the reliability of the launch vehicle before the two piloted launches.[20]

It was still sweltering during the first days of August at Tyura-Tam. All the living quarters were filled with representatives who had flown in from various agencies, services, and enterprises. For the first time the launch schedule called for three days of continuous work at the launch site. The first day was preparation for Nikolayev's launch; the second day was the launch and beginning of preparation for Popovich's launch; and the third day was Popovich's launch. We had three days of work without rest ahead of us. But no one grumbled.

On 7 August, the State Commission solemnly signed off on the crews. Valeriy Bykovskiy was approved as Nikolayev's backup for *Vostok-3* and Vladimir Komarov as Popovich's backup for *Vostok-4*. This was the first time I'd seen Komarov. He was the type of person whom you trust and like from the moment you meet him. His maturity distinguished him from his colleagues.

The *Vostok-3* launch vehicle was scheduled for rollout to the launch site at 7 p.m., so that launch preparation work could be conducted in the cool of the night. However, at the appointed time, instead of giving his consent for rollout, Korolev gave the command to replace a washer, which the assembler sent by Semyon Alekseyev's organization admitted had been installed erroneously somewhere in the mounting of the seat.[21] Everyone who had gathered for rollout got complete satisfaction from the demonstrative reaming out that Korolev gave to the engineer and assembler from Alekseyev's firm. Yet he did not utter a single word of blame to Chief Designer Alekseyev. Turning to him at the end of his tirade, he said only: "Don't let me see them here again."

But everything turned out fine. No one was banished, and the same people prepared the next seat. After that, everything went according to schedule. On 11 August, *Vostok-3* carrying Nikolayev was inserted into space, and one day later on 12 August, *Vostok-4* carrying Popovich successfully went into orbit.

Communication with the cosmonauts was conducted from our control post located next to the MIK in Colonel Kirillov's three-story service building. This was supported primarily by Gagarin. I admired his composure and skill at finding the right words in rather tedious but essential conversations when the cosmonauts' well-being needed to be determined based on the intonation and timbre of their voices. After all, at that time we had no Flight Control Center (TsUP) with data processing and display systems.[22] The cosmonauts were the sources of "real time"

---

20. In the open press, this Zenit-2 was announced as *Kosmos-7*.

21. Semyon Mikhaylovich Alekseyev (1909–93) was chief designer of Factory No. 918 (later MZ Zvezda) responsible for the design and production of cosmonaut spacesuits.

22. TsUP—*Tsentr upravleniya poletom*.

information. It was 3 to 4 hours after the communications session before we were able to ascertain the status of the systems on the two spacecraft after developing tens of meters of telemetry film in the photo laboratory.

There were heated debates about whether to extend Popovich's flight to four days. At first, it was agreed that both cosmonauts would land according to the following program: Nikolayev—during his 65th orbit after four days and Popovich—during his 49th orbit after three days. However, even on such a seemingly simple question as, "Shall we extend Popovich's flight to four days?", Smirnov and Korolev did not accept the responsibility and decided to report to Khrushchev. Khrushchev responded calmly that if there were no problems with hardware or with Popovich's health, then " . . . why should we begrudge him? Ask him, and if he wants to, and if he is able to fly longer, let him fly for four days." Thus, responsibility shifted temporarily to Khrushchev. Marshal Rudenko acted similarly. He reported to Marshal Grechko, who also did not object to four days in space for our fourth cosmonaut.

The State Commission convened one more time. Korolev, Kamanin, and Gagarin were tasked to speak with Popovich and determine how he felt. Pavel Romanovich, mimicking Voskresenskiy, cheerfully responded that he felt superb—"Grade A!" After this, they decided once again to report "upstairs," and Smirnov once again contacted Khrushchev and Kozlov. And once again Khrushchev gave his consent.

However, Popovich deprived himself of his fourth day in space. During his third day of flight, he reported that the temperature and humidity had dropped to the lowest permissible levels. The temperature in the spacecraft was just 10°C (50°F). After this report, the medics began to fret and requested an immediate landing.

What to do? Khrushchev and Grechko had just given their approval for four days and suddenly we can't even complete a three-day flight. Perhaps the flight would have been extended, but suddenly the following report came in from Popovich: "I am observing a thunderstorm." "Thunderstorm" was the code word signifying that nausea had worsened to the point of vomiting. It never occurred to anyone that he was talking about a *real* thunderstorm. Such agitation began at the control post that conversations about a four-day flight ceased despite the fact that when asked again about the "thunderstorm" Popovich responded, "I observed a meteorological storm and lightning."

On 15 August, Major Nikolayev landed after a four-day flight, and 6 minutes later Lieutenant Colonel Popovich made a successful landing in the same area after a three-day flight. Thus, we had outdistanced the Americans by a total of 60 orbits.

Despite preparation for the next Zenit-2 launch and the Venus launches, the first of which was supposed to take place on 25 August, I was granted permission to fly to Moscow for the reception for Nikolayev and Popovich under the condition that I would be back at the firing range no later than 18 August. I cover the Venus and Mars launches that took place during the second half of 1962 in other chapters.[23] Here, I just want to mention that in August 1962, while preparing for the launch "toward Venus," my comrades and I saw for the first time at the MIK a covey of slender young women in military blouses, who, we were told, were future cosmonauts. The young women were undergoing training. They were studying the launch vehicle and even becoming familiar with the layout of our interplanetary station. When they were led over to our hardware—which had almost finished testing—considerably more curious on-lookers crowded around than the work actually required. Which of the women would be the first to fly? Each person who had joined the throng next to the object waiting to be mated with the launch vehicle probably asked themselves that question . . . .

Approaching the curious bystanders, Kirillov, who loved to joke in such situations, said in a stage whisper: "Here comes Korolev!"

The servicemen and civilians vanished into thin air! I quickly and incoherently finished my explanations, and when the young women were led away, I shouted to Kirillov, "And just where is S.P.?"

"I just did that to scare them. It would have been awkward to chase off respectable people by hollering rudely in front of the young ladies."

"But Korolev isn't at the firing range. He's in Moscow. According to my information, he's even in the hospital."

"Precisely! I was checking their reflexes. Sergey Pavlovich isn't here, but the established order is in effect: more than three shall not gather unnecessarily."

On 25 August, an 8K78 launch vehicle lifted off carrying automatic interplanetary station 2MV-1 vehicle No. 3. Five young women, who watched their first *Semyorka* launch from the observation deck of IP-1, left the launch area and departed "for further service." Out of this group of five, Valentina Vladimirovna Tereshkova was destined to become the first woman in the world to make a visit into near-Earth space. The rest would not even fly into space.[24]

---

23. See Chertok, *Rockets and People, Vol. II*, pp. 589–595.

24. The other women were Tatyana Dmitriyevna Kuznetsova, Valentina Leonodovna Ponomareva, Irina Bayanovna Solovyeva, and Zhanna Dmitriyevna Yerkina.

THE CIRCULATION OF SENSATIONAL RUMORS AND JOKES IS A TRUE SIGN OF NATIONAL RECOGNITION OR SIMPLY THE POPULARITY OF SOME NEW UNDERTAKING. The popularity of the first four cosmonauts was very high. The public was even more interested in them than it had been in Chkalov, Gromov, Baydukov, Grizodubova, Raskova, and other famous, record-setting, heroic pilots in their time.[25] Back in the pre-war days, the family affairs of pilots did not attract much attention. People were satisfied with what was written in the newspapers and broadcast over the radio. It was simply risky to pose unnecessary questions.

It was different now. Jokes even circulated about Khrushchev himself. It was clear that the cosmonauts weren't saints either. Women took an emotional interest in Nikolayev's fate. He was the only bachelor among the four. At home, I was suddenly asked point-blank: "Is Andriyan Nikolayev really going to marry Vertinskiy's daughter?"[26]

I was caught off guard because this was the first time I had heard this sensational news. I called Bushuyev. It turned out he was out of the loop too. Voskresenskiy teased me a bit: "As always, we are the last to find out anything interesting, but Alena tells me that all of Moscow has known about this for a week already. It's just a matter of getting Khrushchev's permission."

We had to confess that, in our perpetual rush, we had actually missed this historic event. The next morning Bushuyev dropped in on Korolev with the very same question. First, Korolev burst out laughing.

"That can't be!"

Then he got right on the direct line to the Kremlin and telephoned Kamanin. He found out that someone had started a false rumor, which was eagerly picked up by a public that was thirsty for sensational space-related news.

WHICH WOULD BE THE NEXT PILOTED FLIGHT? As of early 1963, neither we at the bottom at OKB-1, nor those at the top at the VPK, nor the Air Force, nor the Rocket Forces had a clear-cut position on this matter.

We were carried away with designing the Soyuz. For us, this was really an interesting, creative task. Here we had the opportunity to implement ideas

---

25. Valeriy Pavlovich Chkalov (1904–38), Mikhail Mikhaylovich Gromov (1899–1985), Georgiy Filippovich Baydukov (1907–94), Valentina Stepanovna Grizodubova (1911–93), and Marina Mikhaylovna Raskova (1912–43) were all famous Soviet aviators of the 1930s, a time when these pilots achieved extraordinary celebrity status in the popular imagination.

26. Aleksandr Nikolayevich Vertinskiy (1889–1957) was a famous Soviet artist, actor, composer, poet, and singer. His daughters Mariana and Anastasia became famous actresses of the Soviet era.

that had spontaneously cropped up while we analyzed the shortcomings of the Vostok, which was incapable of maneuvering. Feoktistov seized the initiative and among the design engineers he was the primary and rather bold promoter of new ideas.

Despite the fact that my head was about to explode with worries about the Venera, Mars, Zenit, and R-9 launches after the successful flights of Nikolayev and Popovich, I had to delve much deeper into the programs of future piloted flights. Would we continue the Vostok series or shift our remaining resources to the Soyuz, which had a status not yet completely clear? Shortcomings in the Vostok systems were already evident. Much of what we had done for the Zenits could also be used in a piloted version.

But when and how? When Korolev left the hospital, he was told that Malinovskiy had rejected the order of new Vostoks. "Vostok spacecraft have no military value; we're not going to put them into service. It's not our business to order them—let the VPK deal with them," said the minister of defense.

It appeared that Korolev was not very distressed by this announcement. He also understood that it would be difficult to prove the Vostoks' military value. If we were to launch another series of 10 spacecraft, as the Air Force proposed, then Turkov's (the OKB-1 factory director) production capacity would be completely exhausted and any Soyuz programs would be out of the question.

Our former Lead Designer, Oleg Ivanovskiy, the last one from the launch control team to have seen Gagarin off into space, now had a separate office in the VPK suites in the Kremlin. I visited him regarding preparing a resolution about the upcoming Venus and Mars launches in 1963 and 1964. Of course, we got on the subject of the Vostoks. He put the question to me: "It's clear that there will be no Soyuz in 1963, but what about 1964?"

According to inter-organizational rules, I did not have the right to express thoughts that differed radically from those accepted in OKB-1 and authorized by government decrees. Finding myself in the office of Ivanovskiy, with whom I had interviewed when I signed on at NII-88, and then with whom I had shared so many trials and tribulations at the OKB and at the firing range, I told him what I had already been thinking, what I had already sized up, and what I had talked about time and again with my comrades: "Soyuz flights won't be possible in 1963 or in 1964. We won't begin the flight development tests until 1965. We might manage to pull off the first flight with a crew in 1966."

If S.P. had heard my response, he would have clobbered me. But Ivanovskiy did not betray me. I did, however, err in my predictions. The future turned out to be far worse.[27] The Romans had demanded "bread and circuses" during the times of Nero. There was still bread in Moscow. On the outskirts, especially in the villages, the bread situation was a lot worse. Khrushchev's infatuation with corn did not solve the problem.[28] In such a severe economic situation, triumphs in space were the most distracting "circus." Grumblings about agricultural failures were muffled to a certain degree by the commotion surrounding the spaceflights.[29] For political reasons, we could not halt or delay piloted spacecraft launches for long.

For us guidance specialists, the main—and for the time being, very murky—problem was rendezvous and docking on orbit. The primary stimulus was not so much the shortcomings of the Vostok series, as it was a new mission—piloted flight around the Moon. It was obvious that the Vostok was completely unsuitable for that. We did not have a new launch vehicle. Chelomey had already reported his proposals to Khrushchev concerning a lunar flyby using the new UR-500 heavy rocket with his own piloted spacecraft.[30] Our pride in our organization would not allow this. We needed our own version!

And we got one. The lunar fly-by could be executed using a space train—the Soyuz complex—that would be assembled in orbit. Operations on the Vostok series were already somewhat routine compared with those on the Soyuz series. Therefore, the general mood in our systems departments favored new, revolutionary proposals. Conversations about the possibility of ordering another series of eight or nine Vostoks were hardly a cause for enthusiasm among our engineers.

A bunch of us often met with Armen Mnatsakanyan. We included Rauschenbach, his loyal "sword bearers" Bashkin and Legostayev, specially assigned "rendezvous" theoreticians Shmyglevskiy and Shiryayev, radio engineer Nevzorov, and myself. Mnatsakanyan and his coworkers, who had developed the command radio link, were acquainted with us from our joint work on Vostoks and Zenits. But a different group worked on developing the radio

---

27. The first piloted Soyuz flight was accomplished in 1967. The first relatively successful mission was in 1968.
28. Khrushchev became a great enthusiast of corn after his trip to the United States (particularly Iowa) in September 1959, and decided to introduce the crop to Soviet agriculture despite entirely different climate conditions.
29. A series of agricultural crises during the Khrushchev era severely limited Soviet industrial growth. The most serious crisis was in 1963 resulting from a combination of mismanagement, lack of innovation, and very bad weather. Agricultural crises, such as these, had negative effects on the consumer economy in urban areas, and they undercut much of the popular support for Khrushchev.
30. The UR-500 and its various modifications were openly known as the Proton, named after its first satellite payloads launched from 1965 to 1968.

system that measured the spacecraft positions relative to one another as needed on the Soyuz. The linchpin of this new undertaking was Yevgeniy Kandaurov. He already had experience developing on-board aviation radar for the target-seeking warheads of air-to-air missiles, and therefore he considered solving a space problem in a strictly automatic mode to be completely feasible.

News about our work on the in-orbit assembly problem quickly spread around the secret radio facilities. We received several more proposals for collaborating on the development of relative measurement systems. As a safety net for the work of Mnatsakanyan's NII-648, Korolev agreed to sign contracts with NII-158 in Leningrad, with Bogomolov's OKB MEI, and with TsKB Geofizika. The development of automatic hybrid and manual rendezvous systems occupies a special place in recent cosmonautics history and deserves to have a special work devoted to it.

In late January 1963, during the latest discussion of Soyuz problems, S.P. announced that we were supposed to prepare Vostoks for a group flight of women, and then be done with it—let the Air Force deal with Vostoks. If they like it, let them order another series of 10 and put them into service.

"It's time for us to do some serious work on the Moon," he said.

Being in an unusually melancholy, depressed mood, as Bushuyev and I later surmised, S.P. laid out for us his thoughts concerning the general management of space operations. The Ministry of Defense, to whom we were in indentured servitude, lacked a unified point of view as to which of the service commanders-in-chief should be in charge of space. Before the piloted flights, we had only dealt with the artillerymen, and then with the Rocket Forces. After Gagarin's flight, the Air Force took on a large share of the responsibilities. We couldn't manage without them in human spaceflight. At the time, a struggle was going on between the Air Force's Konstantin Vershinin and the Strategic Rocket Forces' Nikolay Krylov.[31] However, it almost didn't matter to these two and to other marshals who were already getting on in years. The main scuffle was going on between lower-level generals. Minister of Defense Malinovskiy generally had little interest in all of this. All he cared about was less upheaval and less reshuffling so that he could live out his days more peacefully. Sure, the marshals had earned a peaceful life. But this appetite for tranquility was slowing down our work. Ustinov didn't want to give up space, even piloted flights, to the Air Force. Smirnov—Ustinov's man—would also adopt that line.

---

31. Nikolay Ivanovich Krylov (1903–72) served as commander-in-chief of the Strategic Rocket Forces from 1963 to 1972. Konstantin Andreyevich Vershinin (1900–73) served as commander-in-chief of the Air Force from 1946 to 1949 and then again from 1957 to 1969.

We couldn't risk spoiling our relationship with the Rocket Forces because the firing range and the Command and Measurement Complex were all theirs. We started out with them, and without them we couldn't launch a single rocket. And then artillery and Rocket Forces people such as Smirnitskiy, Sokolov, Karas, Kerimov, Tyulin, and even Mrykin. Say what you like, they were still our people. We were all tied together with the same rope and inseparable. Mishin, who had always sympathized with the Air Force, asked: "Do we really need to get mixed up in these disputes? Let the marshals sort it out among themselves."

"We absolutely have to get mixed up in this," was Korolev's firm response. "If we let them decide things without taking us into account, the result will always be stupidity. For the time being, no one is coming out against the Soyuz. Let's have a look at the schedule."

We had looked at the Soyuz schedule time and again. And time and again, Turkov had argued that if he got an order for a series of another 10 Vostoks, then he could not produce the Soyuz in 1964. However, even that wasn't the issue now. At that point, Turkov didn't even have any drawings. This was a familiar ploy. It was easy to make Korolev seethe, if you said that it was impossible to have a serious discussion of deadlines for the production schedule of a new spacecraft if the factory had no drawings. But this time things ended without much hubbub.

Despite all the exotic qualities and outward seductiveness of Feoktistov's idea for the lunar spacecraft design, everyone was developing an aversion to this idea. Everything was so convoluted, complex, and uncertain. Korolev sensed this, too.

As chairman of the Interdepartmental Scientific-Technical Council on Space Research (MNTS-KI), Keldysh was also unable to precisely formulate the Soviet cosmonaut's immediate and top-priority missions.[32] He very aggressively insisted on flights to Venus, Mars, and a soft landing on the Moon; but as soon as it came to piloted launches, Keldysh waited for Korolev's initiatives.

Influenced by information about the United States' progressive broadening of their front of operations in preparation for a flight to the Moon, Keldysh convened his Council in late April 1963. S.P. collared almost all of his deputies for this meeting—Mishin, Bushuyev, Okhapkin, Kryukov, and me.

---

32. The MNTS-KI was a special interagency body attached to the Presidium of the USSR Academy of Sciences to provide high-level recommendations on the scientific direction of the Soviet space program. Throughout the 1960s and 1970s, under Keldysh's direction, the Council served as an advisory body arbitrating many conflicts in Soviet space policy. The Council was established in January 1960 by reorganizing an older commission established in 1956 to direct the Object D satellite program.

Korolev's report boiled down primarily to an account of the N1 rocket design.[33] We still had no lunar spacecraft design or even a landing procedure. The scope of the report was the same as the draft plan overview from 1962. Korolev said that the N1 would make it possible to send a man to the Moon and return him to Earth. We would eventually be able to set up a scientific station on the Moon. In terms of military goals, the N1 would make it possible to create an orbital station that could conduct continuous detailed intelligence gathering, could intercept enemy spacecraft, and if necessary, also deliver strikes against Earth. I felt I needed to make a few remarks concerning problems of radio monitoring and data transmission at near-lunar ranges. The Council acknowledged that it was advisable to report to the Central Committee about issuing a special decree to speed up these operations.

I also recall this particular Council meeting in Keldysh's office because it brought annoying aggravation. That same day a parade was being held on Red Square for Fidel Castro.[34] As a result, traffic through the center of

From the author's archives.
Chertok relaxing.

---

33. The N1 was the heavy-lift launch vehicle proposed in the early 1960s for a wide range of ambitious space missions. After 1964, the fate of the N1 was tied to the Soviet human lunar landing program, a competitor to the Apollo lunar landing project. Chertok will address the development of the N1 in detail in Volume IV.

34. Cuban leader Fidel Castro paid his first historic visit to the Soviet Union in 1963, arriving on 27 April and departing on 3 June.

Moscow was practically shut down. As always in such instances, the ring roads were overflowing with automobiles.[35] I had been driving around in a new light gray *Volga* for just a month. I was hurrying home from the Institute of Applied Mathematics on Miusskaya Square so that I could catch a glimpse of the legendary Fidel on television. Along the way, as often happens, I was critically analyzing my presentation to the Council. In hindsight, I found it cowardly, trying to suit Korolev and Keldysh by simplifying the problem of developing a guidance and control system for the lunar vehicles. I had said nothing about an autonomous inertial system and lunar landing problems. This mental self-criticism distracted my attention from the traffic lights and I was punished on the spot. While turning left from Sushchevskiy Val onto Sheremetyevskaya, I didn't wait for the green arrow, and a truck, which clearly had the right of way as it made a right turn, ran into the right side of my car. Motorists must understand me. After a shakeup like that, your whole world falls apart, and space moves to the background for a while.

---

35. The major thoroughfares in the modern city of Moscow emanate from the center like spokes in a wheel. These are criss-crossed by progressively bigger circular loops around the center. The most outer ring is the Moscow Automobile Ring Road (MKAD), opened in 1961 as a loop around the city. It served as the administrative boundary around Moscow until the 1980s.

Chapter 8
Man and Woman

The flight program for *Vostok-5* and *-6* was born amid bitter disputes. Leaders at various levels were fighting primarily about cosmonaut candidates. During these multi-day debates, I either did not participate at all, or I was a passive observer. Korolev, Bushuyev, Feoktistov, and Tsybin put so much passion into these disputes that it seemed ridiculous to me and many of my friends.

The first stumbling block was the matter of simultaneous or separate flight. They agreed that following the example of the preceding flight, this should be a dual flight. If it was a dual flight, should it be two women, two men, or a man and a woman? More than the others, Kamanin fought for a strictly female dual flight. At first, Korolev, Tyulin, and Keldysh thought that in general there was no need to rush into flights with female crews. Spaceflight duration needed to be confirmed. It needed to be extended to six to eight days; experiments in manual attitude control needed to be introduced; the spacecraft should be equipped with additional equipment for observation and photography. In short, we needed to show that there was a benefit from having a human being in space. But the memory of our sensational space feats was fading, and rather than a routine project, political considerations demanded a new space sensation. The flight of a woman might be just that.

In April 1963, we finally agreed upon a dual flight by a man and a woman. An agreement was reached as to a male candidate without particular controversy: Valeriy Fedorovich Bykovskiy, with Boris Valentinovich Volynov as his backup. Passions raged over the female candidates. Joining forces with Gagarin, Korolev talked Tyulin and Mrykin into supporting Valentina Tereshkova. Keldysh and Marshal Rudenko, on behalf of the Academy of Sciences, stood up for Valentina Ponomareva, proposing Tereshkova as her backup.

In May, the chief designers had reported to the State Commission headed by Tyulin that all systems were ready, but that it was as yet unknown for whom the seat in the spacecraft would be designed. Finally, they decided to go to the Cosmonaut Training Center and select a candidate once and for all. Korolev, Bushuyev, Keldysh, Tyulin, Mrykin, Rudenko, and Kamanin gathered at TsPK

From the author's archives.

Cosmonaut Valeriy Fedorovich Bykovskiy (1934–) in a picture dating from the early 1960s.

and decided in favor of Tereshkova. They decided to kill two birds with one stone: Bykovskiy was supposed to set a new spaceflight duration record of eight days and Tereshkova was supposed to fly for no more than three days.

In late May, once again all the space elite gathered at the firing range. We were constantly aware of the inconvenience of our living conditions. We had come to grips with them in the early days, but now they were aggravating: the hotels were cramped, there were lines in the dining halls, transportation was a problem, and there were delays obtaining passes. After a series of complaints, Korolev had some serious talks with firing range chief General Andrey Zakharov. However, their relationship remained shaky, and Korolev complained that he couldn't develop an understanding with Zakharov.

I had just flown back to Moscow from the firing range after a Zenit-2 satellite launch, and Korolev called me right back there to give a readiness report to the review team and State Commission on the emergency rescue system (SAS) and automatic engine unit shutdown (AVDU) system.[1] The preparation of the Vostoks was proceeding normally. This made it possible for us to have a small party on 3 July to celebrate Aleksey Bogomolov's fiftieth birthday.

A business session of the State Commission was held the morning of 4 June, and that evening a film and audio recording were made of a "show" session.[2] Major Bykovskiy and Second Lieutenant Tereshkova were confirmed as spacecraft commanders. Male comments, unsuitable for audio recording, were inevitable.

---

1. AVDU—*Avtomaticheskoye vyklyucheniye dvigatelnoy ustanovki*. A secret Zenit-2 reconnaissance satellite was launched on 24 May 1963 and publicly announced as *Kosmos-18*.

2. In the early days of the Soviet space program, all films of State Commission meetings were usually staged for public consumption. The *actual* deliberations on pre-flight discussions were never filmed.

Second Lieutenant Valentina Vladimirovna Tereshkova (1937–), the world's first woman in space.

"Look how Tereshkova has blossomed. A year ago she was a plain Jane, and now she's a real movie star," remarked Isayev seated next to me. "And imagine what she'll be like once she's flown," I responded, and we both knocked on our wooden chairs.

True, after getting a look at Ponomareva, we decided that she "didn't look half bad" either. But she didn't glow like Tereshkova. Judging by her appearance, she was overly serious, and it seemed to me that she was simply pouting because she remained the backup.

For lack of a flight control center and standard flight directors, competent members of the State Commission took on these duties. Flights lasting several days required shifts. Therefore, the review team was split into four shifts with Ishlinskiy, Kuznetsov, Mrykin, and Kamanin placed in charge of them.

Meeting the day before the launch, on 15 June 1963, are the three leading members of the female cosmonaut corps. From left to right: Valentina Ponomareva, Irina Solovyeva, and Valentina Tereshkova.

Bykovskiy's launch might have taken place on 7 June, but it was postponed due to weather conditions. The wind speed on the ground exceeded 15 meters per second—the maximum permissible speed according to our specifications. The rocket was not erected until the morning of 9 June. Launch preparations proceeded normally with an eye toward launch on 10 June. Late in the evening, Korolev suddenly called Bushuyev and me to his cottage. We received strict instructions to fly to Moscow at dawn and drive directly from Vnukovo Airport to the Academy of Sciences to see Keldysh and find out why he was opposed to the launch.

It turned out that Keldysh, who had stayed on in Moscow, had received a warning from Andrey Borisovich Severnyy, director of the Crimean Observatory, about a dramatic increase in solar activity and, consequently, an increase in the radiation level in near-Earth space. Bushuyev and I were included in Keldysh's commission, which was supposed to give the go-ahead for launch, depending on the sunspot forecast.

An order is an order, and on the afternoon of 10 June at Leninskiy Prospect, we announced to Natalya Leonidovna—aide to the Academy president—that we had arrived "in furtherance of our duty." The president's office was full of scientists, and we had quite a time trying to understand what the heated argument was about. Gradually, helping each other and using hints from Academician Vernov, we surmised that no one really had any idea of the actual radiation situation in space, but what Severnyy (the top solar specialist) was observing provided cause to assume that the intensity of the radiation would increase by a factor of 100 in the next two days.

An hour after our arrival at Keldysh's office, Natalya Leonidovna called us both in for a conversation over the high-frequency line with Korolev. After he had ascertained that we had arrived safely and had already been in conference for an entire hour, he pounced on Bushuyev, and when the latter had "gotten what he deserved," he handed the phone to me; and Korolev pounced on me. S.P. explained that now Bushuyev and I alone were responsible for all the mess on the Sun, and if we did not take the necessary measures, then the scrubbing of the flight, which had been postponed until 12 June, would be our fault. I reiterated to him the forecast for a 100-fold increase in radiation—the most recent report that I had heard as I left the office—and promised that early in the morning on 12 June, solar activity data would come in from the Crimea and other observatories, enabling us to make a decision. After ultimately being ordered not to leave the Academy until we had managed to get Keldysh's agreement, we nevertheless arranged to spend the night at home, since the Presidium building had no facilities for that and the "Solar Service" also rests at night.

Calls from the Central Committee bureaucracy and VPK, from Tyulin and Korolev at the firing range, and conversations with the Crimean observatory had the usually calm Keldysh seething by the end of the day. He did not have the authority to divert calls on the "Kremlin hotline" to Natalya Leonidovna, and with his office full of witnesses, he took the liberty of responding very disrespectfully to Central Committee Secretary Frol Kozlov that he did not have people capable of bringing the Sun to proper order.[3]

In parting, I reminded Keldysh of Kozma Prutkov's observation that, "If they ask you what is more useful, the Moon or the Sun, boldly answer, 'the Moon.' The Sun shines during the day when it's already bright, but the Moon shines at night!"[4] This cheered him up tremendously. Bushuyev and I promised to be back in his office the next day by 7 a.m.

For two days, 11 and 12 June, arguments with the "Solar Service" continued non-stop. As the night of 12 June approached, good judgment began to prevail. An analysis of scientific developments on the life of the Sun showed that, during the past 100 years, no disasters had been attributed to Sun spots. The metallic housing and thermal-protective coating of the Vostok were sufficient radiation protection, even exceeding conventional protection for these altitudes by a factor of 10. Rather than short-term exposure, it was the total radiation exposure that was hazardous, and this would be many times greater for sunbathers on southern beaches and high-altitude mountain climbers than the exposure that a cosmonaut could receive during a week-long flight. Deputy Minister of Health Avetik Burnazyan dealt Keldysh the final blow. Burnazyan was responsible for the health of atomic industry workers. After hearing Keldysh out, he indicated that the worst exposure our cosmonaut could encounter during a week-long flight would total less than the daily dose that, according to the norms, is considered permissible for personnel working on atomic reactors. At 2200 hours Moscow time (at Site No. 2 at Tyura-Tam it was midnight), Keldysh woke up Korolev and then Tyulin and informed them that we were all writing the findings that would grant permission for a launch on 14 June.

---

3. Frol Romanovich Kozlov (1898–1965) was the Secretary of the Central Committee in charge of defense industries and the space program, i.e., the de facto head of the Soviet space program, from 1960 to 1964.

4. Kozma Petrovich Prutkov was a pseudonym collectively used by four authors (including Aleksey Konstantinovich Tolstoy) in the mid-19th century. Under the Prutkov name, they wrote a number of satirical aphorisms, quotes, and fables ridiculing the imperial-era bureaucracy.

In response, Korolev demanded that Bushuyev and I also sign this document, send an official high-frequency telegram, and fly back to the firing range with the original. Bushuyev and I were also instructed to bring Keldysh with us as a "solar hostage." Our flight out of Moscow was delayed and, by the time we arrived at the firing range, Bykovskiy's launch had already taken place.

From the author's archives.

Meeting at Site No. 2 before the launch of *Vostok-6.* In the foreground are, from left to right: Marshal Nikolay Krylov (Commander-in-Chief of the Strategic Rocket Forces), Georgiy Tyulin (Chairman of the State Commission), and Chief Designer Sergey Korolev. Behind them, in civilian clothes on the left, are Chief Designers Andronik Iosifyan and Nikolay Pilyugin.

It turned out that the launch control team—which had taken a break during the period of the solar flares—the State Commission, and Bykovskiy himself, during his last hours on the ground, had been under stress that was a lot more hazardous in terms of its possible consequences than the solar activity.

After T-minus 60 minutes, Chief Designer Alekseyev reported to Korolev that his engineer V. I. Svershchek had failed to cock the ejection mechanism on the seat. Ejection was impossible—"we can't launch!" A 30-minute delay was announced. They had to open the spacecraft hatch and cock the mechanism with the cosmonaut sitting in the seat.

At T-minus 5 minutes, it turned out that a gyro horizon on the third stage had failed. Replacing the main gyro on a fueled rocket with a cosmonaut seated in the vehicle, to boot, was an emergency situation. But there was no way out—either postpone the launch for another 24 hours or take the risk. The chief guidance and control specialists—Pilyugin, Kuznetsov, and Ryazanskiy—convened and talked Korolev into taking a justified risk. This operation required that Bykovskiy be on the ground an additional 3 hours in the closed spacecraft.

In addition to all these headaches, the spacecraft went into an orbit that was lower than the calculated one. The guaranteed orbital lifetime did not exceed 10 to 11 days. The ballistics experts joked that the danger of solar activity was not in the radiation exposure, but in the "thickening" of the upper atmosphere. It could slow down a spacecraft so much that, heaven forbid, it "buries itself" by the eighth day and then lands who knows where. All the Command and Measurement Complex services and ballistics centers were given the strict order to continuously monitor the *Vostok-5* orbit, to compute, forecast, and report . . . report . . . after each measurement session.

Unlike the previous launch, the launch of *Vostok-6* carrying Valentina Tereshkova on 16 June proceeded right on schedule, precisely, without any delays or off-nominal situations. After Levitan's first report and the TASS publications, the press started to refer to Valentina Tereshkova by the code name assigned to her during the flight: "our *Chayka* (Seagull)." This aptly selected code name stuck with Valentina for many years.

The orbit of the *Yastreb* (Hawk), i.e., Bykovskiy, dropped more rapidly than the first prognosis. On 17 June the decision was made to limit the time of his flight to five days, landing on the 82nd orbit. *Chayka*'s flight program had provided for a landing during the 49th orbit, but during day two the flight directors began to complain about her responses, which were not always clear. Whether she was fatigued, or suffering from nausea in weightlessness, she sometimes gave evasive responses to direct questions. After being tasked with manual attitude control of the spacecraft, she failed to execute it on her first attempt. This didn't bother us very much, but it really annoyed Korolev. Unable to tell *Chayka* off himself, Korolev reprimanded those conducting the conversations and demanded that Rauschenbach personally explain how to control the spacecraft for "landing" orientation.

On the morning of 18 June, the attention of the State Commission and all our command post "enthusiasts" switched our attentions from *Chayka* to *Yastreb*. Khabarovsk had received a message from Bykovskiy over the shortwave channel: "At 9 hours 5 minutes there was a space bump (*stuk*)." Korolev and Tyulin immediately began to develop a list of questions to ask Bykovskiy when he entered our communication coverage zone, so that we could understand the magnitude of the danger threatening the spacecraft. Someone was assigned the task of calculating the size of a meteorite large enough to produce a "bump" that the cosmonaut would hear. They also racked their brains over what might happen in the event of a collision that did not result in depressurization. Kamanin was delegated to question Bykovskiy.

At the beginning of the communications session, when asked about the nature and area of the "bump," *Yastreb* answered that he didn't understand what Kamanin was talking about. After being reminded about the radiogram transmitted at 0905 hours and having its text repeated by *Zarya*, through his laughter Bykovskiy answered: "I said stool (*stul*), not bump. There was stool, do you understand?" Everyone listening got a good laugh out of his answer. They wished the cosmonaut further success and told him that despite his valiant deed, he would be returning to the ground at the beginning of day six. The incident involving the "cosmic stool" has gone down in the oral history of cosmonautics as a classic example of the unsuccessful use of medical terminology over a space communications channel.

On 19 June, more anxiety than ever arose on the ground once again over Tereshkova's manual attitude control training sessions. Korolev was beside himself until *Chayka* reported that she had gotten through with her attitude control task. Commands were sent almost simultaneously from the ground to both spacecraft to activate the automatic landing cycle. Bykovskiy sent reports that he was visually monitoring the activation of solar orientation. Then naval ships picked up his report that the braking engine was operating.

From the author's archives.

Valentina Tereshkova after landing, shown here beside equipment from the *Vostok-6* descent module.

Chayka was silent. Either communications were the culprit, or she was so agitated that she forgot about reports. For more than 2 hours, it was unclear what had happened with *Vostok-6*. Korolev darted between Bykov, who was handling communications, Kamanin, who was pestering the aviation search and rescue services, and officers, who were attempting to receive reports from air defense facilities. The commands to land had been issued at 1000 hours, and it wasn't until 1700 hours that explanatory reports saying both cosmonauts were alive and well were finally received.

I have kept a rare photograph. A correspondent had showed up at Tereshkova's landing site before the colonels and generals who were responsible for her safety. Tossing aside all medical instructions, she made short work of the local foodstuffs that had been conveniently laid out on a parachute. During her three days in space, *Chayka* had had no appetite.

Despite the happy ending of Tereshkova's flight, we guidance and control specialists had to make sure that the unsuccessful attempt at manual control was not our system's fault. Could it be that manual control of a spacecraft really did require a pilot with experience flying fighter planes? Tereshkova's experience was a precedent that could be interpreted in two ways. Kamanin, representing the Air Force, obtained evidence supposedly proving that only a pilot could command a spacecraft. We engineers, who had designed the control system, believed that it was much simpler to control a spacecraft than an airplane. All the processes are more spread out over time. One has the opportunity to think. A spacecraft isn't going to go into a tailspin and if a firing of the braking engine is scheduled, then, according to the laws of celestial mechanics, the spacecraft will not depart from its orbit until this deorbit burn takes place. Consequently, any physically and mentally normal individual who has been trained for two to three months can control it—even a woman!

Opponents conclusively argued that people are allowed behind the controls of an airplane or the steering wheel of an automobile after logging many hours of flight or road time with an instructor. One never goes on a solo flight or road trip the first time out. No simulator and no training can replace the experience of flying solo in an airplane or driving an automobile. On a spacecraft, the cosmonaut's attention is distracted by the unfamiliar state of weightlessness, which disturbs vestibular function, and therefore an ordinary mortal is capable of making many mistakes. A pilot who is used to fighting for his life is more reliable anyway.

For some reason, more than his deputies, ourselves, and the other "chiefs," Korolev visibly showed his dissatisfaction: "So now I'll have to agree one more time to launch a broad!" We considered these petulant outbursts from Korolev to be just for show. He simply needed to publicly blow off the steam that had been building up during the stressful days of Bykovskiy's and Tereshkova's flight. In the dining hall, Leonid Voskresenskiy didn't pass up the opportunity to needle Korolev with a historical reference. As we all know, Catherine II had no teacher to train her to govern Russia. She dealt with this problem on her own and better than subsequent emperors. Korolev glared at Leonid, but then the hint of a smile crept over his face, and without saying a word he continued his meal.

We decided to find out what had actually happened from *Chayka* herself in a candid conversation "without the brass." The cosmonauts' first days after returning to Earth were scheduled for medical examinations, meetings, receptions, press conferences, and other obligatory events. We would have to conduct our conversation as soon as possible before the hurly-burly on the ground erased the perceptions of the flight. My appeal to Kamanin requesting that 2 hours be allotted in Tereshkova's schedule to meet with specialists was unsuccessful. I complained to Korolev. A half hour later, he telephoned and notified me: "Tomorrow at noon Tereshkova will be in your office unescorted. Don't call in a lot of people. It's only a talk on business matters; no clamoring for autographs."

From the author's archives.

Chief Designer Sergey Korolev speaking at the wedding reception of cosmonauts Valentina Tereshkova and Andriyan Nikolayev in late 1963. From left to right are Air Force Marshal Konstantin Vershinin, Tereshkova, Nikolayev, Minister of Defense Rodion Malinovskiy and his wife, and Korolev.

There were more people wanting to be at such an extraordinary meeting than could fit in the office. Rauschenbach and I decided to select no more than 12—as many as could comfortably sit around Grabin's conference table.

Everything was going according to schedule. Tereshkova arrived escorted by Lelyanov—Korolev's personal assistant, a former KGB operative. I introduced *Chayka* to those assembled and said that we would like to hear her detailed account of her attempts at manual control, her perceptions, visual observations, and if possible, let her freely criticize everything that had hampered attitude control and activation of communication modes. In general, we very much wanted her to candidly tell us everything that she thought about the spacecraft.

Suddenly Korolev entered the office.

"Excuse me, comrades, I need to talk to Valya. I'll give her back to you in 10 minutes."

I opened the "lounge" behind the office. Instead of 10 minutes, the confidential conversation lasted about 30 minutes. Korolev was the first to appear. After glancing at the assembled crowd, he flashed a devilish smile and quickly left. We waited for Tereshkova for several more minutes. She could not conceal her teary eyes and state of dejection. We realized that now we weren't going to have the conversation that had been arranged. To facilitate the discussion, Rauschenbach started to ask questions himself. Hoping to defuse the situation, Bashkin cut in and recalled his impressions at the command post during the flight.

I had the feeling that she was about to burst into tears. Ultimately, we would figure out this manual control issue, but now we needed to free Tereshkova from our interrogations. After taking *Chayka* to her car, I promised that we would still find time for a serious talk. When I returned, a stormy discussion was raging around the conference table about what had happened. It seems that Kalashnikov had remarked, "A woman, even a cosmonaut, is still a woman. She's easier to offend than a thick-skinned guy." None of us had any idea why S.P. needed to reduce Tereshkova to tears. I saw *Chayka* in tears one more time, after the death of Gagarin.

Years later, not only the members of the commission that had selected Valentina Tereshkova from among the female candidates but everyone could see that she had reached truly cosmic heights in public service. In 1968, Tereshkova became Chairperson of the Committee of Soviet Women, and in 1969, the Vice President of the International Democratic Federation of Women and member of the World Peace Council. Beginning in 1971, she was a member of the CPSU Central Committee. In 1974, she became a deputy and member of the Presidium of the Supreme Soviet of the USSR and Chair of the Society for the Friendship of the Peoples of the Soviet Union (and then Russia) with Other Countries. Six years after her flight, Tereshkova graduated from the N. Ye. Zhukovskiy Air Force Engineering Academy and received the rank of Air Force Colonel. In terms of the number of awards and honorary titles she received in various countries and cities after her flight, in no way did she yield to her predecessor male cosmonauts. She reached the rank of Major General. However, for those who were involved in the creation, preparation, and launch of *Vostok-6*, Tereshkova would always be "Seagull." For my fellow comrades and me, an architectural landmark— a villa on Vozdvizhenka Street—is not associated with the name of industrialist A. A. Morozov. For us, it's "Seagull's" residence.

In the spaceflights of the ensuing years, women have proved that they are capable of working in space as well as men, even performing such vital and complex tasks as piloting and landing the Space Shuttle. *Chayka's* successful landing marked the end of the Vostok flights. Ahead of us were the Soyuz flights, but we had been delayed along the way to them for the sake of the Vostoks.

Chapter 9
The Voskhods and the First Spacewalk

The domestic propaganda system made broad use of the successes of the piloted flights of Gagarin, Titov, Nikolayev, Popovich, Bykovskiy, and Tereshkova to prove the superiority of the Soviet Union, and consequently, of its socialist system. The fact that we had fallen behind absolutely in the number and reliability of ICBMs and submarine-launched ballistic nuclear-tipped missiles was kept secret from the Soviet people. Our space triumphs served as excellent camouflage for our stunted combat missile strength. It seemed to the entire world that the United States had fallen hopelessly behind in space.

The first real U.S. human spaceflight took place on 20 February 1962, after they had successfully modified the Atlas ICBM, converting it into the Atlas D. The Americans still had nothing to counter the Nikolayev/Popovich and Bykovskiy/Tereshkova dual flights. Above all, they trailed behind because they lacked a launch vehicle with a lifting capacity comparable to the R-7, the backbone of our program.

In 1962, information about the U.S. Dyna-Soar project caused a small stir in our space circles.[1] It was an elaboration of the famous X-15 winged rocket-plane design. The Martin Company received the order to modify the Titan ICBM and created the Titan II, which was designed for combat duty carrying a hydrogen bomb with a yield of 10 megatons. The Titan II also was supposed to insert a piloted rocket plane into space. Boeing received the order for the rocket plane.

The Dyna-Soar project entailed returning to Earth "on wings" and landing on an airfield "airplane-style" with a pilot at the controls. Now one can say that this project anticipated the Space Shuttle design implemented 15 years later. The insertion into space of the Dyna-Soar rocket plane was supposed

---

1. The U.S. Air Force proposed the Dyna-Soar (later the X-20) concept in 1957 as a piloted space plane for a variety of military and intelligence tasks in low Earth orbit. The reusable, single-seater vehicle was to be inserted into orbit by a Titan IIIC and glide back to Earth. In 1963, the program was cancelled before any significant hardware was built.

to begin in 1965 using a subsequent modification of the United States' most powerful rocket, the Titan III. According to its design, the mass of the rocket plane reached 7 to 8 metric tons and continued to grow. However, the Apollo project urgently needed a spacecraft similar to it in terms of the flight and return-to-Earth plan for systems development and verification. The Rockwell Corporation was tasked with developing the two-seat Gemini spacecraft and the Dyna-Soar project was cancelled in December 1963.[2]

Thus, from 1959 to 1965, projects for cruise missiles and rocket planes, such as the Burya and Buran in the USSR and Navaho and Dyna-Soar in the U.S., were cancelled absolutely independently by our respective defense ministers.[3] The Spiral project of A. I. Mikoyan's OKB was the last to be cancelled. German Titov was being trained for flights on this rocket plane. The project did not have the Rocket Forces' support and ended during the phase when an analog rocket plane was being test-dropped from a Tu-95 aircraft. Cosmonaut Titov ultimately left the Air Force and transferred to the Main Directorate of Space Assets (GUKOS), which was subordinate to the commander-in-chief of the Strategic Rocket Forces (RVSN).[4]

Meanwhile, information about the Americans' development of the two-seater Gemini and the three-seater Apollo spacecraft reached Khrushchev. Khrushchev had a fully competent private consultant in his son, Sergey.[5] We were getting ready to counter the American projects with two-seater Soyuz vehicles. But when would we have them? Despite all of Korolev's optimism, there was an ample number of sensible people who believed that it was obvious that the Soyuz would not fly in 1964. That is when Khrushchev instructed Korolev to adapt the Vostok for the flight of three cosmonauts at once. At first it seemed completely unrealistic to put three people in a single-seater Vostok. Revolutionary designs needed to be implemented using ideas that had just been conceived for Soyuz vehicles.

---

2. The cancellation of Dyna-Soar had little to do with the Gemini program but rather with questions of cost and usefulness.

3. The Burya and Buran were intercontinental missiles developed by the Lavochkin and Myasishchev design bureaus respectively in the 1950s. Only Burya reached the stage of flight-testing. Both programs were cancelled after test flights of the early ICBMs such as the R-7. Navaho was a similar intercontinental cruise missile funded by the U.S. Air Force. After several flights of a test version in the late 1950s, the program was terminated. Although cancelled, Navaho technology proved extremely useful for a number of follow-on programs in the American missile and space programs.

4. GUKOS—*Glavnoye upravleniye kosmicheskikh sredstv.* GUKOS was name of the Soviet "military space forces" between 1970 and 1986. It was subordinate to the RVSN until 1981, after which it came directly under the purview of the Minister of Defense.

5. Khrushchev's son Sergey Nikitich Khrushchev (1935–) was a missile guidance systems engineer at Chelomey's OKB-52 from 1958 to 1968.

The Voskhod (Rise) vehicle came tumbling down on us rather unexpectedly and clearly put a damper on our work on Soyuz spacecraft.[6] On 3 December 1963, the CPSU Central Committee issued a decree that had been an entire year in the making. According to the decree, flight development tests on the piloted 7K (Soyuz) spacecraft were supposed to begin in 1964, and on the entire Soyuz complex, designed for a lunar flyby, in 1965–66.

The department chiefs under me did not conceal their astonishment at the usual get-togethers: "You are demanding that we speed up our work on the Soyuz. We can't tell Korolev, but you can certainly see that in 1964 not even a hedgehog will be able to fly on it. And now they're gearing up for a three-seater Vostok."

Korolev also understood all of this very well. Khrushchev had nudged him into a game of chance, and he, having secretly moved Soyuz to the back burner, had personally plunged into developing the Voskhod. Here, rapid tactical success was possible and for its sake, the strategic objective was pushed aside.

To accommodate three cosmonauts, it was necessary to do away with the seat ejection landing scenario. All three cosmonauts would land in the descent module itself. This would require the development of special seats with reliable shock-absorption and the introduction of a new system—a soft landing system. It seemed totally impossible to put the three cosmonauts in space suits. They could be squeezed with difficulty for a short period of time in a very uncomfortable position wearing just training suits.

The decision to forgo spacesuits provoked fierce resistance from the Air Force. But under pressure from Korolev, opponents backed off one after another. Kamanin remained the most consistent proponent of spacesuits. But he too had to back down after pressure "from the top," from Marshals Vershinin and Rudenko.

Modifications caused the vehicle mass to increase to 5.7 metric tons, which was 1,000 kilograms more than Vostok. The third stage—the Block Ye on the 8A92 model of the *Semyorka*—was replaced with the more powerful Block I, which had Kosberg's engine with a thrust of 30 metric tons.[7] The 11A57 launch vehicle was the one that would insert the Voskhod spacecraft on orbit.

---

6. More broadly, the Russian word *voskhod* denotes "rising above the horizon," in the sense of a "sunrise."

7. The 8A92 launch vehicle was actually never used to launch the Vostok spacecraft although it was very similar to the 8K72K launch vehicle that did. Both used Kosberg's RD-0109 (8D719) engine for the Block Ye third stage. On the 11A57 launch vehicle which carried the Voskhod spacecraft, the Block I third stage used Kosberg's more powerful RD-0108 (or 8D715P) engine.

THE SECOND SERIOUS CONFLICT WITH THE AIR FORCE AROSE REGARDING THE COSMONAUT CANDIDATES FOR THE FLIGHT OF THE WORLD'S FIRST THREE-SEATER SPACECRAFT. As long as Vostoks had been single-seater vehicles there had been no particular controversy. Only military pilots, specifically fighter pilots, could fly into space. The first group of five cosmonauts had been selected on that principle. An exception had been made for Tereshkova. She was not a fighter pilot, but she was launched into space. It was political policy, and it worked.

But it didn't make sense to have a crew of three cosmonauts who were all pilots. Even Kamanin was forced to agree with that. Korolev proposed a crew with the following makeup: a vehicle commander, who was a military pilot, and two civilians—a flight surgeon and an engineer. A real battle ensued over the second and third spots. Kamanin proposed a military physician and a military engineer. He hoped to somewhat thin out the line of cosmonaut candidates that had formed. The Ministry of Health proposed Boris Yegorov—a physician from outside the Air Force system, while Korolev proposed Feoktistov as flight engineer. Kamanin ultimately agreed to Yegorov, but he fought against Feoktistov's candidacy on all levels. Initially, the Voskhod group included Lazarev, Polyakov, Yegorov, Sorokin, and Katys. For Korolev, Feoktistov's candidacy had become a matter of principle and prestige. He threatened to sever all connections with the Air Force, to set up his own cosmonaut training service, and prove that engineers can control a spacecraft as well as pilots.

At first, Commander-in-Chief of the Air Force Vershinin and his deputy Marshal of Aviation Rudenko supported Kamanin without particular enthusiasm. Ustinov and Serbin maintained neutrality, waiting to hear what Khrushchev would say. When they appealed to Khrushchev, he declared that he was not involved in cosmonaut selection.

The battle with the Air Force over the crew makeup was fundamental in nature. We supported Korolev not just because he was our boss. I, in particular, thought that the degree of automated control that was already on Vostoks was even better on Zenits; and this better degree of control would be absolutely marvelous on the next generation of spacecraft. Therefore I thought that the crew should conduct research, intelligence gathering, and experiments. A good engineer could control a spacecraft as well as a pilot if there were no obvious medical contraindications. Mishin took an even more extreme position. He believed that only engineers and scientists should fly, and that the expensive training at the Cosmonauts Training Center needed to be simplified or dispensed with completely. The flight of Feoktistov threatened the Air Force's monopoly in cosmonaut selection.

During this period there was a change in the management at Factory No. 918—later called Zvezda (Star) Factory—which was our main subcontractor for seats, spacesuits, and life support systems. Gay Severin was appointed chief designer in place of Semyon Alekseyev. Together with Bushuyev and Feoktistov, he would have to solve the difficult problem of putting three seats where only one had been before. In June, after taking the measurements of all the candidates, Severin informed Bushuyev that there were "giants" among them, whose height in the sitting position exceeded the maximum permissible value for the Voskhod. This made it easier for Korolev to push through Feoktistov's candidacy.

The decree calling for the three-seater Voskhod was not issued until 14 June 1964. By that time, Komarov and Volynov had been added to the list of original candidates. On 21 August, a session of the VPK verified the fulfillment of the Voskhod decree. Korolev was essentially reporting on work that had begun in January. Nevertheless, he was in the position to announce that for the most part the enormous project involving numerous OKBs and NIIs was drawing successfully to an end. At this session of the VPK, Korolev first reported about operations on the *Voskhod-2*, which would presumably support a cosmonaut's egress into open space.[8] This was one more tactical ploy to pre-empt U.S. operations.[9] Korolev's report on the *Voskhod-2* was also intended to help new Chief Designer Severin, who within a limited number of months was supposed to develop a spacesuit for a spacewalk.

I was at the firing range at that time; we were about to launch the second Molniya.[10] Bushuyev later told me the details firsthand.

The VPK approved a proposal for the launch of an unpiloted test Voskhod before 5 September. For us guidance and control specialists, this was very important, since it gave us the opportunity to test the ionic attitude control system in telemetry mode and to test the reliability of the new landing system. At the insistence of the Air Force, a test of the new parachute system was conducted in Feodosiya (in the Crimea) by dropping a mockup of the spacecraft from an airplane. This prolonged the overall training cycle, and Korolev was against such complicated and time-consuming experimental operations.

---

8. Russian parlance for spacewalks is "egress into open space." For the sake of clarity, in this volume, we use the term "spacewalk."

9. This is a reference to the American dual-astronaut Gemini program designed to master complex operations in Earth orbit such as rendezvous and docking, long-duration flight, and extravehicular activity.

10. The second Molniya-1 (or 11F67) communications satellite was launched on 22 August 1964 and named *Kosmos-41*. The first failed to reach Earth orbit on 4 June 1964. See Chapter 16 of this volume.

In this case, he quickly realized that he was wrong. On 6 February, during airplane testing, a Voskhod mockup dropped from an altitude of 10,000 meters crashed. According to reports from Feodosiya, after the drop, the parachute system hatch failed to jettison and the parachutes did not come out of the container.

The State Commission and Korolev were at the firing range hoping for a positive report from Feodosiya, after which the Voskhod test model was supposed to be launched. As usual, when it rains, it pours. The next day, a Zenit spacecraft, every bit of it produced in Kuybyshev, failed to lift off into space due to a failure of the "core"—the Block A of the launch vehicle.

From the author's archives.

Nikolay Pilyugin, Aleksandr Voytenko, Nikolay Kamanin, Sergey Korolev, and Yevgeniy Shabarov at Baykonur Cosmodrome in 1964.

I was torn between the Molniya's solar array, which had failed to deploy, the failure of the Voskhod's parachute hatch to jettison, working on the commission investigating the "core's" failure to start up on the Zenit's launch vehicle, and preparing the Ye-6.[11] The worst incident was the parachute hatch. If the airplane drop hadn't been performed, we might have killed three cosmonauts. Unthinkable!

---

11. The Ye-6 was the designation of the second-generation Soviet robotic lunar probes designed to accomplish a survivable landing on the surface of the Moon. Two Ye-6 probes were launched in 1964; both failed to reach Earth orbit. See Chapter 13 of this volume.

True, an experimental flight of a Voskhod test model had already been scheduled. At the second production site, with the help of German Semyonov and Viktor Kalashnikov, we set up an experiment on the mockup to test the jettisoning of the ill-fated hatch. We were forced to witness that the pyro cartridge detonation circuit, which had no backup, was capable of failing. I tormented myself, Kalashnikov, and the circuit developers.

On 16 September, Kalashnikov and I flew out to the firing range to report about our "evil deeds" and went straight from the airfield to the MIK. With Shabarov and Kirillov present, S.P. at first listened calmly to my extremely self-critical report. Then he asked for Kalashnikov's interpretation. Kalashnikov didn't manage to finish his rather convoluted explanation. Such an outburst of indignation followed that even Kirillov and Shabarov, who were completely innocent, cringed. Korolev showed no restraint in his choice of words. Many forgave him for this, I among them. But in the process, he got himself worked up and grew angrier and angrier. It was pointless to object, make excuses, or argue. Especially since, essentially, he was right.

Tyulin stopped by the office where the "execution" was under way. S.P. immediately fell silent. Taking advantage of the hush, Tyulin said that the State Commission needed to be informed of the final makeup of the Voskhod crews. We needed to switch our attention to that. In a dismal tone, S.P. said: "Boris, you will report to the State Commission about that hatch. Now scram, both of you!"

On 18 September, at the State Commission session, I reported that the hatch's failure to jettison during the drop over Feodosiya occurred through the fault of our electrical circuit, which on the airplane-borne mockup differed from the standard flight model! The circuit on the mockup had been simplified for the sake of meeting deadlines. But I didn't report this disgraceful detail. I just assured them that now the error had been cleared up completely; the circuits had been verified through test-jettisoning operations at the factory; and the circuit had been modified for the test spacecraft. I guaranteed the reliability of the landing electrical system.

After my report, Korolev gave a very calm and convincing speech. He confirmed that he had personally studied this incident. The hatch jettisoning circuit had serious defects. Now it had been reworked and the main elements had redundancy. Factory tests had confirmed the reliability of the new circuit. In conclusion, Korolev asked the State Commission to give their approval for the launch of a Voskhod carrying dummies, without waiting for repeat airplane drops of a spacecraft mockup in Feodosiya. The State Commission gave their consent and we were allowed to go back to Moscow, while the "inner circle" stayed behind to squabble over the candidates for the Voskhod flight.

During those days at the firing range, an unprecedented campaign began to prepare for Khrushchev's visit.[12] The centerpiece of the program was supposed to be the launching of combat missiles. The piloted launches were not the center of attention for the time being. I didn't attend the demonstration launches. According to the accounts of eyewitnesses, they all went very well. The featured event of the program was a performance from future cosmonaut Leonov. Dressed in a spacesuit, he demonstrated a "spacewalk" and returned to the spacecraft with the assistance of a crane.

On 25 September, I was back at the firing range. Preparation was under way for the launch of the test Voskhod carrying dummies. The State Commission appointed on 29 September had to hear Bogomolov's report about the *Tral* telemetry link failure. The system needed to be dismantled and replaced and the spacecraft needed to be retested. This would take at least five days. Because of the recent events, S.P. lost his temper at the drop of a hat. "Storm clouds" were still hanging in the air, and for that reason, when the schedule was disrupted, not everyone was fully loaded with work. Under these stressful circumstances, arguments and aggravation sometimes occurred over trivial matters.

The next State Commission meeting didn't convene until 5 October. Bogomolov had very recently been fêted on the occasion of his 50th birthday, first here at the firing range and then at a function at OKB MEI. Now he was being trampled and tormented for the poor quality of the *Tral* system.

Early on the morning of 6 October, preparation began for the launch of the first unpiloted Voskhod. Everything proceeded so normally and according to schedule that even S.P. strolled around the launch pad at a leisurely pace and appeared to be perfectly calm. The launch took place at 1000 hours. The Voskhod test model went into the calculated orbit.[13] A day later, it landed in the proper area. According to the report from the landing site, the soft landing system had functioned superbly. Our sins for the hatch's failure to jettison and Bogomolov's failed *Tral* were forgiven and forgotten.

Now it was time to give the "green light" to launch the piloted *Voskhod*. But suddenly yet another sinner appeared. A report arrived from Voronezh about the occurrence of "high frequency" during firing rig test checks of the third stage engines. Kosberg on the State Commission argued that this phenomenon only occurs during rig tests. It doesn't happen in flight. The Commission trusted Kosberg. They were very tired of waiting for the launch.

---

12. Khrushchev visted Tyura-Tam on 24–25 September 1964.
13. In the open press, the vehicle was known as *Kosmos-47*.

From the author's archives.

Shown here is the crew of the spaceship *Voskhod*: from left to right are Konstantin Feoktistov, Boris Yegorov, and Vladimir Komarov.

The arguments over the makeup of the crew ended in a victory for Korolev: Komarov, Yegorov, and Feoktistov would fly. Test pilot Komarov, young physician Yegorov, and designer Feoktistov would bring back priceless information from space. Such were the hopes.

Compared with the preceding six piloted launches, the upcoming *Voskhod* launch was undoubtedly a greater risk. First, the cosmonauts were flying without spacesuits. If the spacecraft depressurized, death was inevitable. Second, due to volume and mass limitations, the spacecraft was only equipped with a two-day margin of life support resources. There could be no delay in returning to Earth. Third, the new landing system had only been tested once!

On 11 October, a get-together was arranged at the launch site for the *Voskhod* crew and the military contingent participating in the upcoming launch. The rally pulled everyone together. After the "at ease" command was given, all the garrison soldiers standing on the launch pad felt like they had a critical stake in the impending achievement of Soviet science.

Taking advantage of the fact that spacesuits weren't required, Korolev and the three crewmembers rode the elevator up to the spacecraft, supposedly for a briefing. Someone like Feoktistov needed no briefing. He knew the spacecraft and all of its systems better than Korolev. I think that Korolev was very nervous and he was looking for a way to calm down. But technology back then did not tend

From the author's archives.

The *Voskhod* crew of Vladimir Komarov, Konstantin Feoktistov, and Boris Yegorov sets out for Baykonur in October 1964.

to be soothing. During general tests of the launch vehicle, the *Tral* system's on-board transmitter had failed, this time on stage three. Replacement and retesting would take at least an hour—under such stressful circumstances a snafu was unavoidable.

"Aleksey Fedorovich," said Kirillov to Bogomolov, "go find Korolev and Tyulin and brief them yourself. I don't want to get it in the neck for you."

Bogomolov left the launch site and went to the engineering facility. He found Korolev in his office along with Tyulin and several members of the State Commission. After Bogomolov's cheerful report that the *Tral* replacement would be finished in 10 minutes, Korolev exploded and shouted: "Get out of here! I can't even look at you! You lily-livered brat!"

From the author's archives.

Sergey Korolev speaks before the staff of OKB MEI. Chief Designer Aleksey Bogomolov is on the right.

Many other similar expressions came crashing down on the 50-year-old "brat." During this scene, which was quite painful for all those witnessing it, Kirillov reported from the launch site that the *Tral* replacement would be completed in an hour, and that the launch vehicle preparation schedule would not be disrupted because of the available time margin.

"So much for your 10 minutes! I don't want to have anything more to do with you!"

This was the last incident before the launch of the three-seater spacecraft. On the eve of 12 October—the day of the *Voskhod* launch—a freeze set in. During the clear, still night, the temperature fell to minus 10°C [14°F]. At 7 a.m., after grabbing a quick breakfast as Site No. 2, we hurried to the *bankobus* around 200 meters from the launch site.[14] At the last State Commission session, the decision was made to begin fueling and launch. Yegorov and Feoktistov made the first trip up in the elevator; next went Komarov accompanied by lead designer Yevgeniy Frolov. At T-minus 15 minutes I left for IP-1. The launch went off beautifully and normally. After stage separation, I observed the flicker of the glowing columns

Asif Siddiqi

Konstantin Feoktistov before his launch on Voskhod in 1964 saying goodbye to Yuriy Gagarin. Feoktistov was one of the senior designers at OKB-1 in charge of designing the Vostok and Voskhod spacecraft.

on the electronic screens of the *Tral* receiving station along with the telemetry specialists. After 525 seconds of flight, Vorshev triumphantly announced: "We have separation of the object."

Bogomolov, who had been standing with me next to the *Tral* unit, felt completely satisfied. His device had made it possible for everyone on the ground to see that the spacecraft and the three cosmonauts had successfully gone into orbit.

When Bogomolov and I pulled up to Site No. 2 after a leisurely drive from the tracking station, Bratslavets had already set up television coverage of the cosmonauts. Using the high-frequency communications system, Tyulin and Korolev had called Pitsunda where Khrushchev was on vacation and briefed him on the flight.[15] During a second call, they briefed Brezhnev, Smirnov, and Ustinov. At that time those three weren't thinking about space. None of us could imagine that in such joyfully triumphant moments as these seemed to us, Korolev and Tyulin were conversing not only with the head of state, but also with the man who was preparing to replace him the following day.[16]

---

14. The term *bankobus* was left over from the late 1940s during rocket testing at Kapustin Yar and described a vehicle for shelter close to the pad. See Chertok, *Rockets and People*, Vol. II, p. 34.

15. Pitsunda is a village in the Georgian Caucasus.

16. Unbeknownst to Khrushchev, by a secret decision of the Presidium of the Central Committee, he was stripped of all his positions in the Communist Party and the government on 12 October. The decision was affirmed the following day by a meeting of the Central Committee. The anti-Khrushchev faction was led by Leonid Brezhnev.

The Kremlin conspirators did not dare to eliminate direct communications and conversations between Khrushchev and the *Voskhod* crew ahead of time. Communications had been arranged. Komarov reported to Khrushchev that the flight was proceeding normally. Khrushchev wished them a safe return to Earth and said that he looked forward to seeing them soon. Kamanin talked with Vershinin and asked him to approach Malinovskiy with a request that Komarov be awarded the rank of Colonel-Engineer and Yegorov and Feoktistov—the rank of Captain.

According to the control post duty roster, I found myself on the same shift as Gagarin from 3 a.m. until the spacecraft landed. A large portion of our shift was taken up with "dead orbits," when communication with the crew was impossible. I enjoyed observing Gagarin when he was talking with the crew. He clearly could not conceal how much pleasure he derived from what he was doing. Komarov's reports from space were calm and confident.

At 8 a.m. Korolev and Tyulin decided that it would be a good idea to brief Khrushchev before the landing.

"He's still sleeping," warned Tyulin.

"It's OK, he won't chew us out for this," Korolev assured him.

They were connected with Pitsunda via Moscow, and Korolev reported that everything was OK on board. The flight program was designed for one day. Khrushchev knew about that. Nevertheless, "for formality," Korolev requested Khrushchev's permission to land. Both Korolev and Tyulin thought that it wouldn't be excessive to remind Khrushchev in this way of his personal contribution to the creation of *Voskhod*. He gave his permission to land. This was the last conversation with Khrushchev. The Kremlin conspirators had already removed him from authority over our great nation. Before the command was issued to activate the descent cycle, Korolev asked Gagarin to get Komarov on the line.

"This is '20.' How do you feel? Are you ready for final operations?" asked Korolev.[17]

"This is *Rubin* (Ruby). I feel fine. Many interesting developments. I'd like to continue working," responded Komarov.

Korolev glanced at the people around him. The majority were shaking their heads. It was clear to him as well that it was very risky to extend the flight for another day. With the hint of a smile, he pressed the microphone talk-button and responded: "But that wasn't the agreement!"

"You're right, it wasn't, but there's a lot of new stuff."

---

17. *Dvadtsatyy* (20) was Korolev's call-sign during all the early piloted space missions.

"You can't see all of the wonders there are, *Rubin*. Do you read me? This is '20', over!"

"'20,' this is *Rubin*. I read you. We're ready for final operations."

"*Rubin*, this is '20.' All of us, your comrades, are here. We wish you fair winds. This is '20,' over."

Rubin's response was drowned out in noise. NIP-4 in Yeniseysk reported that the command had been issued to start up the descent cycle. NIP-7 in Kamchatka issued a backup command to be on the safe side. These commands started up the sequencer program, according to which, at 1008 hours 56 seconds over the Gulf of Guinea, during the 16th orbit, the braking engine should be started up, if the spacecraft had first been oriented with the engine nozzle in the direction of the spacecraft's motion. The usually agonizing wait for the report from the ships *Dolinsk* and *Krasnodar* was brief this time. At 1025 hours, the report was relayed via Odessa and Moscow that the commands for engine startup and shutdown had passed. Next, the excruciating uncertainty began. There was no communication with the spacecraft.

Everyone assembled stood in strained silence waiting for reports from the search service's chief, General Kutasin. When the latter reported that an Il-14 pilot "sees the object," Korolev couldn't stand it. He took the microphone away from Gagarin and shouted: "This is 20! How many parachutes does pilot Mikhaylov see—one or two?"

If only one canopy had deployed—this was bad. The landing speed might be 8 to 10 meters per second. If, on top of that, the soft landing engine failed to operate, injury to the cosmonauts would be inevitable. After an excruciating pause, Kutasin reported that the spacecraft was descending on two parachutes. Again we waited. Finally: "Pilot Mikhaylov sees the spacecraft on the ground and there are three people near it waving their arms." Dear, unknown pilot Mikhaylov! If you only knew what thundering applause and embraces your brief message produced!

"I never would have believed that it was possible to make a Voskhod out of a Vostok and have three cosmonauts fly in it," said Korolev, radiating a rare glow. In his joy he even embraced the "brat," Bogomolov.

By the decision of the State Commission, the cosmonauts were supposed to fly from Kustanay to Tyura-Tam. They had been flown by helicopter to Kustanay and, according to ritual, they were supposed to report to Khrushchev about their safe return. We all were also expecting to hear Khrushchev's gratitude and congratulations. But time passed, and we heard nothing from Moscow, nothing from Kustanay, nothing from Pitsunda—no calls, no congratulations. We broke for lunch and some rest.

The news soon spread that Smirnov had telephoned State Commission Chairman Tyulin from Moscow and informed him that there would be no conversation with Khrushchev and that the cosmonauts could fly to the firing range from Kustanay. That same evening, under the floodlights, we met all three cosmonauts at the airfield. They came down the airplane stairway without any signs of fatigue. Feoktistov, whom the physicians had categorically barred from flying, looked particularly happy.

The next day the State Commission arranged an expanded, almost open, session attended by around 250 people. Each of the cosmonauts gave an account of his perceptions. After the session there was a banquet with toasts to the health of Korolev, the crew, and all those who had been involved in preparation.

From the author's archives.

Cosmonauts Vladimir Komarov, Konstantin Feoktistov, and Boris Yegorov at their welcome to the OKB-1 premises in Kaliningrad in 1964. In the background, just behind Yegorov (with the papers in hand) is Boris Chertok.

On behalf of the crew, Komarov thanked everyone and pledged that they were ready to fulfill the new assignment of the Party and the government. He saw fit to say that he and the crew were grateful to Nikita Sergeyevich, who had shown particular concern for the development of Soviet space technology.

Next, the cosmonauts met with the launch control team. We had lost ourselves in speculation: Why was Moscow silent? Why had no one saluted and congratulated the cosmonauts or us scientists?

Readers know from other memoirs what happened back then in Moscow and on Cape Pitsunda. Late in the evening of 14 October, it dawned on us that Moscow was not pleased. Moscow rejected Komarov's previously prepared statement addressed to Khrushchev and ordered that the text be changed.

The next morning the stunning news came over the radio. Khrushchev had been relieved of all his duties in the Party and government. Brezhnev and Kosygin had taken his place. We could not conceal our disappointment with this news. We considered Khrushchev to be a devotee of cosmonautics and reasoned that his departure would not work in our favor. On 16 October, the State Commission, followed by almost everyone who had participated in the latest historic flight, departed for Moscow, leaving behind the cosmonauts to languish at the firing range waiting for a special invitation.

Five days after their return to Earth, the *Voskhod* crew finally received the command "welcoming" them to the Great Kremlin Palace. At the reception the practices established under Khrushchev were maintained. After "tanking up," we poured out our hearts singing traditional favorites such as *Bravely we go into battle*, *Over hill and dale*, and even the sorrowful *Why do you stand there swaying, slender mountain ash . . . .*[18]

In the midst of the festivities, an unrecognizably somber Kamanin came up to our table. He was looking for Korolev, but the latter had been taken away to meet with the new leaders. Kamanin told us: "In Yugoslavia an Il-18 airplane crashed—the very one that flew the cosmonauts from Kustanay to Tyura-Tam. Chief of General Staff Marshal Biryuzov was killed."[19]

In November 1964, after the sensational flight of the three-seat *Voskhod*, a small group gathered in Korolev's tiny office to discuss the production schedule for subsequent Voskhods. I recall that Bushuyev, Turkov, and Okhapkin were there. No schedules and precise plans for our future operations had been signed off on yet. There was bickering going on at the bottom and the top. S.P. himself, as we used to say, was in

From the author's archives.

Cosmonaut Yuriy Gagarin shares a light moment with *Voskhod* crew members Boris Yegorov (left) and Konstantin Feoktistov (right).

---

18. The first two are Bolshevik songs dating from the revolutionary era while the third was a popular love song from the Khrushchev era.

19. Before his stint as Chief of General Staff, Marshal Sergey Semenovich Biryuzov (1904–64) was the commander-in-chief of the RVSN from 1962 to 1963 and the Air Defense Troops from 1955 to 1962.

a "dismantled" state. Before beginning our discussion, either as a challenge, or as a rebuke, he brusquely handed me a thin booklet and said, "Kostya and Serega have already read it, now it's your turn! They have gathered all their forces into one fist and are not hiding their plans. And up until now we have been keeping secrets from ourselves. We can't come to an agreement with the minister of defense, or with the Air Force, or with the VPK. Give one a series of Vostoks, another—Voskhods; and we only need the Soyuz. Brezhnev needs something launched, and quickly, to show that after Nikita our projects are doing just fine. Look at how the Americans work!"

Our discussion in Korolev's office was chaotic. S.P. was obviously out of sorts. Despite the traditional Kremlin festivities after the return of the *Voskhod* crew, he still had not established a rapport with the new political leadership. The enthusiasm that Khrushchev had encouraged was missing now. This tormented him; it introduced uncertainty into his thoughts about the future.

"There is hope," said S.P., "that Ustinov will be the new Central Committee Secretary for Defense Matters, instead of Brezhnev.[20] This strengthens our positions. Incidentally, just between us, now there will be a new minister.[21] I don't think either of them will support Chelomey's latest lunar adventure."

At this juncture, Okhapkin felt he needed to put in a word. He had a very resolute command of engineering that he had maintained since the time he had worked with Tupolev.

"One shouldn't underestimate Chelomey. He has amassed a very strong staff in Fili. This is the design school of Tupolev and Myasishchev. If they are given free rein and resources, they will develop designs that are as good as those of the Americans. And their factory, not to offend Roman Anisimovich (Turkov), is magnificent. There's no need for us to fly off the handle. Now is a convenient time to seek common ground with Chelomey."

After returning to my office, I began to study the booklet that Korolev had handed me. The information had been published by the Central Aero-hydrodynamic Institute (TsAGI).[22] It contained speeches by leading American scientists and directors of operations in the field of astronautics. I have held onto this material, and now, after rereading it, I am convinced that one could accuse the Americans of striving for global space hegemony to a greater extent than us.

---

20. Ustinov became Secretary of the Central Committee for defense industries and space in March 1965.

21. This is a reference to the formation of the new Ministry of General Machine Building in March 1965 and the appointment of Sergey Afanasyev as its head.

22. TsAGI—*Tsentralnyy aerogidrodinamicheskiy institut*—was the leading aeronautics R&D institution in the Soviet Union.

I cite excerpts from the "closing remarks" based on the results of NASA's work in 1963:

It is important that space missions be discussed in the government, in industry, in universities, and in scientific communities in order to make the most sensible decisions for the future.

More simply put, the goal of the United States is to achieve superiority in all the most important fields of space exploitation and conduct this space program so that our expertise in scientific, technical, and operational matters related to space become evident to the whole world.

To win supremacy in space we must conduct research over a broad range . . .

For superiority in space we need to have leading edge technology that will make it possible to insert ever greater payloads into orbit around the Earth and to execute flights to the Moon and other planets . . .

We need to perfect technology to transmit larger amounts of data at enormous distances. Moreover, American supremacy in space requires the know-how to launch spacecraft at a precisely designated time. It also requires the ability to increase payloads launched into precise orbits. We need to study the mechanics of maneuvering in space and rendezvousing with other large objects. We need to master the procedure for making a precise landing on the Moon and planets and returning to the Earth at ever-increasing high speeds.

And finally, superiority in space means that we need to learn how to manufacture, assemble, test, and prepare spacecraft and components that are capable of effectively operating in space not for months, but for years . . .

Logic requires that we strive to expand cooperation between governmental scientific research organizations, flight testing centers, industrial organizations, and universities while carrying out space research.

The activity of each of these organizations and their joint efforts must be aimed at achieving a common goal—the superiority of the United States in space . . .

To achieve the superiority of the United States in space, it is necessary that we conduct research on the Moon before the end of this decade. As President Kennedy noted, this will not simply be the flight of American explorers [but of] the entire nation, every American must make a contribution to the successful performance of this flight . . .

It is appropriate to recall that 90% of every dollar is spent in industry or through contracts, or in cooperative programs between industry and universities "Space" dollars penetrate into almost every realm of the U.S. economy through subcontractors. Thus, space exploration is truly a national mission.[23]

---

23. BNTI TsAGI (*Byuro nauchno-tekhnicheskaya informatsiya Tsentralnogo aerogidrodinamicheskogo instituta* [*TsAGI Scientific-Technical Information Bureau*]) Translations, '91, 1964.

We assiduously followed what was going on in the United States not only in the open press, but also in secret communiqués that were provided to Korolev without mention of the source. He, in turn, was authorized to share them with his "inner circle." In our news media it was difficult to track down the history of the first years of American astronautics.

I feel compelled to very briefly recall this history for the reader. American operations had a very strong effect on our plans. American historians of astronautics assert that our successes were the primary reason why the United States converted its space programs into a top-priority, nationwide challenge.

The German roots of American astronautics ran considerably deeper than in Soviet cosmonautics. In May 1945, Wernher von Braun, who had headed the large group of missile specialists evacuated from Peenemünde, surrendered to the Americans. During the operation codenamed "Paperclip," German specialists were interned by the U.S. Army, and then transported to Fort Bliss near El Paso, Texas, on the Mexican border. Meanwhile, in 1945, Aleksey Isayev and I were setting up Institut RABE and gathering up crumb by crumb what was left after the successful American operation to seize the best German specialists.

The Americans first transported abroad 118 leading specialists, headed by Wernher von Braun.[24] After that, "supplemental staffing" continued. According to information from the American press, 492 missile specialists and 644 of their dependents arrived in the United States from Germany. They soon comprised a genuine "brain trust."

The "peacetime prisoners" lived behind barbed wire, free to move about within their living quarters. Many Americans had mixed feelings about them. The "Nazi scientists" were met with skepticism or open hostility. The press published horrible photographs of the Dora camp and said that prisoners of concentration camps had mass produced the V-2 missiles under conditions that took more lives than the 3,500 missiles launched against such targets as London and Antwerp.

The German specialists worked under the supervision of American officers. The main work was preparing and launching captured models of V-2 missiles at the White Sands test range. For months the Germans translated and edited documents from the wealth of Peenemünde archives that had fallen en masse into American hands. The U.S. military finally grew tired of holding the Germans as semi-prisoners. To make them U.S. citizens, they arranged for the Germans to take an excursion into Ciudad Juárez, the neighboring Mexican

---

24. Initially, in 1945, 127 German rocket scientists and engineers accepted the offer to relocate to the United States.

city. They crossed the border on foot, and once they had returned to U.S. territory, having officially become legal immigrants, they had the right to obtain American citizenship.

In April 1950, the U.S. Army decided to set up the development of missile weapons at the Redstone Arsenal in Huntsville, Alabama. Von Braun's entire team was first assigned to the Ordnance Rocket Center, which was part of the Army Ordnance Command.[25] This command was responsible for the entire Army missile program. Soon, the group that von Braun headed in Huntsville obtained department status and was called the Rocket Ordnance Research and Development Division. In 1956 it was renamed the Department of Operations Development in the Army Ballistic Missile Agency (ABMA).

During this time, independent of the Army, the Air Force and Navy were attempting to create their own missile research and development division. However, the Army had advantages over its competitors in the field of missile technology, having the German brains of von Braun's team at its disposal. In Huntsville, this team was rapidly supplemented with U.S. military specialists and more than 100 specialists from General Electric (through contracts).

The first missile developed in the U.S. under von Braun's supervision was the Redstone. The first successful launch of this missile took place on 20 August 1953. I recall that at that time we had put the R-1 and R-2 missiles into service and were getting ready to test the R-5, which had a range of 1,200 kilometers.

Officially, Major General John Bruce Medaris from the Army Ballistic Missile Agency was in charge of the missile program in Huntsville. He increased the missile staff in Huntsville to 5,000. According to directives issued in 1956, ground troops were limited to surface-to-surface missiles with a range up to 200 miles (320 kilometers). The Air Force was responsible for ground-based surface-to-air missiles, while the Navy was responsible for ship-based air defense missile systems.

Officially, the United States had no plans for work on an intercontinental missile before 1955. Von Braun designed the Jupiter C missile, the 1956 version of which was a three-stage updated stage of the Redstone rocket. The Jupiter C missile was developed despite the Pentagon's restrictions on flight range; and on 20 September 1956, the first three-stage Jupiter C missile equipped with a mockup fourth stage, in place of a real stage with a satellite, flew 3,300 miles (5,300 kilometers).

---

25. The Ordnance Rocket Center was formally established at Redstone Arsenal on 1 June 1949. After their transfer on 15 April 1950, von Braun's team was renamed the Ordnance Guided Missile Center. The Army Ordnance Missile Command (AOMC) was formally established much later, on 31 March 1958.

The United States' first plans for an artificial satellite of Earth were announced in December 1948.[26] In 1955, information emerged about U.S. intentions to launch a satellite for the International Geophysical Year (IGY), which lasted 18 months from mid-1957 until the end of 1958.[27]

On 4 October 1957, after hearing about the launch of the Soviet *Sputnik*, Wernher von Braun told Secretary of Defense Neil H. McElroy: "We knew that they were on the verge of producing this. We can launch our satellite in 60 days. Just give us the green light and those 60 days."[28]

The era of the United States' self-confidence had ended. The first Soviet satellite had profoundly shaken millions of Americans. The dogma about U.S. technical superiority had crumbled. Prominent *Washington Post* columnist Chalmers Roberts ranked the importance of the launch of the first satellite on a par with such 20th century events as the October Socialist Revolution in Russia and the splitting of the atom to achieve a nuclear chain reaction.

The Eisenhower Administration was subjected to severe criticism. Senator Lyndon B. Johnson (later President of the United States) issued a statement:

Control of space means control of the world, far more certainly, far more totally than any control that has ever or could ever be achieved by weapons, or by troops of occupation. From space, the masters of infinity would have the power to control the Earth's weather, to cause drought and flood, to change the tides and raise the levels of the sea, to divert the Gulf stream and change temperate climates to frigid.[29]

---

26. Chertok's claim is slightly overstated. At the time, U.S. Secretary of Defense James V. Forrestal, in an annual report to the President, included remarks announcing feasability studies on launching an artificial Earth satellite. Despite the announcement, satellite research in the U.S. remained at a low level until about 1955.

27. The IGY was designated by an international coalition of scientists as a period of global scientific research to coincide with heightened solar activity. Lasting from 1 July 1957 to 31 December 1958, the IGY program involved the use of ground, air, and space-based instrumentation to study an enormous range of geophysical phenomena.

28. The von Braun quote that Medaris gives in his memoirs is "We knew they were going to do it . . . We have the hardware on the shelf. For God's sake, turn us loose and let us do something! We can put up a satellite in sixty days, Mr. McElroy. Just give us the green light and sixty days!" See John B. Medaris with Arthur Gordon, *Countdown for Decision* (New York: Putnam, 1960), p. 155. Neil Hosler McElroy (1904–72) was U.S. Secretary of Defense from 1957 to 1959 under President Dwight D. Eisenhower.

29. The text that Chertok quotes from was published in a number of different places, including "Text of Johnson's Statement of Nation's Defense and Race for Space," *The New York Times*, 8 January 1958, p. 10.

The United States plunged into a battle for space, stoked by jingoistic fervor and military hysteria. On 1 October 1958, President Eisenhower created the National Aeronautics and Space Administration (NASA). Preparation for the launch of the first American satellite was conducted in an atmosphere of hysteria and anxiety. Disputes between the branches of the armed forces escalated sharply. Independently of one another, all three branches of the armed forces were striving to save America's prestige. The first attempt was made with the Vanguard rocket on 6 December 1957. The 72-foot long (22 meters), three-stage rocket lifted 5 feet (1.5 meters) off the ground, sat back down, and exploded.

After the failure of the Vanguard rocket that the Air Force [*sic*] had prepared, von Braun's German team was entrusted to launch the first U.S. satellite.[30] The Jupiter C rocket developed by the Germans was used for this.[31] *Explorer 1*, the first U.S. satellite, was inserted on orbit on 1 February 1958.[32] In the first half of 1958 alone, the U.S attempted to launch a total of six satellites using the Vanguard rocket. Only one was successful.[33]

The Soviet *Sputnik* had provided the impetus for the United States to create a governmental science organization. During the presidential election campaign of 1960, John F. Kennedy skillfully used the U.S. lag behind the Soviet Union in the field of space exploration to criticize the Eisenhower-Nixon Administration.

After the first Soviet *Sputnik*, the flight of Gagarin was the second psychological and political shock for the American public. Immediately after Gagarin's flight on 25 May 1961, President Kennedy spoke before a joint session of Congress and said: " . . . I believe that this nation should commit itself to achieving the goal, before this decade is out, of landing a man on the Moon and returning him safely to the Earth. No single space project in this period will be more impressive to mankind, or more important for the long-range exploration of space; and none will be so difficult or expensive to accomplish."[34]

---

30. The Vanguard project was actually run by the Naval Research Laboratory.

31. The modified Jupiter C launch vehicle in its orbital version was named Juno 1.

32. The launch was on 31 January 1958 Eastern Time.

33. The Vanguard launch attempts were on 5 February, 17 March, 28 April, 27 May, 26 June, and 26 September 1958. *Vanguard 1* was successfully launched into orbit on 17 March 1958.

34. The speech, known as the "Special Message to Congress on Urgent National Needs," was delivered by Kennedy before a joint session of Congress on 25 May 1961. For the complete text, see John F. Kennedy Library and Museum Web site, *http://www.jfklibrary.org/ Historical+Resources/Archives/Reference+Desk/Speeches/JFK/003POF03NationalNeeds05251961 .htm* (accessed 6 May 2007).

The first U.S. piloted space flights seemed primitive to us. However, under the intense press of Party propaganda, we underestimated President Kennedy's famous speech, and did not match it in the near future with an equivalent national goal. The Apollo project focused national efforts on space, and served to a certain degree as the standard for U.S. achievements in this field; the leaders of the U.S. space program acknowledged that this project would help to achieve the capability to carry out a broad range of operations in space to satisfy the potential needs of national policy.

The Mercury project and later the Gemini project were subordinate to the main Apollo program. Beat the Russians in the race to the Moon and at the same time occupy the minds of average Americans with something grandiose! As for expenses—the Americans' and our expenses actually exceeded what had been planned at first. For the Mercury program, $334.5 million was allocated, while $384.1 million was actually spent. The Mercury program was supposed to have been the United States' strong response to the Soviet Union's space successes. Inserting a human being into orbit was a natural sequel after reliably mastering the technology for producing and launching automated satellites.

The Air Force showed the first initiative in the United States for the production of a piloted space system. They were counting on the winged design of the X-15 rocket plane, which they had begun developing as early as 1954. The launch of the first Soviet Sputniks upset their apple cart. In 1958, the U.S. did not have a launch vehicle capable of inserting a piloted spacecraft into space. Striving to eliminate the organizational mess and remove the military from the human spaceflight programs, President Eisenhower delegated leadership to the newly created Federal civilian NASA. NASA did not have a title similar to our "chief designer." Wernher von Braun actually performed these duties, but only for launch vehicles. His team was not involved with the development of spacecraft.

From the very beginning, NASA specialists were geared toward the flight of one individual in a spacecraft of minimum volume and weight. The vehicle, which was named Mercury, was called a capsule. Compared to our Vostoks, it really was a capsule. In February 1959, the McDonnell Aircraft Corporation won the contract to develop it. Therefore, the head of this company, James McDonnell, who personally oversaw this project, can be considered the chief designer of the Mercury capsule.[35] Twelve Redstone

---

35. Max Faget (1921–2004) of NASA's Space Task Group at the Langley Research Center in Hampton, Virginia is considered the main designer of the Mercury spacecraft.

rockets were used just for the flight development tests of the systems for the Mercury program out of a total number of 37 launched for flight development tests on the rockets. The Mercury-Redstone launch vehicle had a launch mass of 29.9 metric tons, a height of 24.4 meters, and a diameter of 1.78 meters. The rocket was an elaboration of the V-2 Liquid oxygen and kerosene served as the propellant. The control system and propulsion system were substantially modified. The combat version had a separable warhead and load-bearing fuel tanks. The thrust of the liquid-oxygen rocket engine was 37.7 metric tons. A solid-propellant rocket engine was used for separation.

In our quest to be the first to send a man into space, we had a fundamental advantage over the Americans—the secrecy of our program. Although brief TASS reports gave American specialists a certain idea about our developments, they were confident that, if not a full-fledged orbital flight, they at least had a "hop"—in the form of a high-altitude ballistic flight with several minutes of weightlessness—"in the bag." The Mercury-Redstone launch vehicle was declared ready for piloted flight on 24 March 1961. The preparation and the last launches of the flight development testing program had been conducted with press and television coverage. The VPK had notified Korolev that the Americans were prepared for a piloted launch no later than April. That is why he was in such a hurry with the *Vostok*.

Alan B. Shepard, Jr. executed the first suborbital flight on 5 May 1961. Despite Gagarin's worldwide triumph, U.S. television and mass media spared no effort or expense to generate enthusiasm among the American public. The flight lasted just 15 minutes 22 seconds. This was the world's first suborbital piloted flight and the first manual control of a spacecraft. The launch was postponed several times for various reasons. Shepard had to sit in the capsule for more than 4 hours. He complained that his bladder might burst. His spacesuit provided no toilet facilities and he was granted permission to relieve himself in his pants. The first American astronaut touched space with wet pants. Before the launch, classes in schools and work in companies were stopped, and traffic on the streets came to a standstill. As many as 70 million viewers watched the live televised coverage.[36] Touchdown took place according to the program in the Atlantic Ocean, but the capsule overshot the touchdown site by 11 kilometers. A rescue helicopter delivered the capsule and astronaut to the deck of the U.S.S. *Lake Champlain* 25 minutes after their launch from Cape Canaveral, Florida.

The second U.S. piloted spaceflight was also suborbital. Liftoff took place on 21 July 1961 after being delayed three times for various reasons for 80 minutes. Pilot Virgil I. "Gus" Grissom was in flight for

---

36. Contemporaneous estimates place the viewer number as closer to 45 million.

15 minutes, 37 seconds. The capsule came down in the Atlantic Ocean 486 kilometers from its launch site, overshooting the design touchdown site by 15 kilometers, 10 kilometers from the aircraft carrier U.S.S. *Randolph*. Hindering each other, teams of rescue helicopters managed to pick up Grissom a second before it seemed he would go under.[37] Miraculously, Grissom was saved. In 1965 he flew as commander of the *Gemini III* spacecraft and might have become the first human being on the Moon. In January 1967, he perished in a fire on the Apollo spacecraft.[38] The Mercury capsule that Grissom had flown was not recovered and sank to a depth of 4,890 meters.

In 1985, Curt Newport, an undersea explorer and remote-controlled submarine expert, began to develop the idea of hunting for and raising the capsule. He succeeded in raising the legendary capsule on 20 July 1999 and delivered it to Port Canaveral on 21 July, the anniversary of Grissom's launch.

Despite the large stockpile of Redstone launch vehicles, the Mercury program managers decided to conduct the third piloted flight using the Atlas D launch vehicle. The Convair Company began developing the Atlas rocket as early as 1946.[39] The Atlas was conceived as a one-and-a-half-stage rocket. During liftoff, five engines fired simultaneously—two liquid-propellant booster engines, one sustainer engine, and two control thrusters. After 2 minutes the side-mounted liquid-propellant booster engines were ejected and the rocket continued the powered portion of flight by dint of the thrust from the sustainer engine and two control thrusters. The engines of the Atlas intercontinental missile operated on oxygen/kerosene propellant components. Its launch weight fully loaded for different modifications was 117 to 120 metric tons. While its launch mass was slightly greater than that of the Redstone, for prospective space plans, the Atlas in its combat version was a light-duty launch vehicle. Flight development testing of the Atlas launch vehicles began in June 1957, just two months after we began similar tests with our R-7 rocket.

The first space modification was the Atlas D for the Mercury program. The hybrid control system combined an autonomous inertial unit and Doppler

---

37. A hatch on the *Liberty Bell 7* spacecraft blew after splashdown, allowing the capsule to quickly fill with water and sink to the bottom of the ocean.

38. On 27 January 1967, during a countdown simulation in preparation for the scheduled launch of the Apollo 1 mission, a flash fire killed the three-man crew of Virgil I. Grissom, Edward H. White II, and Roger B. Chaffee.

39. The development of the Atlas can be traced back to the MX-774 concept proposed by Convair and funded by the Army Air Force in April 1946 to explore long-range air-breathing and rocket propulsion missiles. The MX-774 study was cancelled the following year and ICBM studies in the U.S. were not renewed until 1951 when the Atlas concept was proposed by Convair. However, it was only in 1954 that the Atlas program was put on an urgent footing.

tracking and precision speed measuring radar. In general, the Atlas contained many design concepts that were surprisingly similar to the R-7 rocket. But no one had stolen ideas from anyone. For the intercontinental missile, they were "in the ether," and Soviet and American engineers came up with the designs absolutely independently. The Soviets, headed by Korolev, were the luckier of the two groups. They were ordered to produce an intercontinental missile for a "payload" with a mass of 3.5 metric tons. No one made such a demand on the American engineers. The Atlas D rocket did not go into service in the U.S. Air Force until September 1959.

German Titov's 24-hour flight on 6 August 1961 crushed the American hopes of celebrating the breaking of Yuriy Gagarin's record. It wasn't until 20 February 1962 that John H. Glenn, Jr. lifted off in the Mercury capsule on the Atlas D launch vehicle. His flight lasted 4 hours, 55 minutes, 23 seconds. This was the United States' first full-fledged orbital space flight. During the flight, the automatic attitude control system failed. Glenn controlled the capsule manually for two orbits.

After Glenn, the Atlas D launch vehicle inserted a piloted Mercury capsule three more times. On 24 May 1962, M. Scott Carpenter completed a flight lasting 4 hours, 56 minutes. On 3 October 1962, Walter M. Schirra, Jr. achieved a flight duration of 9 hours, 13 minutes, 11 seconds. L. Gordon Cooper, Jr., who lifted off on 15 May 1963, was the first U.S. astronaut to spend more time in space than German Titov. Cooper's flight lasted 34 hours, 19 minutes, 49 seconds. We received this information when we were at the firing range during the flight of Bykovskiy and Tereshkova. I took the opportunity to go public with my "discovery."

"15 May—that's our lucky date, anyway, and now it's the Americans' lucky day, too!"

Korolev demanded that the ballistics specialists give him a precise quote on the flight duration by which a Vostok surpassed the Americans' latest achievement. It turned out that Bykovskiy set the Vostok duration record on 19 June 1963. He outstripped Cooper's flight duration record by 85 hours, 47 minutes. However, the Americans beat our records for the amount of time a human being had spent in space after they began their Gemini spacecraft missions. In August 1965, astronauts L. Gordon Cooper and Charles "Pete" Conrad completed a flight on *Gemini V* lasting 7 days, 22 hours, 55 minutes, 14 seconds. The advantages of Gemini spacecraft over Vostoks and Voskhods were obvious. Our first Soyuz vehicles had only just appeared in the factory assembly shop.

AFTER KHRUSHCHEV'S FALL, WE WERE ALL WORRIED ABOUT THE ATTITUDE OF THE NEW PARTY AND GOVERNMENT LEADERSHIP toward our previous "space schemes" that Khrushchev had approved. Based on the comments of "our people" in the Central Committee and VPK bureaucracy, new Council of Ministers Chairman Aleksey Kosygin was a humorless individual who had not previously been enamored with propaganda and "space spectacles." Kosygin's front burner concerns were with the national economy in general and its reorganization. It was rumored that he had conceived a reform that would establish new procedures in industry and agriculture. His relations with Ustinov were far from warm. He intended to economize on the army, which since Khrushchev's departure had pinned its hopes on restoring the former might of the air and naval fleets. The All-Union Council of National Economy (*Sovnarkhoz*) was phased out.[40] Its chairman, Ustinov, was transferred to the coveted post of Central Committee Secretary for Defense Matters.

After becoming General Secretary, Brezhnev was obliged to show his gratitude to the Party leaders of the *oblasts* and republics that had put him in power during the plenary session in October 1964. They had not forgiven Khrushchev for his anti-Stalin campaign, which had undermined the influence of Communist Party leaders on the masses and had risked unpredictable intra- and extra-Party complications. At first, the new General Secretary didn't put together his own strong faction of loyal people; he needed to exercise caution.

It was high time that we (i.e., the Korolev design bureau) developed a concise and convincing program of space activity, first and foremost, for defense purposes, and then for scientific and political ones. We needed to come out with this program first, before Chelomey or Yangel got ahead of us. The Central Committee and VPK bureaucracies insisted on developing such a combined long-range program, but they also believed that "like it or not, it was necessary to pay off prior debts."

OKB-1 was left with large debts. Above all, these were the projects for the spacewalk, then the soft-landing on the Moon, and the Venus, Mars, and Molniya. Operations on the N1 launch vehicle were still in a state of disarray. At the same time, we still hadn't gotten the R-9 missile into shape and were already tackling a new solid-propellant missile. The Soyuz program clearly failed to meet its deadlines. This was generally the prevailing mood about us that we found in the offices on Old Square and in the Kremlin, i.e., in the Central Committee and the VPK bureaucracies.

---

40. The All-Union *Sovnarkhoz* (*Sovet narodnogo khozyaystva* or Council of the National Economy) was a body that existed from 1963 to 1965 to coordinate economic and industrial development. The decentralized *Sovnarkhoz* system was abolished in 1965 after Khrushchev's overthrow.

VPK employees received information about the United States' space activities before we did. They could "tease" us with the U.S. programs, asking questions to which they already knew the answers. In addition to open U.S. publications, detailed information was coming in "via special channels" about the development of a series of two-seat Gemini spacecraft that were intended for the experimental development of systems and the training of astronauts for the Apollo program. The first unpiloted Gemini spacecraft had already completed a three-day flight in April 1964. One more automated flight was scheduled for 1965 and then five piloted flights. By placing two astronauts in each spacecraft, the Americans overtook us in terms of the number of piloted launches and the number of cosmonauts.

It seemed that in the propaganda of space achievements our mass media were unrivaled. However, reading the so-called "white TASS" and other materials from the U.S., we found that the Americans had outdone us in terms of fanning the military hysteria and persuasively campaigning to get the American public's backing for space expenditures.[41] They had spent a long time looking for a concept that would integrate pure space science with the ambition "to prevent the Reds from taking over in space." Appeals to develop the perfect space weapon outnumbered publications about the peaceful uses of space. A deliberate campaign was being waged for the development of a government strategy for the military uses of space. One must give credit to influential U.S. politicians—they immediately appreciated the organic connection between fundamental science and the goals of future weapons systems. To the joy of scientists, it was shown that an assessment should be made based not on actual current achievements, but looking far into the future. Support for science was necessary because the knowledge accumulated over a certain period would lead to revolutionary transformations in future weapons systems.

Senators who supported increasing appropriations for space reasoned that, sooner or later, moving forward would pay off. They said, "Today it is impossible to precisely list all the benefits from space exploration; Columbus was looking for a route to India, but discovered America, and this discovery surpassed all expectations." Back in 1959, a well-respected academic journal wrote:

---

41. "White TASS" represented one of three types of TASS news during the Soviet era. Green or Blue Tass was intended for the public and comprised extremely sanitized versions of domestic and international news. White TASS—the equivalent of secret news—included very candid accounts of domestic and international events prepared only for government ministries and Communist Party offices. Red TASS was top secret information, i.e., completely unexpurgated information from foreign news agencies, delivered only to the topmost individuals of the government and Party structure (including the Politburo).

Projects to send a man to the Moon or Mars perhaps do not fit under the heading of purely military endeavors, but from the standpoint of their military implications, the realization of these projects might be by far more important than all the work of the general staffs.[42]

While our propaganda trumpeted each Soviet space success as a new victory in the struggle for world peace and as evidence of the superiority of socialism, "decadent" capitalism was certainly not thinking of surrendering. The arguments for increasing appropriations for space were public and clear to any "average" American. Britain's dominance at sea had enabled her to achieve military superiority and to reign supreme in the world for centuries. Superiority in the air had guaranteed the Allied victory during World War II. If the United States achieved superiority over the Soviet Union in the acquisition of spaceflight technology, then it would confirm its supremacy on Earth. The expansion of the United States' military might required that the U.S. expand its fundamental scientific knowledge and explore the new ocean of space.

It was far-sighted politicians, rather than the chiefs of staff of all the branches of the U.S. armed forces, who argued that fundamental research opening up the boundaries of knowledge would enable the United States to become a world leader through supremacy in space. One should not cut back expenditures on science. The answers that science provides are always broader than the questions asked of her. This is what those politicians—whom our propaganda considered to be lunatic aggressors—were saying and writing.

Thirty years later, in the 1990s, there were no politicians or generals to be found in Russia who grasped these truths. The Russian science that the U.S. senators had so feared had been defeated without using any of their new weapons.

Assessing our first successes in space, American experts assumed that they could be achieved only by freezing the program for the production of combat missiles. It now seems to me that this assessment was not far from the truth.

It is curious that our thoughts on the conservative views of distinguished old marshals sometimes coincided with the initial positions of U.S. senators. They accused their military men of intellectual stagnation that prevented them from

---

42. The original quote in English communicates a subtly different message than the excerpt that Chertok cites, suggesting that in translating the original passage into Russian, much of the meaning was misinterpreted. The original quote was: "To put a man in space, or on the moon, or on Mars may not be a military or a scientific goal in any rigid sense. But it is not a circus stunt, either; it is an adventure comparable to Columbus' non-military, non-scientific maritime probe—which turned out to have more military consequences than all the work ever done by staff planners and more scientific consequences than all the work ever done in laboratories." See "Why the U.S. Isn't On the Moon" (editorial), *Fortune* Vol. 60 no. 6 (December 1959), p. 95.

making sufficient military use of the capabilities acquired through NASA. In this regard, it was proposed that the Pentagon order special military spacecraft that would be in a constant state of combat readiness, whose launch sites would be classified, and whose crews would undergo preliminary training in space.

Having become well known for his anti-Soviet and anti-Communist intransigence, Senator Barry Goldwater made the following axiomatic pronouncement: "He who controls aerospace will rule the world." He called for the United States "to move ahead of Moscow and be the first to occupy the driver's seat in the carriage of future world events." He also called for the United States to put such weapons systems into space that would make it possible to wipe any country off the face of Earth in a matter of seconds! Goldwater believed that all human spaceflight projects should be military and should also be implemented by the military.

Attacks of the American "hawks" on their own government affected both the Apollo program and its first phase [*sic*]—the Gemini spacecraft. By September 1963, the Gemini project was 18 months behind its original schedule and its actual cost threatened to be almost twice what had been anticipated. This time lag enabled us to be the first to perform a spacewalk. Astronaut Edward H. White II performed an egress from Gemini IV in June 1965—two-and-a-half months after Aleksey Leonov.

In the House of Representatives, there were speeches alleging that the U.S. was allowing the Soviets to walk unhindered down the path to attaining military supremacy in near-Earth space, while the U.S. was occupied with a project to send a man to the Moon, which was of no practical importance. In their attacks on the program initiated by President Kennedy, the American militarists intentionally exaggerated our achievements. We surprised ourselves when we were able to launch three at once on the former Vostok. But we were even more surprised when we found out that Goldwater had called our *Voskhod* a prototype of the Soviet "space battleship." If only he had been right!

Studying the materials about the Gemini spacecraft and the program of planned flights, we saw that in the coming year the Americans might bypass us both in terms of technology and in the number of piloted flights. The Gemini spacecraft weighed just 3.8 metric tons. The Vostok weighed almost 1 whole metric ton more, and the *Voskhod-2* weighed a little less than 2 metric tons more than the Gemini. But the Gemini outperformed the Vostoks and Voskhods across the board. It was equipped with a special forward compartment containing radar for search and rendezvous with other spacecraft. Behind the forward compartment was what we guidance and control specialists dreamed of, the attitude control system compartment, and behind it, the instrument compartment, which separated before reentry into the atmosphere. The headlight-shaped descent module was designed for a guided descent taking advantage of its highly aerodynamic features. Soft-landing engines were not needed because it was designed to touch down in the ocean rather than on land. Both astronauts were seated in ejection seats. Our Vostoks had the capability to eject only one cosmonaut. Oxygen-hydrogen fuel cells provided electric power for the Gemini. Water generated in these electrochemical generators was used for drinking. An inertial unit that had its own electronic computer (TsVM) served as the basis for the control system.[43] We had ordered the development of fuel cells and a computer, but no provision was made for them, even for the Soyuz. The Gemini computer made it possible to conduct autonomous navigation using a manual sextant, to process information using new principles, and to warn the pilots of critical modes and errors that they committed.

The Gemini propulsion system made it possible to execute various in-orbit maneuvers. Vostoks and Voskhods did not have such capabilities. It wasn't until the Soyuz vehicles that we provided the widest range of maneuvering capabilities. But Geminis were already flying, and we hadn't even tested our circuitry yet.

After all the modifications, our R-7 launch vehicle made it possible for us to insert the future Soyuz spacecraft weighing 7 metric tons into near-Earth orbit. The Titan II inserted spacecraft half that size. Nevertheless, the Gemini enabled the Americans to do a lot more than we could do on the Voskhods: long-duration flight; various experiments involving rendezvous; and even a spacewalk without using an airlock—directly from the main hatch! This was, in our opinion, a risky decision. But, why did everything come easier for them? More than anybody else, Sergey Okhapkin was outraged by the excessive weight of the systems delivered to us. He was a designer who had been brought up in the weight-conscious aviation culture and he let our excess weight affect him very emotionally. He admired the simplicity and daring that the Americans resorted to when dealing with complex situations. They entrusted a great deal to the individual, whereas we installed heavy trunks full of all sorts of triple-redundant automatic controls. Ten years after the flight of *Voskhod-2*, when the Americans became acquainted with our technology during the joint Apollo-Soyuz Experimental Flight (EPAS), they were astonished at our ability for complete automatic control without using a computer or human intervention.[44]

---

43. TsVM—*Tsifrovaya vychislitelnaya mashina*—literally means "Digital Computing Machine."

44. The Apollo-Soyuz Test Project (ASTP) was known as EPAS—*Eksperimentalnyy polet Apollon-Soyuz*—or Apollo-Soyuz Experimental Flight by the Soviets.

Korolev had promised Khrushchev that we would amaze the world by having a cosmonaut perform a spacewalk from a spacecraft. Now no one at the top was rushing us at all, except for Korolev himself and the U.S. Gemini program. The Americans had announced that the first piloted flight of this program would take place in March 1965. My colleagues and I calculated that we could manage to launch the *Voskhod-2* in February 1965.

It made no sense for us to repeat a 24-hour flight with three cosmonauts, and the Voskhods were not designed for lengthier flights. In addition to the three-seat *Voskhod*, the government decree of 14 June 1964 had called for the solution of one more earthshaking problem—a spacewalk by a human being. The spacecraft designed for this was referred to variously as either Vykhod (Egress) or *Voskhod-2*. The latter name became the accepted one at Korolev's insistence. Lead designer Feoktistov had been busy training for his flight on the *Voskhod* and did not get seriously involved in work on *Voskhod-2* until November 1964. Korolev promised Khrushchev that the flight of *Voskhod-2* and coverage of the human "space walk" could be broadcast to the entire world thanks to television. To speed up the television operations, I had to visit NII-380 in Leningrad a couple of times. The institute director, Igor Rosselevich, and the leading space television specialists already had experience working with us and they tackled this new task enthusiastically. In addition to the narrow band system that had already flown and which transmitted just 10 lines at a speed of 10 frames per second, they developed a 100-line broadband system, also with a 10-frame operating speed, as opposed to the broadcasting standard of 25 frames per second.

The most complex problem proved to be developing a soft air lock chamber and spacesuit for the extravehicular activity. This work fell primarily to Gay Ilyich Severin. Affable, possessing a good sense of humor, and quickly finding his way around in technology, space medicine, and our clannish politics, Severin immediately fit in to our community. He was the only one in

RKK Energiya

The *Voskhod-2* spacecraft during assembly. Note the accordion–like airlock in stowed position affixed on the descent module.

our company who reached a professional level in alpine skiing. His face always had an alpine suntan, even in the off-season.

The first contacts—and the first business-related arguments—that my comrades and I had with Severin were over the umbilical electric power supply technology and the setting up of two-way voice communications with the cosmonaut during his extravehicular activity. All types of support would have to be fed into the spacesuit from the spacecraft via a special tether. Back then, there was no feasible technology in terms of size and mass that would enable a cosmonaut to operate autonomously in a spacesuit. The airlock complicated the egress process and made the spacecraft heavier, but it also made it safer than the Gemini. In its stowed state, it was secured to the external surface of the descent module. During the first orbit, the spacecraft commander activated airlock pressurization and monitored it as it was set in the working position. The airlock had two hatches, one connecting it with the descent module and one for a spacewalk. Unlike the Gemini, the availability of two hatches made it possible to maintain the pressure integrity of the descent module during the cosmonaut's egress and return. After the cosmonaut's return, the airlock was ejected, but the frame to which it had been secured remained on the surface of the descent module. All operations involving pressurization, opening and closing the hatches, releasing pressure, and jettisoning the airlock required a strict sequence. Our electrical engineers developed a special control panel for all of these operations. Backup of the commands controlling airlock pressurization, opening and closing the internal hatch, the subsequent release of pressure, and jettisoning of the airlock was provided from the ground via command radio link.

The reliability of these operations and the airlock's pressure integrity needed to be tested on an unpiloted *Voskhod-2* test model. Its launch took place on 22 February.[45] The flight proceeded without any serious glitches. During the checkout procedure on the new television radio link, an unusually clear (for space television) image of the airlock appeared on the screen of the receiver installed at the command post. Everyone present was thrilled and began to congratulate the television specialists. Only after the spacecraft had left the coverage zone did they shower Severin with congratulations—without telemetry they had seen that the airlock was functioning normally and the "wind" wasn't blowing it away.

Korolev had a knack sometimes for fanning tension when there was no need for it. He demanded that Colonel Bolshoy immediately send explicit instructions to General Karas in Moscow to keep the televised image of the

---

45. The spacecraft was known in the open press as *Kosmos-57*.

airlock classified until a special State Commission decision was issued.[46] The television image from *Voskhod-2* was broadcast only over our territory. There were receiving stations in Moscow, at the firing range, and in Simferopol. Not a single TV viewer was able to receive the image because the radio link did not conform to any broadcasting standard. Nevertheless, the instructions were passed on. The airlock image broadcast during the second orbit was also enthusiastically received in Moscow. The *Voskhod-2* test launch had already been announced as *Kosmos-57*, and Smirnov had briefed Brezhnev on the successful experiment. However, at the beginning of the third orbit, in the coverage zone of NIP-4, NIP-6, and NIP-7, *Kosmos-57* "disappeared" from all communications channels. The *Tral* and *Fakel* systems fell silent, as did the television and telephone simulation. When anti-ballistic missile assets were called on for assistance and provided with target designations, they were unable to find anything similar to our spacecraft. On the fourth orbit no facilities were able to detect the spacecraft in space. The State Commission and "enthusiasts" shaken by the loss of the spacecraft gathered at the command post. I ventured a guess that such a thing could only have happened if the spacecraft had been blown up by the APO system. But what would have triggered it? Korolev seized on this idea and pounced on me: "Most likely, you overlooked something there. Tell your people to look into this immediately."

Tyulin disagreed. He proposed that an official commission be appointed under the chairmanship of Kerim Kerimov. The commission began working to determine the operating logic of the APO, the reliability of its protection against false actuation, and the possibility of receiving false commands. They quickly learned that the APO actuates in the event of an off-nominal descent when there is the danger of landing on foreign territory.

Mnatsakanyan rescued us. After analyzing what commands were used for transmission "on board" via his command radio link, he and his specialists determined that if one of the airlock control commands was sent simultaneously from two ground stations, it was converted into a descent cycle command. If this happened, it meant that the TDU had fired, and the APO detonated during the unauthorized descent and destroyed the spacecraft. We very quickly found the guilty parties.

By midday on 25 February, the accident investigation commission had unequivocally determined that the triggering of the APO had been caused

---

46. Amos Aleksandrovich Bolshoy was the deputy chief of the control center of the Command and Measurement Complex (KIK). Andrey Grigoryevich Karas (1918–79) was the Commander-in-Chief of the Central Directorate of Space Assets (TsUKOS), the predecessor to the "military space forces."

when Kamchatka stations NIP-6 and NIP-7 issued two identical overlapping commands No. 42. The on-board decoder interpreted the two superimposed commands as a single command No. 5—"descent." Mnatsakanyan and Bolshoy briefed the commission. Both confirmed that only NIP-6 was supposed to have issued command No. 42. NIP-7 was supposed to have been silent, but took the initiative because it had no explicit prohibition.

Tapes from experiments at NII-648 and tapes—material evidence— delivered from the (command) posts confirmed the validity of the theory involving the superimposed commands and the transformation of two harmless commands into one catastrophic one. Now Korolev criticized Mnatsakanyan and me, and I must admit, rightfully, for the failure to protect such a crucial command as "descent." Besides determining the causes, the commission made a special ruling demanding the protection of particularly crucial commands in Mnatsakanyan's radio link.

MARCH AT THE COSMODROME BEGAN WITH DAMP, DREARY WEATHER. The low clouds were sometimes loaded with wet snow and more often with fine drizzling rain. The steppe was covered with patches of dirty snow living out its last days. The launch of (lunar probe) Ye-6 vehicle No. 9 was scheduled for 12 March. This time I ended up being an observer at IP-1 along with former and future cosmonauts. My perpetual preoccupation with information and reports after the spacecraft entered its intermediate orbit kept me from getting a look at the new cosmonauts.

The State Commission's business session began on 16 March at 1 p.m. Korolev gave a very detailed readiness report. Next, as tradition had it, all the chiefs and firing range chief Zakharov gave presentations, and Kirillov reported on the spacecraft and rocket test results. On behalf of the Air Force, Kamanin reported on the readiness of all the post-landing crew search and rescue services. The State Commission accepted Korolev's proposal for rollout on 17 March and launch on 18 March. After the official portion of the meeting I reminded everyone that on Paris Commune Day the launch should be successful.[47] That evening, also in keeping with tradition, the State Commission approved the Air Force's proposal for the crew. The primary crew: Lieutenant Colonel Pavel Ivanovich Belyayev with Major Aleksey Arkhipovich Leonov performing the spacewalk, and backup crew: Major

---

47. "Paris Commune" refers to the socialist government that ruled Paris for a little more than two months (from 18 March to 28 May) in the spring of 1871 during a national crisis after the Franco-Prussian War. The term dates back to the Paris uprising during the French Revolution of 1789.

Dmitriy Alekseyevich Zaikin with Major Yevgeniy Vasiliyevich Khrunov performing the egress.

A meeting took place at the launch site on 17 March—a get-together for the crew, launch control team, and everyone involved in launch preparation. About 500 people gathered. During these occasional meetings, when we had the opportunity to observe the four new aspiring cosmonauts, Leonov stuck in my memory more than the others.

"He has some of Gagarin's daring," I confided in Ryazanskiy. "He has observant eyes; they don't skim over the surface—they linger!"

Ryazanskiy agreed with me and added that, besides everything else, Leonov stood out by dint of some natural refinement. The future showed that we had been right. Among other things, Leonov had an artist's eye.

On launch day, 18 March, the weather was lousy. There was a low cloud cover and intermittent drizzling rain. This was the second year that Voskresenskiy wasn't at the launch site due to heart disease. Voskresenskiy had a talent for making the difficult hours at the launch site pass by with witty jokes that were always apropos. It was like adding a spicy sauce to bland food. Korolev entrusted technical supervision of operations at the launch site to Shabarov and according to all indicators he coped.

Yuriy Gagarin, Aleksey Leonov, and Pavel Belyayev at the pad area in March 1965 before their flight on the *Voskhod-2* spacecraft.

From the author's archives.

When the time comes for the fueling process, all bystanders leave the launch pad in front of the rocket. In the white clouds of vapor that formed as the oxygen vented, more than anyone else, one caught glimpses of Barmin's hulking figure. During the fueling of the rocket—with the spacecraft and cosmonauts at the top—Barmin did not leave the launch pad. Again and again, his controllers in charge of various operations came up to him with their reports. Korolev was also usually strolling about, inhaling the oxygen-enriched air, and did not retreat to the bunker with Shabarov and Kirillov until T-minus 15 minutes.

Asif Siddiqi

Korolev bidding farewell to *Voskhod-2* commander Pavel Belyayev on 18 March 1965.

From the author's archives.

Meeting with cosmonaut Aleksey Leonov (with his back to the camera) at the launch site on 15 March 1965 are, from left to right: I. Yu. Luchko, Yuriy Gagarin, I. G. Kurmanov, Anatoliy Kirillov, V. S. Belyayev, Yevgeniy Shabarov, and A. I. Yatsushko.

The bunker was my "work station" during all the pre-launch tests. I was involved with the analysis of the results and monitored the coordination of the preparation of all the intricate interface between the on-board and ground-based spacecraft systems. At T-minus 30 minutes I climbed out of the bunker, reported to Korolev, and then drove to IP-1. This time I set out earlier to stop by at the MIK; I wanted to see with my own eyes whether the television picture that Bratslavets had reported about from the command post to the bunker was really going via the new broadband link from the spacecraft.

After emerging from the bunker onto the launch pad, I saw Korolev standing in the distance on the access road to the parking lot rather than in his usual position by the rocket. As I met up with Shabarov, he said, "It's better not to approach S.P. right now!"

Keldysh and Barmin were standing with Korolev. Korolev was saying something and gesticulating dramatically, which he rarely did. Keldysh stood hanging his head like a delinquent little boy. Barmin was smiling. When I drew closer, Barmin was walking toward me and also warned me: "Don't approach them now."

I went to my car and drove away. I learned from Barmin what the three of them were talking about after Korolev had passed away. After the cosmonauts had been strapped into the spacecraft during the fueling operation, Keldysh was suddenly called away from the launch site to Site No. 2 to talk to Moscow. Mikhail Suslov was waiting for him on the high-frequency line to the Central Committee. Expressing no interest in the status of affairs at the cosmodrome, Suslov ordered Keldysh to fly to Moscow immediately to conduct a general assembly of the Academy of Sciences during which they were supposed to condemn the behavior of Academician Andrey Sakharov and perhaps even rescind his membership in the Academy.[48] Instead of objecting, Keldysh returned to the launch site and requested that Korolev quickly let him have our airplane to depart for Moscow. When Keldysh explained what was going on, as Barmin recounted, Korolev flew into an absolute fit of rage. At that time, Barmin was the third Academician on the launch pad. Korolev and Keldysh stepped out of the way of the scurrying launch preparation teams

---

48. Andrey Dmitriyevich Sakharov (1921–89) was a preeminent Soviet physicist who is considered the "father" of the Soviet hydrogen bomb. During his later life, he became a prominent dissident, openly criticizing not only Soviet nuclear strategy but also the repressive nature of political and civic life within the Soviet Union. He was awarded the Nobel Peace Prize in 1973. In 1980 the Soviet government stripped him of his various honors and exiled him to the city of Gorky. Chertok is referring here to a movement to strip Sakharov of his full membership in the Academy of Sciences.

and Korolev shouted at him that he was not going to hand over any airplane even for the Academy. Such a question is a disgrace! He would not allow Keldysh to leave the firing range. He was prepared to talk with Suslov himself and tell him that for the next two days it would be impossible for the president of the Academy of Sciences to leave. It all ended when Korolev asked the "Theoretician of Cosmonautics" to withdraw to the bunker, stay there until the spacecraft was in orbit, and check the text of the communiqué that Ishlinskiy prepared one more time.

As far as I know, 1965 was only the beginning of Sakharov's dissident activity. In view of the subsequent space triumph, with which Keldysh was so directly involved that he couldn't even leave the cosmodrome, he was forgiven for his failure to follow Suslov's instructions. Sakharov, however, was not forgiven. I grant that the episode at the launch site served as a warning for Keldysh, after which Sakharov's expulsion was never questioned at the Academy's general assemblies.

The rocket carrying *Voskhod-2* lifted off normally and quickly disappeared, leaving behind a perceptible hole in the blanket of clouds. On the *Tral* scopes, all the green columns quivered within the acceptable ranges—reports came into the bunker over the loudspeaker that everything was normal. Just when the reading "530" appeared on the screens, everything began to jump, but the spacecraft had already become an independently monitored object.

The flight of *Voskhod-2* went down in history twice. In the first official, open version, everything was said to have gone splendidly. In the second version, which unfolded gradually and the details of which weren't published at all, there were at least three "incidents."

Leonov was observed via television, and the image was relayed to Moscow. During his egress from the spacecraft to a distance of 5 meters, he waved his hand in open space. Leonov was outside the airlock for 12 minutes, 9 seconds. But it turned out that it was easier for him to exit than to come back. His spacesuit inflated in space and refused to squeeze back into the airlock. Leonov was forced to release pressure to get "skinnier" and make the spacesuit

These three stills are from the external movie camera on the Soviet *Voskhod-2*, which recorded Aleksey Leonov's historic spacewalk on 18 March 1965. Leonov's EVA made him the first human ever to walk in space, giving the Soviet Union yet another space "first."

Asif Siddiqi

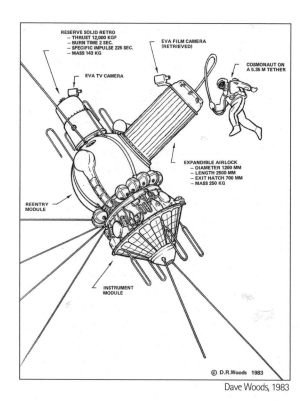

RESERVE SOLID RETRO
— THRUST 12,000 KGF
— BURN TIME 2 SEC.
— SPECIFIC IMPULSE 225 SEC.
— MASS 143 KG

EVA FILM CAMERA
(RETRIEVED)

COSMONAUT ON
A 5.35 M TETHER

EVA TV CAMERA

EXPANDIBLE AIRLOCK
— DIAMETER 1200 MM
— LENGTH 2500 MM
— EXIT HATCH 700 MM
— MASS 250 KG

REENTRY
MODULE

INSTRUMENT
MODULE

© D.R.Woods 1983

Dave Woods, 1983

Diagram of Aleksey Leonov's historic spacewalk from *Voskhod-2* in March 1965. Note the extended airlock, which was jettisoned before reentry.

softer. Nevertheless, he had to crawl back in headfirst rather than feet first as stipulated. We didn't learn about all the vagaries of the return to the spacecraft until after the cosmonauts had landed.

The second incident was an inexplicable drop in pressure in the cabin pressurization tanks from 75 to 25 atmospheres after Leonov's return back into the spacecraft. They would have to land no later than the 17th orbit, although Grigoriy Voronin—chief designer of this part of the life support system—assured everyone that there was enough oxygen for another 24 hours.

The third incident caught us completely by surprise. The automatic solar orientation system malfunctioned, and consequently the braking engine unit (TDU) did not fire. The spacecraft "waved at us," stayed in orbit, and entered its 18th orbit. It was recommended that the crew land the spacecraft manually during the 18th or 21th orbit.

For 4 hours at the command post it was unclear what had happened in space. Korolev completely wore out Bykov, Kamanin, and Bolshoy, asking for communications. There was no direct contact with the spacecraft. There was only indirect data. Radars of the Air Defense Troops pinpointed the spacecraft's entry into the atmosphere and its descent over Central Russia. One of the shortwave receiving stations supposedly picked up the telegraph dispatch "VN . . . VN . . . VN," which stood for "all normal."[49]

---

49. VN—*Vse normalno.*

Finally a report came in from a search helicopter. It had discovered a red parachute and two cosmonauts 30 kilometers southwest of the town of Bereznyaka. The dense forest and deep snow prevented the helicopters from landing near the cosmonauts and there were no populated areas nearby.

The landing in the remote taiga was the last incident in the saga of *Voskhod-2*. The cosmonauts spent the night in a Northern Ural forest. All the helicopters could do was fly overhead and report that "one is chopping wood and the other is making a campfire." The helicopters dropped supplies to keep the cosmonauts warm and provide them with food, but they weren't able to pluck Belyayev and Leonov out of the taiga. A group of cross-country skiers including a physician made a landing about 1,5 kilometers away. They reached the cosmonauts over the snow 4 hours later, but did not take it upon themselves to bring them out of the taiga.

A real competition developed for the rescue of the cosmonauts. The firing range service, warmed up by Tyulin and Korolev, sent its own rescue mission, headed by Lieutenant Colonel Vladimir Belyayev and foreman of our factory Yuriy Lygin, to Perm. From Perm they flew by helicopter to an area 2 kilometers from *Voskhod-2* and were soon exchanging embraces with the cosmonauts. Air Force Marshal Rudenko forbade his rescue service from evacuating the cosmonauts from the ground to a hovering helicopter. They remained in the taiga for a second cold night. True, now they had a tent, warm fur clothing, and sufficient food. The matter had gone to Brezhnev. They had convinced him that it was dangerous to lift the cosmonauts up into a helicopter hovering above the ground. Brezhnev agreed and endorsed the proposal to cut down the trees in the vicinity to prepare a landing area.

On 21 March, cosmonauts Pavel Belyayev and Aleksey Leonov made their way to an Mi-4 helicopter over a ski track laid down by Lieutenant Colonel Vladimir Belyayev from the firing range. They transferred from that helicopter to a heavy Mi-6, which took them to Perm. Two days later, after landing 70 kilometers from the *oblast* center, the cosmonauts had the opportunity to

The *Voskhod-2* cosmonauts debriefing in the cabin of an airplane following their mission. From left to right are Sergey Anokhin, Aleksey Leonov, I. G. Borisenko, Pavel Belyayev, and Yevgeniy Karpov.

From the author's archives.

report to the General Secretary that they had fulfilled their assignment. That's the kind of communications technology we had in those days! Standing in the taiga next to the descent module, it was impossible to communicate even with a hovering helicopter. This was our major defect.

We learned a good lesson, but subsequent events during the Soyuz flights showed us once again that setting up reliable communications with the crew once it is on the ground—a problem that even the technology of the 1930s could handle—was even more difficult 30 years later than it had been during the time of the Papanin expedition.[50] Otherwise, everything went according to the best traditions.

On the evening of 21 March, all the active functionaries of the town of Leninsk, including the Young Pioneers, gathered at the firing range airfield for a celebratory reception.[51] On that warm evening, it was strange to see Belyayev and Leonov emerge from the airplane in winter boots and fur coats. They were given 24 hours to rest; and on the morning of 23 March, they flew to Moscow on an Il-18 for a jubilant reception there.

From the author's archives.

Cosmonauts Pavel Belyayev and Aleksey Leonov return to Baykonur after their spaceflight.

50. For the Papanin expedition, see Chertok, *Rockets and People, Vol. I*, Chapter 7.

51. Between January 1958 and December 1995, the residential town at the Tyura-Tam firing range was officially known as Leninsk. In December 1995, the town was renamed Baykonur. Young Pioneers were the Soviet-era mass youth organization for 10 to 15 year olds, somewhat analogous to Boy Scouts and Girl Scouts in the United States.

The newspapers were filled with the traditional proclamations to the peoples and governments throughout the world, with congratulations to the scientists and designers, engineers, technicians, and workers, and with news stories about the conversation of the leaders of the Communist Party and Soviet government with the *Voskhod-2* crew. During the flight, a direct radio communications session had been held between the cosmonauts and the Party and government leaders assembled in the Sverdlovskiy Hall at the Kremlin. Up until the flight of *Voskhod-2*, all cosmonauts had reported from space directly to Khrushchev. Now they addressed a "collective leadership."

From the author's archives.

Cosmonauts Pavel Belyayev, Vladimir Komarov, Aleksey Leonov, and Yuriy Gagarin.

The reception at Vnukovo airport was every bit as good as in the past. Just the list of those in attendance took up a newspaper column. The last name on that list was Academy of Sciences President M. V. Keldysh. Foreign diplomats were mentioned but the scientists whom the Central Committee and Council of Ministers had so triumphantly referred to—not a single one of them were mentioned.

It was March 1965, not April 1961. Nevertheless, once again Red Square was filled with thousands of Muscovites. In old photographs of that day, I see on the people's faces sincerely joyous smiles, nothing false, no feigned elation. That's how it really was. After their flight, Belyayev and, especially, Leonov gave speeches dozens if not hundreds of times. But that day, 23 March 1965, on Red Square, Leonov really spoke from the bottom of his heart, in the true sense of these words:

> *I want to tell you that I was simply astounded and spellbound by the picture of the depths of space that I saw—its grandeur, its immensity, the clarity of its colors and the sharp contrasts between the velvety darkness and the blinding radiance of the stars. To complete this picture, imagine that against this background I see our Soviet spacecraft illuminated by the bright light of the Sun's rays. When I went out of the airlock, I felt a powerful stream of light and heat that reminded me of a welder's torch. Above me was black space and bright, unblinking stars. The Sun appeared to me like an incandescent, fiery disc. I felt a sense of vastness and lightness; it was bright and good . . . .*

The evening reception at the Great Kremlin Palace was in keeping with the best traditions. At this reception, Katya (my wife) set out on an "unescorted voyage" and hit the jackpot. At home now, as a precious relic, we have kept a brochure—news coverage of the flight signed by Gagarin *"To Yekaterina Semyonovna ~~Chertok~~ Golubkina"* and including the autographs

From the author's archives.

of Andrey Tupolev, Dmitriy Ustinov, Sergey Korolev, Aleksey Leonov, Pavel Belyayev, *and* "Kolki from Arbat"—as Nikolay Golunskiy signed his autograph.[52]

Before the obligatory postflight press conference, a heated argument flared up: should the cosmonauts tell the truth about the flight? Was it necessary to say that there were difficulties during the return into the airlock, that there was a failure

An official portrait of the *Voskhod-2* crew taken after their mission shows Pavel Belyayev (left) and Aleksey Leonov (right).

---

52. Nikolay Pavlovich Golunskiy was a telemetry specialist who worked in Chertok's department at OKB-1.

of the automatic orientation system and consequently an emergency landing in the taiga, overshooting the calculated landing site by 368 kilometers? It is astounding and incomprehensible why, but Keldysh demanded that the cosmonauts say nothing about the failure of the automatic system, and they asserted that the spacecraft had landed at the calculated site and they had not spent two days in the taiga, but had been resting at the cosmodrome under the observation of doctors. Korolev strongly objected to such lies and said that he would speak with Brezhnev. Kamanin, who also felt that they should tell it the way it happened, supported him.

More than 1,000 people were assembled for the press conference in the auditorium at Moscow State University. I was not at this press conference. But after hearing the accounts of friends who were there and reading the report, I understood that they were not given permission to tell the whole truth. Keldysh's introductory speech was unusually short and ended with him awarding the cosmonauts Tsiolkovskiy Gold Medals on behalf of the Academy of Sciences.

In his speech, Belyayev, who had been promoted to the rank of Colonel, stated that the spacecraft had landed successfully in the vicinity of the town of Perm on 19 March at 1202 hours. Poor Belyayev! They forbade him from telling the real truth and made him say something that sounded plausible.

From the author's archives.

Cosmonaut Aleksey Leonov making his postflight report to the members of the State Commission in 1965. Sitting on the left is Major General Nikolay Kuznetsov, the director of the Cosmonaut Training Center.

It turned out that the cosmonauts had longed to use manual attitude control to execute the de-orbit maneuver. And when, during the process of preparing for landing using the automatic descent cycle, they noticed certain irregularities in the operation of the solar orientation system, this made them very happy. Now they had the opportunity to execute a landing manually and thus discover one more remarkable capability of Soviet piloted—now in the full sense of the word—spacecraft. It turned out that the cosmonauts had asked the "ground" for permission and had feared that they would not be permitted. The manual landing system operated flawlessly, "and we landed approximately where they had calculated we would, but we overshot the landing site slightly due to the newness of this type of landing." In conclusion, Belyayev congratulated American astronauts Grissom and Young, who had completed a flight on the Gemini spacecraft on the day our cosmonauts returned to Moscow.[53]

Leonov devoted his speech entirely to the colorful description of the novelty of his impressions during his spacewalk. He happily alluded to his painting hobby, saying that "for those who are familiar with a paint brush and easel, it is difficult to find a more magnificent picture than the one that opened up before me."

Thus ended the phase of Vostoks and Voskhods, which took up five years of truly heroic labor in our history. With all the new problems, and despite the imperfection of the technology and the risky decisions, all eight piloted launches had a happy, and one might say, triumphant ending. A two-year break in our piloted launches was coming.

In September 1965, the 16th Congress of the International Astronautical Federation (IAF) took place in Athens. The chief scientific representative of the Soviet Union at this congress was the indefatigable Academician Leonid Sedov. The main attractions of the Soviet delegation were cosmonauts Belyayev and Leonov.

The Americans were very open, and they were not afraid to send, as a delegate to the congress, the technical director—or in our terms, chief designer—of the not yet flying, but already famous, Saturn rocket series, Wernher von Braun.

The European press had published a lengthy article by the "father of the V-2 missile, German scientist, naturalized U.S. citizen, Wernher von Braun." One must give credit to von Braun. The article, which described the state of

---

53. Astronauts Virgil I. "Gus" Grissom and John W. Young flew the 5-hour *Gemini III* mission on 23 March 1965.

cosmonautics at that time, the immediate plans of the United States and a prognosis for the future, is still a very interesting read 30 years later. Von Braun describes the primary features of the Saturn IB and Saturn V rockets, compares their capabilities for delivering payloads into near-Earth orbit and the costs of possible piloted flights into near-Earth orbit and to the Moon. In this article, and in subsequent interviews, he stated that a flight to the Moon will take place no earlier than 1969 and no later than 1975. Von Braun's prognosis turned out to be accurate. Under his supervision, all six American expeditions to the Moon were carried out during the period from 1969 to 1972. But with regard to other expeditions, von Braun was too optimistic. He wrote:

The first manned missions to Venus could take place in 1975, if enough funds are allocated and if all the industrial facilities are used to this end. We'll need to wait until the 1980s for the first landing on Mars or on one of its moons . . .

. . . Traveling to the Moon under the Apollo program is simply a reconnaissance mission, similar to what an army does before it penetrates into unknown territory. After the Apollo program the real conquest of space will begin.

Von Braun went on to discuss the problems of nuclear power sources for space transport, reusable systems, astronomical observatories, and orbital "hotels."

In the Soviet Union, research and projects on that subject were classified "Top Secret." The things that were published in popular literature by incompetent journalists were so watered-down and far from the actual technology and its problems that specialists considered them to be the kind of pap written for elementary school children.

Von Braun didn't just write and tell stories. He brought to the congress a model of the lunar vehicle, a 360-foot (110 meter) Saturn V rocket, and dozens of various slides. The newspapers wrote that "the still youthful, but gray-haired" Doctor von Braun leads an army of 300,000 men.

"Only by laying all of our cards on the table can we hope to motivate the Russians to tell us what they are doing," von Braun told the correspondents.

The Americans were not bluffing when they spoke about their space plans. They really did "lay their cards on the table." This placed our delegation in an awkward position. Other than cosmonauts Belyayev and Leonov, our delegation did not include any genuine qualified rocket scientists or the creators of spacecraft. But we could shine, having shown the design of three *Semyorka* models and at least a general view of the UR-500.[54]

Assessing the closed nature of our society at that time after 30 years, one can assert that it could not be explained by common sense, or by ideology, or by real concern for security. Even Khrushchev had been unable to overcome the stagnant bureaucratic system of the most powerful Party administration in the world, which operated on the principle "hold on and don't let go."

In response to the numerous questions regarding a landing expedition on the Moon, the Soviet scientists at the congress refused to confirm or deny reports that they were getting ready to send a piloted flight around the Moon. According to the instructions received in the Central Committee before departing for Athens, the members of our delegation answered, "Wait and you'll see for yourselves."[55]

But our delegates weren't able to avoid meeting with von Braun. Sedov met and talked with him, and Belyayev and Leonov had an hour-long conversation with him. In this regard, a secret scandal flared up in our midst, which I indiscreetly referred to as a "tempest in a teapot" in a conversation with Tyulin over the direct line to the Kremlin. After reading a TASS news flash about the events in Athens, Pilyugin called me on the Kremlin line. He was agitated.

"I'm perturbed that Sedov is mixing and mingling with that fascist, SS officer von Braun. I want to persuade Keldysh and Sergey to investigate his behavior in the Academy Presidium. You gathered materials in Germany about the atrocities in the underground V-2 factory.[56] Thousands of people died there with von Braun's knowledge. You've got to help me!"

"How am I going to help you? Punish innocent Sedov for von Braun's sins? In my opinion, it's more than unjust; if anything, it's even ridiculous."

"He did not have the right to talk with von Braun!"

---

54. This is not strictly true since the design of the Semyorka (R-7) rocket was first revealed to the West during the Paris Air Show in June 1967. A full image of the UR-500 Proton rocket was not revealed until the launches of the Vega interplanetary probes in 1984.

55. BPI [Press Information Bulletin] no. 223, TASS, 21 September 1965.

56. The production of the V-2 during the latter part of World War II was carried out at the Mittelbau complex near Nordhausen, which included factories, depots, and prisoner camps. Nearly 25,000 concentration camp workers who were housed in the Dora prisoner camp died under horrific conditions while working on the production of the V-2 at the adjacent Mittelwerk assembly factory. For a recent investigation of von Braun's possible complicity in the conditions and work at Mittelbau, see Michael J. Neufeld, "Wernher von Braun, the SS, and Concentration Camp Labor: Questions of Moral, Political, and Criminal Responsibility." *German Studies Review* 25/1 (2002): 57–78.

"Remember, Nikolay! In 1945, when we were in Bleicherode, we tried to arrange the kidnapping or enticement of this same von Braun from the American zone to our Institute RABE. And we certainly didn't want to get our hands on him to punish him for being in the Nazi party, but rather to make use of his experience and knowledge for our country. If we had succeeded, I am certain that not just we, but also ministers would be associating with him. Incidentally, you've sat at the same table with former Nazis like Dr. Ruhl more than once. You've worked successfully with them, drank coffee and even 60-proof corn liquor."

Realizing that we didn't see eye to eye and that I did not share his indignation with Sedov's behavior, Pilyugin slammed down the phone in a fit of anger.

I was also perturbed, but for a completely different reason. Nine years had passed since the 20th Party Congress.[57] Everything that had come to light then after Khrushchev's secret report was horrible. After this, the rehabilitation of the many tens of thousands of ordinary people, very famous military figures, Party workers, writers, and scientists had raised hopes for political change; people hoped that no one would ever have to fear falling into the category of "enemies of the people," being accused of "worshipping" Western science, or simply of being acquainted with a foreign scientist. After Khrushchev's visits to the United States, there was a confidence that efforts would be made on our side to lift the heavy iron curtain.

The Americans had given von Braun U.S. citizenship and trusted him to manage the enormous national space program.[58] They permitted him not only to fly to Athens, but also to report on projects that, according to our conventions, would have to be classified "Top Secret."

Many of us naively assumed that now we would have the derogatory stamp "prohibited from travel abroad" lifted from our passports, especially for those of us who had never been repressed, been "on trial and under investigation," or had not lived during the war "in territories that were temporarily occupied by the Germans."[59] But the Party bureaucrats schooled under Stalin could not bring themselves to such liberalization, even under Khrushchev.

---

57. At the famous 20th Party Congress in February 1956, Khrushchev, in a "secret" speech, denounced Stalin and his crimes, causing shockwaves through Soviet society. The speech set the stage for a brief period of de-Stalinization and openness in Soviet society known as the "Khrushchev Thaw."

58. Von Braun became a naturalized U.S. citizen in 1955.

59. In Russian parlance, "repressed" meant that you had been arrested, imprisoned, or executed during the Stalinist era. One of the possible charges for arrest was living in German-occupied territory during World War II.

Aleksey Adzhubey, in his memoirs *Those Ten Years*, writes:

Korolev, Glushko, Keldysh, and Kurchatov, together and separately, were often at Nikita Sergeyevich's dacha. His vast assortment of business did not deter Khrushchev on his day off from waiting for them with delighted impatience to join him for dinner. In general, he valued people of science and engineering work, and placed them, so to speak, higher than scholars . . . . Behind scientific and technical discoveries, his mind instantly searched for the material benefit, the way to move forward and, most important, the social effect.[60]

Only Kurchatov was permitted once to travel to a scientific symposium in England.[61] Who kept Khrushchev from taking Korolev out of the shadows and putting him next to himself on the stand during the reception for Gagarin or the other cosmonauts? Why didn't it even occur to Korolev or a single one of the chief designers to request permission to speak or simply attend the international space congresses? The cleansing wind of the 20th Party Congress was still unable to dispel the stagnant atmosphere of fear that Soviet citizens who traveled abroad would see something that would make them waver, change their convictions, or who knows, on top of that, some enemy intelligence service might recruit them. Khrushchev "discovered America" for himself and his family. Nevertheless, the scientists that he really valued so much were unable to participate in this discovery.

After Khrushchev's removal (in 1964), the bureaucracy of the administrative system of power formed by Stalin breathed a sigh of relief and undertook a complete overhaul, patching up breaches that had been made in the iron curtain. Barely 1 hour after my conversation on the Kremlin line with Pilyugin, Korolev called me directly from his office.

"Boris, what materials did Nikolay ask you for about von Braun?"

I told Korolev that I had no compromising materials on von Braun. There's a book called *Secret of Huntsville*, in which the author, a German journalist, tells the story of Mittelwerk, the anti-Nazi underground, and very briefly writes about how the Americans pampered the German rocket specialists, including

60. A. Adzhubey, "Te desyat let" ["Those Ten Years"], *Znamya* [*Banner*] no. 7 (1988). His complete memoirs were published as Aleksey Adzhubey, *Te desyat let* [*Those Ten Years*] (Moscow: Sovetskaya rossiya, 1989).

61. Kurchatov visited England in April 1956.

von Braun, who in the author's opinion should be considered a war criminal.[62] The author's description of the V-2 in this book is very brief and lacking technical details.

"Let Nikolay have that book. He wants to talk Keldysh into bringing a vote of censure (in the Academy) against Sedov for associating with von Braun. I'm not about to get mixed up in that game and advised him not to."

I was annoyed with Keldysh, Korolev, Pilyugin, and myself. To vent my feelings, I called Tyulin (also on the Kremlin line). Now he was our deputy minister, but I had remained on good terms with him.[63] I asked him to advise Pilyugin to forgo his aggressive actions against Sedov. At the same time, pretending to be a naïve simpleton, I chastised Tyulin:

"As the new deputy minister, you should have raised the question in the Central Committee, or wherever, so that they would send the real creators of our technology to those congresses, and not second-rate bureaucrats from the VPK who have nothing to do with the work of prominent academicians. Then we wouldn't have this 'tempest in a teapot'."

Tyulin would not accept my admonishments.

"You understand full well that I can't change anything in this matter. Especially now. I advise you not to get involved in unnecessary conversations on this subject anywhere; go enjoy your good health!"

Many years later, when Korolev, Pilyugin, and von Braun were no longer alive, I met up with Leonid Ivanovich Sedov at one of our academic gatherings. I asked his opinion about von Braun.

"He was a very pleasant, intelligent, and even charming individual and conversationalist. Of course, he was fanatically devoted to his work."

DURING 1988–89, OUR NPO ENERGIYA, TOGETHER WITH THE ACADEMY OF SCIENCES PHYSICS INSTITUTE AND THE EUROPEAN SPACE AGENCY (ESA), developed the design for a spacecraft equipped with a 30-meter parabolic antenna that was supposed to be inserted in orbit more than 1 million kilometers from Earth by the super heavy Energiya rocket. A radio interferometer would

---

62. The book that Chertok cites here was originally published in German as Julius Mader, *Geheimnis von Huntsville: die wahre Karriere des Raketenbarons Wernher von Braun* [*The Secret of Huntsville—The Real Career of Rocketbaron Wernher von Braun*] (Berlin: Deutscher Militärverlag, 1963). An updated version was published in 1967. An abridged translation from German to Russian was published in two different versions, both by Julius Mader: *Tayna Khantsvilla* [*The Secret of Huntsville*] (Moscow: Politizdat, 1964); *Tayna Khantsvilla: dokumentalnyy rasskaz o karerye 'raketnogo barona' Vernera fon Brauna* [*The Secret of Huntsville: A Documentary Story on the Career of the 'Rocket Baron' Wernher von Braun*] (Moscow: Politicheskoy literatury, 1965).

63. At the time (in 1965), Tyulin was first deputy minister of general machine building.

be formed in conjunction with large ground-based antennas. This made it possible to study the most remote corners of our universe.

The design was daring and in terms of engineering—beautiful. Once the spacecraft was just 1,000,000 kilometers from Earth, humankind would be able to find out what was happening "at the last frontiers of creation"—on the very edge of the universe. Our leading specialists, designers Yuriy Denisov and Yakov Kolyako, and one of the authors of the project, astrophysicist Nikolay Kardashev, and I participated in defending this project at the international symposium in the Netherlands in the town of Noordwijk—a research center of ESA. According to tradition, after three days of arguing and discussing, the director of the center arranged for a friendly dinner that was held in a restaurant with authentic Indonesian cuisine. Although the restaurant was on the shore of the Atlantic Ocean rather than on the fabulous islands of Indonesia—the Netherlands' former colony—everything, even the wait staff, acted as though we were "there."

By European standards, the dinner was absolutely exotic. The rocket scientists, astronomers, and physicists of Europe, the United States, and Japan felt free and easy. An elderly American physicist took a particular liking to me. I convinced him that I could match the pirates of *Treasure Island* when it came to drinking Dutch gin. He became completely relaxed and expressed thoughts that also tortured me: "Now we are friends and like-minded people. Together America, Europe, and Russia are capable of creating a system for great scientific discoveries. Separately, each of us had already created part of this system. You built an outstanding rocket; we are working on supersensitive instrumentation; and here in Europe, they are creating the antenna to be deployed in space that is unique in terms of the precision of its surface. Together we have yet to develop the unique electronics to control and transmit the measurement data. We meet freely, exchange ideas, and even sit together in a restaurant and understand that each of us can't solve this problem in isolation in our own country.

"Six or seven years ago we couldn't even dream about associating freely. The Cold War separated us. Now these barriers have been removed. But it's too late! If we had begun this work 10 years ago, together we would have made a system that would have surpassed the Moon landing in terms of scientific value. But then we were separated by distrust. Both we and you received large amounts of money so that we scientists could create systems of mutual annihilation. We were very successful. So were you in Russia. Now they aren't necessary. But we still won't achieve anything together. Science unites us, but commerce separates us. Neither the United States nor Europe will give as much money for this work as is needed. This isn't a weapon, and it's not a machine for packaging cigarettes. And you and I are cut off once again, remaining friends in our common poverty—we don't yield profits."

The American was right. We had attained the freedom to travel abroad and to associate with scientists from any nations. It turned out that while money doesn't buy happiness, it does give you a sense of freedom. But freedom without money doesn't give us much.

In the early 1990s, after receiving, as it were, all the types of "freedom" that we had dreamed of, we had lost the freedom of the "Russian revolutionary élan" in scientific work. "Russian revolutionary élan" did not get along with the ideological principles of the new Russian regime. Concerning the other side of the Leninist slogan about "Americans' practicality in business," it only got in the way of the new powers.

The project that we dreamed about that evening required about 1 billion dollars to be implemented, split roughly evenly among Russia, Europe, and the United States. It turned out that this was beyond the power of both Russia and Europe. American scientists obtained funding for the very expensive Hubble Space Telescope project. They no longer cared about our concerns. But all of that was a long time after the Vostoks and Voskhods. Belyayev and Leonov were the last cosmonauts that Korolev managed to monitor in space and meet on Earth. The Soyuz series came next; but that happened without Korolev.

Chapter 10
Radio Engineering Digression

Radio engineering systems and equipment played a decisive role in the development of 20th century missile technology and cosmonautics. I believe that without introducing the newest achievements of radio engineering into the rocket and space endeavors of humanity, these efforts simply make no sense. The snowballing process of space exploration stimulated new radio engineering developments. The achievements of radio engineering accelerated the process of rocket testing and experimental development. Intertwining a purely rocket system (particularly a rocket-space system) with various radio engineering systems at all stages of its life cycle not only increases its effectiveness, but often leads to discoveries that were not always predictable.

For many years I have been teaching a course to students at the Moscow Physical Technical Institute and the Bauman Moscow State Technical University (formerly MVTU, the Moscow Higher Technical School) that I provisionally call "Large rocket-space systems."[1] In one of the first lectures, while describing the structure and makeup of one such large system, I list the primary functions performed by the radio engineering systems in present-day rocket technology and cosmonautics.

During my more than 60 years of work in the rocket-space field, I have found that a good radio engineer working in the rocket-space industry soon grasps the nuts and bolts of rocket science. He begins to understand the principle of rocket flight, the structure of the rocket engines, the rocket production and operating process, and the fundamentals of cosmonautics. By contrast, the mechanical engineer who creates the rocket itself, or the spacecraft that we refer to as the "article as a whole," simply does not grasp the basic fundamentals of radio engineering.

---

1. MVTU—*Moskovskiy vyssheye tekhnicheskoye uchilishche.* The institute was known as MVTU from 1918 to 1930 and from 1943 to 1989.

That is why I thought it would be a good idea to begin this chapter by providing a list of the basic functions that radio systems fulfilled during the creation of various rocket-space systems, adhering to a historical stage-by-stage approach to the greatest extent possible. Radio systems have allowed us to:

1. Measure flight speed and issue commands for rocket engine shutdown to ensure a precise target hit in terms of range.

2. Monitor deviations of the rocket from the assigned direction and introduce corrections into the control system to reduce lateral dispersion.

3. Provide radio telemetry, i.e., monitor parameters characterizing the processes on board the missile or spacecraft and transmitting them to the ground.

4. Provide radio monitoring of the missile flight trajectory and determine spacecraft orbital parameters.

5. Provide command radio link, making it possible to transmit flight control and operational commands for various on-board systems from the ground to the spacecraft.

6. Provide two-way radio telephone communications for piloted spacecraft systems.

7. Provide one-way or two-way television communications.

8. Create systems for establishing mutual coordinates for spacecraft rendezvous and docking systems.

9. Provide radar and television observation of Earth's surface and transmission of results to the ground via radio link.

10. Provide relay of radio-telephone, television, and other forms of information (communications satellite systems).

11. Create special radio systems for planetary and deep space exploration.

During the first decade of the missile era, no more than two organizations were commissioned to develop systems to perform these various functions. As demands grew in scope and complexity, many tasks required that new organizations be brought in. A creative "radio-rocket" competition developed.

The history of Russia's domestic rocket-space radio engineering is very closely linked to the history of international radio engineering and radio electronics, the progress of which in the 20th century has largely determined the level of modern civilization.

In the history of technology, it is usually considered that humankind's practical use of radio began with the inventions of Aleksandr Stepanovich Popov (1859–1906) and Guglielmo Marconi (1874–1937). In my lectures to students, I contend that the birth of radio should begin with Michael

Faraday's (1791–1867) discoveries and the fundamental works of prominent mathematician James Clerk Maxwell (1831–79).

In the 19th century, science developed very slowly. Maxwell published "A Dynamical Theory of the Electromagnetic Field" in 1864. But it wasn't until 1887 that Heinrich Rudolph Hertz experimentally proved Maxwell's theoretical conclusions. It is not surprising that the first message Popov transmitted "over the radio" was Hertz's name. The first transatlantic radio link was not established until 1901. One hundred years later, the data transmission range had been increased by a factor of 1,000,000.

At the beginning of the 20th century, the rate of progress in radio communications technology sped up thanks to the development of a fundamentally new invention, vacuum tubes. But it would require another six years to successfully transmit speech using vacuum-tube radio transmitters. From that moment on, radio technology began to develop by leaps and bounds.

It is surprising that Marconi did not see the need for some other use of radio besides the wireless telegraph using Morse code, particularly during the time when technology for transmitting and receiving speech and music had surged forward at irrepressible speed. In all sectors of technology involving transmission and reception of radio signals, success was achieved by further enhancing electrical circuits using vacuum tubes: small ones (power-wise) for receivers and kilowatt-power ones for transmitters.

Thus, Lee de Forest's triode was followed by the tetrode, pentode, and even more complex tubes.[2] The field of radio broadcasting absorbed the long-wave and then the medium-wave range one after the other.

The development of the shortwave ham radio movement led to fundamental discoveries in the field of radio wave propagation. Both amateur and government radio operators began to exploit the 10- to 100-meter shortwave range, thanks to the waves' reflection off the ionized layers of the upper atmosphere. Inventors of military communications and remote control systems also mastered the 1- to 10-meter very-high frequency range. This range of radio waves opened the way for an avalanche of developments in the technologies of radar and television.

Of the aforementioned radio systems, which are integral parts of a rocket-space complex, motion control systems were the first to be created. The Germans had developed them to reduce the dispersal of V-2 missiles. Lateral radio correction and radio range control were part of the R-1, R-2, R-5, R-5M, R-7, and R-9 missiles. These systems were much improved and, in their latest edition, were

---

2. Lee de Forest (1873–1961) invented the first triode vacuum tube, known as the Audion.

a far cry from their German ancestors. The R-9 rocket was the last to require radio control. For all subsequent generations of ICBMs, autonomous inertial control systems alone provided the requisite target striking accuracy.

During the first years designing the "zero" generation of ICBMs—the R-7 and R-7A in the USSR and the Atlas and Titan I in the U.S.—no particular importance was attached to problems of launch site vulnerability and the amount of time a missile stood on duty after being fueled. The main criteria that generated the fiercest debates were maximum range, warhead yield, and firing accuracy.

For the R-7 and R-7A, the developers and military customers somehow quickly came to an agreement on the first two criteria, because in this design the missile carried a warhead with a maximum yield, the mass of which had been determined at Arzamas-16, and that was as good as if God had pronounced it. The mass of the entire warhead was the factor that determined range. The control system determined 90 percent of the accuracy, or Circular Error Probability (CEP). Here, room opened up for many alternative variations.

Ever since our work in Germany, we had been aware of the fact that autonomous inertial systems without radio correction were not capable of providing a high degree of accuracy. Both calculations and the experience gained on medium-range missiles showed that for the R-7, without radio control, the CEP might reach tens of kilometers.

During the 1950s, the developer of the radio control systems for our missiles was NII-885 of the Ministry of Communications Equipment Industry (MPSS).[3] However, for the sake of historical accuracy, NII-20 of the MPSS should be considered the cradle of Soviet radio control systems. Mikhail Ryazanskiy, Yevgeniy Boguslavskiy, and a number of other radio engineers from that institute joined me in Germany at the Institute RABE.

In 1946, Ryazanskiy headed a group of specialists in radio engineering, radar, and radio navigation who were transferred from NII-20 to NII-885. Ryazanskiy was named chief engineer of NII-885. That same year, Boris Konoplev, who had remained at NII-20, received the assignment to develop a radio control system for the R-3 missile with a range of 3,000 kilometers and began studying control systems for missiles with ranges up to 10,000 kilometers as part of the

---

3. MPSS—*Ministerstvo promyshlennosti sredstv svyazi.*

N3 scientific-research project (NIR).[4] The harsh lessons of World War II were not lost on the nation's top political leadership. Even at the level of a single ministry, it was considered beneficial to conduct backup studies on complex systems. They did not try to cut the budget on that.

I had not seen Boris Konoplev since 1937, when the search for Levanevskiy's crew had ended.[5] After that he had been a student in the physics department at Moscow State University and laboratory manager at the Academy of Sciences Institute of Theoretical Geophysics under Academician Otto Yulyevich Schmidt. During the war, Konoplev had developed automatic radio meteorological stations for the Arctic. In 1943, he was transferred to NII-20, where he developed a precise aircraft radio navigation system, which was put into service. For this he received the Stalin Prize in 1946.[6]

At NII-20 Konoplev organized research work on a large scale on precise radio guidance systems for missiles. I ran into him at Kapustin Yar in 1948. At that time, Pavel Tsybin came out with a ditty aimed at Konoplev that began:

To analyze the plume effect
To us has come Mr. Konoplect

Despite of all the fun poked at him, for several years Konoplev conducted investigations that helped in the selection of the optimal radio wave ranges for our future radio systems and in determining the sites for the placement of ground radio facilities. In 1950, Konoplev and a group of co-workers transferred from NII-20 to NII-885. There he took on the management of all developments for long-range missile radio control systems. Two warring teams—those of Yevgeniy Boguslavskiy and Mikhail Borisenko—were transferred to his management. They had been fighting for the exclusive right to do the most prestigious work. With his will and authority, Konoplev quashed both, briefly easing the situation for Ryazanskiy, who had taken on the role of peacemaker.

---

4. NIR—*Nauchno-issledovatelskaya rabota* (literally, Scientific-Research Work). The N3 was one of three major research and development projects conducted by the Soviet missile industry in the late 1940s and early 1950s. N1 involved a single-stage missile with a range of 3,000 kilometers; N2 encompassed a missile using storable propellant components; and N3 involved exploratory research on a Soviet ICBM. See Asif A. Siddiqi, *Challenge to Apollo: The Soviet Union and the Space Race, 1945–1974* (Washington, DC: NASA SP-2000-4408), pp. 115–127.

5. See Chertok, *Rockets and People, Vol. I*, Chapter 7.

6. The Stalin Prize was awarded between 1940 and 1952 to Soviet individuals for significant achievements in the sciences, arts, and architecture.

In 1953, Konoplev successfully completed the development of a radio control system for the R-5, updated it in 1954 for the R-5M, and then set about his main task—developing a draft plan for the R-7 radio control system. This design involved a series of fundamentally new concepts: multifunctional pulsed radar for trajectory measurements and transmission of commands; lateral radio correction using differential distance measurement; and command coding. The scale of operations was such that the NII-885 pilot plant was completely loaded down with Konoplev's orders.

At the height of equipment manufacturing operations, fundamental disagreements arose between Konoplev and Pilyugin (and Pilyugin's ally, Ryazanskiy). I was on good terms with each of them. Through me or directly, each of them tried to convince Korolev that they were right. When I had to find out how much work NII-885 would be faced with at Tyura-Tam in order to bring the R-7 radio control system up to par, I came to the conclusion that with these complex, although very smart radio systems, we would have no time to deal with the missile itself.

I voiced my misgivings to Korolev. His response was calm: "It may be that we will have to decide to launch regardless of whether the radio system is ready. If the radio experts are going to be 'last in line,' we will start flying without them. We have enough problems of our own. And don't even think about talking about this. We're going to require that everybody be ready exactly according to the schedule. And you and Pilyugin keep thinking how to make the autonomous system even more accurate."

At NII-885 itself, a fight had heated up over priorities in research and production between Pilyugin (focused on inertial guidance) and Konoplev (adopting radio control guidance). Ryazanskiy, who had consulted with Korolev, sided with Pilyugin. Konoplev took offense and threatened to leave NII-885. But it wasn't a matter of hard feelings. Konoplev was smart enough to understand that in the future, ballistic missile control systems should move away from unwieldy and complicated radio systems. He first expressed these thoughts in Leningrad, when we visited NII-49 together. We had commissioned this shipbuilding industry institute to develop a gyro-stabilizing platform with "air suspension." Vyacheslav Arefyev was in charge of a young staff. He made the case that if he succeeded in achieving the parameters set forth in the inertial system design for the platform and for the accelerometers installed on it, then theoretically, it would be possible to achieve the required CEP *without* a radio system. At that time, the manufacturing process for our gyroscopic technology would not yet allow us to take advantage of this simple approach. Besides improvement of the gyroscope electromechanics, fundamentally new electronics were also needed. The time of on-board digital computers had not yet come for our rocket technology. The Americans fully realized the advantages of an inertial system combined with an on-board digital computer to the full extent on the Titan II and Minuteman I missiles in 1962, i.e., before we did. On the earlier Atlas and Titan I, striving to attain a high degree of accuracy, the Americans used radio systems, just as we did.

One time Konoplev let it slip to me that, if it had been up to him, he would have completely redone the system developed and put into production for the R-7. In 1955, Konoplev did leave NII-885 after all, "slamming the door behind him." Then in 1959, he was named head and chief designer of OKB-692 (later called Elektropribor, and then Khartron) established in Kharkov. By that time, a logically explicable metamorphosis of Konoplev's world view had taken place. Having become the chief designer responsible for the entire R-16 missile control system, he abandoned radio control and developed a purely autonomous system. Konoplev was killed on 24 October 1960 in the Nedelin disaster.[7] He would never know that the R-16—the first domestic ICBM without a radio control system—had a CEP of 2,700 meters at a range of 12,500 kilometers. This was just 700 meters more than the CEP of the R-7 missile, which had a very complex radio control system developed by his successors at NII-885, chief among whom was Yevgeniy Boguslavskiy.

By early summer 1956, the development of the radio flight control system for the first R-7 intercontinental missiles had for the most part been completed and the agonizing stage of manufacturing the on-board equipment and the very bulky ground-based equipment had begun. The on-board portion of the system, including the antennas, weighed more than 200 kilograms.

Despite its great mass, the on-board equipment was not very reliable. During launch preparation of the first R-7 missiles, the reliability of the radio system created a lot of headaches for the developers, testers, and State Commission. The on-board radio transmitter was particularly notorious. At that time, a powerful transmitter could be built only using radio tubes, but the tubes got warm, as they were supposed to. The transmitter turned into a powerful electric oven. There was no external cooling; and during prolonged testing at the engineering facility or launch site, it often malfunctioned. Replacing the transmitter turned into a separate problem. Considerable precautions had to be taken to remove the black box, which weighed 60 kilograms, from the instrument compartment. Four soldiers transferred it to the laboratory for subsequent diagnosis and repair. Local quipsters badmouthed the transmitter and nicknamed it the "radio coffin."

---

7. See Chertok, *Rockets and People, Vol. II*, Chapter 32.

A multi-channel pulse communications line operating in the 3-centimeter wave range used coded signals to measure the motion parameters and transmit control commands to the missile. The system had two control posts—a primary and a "mirror," located 500 kilometers from one another. A straight line connecting the two posts was supposed to be perpendicular to the direction of flight.

It was precisely this "dispersion" by 250 kilometers to the right and left of the launch pad that proved decisive in selecting the construction site for the pad and the entire NIIP-5 firing range—the future Baikonur cosmodrome.

For lateral correction, the difference between the distances from the missile to the primary and mirror control posts was measured, and signals corresponding to the missile's lateral displacement and lateral velocity relative to the firing plane were generated. These signals entered the stabilization controller developed in Pilyugin's department; and after being processed, differentiated, amplified, *and* added to commands from the gyroscopes, the signals went to the control surface actuators, turning the control thrusters to the required angle. That is how yaw control was carried out. Range control was executed using a special computer located at the primary control post. When the missile attained both the calculated value for the final velocity *and* the coordinates that satisfied the function determining the flight range, the computer uplinked the preliminary and main commands to shut down the second stage engine.

In 1957, the radio control system ground equipment was contained in 15 military trucks. Soon thereafter, brick buildings called "control stations" were built. The equipment and trucks were moved into permanent buildings. Hundreds of kilometers from the launch site, in the barren steppe, it was necessary to build not only buildings to house equipment, but also barracks, dining halls, and power plants. Creature comforts needed to be provided for the military contingent and the civilian radio industry specialists. The unwieldiness and vulnerability of the radio control system were obvious.

It wasn't easy living and working at Site No. 2, but it was much more difficult at the garrisons lost in the desert by the control stations. Unlike the "industry guys," who were absorbed with straightening out their complex and unpredictable radio equipment, the soldiers and young officers really suffered. From those times, soldiers' folk poems remain, describing the mood at those small garrisons forsaken in the semi-desert:

In summer no clouds grace the sky,
Not a blade of green grass do I see,
Nature has only created
Dry thorns and the saxaul tree.
Plants wither and writhe in the sun,

Beseeching the heavens for rain,
But heaven does not hear their prayers,
They wait for cool droplets in vain.
A wicked wind roams o'er the dunes,
The whirlwinds howl as if insane,
And sand drifts engulf our youth,
Which we never shall regain.[8]

However, the semi-desert of Kazakhstan would not be the only place from which our *Semyorkas* would lift off. The "great missile pad construction" had also begun in the north in Arkhangelsk *oblast*. By 1962, the Plesetsk firing range was supposed to have four *Semyorka* launch sites.

The 500 kilometer long "radio moustache" for four launch sites did not fit in the "green taiga sea" of Arkhangelsk *oblast*.[9] There were intensive operations to simplify and increase the reliability of the radio control system. Pilyugin's and Kuznetsov's teams labored to increase the accuracy of the primary command instrumentation and entire autonomous system. The radio specialists looked for ways to do away with the two equidistant stations and convert the system into one that relied on a single control post.

When we made the transition from the R-7 to the R-7A, a new system had already been developed. The autonomous system, which we called "the Pilyugin system," ensured that the lateral dispersion met the operational requirements. Meanwhile, the radio system was relieved of its lateral correction duty, retaining only the accurate range control function. True, for autonomous range control, instead of one longitudinal accelerations integrator, the "Pilyuginists" put in three, and they introduced an apparent velocity control (RKS) system.[10] We and they tried to demonstrate that "pretty soon" we would be able to do without radio control. But the flight tests showed that in measuring the main parameter determining flight range—velocity at the end of the powered flight phase—the radio system was an order of magnitude more accurate than the autonomous instruments of that day.

A new, substantially simplified radio system was created for the R-7A missile. All ground-based range control facilities were concentrated at one post located near the launch site. Instead of two preprogrammed rotating antennas, a single

---

8. Verse translated by Cynthia Reiser and Laurel Nolen.
9. "Radio moustache" or "radio whiskers" (*radiousy*) was the nickname given to part of the radio footprint necessary for the R-7's radio control system. The other part was the "radio tail" [*radiokhvost*]). See Chertok, *Rockets and People, Vol. II*, p. 306.
10. RKS—*Regulirovaniye kazhushcheysya skorosti*.

non-rotating antenna was installed on board the missile. But the upgrading of radio control methods did not stop there.

While developing the pulse differential ranging method, which limited the ability to shift the direction of flight in real time, NII-885 was also developing a single-command-post phased system. The control station was supposed to be located near the launch site and provide unlimited target selection in terms of direction. Work on the system began in 1959 for its use on the R-9 missile. Mikhail Borisenko was the direct head of operations for the development of phased radio control systems.

My memories of Borisenko in his white sheepskin coat went all the way back to the cold autumn days of 1947 in Kapustin Yar. He had come to missile technology from the airborne troops, having passed through the grueling school of the war. A rather coarse wisecracker, not intimidated by the brass, he mastered the technology of lateral radio correction (BRK).[11] First, he worked on the German *Viktoria* system, and then he managed the team that developed its own BRK system for the R-1 and R-2.

Building on the experience of developing lateral radio correction systems for the R-1, R-2, and R-5 missiles, Borisenko came out with a proposal in 1956 to develop an alternative system in the centimeter radio wave range using phased measurement methods. This system would have broader applications and be more compact, more accurate, and simpler to operate than the pulse system developed for the R-7 and R-7A. In the system that Borisenko proposed, for the first time, transistors were used in place of vacuum tubes in flight equipment. This was a qualitative leap, but the reliability of transistors in those days was not much different from that of vacuum tubes.

During the R-9's first flight tests, the ground control post of the missile radio control system was located at Site No. 1A at a distance of 1.2 kilometers from the launch site. The ground unit included five standard all-purpose truck trailers (KUNGs) packed with equipment.[12] The control post also had 12 parabolic antennas—two for transmitting and 10 for receiving. To simulate the operation of the on-board equipment, a remote check point with a horn antenna was set up 150 meters from the launch site in a dugout. This facility was intended to perform the prelaunch adjustment of the antenna systems, i.e., check the electrical and geometrical axial alignment of the antenna patterns.

---

11. BRK—*Bokovaya radiokorrektsiya.*
12. KUNG—*Kuzov universalnyy normalnykh gabaritov.*

At Borisenko's request, when the equipment was put into operation, antenna adjustment was conducted from a tethered aerostat. It had equipment simulating on-board equipment, electric power sources, and a radio for communication with the pilot installed in its basket. This operation seemed so exotic that jokesters suggested that we refer to the aerostat pilot as the "tethered cosmonaut."

The new, single-post radio control system developed for the R-9 undoubtedly represented cutting-edge radio engineering achievements. However, it made the operation of the combat missile substantially more complicated. The aspirations of the new chief designers to produce combat missiles with purely autonomous control systems were understandable.

Borisenko openly criticized the Konoplev-Boguslavskiy system, the development of which Ryazanskiy had supported. Technical disagreements worsened the personal relationships between Borisenko and the Ryazanskiy-Boguslavskiy duo. Borisenko capitalized on the failures that had occurred due to the radio systems during *Semyorka* launches to demonstrate the flawed nature of the path that Boguslavskiy had chosen and the advantages of the system that he was proposing.

On 2 January 1959, the radio control system failed to issue the primary command to shut down the engine of our Moon shot rocket. An investigation determined that the antenna of the main control post radio direction finder had been erroneously set to communicate with the launch vehicle using one of the secondary lobes of the antenna pattern rather than the main beam. During powered flight, the rocket deviated from the design trajectory. *Luna-1* flew past the Moon and was declared a "new planet" of the solar system and named *Mechta* (Dream).

Two years later, during the first piloted launch on 12 April 1961, due to the erratic operation of the radio complex power converter, the radio command to shut down the engine was not issued at all. The engine was shut down by a command from the autonomous system integrator, which had been set for a velocity exceeding the design velocity for the radio system. That is why the first *Vostok* overshot the predetermined landing area and landed in Saratov *oblast*. As a result of the failure of the radio control system during Gagarin's flight, the decision was made not to use it during piloted flights. This was a severe blow to the prestige of NII-885 and to the standing of Ryazanskiy and Boguslavskiy.

But competition bore fruit. Two radio links were used in the R-9 radio control system: a continuous-wave link to measure the radial range and a pulsed one to transmit control commands on board. When the launch silo was built for the R-9, the radio control system was also sheltered under ground. The antennas were placed in special silos and required special remote-controlled systems to raise them and point them after the protective roof rolled back.

From 1961 to 1965, a standardized radio control system was produced for Chelomey's UR-100 and Yangel's R-36 missiles. But both missiles were put into service just with inertial guidance and control systems. An on-board computer processed the information from sensitive command instrumentation installed on gyro-stabilized platforms. This made it possible to take procedural errors into consideration and implement optimal control algorithms. In the mid-1960s, the development of radio control systems was halted. A new explosion of ideas and developments for radio control systems occurred in the late 1970s. These were autonomous radar systems for guiding multiple warheads using digital ground maps. By that time, completely different people were developing these systems.

Boguslavskiy switched his energy to producing radio telemetry systems and spacecraft control systems. He died suddenly on 18 May 1969. For several more years, Borisenko attempted to develop compact all-purpose radio control systems, but soon realized that it was useless to compete with inertial systems in terms of simplicity and reliability. In 1974 he left NII-885 for an institute that was developing special communications systems. NII-885 was subsequently renamed the Scientific-Research Institute of Space Device Engineering (NII KP). The radio systems developed in recent years in this NII comprehensively solve problems of radio control, telemetry, television, and radio-telephone communications.

RADIO FLIGHT CONTROL SYSTEMS FOR COMBAT MISSILES AND SPACE ROCKETS LEFT BEHIND A BODY OF SCIENTIFIC WORK THAT WAS USED IN THE DEVELOPMENT OF RADIO FLIGHT TRAJECTORY MONITORING SYSTEMS. Fierce competition continued in this field. The young OKB MEI, with its new original developments, was crowding out the main institute, NII-885.

During flight tests on the R-2 missile during the summer of 1952, all the missiles were equipped with *Indikator-D* equipment. The idea behind its development belonged to Vladimir Kotelnikov. At that time, future Academician and OKB MEI Director Aleksey Fedorovich Bogomolov served as radio control system deputy chief designer for the first time. Beginning in early 1953, series production of ground and on-board equipment got underway at radio Factories No. 304 in Kuntsevo and No. 567 in Kazan. In late 1954, the radio trajectory monitoring (RKT) system developed at MEI went into service.[13] Since that time, virtually no long-range ballistic missile has passed through flight tests without this system.

---

13. RKT—*Radiokontrol trayektorii.*

During 1955–56, test models were developed and manufactured, and in early 1956, during launches of M5RD missiles from the Kapustin Yar firing range, a precise angle-measuring phase direction-finding system was tested as part of the *Irtysh-D* station and *Fakel* on-board responder.

The OKB MEI team surpassed other domestic organizations in the mastery of the new element base via the practical use of semi-conductors, or transistors. Series production of *Rubin* transponders at the Kazan radio factory began in the summer of 1959.

One after the other, Command and Measurement Complex ground telemetry stations were equipped with the *Kama-IK* radar, created using the latest radar achievements (complex pulsed coherent signals; precise range, radial velocity, and direction-to-signal-source measuring instruments; and real-time data processing with communications line output). Not only missiles, but also spacecraft were equipped with new special coherent on-board transceivers.

In view of the refusal to use NII-885's radio control systems as standard equipment on combat missiles and launch vehicles, OKB MEI monopolized trajectory data output for each launch virtually "without a fight." But the real competitive battle for a place on missiles and spacecraft was flaring up in the field of radio telemetry systems.

I have already written that in 1948, future Academician Vladimir Kotelnikov headed a small group of scientists and engineers at MEI. Independently from NII-20 and NII-885, this group very boldly tackled work on trajectory measurement and telemetry systems for the first ballistic missiles. Kotelnikov's successor, Aleksey Bogomolov, transformed the group

From the author's archives.

Academy of Sciences President Mstislav Keldysh and Chief Designer Aleksey Bogomolov.

of enthusiasts into a powerful modern OKB. He headed OKB MEI for 30 years and became a full member of the USSR Academy of Sciences. At times Bogomolov irritated radio system "fundamentalists" at NII-885 and other "reputable" firms by the fact that he took on work that distinguished firms in respected ministries had rejected.

OKB MEI began a veritable assault in the battle for radio flight monitoring and telemetry systems in 1955. Beginning with the third Sputnik—the first space laboratory—it became essential to uplink radio commands from the ground to execute the flight program. In essence, precisely such a command radio link system enables a human being to participate in the control loop. That is why modern-day flight control centers are called "control centers," in that they are actively involved in the processes on board spacecraft.

Despite the general crisis in the Russian radio electronics industry, the OKB MEI staff has retained its performance capacity and to this day its developments cover almost the entire spectrum of radio-related space projects. Despite its "collegiate" origins, in terms of the importance of its contribution to domestic rocket-space radio engineering, OKB MEI is undoubtedly second only to NII-885.

THE FIRST TWO DECADES OF THE MISSILE ERA WERE MARKED BY THE TURBULENT DEVELOPMENT AND SUBSEQUENT COMPLETE TERMINATION OF DEVELOPMENTAL OPERATIONS on radio flight control systems for all types of long-range ballistic missiles. However, at the same time there was a sharp increase in the need to enhance the data transmission capacity of telemetry monitoring systems and passive trajectory monitoring systems (those not affecting the missile).

A government decree dated 13 May 1946 specified a single head organization responsible for the development of radio systems—NII-885. However, this institute did not hold onto its monopoly in the radio field for long. During the second decade of the missile era, by 1967, there were already seven scientific-research institutes developing radio engineering systems for the new industry. This is not counting the series production factories. Each chief designer of rocket-space systems—Korolev, Yangel, Chelomey, and later Makeyev, Reshetnev, Kozlov, and Polukhin started up their own departments of radio engineering and antenna feeder systems.[14]

---

14. These were chief designers of the following design bureaus: S. P. Korolev (OKB-1), M. K. Yangel (OKB-586), V. N. Chelomey (OKB-52), V. P. Makeyev (SKB-385), M. F. Reshetnev (OKB-10), D. I. Kozlov (TsSKB), and D. A. Polukhin (KB Salyut).

As a rule, the Ministry of Defense operated the entire ground portion of the following three systems: radio telemetry; trajectory measurement; and command radio link systems. The Command and Measurement Complex, which was a primary component of Military Unit 37103, was a very powerful organization with hundreds of highly qualified radio engineers. This body of military radio specialists worked closely with developers from industrial organizations. Located on the border between "Earth and space," military engineers, with their operational experience, made a great contribution to improving the reliability of radio systems.

From its very beginning, World War II stimulated the development of radar and radio navigation. These areas of radio engineering developed independently of rocket technology, which had yet to gain recognition. Unlike the areas listed above, radio telemetry was born and developed in the form of a radio electronic system that was essential for rocket technology. The firing rig is essential to the creator of a rocket engine. The radio telemetry system is essential to the developer of a rocket, its engine, and the control system. And that's not all.

A test pilot can observe everything that takes place during an airplane's flight tests. If he returned alive after a flight, he could report all requisite information in detail to the airplane's creators. The test pilot not only controls the airplane, he is also a sort of medium for stored information. In case the pilot dies, "black boxes" were invented in the late 20th century. They need to be retrieved from among the airplane's wreckage, and after the recorded flight data is analyzed, the cause of the catastrophe is learned.

Since the early days of its creation, a missile has been an automatically controlled unpiloted flying vehicle. During missile flight tests, observing its flight from the ground did not provide information capable of elucidating what was happening on board during deviations from the flight program.

A system to track the processes taking place on board was essential. While developing the first long-range ballistic missile—the V-2—the Germans also developed and broadly used the first radio telemetry system in the history of radio engineering, the *Messina-1*. This system was developed simultaneously with all systems necessary for control and flight. The *Messina-1* was not involved with flight control; it monitored in real time and transmitted to the ground only four of the most important parameters describing the operation of the engine and missile control system.

Soon after the organization of the Institute RABE in 1945, while hunting for missile instrumentation left behind in Thuringia after the Americans departed, we found several radio instruments with vacuum tubes, the purpose of which we could not immediately discern. Wernher von Braun's former deputy in the field of radio electronics, Helmut Gröttrup, who had crossed over from the American zone, helped us out. Right away he determined that these were instruments from the *Messina-1* telemetry system. Considering the importance of the discovery, I quickly set up a laboratory in the semi-basement of the military commandant's office in Bleicherode. I put Captain K. A. Kerimov, who had just arrived from Moscow, in charge of it under the scientific supervision of Major-Engineer Boguslavskiy. One could say that that was the beginning of the victorious march of future General Kerimov, who for many years was chairman of the State Commission on Space Flight Programs.[15] Experienced radio engineer G. N. Degtyarenko, who had come from NII-20, rendered real assistance to Kerimov in studying the German *Messina*. According to the unwritten laws of missile technology, no useful initiative goes unpunished, therefore after the historic decree of 13 May 1946 was issued, NII-20 received the assignment to reproduce the *Messina* system using domestic parts. The *Messina* used the principle of frequency-division multiplexing, i.e., a separate sub-range of frequencies was allocated for each channel, which enabled information to be transmitted continuously from sensors that converted physical processes into electrical signals. The bandwidth capability made it possible to transmit changes in the monitored parameter as frequently as every two kHz. The first telemetry system had a total of four channels. On the ground, the information received from the on-board transmitter was recorded onto the thermal paper of an oscillograph.

The domestic eight-channel system was given the name *Brazilionit*. Flight tests of the German V-2 missiles in 1947 and of their domestic analog R-1 missile in 1948 showed that *Brazilionit*'s limited capabilities did not allow for a full-fledged analysis of the processes that occur in flight. Chief Designer of engines Glushko demanded that he be given four channels to monitor the propulsion system parameters. Pilyugin required at least nine channels. During those first years as deputy chief engineer at NII-88 and chief of the guidance and control department, I was responsible for channel allocation, the measurement circuitry, the antenna feeder system, and development and actuation of the primary sensors of the telemetry system. At that time a decision was made quickly. As early as May 1948, a telemetry laboratory was set up in my department No. 16 and given responsibility "for everything": ordering new systems; developing and ordering sensors; developing a measurement program; overseeing the production of ground-based recording equipment; and processing measurement results.

---

15. Kerimov served as chairman of the State Commission for piloted flights from 1966 to 1991.

In late 1948, NII-88 issued a design specification for the development of a new telemetry system to two organizations at once, NII-885, where Ye. Ya. Boguslavskiy was in charge of the project, and the special operations sector (later OKB MEI), where V. A. Kotelnikov was in charge at that time. Both organizations, as well as other missile specialists, were working at a wartime pace. In 1949, just a year later, Boguslavskiy proposed a new STK-1 radio telemetry system under the code name *Don*. The equipment was ready for delivery in time for the second phase of flight tests on the R-1 missile. The *Don* system used the principle of time-pulse modulation. The channels were divided according to time rather than frequency. Each parameter received its own "window" when it was polled. A poll was conducted at a frequency of 62.5 times per second for each channel. The system had 16 measurement channels with a total data transmission capacity of 1,000 measurements per second. An electronic switch switched the channels. The system's on-board complex consisted of four units. Power supply and output signal lines comprising the telemetry system's on-board cable network extended from the distributor unit to each sensor in every nook of the missile.

The *Don* receiver station was located in a truck trailer. Information was recorded onto movie film directly from the cathode ray tube. At the center of the main console was a visual monitoring screen: all 16 channels were displayed as columns, whose height corresponded to the current amplitude of the parameter being measured. The *Don* system became the backbone for missile technology up until development of the R-7 ICBM began.

Korolev could not come to grips with the fact that telemetry information, the "eyes and ears" of a missile tester, was organizationally outside his administrative power. In 1949, with my consent, he set up a telemetry data processing and analysis group in my department at NII-88. Young specialists N. N. Zhukov, A. I. Ostashov, V. V. Chernov, and O. A. Nevskaya comprised the group and were later joined by N. P. Golunskiy. I feel compelled to name each of them because these young engineers, who had just arrived from Moscow Aviation Institute (MAI), made an invaluable contribution to the procedure for testing missiles and telemetry systems.[16]

The front of operations on telemetry systems was expanding by leaps and bounds. New missile organizations wouldn't even think of testing and verifying the reliability of a system without using the latest advances in radio telemetry technology. In late 1952, a government decree created a new specialized organization for the development of high-speed telemetry

---

16. MAI—*Moskovskiy aviatsionnyy institut.*

systems—SKB-567. Ye. S. Gubenko was named chief designer. In 1954, having studied the experience of all preceding radio telemetry systems and in an effort to consciously meet the requirements of the chief designers of missiles, the young team at OKB MEI, which A. F. Bogomolov headed after Kotelnikov's departure, began to develop its own *Tral* radio telemetry system. They made skillful use of the revolution in electrical engineering: transistors replaced vacuum tubes.

The *Tral* system used time-pulse code, which ensured maximum interference immunity. Successful circuitry designs ensured a high degree of reliability and accuracy of measurements. The *Tral* had a data transmission capacity of 6,000 measurements per second, which was six times greater than the *Don* in service at that time. Each of the 48 channels had a polling rate of 125 per second. The *Don* and *Tral* systems could not be used to measure the vibration process. This "blind spot" was covered by the RTS-5 system that SKB-567 developed in 1955.

Flight tests on the R-7 missile required such a volume of measurements that a single set of the new *Tral* system was not enough. Considering the decisive importance of radio telemetry during flight tests on the first ICBM carrying a thermonuclear warhead, the VPK created a special commission to select a system. Two systems were competing, the *Tral*, developed by OKB MEI, and the RTS-7, developed at SKB-567. After heated arguments, the commission recommended the *Tral* system. For many years it was the backbone not only for the *Semyorka* series of combat missile and space launch vehicles, but for other missiles as well. The Ministry of the Radio Industry lost the competition for the telemetry spot on missiles. After modifications, the RTS-7 system began to be used extensively in the aviation industry for airplanes and anti-aircraft guided missiles.

The R-7 required the installation of at least three sets of *Tral* telemetry equipment to monitor almost 1,000 parameters. The total amount of labor required for missile tests rose sharply because of telemetry systems.

As on-board equipment became more sophisticated, the ground telemetry system followed suit and became many times more complicated. Each tracking station (IP) required a ground installation capable of simultaneously receiving data from three on-board telemetry transmitters with proper backup equipment (for reliability). The Ministry of Defense used its resources to build and put into operation two tracking stations at the firing range and four tracking stations along the flight path all the way to Kamchatka for the first R-7 missile launches. Each tracking station required not only technology, but also highly qualified specialists to operate the sophisticated radio complexes, perform real-time analysis of the data, and transmit it to the control center, which back then

From the author's archives.

Veterans of the Soviet ground tracking network, known as the Command and Measurement Complex (KIK). Seated from left to right are: V. F. Shtamburg, Georgiy Tyulin, and Armen Mnatsakanyan. Standing are Boris Pokrovskiy, Pavel Agadzhanov, and E. M. Kogan. Mnatsakanyan was chief designer of rendezvous systems for Soviet piloted spacecraft.

was located right at the NIIP-5 firing range and at NII-4. Military manpower constructed a Command and Measurement Complex (KIK) that extended over the entire country. Almost all the officers commanding the tracking stations were war veterans and approached their new duties with the same sense of accountability with which they had carried out combat missions.

After the Ministry of General Machine Building was formed in 1965, the organizations that developed radio electronic systems for missiles and space technology were transferred under its authority. OKB MEI was an exception. Bogomolov decided to remain independent, figuring that the trajectory, orbit, and telemetry monitoring systems that his staff had developed had found such broad application that missile technology and aviation could not endure without OKB MEI. However, officials from the Ministry of General Machine Building skillfully used revolutionary foreign breakthroughs, especially in American electronics, to limit the opportunities of the intractable OKB MEI in the competitive war with NII-885, SKB-567, and NII-648.

Despite the improved reliability of vacuum-tube instruments, their size and the demand for electronics during the late 1960s made it impossible to meet the avalanche of requirements in terms of the radio telemetry systems' data transmission capacity, their weight, size, and reliability.

The new designs of the super heavy rockets Proton and N1, and of new spacecraft, required a data transmission capacity of up to 100,000 measurements per second. After our specialists met with Americans during the implementation of the Apollo-Soyuz program and had learned about the Space Shuttle project, there was talk of systems that had a data transmission capacity of up to 1,000,000 measurements per second.

The revolution in electronics demanded brand-new, enormous capital investments in new production technologies. The transistors that had replaced vacuum tubes turned out to be just the beginning of an historic electronic revolution for humankind. Single-body transistors were followed by integrated circuits and very-large-scale integrated circuits; in lieu of being wired, instruments were now produced from printed circuit boards. The brand new mass production of super-miniature components was emerging.

In the 1970s, an event took place that brought significant changes to all areas of radio electronics, without exception: the microprocessor was born. The age of digital electronics had begun. Developers of radio electronics systems for rocket-space technology, competing in terms of basic performance indices, were eager to make broad use of integrated circuits. Unfortunately, our electronics industry had fallen far behind the U.S. in terms of rates of integration and methods for manufacturing micro- and hybrid circuits. The U.S. mastered the mass production of highly integrated circuits, mainly for consumer goods: televisions, communications devices, and portable computers. Elements that showed the greatest degree of reliability in consumer goods were used in military programs. For us it was just the opposite. The most cutting-edge developments were classified and used above all for defense purposes. The result of this was that sometimes military acceptance cleared no more than 5 to 10 percent of the total number of manufactured integrated circuits for use. Manufacturers encountered great difficulties in obtaining permission to freely sell to the civilian sector the micro-electronics parts that the military had rejected.

Though belatedly, the revolutionary process in technology took hold of our design bureaus and factories one after the other. Instrument designs changed radically thanks to the development of film and ceramic media for chips. Large and super-large fully-digital multilayer hybrid circuits came on the scene. Instead of being wired, extremely complex circuits were "printed" on multilayer boards that provided a high degree of packing density that couldn't even have been anticipated in the 1960s. In the production of integrated circuits, integration lines three to four microns wide became common, and in prototype models they were as small as one to two microns. Electron-beam lithography became a primary factor in the manufacture of photo templates with micron-linewidth for the manufacture of large-scale integrated circuits. The newest advances in microelectronics technology were classified in the Soviet Union. The various branches of the defense industry were not able to order that same quantity of components, whose production in the United States had ensured a high degree of reliability. This was the Achilles heel of our radio electronics industry.

Chapter 11
Star Wars

Young people of the early 21st century (in Russia) associate the concept of "Star Wars" with televised science fiction "thrillers" in which a superhero actor, with the help of a beautiful starlet, saves earthly civilization from an extraterrestrial invasion. For this, usually, they used a wonder weapon that was a modification of Engineer Garin's "death ray," the brainchild of writer Aleksey Tolstoy back in the 1920s.[1] The concept and term "Star Wars" appeared in the 1960s outside the realm of science fiction. The U.S. mass media referred to the scientific, technical, political, and diplomatic competition between the Soviet Union and the United States to produce space-based missile-defense systems (RKO) as "Star Wars."[2]

Air defense (PVO) or anti-aircraft weapons appeared in the early 20th century, and during World War II they were of vital strategic importance.[3] By the 1960s, using the achievements of missile, radar, and electronic technology, domestic air defense had achieved outstanding results. Beginning in 1955, if a massive bomber attack had taken place from any direction, not one of them would have managed to drop a bomb on Moscow or Leningrad. However, air-defense systems were powerless against ballistic missiles flying through space. Powerful nuclear-tipped missiles as an independent branch of the strategic forces appeared in the 1960s.

---

1. Aleksey Nikolayevich Tolstoy (1883–1945) was an influential Soviet science fiction writer whose works included *Aelita: Sunset on Mars* (1923) and *The Hyperboloid of Engineer Garin* (1927).

2. RKO—*Raketno-kosmicheskaya oborona*—literally stands for "missile-space defense." Chertok's elaboration of the "Star Wars" concept as being from the 1960s is somewhat anachronistic and misapplied. The phrase "Star Wars" as a reference to space-based warfare came into usage only after U.S. President Ronald Reagan's famous speech of 23 March 1983 when he proposed the Strategic Defense Initiative (SDI), i.e., a national missile defense shield to protect against Soviet nuclear attack. Senator Edward M. Kennedy (D-Mass) was probably the first major public figure to refer to Reagan's proposal as "Star Wars," an allusion to the enormously popular science fiction movie released in 1977. Since then, "Star Wars" has been typically used as a pejorative term to describe space-based missile defense systems, not as a more general descriptor for military competition in space between the United States and its opponents.

3. PVO—*Protivovozdushnaya oborona*—literally stands for "anti-air defense."

Scientific schools established during the development of anti-aircraft missile systems were the basis for the development of space-based missile-defense systems. The solution of missile-defense problems required the creation of three different systems: a space monitoring system (KKP), a missile attack early warning system (SRPN), and even a space defense system (PKO)—the latter to defend against an all-out attack, not by missiles, but by potential enemy spacecraft.[4] All of these systems were supposed to be developed regardless of the type of physical principle used to destroy ballistic missiles launched from the ground or from submarines.

Wide-ranging missile-defense activity was under way in parallel with larger projects for the creation of a "nuclear missile shield" and diverse rocket-space technology. Projects dealing with all types of missile defense were more highly classified and secretive than everything associated with strategic missile technology and cosmonautics. Many thousands of teams were involved in producing the most sophisticated military technology systems and putting them into service. Truly outstanding results were achieved in radar, computer, optical, and laser technology, as well as in programming, data transmission, and processing. Despite every imaginable international agreement between the United States and Soviet Union calling for the reduction of activity on missile-defense systems, work continued in various areas on anti-ballistic missiles, their control systems, and special space-based surveillance and monitoring technology. During his presidency, Ronald Reagan took the initiative, replacing the concept of "Star Wars" with his Strategic Defense Initiative (SDI).[5] This did not change the substance of the matter.

I was not directly involved with the development of air-defense and anti-ballistic missile systems. I do not pretend to pass on to future generations the invaluable experience of the creators of these systems. Their path was no easier, and was at times more difficult than the one we were on as creators of rocket-space technology and the notorious "nuclear missile shield." I recommend that those who are interested in the heroic story of the creation of modern air defense and missile-defense systems look into the following publicly available works written by individuals directly involved in the development of air-defense and missile-defense systems: K. S. Alperovich's

---

4. KKP—*Kontrol kosmicheskogo prostranstva;* SRPN—*Sistema rannego preduprezhdeniya o raketnom napadenii;* PKO—*Protivokosmicheskaya oborona.*

5. Here again, Chertok misinterprets the root of the "Star Wars" term. Space-based defensive weapons had been conceived by military strategists since the 1950s although there had been no major American initiative on the topic until Reagan's 1983 speech. Soon after, the phrase "Star Wars" came into common usage as a descriptor for space-based missile defense.

How a New Weapon Was Born, G. V. Kisunko's *Secret Zone*, and military journalist and writer Mikhail Pervov's major work, *How Space-Based Defense Systems were Created.*[6]

And nevertheless, I thought it necessary to include a "Star Wars" chapter in my work. Why? I was directly involved with the establishment of rocket-space science and technology in the Soviet Union. Casting aside the remnants of modesty, I shall venture to include myself among its founding fathers. With regard to air-defense and missile-defense systems, I consider myself to be an eyewitness, whose testimony might be of interest as far as the history of the times and the personality traits of the leaders who courageously took responsibility for solving what was seemingly impossible.

I shall begin with evidence of "complicity."

FROM 1940 TO 1945, I WORKED AT AVIATION FACTORY NO. 293, which should be considered the alma mater of air defense jet fighter planes. I wrote about the development of the BI aircraft, its creators, and my own involvement in that project in Volume I of *Rockets and People.*[7]

Isayev, Bereznyak, Mishin, Melnikov, Bushuyev, and I—future Heroes of Socialist Labor and winners of numerous awards were graduates of Bolkhovitinov's "school"—Factory No. 293, located on a small piece of land in Khimki. This piece of land occupied by Factory No. 293 suffered through a period of obscurity while awaiting the advent of the missile age. In 1953, Petr Dmitriyevich Grushin arrived at the factory to serve as chief and chief designer.

On 20 November 1953, Special Design Bureau No. 2 (OKB-2) was set up using the facilities of the pilot production plant of Factory No. 293. Incomprehensibly, for a short time, it became part of the atomic Ministry of Medium Machine Building. Petr Dmitriyevich Grushin was appointed chief and chief designer of OKB-2. After many changes in affiliation to various high-ranking government agencies, Factory No. 293 came to be referred to as the Fakel Machine Building Design Bureau (MKB).[8] From 1953 through 1991, a period of 38 years (!), the chief and general designer

6. K. S. Alperovich, *Tak rozhdalos novoye oruzhiye: zapiski o zenitnykh raketnykh kompleksakh i ikh sozdatelyakh* [*How a New Weapon Was Born: Notes on Anti-Aircraft Missile Complexes and Their Creators*] (Moscow: OAO 'TsKB Almaz,' 1999); G. V. Kisunko, *Sekretnaya zona: ispoved generalnogo konstruktora* [*Secret Zone: Confessions of a General Designer*] (Moscow: Sovremennik, 1996); Mikhail Pervov, *Sistemy raketno-kosmicheskoy oborony sozdavalis tak* [*How Space-Based Defense Systems were Created*] (Moscow: Aviarus-XXI, 2004).

7. Chertok, *Rockets and People, Vol. I*, pp. 160–163, 173–178, 187–199.

8. MKB—*Mashinostroitelnoye konstruktorskoye byuro.*

of Fakel was future academician and twice-decorated Hero of Socialist Labor Petr Dmitriyevich Grushin.[9]

OKB-2's first anti-aircraft missile was put into service for the Air Defense Troops in 1957. OKB-2 (Isayev's, Mishin's, and my alma mater) was awarded the Order of Lenin for producing this family of missiles. Grushin was awarded the title Hero of Socialist Labor, and in 1958 Nikita Khrushchev and Leonid Brezhnev both came to the factory to present him the award. Grushin's 1D (or V-750) missile was the first in the world to strike an actual enemy aircraft, the U.S. Lockheed U-2 reconnaissance aircraft piloted by Francis Gary Powers. He was shot down near Sverdlovsk on 1 May 1960. While giving Grushin his due, it bears mentioning that his missiles were part of the larger S-75 system, the general designer of which was Aleksandr Andreyevich Raspletin.[10] Both Grushin and Raspletin became academicians that year. Having interacted with them in various academic and semi-academic situations, I must say that they were distinguished, extremely talented engineers, gifted with organizational skills and trailblazing enthusiasm.

Grushin's missiles in Raspletin's S-75 system destroyed hundreds of American airplanes during the Vietnam War. They were an important factor in the U.S. withdrawal from Vietnam.

In 1961, a unique problem was solved for the first time: Grushin's V-1000 missile, guided by Kisunko's radar and computer system, struck the warhead of an R-12 ballistic missile launched from Kapustin Yar.[11]

Not long before Gagarin's launch, after a hot day at the firing range in Tyura-Tam as we were taking a stroll, Isayev remarked apropos Grushin's missiles: "Just think, those Americans were shot down by missiles that were developed at the place where we first began to understand what a liquid-propellant engine was." Isayev had every right to be proud. Grushin's missiles,

---

9. Although little known among Western space historians, Petr Dmitriyevich Grushin (1906–93) was one of the most influential missile designers in the Soviet Union during the Cold War. Because he oversaw the development of several generations of air defense missiles, Grushin was promoted in 1966 to membership in the Central Committee of the Communist Party, a rare honor bestowed not even to Korolev during his life.

10. The S-75 system (known in the West as the SA-2 Guideline) was officially declared operational on 11 December 1957. Aleksandr Andreyevich Raspletin (1908–67) served as a chief designer (1953–63) and then general designer (1963–67) of KB-1. He, along with Grushin, was a giant in the Soviet air defense missile sector.

11. The V-1000 missile was part of System A, the first experimental anti-ballistic missile system developed by the Soviet Union. Grigoriy Vasilyevich Kisunko (1918–98) was the "patriarch" of the Soviet anti-ballistic missile program. In the 1960s and early 1970s, he was chief designer at OKB-30, which later became the heart of the large Vympel Central Scientific-Production Association (TsNPO Vympel) that conducted research on anti-ballistic missile and early warning systems during the Soviet era.

which had downed American airplanes, had Isayev's engines. Grushin passed away in 1993. He can be credited with producing more than 20 models of anti-aircraft and anti-ballistic missiles.

Rafail Borisovich Vannikov, whom I met in Germany, was the most actively involved in the development of anti-ballistic missiles. At the rank of major, he was one of the first organizers of the Special Purpose Rocket Brigade.[12] Beginning in 1947, Vannikov, the NII-88 senior military representative, was in charge of the military acceptance of missiles in Podlipki, and in 1953, he was appointed regional engineer at OKB-2 and managed the acceptance of anti-aircraft and anti-ballistic missiles.

Now, retired Colonel Rafail Vannikov, Deputy Chief Designer of MKB Fakel, never fails to remind me about all the rocket-space anniversary dates and continually extends invitations to visit the former Factory No. 293. But I keep putting it off. Alas, with age a loss of curiosity is already showing.

The contribution of the Moscow suburb Khimki to missile-defense systems was not limited to the development of anti-ballistic missiles by Grushin's firm. Georgiy Nikolayevich Babakin, who at my suggestion in 1949 began working on control systems for surface-to-air guided missiles, began developing, in 1967, the US-K spacecraft for the missile attack early warning system as the chief designer of the OKB at the Lavochkin Factory. The first experimental spacecraft produced at Lavochkin Scientific-Production Association (NPO Lavochkin) for the developmental testing of the US-K early warning system, *Kosmos-520*, was not inserted into a high elliptical orbit until 19 September 1972.[13] Academy of Sciences Corresponding Member and Hero of Socialist Labor Babakin died suddenly a year-and-a-half before this event.[14]

Sergey Kryukov, the former deputy of Korolev and later of Mishin, became the chief of the Lavochkin OKB, responsible for developing the spacecraft for the missile early warning system. For seven years Kryukov managed the factory's broad spectrum of space-related projects. The Lavochkin Factory became NPO Lavochkin in 1974. Kryukov's departure from Korolev's old OKB-1 was not a career move. Differences of opinion with Chief Designer Mishin made it impossible for him to continue working there. In 1978, Kryukov left the

---

12. Rafail Borisovich Vannikov (1922–) is the son of Boris Lvovich Vannikov (1897–1962), one of the top managers of the Soviet nuclear weapons industry in the postwar era.

13. NPO—*Nauchno-proizvodstvennoye obyedineniye*. Scientific-Production Associations were industrial conglomerates that were introduced into the Soviet defense industry in the early 1970s as a means to consolidate R&D and series production under one roof. In the case of Lavochkin, for example, in 1974, its design bureau was attached to its factory to form NPO Lavochkin.

14. Babakin died on 3 August 1971.

anti-ballistic missile field and returned to missile projects. On 1 August 2005, leaning on my cane, I plodded home from his funeral at Ostankino Cemetery to Academician Korolev Street. It wasn't so easy to arrange with the Moscow cemetery authorities to set aside a patch of ground in the Ostankino Cemetery for Sergey Kryukov's burial. Of course, this was beyond his family's power and pocketbook. His former organization, now named for S. P. Korolev, helped.

A GREAT DEAL HAS BEEN WRITTEN AND SAID ABOUT THE DEVELOPMENT OF SOVIET ATOMIC WEAPONRY AND MISSILE AND SPACE TECHNOLOGY. Considerably less is known about the work of our scientists in the field of defending against air and missile attack. In this regard, the development of Moscow's unique air defense system, which was unequalled in the world at that time, is very illustrative. I am trying to fill this gap with brief episodes from this great story.

In the spring of 1953, the Korolevian Council of Chief Designers and the entire body (which was small by today's standards) of the missile elite had gathered at the State Central Firing Range (GTsP) in Kapustin Yar.[15] We were conducting flight tests on R-5 and R-11 missiles. We considered everything that happened in the area from Kapustin Yar to Vladimirovka, which had a major Air Force NII located nearby, to be the sphere of interest of the missile industry and the Ministry of Defense.

Colonel-General Vasiliy Ivanovich Voznyuk was the generally recognized and actual boss of the firing range and all the "surrounding areas." Contending with his hard-nosed personality was an issue not just for the servicemen directly subordinate to him, but also for the chief designers, including Korolev. The staff of testers, especially the launch control team, consisted almost equally of military and civilian specialists. Korolev was very particular when it came to assigning these staffers to the most responsible positions and operations. This sometimes led to conflicts with Voznyuk, who justly believed that the placement of military personnel was his prerogative, and certainly not that of civilian chief designers, even if it was a chief such as Korolev.

This time it became apparent that the reshuffling among military personnel took place against, rather than at, the will of the all-powerful Voznyuk. One major point of contention was when Major Yakov Tregub, the officer in charge of the launch control team, was transferred from the tight-knit group of testers to the Third Main Directorate (TGU) of the Council of Ministers.[16]

---

15. GTsP—*Gosudarstvennyy tsentralnyy poligon.*

16. The TGU—*Tretye glavnoye upravleniye* (Third Main Directorate) was a special top secret organization established in 1951 to manage the Moscow air defense project known as Berkut (or S-25).

Back in 1952, during one of the meetings of the Council of Chiefs, Korolev had complained: "Our former chiefs are in charge at the Third Main Directorate now. They know everybody like the back of their hands and will lure away good people."

I did not learn what the Third Main Directorate was until later, but I will tell the reader the history of how it came to be now.[17] In 1948, Stalin personally demanded that reliable air defense be set up over Moscow using up-and-coming anti-aircraft systems. He gave supervisory responsibility over all the projects involving this problem to Lavrentiy Beriya, who had already proven himself to be an excellent project organizer in the nuclear field. The First Main Directorate (PGU) carried out the nuts-and-bolts organization of the work for the development of atomic weaponry.[18] Boris Lvovich Vannikov, who reported directly to Beriya, was in charge of the PGU.

Stalin signed a decree of the USSR Council of Ministers (in August 1950) calling for the development of an anti-aircraft missile system to protect Moscow against an air attack with the resolution that "We must have an air defense missile within a year." The decree spelled out the main requirements stating that the system must be capable of simultaneously destroying up to 1,000 aircraft flying from different directions. Placed in charge of developing the enormous and highly complex system was KB-1, the chief designers of which were Professor Pavel Kuksenko and Lavrentiy Beriya's son Sergey.

A new governmental agency, dubbed the Third Main Directorate (TGU) of the USSR Council of Ministers was created to manage the development of the system. The TGU was set up using an organizational structure that had already been used for the atomic project. Ustinov's former deputy, Vasiliy Ryabikov, was appointed TGU chief. Lavrentiy Beriya instructed Chief of the First Main Directorate Vannikov to render Ryabikov all the help he could—first and foremost, with staffing. Vetoshkin was named Ryabikov's deputy. Both had a very good idea of which of the missile specialists, both civilian and military, it made sense to recruit for the new assignment.

Before his transfer to TGU, Vetoshkin had been in charge of the Seventh Main Directorate of the Ministry of Armaments, i.e., the ministry's missile directorate. We were all annoyed that we had been deprived of two good bosses (Ryabikov and Vetoshkin). Not only had both been competent specialists, but they had been very decent, kind human beings and wise managers. It is extremely rare that one encounters people of integrity among high-ranking officials.

---

17. Also see Chertok, *Rockets and People, Vol. II*, pp. 208–209.
18. PGU—*Pervoye glavnoye upravleniye.*

Now we found out that on Vetoshkin's advice, Major Tregub had been promoted to the rank of lieutenant colonel and transferred to TGU as firing range chief engineer for tests on anti-aircraft missile complexes. This firing range was located in the vicinity of Kapustin Yar. Military servicemen at this new firing range were not subordinate to Voznyuk and outside the sphere of authority of the Ministry of Defense.

We used our own "home" airfield, *Konstitutsiya* (Constitution), to fly into and out of Kapustin Yar. No one could ever explain how it got that name. Now this small airfield on the steppe had become overloaded, serving not just us, but also the anti-aircraft missile firing range that the top brass frequented.

During April and May 1953, the first full-scale *Berkut* (Golden Eagle) anti-aircraft missile complex (soon thereafter renamed System-25 or S-25) was tested at the new firing range. The system, which included V-300 missiles, was designed to create an impenetrable missile shield protecting Moscow against an attack of aircraft from any direction.

For firing tests against real targets, two Tu-4 airplanes (exact copies of the U.S. B-29 Superfortress) took off from the Vladimirovka airfield. One airplane was supposed to be the target, while the other went along as an escort. When the aircraft reached the turning point for its bombing run, the crew of the target aircraft parachuted out of the airplane leaving it on autopilot. The escort aircraft reported that "The crew has bailed out of the target" and departed from the path of the bombing run. In all, five airplanes were fired on and brought down by missiles. From our firing range, I observed this operation, which involved the destruction of the best bombers of World War II, with mixed emotions. On the one hand, I felt sorry for those good airplanes, and on the other, I was pleased with the synthesis of radio engineering, automatics, and missile technology achievements. I couldn't help recalling our naive attempts at solving a similar problem during the war using makeshift means.

After Lavrentiy Beriya's arrest, his son Sergey was barred from working at KB-1. Aleksandr Raspletin was named chief designer of the S-25 system. Semyon Lavochkin was the chief designer of the anti-aircraft guided missile for this system. I wrote about my first acquaintance with the V-300 missile in an earlier volume.[19] At that time, I did not yet know that the system that included Lavochkin's missile was called *Berkut*, from the initial syllables of its chief designers Beriya and Kuksenko. That is why the *Berkut* system was renamed S-25 after Beriya's arrest.

---

19. See Chertok, *Rockets and People, Vol. II*, Chapter 11.

Military missile specialist Yakov Tregub was appointed chief engineer of the test firing range; Petr Kirillov was chief designer of the missile's autopilot; Vladimir Barmin was chief designer of the ground-based launcher; and TGU deputy chiefs Valeriy Kalmykov and Sergey Vetoshkin were the general leaders of the testing process.[20] Ryabikov and Vetoshkin dropped in on us at the firing range as guests and together enjoyed observing one of the R-5 launches. Vetoshkin was in excellent spirits and told us that even before tests were concluded on the *Berkut*, the construction of two missile-defense rings around Moscow would be completed.

Stalin's assignment calling for the creation of an impenetrable air defense system around Moscow was fulfilled.[21] Under Cold War conditions, a system had been created that was extremely powerful for those times and had unique combat performance characteristics unequaled in the world. Aleksandr Raspletin deserves a great deal of credit for this. Sixty-six multi-channel anti-aircraft missile systems were deployed in two rings placed 48 and 90 kilometers from Moscow. Each system could fire 20 missiles against 20 targets. Target acquisition radars located the airplanes and were subsequently switched to target tracking and anti-aircraft missile control mode. In the vicinity of each missile emplacement were missile preparation facilities, residential areas, and all the necessary auxiliary facilities.

Moscow's air defense anti-aircraft missile systems were interconnected by two concrete ring roads. Access to these roads was permitted only with special passes. Forbidden fruit is always sweet. Korolev instructed his deputy for security, KGB Colonel Grigoriy Mikhaylovich Yakovenko, to provide the managerial staff of his OKB-1 with passes to these forbidden roads. Thus, we obtained access via excellent roads to the best forests for mushroom-picking in the area surrounding Moscow.

System-25 was in service for more than 30 years. The potential capabilities built into the S-25 made it possible within a brief period of time to develop the S-75 and S-125 anti-aircraft missile systems. Chief Designer Petr Grushin's OKB developed missiles for these and many subsequent air defense systems, and then for anti-ballistic missile systems as well. On 1 May 1960 near Sverdlovsk, the S-75 system shot down the high-altitude U-2 reconnaissance airplane piloted by Gary Powers. The Americans had considered this aircraft

---

20. *Author's footnote*: At that time, by agreement with Ustinov, all operations at NII-88 dealing with the reproduction of German *Wasserfall* anti-aircraft missiles were halted and a team of control system developers headed by Babakin was transferred from NII-88 to Lavochkin's OKB-301.

21. The first S-25 systems were put on service duty in June 1956.

to be invulnerable. During the Vietnam War, S-75 systems posed a threat to the Americans' newest B-52 bombers.

The successes of microelectronics and computer technology and the development of the theory and design of phased antenna arrays made it possible to solve problems on a new technical level for the development of effective anti-aircraft missile systems. A new generation of systems appeared in the 1980s. The S-300 (or SA-10 Grumble) and S-300PMU (or SA-20 Gargoyle) anti-aircraft missile systems were demonstrated at exhibitions of the latest weapons systems. The chief designers of these systems had long ago abandoned the liquid-propellant rocket engine and switched to using only solid-propellant missiles.

System-25 missiles had a range of just 25 kilometers and could shoot down airplanes at altitudes no greater than 18 kilometers. S-300 systems have a kill zone as great as 100 to 150 kilometers. Within this zone, they are capable not only of shooting down aircraft, but they also can strike ballistic missiles during the descent portion of their trajectory. The historical irony of the super secret S-300 system is that, after proving itself to be one of the best such systems in the world, it has now gone on sale on the international open market. The Russian Ministry of Defense has no funds to pay for the further development of such costly systems. The creative teams established by Academicians Raspletin, Bunkin, and Mints, which were unique in terms of their intellectual potential, are now struggling to survive.[22]

A German company that deals in plants and flowers, toilets, construction materials, and kitchen appliances is comfortably ensconced in the once top-secret, legendary building of KB-1 located at the junction of the Leningrad and Volokolamsk highways. The signboards of banks and some corporations adorn the main entrance on the other side. Companies from those same rich and prosperous nations with whom we waged the Cold War for so many years now lease thousands of square meters here. The "new Russians," acquiring custom fixtures for their bathrooms in these once top secret premises, have no idea what their fellow countrymen created within these walls several years ago.[23]

When I became familiar with the technology of System-25 and later with its modifications, I couldn't help recalling wartime projects such as Germany's

---

22. Boris Vasilyevich Bunkin (1922–) succeeded Raspletin at KB-1, where he served as general designer until 1998. KB-1 is now known as NPO Almaz Named After Academician A. A. Raspletin. Aleksandr Lvovich Mints (1895–1974) was directly involved in a variety of air defense and missile defense projects in the 1950s and 1960s. He officially headed the Radio Engineering Institute of the Academy of Sciences from 1957 to 1970.

23. Here, "new Russians" is a pejorative reference to the *nouveau riche* who gained wealth despite the near-economic collapse of the Russian economy in the 1990s.

attempt to develop an aviation-impermeable missile defense using *Wasserfall* missiles. It took humankind 2,000 years to go from a wooden shield covered with leather to the concept of *Wasserfall*. And to go from the *Wasserfall* concept, which was still imaginary in 1945, to System-25—an actual missile shield protecting an enormous city—took just 10 years.

A missile shield against aircraft had been created for Moscow. Now the time had come to protect the nation, or at least the capital, against *ballistic* missiles. The anti-aircraft system had to be supplemented with an anti-missile missile system. Surface-to-air missile systems had to be modified to practically solve the problem of a "missile-to-missile" hit. At first, this seemed completely unrealistic, even to air defense specialists. But the Cold War allowed scientists to be presented with tasks that, until recently, had seemed fantastic.

During the "hot war" years, it had been impractical to propose to heads of state inventions or ideas that required several years to be realized. During the Cold War, it was possible to "relax" and allow scientists to work on projects with a production cycle of five to seven years or more. Usually, to begin with, the producers promised it would take three to four years, and then they requested at least two more years. Two years later, it became clear that if funding were increased two- to threefold, then the promised problem would be solved in five years. Thus, an 8 to 10-year production cycle for large complex systems was gradually established.

It was the General Staff that took the initiative for the development of a missile-defense system. Rather than turning to scientists, seven Marshals turned to the Central Committee of the VKP(b).[24] Upon receiving the Marshals' letter, the Central Committee assigned to the very same KB-1 to develop a draft design of the missile-defense system.

In 1955, at KB-1, a division was created to work on the missile-defense system. Grigoriy Vasilyevich Kisunko, a 36-year-old doctor of science and electrodynamics specialist, was appointed division head. He began by setting up proof-of-concept experiments. They were necessary to prove that, at distances of 1,000 kilometers and more, it was possible to "see" a target—a warhead that had separated from a missile—and distinguish it from the missile hull.

I met future missile-defense system Chief Designer Grigoriy Kisunko for the first time in Kapustin Yar. He flew out to the firing range in 1954 to familiarize himself with the warhead container of the R-5 missile as a radar

---

24. VKP(b)—*Vsesoyuznaya kommunisticheskaya partiya (bolshevikov)* (All-Union Communist Party [Bolsheviks])—was the official name of the Communist Party during the early years of the Soviet era. In later years, the phrase was shortened to simply the Communist Party of the Soviet Union.

search object. Korolev assigned Voskresenskiy to deal with Kisunko, and he told Kisunko that radar issues were my purview.

On 2 February 1956, full-scale tests were conducted on our R-5M missile carrying a real nuclear warhead.[25] Now the opposite problem was posed. They needed to prove that a missile warhead could be struck and knocked out of action with high-velocity fragments from an anti-ballistic missile, which would explode in its path before or during entry into the atmosphere. Our Factory No. 88 manufactured R-5M warhead containers ordered by Samvel Kocheryants, who was chief of KB-11 at Arzamas-16 and chief designer of the nuclear device.

The experiments were conducted under the supervision of Academician Yuliy Khariton. Khariton was surprised that it wasn't so easy to damage an atomic bomb with fragments. The hull of the R-5M missile warhead container proved to be surprisingly sturdy. It was necessary to develop a more effective warhead for the anti-missile. The atomic scientists proposed testing our warhead container at the nuclear firing range near Semipalatinsk. It was interesting to test the effects of the shock wave of a nuclear explosion. What we had in mind was a scenario on the principle that "one atomic projectile needs to hit another atomic projectile."

A Council of Ministers decree calling for the development of a missile-defense system came out on 17 August 1956, a month after the first domestic strategic nuclear missile went into service![26] In our circle, Korolev was the first to find out about top secret ("special file") decrees being issued. Over lunch in the small cafeteria where we tried to learn from him the latest news that we were not officially supposed to know, he leaked the following: "I'm afraid that these boys are going to fall flat on their faces. Sergey Ivanovich was complaining that out there at Sokol they haven't come up with the kind of tight-knit team that we have."

"Sergey Ivanovich" meant Vetoshkin, and "Sokol" (Falcon) meant KB-1, which Korolev referred to by the name of the nearest metro station. The station, in turn, had gotten the name Sokol from the small pre-war town in this area, half of which was occupied by dachas.[27]

The intensive development of anti-ballistic missiles in the USSR began concurrently with the development of offensive strategic missiles, i.e., missiles armed with nuclear warheads. In this field, we were ahead of the Americans.

---

25. See Chertok, *Rockets and People, Vol. II*, Chapter 15.

26. The R-5M nuclear-capable intermediate range ballistic missile system was officially declared operational on 21 June 1956.

27. The region was originally used for falcon hunting.

The team located in Khimki on the premises of what had once been Factory No. 293 developed the first V-1000 anti-ballistic missile. As I have mentioned earlier, I began working at aviation Factory No. 293 in 1940. The factory's chief designer at that time was Viktor Bolkhovitinov. Now the chief designer of the first anti-ballistic missiles here was Petr Grushin, the former deputy of Semyon Lavochkin. He was involved in the development of the anti-aircraft missiles for System-25 and System-75. The great amount of experience from developing anti-aircraft guided missiles was used in designing anti-ballistic missiles. There was a single Chief Designer—Aleksey Isayev—for the engines of all of the following: anti-aircraft guided missiles, the new anti-ballistic missiles, our R-11 short-range ballistic missiles, and the submarine-launched strategic ballistic missiles.

In 1962, we got together once again, celebrating the 20th anniversary of Bakhchivandzhi's first flight in the BI-1.[28]

He asked me, "Do you remember how Sasha Bereznyak and I talked you into developing a search and guidance system for our plywood BI to use against German bombers back in 1941? What a bunch of naive blockheads we were! Back then we thought that a guidance system was a trifling problem— that the main thing was the engine. Now, when conferences are held on this new 'System A,' they don't even remember me, my engine, and even Grushin's missile—it turns out to be a minor detail. The crux of the matter is the enormous guidance system."

With his typical razor-sharp wit, Isayev very passionately told me about his participation in various conferences and scientific councils on the missile-defense problem: "When it comes to the guidance and control system—or in essence, the whole eternal problem of interception—passions always flare up. Sometimes it seems to me that the anti-ballistic missile itself simply gets in their way. The main issues here are over-the-horizon radar, computing the enemy missile's trajectory, and prolonging our own missile's trajectory. If you could see what they're building at Sary-Shagan!"[29]

Indeed, the interception and destruction of a ballistic missile's warhead by an anti-ballistic missile was an extremely complex problem. It required the creation of a special organization to design and test a system that included a ballistic target early warning radar; an anti-ballistic missile pinpoint navigation radar; an anti-ballistic missile control radar; stations to transmit

---

28. Grigoriy Yakovlevich Bakhchivandzhi (1909–43) conducted the first flight of the Soviet BI-1 rocket-plane on 15 May 1942. See Chertok, *Rockets and People, Vol. I*, Chapter 13.

29. Sary-Shagan was the location of the anti-ballistic missile testing grounds, later renamed the Scientific-Research and Test-Firing Range No. 10 (NIIP-10).

control commands to the anti-ballistic missile; a command-computer center; radio relay communication lines between system facilities; and, finally, the anti-ballistic missile launch facilities.

Our nation's first experimental system was called System A. The design, production, and testing of this system was conducted under the management of General Designer Kisunko. The chief designer of the ground-based launcher was Vladimir Barmin. Radar systems were produced jointly with Academician Aleksandr Mints' Radio Engineering Institute, and one of the first Soviet computers—the system's electronic·brain—was developed at Academician Sergey Lebedev's institute.[30]

On 4 March 1961, for the first time in the world, System A intercepted and destroyed the payload of one of Yangel's R-12 missiles. Subsequently, V-1000 anti-ballistic missiles were launched against the payloads of R-5M and R-12 missiles. In so doing, we verified the destruction reliability of missile warheads. A modified V-1000 anti-ballistic missile with a heat-seeking homing head and on-board computer was developed to increase accuracy and effectiveness. Full-scale tests showed that it was possible, in principle, to intercept and destroy ballistic missile warheads during the descent segment of their trajectory. System A had great scientific, technical, and strategic political significance. It cleared the way for the production of missile-defense systems and, most of all, for the creation of a missile shield to protect Moscow, much like the earlier System-25.

But, oh, what a difficult journey it would be. In his previously mentioned memoirs, Corresponding Member of the Russian Academy of Sciences G. V. Kisunko describes some of the peripeteia of this grandiose project, which our Cold War opponents subsequently rated very highly. The author of *The Secret Zone* gives a very harsh and subjective assessment of the actions of certain scientists and leaders. I cannot agree with his descriptions of Raspletin, Mints, Kalmykov, and Chelomey. Nevertheless, *Confessions of a General Designer* (as it was subtitled) confirms that in the early 1960s, we were ahead of our Cold War opponents in realizing the principles of a missile-defense system.

Developments in air- and missile -defense systems in the first decade of the Cold War were taking place at such a rapid rate that in this radio engineering field we were leaving the Americans in the dust. We had in fact been the first to begin implementing a program in the 1950s that the Americans later

---

30. This institute was the Institute of Precision Mechanics and Computing Technology (*Institut tochnoy mekhaniki i vychislitelnoy tekhniki*—ITMiVT).

hailed as the SDI in the 1980s.[31] We had examined many of the ideas widely touted in the Americans' SDI program quite some time before. It is illustrative that the Americans did not succeed in repeating our attempt at destroying a ballistic missile warhead with an anti-ballistic missile until 1984, 23 years after the Kisunko-Grushin experiment![32]

However, on no account should we underestimate those American accomplishments which used new scientific principles for systems which were part of the extremely broad concept of SDI. There was a reason, after all, why the American propaganda machine referred to SDI as the "Star Wars" program.

The anti-ballistic missile boom started in the U.S. after the first Soviet satellites appeared. The first American missile-defense project was the Nike-Zeus system. It was supposed to involve approximately the same set of elements as our analogous system (System A). However, a fundamental difference from the Soviet system was the use of a three-stage solid-propellant Nike-Zeus missile with a thermonuclear warhead. The use of an anti-ballistic missile with a "fashionable" thermonuclear warhead substantially reduced the demand for guidance precision. It was assumed that the effects of the nuclear explosion of the anti-ballistic missile would neutralize the warhead of a ballistic missile even if it was 2 to 3 kilometers away from the epicenter. In 1962, to determine the effects of a nuclear burst, the Americans conducted a series of high-altitude nuclear tests, but soon thereafter work on the Nike-Zeus system was halted.

However, in 1963 development began on the next generation missile-defense system—the Nike X. It was necessary to develop a missile-defense system that was capable of protecting an entire region, rather than a single installation, against Soviet missiles. The Spartan missile, with a range of 650 kilometers and a 1-megaton-yield nuclear warhead, was developed to destroy enemy warheads while on long-range approaches.[33] A warhead with such an

---

31. The Russian abbreviation for SDI is SOI—*Strategicheskaya oboronnaya initsiativa* (Strategic Defense Initiative).

32. The first U.S. intercept was a year after the Soviet one. The Cold War-era U.S. anti-ballistic missile effort encompassed a number of successive programs beginning the mid-1950s with the Nike-Zeus system. The system (using a Nike-Zeus B missile) was tested on 19 July 1962 with a successful intercept of an inert target launched on an Atlas D from Vandenberg Air Force Base. Nike-Zeus was succeeded by the Nike X system whose missiles (Spartan and Sprint) were tested a number of times during 1965–73. These missions included a successful intercept of a Minuteman reentry vehicle by a Spartan anti-ballistic missile on 28 August 1970.

33. The Spartan missile was basically an upgraded Nike-Zeus B missile and supported the exoatmospheric component of the Nike X system. The original model was designed for use with a single 5 megaton warhead and have a range of 740 kilometers. An improved version was projected to carry a 1 megaton warhead.

enormous yield would create a guaranteed intercept zone of several warheads and possible decoy targets in space. Testing of this anti-ballistic missile began in 1968 and continued for three years.

In case part of the enemy's missile warheads managed to penetrate the space protected by Spartan missiles, the missile-defense system contained shorter-range Sprint anti-ballistic missiles. These missiles were designed for use as the main instrument to protect a limited number of installations. It was supposed to hit targets at altitudes up to 50 kilometers.

Unlike their Soviet counterparts, American designs in the 1960s used powerful nuclear weapons installed on anti-missile missiles as a feasible way of destroying enemy warheads. The abundance of anti-ballistic missiles did not guarantee the protection of all the areas to be defended, and if they were used, the entire territory of the United States was threatened with radioactive contamination. By the late 1960s, the United States had deployed an experimental system based on the Thor missile. These missiles were also equipped with powerful thermonuclear warheads. It was assumed that in the event of a conflict, these anti-ballistic missiles would destroy Soviet reconnaissance satellites.[34]

Despite using the most state-of-the-art radio systems for detection, tracking, and guidance that were available at that time, all anti-ballistic missile weapons systems developed in the U.S. in the late 1960s were designed to cover only small areas. The interceptions and destructions of enemy ballistic missile warheads were supposed to take place during the descent trajectory at a relatively low altitude over U.S. territory.

Systems developed in the Soviet Union at that time were also able to detect and destroy warheads only during the final descent trajectory. The [subsequent] A-35 system was supposed to carry out these tasks. The system's general designer, Kisunko, had demonstrated the possibility of striking a target missile with an interceptor missile back in 1961. Kisunko's team separated from KB-1 and became the independent OKB-30. This organization was also supposed to protect Moscow against a nuclear missile attack much the same as System-25 had protected it against an air attack. However, this task was thousands of times more complex.

Chelomey came up with a sensational anti-ballistic missile project, the *Taran* (Battering Ram). The first time I, and most likely also Korolev's other

---

34. Chertok is probably referring here to the U.S. Air Force's Project 437 anti-satellite (ASAT) program which was declared operational in 1964. The system was tested 16 times between 1964 and 1970.

deputies, heard about *Taran* was when Korolev came back very agitated from one of the conferences that Khrushchev held rather frequently on various types of weaponry. Usually Korolev spoke respectfully about Khrushchev's conduct at such conferences, being careful not to joke or be critical toward him, even when he was not happy with the decisions that were made. This time Korolev was indignant that Khrushchev had approved Chelomey's proposals without any serious examination. In May 1963, a Central Committee and Council of Ministers decree was issued calling for the development of a pre-draft plan for the *Taran* system.[35] To assist General Designer Chelomey, Academician Mints had been appointed head of development of the pre-draft plan.

Korolev seethed as he told us this.

"Chelomey wants to be the missile-defense general designer. That's why he's proposing that they use his UR-100 missile. Great! But the system is supposed to be designed by Mints, and for missile defense, the system's the main thing! Is the missile for the system or is the system being designed around a ready-made missile? After all there's also General Designer Kisunko. He's making a system, and Grushin is making the missile for his system. That's how it should be! Vladimir Nikolayevich hasn't yet proven that his *Sotka* (UR-100) is the best intercontinental missile in the world, but now he's found another job for it. Don't you see, he's claiming that it can also be an anti-ballistic missile. If a 10-megaton warhead is put on it, and it's launched toward American warheads, then when they meet at the trajectory apogee, so as not to damage the eyesight of the Soviet people, 100 *Sotki* will make such a nuclear bonfire that supposedly not a single warhead will break through it.[36] It turns out that when all American missiles have decided to attack the Soviet Union, their trajectories pass through (as if by request!) a small area of space that is a convenient place to intercept them. Super-powerful nuclear charges detonated in this area are capable of destroying tens, and perhaps hundreds of warheads flying toward us along with all the decoy targets."

"I was talking with Keldysh," continued Korolev, "and his guys estimated that, given that the Americans are not the fools that Nikita Sergeyevich [Khrushchev] is being told they are, it will be necessary to use up at least 200 *Taran* anti-ballistic missiles at 10 megatons each to destroy 100 Minuteman warheads of 1 megaton each—a total nuclear irradiance of 2,000 megatons! But Chelomey has already solved the main problem. The first *Sotka* has yet to

---

35. The "pre-draft plan" (*avanproyekt*) in the Soviet R&D process preceded the preparation of a "draft plan" (*eskiznyy proyekt*) and involved the compilation of a set of detailed technical proposals outlining the design and operation of a particular weapons system.

36. *Sotki* is the plural of *Sotka*.

fly, and already plans need to be made to produce thousands! [This missile] is indeed 'universal'!"[37]

To calm S.P. down, I remarked that in the case of a nuclear missile attack, all problems could be solved a lot more simply than Chelomey, Mints, Kisunko, and the other anti-missile specialists suggest.

"Okay, Boris, let's hear it—what have you got?" S.P. coaxed.

"Armenian Radio is asked: 'What should one do in the event of a nuclear missile attack?' Armenian Radio answers: 'Quickly roll yourself up in a white sheet and slowly walk to the cemetery.' 'But why slowly?' 'So that you don't cause a panic'."[38]

My joke did not prompt the merry laughter that I expected. Korolev himself rarely told jokes, but he encouraged comrades who knew how to tell them. However, I noticed that during any conversations in which "megatons" were mentioned, his mood became somber and he didn't react to those jokes. When conversation turned to the prospects for nuclear missile war, many lost their sense of humor in Korolev's presence. This time Bushuyev commented on the situation citing the "top secret" analyses of atomic medical specialists, with whom he often met regarding piloted programs.

"So," he said, "they've calculated that if just one fifth of the 50 Titan II missiles that the Americans are putting on duty now break through to the Soviet Union, then a minimum of 35 million people will die!"

The *Taran* system caused quite a stir in the offices of ministers, the Central Committee's Defense Department, and in the General Staff. Despite the support of First Deputy Minister of Defense Marshal Andrey Grechko and Chief of General Staff Marshal Matvey Zakharov, the office of the Central Committee (including Ivan "The Terrible" Serbin) started to forget about it after 1964.[39] The matter did not go beyond a pre-draft plan.

In 1973, when Chelomey was no longer so all-powerful, at his initiative, as he described it, three "highly confidential" meetings took place in Reutov. Three people participated in the conversation that I am recounting: Chelomey, Bushuyev, and I. For some reason, Chelomey's first deputy, Gerbert Yefremov, who had met us and flung open the door to his boss' office, was not invited

---

37. The abbreviation "UR" stood for *universalnaya raketa* (Universal Missile), i.e., "universal" as in "all-purpose."

38. A vast variety of political jokes were in circulation during the Soviet era. One popular genre was an Armenian Radio or Radio Yerevan joke, which had a format consisting of a question posed to Armenian Radio followed by an irreverent answer poking fun at the system.

39. Chelomey lost much of his standing in the top echelons of the Soviet defense industry after the fall of Nikita Khrushchev in 1964.

to join our conversation. I had the opportunity to ask Chelomey why it had been so easy for him to abandon the *Taran* in 1964.

"Well, you see, the radiomen let me down. They had assured me that it would be very simple to detect each individual missile coming at us, to extrapolate its trajectory, and calculate the place and time that the UR-100 anti-ballistic missile would intercept it. But it turned out that it was child's play for the enemy to blind or even destroy all these massive early warning antenna systems beforehand. And without them, the problem could not be solved. Ten years later, the Americans were also studying the feasibility of such a defense, but they rejected it and now they are designing a three-echelon missile-defense system. They correctly believe that interception needs to be initiated during the boost phase with strikes from space, using space-based systems to destroy warheads that have survived during the midcourse phase, and if one of them breaks through, finishing it off during the terminal phase. But all of these systems are a thing of the future. And it's not just us who will need decades—they will, too. That's why we need to collaborate to create space systems that make those systems simply unnecessary. They stole my satellite destroyer, the IS, but never mind.[40] It's already obsolete now. I hope that if we keep good track of each other, then we'll only be launching rockets into space rather than at Earth."

"Before we wage war in space with the Americans," concluded Chelomey, "we need to make cosmic peace among ourselves at home."

After establishing parity between the U.S. and the USSR in strategic weaponry, the threat of the nuclear militarization of space resulted in a certain easing of the arms race in the field of missile-defense systems. The Anti-Ballistic Missile (ABM) Treaty was signed on 26 May 1972, and a protocol to that treaty reducing the number of anti-ballistic missile system deployments was signed on 3 July 1974.

However, the prospect of a thaw in the Cold War conditions did not dovetail with the interests of the most reactionary circles and the true masters of the military-industrial complex. Along both sides of the Cold War front, ideas began to accumulate and be realized in two directions. The developers of ballistic missiles took into account the possible threat to their formerly "absolute" weapon from anti-ballistic missiles and began to forge a missile "sword" capable of breaking through a future missile "shield." "Shield"

---

40. Chelomey's co-orbital anti-satellite program was named IS—*Istrebitel sputnikov* (Satellite Destroyer). Although Chelomey originated the program, in 1965, after Khrushchev's fall, overall management of the project was transferred to another organization, KB-1.

ideologues switched to developing super systems consisting of several independent subsystems to destroy enemy missiles during various trajectory segments, from the moment of launch until entry into the atmosphere in the target area.

For the Soviet economy, this next turn in the Cold War, during which it was necessary to concurrently produce new generations of strategic missiles *and* anti-ballistic missile systems using new scientific principles, was beyond its strength.

However, the scientists and engineers drawn into this race were enticed by the possibility of implementing what seemed to be absolutely fantastic ideas. This maelstrom sucked in the leading rocket-space organizations. Our TsKBEM under Mishin's leadership, Chelomey's TsKBM, and Yangel's KB Yuzhnoye plunged into it.[41] In the field of anti-ballistic missiles, projects in the Soviet Union were running more or less on par with the achievements of the missile specialists in the United States. In the field of anti-ballistic missiles, we had more than enough ideas, but we had far fewer resources than the Americans to convert those ideas into real systems. The Americans began to spend actual billions of dollars. We could squeeze out billions for anti-ballistic missile systems only if we cut back on modernization programs and programs to develop new strategic missile systems.

Intelligence about American methods for overcoming our anti-ballistic missile systems determined the structure of our systems. On the other hand, the internal "civil war" between the schools of Yangel and Chelomey resulted in a competition to produce missile system structures capable of withstanding an enemy nuclear strike on the ground and overcoming the enemy's missile-defense system while en route to the U.S.[42] Due to these internal and external factors, the debates over the development of land-based and submarine-launched ballistic missiles resulted in similar flight pattern structures for almost all long-range ballistic missiles.

In the late 1960s, the U.S. put missiles with multiple independently reentry vehicles (MIRVs) into service. In the Soviet Union, Yangel was the first to respond to this challenge and began to update the R-36 missile, converting it into a multiple-warhead missile, while at the same time improving the launching silos, introducing the so-called mortar launch, and equipping the missiles with means to overcome anti-ballistic missiles. Chelomey was not to be outdone by him. The same factors were in effect when it came to Makeyev's

---

41. In 1966, design bureaus across the Soviet defense industry were renamed. Korolev's OKB-1 became TsKBEM (Central Design Bureau of Experimental Machine Building) while Chelomey's OKB-52 became TsKBM (Central Design Bureau of Machine Building).

42. For the internal "civil war" in the Soviet missile program, see Chapter 5 of this volume.

submarine-launched missiles. At times we had successes of our own, updating our RT-2 solid-propellant missile.

In the end, a sort of missile trajectory "world standard" from launch to target was formulated. A missile's flight pattern can be divided into four phases in time and space. The **first** phase is an initial boost phase that begins the moment a land-based missile emerges from the silo under the effect of the powder gases, or the moment a submarine-launched missile breaks the surface of the ocean.

The **second** phase is a boost phase when the engines of the first, second, and—if there is one—third stages operate in sequence. Characteristic of this trajectory phase is the intense radiation of the engines' flames in the visible and infrared portions of the spectrum. Space-based monitoring systems are able to verify missile launch based on this radiation.

The **third** phase begins after the engines of the last stage shut down and the reentry vehicle containing independently targetable warheads and decoy targets separates from the launch vehicle. The control system in this vehicle consecutively launches each warhead on its own target according to a program loaded into the on-board computer. As Pilyugin once expressed, bragging about the control system developed at his Scientific-Research Institute of Automatics and Instrument Making (NIIAP): "We're delivering payloads to their individual addresses." The final verification of the "address" was loaded into the memory of the warhead's guidance system in the form of a digital map of the terrain—a sort of "portrait" of the target.

The **fourth** and last (terminal) phase is the re-entry into the atmosphere and flight of the warheads to their individual "addresses." Comparing the actual radar "picture" with its digital map of the terrain, each warhead rushes toward its target over its own trajectory. As the warheads home in on their targets, decoy warheads and infrared radiation sources fly toward various targets, hampering the work of the missile-defense system, if it is getting ready to counter them.

Khariton, Kocheryants, and Zababakhin each formed his own school in nuclear technology, as did Korolev, Yangel, Chelomey, Makeyev, and Nadiradze in the field of missile technology.[43] One can argue at great length, tracking down the achievements and mistakes in the development of missile systems made by each of these chief designers during their lives and posthumously by their schools, but one thing is beyond dispute, that they were in charge

---

43. Samvel Grigoryevich Kocheryants and Yevgeniy Ivanovich Zababakhin were leading nuclear physicists in the Soviet atomic bomb program.

of scientific and production teams that, together with hundreds of other organizations in the late 1970s and early 1980s, countered American science, technology and industry with a nuclear missile force in no way inferior to the one that was hanging over us.

All the major missile specialists, with the exception of Nadiradze, were merged into a single Ministry of General Machine Building and single department of the Academy of Sciences. They relied on the same chief developers of engines, control systems, and land-based launching systems. Looking back, I can say with a clear conscience that there was a lot more that united them than separated them. When they created and produced new systems, they possessed real power that no ministry or high-ranking official standing over them would dispute or attempt to take away from them.

It was quite a different case with our anti-ballistic missile technology. Our anti-ballistic missile system succeeded in creating a "shield" that only partially covered Moscow during the terminal flight phase of nuclear warheads. In the USSR, the Vympel Central Scientific-Production Association (TsNPO Vympel), subordinate to the Ministry of the Radio Industry, was established in 1970 to bring organizational order into this field. This organization brought together dozens of the nation's most powerful radio engineering teams, who had experience developing large-scale radar systems for various applications. The development of the "hard kill" systems themselves—the anti-ballistic missiles, their warheads, and possible space-based systems to kill enemy missiles using particle beam or laser beam weaponry—was not part of Vympel's scope of work. As before, the Ministry of the Aviation Industry was in charge of the missiles for air defense and missile-defense systems, the Ministry of the Defense Industry was in charge of time fuses and kinetic energy weapons, and the Ministry of Medium Machine Building was in charge of all types of nuclear warheads and problems related to particle beam weapons. The new NPO Astrofizika, headed by the son of Minister of Defense Ustinov, was created in the Ministry of the Defense Industry to develop a laser "death ray" weapon.[44] Teams of top-notch specialists worked on dozens of different Central Committee and Council of Ministers decrees, VPK resolutions, Ministers' orders, and also on their own plans.

When the issue at hand is the development of new systems using all the scientific discoveries to date, the freedom of creativity that a scientist needs

---

44. Ustinov's son, Nikolay Dmitriyevich Ustinov (1931–), was a leading scientist at the Luch Central Design Bureau (TsKB Luch), which produced military laser weapons. He became a corresponding member of the Academy of Sciences in 1981.

has to be severely restricted, especially in terms of fundamental scientific research. But no matter who had authority over the scientific team, these restrictions cannot be placed in the hands of an administrator, even if he is well known as a "good organizer," or has access to the offices of ministers and support from the supreme powers of the bureaucracy.

I cannot pretend to be a judge in the history of the competition between the Soviet and U.S. missile-defense systems. Too often I feel like a dilettante. Let's turn once again to Grigoriy Kisunko. He composed an epigraph for the last chapter of his memoirs *The Secret Zone*: "There is no sadder tale on this Earth than that of the Soviet anti-ballistic missile."

The term "Star Wars" in the context of missile-defense systems acquired a special ring after President Reagan proposed the SDI program in March 1983. The American SDI program called for the development of systems that varied in terms of structure and operating principle, each designed to knock out a ballistic missile during one of the four characteristic phases of its trajectory. The Americans had clearly beaten us. We were no longer capable of funding systems on a scale comparable to Americans. After many years of high-level negotiations, the hype surrounding "Star Wars" quieted down. Despite all manner of international agreements calling for the suspension of large-scale operations on missile-defense systems, scientists were allowed to work on this problem.

An effective system making it possible to destroy the missiles themselves or their warheads before they reached Earth was first developed in the Soviet Union. In any event, through the efforts of Soviet scientists and hundreds of scientific and industrial enterprises in the late 1970s, Moscow became the only capital in the world to be surrounded not only by an anti-aircraft, but also by an anti-ballistic missile "shield." Moscow's first A-35 anti-ballistic missile system, produced under Kisunko's leadership, was put into service in 1972. The updated A-35M system was produced after Kisunko's removal when the "anti-ballistic missile" field was transferred from scientists to frequently replaced administrators. After various organizational experiments, General Designer Anatoliy Basistov began to take hold of the field. By the early 1990s, the new generation A-135 missile-defense system had been set up around Moscow.[45] It seemed that now the "beloved city may sleep soundly."[46] Neither airplanes, nor even missile warheads could penetrate its defenses. The

45. The A-135 system was officially declared operational on 17 February 1995.
46. This is a line from a song from the popular World War II-era film *Istrebiteli* (Fighter Pilots).

Moscow A-135 missile-defense system was designed to intercept and destroy warheads during trajectory phases three and four. Our scientists were ahead of their time and the Americans in these developments. The Americans did not yet have a similar system.

Anti-ballistic missile firing complexes were positioned around the capital along a radius of about 100 kilometers. Each complex was a small, heavily guarded town with operational, technical, and residential zones. The technical zone consisted primarily of powerful radar, a super-high-speed computing center that executed rendezvous and control tasks, anti-ballistic missile launch systems, and the anti-ballistic missiles themselves, capable of destroying an enemy nuclear warhead with its own nuclear charge. Hundreds of engineers in officers' uniforms were on duty around Moscow in the missile-defense bastions that were packed with special-purpose radio engineering complexes. The very well-furnished residential zones of these towns had all the necessities for the tranquil life of the servicemen and their families.

Ballistic missile early warning centers using gigantic antenna systems, or "arrays," were erected in the Moscow region, Latvia, and Armenia. They were also erected near Krasnoyarsk but never went into service. All the launch complexes and early warning centers were linked by the most state-of-the-art communications system, over which a single command center controlled this massive and once-unique system.

I WAS BRIEFLY INVOLVED IN MISSILE-DEFENSE DEVELOPMENTS FOR NPO ENERGIYA IN THE EARLY 1980S. In 1981, Yuriy Semyonov was named both first deputy to General Designer Valentin Glushko and chief designer of the Buran spacecraft. He replaced Igor Sadovskiy, who for some reason did not always see eye to eye with Glushko on prospective programs. One such program was the payloads for the super heavy-lift Energiya rockets. Boris Gubanov had been named chief designer of the already conceived *Energiya* rocket.[47] Neither Glushko, nor Semyonov, nor Gubanov had recommended using the *Energiya* rocket for our "Star Wars" program. As I recall, Sadovskiy was the first to come up with this initiative. Having been relieved of responsibility for the future Buran spacecraft, Sadovskiy, with his inherent passion for cutting-edge and challenging problems, attempted to breathe new life into strategic space systems projects.

---

47. Yuriy Pavlovich Semyonov (1935–) served as first deputy general designer of NPO Energiya from 1981 to 1989 before succeeding Glushko as chief and general designer of the organization. During his tenure under Glushko, he served as chief designer of the Buran spacecraft. Igor Nikolayevich Sadovskiy (1919–93) and Boris Ivanovich Gubanov (1930–99) were, at various points, chief designers of the Buran and Energiya vehicles respectively.

Mishin had started feasibility studies for the development of space systems capable of striking ballistic missiles and military spacecraft at NPO Energiya back in the early 1970s. However, the organization's primary subject matter had very much overshadowed these projects. Sadovskiy proposed involving the various departments and uniquely specialized control systems and computer science experts subordinate to me in the process of designing the military spacecraft.

When I began to go over the proposals of Sadovskiy's conceptual designers, it turned out that he had already personally contacted Academician Anatoliy Savin, who was in charge of TsNPO Kometa, one of the leading organizations involved with missile-defense programs.[48]

Military space stations inserted into various orbits, including geosynchronous ones, formed the basis of the new programs. The stations would be equipped with facilities for round-the-clock surveillance of Earth's entire surface. One station or another was supposed to detect the launch of any missile or the insertion into space of any spacecraft and immediately transmit this information to the command post. In the event of the outbreak of "Star Wars," the stations were equipped with various means of knocking out missiles during the boost phase, and if they didn't succeed, then they started hunting for warheads. When commands were received from the ground, military spacecraft were also supposed to be destroyed. It was proposed that platforms be equipped with powerful laser weapons and self-guided rocket projectiles to strike ballistic missiles and spacecraft.[49]

Rough calculations showed that the guaranteed destruction of the enemy's primary missile forces, after they had been successfully launched from silos and submarines, would require the continuous alert status in space of as many as 20 heavy-duty military stations. In addition to the ground control complex for these stations, the launch vehicles, and all the miscellaneous equipment, a missile-defense system would be needed that would be many times more expensive than the entire nuclear missile shield in terms of volume of production and potential expenditures.

Realizing the lack of economic feasibility in creating such a system in the next 20 years, I switched my "foreign policy" activity to the development of

---

48. TsNPO Kometa was the systems integrator for a number of high profile Soviet military space systems including the co-orbital anti-satellite system, the missile early warning system, and the ocean reconnaissance satellite system. Anatoliy Ivanovich Savin (1920–) served as head of TsNPO Kometa (in its various incarnations) from 1958 to 1999.

49. For a detailed account of these programs, see Konstantin Lantratov, "The 'Star Wars' That Never Happened: The True Story of the Soviet Union's Polyus (Skif-DM) Space-Based Laser Battle Stations: Part I)," *Quest: The History of Spaceflight Quarterly* 14 no. 1 (2007): 5–14 and Part II, *Quest: The History of Spaceflight Quarterly* 14 no. 2 (2007): 5–18.

super-powerful space communications systems as a payload option for the *Energiya* rocket.

While General Designer Glushko was still alive, I proposed using the new Energiya launch vehicle to create global space communications and surveillance systems.[50] Only the Energiya was capable of inserting into geosynchronous orbit the heavy space platform required for such a system. Yuriy Semyonov, Glushko's first deputy at that time, actively supported this project. After Glushko's death, Semyonov, having become general designer, pushed the project through to obtain a positive decision from the Soviet Union's Defense Council.

With the support of the Ministry of General Machine Building and the Ministry of Communications, a decree of the USSR Council of Ministers was prepared and cleared with all interested agencies and union republics. Chairman of the Council of Ministers Nikolay Ryzhkov did not manage to sign it—the Soviet Union collapsed, burying many promising projects under its own rubble. The cabinet of ministers of the new Russia showed no interest in such projects.

We had spent three years developing the space communications system project, which should have more than met the needs of every human being for fully integrated telephone communications—in any populated area, on an airplane, in an automobile, on a seagoing vessel, in the desert, or in a megalopolis. We managed to interest the leading missile-defense development NIIs and KBs in this project. Economic reforms forced the managers of these organizations to seek work for their uniquely qualified personnel in keeping with the well-known slogan from satirists Ilf and Petrov: "Saving drowning men is up to the drowning men themselves."[51]

During this period I became closely acquainted with many of the people who had developed the now legendary missile-defense systems. The head organization that developed missile-defense systems was the Scientific-Research Institute of Radio Instrumentation (NIIRP).[52] NIIRP Chief Engineer Boris Grebenshchikov proposed that I visit one of the anti-ballistic missile launching complexes on the outskirts of Moscow to discuss the feasibility of using its large antennas and all of the radio engineering equipment as a control post for the future global communications system. NIIRP was actively involved in this system's design process.

---

50. Glushko died on 10 January 1989.

51. Russians often invoke this phrase from the novel *Dvenadtsat stulyev* (*The Twelve Chairs*) for situations when it is clear that no one is going to help you and you must rely on your own devices.

52. NIIRP—*Nauchno-issledovatelskiy institut radiopriborostroyeniya*—is the current name of Kisunko's OKB-30, originally established in December 1961.

What I saw in the approximately 20 square kilometers of forested restricted area made a double impression on me. As a specialist, I was delighted with the facilities, which were the result of the colossal, intensive labor of many thousands of Soviet people. As a citizen and patriot of this nation, the senseless, unpunished devastation of this unique system depressed and humiliated me. The residents of neighboring villages could now drive into this once highly restricted area and select "home furnishings" to their liking—from toilets to parts from large computer systems. The missile launching units still remained, now overgrown with weeds.

Five years have passed since then. The grandiose missile-defense systems, capable of providing protection with information and force against ballistic missiles for Moscow, and in theory any other regions of the country as well, had been destroyed without a single shot being fired. There's an old saying: "Once your head is off, it's no use crying about losing your hair." If we suffered an overwhelming defeat in the Cold War, is it worth it to grieve over the missile-defense system?

History teaches that in all times there has been a close dialectical link between the offensive and defensive weapons of opposing sides. In the process of developing strategic nuclear forces, a balance was established between the two superpowers, referred to as "mutually assured destruction" (MAD). Maintaining the offensive strategic assets of both sides, the U.S. and Russia at this point would have guaranteed peace if both sides had not begun broad-scale operations on missile-defense systems. Now the balance has been disrupted. We have virtually stopped developing and testing (much less deploying) new assets. Russia runs the risk of irretrievably losing the very rich intellectual potential capable of developing such unique systems.

By ruining the missile-defense system that was created by the self-sacrificing labor of the entire nation, we are losing our "shield" and now must pin our hopes on the retaliatory strike of our missile "sword." According to information available from overt and covert sources, the Americans have not halted operations on "Star Wars" systems.

The standoff between Russia and NATO is no longer as acute as it once was. But where is the guarantee that 10 to 15 years from now, effective offensive nuclear missiles will not end up in the hands of some other aggressive nations? In 1933, in one of the most "democratic" European nations, authority ended up in the hands of the criminal Nazi regime that unleashed World War II. No systems analysis using the most state-of-the-art methods for modeling and forecasting human social development can prove that something similar will not happen in the 21st century somewhere in Asia or South America. Whether we like it or not, the United States continues to develop strategies and technical systems for the direct protection of its own territory against nuclear weaponry. Ballistic warheads with target-seeking guidance systems and beam weapons developed using new energy principles (lasers, particle beams, and electromagnetic weapons) remain the basis of these strategies. The term "Star Wars" gradually disappeared from the vernacular of politicians, diplomats, and journalists. But work on the actual armaments continues.

I must confess to the readers of my *Rockets and People* book series that Russia's termination of broad-scale operations on missile-defense systems has indirectly contributed to the original publication of these books. The fact is that young, enterprising specialists at NIIRP, which had been left without government funding, began to seek out other fields in which to apply their expertise. The creation of Space Communications Systems (KOSS) Closed Joint Stock Company was the result of this activity.[53] KOSS used its wealth of experience in producing large radio engineering systems to develop a unique communications system using small low-orbit satellites.

The chief designer of this communications system, Igor Dunayev, and his close colleagues took the risk of providing the financial backing to the Mashinostroyeniye Publishing House so that my memoirs and musings might see the light. It seems to me that this kind of conduct on the part of the KOSS directors—unusual in our time—is a testimony to the fact that the people who toiled on the domestic "Star Wars" problems were not only talented, highly-specialized individuals, but they were also people who "thought big" and held on to their dreams and optimism.

NINE YEARS HAVE PASSED SINCE [1997 WHEN] I HANDED OVER TO THE MASHINOSTROYENIYE PUBLISHING HOUSE THE (RUSSIAN) MANUSCRIPT of my third volume of *Rockets and People*, which contained a chapter called "Star Wars." During this period, a semblance of warm or "partnership" relations has developed between the presidents of the United States and Russia, and business-like, and even amicable cooperation, has been established between their space agencies. A burst of controversy reminiscent of the supposedly extinguished Cold War proved to be all the more surprising. In its March 2006 issue, the influential American journal *Foreign Affairs* published an article by two respected experts comparing Russian and U.S. nuclear potentials and missile-defense technology.[54] The authors compared the latest data on the status of Russian strategic nuclear assets with the near-term plans for the

---

53. KOSS—*Kosmicheskiye sistemy svyazi.*
54. Keir A. Lieber and Daryl G. Press, "The Rise of U.S. Nuclear Primacy," *Foreign Affairs* 85 no. 2 (March/April 2006): 42–54.

development of U.S. strategic programs. The American scientists' analysis shows that the U.S. continues to successfully improve its strategic nuclear triad at the same time that Russian assets are rapidly degrading. In view of the obvious disruption of parity, the nuclear deterrent policy will soon play no role. If the U.S. is the first to launch a nuclear strike, then in the best case Russia will be able to respond to it with individual ballistic missiles, which the U.S. missile-defense system will be able to intercept.

Regarding the articles of the American scientists, the Russian press published the thoughts of respected Russian specialists. The thrust of the response was to confirm that if the U.S. destroyed almost all of Russia with a first strike, then the intact remains of the nuclear arsenal would still be sufficient for a worthy retaliatory strike. In this very grim discussion, there was no answer to the question that I would ask the American and Russian strategists. Why would the Americans take this risk? In terms of scale, the collapse of the Russian economy and the degradation of Russia's high-tech industries, and consequently, of the entire military-industrial complex, already correspond to the losses that the nation would suffer in the event of a nuclear strike. True, in the latter case there are not millions of fatalities and future generations will not be able to accuse the U.S. strategists of committing the most heinous crimes in the history of humanity. The Russian elite has already committed a crime against its people, having consistently conducted a policy of bringing Russia down to the role of no more than a source of raw materials for other industrialized nations. Such a source has no need for either fundamental or applied science. The military power of the strategic nuclear armaments that rely on those sciences will fade away of its own accord.

Chapter 12
Spying from Space

The USSR and the U.S. came up with the idea of using space for the purposes of reconnaissance almost simultaneously, just two years after the launch of the first *Sputnik*. In 1959, an automated automatic spacecraft obtained photographs of the far side of the Moon for the first time. It would seem that photographing any area of Earth's surface would be far easier.

It turned out that taking a photograph of Earth's surface that would have practical military, scientific, or economic value was not all that simple. During the years of the bitter Cold War, there were periods when it verged on a hot one—World War III. The lessons that we, our military, and political leaders should have drawn from the experience of World War II called for us to obtain credible, objective information about a potential enemy's readiness for war. It is astonishing that neither the minister of defense nor the General Staff was the first to demand that space be used for strategic reconnaissance. The initiative originated, if not from Korolev personally, then, at any rate, from his entourage. At this point, it is difficult to say with any certainty who came up with the idea first.

Even before World War II, the development of military technology gave rise to new service branches and a new type of armed forces—aviation. After World War II, nuclear weaponry on missiles and aircraft carriers fundamentally altered the scale of warfare in time and space.

The scientists, designers, and military-industrial leaders who had actually created this technology had overtaken career military leaders in the development of a strategy for using the newest scientific achievements for the purposes of war. Scientists found support, first and foremost, among the country's top political leadership. And only after career military leaders—heroes of the past war—had received instructions "from the top" would they get involved in the organization (but not in the development) of the new armed services.

Theoretical works of military science, citing classic statements of Clausewitz, maintain that strategy is the domain of the armed forces' supreme command.[1] The history of atomic and then missile and rocket-space technology makes it possible to contend that the development of post-World War II strategy has been defined by the leaders of science and technology rather than great military leaders. They have borne the responsibility for creating new types of weaponry, the very emergence of which required them to be directly involved in the development of a strategy.

Having been skeptical about rockets, many military strategists cited convincing examples of the Germans' use of V-2 long-range ballistic missiles which had the strategic objective of knocking Britain out of the war. Five hundred thirty-seven missiles reached London, killing 2,700 people. The German missile attacks did not demoralize the British. They withstood them. The new "vengeance weapon" did not save Nazi Germany from a crushing defeat by the Red Army.[2] Consequently, many generals believed that we should not repeat the Germans' mistake by giving priority to missiles at the expense of other types of weaponry.

The demands made in past wars for leadership qualities were now made to the fullest extent on those who were at the leading edge in creating military technology. These qualities included: a wealth of ideas; the ability to intuitively interpret the situation; mental flexibility; the ability to penetrate into uncertainty; a talent for making decisions despite a lack of basic data; erudition enabling one to make a sober assessment of internal and external circumstances; and the ability to remain steadfast in the face of setbacks, catastrophes, and the attacks of hardliners during the Cold War. More was required from them than from the great commanders. On their shoulders, they carried not only the burden of developing the principles behind the new types of weaponry but also of creating fundamentally new branches of industry.

Based on the experience of World War II, the highest-ranking officers of the armed forces of the Soviet Union and the United States were unenthusiastic about missile technology until a missile was capable of carrying an atomic warhead. Their attitude changed radically after 1956. It would take the two missile powers' defense departments, another two years to realize that a missile

---

1. Carl von Clausewitz (1780–1831) was a Prussian general and military strategist most widely known for his work *On War* (*Vom Krieg*), which advocated the concept of total war on an enemy's territory, property, and population.

2. For the purposes of wartime propaganda, the Nazis used the designation *Vergelstungwaffe* 2 (Vengeance Weapon 2 or V-2) to refer to the A4 ballistic missile.

was not just capable of carrying strategic weapons of mass destruction, but it could also serve as the launch vehicle for a strategic reconnaissance spacecraft.

Just two years after the launch of the world's first artificial satellite, it was rocket scientists in the Soviet Union and in the United States, not the ministries of defense, who proposed using space for strategic reconnaissance. Enthusiastic professionals at the General Staff's Main Intelligence Directorate (GRU) supported the proposals.[3] Reconnaissance from space promised many advantages. The violation of national borders on land, at sea, and in air space for the purpose of intelligence gathering was called espionage. Spies could be killed on entirely legal grounds. There were, however, no international agreements concerning national borders in space, and there are none to this day. Any vehicle whose speed exceeds 7.9 kilometers per second becomes an artificial satellite, and according to the laws of celestial mechanics, is not capable of honoring national borders.

By its nature, space activity is global. Depending on its inclination angle, a satellite is capable of flying over any nation and over any top secret area on the territory of any country. Artificial satellites opened up fundamentally new opportunities for legal strategic reconnaissance. The impunity of this activity was guaranteed.

The GRU chiefs appreciated these qualities. Korolev skillfully combined the military's interest in developing reconnaissance capability from space with his dream of a piloted spacecraft. With the aid of Keldysh, Rudnev, Ustinov, the Ministry of Defense's "missile faction" (Mrykin and Smirnitskiy), and the individual most interested in the newly discovered capabilities of the GRU, General Kostin, the objectives for the development of reconnaissance spacecraft and a piloted vehicle were included in the same space program decree.[4]

In the early 1960s, when the U.S. was overtaking the USSR in terms of spy satellite launches, our U.N. representative announced a proposal to prohibit the insertion of spy satellites into space. However, after we had developed our own Kosmos spy satellites, we made no more attempts to place restrictions on space surveillance and monitoring technology. Khrushchev, who had initiated the prohibition campaign, wrote the following statement in his memoirs on that subject: "I recall the uproar in 1960 when the United States' U-2 spy plane flew into our air space and we shot it down. We were outraged that our sovereignty had been violated. Times have changed.

---

3. The GRU—*Glavnoye razvedyvatelnoye upravleniye*—was the primary military intelligence agency of the Soviet Union.

4. General Pavel Trofimovich Kostin headed the space reconnaissance directorate of the GRU.

Now this sovereignty is not violated, even though U.S. satellites trace the globe, photographing anything they want. These are also intelligence gatherers, even better intelligence gatherers than the U-2 airplanes, but we don't protest. We have satellites just like theirs flying over the U.S., also taking photographs, and reporting to the military. America doesn't protest either because for the time being there is no way of combating these satellites and there's no sense in protesting. Moreover, this is a reciprocal opportunity, which each side is getting. For the time being there is no agreement on sovereignty in space."[5]

By the beginning of the 21st century, in addition to the U.S. and Russia, at least 10 more nations had managed to get spy satellites.[6] After the destructive reforms of the 1990s, Russia has been forced to rebuild its space-based remote sensing, surveillance, monitoring, and reconnaissance systems using the latest achievements of radio engineering and optical electronics—it has to if it wants to be considered a great power.

To realize his dream of piloted spaceflight, Korolev needed the support of the military. And not just for piloted spaceflight. During the 1950s and 1960s, any space program required decrees from the Central Committee and Council of Ministers. But such a decree could not be issued without the approval of the Ministry of Defense.

The decree issued in May 1959, coordinated with Korolev, made the production of a photoreconnaissance satellite an urgent defense task. The launch of a spacecraft "with a human being on board" at that time was still not considered a priority, and the Ministry of Defense showed no initiative at all on that front. At the initiative of the GRU, on 10 December 1959, a decree was issued calling for the separate development of satellites to perform photo- and radio-reconnaissance, navigation, and meteorological missions. On 4 June 1960, at the insistence of the GRU, but prompted by our conceptual designers, who were photoreconnaissance enthusiasts, one more decree was issued calling for flight development testing of photo- and radio-reconnaissance systems.

In contrast to our "hyped up" peace propaganda and our silent military, American strategists did not hide the fact that, in the military use of space, priority should be given to intelligence-gathering systems. They developed MIDAS satellites for photoreconnaissance and SAMOS satellites for radio

---

5. N. S. Khrushchev, *Vospominaniya: vremya, lyudi, vlast, t. 4* [*Memoirs—Time, People, Power, Vol. 4*] (Moscow: Moskovskiye Novosti, 1999), p. 220.
6. These nations include China, France, Germany, Italy, Israel, Japan, India, South Korea, and Iran.

reconnaissance.[7] According to the sketchy information that we received, the Americans were not satisfied with the initial results and were designing new satellites for all types of intelligence gathering from space.[8]

The Krasnogorsk Optico-Mechanical Factory was designated the main developer of photographic equipment for our future reconnaissance systems. This factory had a lot of experience in producing equipment for aerial photography. Our requirements called for the photographic equipment to be installed in a descent module whose window guaranteed an airtight seal without frame distortion. The filming and film advance process needed to be fully automated and the film needed to be preserved in a special cassette for descent and landing that could withstand up to 20 g's. The first reconnaissance camera produced at the Krasnogorsk factory was called *Ftor* (Fluorine). This provisional name stuck for years.

Camera Chief Designer Beshenov and his coworkers required that our conceptual designers and designers create "special conditions" for the optics of the cameras. One of the most difficult conditions was that the lens be maintained at a constant temperature deviating by no more than 1°C, while the rate of change was not supposed to exceed 0.1°C per hour. A slight difference in the temperatures on the panes of the window altered their curvature. This caused distortion of the image for the camera's telephoto lens. We were supposed to input flight velocity and altitude data into the camera. They were used in the film advance mechanism to compensate for the shifting of the image. The specified photo resolution could be ensured only if the deviation from the specified speed of the film's compensating motion did not cause the "stopped" image to shift by more than 0.01 millimeter during exposure.

All of these problems, and a whole lot of other ones, seemed to be solved in calculations and ground tests, but what would happen in flight? The ground would issue commands to control the selection of light filters

---

7. SAMOS was a U.S. Air Force program to develop (at least initially) a reconnaissance satellite capable of real time radio readout of scanned photographs. After such a system was shown to have significant limitations, later SAMOS missions focused on film recovery techniques. SAMOS vehicles were launched between 1960 and 1963. Some satellites also tested experimental ferret electronic intelligence systems. The MIDAS program was focused on missile early warning by using infrared sensors. The first MIDAS launch was in 1960. Both SAMOS and MIDAS emerged from the WS-117L intelligence satellite program from the late 1950s.

8. Although Chertok does not mention it, the main American photo-reconnaissance system was the top-secret CORONA program, whose satellites were initially launched under the Discoverer moniker to disguise their true missions. U.S. President Dwight D. Eisenhower approved the CORONA program in February 1958. It was the first operational reconnaissance system deployed by any nation. At least 144 CORONA satellites were launched between 1959 and 1972.

and the exposure and would select the coordinates where the photo session would begin and end as well as the number of frames. The *Granit* (Granite) on-board sequencer was loaded with a detailed program controlling the photographic process via radio link. Isaak Sosnovik and Nina Kvyatkovskaya labored over its development in Shustov's radio electronics department. Conceptual designers Yevgeniy Ryazanov and Yuriy Frumkin coordinated the "ideology" of *Ftor's* operation with *Granit's* capabilities. There was also a real-time photoreconnaissance version, making it possible to receive information immediately in flight without having to wait for the descent module to land, hunt for it, and then recover, deliver, and develop the film. The special *Baykal* photo-television complex was developed for this. Immediately after a photo session, the film went to a developing unit right on board. After development, fixation, and the drying process, the film was pulled frame-by-frame in front of a video camera and transmitted via the *Kalina* (Guelder Rose) television channel to the ground. NII-380 in Leningrad developed this complex device.

When I visited NII-380, publicly known as the All-Union Scientific-Research Institute of Television (VNIIT), I received a very warm and responsive reception, and not only because I came as a wealthy customer armed with decrees.[9] NII-

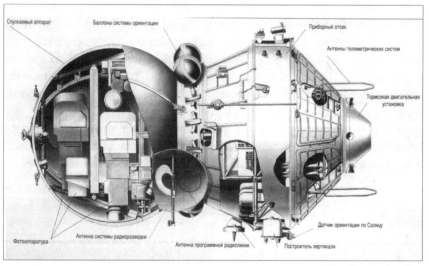

From the author's archives.

Layout of the Zenit–2 reconnaissance spacecraft. The legend from top left (going clockwise) is: descent module, balls for the attitude control system, instrument compartment, telemetry system antennas, braking engine unit (TDU), solar orientation sensor, vertical, radio–link antenna, radio reconnaissance system antenna, imaging equipment.

---

9. VNIIT—*Vsesoyuznyy nauchno-issledovatelskiy institut televideniya.*

380 was the country's most qualified and prestigious organization in the field of television. Institute Director Igor Aleksandrovich Rosselevich and the young enthusiasts of television technology who worked under his leadership did not need to be persuaded. They themselves had been striving to bring television technology into space. The first piloted *Vostok* spacecraft and Zenit-2 reconnaissance satellite were equipped with television equipment and a radio link that transmitted video from on board that were developed at NII-380.

Right off the bat, the testers at the firing range renamed the *Baykal* system, which occupied a lot of space in the descent module, the "Bath & Laundry Trust Co." or BPK for short.[10] And for good reason. During the very first tests, it leaked fluids and spewed out steam. All in all, its flight preparation caused a lot of headaches; its young developers were forced to endure many criticisms and wisecracks.

For radio reconnaissance, i.e., detecting the locations of radar systems and searching for the opportunity to intercept radio communications, Institute No. 108 (Aksel Berg directed it in his time) developed a complex radio system called *Kust* (Bush). The *Kust* system collected its information on a special recorder that was also supposed to return to the ground in the descent module.

It was much more complicated to control the Zenit-2 space reconnaissance satellite than the Vostoks. A rather complex program for control from the ground via special command radio link was devised to ensure that the necessary targets entered the camera's field of vision. The Vostoks and the Luna, Venera, and Mars spacecraft used one-time commands and settings to perform control functions, i.e., they used a specific numerical value for a limited number of parameters. In comparison, the amount of information that needed to be transmitted to the Zenit was 10 times as great. Each photo session required its own individual program.

Two program command radio links were developed simultaneously for the Zenits: the first at NII-648 and the second at NII-10. Yuriy Kozko's development at NII-648 came out ahead. It ended up on the first Zenit.

The story behind the selection of the command radio links and the entire ground-based Command and Measurement Complex is quite interesting. While preparing the government resolution for the beginning of operations on the first artificial satellite, the Ministry of Defense was against being saddled with work outside its purview, namely the development of a ground-based Command and Measurement Complex. The need to be involved in space exploration never dawned on the generals who were still analyzing the

---

10. BPK—*Banno-prachechnyy kombinat.*

triumphant end of World War II. In their opinion, this was purely the business of the Academy of Sciences. You've got to give credit to the directors of NII-4, especially A. I. Sokolov, G. A. Tyulin, and Yu. A. Mozzhorin. Risking their military careers, they stubbornly insisted on the Ministry of Defense being directly involved in developing satellite monitoring and control systems. Disputes over these issues were heated and continuous until Minister of the Defense Industry D. F. Ustinov appealed directly to Minister of Defense Marshal G. K. Zhukov. Marshal Zhukov appreciated the potential for using space on behalf of the nation's defense earlier than the generals in the General Staff. He gave all the necessary instructions. I recall that Zhukov was named minister of defense in February 1955 and relieved of this post on 27 October 1957. During Zhukov's tenure as minister of defense, organizational and personnel reforms took place that fostered very close collaboration [between the military and] scientists on the cutting edge of the scientific-technological revolution. If you sum up the comments of my military friends, they viewed Marshal R. Ya. Malinovskiy's appointment to the post of minister of defense as a transfer of initiative for military advancement from the Ministry of Defense directly to Khrushchev.[11]

It wasn't until September 1960 that the Third Directorate of "Artificial Satellites and Spacecraft" was formed within GURVO (Main Directorate of Reactive Armaments). Colonel and Engineer K. A. Kerimov, whom I knew very well from Germany and Kapustin Yar, was appointed chief of this directorate; Colonel and Engineer V. I. Shcheulov, also a close acquaintance from Kapustin Yar, was soon appointed as his deputy. The actual formation of the first specialized military-space units did not begin until 1961. Strategic Rocket Forces officers comprised the command and engineering staff of the space units. They had undergone good missile training at all levels to the point of launching ICBMs. In addition, Generals and officers from GURVO, scientists from NII-4, officers from the Space Communications Telemetry Center, and officers from military missions and the testing directorates of firing ranges passed through the good training of working jointly with industry.

The military space forces had been obliged to analyze the potential capabilities for the practical use of space for military purposes before we industrial engineers did. However, the priority that Khrushchev personally gave to the prestigious scientific space goals compared with the secret defense projects in the 1960s meant that the Soviet Union developed space-based

---

11. Marshal Rodion Yakovlevich Malinovskiy (1898–1967) was appointed to replace Zhukov as minister of defense in October 1957.

reconnaissance, remote sensing, meteorological, and communications systems at a slower rate than the U.S. The Americans were the first to launch the Atlas-SCORE military communications satellite (on 18 December 1958), the TIROS weather satellite (on 1 April 1960), the Transit navigation satellite (on 13 April 1960), and in late 1960, beat our Zenit-2-based photoreconnaissance and cartography program by more than a year-and-a-half.[12] OKB-1 began to develop the first generation of reconnaissance and cartographic systems working in direct contact with GRU specialists rather than with the official space technology customer in the Rocket Forces system.

At Korolev's own initiative, the Zenit-2 borrowed the Vostok descent module to bring back to Earth the photographic equipment and film with the results of seven to eight days of observations. At OKB-1, Sector Chief Ye. F. Ryazanov was responsible for both the actual conceptual design of the spacecraft, and for selecting and ordering the photo surveillance equipment. Initially, his sector had been part of Department No. 9, headed by M. K. Tikhonravov. As the project developed, the sector was converted into Department No. 29, which Ye. F. Ryazanov headed, while Yu. M. Frumkin became chief of the Zenit-2 conceptual design sector. I managed, with great difficulty, to break away from the turbulent

stream of Vostok and R-9 problems to devote the necessary attention and guidance to designing the systems for Zenit-2. The managers directly involved with the development of the on-board systems—B. V. Rauschenbach, O. I. Babkov, I. Ye. Yurasov, V. P. Legostayev, Ye. A. Bashkin, D. A. Knyazev, Yu. S. Karpov, B. V. Nikitin, V. V. Krayushkin, and A. G. Melikova—found themselves in the same situation, hemmed in by problems.

Yevgeniy Fedorovich Ryazanov was a department chief at OKB-1 who played a major role in the development of the Zenit-2 reconnaissance satellite.

From the author's archives.

---

12. SCORE—Signal Communications by Orbiting Relay Equipment; TIROS—Television InfraRed Observation Satellite. The first launch attempt in the CORONA photo-reconnaissance program took place on 28 February 1959. The first successful return of images was in August 1960 (with *Discoverer 14*), about one-and-a-half years ahead of a similar Soviet achievement.

Intelligence officers who had experience interpreting aerial photography explained that the information would be of value if we could ensure a resolution of 10 to 15 meters. As a result, we needed to determine performance data of the photographic equipment during the first launches.

We also needed to develop a new command radio link from the ground, as opposed to the very basic one we had used on the Vostoks. A new command radio link (KRL), code-named *Kub* (Cube), was developed per our specifications at NII-648, the director and chief designer of which at that time was Armen Sergeyevich Mnatsakanyan.

Independent of our order, the military had also ordered a new KRL for their space projects from NII-10, the director of which was Mikhail Pavlovich Petelin. After the first Zenit-2 was assembled and sent to the test facility for electrical tests, I received an unexpected phone call from Yevgeniy Ivanovich Panchenko at GURVO, who gave me a friendly "heads up," letting me know that the military was categorically opposed to using *Kub* and demanded that *Podsnezhnik* (Snowdrop) radio systems—developed per their specifications at NII-10—be installed on the Zenits. I was often at NII-10 on gyroscope business and Director Petelin had alerted me that they were developing the *Podsnezhnik* radio system, but they wouldn't be ready for delivery until late 1961. And we were planning to launch the first Zenit-2 at the end of 1961. Replacing the radio equipment of one command radio link with another required a large number of changes in the on-board electric systems and the instrumentation of the on-board complex control system (SUBK).[13] Consequently, all the ground tracking and control facilities would have to be retrofitted with new control stations. I responded that replacing the radio complex would require that the dates for the beginning of the Zenit flight test be pushed back by at least one year!

About three days after my talk with Panchenko, Korolev telephoned me and demanded a detailed report. He said that the next day he and I were supposed to be at a meeting in the office of Commander-in-Chief of the Rocket Forces Marshal Kirill Moskalenko. To defend the military's stance, some "heavy artillery" were attending the marshal's meeting: General A. I. Sokolov, Chief of NII-4; General A. I. Semyonov, Chief of GURVO; his Deputy A. G. Mrykin; and Ye. I. Panchenko, who was the most knowledgeable of them all about radio engineering. To put it mildly, Marshal Moskalenko could not understand our differences with the military.

---

13. SUBK—*Sistema upravleniya bortovym kompleksom.*

Korolev took the floor first and began to explain in layman's terms the tasks of the radio link and how things stood with the preparation for flight tests of the first reconnaissance satellite. The marshal listened attentively, but was puzzled as to what they wanted from him. Why were we all there?

Mrykin suggested that we listen to the GURVO proposals that Major Panchenko was presenting. Korolev knew all of our opponents quite well and did not understand why they were putting on such a show in front of a distinguished combined-arms marshal, who in recognition for all his service during World War II, had been named commander-in-chief of the Rocket Forces for inexplicable reasons and was supposed to know his way around radio engineering, to boot.

General Semyonov said that Major Panchenko would present all the technical arguments. I was feeling intensely uneasy. Of all people, Yevgeniy Ivanovich Panchenko understood full well that if Korolev was promising to deliver the first reconnaissance satellite to the firing range in six months, he needed to mention that installing it [*Podsnezhnik*] instead of the *Kub* would require pushing back the beginning of flight development tests. Considering the advantages of the *Podsnezhnik* radio link, in particular its resistance to jamming and the number of commands, they, the customers were in agreement with that. But Panchenko, who had received a preparatory "pep talk" from his commanding officers, raved about the *Podsnezhnik*, but kept quiet about the dates and the amount of work required to modify the vehicle and the ground facilities.

Korolev nudged me. I took the floor and said that we were in complete agreement with the arguments about the advantages of *Podsnezhnik*, and moreover, we had already begun developing documentation for a version of the Zenit with the new radio link. However, it was advisable to begin flight tests on the first Zenits using a radio system that has already been proven in service in order to check out the primary objective—the descent module's photographic and recovery system.

Mrykin asked the following question: "And who is going to guarantee that the Americans don't interfere with control and issue a command for the module to land on their territory?"

"The Americans don't have a *Kub* ground system capable of transmitting the whole series of landing commands in the proper code. If they manage to affect the on-board receivers with their interference generators and any landing commands are issued outside our territory, then the emergency vehicle destruction (APO) system will actuate and the spacecraft will be destroyed."

Korolev couldn't stand it and turned to the marshal: "Kirill Semyonovich, I promise to install the *Podsnezhnik* system beginning with one of the next

Zenits. Please instruct GURVO to coordinate this proposal with us and submit it to you for approval, but the beginning of flight tests should not be delayed. And moreover, we will work on and preclude the possibility of a safe landing on U.S. territory." Moskalenko listened appreciatively and looked at Semyonov. Without consulting his advisors, the latter hastened to announce: "I believe that we can agree with Sergey Pavlovich's proposal. We will work out a specific decision and report to you for approval."

"So, we have an agreement. I thank you all."

When we had left the meeting, Korolev could not stand it any longer and vented at Mrykin and Sokolov: "Why was it necessary to get the marshal involved in this particularly technical matter? It's our duty to find a solution. Did you want to shift the responsibility from yourselves to the commander-in-chief? In the future, please arrange it so that we can resolve such issues without getting the brass involved."

As we drove back, Korolev told me, "And your friend Panchenko is a fine fellow too. Why couldn't you have made some arrangement with him beforehand?"

Over the course of two years, 13 Zenit-2 spacecrafts were launched. In 1964 the Zenit-2 complex was put into service with the *Podsnezhnik* radio link. I have to say that during two years, we and all those assisting us had made so many changes that the Zenit-2 that was put into service was quite different from the first Zenit-2, which was launched on 28 April 1962, but flew just three days instead of the planned seven. The early return to the ground was not the Americans' fault, but the result of a large number of glitches in the operation of the primary photographic equipment, the photo-television system, and the attitude control system. Now that I've provided this long introduction, let's get back to my instructive history of the launch of the first Zenit-2.

THROUGH 1960, RAUSCHENBACH'S ENTIRE "TEAM" TRANSFERRED TO US FROM NII-1.[14] This group was significantly strengthened by our regulars and the former "Grabinite" electronics engineers, circuitry specialists, and designers. After this, early 1961 saw the production of the instruments they had developed, which constituted the motion control complex that has been called *Chayka* since the days of the system used to photograph the far side of the Moon. For newer spacecraft they developed new motion control and navigation systems called *Chayka-1*, *Chayka-2*, and so on. Modern Soyuz spacecraft carry the *Chayka-5*.

The innovations that went into subsequent versions of the *Chayka* were given their first trial run on the Zenit-2. First and foremost, this included the

---

14. See Chertok, *Rockets and People, Vol. I*, pp. 474–476.

Yuriy Stepanovich Karpov was one of the engineers responsible for development Zenit-2's on-board systems.

From the author's archives.

infrared vertical (IKV), which kicked off the whole orientation process, the *giroorbitant*, and the not always reliable ionic course orientation system that replaced it.[15]

The requirements for orientation accuracy were very high compared with the Vostoks. Equipment left over from the Vostoks was of no help here. Rauschenbach's staff members Tokar, Legostayev, Knyazev, Branets, and Komarova were occupied with these new problems. The motion control system was supposed to maintain tri-axial orientation in the orbital coordinate system for practically the entire flight. In this system, one of the axes points along the local vertical while another would do the same along the spacecraft's line of movement. Deviation from the assigned orientation by more than one degree drastically reduced the image quality.

The operational lifetime in orbit, based on the reserves of film and propellant for orientation, was eight days. The only way to maintain operation of all the systems during this time was to use the Sun to replenish the electric power reserves. Swiveling solar arrays needed to be developed.

The full automation of all on-board processes with continuous ground monitoring and intervention by a program-command radio link required the development of an on-board complex control system operating on new principles. Karpov and Shevelev, the main gurus behind the on-board complex, were loaded down with Vostoks. They assigned the actual development of the system to the two "electrician brothers" Aleksandr and Nikolay Petrosyan. Since they weren't constrained by the constant micromanagement of supervisors, they really became absorbed in their work and developed a centralized power distribution and program logic control system for the entire on-board complex. The "Petrosyan Duo" could quickly trace all the intricate paths of electrical commands through a multi-page schematic electrical diagram and answer the question: "What will happen if there's a break here, and a "minus" to the frame there?"

---

15. *Giroorbitant* is the Russian term for a gyroscopic instrument used to determine the angle of the velocity vector of a spacecraft relative to the orbital plane.

The Zenit's return to Earth also differed from that of the Vostoks. We couldn't rely only on solar orientation, which provided a very limited number of dates and times for landing at a specific firing range. Before executing the deorbit burn, the IKV helped orient the spacecraft with one axis pointing toward Earth, and using the *giroorbitant,* the vehicle turned about this axis so that the nozzle of Isayev's TDU was pointing along the orbital velocity vector. After selecting the most suitable orbit for executing a landing on the given day, a program was uplinked in one of the coverage zones. Next, the *Granit* ran it, issuing all the necessary commands for orientation and engine startup.

All the on-board processes were not only strictly regulated in time, but they also had to be precisely synchronized with Earth-time. A special high-precision clock, the *Liana,* was developed for the Zenit-2. A man by the name of Begun, the chief designer at one of the electronics NIIs in Leningrad, produced the *Liana.* Consequently, when the *Liana* on board started acting up, there was a spate of jokes that *Liana* had "Begun to run fast, or Begun to run down."[16]

Besides ours, more than 20 organizations, each with their own chief designers, were involved in producing the Zenit-2. And there were some who got offended. Ryazanskiy got miffed at me and then complained to Korolev that the orbital monitoring equipment from NII-885 wasn't being installed on the Zenit-2. Actually, we had made a collegial decision to install the *Rubin* responder for orbital monitoring and the *Tral* to transmit telemetry data. Bogomolov's team at OKB MEI had developed both systems. Korolev was annoyed by the failures of Bogomolov's equipment on the launch vehicles, but his response to Ryazanskiy was, "If you want to propose something better, I will immediately give the order to 'throw Bogomolov overboard'."

But at that time NII-885 had nothing better to offer. Having received no assignments for the Zenit-2, Pilyugin viewed this as a potential threat to his monopoly on the development of inertial systems. This was actually the truth. What's more, a very large portion of the work had been handed over to Viktor Kuznetsov, with whom we had started working directly on developing motion control systems.

The draft plan of the Zenit-2 had been drawn up, discussed, and approved during the gap between the flights of Gagarin and Titov. Working drawings of the entire structure and instruments were released into production even before the draft plan had been approved.

---

16. The original pun in Russian is based on the designer's name, which means "runner" in Russian. The joke was that *Liana* ran from the Runner or fell behind the Runner.

After returning from the firing range in August [1961] following Titov's [*Vostok-2*] flight, I discovered that [our factory director] Turkov, having been raked over the coals by Korolev, was sending the entire [Zenit] spacecraft to the monitoring and testing station (KIS) for tests without waiting for it to be completely integrated. It wasn't until we were at the KIS and had connected all the instruments with hundreds of cables into a single system that we realized that we should not be reassured by the vows of the most reliable electricians that all the circuits had been reviewed dozens of times on paper and checked by independent inspectors. The very first activation at the KIS showed such a number of errors and absolutely inconceivable circuit tangles that we felt like starting the design process all over. But we shook off that idea and began to forge our way through the morass of our own errors.

Having hurriedly tested the Zenit-2 at the KIS, Turkov sent it to the engineering facility at Tyura-Tam. After Korolev assigned me to be technical director of testing at the engineering facility, I sent lead designer Boris Rublev to the firing range, and stayed behind to make sure that everyone involved in the operations from my organization and cooperating organizations managed to fly out. Korolev sent Pavel Tsybin to the firing range as the main representative on vehicle design.[17] The decision to transfer all further Zenit projects to Kuybyshev had not yet come to fruition, and various signs seemed to indicate to us that S.P. intended to appoint Tsybin Chief Designer of Zenits in the event the project stayed with us for a long time.

Korolev at a reunion with his friends from his glider–flying days in the late 1920s. From left to right are Korolev, unknown, Khalutin, Sergey Anokhin (in uniform), and Pavel Tsybin (smiling).

From the author's archives.

17. Pavel Vladimirovich Tsybin (1905–92) was a deputy chief designer at OKB-1 with a long career in the Soviet aviation industry. He designed several experimental high-speed aircraft in the 1940s and 1950s before joining Korolev's organization in 1960. Tsybin first met Korolev in the 1920s while both were glider enthusiasts.

347

I arrived at the engineering facility on the cold, dreary Saturday of 11 November, and settled into the well lived-in cottage No. 3. The next day, despite its being Sunday, a cycle of so-called "test firings" and "partial programs" was already under way at the MIK. Only here did we discover how much various and sundry electrical "trash" in the form of all kinds of interference was wandering through the common power busses. Interference was penetrating from the *Tral* radio transmitters to the input of the command radio link receivers and was blocking the passage of commands. Yuriy Kozko, the command radio link developer from NII-648, and Mikhail Novikov from OKB MEI had set up a study of systems' interference using a double-beam oscillograph and were experimenting with various filters. Bashkin realized that the amplifier in the ionic system needed to be changed. Kaznacheyev found a wiring error in the *Ritm* (Rhythm) event controller. There was a "minus" in contact with the frame in the command radio link transmitter. In addition to exact time tags, the *Liana* was feeding some sort of inexplicable noise into the power supply network. The guaranteed service life of the infrared vertical instruments had expired, and they needed to be completely rechecked for flight clearance. *Baykal* was interfering with *Kust*, and so on. In one day my notebook was filled with dozens of glitch notations. On 19 November, we began to see the light at the end of the tunnel; we fixed *Baykal* by modifying *Kalina*; the noise in the *Liana* disappeared after we installed feed-through capacitors. We re-soldered dozens of cables to correct mistakes, and we put together the entire electrical system and moved onto the next task.

On 21 November, we began integrated test cycles. Two more days were fraught with dozens of glitches. Once again we found instruments that needed to be modified or replaced. By the morning of 25 November, we had assembled and closed the instrument compartment and moved the entire object on to a wooden rig so that foreign metal would not affect the tuning of the antennas.

During the most stressful period of the Zenit-2 launch preparation, we were notified at the firing range that the Minister of Defense and Marshal of the Soviet Union Rodion Yakovlevich Malinovskiy would be coming for a review of rocket-space achievements. From 1956 to 1957 he had been commander-in-chief of the Ground Forces. In military circles it was said that he was not ambitious and hadn't at all been striving to become the minister of defense of a nuclear superpower. Nevertheless, in late 1957, at Khrushchev's initiative, his candidacy for that high post was approved. The time had come for Malinovskiy to become familiar with rocket-space technology.

In the autumn of 1961, a rumor swept through the firing range that Nikita Sergeyevich [Khrushchev] would accompany Malinovskiy, but soon it came to light that Khrushchev would not be coming. In view of the VIP visit, almost all officers and soldiers were taken away from testing operations and thrown into a rush clean-up job.

During this time, I was responsible for Zenit deadlines and preparation quality and tried to protest against this diversion of personnel, having voiced my indignation to State Commission Chairman Kerim Kerimov and Chief of the firing range's First Directorate Colonel Anatoliy Kirillov. Smiling, Kerimov had replied that the firing range command knows "which side its bread is buttered on" and said that they would not listen to him anyway. Kirillov didn't pass up the opportunity to give a sermon: "You civilians respect your ministers, but you aren't afraid of them. If one offends you—you go work for another under even better conditions. Your ministers have primarily a psychological authority over their people. It's a completely different matter with us military types. In the army, the authority of the minister of defense is barefaced and absolute. One can be booted out without recourse for dirt on the road and disorder in the barracks and even be evicted from his military lodgings. We can have no respect for leadership; we can despise it; but we simply must fear it."

Korolev was preparing to fly to the firing range, figuring that Khrushchev would show up there. When it became clear that only Malinovskiy would be coming, he sent Mishin there with instructions to set up an exhibition of our achievements and prospective developments in the MIK. S.P. called me and ordered me to help Mishin in every way possible in this undertaking and not to balk at meeting with the marshal.

Malinovskiy's appointment replacing Zhukov as minister of defense in 1957 happened so suddenly that it shook up not only military circles, but also all of Soviet society at that time. As recently as December 1956, Zhukov had been awarded his fourth Hero of the Soviet Union Golden Star medal on the occasion of his 60th birthday, and in early 1957 he was made a member of the Central Committee Presidium. For the people of the Soviet Union, the name Zhukov was directly associated with victory in the Great Patriotic War. He was universally recognized as a great military leader. He commanded the highest authority not only in his own country but also throughout the world. After he became minister of defense, he established working relationships with the scientific and technical circles of the military-industrial complex; he demanded that military cadres pay attention to scientific-technical progress, the development of military science, and, consequently, the restructuring of the armed forces. Zhukov's growing authority and popularity had frightened Khrushchev. He insisted on convening a Plenum of the Central Committee, which accused Zhukov of "violating the [Armed Forces] leadership principles of Lenin's Party." Zhukov was charged *in absentia*. At that time he was on board the cruiser *Kirov* conducting an official visit to Yugoslavia and Albania. The central press and radio regularly reported on the enthusiastic receptions for the "great marshal of the Soviet Union, the greatest military leader of World War II" being given abroad. And suddenly! . . . [18] Military officers whom I knew from work and all the civilians involved in the military construction work were stunned. They all had the same response: "Nikita got scared!"

At the 20th Communist Party Congress, Khrushchev had showed courage and dispelled Stalin's cult of personality, but just four years later he decided that the Soviet people must have only one idol: himself, Khrushchev. What if Zhukov got it into his head to carry out a military coup? Zhukov had every chance of becoming a second idol. He needed to be removed before it was too late. Docile and harmless Malinovskiy was named to the post of minister of defense. Afterward, if we are to believe Khrushchev's memoirs, he felt very sorry for Zhukov, but held him in exile at a dacha.

ABOUT THREE DAYS BEFORE THE VIP VISIT, A PERSONAL COOK AND SERVICE STAFF ARRIVED. The latter prepared the dining hall at Site No. 10, where the marshal's residence was, and our "deluxe" dining hall for the chief designers at Site No. 2.

Malinovskiy arrived on the evening of 27 November together with Strategic Rocket Forces Commander-in-Chief Marshal Moskalenko. After meeting with them, firing range chief Zakharov and GURVO chief Smirnitskiy tried to persuade them to go rest for a while. However, contrary to

A detailed model of a typical R-7-type launch pad shows a mockup of a launch vehicle being moved on rails to the pad. This model is on display at the Tsiolkovskiy State Museum of the History of Cosmonautics in Kaluga.

Asif Siddiqi

---

18. In October 1957, Khrushchev forcibly removed Zhukov from his post as minister of defense and also deprived him of his membership in the Central Committee of the Communist Party.

what one might expect, the 60-year-old marshal declined to rest and demanded that we show him the launch of the R-7A (or 8K74) intercontinental missile that had been put in service. The launch had been in preparation for three days at Site No. 31. Everything was going well until T-minus 30 minutes when the cavalcade of vehicles escorting the minister of defense appeared.

Mishin was at Site No. 31 at that time representing the chief designer. I had stayed at Site No. 2 and was relaying reports from the launch site to Korolev in Podlipki via the high-frequency communications channel.

They explained to Malinovskiy that they were at T-minus 20 minutes in the countdown. He did not get close to the missile; he withdrew to a specially dug reinforced clean trench on the crest of a hill where an icy wind was blowing. Right away the *bobiks*, well-known manifestations of the "visit effect" started. The valve assembly feeding liquid oxygen had been installed askew and an oxygen leak appeared. A 30 minute delay was announced. When the time came to remove the trusses, through an oversight, the cable of the tank fueling gauge broke. Once again a delay was announced—this time for 1 hour! When Korolev heard about this, he asked me to pass on the order to "Drain the oxygen! Make note of the team's disgraceful work! Replace the cable!"

The marshal had been freezing in his trench for 2 hours now. He became incensed and said, "It's more like T-minus 12 hours. You have already wasted three days. Now it's 2300 hours. Tomorrow at 1100 hours, I want to see you people work!"

On the way back, we wanted to bring him to our dining hall for a bite to eat. The best restaurants in Moscow would be proud of the dinner they served there. Malinovskiy turned down dinner at Site No. 2 and departed for his residence, having given warning that early in the morning he would be at Site No. 41 where preparation was under way for the launch of Yangel's R-16. But the streak of bad luck also extended to Yangel's launch site. The on-board self-activating batteries produced reduced voltage and needed to be replaced. The launch was postponed for several hours.

Malinovskiy was livid and ordered the firing range supervisors and directorate chiefs to stay at Sites Nos. 41 and 31 to monitor launch preparations. He promised to return soon to verify. Then, accompanied by Moskalenko, he drove over to our area to see Site No. 1, the exhibition in the MIK, and an R-9 launch.

At Site No. 1 Malinovskiy walked up to the edge overlooking the moat-like gas-venting ditch. Leaning heavily on the railing, he asked Moskalenko, "How high does the gas come up?"

Marshal Moskalenko didn't know what to say. One of Barmin's colleagues, who happened to be standing nearby, kept his wits about him: "Up to that notch, Rodion Yakovlevich."

The minister of defense looked around disgruntled: "Why did some random civilian answer me instead of an officer?"

For show they rolled out the 8K72 three-stage launch vehicle with Vostok No. 5 to the launch site and demonstrated how it was erected in the launcher. They wanted to show one of the future cosmonauts in a spacesuit—they rehearsed the report for this "show" for about three days. But all Malinovskiy said was, "Why all this fuss when it's freezing cold?"

Without lingering, without a word of thanks, he climbed back into his car to drive down to the lower-lying R-9 launch site. Mishin just managed to meet him there and begin his report explaining what a great future intercontinental missile this was and what super-cooled oxygen was, when Malinovskiy interrupted his report: "Colonel-General Ivanov was here; he sized everything up and gave me a report. Don't waste time. I've been briefed on everything."

The R-9 launch was postponed, and the minister of defense departed for the military builders' base. There, when all available soldiers had been hastily assembled and were standing in formation, he asked an unexpected question: "Men! You are pioneers, volunteers. How is your work going here?"

The soldiers fell out of formation and crowded around the minister, showering him with complaints. The marshal's escorts managed with some difficulty to extricate him from the crowd and drove him back to Yangel's Site No. 41. There, it turned out that today's launch had been postponed. Another 24 hours was needed for preparation and the correction of glitches.

"You've covered yourselves in shit," said Malinovskiy in the heat of the moment. "If this is the way it is with you, what in the world is going on in the combat units? Where is your head at, Kirill Semyonovich [Moskalenko]?"

This question was directed at Commander-in-Chief Moskalenko. But what could a dyed-in-the-wool infantry marshal say? It was agreed that the following day two launches would take place: the R-16 from Site No. 41 at 1000 hours and the 8K74 an hour later from Site No. 31.

On the frosty, overcast morning of 29 November at Site No. 41, once again there was a delay. Malinovskiy drove over to Site No. 31. This time our *Semyorka* and the launch crew did not let us down. They worked precisely. The flight proceeded normally. Wanting to please their VIP guest, the local command overdid it. During the flight, they raised the volume on communications so high that the immediate vicinity resounded not only with information about the flight status, but also with top secret performance data.

"You would have invited foreigners to the show. You should only launch cosmonauts, not missiles."

After the successful launch of the 8K74, the launch crew was assembled in formation on the launch pad. It was the tradition to give a brief report on the results of the launch and for the command to express its gratitude—Nedelin had started this practice, and it had been continued under Moskalenko. The formation tensely waited for at least two or three words from the minister of defense. However, the frowning Malinovskiy began to walk behind the formation rather than along its front line where a colonel was standing erect ready to give his report. Next came the command "about face!" Now Malinovskiy was before the front line of the formation. But he continued to slowly amble to the parking lot without raising his downcast head. The "dismissed!" command was not given, but the formation fell out of its own accord. The insulted officers watched in bewilderment as the marshals departed.

The resilient Anatoliy Kirillov cheered up the dispirited launch crew officers: "The most pleasant sight at any military review is the dust kicked up as the brass drives away. The same is true for a rocket technology review."

After this first successful launch, Malinovskiy drove over to us in the MIK. The most senior among the officers he met was Kirillov's deputy, Colonel Bobylev. Standing bolt upright, he began his report: "Comrade Marshal of the Soviet Union! This is the Assembly and Testing Building. Preparations are carried out here . . . ."

With the wave of his hand, the marshal halted the colonel's subsequent oratory: "I don't need you to tell me what's what. You already take me for a complete fool. Instead, why don't you tell me where the latrine is around here."

The colonel was crestfallen, but obediently led the marshal to the restroom, where providently for the last three days a lieutenant had been specially posted to fend off unauthorized personnel at all costs. Relieved, the minister walked around looking at the posters on display, clearly having no desire to watch and listen. Paying no attention to the behavior of this high-ranking guest, Mishin went on the offensive and began a loud and effusive briefing on the global missile: "In two to three years, conventional missiles will be ineffective. Only our global missile, developed on the basis of the R-9, will be capable of performing any missions. The Americans do not yet have any similar development."[19]

At these words, it seemed that Malinovskiy emerged from his funk, pricked up his ears, and started to listen. Observing him, I decided that he was tired

---

19. This is a reference to OKB-1's proposal for the GR-1 "global missile" designed to orbit nuclear warheads and then de-orbit and deliver them at the appropriate time to enemy targets. The GR-1 never got beyond the mockup stage, although a parallel global missile program, Yangel's R-36-O, reached flight-testing stage and was on service duty from 1968 to 1983.

or ill and that it was difficult for him to stand in his marshal's overcoat and sheepskin hat and listen to a briefing on the new missile's flight profile.

We had placed tables and chairs by the display stands figuring that both elderly marshals would take a load off their feet and attentively listen to the briefings, quenching their thirst with mineral water. We even remembered to have a coat rack with hangers. But Malinovskiy did not take off his coat and sit down. This was his way of showing his displeasure not only with current missile technology, but also with future technology. However, during the briefing he asked two questions: Why were there three stages, and where would the first two come down? Then, he asked why it was necessary to brake during the third stage, where this would happen, and how. Mishin gave a very straightforward explanation, and the marshal seemed to understand. Posters also displayed pictures of *Desna* and *Dolina*—the future silo-based missile launching sites for R-9 missile complexes. I had just begun my very eloquent presentation about the Zenit reconnaissance spacecraft when an adjutant arrived and informed us that T-minus 1 hour had been announced at Site No. 41 and it was time to drive over there for the R-16 launch.

Malinovskiy snapped back, "Never mind, they'll have another delay of 2 hours or so, and then it will be called off again."

I even managed to describe the photographic equipment [of Zenit] and suddenly he asked, "Do you change the focus in flight?"

He clearly liked Zenit and my presentation about its prospects. But when I said that we had the capability of detecting the concentration and movement of armored formations, the minister of defense remarked reasonably: "While you develop the film, figure out where your and the enemy's troops are, or else while you're giving your report, the tanks will already be long gone."

He was right. More than 30 years have passed since that time. Our television technology enables us to transmit intelligence data directly from space in real time, without returning film cassettes to the ground. The marshal left Zenit and returned once again to the R-9 lying on the transport trailer. Pointing to the engine nozzle, Malinovskiy asked, "Is this where it burns?"

Mishin confirmed this, but he was careful to say that the burning takes place higher up in the combustion chamber and the gases produced by combustion exit through those nozzles.

"Why did they take a new motor? That one next to it has been tested. Glushko told me that it's a very reliable motor."

The marshal was pointing for comparison at the 8K74 cluster lying nearby. Evidently the morning launch of this rocket had made an impact. Without waiting for Mishin to finish his long response, the marshal, smiling for the first time, thanked us and shook everyone's hand.

"So, you're staying after all. And you, the military intelligentsia? Or is there another chief designer there and that launch doesn't concern you?"

Finally, after many rechecks, the R-16 launch also proceeded normally. Considering the experience of the preceding launch at Site No. 31 and to avoid an embarrassing situation, the decision was made not to have the launch crew form up. Having received word of the successful R-16 launch, we headed to the dining hall. After two such difficult days, the marshal really should also check out the operation of the "marshal's" dining hall. That morning at breakfast, we had been notified that if the marshal stopped in for lunch, it would be a special one.

Despite this psychological preparation, we were stunned. In front of every place setting lay a leather-bound menu. To us it read like a fairy tale: "Fresh red salmon caviar, sturgeon in aspic, cold-smoked barbel, cured sturgeon fillets, back of Siberian salmon with lemon, sprat with lemon, crab with peas in mayonnaise, roast beef, ham with horseradish, chicken salad, mushrooms with onions, radishes with sour cream, farmer's cheese, sour cream, borsch Moscow-style, home-made noodles, sturgeon soup, Polish-style pike perch, braised veal, stewed chicken, steak with French fries and cornichons, chicken Kiev, crepes with jam, crepes with sour cream, black coffee, coffee with milk, tea with lemon, tea with jam." Crystal bowls sat on the table, mounded with fruit—oranges, apples, and grapes. There were all sorts of mineral water, including Yessentuki, known for its healing properties.

We began a competition to see who could try the most dishes. Someone expressed regret that the menu did not contain certain items [i.e., alcohol] beneficial to our work.

"While the marshal is here—no way! We are under a strict prohibition policy for the time being," warned the military exchange officer.

Marshal Malinovskiy never graced our dining hall with his presence. I tried to do justice to the most enticing *zakuski* and dishes by "squirreling" some away, but I simply couldn't spend more than three rubles on the "whole kit and caboodle." Someone boasted that he had eaten an entire five rubles' worth of dishes. By today's standards, the prices were really fantastic: the price of the most expensive appetizer—caviar—served as a standard of comparison. A decent-sized portion cost just 47 kopeks.[20]

After such a lunch, it was impossible to eat dinner. We didn't show up at the dining hall until the next day for breakfast. The fabulous splendor had vanished. Instead of Muscovite beauties, we were met by our old, familiar waitresses.

---

20. In the official exchange rate at this time, this was equivalent to about 50 cents.

Nevertheless, we believed that the nourishing diet of the last two days was fine compensation for the high anxiety that came with the inspection.

Sorting through my archives, I found a yellowed document, which upon closer inspection turned out to be the menu for 29 November 1961 of "deluxe" dining hall No. 5 at Site No. 2. I had snatched the menu as a souvenir 36 years before. My younger comrades felt that the document was of great historic value because it was authenticated by the signatures of the dining hall manager, the accountant, and production manager.

MARSHAL MALINOVSKIY'S DEPARTURE BROUGHT NO RELIEF FOR THE TESTERS. They had to dismantle the instrument compartment to replace a failed memory unit (ZU), look for a fault in a temperature sensor, and correct five more glitches.[21] They corrected the glitches, assembled the compartment, and reconnected it to the test panels. While they "tinkered" with the instrument compartment, we sent the descent module sphere into the pressure chamber.

On 3 December, we finally performed the mating operations, and for the first time we fully assembled the entire reconnaissance spacecraft. After assembly, we began to activate the systems and discovered the power in the command radio link transmitter dropped catastrophically. Again we began dismantling, replacing, and repeating pressure chamber tests. On 8 December, all the retesting, resoldering, and replacement operations were completed. The braking engine unit (TDU) was mated and attached. The *Chayka* tanks were filled with nitrogen.

I had known General Kerimov, Chairman of the State Commission for Zenit-2, for a long time. In 1945, when he was a captain, he had been my subordinate at the Institute RABE. We easily found common ground, especially since there were no higher-ups capable of driving a wedge between us at the firing range at that time.

Having "sized things up" in a very serene State Commission, we scheduled the launch for 11 December. That morning there was a heavy fog, a rarity for that area. From the bunker, the launch site 150 meters away could not be seen. The periscopes were useless. All hope was pinned on information from Arkadiy Ostashev, who had been sent to the first tracking station (IP) with instructions to plant "four eyes" at each telemetry screen and report everything to us in the bunker in real time.

---

21. ZU—*Zapominayushcheye ustroystvo.*

At 1240 hours the fog surrounding the periscopes filled with a thunderous roar and went from gray to pink. The launch proceeded normally. We tensely listened to the reports from the ground tracking stations (NIPs). The first one had already reported that everything was going normally. Yeniseysk and Saryshagan made contact. The third stage engine fired and was operating. Now if it could only hold on for 690 seconds. That's when separation was calculated to take place. Ostashev's calm, confident voice broke almost into a scream 410 seconds into the flight: "Pressure drop in the chamber!"

"Pressure at zero!"

There were 2 seconds of tense silence. The bunker packed with testers and developers was still.

And then came a report that dashed all hope: "Reception terminated over all lines."

So much work had been invested and so many hopes had been tied to this first reconnaissance spacecraft; and we had been betrayed by the very same *Semyorka* (with the designation 8A92) to which we were entrusting the lives of cosmonauts.[22] We gave the usual commands to immediately deliver the *Tral* films to the MIK for development. We would have to immediately brief Moscow on the causes of the failure in Block Ye. A silent throng gathered at the exit to the surface. Nikolay Petrosyan slowly climbed up the steps ahead of me. At the very top of the steps exiting the bunker, he covered his face with his hands, and leaning up against the cold cement wall, began to sob bitterly. No one tried to console him.

Three hours later we heard the results of the film analysis. The "trousers"— the gas lines from the gas generator to the turbo-pump assembly—burned through. Thrust dropped; the stage lost control; and the emergency engine shutdown (AVD) end contact tripped. This signal was transmitted on board. There, the emergency vehicle destruction (APO) system received and analyzed it. The logic programmed into the system prohibited landing outside authorized territory. The APO operated correctly and our first reconnaissance spacecraft self-destructed, blown up by 3 kilograms of TNT. The APO system developers were the only ones to gain satisfaction. Its failure-free performance was confirmed right off the bat.

My conversation via the high-frequency communications channel with Korolev was very difficult. The failure was a double whammy: first, the loss of our new spacecraft, and second, the shattering of our confidence in the

---

22. Strictly speaking, the 8A92 never launched cosmonauts. The launch vehicle for the Vostoks was the slightly different 8K82K variant of the *Semyorka*.

reliability of the launch vehicle while preparing for the piloted launches scheduled for 1962. He ordered us to forget about flying home and immediately start working around the clock to prepare the next Zenit and launch it before the New Year. Tsyganov grabbed the telephone away from me. He tried to convince Korolev that the "working class"—the entire factory brigade—had been on site continuously for 12 to 16 hours a day, without a day off for 60 days straight. He requested at least one week's respite in Podlipki.

"We asked Turkov to send us a replacement team, but he refused," complained Tsyganov to Korolev.

Tsyganov picked the wrong time to appeal to Korolev. The latter exploded: "You want to go home! You want to spit on the government decree? If necessary, you won't just work 60 days, you'll work an entire 80 days. If you don't feel like it, get the hell out!"

Tsyganov's face fell; he threw down the phone and left. It was an awkward moment for everyone who witnessed the exchange. Why such threats toward the work crew, who were in no way to blame for what had happened and were really working on all the projects selflessly and unfailingly? Here at the firing range, on more than one occasion, my temper had flared up under stress, and I resorted to strong language with my coworkers and even "outsider" engineers. But never once did I raise my voice to factory machinists and assemblers.

About 5 minutes later, while we were silently digesting what had happened, Korolev telephoned back. Now he wanted to talk to me: "Why are you letting everyone there slack off? Stop all the talk about flying home! If necessary, you'll ring in the New Year there! Is there something you're lacking? Tell me, and I'll send everything you need. If you can't reach an agreement with the people, I'll send Voskresenskiy to you! We really need a good launch, understand?"

I promised to straighten out the conflict and report in three days about the normal course of operations with the next Zenit.

After convening our technical council, we decided to restructure the work. We divided all testers and systems specialists into two groups. Each would work 24 hours, receiving an assignment for a realistic volume of work to be completed in that time. As early as five days later, it was clear that round-the-clock continuous work, plus our previous experience, infused confidence in the feasibility of a launch before 30 December. We received affirmations, final statements, and signatures for each system; and the opportunity arose to move the launch to 27 or 28 December, so that even Leningraders would have time to get home for the New Year. Then, along came a new order: the spacecraft must be prepared for long-term storage, put under guard, and everyone must return home.

From the author's archives.

Preparing a Zenit-2 spacecraft for mating with its launch vehicle.

It turned out that the investigation of the incident in Block Ye had resulted in such modifications of the propulsion system that the third stage of the launch vehicle would not be ready until January. That set off a hubbub of departures. The flight detachment did not let us down, and after bidding farewell to the firing range military testers who had nowhere to rush off to, we boarded our Il-14. This time we were not returning as victors.

The first Zenit-2 had perished without going into orbit. But during the process of developing and preparing it, we gained valuable experience in systems design, production, and testing. An interdepartmental team of like-minded people had formed. Among them, over the course of two months toiling at the firing range, that informal sense of camaraderie and kinship of interests had emerged that was so necessary when working on large and complex systems.

Surprisingly, the group departing the firing range for Moscow to ring in 1962 was not dispirited. On the airplane, under the din of the engines, even Kerim Kerimov, whom one would expect to be taking the unsuccessful debut harder than the others, gladly accepted my offer to have an off-the-record discussion of our approach to the upcoming Zenit operations on the airplane.

As technical director, I laid out my personal testament. The demise of the first Zenit-2 through the fault of the launch vehicle was a disaster; but if we honestly analyzed the probability of the equipment's reliable operation after orbital insertion, then we had to be grateful to the launch vehicle for not hauling that Zenit into space. The probability of delivering quality photographs to Earth had been very low. The chances of receiving information from the *Baykal* via the television channel had been even lower.

"And after all, this is Earth! America is not the far side of the Moon. And if we show cloudy pictures instead of clear photographs, no references to the fact that this is a "world's first" event would absolve us. The calculations show that if everything is executed per design, the resolution of *Ftor* with its 1-meter focal length will be at least 10 meters. We must confirm this! We cannot allow unjustified risk in the Zenit program. We have been in too big a rush and wanted to shine with one more achievement after the triumph of Gagarin and Titov. If there had been failures in the first flight preventing the execution of the main mission, the very idea of space reconnaissance could have been compromised. We are grateful to Petr Timofeyevich (I was referring to General Kostin) for his constant support, but we understand that not all generals in the Ministry of Defense are such enthusiasts as he. Just think—we really lucked out that they decided against launching the second Zenit at New Year's. You all felt how rushed our work was and how many undiscovered *bobiks* remained on board. Now each one of us has the opportunity to analyze everything in each system, to think things through, and check things out. Judging by the launch vehicle, we will have about two more months for thorough preparation."

There were about 15 people attending this impromptu conference on the unheated Il-14. Bushuyev, for some unknown reason, had been temporarily removed from piloted launch preparation by Korolev, and didn't feel very confident in our group. Tsybin had been appointed deputy technical chief, that is to say, my deputy. He carried out all of Korolev's orders without complaining, but sometimes he suddenly "bucked" and allowed himself to take exception. He wanted more time to devote to "human" flights. Both Bushuyev and Tsybin supported me. Only my "second deputy"—Arkadiy Ostashev—disagreed with my thesis that we had supposedly lucked out.

"We would have gained, even in the event of the most serious failures, the kind of experience that we could never gain on the ground."

However, all the chiefs of the primary special-purpose systems agreed with me. Phrasing it in various ways, they acknowledged that they had given in to the general hype about the deadlines and signed the certificates for the systems and equipment, knowing full well that they contained defects. Our in-flight conference once again clearly showed the unity of the views and interests of the actual creators of the hardware. Even our "general" customers, Colonels Viktor Shcheulov and Yuriy Kravtsov—both endowed with a sense of humor atypical of their ranks—promised to do everything in their power to delay the acceptance of the launch vehicles until we were confident of the reliability of the spacecraft and all of its systems.

We assured Zenit lead designer Boris Rublev that, within two weeks of our landing at Vnukovo airport, each of us would give him a list of operations that absolutely had to be performed on each system before the next launch. When everyone had given some thought to it and worked on it for a while, it turned out that these "operations," "modifications," and "arrangements" amounted to an entire three months' worth of work.

AFTER HOLDING THE FINAL SESSION OF THE REVIEW TEAM AND STATE COMMISSION IN LATE MARCH, it was decided to conduct the second Zenit-2 launch no later than 25 April [1962] "when ready." In mid-April, everyone involved with the Zenit project gathered once again at the firing range. April at the firing range is a marvelous month in terms of weather and a lucky month in terms of events. We remembered 12 April of the preceding year—now declared Cosmonautics Day.

This time, despite the gathering of the brass, everything was proceeding rather calmly until we started sorting things out in the "Bath & Laundry Trust Co." The chemicals for the *Baykal's* photo developing and fixing process should have been delivered from Leningrad before the very last flight-ready equipment was fueled. By the designated date, all they found on the airplane were waybills that had arrived on the right day at the right time confirming that the photo processing chemicals were in Moscow, received from Leningrad, and forwarded to Vnukovo-1 airport. Further investigation showed that the case containing the chemicals had been loaded into a vehicle headed to Vnukovo-1. But the vehicle had stopped by one more shop at Factory No. 88, where it finished loading up. When they were loading the last cargo items, the case containing the chemicals was in the way. They removed it temporarily and forgot to put it back. The waybills, however, arrived on time. With the Air Force's assistance, Korolev arranged for a special express flight for the *Baykal,* and the case arrived a day late. There were more reprimands given out for that incident than during the Zenits' entire history.

The second launch of Zenit-2 took place on 26 April 1962. The three-day flight of the second Zenit-2 was not publicized. However, this was the very satellite, announced as *Kosmos-4*, that opened the most important era of space activity, effective strategic reconnaissance, and the multi-purpose study of Earth. After the first pictures were developed, General Kostin invited Korolev into the GRU laboratory where the pictures were being processed and interpreted. Korolev was pleased and, right then and there, arranged for Tsybin, Ostashev, and me to also admire "the landscapes of America" provided by our *Kosmos-4* so that we might be "inspired." In confidence, S.P. told us that the pictures had already been shown to Khrushchev, who expressed the desire for a few more reconnaissance spacecraft that would keep track of "our enemies" a bit longer.

We reveled in our leaders' delight but still noted our own errors. Due to glitches in the attitude control system and program discrepancy, the cameras sometimes took pictures of the sky. There were also simple malfunctions in the mechanics. There was no comparison between the quality of the *Baykal's* photo-television information and that of the images on film developed after returning to the ground.

GRU operators scanning the film with a beam spot maintained that, by their estimate, the resolution of the camera with a 1-meter focus approached 5 to 7 meters. And our assignment had been a resolution of 10 meters. We surpassed the requirement twofold. It was still difficult to distinguish a truck from a railroad car, but a good beginning is half the battle.

The next Zenit-2, or *Kosmos-7*, went out on a four-day cycle on 28 July 1962. Now we already had experience. Various photographic modes, including short series and extended runs under various degrees of illumination and positioning of the Sun, were tested.

The GRU of the General Staff turned directly to Korolev, arguing the need to increase the frequency of Zenit-2 launches and correct noted defects. At one of the regular State Commission meetings while discussing flight test programs after the successful *Kosmos-7* launch, General Kostin said that, although the current program was called "flight development tests," in actual fact during each launch we obtained intelligence material of exceptional value for the nation's defense capability. We assured the general and his specialists that we and our subcontractors could make everything much better if another decree were issued. But he was a level-headed man. He thanked us for our efforts in upgrading and asked us to quickly prepare and launch what was on hand. When we pointed out defects in the equipment and systems that required substantial modifications, the officers reassured us.

During one four-day flight, as part of the "flight testing" program, we obtained photographs of regions with a total area of 10 million square kilometers with an average resolution of at least 10 meters. The entire area of the United States is 9.36 million square kilometers. We already had the effective means to keep America under observation. True, due to cloud cover and not always choosing the right time for photos, we sometimes had annoying losses of information. But that was all the more reason to increase the frequency of launches.

THE SUMMER OF 1962 WAS A HOT ONE, AND NOT JUST IN TYURA-TAM. With every passing day, the international situation heated up. Space reconnaissance was a very good thing, but we also needed ICBMs. We had not managed to put either the R-9 or the R-16 into service. Preparations were underway to launch two piloted Vostoks in August. After these, launches to Venus were scheduled in late August and Mars in October. Flight-testing of the solid-propellant RT-1 began in Kapustin Yar after a substantial delay, but for the time being the result was not good. In July Keldysh's expert commission completed its work on the N1 launch vehicle, and the VPK prepared a decree calling for flight-testing to begin on it in 1965.

At the firing range, we kept a close watch on the construction of the assembly plant and the future massive launch facilities for the N1. Unable to conceal his skepticism in my presence, Voskresenskiy told Korolev, "Sergey, you shouldn't agree to begin N1 flight tests in 1965. They're only just excavating for the big MIK, and they haven't even thought about the launch site yet."

Korolev was in a foul mood. The situation involving the N1 was such that, in meetings with Khrushchev and in the expert commissions, it was impossible to back out of old, ill-considered promises. Voskresenskiy had already been guilty of pouring salt in the wound.[23]

Now, analyzing Korolev's behavior in this complex situation many years later, I consider his decision to speed up the transfer of the reconnaissance satellite projects to Dmitriy Kozlov at our Kuybyshev-based Branch No. 3 to be the correct one. But in 1962, carried away with our first successes and with the praise of the military officers, all the Zenit creators at our KB and at the factory felt this idea was like demanding that a family give its favorite child to a faraway orphanage. After several "educational" explanations with Korolev, I realized that further wrangling was useless. He made the decision once and for all. He was too carried away with the N1 and R-9. Bushuyev was up to his ears in work with the piloted flights. Tsybin said that "honey is sweet, but the bee stings," and it's just no use arguing with S.P. Feoktistov had a great influence on Korolev during that time. He was a vocal advocate of unburdening OKB-1, so that it could increase its activity on piloted flights.

As a result of the arguments over whether to transfer the Zenit projects to Branch No. 3 in Kuybyshev, Korolev agreed to retain the development of the new Zenit-4 reconnaissance spacecraft for us [at OKB-1]. He gave in to

---

23. Around 1962–63, Korolev and Voskresenskiy had had a major falling out over the issue of ground-testing of the N1 rocket. Chertok will describe the history of the N1 booster in Volume IV.

Rauschenbach, Ryazanov, and me who had insisted that we keep the program. After three successful Zenit-2 launches, we had gained experience showing that the actual capabilities of intelligence gathering from space exceeded the initial expectations.

The photographic film removed from the descent modules was considered top secret. The film cassettes were delivered from the landing site under armed escort directly to the GRU laboratory. There I had the opportunity for the first time in my life to see how a real map of Earth, not one that has been drawn, looks from space.

The frames varied in format and quality. The short-focal-length camera held 500 frames of 18x18-centimeter format film. According to the assessments of the military officers, who very painstakingly and meticulously studied the photos and compared them with maps and other materials of theirs, the resolution was from 30 to 50 meters. If cloud cover didn't interfere, the pictures contained details that geographical maps did not. The long-focal-length camera held 1,500 frames of 30x30-centimeter format film. It aggravated me when they pointed out a highway and told me that it was possible to make out the cars on it. I couldn't count the cars, try as I might. But the specialists assured me that when the pictures were developed properly the resolution was as high as 7 to 8 meters.

"And what will it take to get the resolution down to 1 meter?"

In response came a long to-do list. First, it would be nice to have fine-grain film that was more sensitive, like what the Americans have. If we had good film suitable for the lens, and then if the focal length were increased, right off the bat this would give us a resolution of 3 to 5 meters.

"That issue's not our concern," I told the officers. "Tell the VPK to issue special decrees for the photo-chemists and opticians at the Krasnogorsk Factory. What else do you need?"

Second, it turned out that orientation needed to be more accurate. Angular oscillations needed to be eliminated and we needed perfect synchronization with Earth's motion so that there would be no blurring.

This was now our concern. We had to develop a more accurate control system that precluded angular shifting. Korolev signed off on the design of this new reconnaissance spacecraft, which was called Zenit-4, under one condition, that the military develop detailed operational requirements taking into consideration the experience of the Zenit-2 flight tests.

In September 1962, we finished preparing the fourth Zenit-2 at the engineering facility. It was difficult for my comrades to come to terms with having to relinquish a piece of engineering work that they had loved so much, and to which we had all become so close. Not only the high-ranking managers, but also the young fledgling engineers understood that we were developing a completely new field of cosmonautics. Perhaps these satellites, code-named Kosmos, were much more necessary at the time than the piloted flights that our propaganda widely publicized. Not in the distant future, but right then, that day, the day after, in the next few months and years, we would be capable of providing the superpower's top decision-makers with ever-more-accurate information needed to make military, political, and economic decisions, to conduct long-term planning, and to rapidly react to critical situations in any region of the planet.

"We finally caught the firebird, and S.P. is making us give it up," said Yurasov, expressing what everyone felt.

While I privately agreed with Yurasov's position, as one of the first deputy chief designers, I did not have the right to dispute a decision that Korolev had already made. To boost morale, I took advantage of an opportunity to hold a "closed stag party," so that my comrades and I could speak our minds. Yurasov, Ostashev, Kozko, Bashkin, Karpov, and another three or four guys (I can't remember exactly) who were vehemently against transferring the Zenits to Kuybyshev accepted my invitation with unconcealed enthusiasm.

I was "temporarily registered" in cottage No. 3 at Site No. 2 with Viktor Kuznetsov, Vasiliy Mishin, and Leonid Voskresenskiy. None of them had flown in this time. Shabarov had been put in charge of launch preparations. Since I was the only lodger, I invited my comrades to spend the evening with me at the cottage for a festive dinner. The table was set with cans of sprats, tomatoes, a huge watermelon, and bottles of Narzan mineral water. The carafe of pure alcohol looked quite humble.

When we had had a good long talk, discussing our immediate prospects and draining the carafe with Narzan chasers, Yuriy Kozko asked for permission to play Voskresenskiy's guitar.

"Let's switch from physics to lyrics," said Kozko. It turned out that he had a pure and powerful voice.

> There's a reason that the wind blew,
> And the storm clouds filled the skies,
> With a peaceful, secret light
> Someone slaked my thirsting eyes.
> The blue haze of my grief dispelled
> By a gentle touch of spring;
> No more sorrow for the Earth,
> This foreign, enigmatic thing.
> The Milky Way does not oppress,
> The mute stars cause no fear,

> For I love the world and endlessness
> As my parents' hearth so dear . . . [24]

The words, the masterful performance, the alcohol—the combination gave rise to rousing applause.

"Yesenin had no fear of stars and eternity. Do you think it's worth it to get so upset about handing over the Zenits?" said Kozko and switched to another repertoire.

Back then, Kozko was irrepressible in his creative research. He didn't get along very well with his official superiors and openly disputed official bureaucratic procedures controlling how scientific research was conducted. His subordinates adored him; and his superiors were rather afraid of his talents, while envious people schemed against him.

Thirty-three years later, as I stood by the coffin of Doctor of Technical Science, Professor Yuriy Anatolyevich Kozko, I recalled that evening at the firing range. Besides command radio links, Kozko developed original methods of radar reconnaissance. He was one of the first to utilize the very broad capabilities that were revealed during the computer processing of radar images. The pinnacle of his creativity was a precision guidance system for MIRVed warheads. He developed a technique for creating electronic digital ground maps. These unusual maps were stored in the memory of on-board computers. The on-board computer, like a navigator, compared the digital map loaded into the memory with the terrain that the on-board radar was scanning below.

Everything did not pan out immediately. And more than once, Kozko reported to the Ministry collegium, explaining that something else needed to be done and some very difficult problems remained to be solved for this navigation system to be just as standard for strategic missiles as an optical sight is for guns.[25] He eventually achieved full success and recognition. But under the terms of the SALT II treaty, the Americans demanded the destruction of specifically those MIRVed warheads that were equipped with this system.[26]

---

24. Untitled poem (1917) by Sergey Aleksandrovich Yesenin (1895–1925). Yesenin was a Russian poet who still remains very popular in Russia today. His poems about love and everyday life resonated with many who saw his verse as embodying the day-to-day struggles of the Revolutionary era. The passage was translated by Laurel Nolen.

25. All Soviet defense industry ministries had a "collegium" or "board" which was the top administrative body within the ministry.

26. The Strategic Arms Limitation Talks (SALT) II agreement was signed by the United States and the Soviet Union in 1979 but the treaty was never officially ratified by the U.S. Senate following the Soviet invasion of Afghanistan. SALT II was eventually succeeded by the Strategic Arms Reduction Treaty (START) talks. In Russian nomenclature, SALT was known as OSV—*Ogranicheniye strategicheskikh vooruzheniy* (Strategic Arms Limitation).

Then the reforms began.[27] He feverishly sought ways to keep his unique staff, which had mastered radio electronic technology and ensured the most critical part of nuclear missile parity. This proved to be much more difficult than creating the most complex systems in the not-so-distant past. His heart stopped without giving multiple prior warnings.

BUT LET'S RETURN TO TYURA-TAM. BEGINNING IN SEPTEMBER 1962, THE ZENIT-2 TESTS WERE CONDUCTED with the active participation of the young specialists of Branch No. 3. They had plunged into this brand new field with such enthusiasm that the actual developers of the Zenit-2 were convinced that our Zenit-2 would also be a favorite child in Kuybyshev. But bitter disappointments could not be avoided. After the *Baykal* complex processed the film and transmitted the images via radio link to the ground, the image quality was poor. At the recommendation of the military officers, beginning with the fourth Zenit-2 flight, we removed the *Baykal* and in its place installed two upgraded cameras, each with a 1-meter focal length. Thus, beginning with *Kosmos-10*, there were four cameras installed on board each reconnaissance spacecraft.[28] Three of them with a 1-meter focal length photographed a path 180 kilometers wide. It was possible to photograph the path in a series of varying lengths. Using programmed turns, it was possible to photograph regions located to the side of the flight path. The joint processing of the photographs made it possible to obtain a three-dimensional image of the terrain, mapping, and precise photographic referencing.

In 1962 there were five successful Zenit-2 launches. The results of reconnaissance photo interpretation were so positive that the wise military officers at the General Staff, ignoring our understanding that those were still test flights, demanded that launch frequency be increased. In the factories and at the firing range, preparation and launches were no longer considered experimental. This was work.

In early 1963, [our] factory director Turkov asked me to speak at a meeting of his shop foremen and to tell them—"to the extent allowed"—about the results of the launches in 1962 and about the strategic importance of the Zenit-2 "to boost morale at the factory." After my brief report, addressing those assembled, Turkov said, "You have not forgotten how we all worked

---

27. This is a reference to the time of *perestroika* (restructuring) and *glasnost* (openness) in the late 1980s, the polices of new Communist Party General Secretary Mikhail Gorbachev, which led to an unprecedented but brief era of liberalism in the Soviet Union culminating in the eventual collapse of the old order in 1991.

28. *Kosmos-10* was launched on 17 October 1962.

during the war. Now the Cold War is going on. Zenits are now more important than cannons. The plan for 1963 calls for the production of five units. The Progress Factory in Kuybyshev is already setting up series production. We must honorably complete this work, which is vitally important for our nation. Keep in mind that for each of you no assignment is more critical until we deliver these articles to the firing range right on time according to schedule."

Over a three-month period in 1963, from March through May, we conducted four successful launches.[29] The 12-day flight of *Kosmos-20*, launched on 18 October 1963, was the last in the experimental series. The reliability of the reconnaissance spacecraft and all of its systems had been proven.

Thirteen Zenit-2 spacecraft had been launched since December 1961. Of these, 10 had fulfilled their missions and three had been lost through the fault of the

From the author's archives.

launch vehicles. In TASS reports our reconnaissance spacecraft were referred to as *Kosmos-4, -7, -9, -10, -12, -13, -15, -16, -18,* and *-20*. The flight development test phase was over. On 10 March 1964, by government decree, the Zenit-2, the launch vehicle, and all the experimental equipment went into service.[30]

Assembly and checkout of a Zenit-2 spy satellite before launch. The robotic Zenit-2 and the piloted Vostok-3A shared the same basic design, with a spherical Descent Module and a double-cone shaped Instrument Compartment. The latter was equipped with a retro-rocket engine for deorbit.

---

29. These launches were *Kosmos-13* (on 21 March), *Kosmos-15* (on 22 April), *Kosmos-16* (on 28 April), and *Kosmos-18* (on 24 May).

30. For the first full revelations on the Zenit-2 flight-program, see Yu. M. Frumkin, "Pervyy sputnik razvedchik" ["The First Spy Satellite"], *Aviatsiya i kosmonavtika* [*Aviation and Cosmonautics*] no. 3 (1993): 41–42; V. Agapov, "Zapuski kosmicheskikh apparatov 'Zenit-2' " ["Launches of the Zenit-2 Space Vehicle"], *Novosti kosmonavtiki* [*News of Cosmonautics*] no. 10 (1996): 65–77.

This was the first instance of a complex spacecraft being accepted into service. The flight development test cycle had lasted less than two years, and just five years had passed since the launch of the world's first primitive satellite. Our mass media had labeled the American reconnaissance spacecraft "spy satellites." We called our spacecraft for all types of reconnaissance "Kosmos" spacecraft. Our interplanetary spacecraft and automated Soyuz spacecraft which were unsuccessfully inserted [into orbit] also went by the same name. Over time, the number of satellites in the Kosmos program, which first began in 1962, surpassed 3,000.[31] The huge number was the result of artificially combining various unrelated endeavors for the purposes of secrecy.

While a satellite is in space, it does not violate anyone's sovereignty or criminal code. Consequently, in principle it cannot be a spy. A spy who has penetrated the territory of a foreign country with the intention of gathering intelligence can be arrested and tried. An aircraft that has violated a country's airspace can be shot down, and a ship that has entered foreign territorial waters can be sunk. But it is against international law to harm or destroy a satellite! In space there are no state boundaries. The very same satellite has the right to conduct scientific photography of a volcanic eruption, a missile base, forest fires, urban plans, and to conduct surveillance of hundreds of natural and economically important objects.

IN 1963, KEEPING KOROLEV'S PROMISE THAT WE WOULD REMAIN THE CREATORS OF FUTURE RECONNAISSANCE SPACECRAFT, we began intensive development of the Zenit-4 design. It was assumed that cameras with a focal length of up to 3 meters would be installed on the new reconnaissance spacecraft series. An autonomous orbital stay time of at least one month was specified. This significant broadening of tactical capabilities required the development of a fundamentally new energy-conserving attitude control and navigation system. Calculations conducted jointly by the conceptual designers and photoreconnaissance specialists resulted in very stringent requirements for precision in these systems. The new control system would have to ensure the following: accelerated attitude maneuvers in the roll plane (perpendicular to the orbital plane) at angles up to more or less than 35 percent, subsequent tri-axial stabilization in this position while maintaining the specified accuracy, and feathering the solar arrays at optimal angles for the most effective illumination of their entire area during flight outside Earth's shadow.

---

31. By the end of 2007, the latest Kosmos was *Kosmos-2,437*, launched along with *Kosmos-2,435* and *Kosmos-2,436* on 25 December 2007. These Uragan-M spacecraft were part of the Glonass-M constellation, the Russian version of the U.S. Global Positioning System (GPS).

The main ideas and assignments for the subcontracting organizations came from Yevgeniy Tokar, Vladimir Branets, Larisa Komarova, and Stanislav Savchenko. Our traditional subcontractor for gyroscopic technology, Chief Designer Viktor Kuznetsov, had a glut of work on ballistic missiles, and I was unable to entice him with the prospects of developing a fantastic gyro complex.

"Try to persuade Gordeyev and Farmakovskiy in Leningrad. The orders for the Navy are getting scarce these days. Maybe you can tempt them with space issues."

I telephoned Farmakovskiy in Leningrad and reminded him of our pre-war activity. In 1936, at the Elektropribor Factory he had developed a vector sight for the long-range DB-A bomber. Back then I had convinced the airplane's Chief Designer, Viktor Bolkhovitinov, to visit Leningrad in person and familiarize himself with the unusual sight. This sight was developed and installed on the first experimental DB-A airplane. It had to be removed when the airplane was modified in an Artic version for a transpolar flight to the United States. I wrote about this tragic flight in a previous volume.[32]

The space debut of Vladimir Gordeyev and Sergey Farmakovskiy culminated with the development of the *Sfinks* (Sphinx) gyroscopic system. This was a system that could be corrected using infrared signals. It contained a dual-rotor *giroorbitant* and gyro horizon. Elektropribor delivered the first *Sfinks* to us in 1965 along with a triaxial stand for simulation and testing of the attitude control system. Unfortunately, the stand was soon abandoned. It wasn't until 15 years later that we managed to reproduce this unique work of gyroscopic technology when the necessity arose. Boris Medvedev at OKB Geofizika developed a scanning infrared vertical with a changing cone-apex scanning angle for an orbital altitude range of 200 to 400 kilometers.

But the most revolutionary innovations for that time were the electric motor flywheels used as precision attitude control actuators and the electric motor for programmed rotations. This continuously operating system generated control moment, replacing low-thrust reactive gas engines. The latter were used only during the spacecraft's initial settling phase and to desaturate the electric motor flywheels. Electric motor flywheels did not exist before the development of the Zenit-4. Andronik Iosifyan, Nikolay Sheremetyevskiy, and Naum Alper, my old friends from the pre-war days of electric engineering, had been enthralled with the problem of powered gyro stabilization. The work was not developed to the point of flight units, but a leap was made from daydreams to the execution of one of cosmonautics' most critical tasks. At

---

32. See Chertok, *Rockets and People, Vol. I*, Chapter 7.

the All-Union Scientific-Research Institute of Electromechanics (VNIIEM), a group was formed that soon developed a control moment gyroscope for the Molniya-1 communications satellite, a spherical gyroscope for the Almaz [piloted space station], and then the most world-famous control moment gyroscopes—gyrodynes—which provided non-propulsive attitude control for the *Mir* orbital complex for 15 years.[33]

Work undertaken to develop the Zenit-4 control system at OKB-1 ended in 1964 when the technical documentation was issued and the test units were manufactured. Korolev insisted on handing over all the accumulated equipment to Branch No. 3. You have to give credit to the team at that branch, which later became the Central Specialized Design Bureau (TsSKB), the primary developer of reconnaissance spacecraft. They did not waste our equipment and nurtured the friendship that we had developed over the years with the main subcontracting groups of the creators of the first reconnaissance spacecraft. Thus, many years later, Korolev's strategy for freeing up OKB-1 for the sake of piloted flights had worked out.

Larisa Komarova's Candidate's Dissertation on control dynamics was a notable event in the history of the Zenit-4. She defended it brilliantly. The first woman to have defended a dissertation in the academic council of our Central Design Bureau of Experimental Machine Building (TsKBEM)—as OKB-1 was called at that time—had to put the finishing touches on our work on automated reconnaissance satellites.[34] Komarova attained the academic degree Doctor of Science and the title of professor for work on piloted spacecraft systems.

MODERN RECONNAISSANCE SPACECRAFT, BOTH OURS AND THE AMERICANS', CAN STUDY EARTH WITH A RESOLUTION THAT MAKES IT POSSIBLE TO REGULATE STREET TRAFFIC. At the height of the Cold War, defending the assertion that supremacy in space was essential, the American "hawks" boasted that the United States' cutting-edge reconnaissance spacecraft already made it possible to determine the number and size of the stars on our officers' shoulder boards. The most revolutionary development in space reconnaissance technology in recent years is the capability to transmit color images in real time. Today, the great achievements of modern video technology are used to monitor the airplanes on the deck of an aircraft carrier, the movement of tanks in local military conflicts, and the position of the heavy hatches on strategic missile launching silos.

---

33. VNIIEM—*Vsyesoyuznyy nauchno-issledovatelskiy institut elektromekhaniki*—was the organization that developed the Meteor weather satellites in the 1960s under Chief Designer Andronik Iosifyan.

34. *Author's note*: Later, TsKBEM became the famous NPO Energiya.

In the office of Vladimir Peshekhonov, the General Director of the Russian Scientific-Production Center of NII Elektropribor and a Corresponding Member of the Russian Academy of Sciences, Professor Farmakovskiy directed my attention to the enormous map of St. Petersburg that was hanging on the wall. This had been a gift from the TsSKB in Samara [formerly Kuybushev] to NII Elektropribor in St. Petersburg. The map was a snapshot of St. Petersburg from space. On the map, you can literally make out each individual building.

I knew the actual state of affairs in Russian scientific organizations that had fostered the creation of true technological wizardry making it possible to see any nook on the globe; it was frightening to think, "is it possible that all of this will be devoured by the criminal chaos of the Russian reforms [of the 1990s]?" At the general assembly of the Russian Academy of Sciences on 29 October 1996, one of the academicians stood at the podium and read the draft of an appeal to the President and government of Russia. I shall quote the last two paragraphs of that appeal:

One must clearly understand that in the 21st century only those nations that develop and use their own high technology on the basis of strong fundamental and applied science will enjoy real independence and security.

If the policy regarding science is not changed now, then the judgment of history will be plain and simple—this policy will be branded as criminal.

As if in confirmation of this gloomy prognosis, in 1996 the press reported:

On 28 September, over the southern Pacific Ocean the Russian Kosmos-2320 *reconnaissance satellite burned up in the atmosphere. It burned up having completed its service life. Russia has no more electro-optical reconnaissance satellites. Meanwhile, this is the only means for monitoring compliance with the strategic arms agreements . . . The satellite was developed at TsSKB Progress in Samara. Its camera resolution was several dozen centimeters . . . It obtained digital images . . . Due to financial difficulties, beginning in the summer of this year Russia has been forced to resort to the unprecedented sale to the CIA of its films from the GRU archives . . . .*[35]

Reading such reports, veterans of space reconnaissance have every right to weep just as the developers of Zenit-2 cried after the loss of their first spacecraft in 1961.

---

35. Sergey Leskov, "Poslednyy rossiyskiy sputnik-shpion sgorel . . . ." [The Last Russian Spy Satellite Burns Up . . . ."], *Izvestiya,* 15 November 1996.

Chapter 13
The Hard Road to a Soft Landing

On 4 February 1966, the whole mass media of the Soviet Union broadcast the latest sensational TASS report about the extraordinary new achievement of Soviet science and technology:

On 3 February 1966, at 21 hours, 45 minutes, 30 seconds Moscow time, the automatic station Luna-9, *launched on 31 January, performed a soft landing on the surface of the Moon in the vicinity of the Ocean of Storms, west of the Reiner and Marius craters.*

The next day, a salute from the Central Committee, the Presidium of the Supreme Soviet, and the USSR Council of Ministers to the scientists and designers, engineers, technicians, and workers, and to all the employees and organizations involved in the creation of the *Luna-9* automatic station, was published. The salute stated that the execution of a soft landing on the Moon was a notable victory of Soviet science and technology, which, after the launch of the first artificial satellite, the first piloted spaceflight, and the first spacewalk by a cosmonaut, was a crucial phase in space exploration.

From the author's archives.

The first panorama of the Moon transmitted by *Luna-9* on 3 February 1966, received at Simferopol (NIP–16).

From the author's archives.

Press conference in 1966 devoted to the first panoramic photos of the lunar surface. Speaking is USSR Academy of Sciences President Mstislav Keldysh.

After that, "Moon Week" began, which swept over the entire world and ended with a press conference in the House of Scientists in Moscow, at which Academy of Sciences President Keldysh, Professor Lebedinskiy, and Academicians Vinogradov and Mikhaylov delivered speeches.[1] As usual, the real designers and creators of *Luna-9* were not included among the speakers at the press conference.

Along with a sense of triumph, my comrades and I also experienced grief and hurt. Grief, because Korolev was no longer with us. He had missed seeing the panorama of the lunar surface by less than a month. He had waited so long for a successful soft landing, had pinned so many hopes on this work, and had invested so much passion in the Ye-6 program (which was now called Luna); and yet his name wasn't mentioned at the press conference or in reference to the flight program, lunar station, and launch vehicle.

---

1. Academician Andrey Vladimirovich Lebedinskiy (1892–1965) was a leading Soviet aerospace medicine specialist. From 1963 to 1965, he was the director of the Institute of Biomedical Problems (IMBP), which directed space medicine research in the Soviet Union. Academician Aleksandr Pavlovich Vinogradov (1895–1975) was a leading Soviet geochemist who contributed to the Soviet nuclear and space programs. From 1947 to 1975, he was director of the Institute of Geochemistry and Analytical Chemistry, which is now named after him. Academician Aleksandr Aleksandrovich Mikhaylov (1888–1983) was a famous Russian astronomer who headed the Main Astronomical Observatory at Pulkovo from 1947 to 1964.

As soon as those of us who had been involved in this work realized that, after many failures, the five-year endeavor had ended in a brilliant success, there was an impromptu proposal to send Moscow a message about dedicating this work to the memory of the Chief Designer, Academician Korolev. Georgiy Babakin (who had taken over the Ye-6 soft landing lunar program from Korolev), Keldysh, and I came out with this idea. True, right off the bat Keldysh had his doubts as to whether our proposal would find its way into the text of official communiqués.

Babakin and I stood our ground and persuaded Georgiy Tyulin (he was the first deputy minister at MOM) to telephone VPK Chairman Leonid Smirnov in Moscow, hoping that this proposal would appear in the official TASS announcement. However, the offices of the Central Committee, which had composed for us the phrase "thanks on behalf of the workers for the congratulations," had their own idea. Rather than dedicate the success of *Luna-9* to the memory of Korolev, it was dedicated, on behalf of "all the organizations involved," to the 23rd Congress of the Communist Party of the Soviet Union.

The Party ideologues reasoned that Korolev's funeral in January had been on such a grand scale that it was quite sufficient. And if *Luna-9* were to be dedicated to a declassified Korolev, who was no longer around, then for the sake of fairness it would be necessary to report who actually had accomplished this extraordinary new triumph of Soviet cosmonautics.

Luna-9, which we referred to as Ye-6 vehicle No. 12 or Ye-6M vehicle No. 202, was a project of considerably smaller scale than the grandiose N1-L3 program.[2] But such an experiment was essential before conducting a landing expedition on the Moon.

Photographing the lunar surface from the closest possible range during an approach or from a low-orbit lunar satellite could not provide adequate information about the properties of the ground or reliable data about the surface, which were needed to design future vehicles executing a piloted landing. In 1959, both we and the U.S. almost simultaneously came up with the idea of developing an automatic station to execute a soft landing, survey the lunar environment, and transmit to Earth television images of "Moon rocks," data about the temperature, radioactivity, and makeup of the lunar strata. Despite our enthusiasm in our quest to be the first to accomplish

---

2. The N1-L3 was the Soviet piloted lunar landing project, comprising the N1 heavy-lift launch vehicle and the L3 lunar landing payload. The L3 mission was officially approved in 1964 to compete with the U.S. Apollo program. Chertok will devote Vol. IV of his memoirs to the Soviet piloted lunar landing project.

From the author's archives.

Partial panorama of the lunar surface, transmitted for the first time by *Luna-9* in 1966.

difficult things, we spent six years from the practical beginning of the design work until the sensational flight of *Luna-9*, which for the first time showed humankind "Moon rocks" from a distance of several meters. We were prepared to overcome difficulties, but did not imagine that they would prove to be many times worse than the greatest skeptics had predicted. Nevertheless, we performed the first soft landing before the Americans.

At Korolev's initiative, on 10 December 1959, Khrushchev signed a Central Committee and Council of Ministers resolution calling for the soft landing on the Moon of an automatic station equipped with special television equipment and scientific instrumentation making it possible to determine whether it would be feasible to travel over the lunar surface. One hypothesis held that there was a thick layer of fluffy dust on the Moon, in which any Earthly structure might sink. Well-known astronomer and science fiction writer Arthur C. Clarke in his short novel *A Fall of Moondust*, published in London in 1961, very graphically described the process of a lunar rover carrying tourists sinking in a sea of Moon dust.[3]

In Tikhonravov's department, the spacecraft design was entrusted to Gleb Maksimov's section, where Nikolay Beresnev's group performed the main

---

3. Arthur C. Clarke, *A Fall of Moondust* (London: Orion Publishing Group, 1961). Published later in Russian as Artur Klark, *Lunnaya pyl* (Moscow: Znaniye, 1969).

design work. The spacecraft and the entire soft landing program were assigned the working designation Ye-6.[4]

Korolev himself exercised conceptual leadership during all the initial stages. As the projects evolved, he handed over supervision of this project to Bushuyev, his deputy for design matters. Due to the crucial nature of control problems in this program, Korolev saddled me with the real responsibility for its implementation.

Control issues were so organically intertwined with ballistics, the selection of landing principles, spacecraft structure, and the spacecraft-launch vehicle interface that very close cooperation with conceptual designers, design engineers, Korolev's main deputies, and at least 10 subcontracting organizations was inevitable. Korolev appointed his old pal from the RNII, Arvid Pallo, as lead spacecraft designer and appointed junior engineer Aleksandr Lugovoy to assist him.

At first, this work was not a priority. The design engineers who had issued the working drawings, and then the production engineers at our factory, brushed off the Ye-6 team leads like pesky flies. Korabl-Sputniks, then the Vostok, Mars, and Venera vehicles, and combat missiles—all these hot items pushed Ye-6 to the end of the line.

When it came to the selection of design concepts and the development of a control system, I found no eager beavers among our specialists. New decrees were coming out one after another. Everyone was so caught up with other programs and occupied with day-to-day troubles that no one wanted to get saddled with one more complex problem. The spacecraft design, the control concepts, and flight program weren't determined until late 1961.

A soft landing on the Moon was one of cosmonautics' most difficult technical problems. A small and relatively simple automatic station would have to conduct the experiment. To avoid crashing into the lunar surface, braking would have to be accomplished before landing using a single rocket engine. This made it necessary to have a supply of fuel on board that comprised more than half the mass of the spacecraft before braking. Before the braking session, the control system had to orient the spacecraft along the lunar vertical, determine the moment to begin braking, and control engine thrust so that descent speed was reduced to zero immediately before contact with the lunar surface. Before the beginning of the braking session, the control system was

---

4. The earlier first generation of robotic lunar probes, launched between 1958 and 1960, were called Ye-1, Ye-2, and Ye-3. The Ye-4 and Ye-5 were never flown. "Ye" is the sixth letter of the Cyrillic alphabet (A, B, V, G, D, and Ye). Objects A, B, V, and G were nuclear warheads and Object D was *Sputnik-3*, launched in 1958.

supposed to maintain spacecraft orientation using optics; after this, the lunar vertical was established, and the distance to the Moon was measured using a radio-altimeter to determine the moment to fire the braking engine.

In addition to technical problems, organizing the joint work of the myriad of project participants demanded a lot of attention. From the first to the last days of the Ye-6 program, keeping the weight of all the hardware and the structure to a minimum was a determining factor in the selection of concepts and layouts. Everyone involved was obliged to develop their designs so that there would be enough mass to fulfill the main objective—for the automatic equipment left on the Moon to transmit images for at least 24 hours.

The three-stage 8K72 launch vehicle, which we used for the lunar expeditions in 1959, was capable of delivering no more than 325 kilograms to the Moon. According to the most stringent calculations, the Ye-6 required at least 1,500 kilograms. Right off the bat, this pointed to the need to use the new four-stage 8K78 launch vehicle, which had already begun flying for the Mars and Venus exploration programs and was expected to be used for the Molniya-1 communications satellite. However, its capabilities were also insufficient for the Venus-Mars program. The search was on for new ideas as to how to reduce the weight of the service systems.

The Ye-6 vehicle called for a radio-controlled trajectory system, an astronavigation system, and an on-board equipment control system, which at the proper moment, upon receiving a command from Earth, could activate the correcting braking engine unit (KTDU) to adjust the trajectory.[5] The radio engineers were given the opportunity to simulate the main concepts under actual conditions before landing a man on the Moon. NII-885 was selected as the main designer of radio systems, but in practice, the new orders fell to Ryazanskiy and Boguslavskiy.

Boguslavskiy proposed developing a single multi-channel radio complex that would receive radio commands from Earth, exchange trajectory information, transmit telemetry parameters, and carry out the main task— transmit images of the local environment after landing. However, the meter-range radio system proposed by Boguslavskiy did not provide a high degree of precision in the measurement of motion parameters. For the interplanetary automatic stations created for the Mars and Venus exploratory missions, SKB-567 developed a decimeter-range radio system. But its workers missed their deadlines so badly that entrusting them with the lunar project in addition to the other one was out of the question.

---

5. KTDU—*Korrektirovochno-tormoznaya dvigatelnaya ustanovka.*

While searching for precision navigation systems, the idea arose (not without my input) to resurrect the astronavigation principles that we had begun developing at NII-88 with Izrael Lisovich back in 1947. The astronavigation system after this was developed and implemented on Semyon Lavochkin's *Burya*. NII-923 Department No. 1, which started out as Lisovich's laboratory, which I had created, now worked within the Ministry of the Aviation Industry system. Aircraft astronavigation systems (SAN) were being developed there.[6] The chief designer of that project, Valentin Morachevskiy, enthusiastically received my proposal to develop a precision astronavigation system for a Moon landing.

Disputes over who would be responsible for developing the autonomous control system, including selecting the gyroscopic system, ensuring stability during engine operation, and developing the logic for interaction with the astronavigation instruments, were resolved in favor of assigning the work to Pilyugin's group. He was named chief designer of the control system. Pilyugin also took on the development of the gyroscopic complex, assigning this crucial and new problem to Vladimir Lapygin. For Pilyugin's specialists, this was their first spacecraft control project. Until now, they had only developed control systems for combat missiles or launch vehicles.

"It can't be that hard," Pilyugin's deputies Finogeyev and Khitrik assured me. They "packed" all the control elements and logic into a device weighing no more than 80 kilograms, which they called the I-100. To achieve this, they eliminated the individual housings for each unit, the interblock cables, and the connectors. Striving to improve the weight margin, Pilyugin's developers combined in this unit the control tasks of the launch vehicle's third and fourth stages and of the spacecraft. The experience that we had gained at a heavy cost on the 8K78 rockets for the Mars and Venus exploration programs was not used.

By the time the Ye-6 launches began, the 8K78 launch vehicle had already performed 10 launches for the Mars-Venera (MV) programs.[7] During the preceding two years of those 10 launches, only during one attempt did all four stages operate normally; this was during the Mars launch in 1962.[8] The other nine launches suffered five failures through the fault of the fourth stage (Block L), two through the fault of the third stage, and two failures during the first and second stage boost segments.

---

6. SAN—*Sistema astronavigatsii.*
7. These launches were performed between October 1960 and November 1962.
8. This is a reference to *Mars-1*, launched on 1 November 1962.

Pilyugin's team would also have to gain experience on a new integrated system that combined the tasks of controlling the third stage, the fourth stage (Block L), and the flight path all the way to the Moon. They really improved the weight in the combined system, but they would have to start all the experimental development and verification from square one. According to the last design weight reports, the total mass of the vehicle racing toward the Moon was 1,580 kilograms. We had been forced to relinquish weight to the television system and complex radio technology. In those days, on piloted spacecraft, wherever we could, we increased reliability at the expense of instrument or system redundancy. In the Ye-6 program, due to rigid weight limitations, it was virtually impossible to use the principle of redundancy and backup. From the standpoint of modern design concepts, this was a fundamental design flaw, but our lack of experience made us more daring.

It remained for the specialists of my departments to develop antenna-feeder units, to handle the overall coordination of the flight program, and to ensure the compatibility of all the systems. This was where I had my fill of trouble. None of our department supervisors wanted to delve into the fine points of the Ye-6's electronic equipment. The most "patriotically" inclined department supervisors frankly said: "You and S.P. gave this whole project to Pilyugin; now you need to pay the piper."

With the active assistance of Yurasov and Ostashev, I began to sort out what exactly the staffs of the various enterprises that had first worked on the project had done, and "to pay the piper" for the mistakes of others.

Keldysh considered the Ye-6 to be a very important project. To assist our ballistics experts, he involved Dmitriy Okhotsimskiy's group from the Academy of Sciences Department of Applied Mathematics (OPM).[9] At OPM, this problem was assigned to the young theoreticians Mikhail Lidov, Efraim Akim, and Aleksandr Platonov, who later became well-known scientists. Keldysh also did a thorough study of the problems associated with determining the flight trajectory, lunar landing precision, and astronavigation. The optimal trajectory and flight program were calculated through the joint efforts of OPM and OKB-1 theoreticians.

During the **first** flight phase, the launch vehicle, with its three stages, inserted the Ye-6 together with the fourth acceleration stage—Block L—into Earth orbit. Out of coverage from Soviet territory, during the **second** phase over the Gulf of Guinea, the Block L fired, bringing the spacecraft to escape velocity and aiming it at the calculated rendezvous point with the Moon.

---

9. OPM—*Otdel prikladnoy matematiki*—was part of the Mathematics Institute of the Academy of Sciences (MIAN).

On the way to the Moon, during the **third** phase, the trajectory had to be thoroughly monitored and corrected using the KTDU to ensure a landing in the previously selected area. During the **fourth** phase, braking occurred and a soft landing was executed. A vehicle with a mass of 500 kilograms, including the mass of the KTDU, descended to the Moon.

Isayev developed a special propulsion system with pump feed and three operating modes. The KTDU was one of the systems whose reliability I had faith in. Isayev jokingly pretended to be offended that, before the actual landing, his KTDU was cast away and ended up out of the station's field of view. He so wanted to have a look at his group's creation lying on the Moon's surface! But the station was supposed to separate and descend clear of the spot where the still-hot KTDU landed.

The shock absorbers—rubber bladders that were pressurized before landing and softened the impact at the moment of contact with the lunar surface—were an ingenious innovation on the spacecraft.

At our factory, four Ye-6 stations were put into production. The first of these was a developmental mockup; the flight program began with Ye-6 No. 2.

On 23 March 1962, the Central Committee and Council of Ministers issued a new decree regarding the Ye-6, which authorized the already established cooperation and set the date for the beginning of launches to 1963! The decree and ministerial order that followed accelerated activity on the Ye-6.

At first, I was only involved with control problems. But in 1963, a vacuum formed in the program's management—Korolev himself withdrew almost completely into piloted spaceflight matters. After a brief harangue, S.P. obliged me to head the entire complex of MV and Ye-6 operations. At the firing range and in the State Commissions, I acted as deputy technical chief.

Arkadiy Ostashev was my *de facto* deputy for testing at the firing range. Together with Maksimov's lead conceptual designers, ballistics experts, and representatives of the main subcontractors, we formed a "small administration," which tried to make up for the two years that had been lost according to the official schedules.

The first flight version of the Ye-6 wasn't fabricated until December 1962. All the participants in the first Ye-6 launch rang in the New Year of 1963 at the firing range by conducting the last phase of testing. During those sleepless nights, we never gave any thought to the low probability for complete success on the first attempt at a soft landing on the Moon.

Using the fundamental concepts of probability theory, it was easy to calculate that the probability of a successful landing and subsequent transmission of images to Earth was no greater than 10 percent. Our optimism was based on the reliability development method over the course of a series of launches.

The more rockets we launch, the more defects are revealed, and ultimately, after 20 to 30 launches, the rocket "begins to fly." This was the reliability development procedure inherited from the artillerymen.

THE FIRST LAUNCH OF A YE-6, VEHICLE NO. 2, TOOK PLACE ON 4 JANUARY 1963. The two new stages of rocket 8K78 and the spacecraft to be delivered to the Moon had been manufactured and tested in less than one year using documentation that was still raw and hadn't been debugged. During the launch, the three stages of 8K78 launch vehicle worked normally, but the engine of the Block L acceleration stage failed to start up in the southern latitudes over the ocean. This was not the first mysterious instance in which the Block L engine failed to start up outside the coverage area of telemetry reception stations from Soviet territory. These failures had been one of the reasons why the Mars and Venus launch program had been disrupted during the previous two years.

The numerous accident investigation commissions created after each latest Block L failure developed all kinds of theories. Among them was one theory that removed the blame from all designers and guidance specialists in advance. Someone from the Academic Institute of Chemical Physics proposed the *oksilikvid* theory. Supposedly, during flight in intermediate orbit, oxygen vapors accumulate in the compartments of Block L. If these vapors mix with fuel vapors when the engine fires, it causes an explosion and the destruction of the Block.

After the first two Blocks L stages "went missing" over the ocean, the Council of Chiefs announced at the State Commission that, "It can't go on like this! We must have information. We need to send a ship with a telemetry receiving station into the area of the Gulf of Guinea." The *Dolinsk* was immediately equipped with a telemetry station and set out from Odessa after taking on-board engine specialist Ivan Raykov and experienced telemetry interpretation specialist Konstantin Semagin. If Block L flew over the Gulf of Guinea, they would transmit information via radio telegraph to the Black Sea Fleet's Odessa Communications Center. It was forbidden to communicate in clear text, so they were limited only to indicating the causes of the failure in heavily encrypted form. Receiving these telegrams, once again we racked our brains; but nevertheless, we learned to guess something or other. It was a month or two before the tape of the telemetry recording, which they had on the *Dolinsk*, reached us by diplomatic pouch.

During the launch attempt on 4 January 1963, a Block L failed for the sixth time. The *Dolinsk*, which had been on duty in the Gulf of Guinea, finally helped unravel the mystery of the Block L failure. There hadn't been any explosion of *oksilikvids*.

This time telemetry identified a failure in passage of the electrical commands from the control system to start up the Block L engine. After long and heated arguments, developer Andronik Iosifyan's PT-500 current converter was named as the root cause. It was installed in the I-100 instrument container, which had been filled with dry nitrogen back at the engineering facility. In such an atmosphere, the carbon brushes of the manifold can rapidly deteriorate. Laboratory tests confirmed this. Iosifyan was indignant; but in the absence of other viable theories, the failure was chalked up to flashover on the manifold, resulting in a short in the power circuit.

To verify the "dry nitrogen" theory as the cause of the PT-500 converter failure, Iosifyan set up round-the-clock experiments at his institute, which used up the entire supply of flight-ready converters. They were not able to prove the complete validity of the theory. The converters malfunctioned after a considerably longer period of time than they operated in flight. Nevertheless, the State Commission accepted this model as the most reliable one. The main measures taken to increase reliability were to humidify the nitrogen and add a small amount of oxygen to the container where the PT-500 was located.

On 3 February 1963, a second attempt was made to send the Ye-6 spacecraft to the Moon. This time it didn't even make it to the fourth stage startup flight phase. The telemetry recordings and trajectory monitoring system indicated that the rocket had veered sharply, pitching toward Earth. After tracking the trajectory further, we realized that the third stage, the fourth stage (which had failed to separate), and the lunar spacecraft had re-entered the atmosphere in the vicinity of the Hawaiian Islands.

Usually during testing of the R-7A and R-9 intercontinental missiles at full range, TASS issued official warnings about upcoming launch vehicle launches in such-and-such an area of the ocean. During space launches, there were no preliminary announcements. Only through indirect signs, and from radio- and photoreconnaissance information, did the Americans guess that we were preparing for space launches. The Americans had learned to pinpoint our launches if the rocket stages fell into their radar coverage zone. When something inexplicable again entered into the atmosphere of the Americans' beloved Hawaiian Islands, a mild fuss was raised in the foreign press, and we stubbornly kept quiet. In the State Commission, even Keldysh was incensed by the categorical prohibition against reporting failures. This restriction came from the Central Committee: "We cannot have space failures." And that was that.

The causes of the second Ye-6 failure were ascertained rather quickly. The new I-100 installed on the spacecraft also controlled the third stage of the launch vehicle. It turned out that the initial orientation of the gyroscopes in the I-100 was set up with a large error relative to the base coordinate system

From the author's archives.

Engineers poring over the first photographs of the lunar surface transmitted by *Luna-9*, on 4 February 1966 at Simferopol. Sitting (with the silver hair) is Academician Mstislav Keldysh. The two men standing on the right are Boris Chertok and Chief Designer Georgiy Babakin.

of the third stage. Right from the start, the third stage was pointed at the ocean. True, a large error in the floating gyroscope also caused the trajectory of the third stage to diverge from the design trajectory in terms of altitude and velocity. As hard to swallow as it was for Pilyugin and his deputies, they accepted the blame.

During those dark days, I watched the hostility between two chiefs, Viktor Kuznetsov and Nikolay Pilyugin, who had once been on friendly terms, with great sadness. Kuznetsov was very pained by Pilyugin's attempts to organize his own production of gyroscopic instruments. Kuznetsov chalked up the failure of the Ye-6, for which Pilyugin had taken the blame, to the incompetence of Pilyugin's gyroscope specialists. And in a private conversation, Viktor hit me with this admonition: "It's your fault that Sergey agreed to this cooperation. This isn't the last spacecraft you're going to mess up. But, thank God, I don't have anything to do with this program any more."

Once again, for two months, we worked day and night to prepare the next Ye-6 launch. In the State Commission for the Ye-6 program, which Georgiy Tyulin had chaired from the very beginning, Pilyugin delivered a speech with a solemn declaration that all the trouble spots in the I-100 had been thoroughly analyzed and the necessary measures had been taken.[10] All the wiring had been coated with a special lacquer, and the tests were made more stringent to reveal possible defects back at the factory.

On 2 April 1963, we made a third attempt to go to the Moon. At last, on launch vehicle 8K78 (lucky number 13), all four stages functioned properly. We were flying to the Moon! We found out about this at Tyura-Tam via high-frequency communication from Moscow 2 hours after the launch.

In its communiqué, TASS announced to the whole world that, "on 2 April 1963 in the Soviet Union, a space vehicle was launched toward the Moon . . . . The launch vehicle was carrying automatic station *Luna-4*, weighing 1,422 kilograms. The automatic station *Luna-4* will reach the Moon in three-and-a-half days."

No mention was made in the official report about the station's actual purpose—to make a soft landing. There had been very heated arguments over this. The majority of those involved with the Ye-6 project, who had devoted not only their intellect, but also their heart and soul to this program, grumbled and at times sounded off vociferously at meetings over such communiqués. Having joined forces with Ryazanskiy, I expressed our shared opinion to Tyulin and Keldysh that this kind of information belittles the real significance of the space program and generates doubt among the public as to its practicality. We also sniped about the last words in the official statement: "A special telemetry complex on the territory of the Soviet Union is tracking the station's flight, determining its trajectory parameters, and receiving scientific information."

We had every reason to be proud of the large 32-meter parabolic antenna and the entire Simferopol NIP-10, which performed the tracking and control functions during the entire route to the Moon.[11] American surveillance assets had precisely determined the location of the large antenna near Simferopol, as well as the large antennas at NIP-16, the Yevpatoriya center for deep space communications. Why did the American press, and later the European press, write more good things about us than we did ourselves? We didn't understand and we were outraged. Tyulin, who was very well versed in the entire process of signing off on the texts of official communiqués, chuckled over our emotions and retorted, "Why report about the Simferopol center if we are forbidden to report even about the launch site?"

It wasn't the first time that, in response to our indignation, Keldysh had to make the excuse that it was out of his hands. The office of the Central

---

10. *Author's note*: A. G. Mrykin replaced Tyulin in 1965.
11. Simferopol is the capital of Crimea in present-day Ukraine.

Committee is invoking the higher interests of the State and, in so doing, is using the argument: "We shouldn't report anything about our failures. If the Americans guess what's going on and know about a lot of other things, then that's their business, their worries. We should not be the source of such information." More succinctly put—let the whole world find out about the launch site, the control center, and our hardships and failures, just not from us.

Tyulin told us that he had asked Korolev to allude to the harm done by this kind of half-truth in publications about the space program during his next personal meeting with Khrushchev or conversation with him over the Kremlin phone line. "But," Tyulin added, "Sergey refused to bring up this subject."

Meanwhile, we had lost so much faith in the reliability of the four-stage 8K78 launch vehicle that all of us, plus the State Commission, were psychologically unprepared for a successful launch. As soon as the *Dolinsk*, which monitored the operation of Block L via the *Tral* telemetry system from the Gulf of Guinea, reported the nominal operation of all systems and that the integrator had shut down the engine, we dashed to the high-frequency communications system to ask the Moscow ballistics center. They weren't ready to answer the question, "where are we flying to?" Subsequent flight control had been transferred to NIP-10 in Simferopol.

At Site No. 2, the technical review team and the entire State Commission gathered for a brief meeting. Korolev asked me who was now personally responsible for flight control. Without hesitation, I named Yevgeniy Boguslavskiy, who, according to the duty schedule, was supposed to be at NIP-10.[12] Boguslavskiy had assigned his deputy, Izrail Pikovskiy, to prepare the Ye-6 on-board radio complex at the firing range, and he personally supervised the adjustment of the NIP-10 ground equipment with the faint hope of finally seeing the then-largest parabolic antenna operate.

Korolev didn't trust anyone to transmit instructions to NIP-10. Communications via secure high-frequency channels from Tyura-Tam to the Crimea went through Moscow. Korolev demanded that Boguslavskiy go to the high-frequency telephone at NIP-10. Despite a very poor connection, S.P. explained to him in detail how crucial the NIP-10 task was now. He said that the State Commission would fly out to Simferopol, and that he, Korolev, would fly to Moscow, and he asked Boguslavskiy to personally see to it that Korolev had a detailed report about the status of the spacecraft sent to him in

---

12. Yevgeniy Yakovlevich Boguslavskiy (1917–69) was one of the leading designers of command and control systems for Soviet missiles and spacecraft. He was a deputy chief designer at NII-885 from 1955 to 1967 and first deputy chief designer from 1967 to 1969.

Podlipki in 6 hours.[13] If everything went according to plan, he promised to be in Simferopol the very next day.

On the spot, lists quickly materialized indicating who was to be on what airplane to which destination. Korolev personally signed off on each list, and the expedition leader responsible for order among "all the civilians" was given the strictest order: seating on the airplanes will take place only in accordance with these lists. One hour later we were already in the air over the endless steppe of Kazakhstan. Along the banks of the Syr Darya River, fresh pools of water from the spring thaw glistened in the sun. The steppe had turned green only to become parched and scorched once again in a couple of months.

Every flight from Tyura-Tam to the Crimea was truly delightful for me. Back then the airplanes flew at altitudes no higher than 6,000 to 7,000 meters. Usually, they landed to refuel in Astrakhan. The route passed over the Caspian Sea and the Caucasus Mountains. I was glued to the window the entire time, trying to recognize the familiar peaks of the Central Caucasus, as well as Dombay and Svanetiya. Katya and I had not hiked through the Caucasus since before the war. In 1935, Katya was even awarded a "Mountain Climber of the USSR" badge for climbing Mount Elbrus. I was not a real mountain climber, but even touring in the mountains gives one an undying appreciation for their grandeur.

From the author's archives.

Chertok and his wife, Yekaterina (Katya) Golubkina, on holiday at Lake Balday.

---

13. Although the Moscow suburb where OKB-1 was located was officially established as Kaliningrad in 1938, residents continued to call the area by the name of the local train station, Podlipki. In July 1996, Kaliningrad was officially renamed Korolev.

Vysotskiy put it very aptly, "Only mountains can be better than mountains."[14] When one has neither the strength nor the time for mountain climbing, merely gazing at the mountains improves a person's emotional state.

From the airplane under a clear sky, a panorama of magnificent beauty opened up. I started to orient myself by searching for Elbrus. It was impossible to confuse its two sparkling snow-white peaks with any other mountain. During these brief minutes of flight over the Caucasus, the thought inadvertently arose—why squander my life on this ghastly technology and fly to the Crimea to some huge antenna to figure out whether there is dust on the Moon? Who needs this if, right here, just 1 kilometer below me, there is such earthly splendor?

But we quickly left the mountains, and below us now was the Black Sea resort coast. We were flying over the sea, and soon we would land at the concrete runway of the airport in Simferopol. Here, NIP-10 Chief Colonel Nikolay Bugayev is already there to meet us. We climb into automobiles and arrive at NIP-10 in 30 minutes.

Despite the scrupulous execution of all the ballistics specialists' demands regarding trajectory correction, we did not hit the Moon. *Luna-4*, which our institute referred to as Ye-6 vehicle No. 4, passed over the lunar surface at 0426 hours on 6 April 1963 at a distance of 8,500 kilometers. With that, releases of official communiqués ended.

For us began the tedious job of searching for the reasons behind the missed Moon shot. It was determined that the primary cause was an error in the SAN. Morachevskiy's OKB was named the culprit. For several reasons—including the fact that this OKB had been created at Keldysh's initiative (it was previously a branch of his institute), and in order to attach a great deal of importance to the program's reliability problem—the State Commission appointed Keldysh as chairman of the accident investigation commission to determine the causes of the failure. I was included in Keldysh's commission and devoted a great deal of time searching for possible causes for the failure of the SAN equipment. Despite laboratory investigations, which were wide-ranging and quite objective, no unequivocal cause could be determined. An analysis of the SAN circuitry, construction, and debugging methods helped to find so many vulnerabilities that the commission was astonished: "Where were you all before? It's hardly necessary to fly to the Moon in order to find dozens of defects in such a complex system. All of this is evident during inexpensive laboratory and factory inspections."

---

14. Vladimir Semyonovich Vysotskiy (1938–80) was a highly influential Soviet singer, songwriter, actor, and poet who still retains in inconic status in Russian society and culture.

Keldysh, who was burdened with a multitude of duties as Academy of Sciences president, surprisingly devoted a great deal of time to the commission's work and delved into such technical details that he often embarrassed the system developers. He posed questions to the developers, Pilyugin (the chief designer of the control complex), me (the head supervisor and system customer), and even military acceptance. There were no excuses, other than references to the constant rush and fear of missing deadlines.

During the uproarious proceedings, dozens of measures to increase the reliability of the SAN were proposed to the developers. The most labor-intensive of these was to establish a normal thermal mode for the instrumentation and all the system elements. Despite the perpetual fear of being overweight, the developers decided to provide a backup for angular trajectory measurements, installing a *Mayak* (Beacon) radio direction finding system.

Korolev, whom I briefed on the results of the commission's work, was outraged. "These are going to be completely new instruments! You and Morachevskiy are going to throw the program off schedule!" And sure enough, we spent an entire year manufacturing the new series and retesting it.

And so once again we were flying to the firing range to prepare and launch spacecraft No. 6 in the Ye-6 program. February, March, and April 1964 was a time of "interplanetary pandemonium" at the firing range. Preparations were under way to launch four 8K78 vehicles, one after the other.

On 19 February, we were scheduled to launch an automatic interplanetary station to Venus; on 21 March, the Ye-6 for a soft landing on the Moon; and on 27 March and 2 April, Venus launches again. In the event that the latest Ye-6 failed, there was one more launch in reserve, scheduled for 20 April. Work at the MIK continued around the clock. The State Commission alternated sessions regarding Venus with those regarding the Moon, discussing both minor glitches and serious defects. They listened to the excuses of chief designers and factory directors.

Under conditions of extreme stress, and with the testers at the engineering facility suffering continuous sleep deprivation, there were exasperating incidents. One time, through the error of the testers, the pyro cartridges of the compartment and KTDU ejection mechanisms were detonated. Another incident was the destruction of an automatic lunar station (ALS) in the pressure chamber due to the external overpressure because they forgot to equalize the pressure in the ALS pressurized housing after leak checks were performed.[15]

---

15. ALS—*Avtomaticheskaya lunnaya stantsiya* (Automatic Lunar Station)—was the generic term used by the Soviets to refer to the Ye-6 soft landing probes.

Tyulin signed telegrams to the ministers demanding punishments for the low quality of the spacecraft. Nevertheless, the four-stage launch vehicles carrying interplanetary stations were rolled out to the launch site and launched on dates within the launch window with delays no greater than 24 hours. Despite a multitude of defects in the electronic equipment of the interplanetary stations, the weakest link of the entire system continued to be the four-stage 8K78 launch vehicle itself.

On 19 February, a very bizarre incident occurred. During the launch of a Venus probe, the engine of stage three—Block I—failed to start up. An investigation showed that the failure was the result of a leak in the kerosene feed line, which passed through a conduit in the liquid oxygen tank. The distribution valve leaked and kerosene froze in the conduit line. The third stage failed to engage altogether, and the probe was lost.

On 10 March at the firing range, the State Commission heard Yevgeniy Shabarov's report on this matter and accepted the program of interplanetary launches for the next two months. On 11 March, the State Commission summoned all the chief designers to Moscow to brief a meeting of the VPK at the Kremlin. "It seems that Khrushchev's patience has run out, and he has ordered his people to mete out the punishment we deserve," Tyulin speculated. We flew to Moscow en masse, leaving Shabarov and Ostashev at the firing range with the instruction not to halt preparations on all the spacecraft for even 1 hour.

On 13 March 1964, at 1700 hours everyone who had just arrived, as well as Korolev and Keldysh, who had remained in Moscow, appeared at the Kremlin before the USSR Council of Ministers Commission on Military-Industrial Matters (VPK). VPK Chairman Leonid Smirnov conducted the meeting in the Oval Hall of the building designed by famous Architect Kazakov.[16] All the commission members were present: ministers, State Committee chairmen, and the leadership of the Central Committee Defense Department.

Korolev reported first. He began by saying that Comrade Brezhnev had called him in for a briefing and that all of us would need to fly back to the firing range that night; and, therefore, he strongly urged that the discussion not be drawn out. As far as the last failure was concerned, this was an aggravating misstep that would not affect subsequent spacecraft. The guilty parties had already been punished for their negligence; everything had been rechecked; and all that was needed was approval for the upcoming launches.

---

16. Matvey Fedorovich Kazakov (1738–1812) was a highly influential Muscovite architect who gained fame during the era of Catherine II. His designs include buildings of the Moscow State University campus and the Kremlin Senate.

Despite the pace that Korolev had set, Keldysh deliberately read the findings of all the expert commissions on the Ye-6 and MV programs slowly and tediously. Then he said that the following day he would be flying to Tokyo and would not be able to participate in the project for the next two weeks.

After very brief reports from the chief designers and Tyulin's conclusions, without further discussion the VPK amicably resolved:

To take into consideration the reports of the chief designers—comrades Korolev, Ryazanskiy, Pilyugin, Khrustalev, and Morachevskiy—that during preparation of spacecraft 3MV and Ye-6, the necessary measures were taken to ensure a high degree of reliability of all the spacecraft and launch vehicle systems taking into consideration the recommendations of the expert commissions and the experience of previous launches.

Next, the VPK changed the subject to a discussion of the status of preparation for the piloted *Voskhod* launch.

That's when I was handed a note: "B. Ye.! Go to the reception area immediately. Iren." Obviously something terrible had happened, because Iren, the disciplined and highly-experienced secretary in the VPK reception area would not have ventured, without extreme necessity, to call me away from such a meeting. When I slipped out of the hall, Iren told me that I was to call my office immediately using the Kremlin line—"Shabarov has put a high-frequency communications call through there from the firing range." Shabarov informed me that 1 or 2 hours before, while testing the Ye-6 ALS, Pikovskiy had mixed up the cables and started a fire "on board." The batteries and instruments had been knocked out of order. The spacecraft was not fit for launch and the schedule was totally disrupted.

When I emerged from the phone booth, Iren remarked, "The look on your face has changed quite a bit." How could it not? At that moment, the doors swung open and the swarm of meeting participants poured out of the hall, making arrangements as they walked for who would go where in what car. Korolev waved me off without stopping to listen to me and dashed over to his meeting with Brezhnev.

I told Tyulin what had happened. We drove to his office and began to call around over the Kremlin line, the high-frequency communications line, and ordinary telephones. The first task was to order new batteries. At Lidorenko's institute we found out that Irina Yablokova, the person in charge of all space-related batteries, had already received the news from the firing range and was getting a new set ready. At our factory, they were already removing the cables from the next ALS to send them on an overnight flight to replace the burned ones. At NII-885, they were rechecking two new instruments to replace the supposedly burned-up units and preparing to dispatch them by airplane. Without any orders "from the top," in response to a call for help from the firing range, all sorts of people had responded and, sure enough, by 3 a.m. everything that was needed to remedy the consequences of the mishap with the ALS had been delivered to the airfield.

Before driving out to the airfield, I managed to call Boguslavskiy in Simferopol. He was sound asleep at NIP-10. For starters, he swore, but then he said that his deputy Pikovskiy really had "burned up" the ALS, and he added, "But if you think that I'm going to lose any sleep over that, you're mistaken. I'm sure that you'll botch up on the trajectory before it even gets as far as [operating] the ALS. Shabarov called me just 1 hour ago and said that he had gathered 'people.' 'People' had promised him that everything would be done by the time the State Commission arrives."

That frosty night at 3 a.m., Tyulin, Khodarev, Morachevskiy, Khlebnikov (Barmin's deputy), and I took off from Vnukovo on our An-12 airplane.[17] We immediately dozed off. The pilot woke us up and informed us that Tyura-Tam was fogged in; the airfield was closed; and we were going to land in Tashkent. Tyulin flew into a rage, but the pilot was adamant. We had left Moscow, where the temperature was minus 19°C [-2.2°F], and landed in Tashkent on a blindingly sunny, hot day. While we snacked on shish kebab in the bazaar, Tyura-Tam cleared the airplane to land, and we were back in the air. After a 12-hour journey we were "home" again.

The "people" that Shabarov had mentioned the night before over the high-frequency communications line with Boguslavskiy, kept their promise. They managed to restore the ALS even without the new cables and instrumentation that had arrived with us. Now they were trying to make up for lost time and keep on schedule.

Meanwhile, the MIK was humming like a beehive. Launch vehicles were undergoing testing in shifts, and at the same time, so were three spacecraft. Piles of cigarette butts cropped up overnight in areas where smoking was prohibited—there was no time to get away to the designated areas for smoke breaks. In the testing hall, the testers were discussing circuitry, calling around to the hotels looking for someone and urgently summoning them, arguing about errors in the instructions, resoldering something, and feverishly operating control panels where multi-colored indicator lights were flashing on and off.

---

17. Vnukovo was the first major international airport in greater Moscow, about 30 kilometers from the city center. It opened for operations just after the Nazi invasion, in 1941.

On 21 March 1964, the fourth launch of the program ended very quickly. The main oxygen valve on the launch vehicle's third stage—Block I—failed to open. This was the landmark 100th launch of a *Semyorka* from the now historic Site No. 1.

Voskresenskiy, who had left his fast-paced testing job and taken on the role of a consultant, gave us a good-natured cussing out about this launch: "Here we have this *Semyorka*, our most reliable launch vehicle, which we use for piloted launches, and it pulls this kind of stunt. And you, Sergey, and Mishin, want to launch a crewed N1 to the Moon without rig testing."[18]

Kosberg's engine specialists were to blame for the failure of the main oxygen valve. "The valve pin broke off,"—such was the finding of the accident investigation commission. All the valves for the Block I stages were to be modified.

After the Ye-6 launch on 21 March, we did not fly home. On 27 March, there was a launch to Venus. All three stages of the launch vehicle operated without any glitches, but once again the Block L shattered our hopes: the automatic interplanetary station remained in a transfer [Earth] orbit. But this time the cause was determined unequivocally. When it passed for a second time over Tyura-Tam, the new *Yakhont* flight recorder—the pride of Bogomolov's telemetry specialists—downlinked information about the sequence of events from the attempt to start up the Block L engine during the preceding first orbit. It turned out that the circuitry developed by Pilyugin's electrical engineers did not take into consideration the built-in time for the actuation of the power switches. The system "broke down," and power was not fed to the valves of the Block L orientation and stabilization system at the time of engine startup and operation. In an unoriented mode, Block L tumbled in its transfer orbit. This fundamental defect in the circuitry, puzzled out thanks only to the new flight recorder, might have been the cause of the previous Block L failures.

As soon as the cause of the latest failure had been determined, Korolev and Tyulin, both furious, started looking for Pilyugin. He turned out to be at the series-production factory in Kharkov. They demanded that he fly to Tyura-Tam immediately.

Meanwhile, for the next Venus launch, this fatal flaw in the circuitry was corrected using a soldering iron in an operation that took 20 minutes. Twenty minutes to correct an error that had resulted in the loss of at least three rocket complexes!

---

18. This is a reference to the risky decision taken by Korolev and Mishin to dispense with static testing of the complete first stage of the heavy-lift N1 lunar rocket, a decision that created an enormous amount of dissension within OKB-1.

Anatoliy Kirillov could not pass up an opportunity to rebuke Pilyugin's staff: "You're nuts! How can you report to the State Commission that it took you 20 minutes to correct an error that may have caused the loss of three launch vehicles? You should have said that after investigations that lasted all night, substantial changes were made to the circuitry. Its reliability is guaranteed! Repeated testing has confirmed that the measures taken were correct. After a report like this, the brass wouldn't be so angry!"

ON 2 APRIL, AUTOMATIC STATION *ZOND-1* WAS LAUNCHED TOWARD VENUS. All four stages operated normally, Zond was on its way, and we switched back to launch preparation for Ye-6 No. 5. It turned out there was a gap of time between the Venus shots and the latest Ye-6 launch, which was scheduled for 20 April 1964. This gave me the opportunity to leave the firing range to check on how things were going with the preparation of the Molniya-1 and bare my soul to my comrades about all the defects discovered during the most recent launches.

During this period, Korolev had many meetings and conversations "at the highest level." Relations between the Strategic Rocket Forces and the Air Force had become quite strained. General Kamanin, who was in charge of the cosmonauts, was campaigning to draw up a resolution transferring piloted cosmonautics to the Air Force. In his opinion, the numerous failures were caused by the incompetence of the engineering staff of the Rocket Forces; the missile officers responsible for the acceptance and quality control of the technology in the industry did not have the experience that aviation specialists had accumulated.

Kamanin, with whom I had had the occasion to meet regarding Vostok and Voskhod control systems as well as cosmonaut training with control technology, once let it slip that Marshal Nikolay Krylov was not at all suited for the post of commander-in-chief of the Strategic Rocket Forces: "He is strictly a joint-services commander. If the minister of defense entrusted strategic weaponry to him, that's not so bad. But Krylov doesn't understand the objectives and problems of cosmonautics. What's more, it gives him too much trouble. He can't grasp, much less support, anything new." That was the opinion of a hero of the Chelyuskin epic. But from the hints of actual missile officers, I understood that General Kamanin wasn't the only one who thought that way.

Korolev invited Marshal Krylov to visit OKB-1, hear our long-range proposals, discuss the N1 program, and have a look at our organization. The visit took place on 16 April. That morning I flew out to the firing range—the scheduled fifth attempt at a Moon shot was just four days away.

The following morning, under heavy fog, we rolled out the latest rocket train from the MIK to the launch site. One hour later we were notified that Korolev was flying out from Vnukovo to join us. Would he be able to land in such fog? I decided to drive to the airfield to meet him. Shabarov, Ostashev, and Kirillov were supposed to show that everything was "shipshape" at the launch site by the time he arrived. When I arrived at the airfield, the fog had dissipated. Our squadron commander, Hero of the Soviet Union Khvastunov landed the brand-new Tu-134 beautifully.

After taking a seat next to me in the car, on the way "home," Korolev immediately started to tell me about Krylov's visit. What S.P. had to say jibed to a great extent with what Kamanin had told me. He gave us the essence of the facts; in his words, Nedelin's death had been a much greater tragedy for our cosmonautics that we had thought.

After Nedelin, celebrated marshals ended up leading the Rocket Forces. Perhaps each of them was a great commander and good military leader, but they had no sense of the future prospects for cosmonautics. Moskalenko, Biryuzov, and now Krylov had absolutely no interest in what was happening on the Moon, Mars, and Venus.[19]

Thank God, Air Force Commander-in-Chief Konstantin Vershinin handled the cosmonauts, but the Rocket Forces were responsible for the problem of reliability as a whole, and this obliged their senior officers to delve into a lot of problems that were incomprehensible and unfamiliar to them. So S.P. tried to tell Krylov about the N1 and the lunar mission profile. But this did not interest the marshal. He perked up only when the subject of the solid-fuel RT-2 combat missile came up. And once again, what interested him was how this missile would affect the "civil war"—the dispute between Yangel and Chelomey for a place in the ICBM system.[20] It wasn't just commanders-in-chief who had no desire to gain a deep understanding of our problem; Minister of Defense Marshal Malinovskiy didn't either. When we drove up to Site No. 2, S.P. changed the subject to everyday matters.

On 20 April 1964, we made our fifth Moon launch attempt. This time all three stages of our 8K78 launch vehicle operated normally. The fourth—the Block L with the ALS—entered a transfer orbit around Earth. Once again, we were in suspense waiting for reports from the *Dolinsk* in the Gulf of Guinea. The report was delayed, and we were already accustomed to consider this a

---

19. The following headed the Rocket Forces after Nedelin's death: Kirill Semyonovich Moskalenko (from 1960 to 1962), Sergey Semyonovich Biryuzov (1962–63), and Nikolay Ivanovich Krylov (1963–72).

20. For more on the "civil war," see Chapter 5 of this volume.

bad omen. And sure enough, the omen proved to be true—the engine of the Block L had again failed to start up: the startup support system had not received the command. This was already the third failure in the operation of the electrical circuitry of Pilyugin's I-100 instrument. According to established tradition, the presumed culprit is named chairman of the accident investigation commission. The State Commission named Pilyugin chairman, and I was once again a member of the latest accident investigation commission.

The *Tral* tapes that came to us two weeks later, by special courier from the *Dolinsk*, did not reveal an obvious, clear cause of the failure. Extensive thermal testing began on Iosifyan's PT-500 current converter and thermal and vibration testing on Pilyugin's I-100. Despite the experience of Pilyugin's designers, it was discovered that many elements of the I-100 had been exposed to unacceptable levels of local overheating. Iosifyan had every reason to exact revenge on Pilyugin for the accusations about "dry nitrogen." Relations between the two enterprises became so strained that I had to take on the role of mediator and peacemaker. Benefiting from the bitter experience on Keldysh's commission, I insisted on modifications to the I-100. To alleviate the thermal mode, Finogeyev proposed introducing pre-launch cooling of the instrument. Barmin agreed to air cooling before blastoff, despite the complication this entailed for its launching equipment.

It took almost another year to implement all the measures that our commission had come up with to increase the reliability of the I-100's circuitry. So many events had taken place during this year that sometimes our worries about a soft landing on the Moon receded into the background and seemed inconsequential.

The spacecraft launched on 2 April 1964 continued its flight to Venus.[21] Zenit launches continued and development of the first Molniya-1 was stepped up. From time to time, I completely dropped what I was doing with the Ye-6 and immersed myself in the problems of my pet project, Molniya-1. In June, it began to fly in its unusual orbit.[22] Molniya-1 was charming by dint of its practical orientation. In June, the purely scientific Elektron satellites were launched.

During this same hot month, June 1964, someone brought a TASS synopsis of a von Braun interview about possible dates for a flight to the Moon. Judging by the text, von Braun was certain that the Americans would land a man on the surface of the Moon before 1970.

---

21. This spacecraft was publicly known as *Zond-1* (Probe-1).

22. The first Molniya-1 (or 11F67) satellite was launched on 4 June 1964, but it failed to reach Earth orbit. A second Molniya-1 successfully reached orbit on 22 August 1964 and was publicly named *Kosmos-41*. See Chapter 16 of this volume.

The chief developer of the German "vengeance weapon"—the V-2 missile—asserted that the scope of operations achieved in the U.S. on the space programs would secure them the leading role in the world. It was hard for us to read this interview, realizing that the N1 program was clearly lagging behind the Saturn V. It seemed that we couldn't possibly master such a "simple" program as the soft landings of the small automatic Ye-6 station—"simple," at least, compared to the N1-L3 program. However, it was comforting that the Americans hadn't yet succeeded in their program for the soft landing of an automatic spacecraft either.[23]

Meanwhile, Khrushchev had not forgotten about the Moon. Korolev, Chelomey, Keldysh, the leaders of the rocket-space industry, and commanders-in-chief of the Air Force and Strategic Rocket Forces were summoned to appear at the Presidium of the Communist Party Central Committee. As a result, on 3 August 1964, Central Committee and USSR Council of Ministers Resolution No. 655-268 "On Work for Lunar and Space Research" was issued. This document ordered Chelomey to prepare for and execute a piloted flight around the Moon in 1967 using the UR-500 launch vehicle. Previous resolutions committing Korolev to prepare an expedition to land a man on the Moon and return him to Earth by 1967–68 were reconfirmed.[24] This same resolution approved the schedule of operations for the production of spacecraft by the end of 1968.

The Mars and Venus programs were not forgotten in this five year plan. Reading this resolution, the abundance and grandiose scale of the tasks was breathtaking. At the same time, the disparity between what was desired and what was possible became abundantly clear.

When Korolev shared his impressions about the discussion of the space plans at the Central Committee Presidium, I asked him whether the Ye-6 program had been mentioned. He replied that no one had raised this subject, and this was for the best. There was no need to aggravate the situation around the failures and invite increased tension from the brass.

Korolev, Keldysh, Tyulin, Bushuyev, and many of our primary subcontractors immersed themselves ever deeper in preparation for launches of the three-seat *Voskhod*. This was also good in the sense that those involved

---

23. NASA began developmental work on the Surveyor program in 1963, which was originally designed for both lunar orbiting (soon abandoned) and lunar landing by robotic probes. *Surveyor-1* successfully carried out the first American soft landing on the Moon in June 1966, about five months after the first Soviet success.

24. For a detailed look at this historic resolution, see Asif A. Siddiqi, "A Secret Uncovered: The Soviet Decision to Land Cosmonauts on the Moon," *Spaceflight* 46 (2004): 205–213.

in the Molniya-1, MV, and Ye-6 projects were not so persistently bothered by the continuous and nerve-racking scrutiny by the brass.

As before, the intense activity on combat missiles continued unabated. The competition with Yangel no longer bothered Korolev. He was much more occupied with relations with Chelomey, who, bolstered by Khrushchev's support, refused to cooperate with us on the N1 program, and set to work on flight around the Moon using the UR-500K. Korolev said that he had a meeting with Chelomey. He proposed supplementing the UR-500K fly-around program with Earth-orbit docking. But Chelomey asserted that he didn't need docking and would create a new heavy UR-700 launch vehicle.

Until this time, all flights of Soviet piloted spacecraft had been accident-free. But we had one failure after the other with the automatic stations intended to explore the Moon, Mars, and Venus. This was depressing and made us think that the cause, after all, wasn't the complexity of the automatic stations, but a lack of attention on the part of the developers during the acceptance process. For piloted spacecraft, the so-called "3KA accountability system" was in effect, i.e., a completely different sense of accountability was evident.

Consequently, it remained only to lower the very strict discipline in the production of piloted spacecraft to the level of automatic stations, and it might have resulted in a catastrophe. What was needed was an administrative, ethical, and formalized responsibility at a single and highest level for both piloted and automated programs. Korolev summed it up saying, "No one should be cut any slack for any glitch."

At the very beginning of 1965, news of the death of Semyon Ariyevich Kosberg in an automobile accident stunned us all.[25] The third stages of all *Semyorkas* were equipped with his engines, and to this day, not a single space project can do without them.

In early March 1965, I once again returned to Tyura-Tam and completely immersed myself in the preparation of the latest Ye-6. The measures recommended by all the accident review commissions were implemented on the ALS. Now we thought that it was just a matter of launch vehicle reliability.

At the firing range, the work was being performed under particularly difficult conditions due to the great convergence of supervisors of all ranks. At that time preparation was under way for the launch of *Voskhod-2*, carrying cosmonauts Belyayev and Leonov. Priority was given to this launch, but due to all sorts of delays on the *Voskhod-2,* the lunar program got ahead of it, and

---

25. Kosberg died on 3 January 1965.

on 12 March 1965, the sixth Ye-6 launch took place with ALS No. 9. The 8K78 launch vehicle lifted off normally. After following it visually until it disappeared in the high clouds, I went over to the telemetry experts and found out that all three stages had operated without glitches. The lunar portion of the missile was inserted into Earth orbit. But then, 1 hour later as expected, we received one more updated report about the latest failure of the Block L engine to start up!

The entire "lunar train" remained in Earth orbit. While we mulled over the reports from the *Dolinsk* and developed hypotheses, Moscow demanded that the State Commission send the text of a communiqué because it was impossible to hide the presence of such a large satellite from U.S. space surveillance systems. Korolev, quite crushed by the latest failure and preoccupied with preparations for the launch of Belyayev and Leonov, brushed off this problem. Keldysh proposed writing at least part of the truth. Tyulin accepted the responsibility and arranged with TASS, which reported about the launch of the latest Kosmos satellite. Thus, a brief report appeared about *Kosmos-60* in high orbit with a perigee of 201 and an apogee of 287 kilometers.

During a discussion of the most probable causes of the failure, Pilyugin once again blamed Iosifyan for the PT-500's lack of reliability. Iosifyan demanded that Pilyugin look for the cause in his own circuitry. This resulted in an actual breakdown in previously normal personal relations. Pilyugin telephoned his institute from the firing range and assigned them the task of setting up PT-500 bench tests with "extra vigilance."

They carried out their boss' instructions with particular zeal, and as a result, after a run lasting many hours, they discovered a balancing washer on the PT-500 rotor that could catch on the fastening screw of the housing lid. Such hampering interference supposedly would cause the PT-500 to malfunction: its windings would burn, followed by a short circuit and general failure of the entire power system.

Whether that happened in flight or not, they didn't find any other credible causes, and the State Commission gave its approval for the next launch under the condition that the PT-500 be replaced with another converter. Iosifyan accused me of favoritism: "Boris, you're a former electrician; you understand that we're glossing over Pilyugin's sins with a decision like this."

I explained myself, arguing that the "presumption of innocence" in the struggle for the reliability of our technology is unacceptable. We must proceed from the principle that "for any glitch there must be an action removing all suspicion of unreliability." You can't pass judgment on and punish an individual based solely on suspicion. Circuits, instruments, cables, and assemblies in our systems need to be changed or modified, without demanding full evidence of their culpability in one accident or another. The PT-500 had not been proven guilty, but suspicion had been voiced against it. Consequently, it had to be replaced and the Block L modified.

Sending fully manufactured Block L stages and spacecraft back to Podlipki from Tyura-Tam for modification was considered an inexcusable waste of time. A brigade from our factory performed the modifications right there at the MIK at the firing range.

This time, instead of one PT-500 current converter, two PT-200 converters were going to be installed. Mechanically, this was not such a complicated thing to do, but it required serious changes in the entire electrical power circuit, in the testing procedures, and at the same time, verification of the thermal mode. It was amazing that after so many failures, people were working with exceptional effort. The example of successful, accident-free piloted launches boosted morale. After all, if it was possible to fly accident-free with a human being on board, this stroke of luck was bound to reach the lunar program at some point. The description of the sixth Ye-6 launch in official reports ended with the brief statement: "The accident was caused by the failure of the PT-500 converter." I left for Moscow for several days to familiarize myself with the experiments that were being conducted at Iosifyan's institute. The experiments could be interpreted in two ways and once again I was convinced that the correct decision had been made.

Incidentally, Korolev's lunar program wasn't the only one that was plagued with failures. On 24 March, I departed for Tyura-Tam on the An-12 with Kerim Kerimov, Chairman of the State Commission for Molniya-1, and Murad Kaplanov and Nikolay Sheremetyevskiy, the chief designers of the Molniya-1 on-board systems, to prepare and launch the Molniya-1 communications satellite. Before our departure from Vnukovo-3, the crew of the aircraft warned that our landing time would coincide with the launch of Chelomey's new heavy UR-500K launch vehicle, which was scheduled for this same hour. This uprated three-stage launch vehicle is now called the Proton. Back then, it was called the *Pyatisotka* since it was assumed that its launch mass would be around 500 tons.[26] As a matter of fact, it reached a weight of 700 tons; and later in a three-stage version, and then in a four-stage version, it reached a weight of 900 tons.

On our approach to Tyura-Tam, we pressed up to the windows, hoping to feast our eyes on the liftoff of the new rocket from the airplane. But the launch was postponed, and we were given the green light to land. That evening,

---

26. *Pyatisotka* is Russian slang form for the number 500.

comfortably ensconced on the roof of our MIK, we observed the *Pyatisotka* lift off. The Sun had already set. The distance, as the crow flies, was no more than 30 kilometers. The majesty of our *Semyorka* launches always set them apart from combat missile launches. Nighttime launches were particularly exotic. Suddenly the steppe was flooded with a sea of light. The spectacle of a launch as daylight turned to dusk promised to be breathtaking, even for us rocket engineers who were already accustomed to beautiful launches.

The liftoff from Site No. 95 appeared as a fiery explosion on the horizon. There was a spray of fire and red clouds of gas. As the rocket gained altitude, resting on its fiery plume illuminated from below the horizon by the sun, its new contours gradually became more clearly defined for us and it settled into its assigned course. Everything pointed to a confident beginning, but . . . a few minutes later our admiration of the majestic launch image suddenly came to a halt. The long tongue of flames from the working engines shortened, became ragged as if they had begun to cough, and the rocket, which had just been glistening in the rays of the setting sun like a Christmas ornament, suddenly fell to pieces. Hundreds of sparkling fragments against the backdrop of the dark evening sky—and internal emptiness. A failed launch, like a sudden death, is a devastating sight.[27]

As if to compensate for this otherworldly but soul-shaking spectacle, two days later we feasted our eyes on the latest beautiful launch of our now tried-and-true combat R-9A.

> Violet hues
> Rise up with the dawn.
> To fire at such a time,
> One needs to be a poet.

In our rocket experience, launches very often took place at dawn or dusk. If Nikolay Tikhonov, who wrote those lines during our Civil War, could have participated in rocket launches, he would have to say we were all part of a poetry workshop.[28]

A distinct fiery plume pierced the violet hues of the sky. Leaving behind a white contrail, the latest *Devyatka* departed for Kamchatka. One hour later,

---

27. This appears to be an error on Chertok's part, since the UR-500 launch failure in question occurred on 24 March 1966, not 24 March 1965.

28. Nikolay Semenyovich Tikhonov (1896–1979) was a popular Soviet-era poet, part of a new generation in the post-Revolutionary period who favored a non-conformist aesthetic. Later in his life, Tikhonov fully embraced the Stalinist cause and remained a favorite of various Soviet leaders until his death.

we received the report that the warhead had reached the target with a deviation of just 500 meters. If this *Devyatka* had had a standard 2 megaton payload instead of a dummy warhead, it's unlikely that such a deviation would have been noticeable.

Just 24 hours later there was one more incident during a combat launch of the latest Yangel missile. During launch preparations, a launch control team operator, who was scurrying around in the bunker, caught the sleeve of his service jacket on a toggle switch that controlled the electrical clamps holding the fueling hoses. The fueling hoses were disconnected and oxidizer poured onto the launch pad. At that time, the tanks were also being filled with propellant. Propellant vapors mixed with vapors from the spilled oxidizer and ignited. The first stage of the missile was already completely filled and it burst into flames. There were no casualties, not counting the bruises that people received jumping out of the windows of the service building in a panic.

At that same time, at Kosberg's enterprise in Voronezh, a large engine-firing test rig was destroyed in an explosion. The explosion was caused by a spark in the oxygen pump inducer. It all would have ended with the local destruction of yet another turbopump assembly; but at that same time, the oxygen line ruptured. Five metric tons of liquid oxygen streamed out onto the small flicker of flame, causing a fire the likes of which Voronezh had not seen since the war. However, these incidents did not delay the preparation and launches of the next Ye-6s.

On 10 April 1965, the seventh Ye-6 launch took place. This was ALS No. 8. "No. 8 won't hit the Moon anytime soon," the now superstitious officers of the launch control team gloomily joked. They were justified in taking caustic jabs at the "civilians"—the developers of the missiles and spacecraft. There hadn't been a single failure through the fault of the launch control team or military detail at the engineering facility at Sites 1 and 2. In all, the calamities, the design, circuitry, or production process had been the culprits.

The life of officers at the firing range was by far more difficult than service in aviation and artillery troop units. There, everything was strictly regimented and on a very tight schedule. There was a time for training, sport, family, and rest. At the Tyura-Tam firing range, the officers working with us always had their daily routine disrupted depending on the number of glitches and defects that were found, changes in launch times, and instructions from the accident investigation commissions.

"Number eight" required yet another accident investigation commission. The third-stage engine did not build up to its operating mode due to a leak in the pressurization system of the propellant and oxidizer tanks. Another possibility was a case of engine burnout in the Block I. The remains of the "lunar train," destroyed during entry into the atmosphere, fell into the waters of the Pacific Ocean.

The next day we drew up a new preparation and launch schedule for the Ye-6 ALS vehicle No. 10.[29] After all the shifts and resolute shortening of the testing cycles, launch day fell on a sacred date—9 May 1965. Numerous war veterans were about to celebrate the twentieth anniversary of the great victory with a real fiery salute at the launch site. Hundreds of soldiers and officers would be on duty at their work stations at the tracking stations along the flight route. Crews on Pacific Ocean expedition ships and those in the Atlantic were also obliged to track the flight on that holiday.

In the interval between the seventh and eighth Ye-6 launches, the latest Molniya-1 communications satellite had to be launched. Everyone involved with the Molniya-1 program was extremely enthusiastic about launching before the May holidays.

On 9 May, I was faced with having to "celebrate" in two locations: at night, in Shchelkovo on the occasion of the first television broadcast of a newsreel of the holiday festivities over the Vladivostok-Moscow line, and in the morning, at the General Staff building, where reports about the Ye-6 launch would be relayed from the firing range.

On Frunze Street, in the General Staff building, there was a communications center where information was gathered from all the tracking stations. General Andrey Karas, who was in charge of the Command and Measurement Complex, took advantage of the holiday and arranged for all the information about the Ye-6 launch and flight to be transmitted to the office of the Chief of the General Staff. On the morning of 9 May, a small group of lunar project planners and I walked through the deserted hallways of the Soviet Army General Staff building. The officer who met us led us into the office, which, he said, was the main site for the planning of military operations during the last two years of the war. General Karas was clearly pleased with his initiative and predicted that on such a festive day, and monitored by the General Staff to boot, the Ye-6 launch would have to be successful. It was an unusual honor to celebrate the twentieth anniversary of Victory in this office. We watched the military parade on television and listened to reports from Vladivostok about the quality of the transmission via Molniya-1. The broadcast was successful, and I even forgot about Ye-6 for the time being.

After NIP-15 reported from Ussuriysk, it became clear that all three stages of the launch vehicle had functioned normally. One hour-and-a-half later,

---

29. Ye-6 vehicle no. 9 had already been launched as *Kosmos-60* in March 1965.

we got the most exciting report from Odessa—the *Krasnodar* in the Mediterranean Sea and the *Dolinsk* in the Gulf of Guinea transmitted telegrams indicating that Block L had performed its task. Karas' officers, who were at the station for communicating with the firing range, were instructed by Korolev to force their way into the General Staff office and deliver the following order: "Chertok and his group are to immediately depart for Simferopol."

On the morning of 10 May, I was already in Simferopol. Korolev, Keldysh, and the primary specialists involved in the spacecraft preparation flew here from the firing range. May in the Crimea is a beautiful time. At NIP-10, Colonel Bugayev, Chief of the Tracking Station, attempted to create a relaxing atmosphere for the honored guests and to arrange for special food. But events en route to the Moon made it impossible to appreciate the wonderful meals prepared by the cook who had been specially called in for this occasion.

Day and night, the trajectory parameters were evaluated and compared with the design value; course correction settings were determined and transmitted to the spacecraft; and confirmation signals from the spacecraft were verified. From time to time, glitches occurred in the reception of telemetry or confirmation signals. The entire radio engineering complex was plagued by operational lapses.

During one of the communications sessions, Boguslavskiy reported that there was strong external interference—evidently, the Crimean or Black Sea Fleet Air Defense Troops system was interfering with us. Korolev immediately contacted the commander of the Black Sea Fleet, and he personally, on behalf of the president of the Academy of Sciences, requested that an order be given to shut down all radio facilities. The admiral reasonably replied that he could do this on the ships, but that he was not authorized to deprive the air traffic control service of communications. Moreover, he added, the radio facilities of the Simferopol airport were not subordinate to him.

Right from the start, I had my doubts as to whether external forces were contributing to unusual interference, which had given us no cause for complaint before.

While S.P., who had summoned Boguslavskiy for consultation, tried to shut down the air defense, naval, and air force radar stations via military communication channels, Pikovskiy and I set off on our own fact-finding mission. We began by checking the cooling system of the maser—the input loop of the large antenna. After climbing up to the base of the 32-meter parabolic mirror, we caught an officer and an engineer who worked for Boguslavskiy in the act of pouring liquid nitrogen from vacuum flasks into the maser's cooling system. We were joined by another of the specialists who had measured the level of the interfering signal. Turning red and stammering, he explained that the interference had suddenly disappeared. "The truth is that the liquid nitrogen delivery was late, and evidently, a self-excitation occurred in the first heterodyne," he said.

Pikovskiy and I quickly descended to the control room, pulled Boguslavskiy away from Korolev, and explained the situation to him. He turned to Korolev and calmly said, "Sergey Pavlovich, it's all my fault. There wasn't any interference. It was a malfunction that my comrades just discovered and corrected. Please pass on my apologies to all the admirals and generals that you had to contact through my fault alone."

Everyone who was present in the large control room braced themselves for an outburst of indignation and a threat such as "you'll walk to Moscow on the railroad tracks!"[30] But S.P. chuckled and right then and there started to make phone calls apologizing to the "admirals and generals." By all appearances, not only were they not offended, they were flattered to have had even this unusual encounter with cosmonautics and a conversation with the mysterious chief designer.

However, getting rid of the false interference did not help. Ye-6 No. 10, which had been called *Luna-5* in the TASS report on the evening of 9 May, lost attitude control during the first correction attempt. We managed, with some difficulty, to get *Luna-5* to settle down, and we puzzled over what had caused its "somersaulting." It turned out that the floating gyroscopes in their settings had been given too little time to warm up. When the engine fired for a correction maneuver, the gyroscopes were not yet ready to support attitude control.

During a second attempt, it was the ballistics specialists' fault that they were unable to execute the correction maneuver. They made an error in calculating the setpoints. Korolev sent Vitaliy Bezverbyy, our man responsible for the calculations, "along the railroad tracks back to Moscow" and demanded that Keldysh apply the very same sanctions against his OPM employee Mikhail Lidov. These two future professors and doctors of science got out of sight, but neither of them set foot "on the railroad track."

S.P., who had accepted Boguslavskiy's apology with a chuckle, was absolutely livid after the mistakes that the "ground" had committed. He called up Mishin in Podlipki, and accused him of losing his sense of responsibility, demanding that he personally supervise all ballistics calculations.

For three days, we tried to make a correction in order to land anywhere on the Moon. We took 1- or 2-hour catnaps. We managed to hit the Moon, but the braking session went awry, again due to the cold gyroscopes. Because of a lack of attitude control, the spacecraft deviated from the calculated landing

---

30. This was one of Korolev's most common admonitions when he lost his temper.

point by 700 kilometers. On 12 May at 2210 hours, having made a splendid salute to Victory Day, Ye-6 No. 10 crashed into the surface of the Moon. For us this was a painful, but very instructive failure.

The world-famous voice of Levitan broadcast the latest TASS communiqué:

Luna-5 reached the surface of the Moon in the vicinity of the Sea of Clouds. During the station's flight and approach to the Moon, a large volume of information was received, which will be essential for the further development of a system for a soft landing on the Moon's surface.

For the first time, it had been officially announced that work was being conducted on the soft landing problem. This was Keldysh's doing. He had finally broken through the wall of harmful ideological censorship with the simple argument: "If we launch ALSs to the Moon so often, then we need to show the scientific results to the world's scientific community. We don't have any yet because they will appear only after a landing. At the end of the day, we need to explain why we are launching one-and-a-half-ton automatic stations, one after the other."

The latest accident investigation commission analyzed all the information and demanded that the control system be modified to guarantee that the liquid of the floating gyroscopes would warm up before correction and braking sessions. In addition, they discovered unacceptable disturbances when the pressurized container of the impact bags was released and jettisoned. Once again, all hands rushed to perform emergency modifications, retest the unit, and make reports to the State Commission.

Just a month later—on 8 June 1965—the ninth launch of a Ye-6 (with factory serial number 7) took place. The launch vehicle performed its mission completely, and the automatic station, which was given the official name *Luna-6*, departed for the Moon.

For this launch, Korolev assembled all the chiefs "in person" at the firing range. After confirming the preliminary reports that *Luna-6* was moving on a trajectory close to normal, Tyulin and Korolev announced that we would depart immediately for Simferopol. The launch took place at 1040 hours Moscow time, and at 1400 hours a team made up of Korolev, Tyulin, Pilyugin, Ryazanskiy, Pikovskiy, Morachevskiy, Bezverbyy, and I had already departed for Simferopol.

During the flight to the Moon of Ye-6 No. 7, we were supposed to demonstrate once and for all that control from Earth was possible and reliable. Before the next correction session, I had the indiscretion to tell Korolev and Tyulin that NIP-10 personnel had gained so much practice in the preceding failures, that now everything should go well. S.P. suddenly flew into a rage, rapped his knuckles on the wooden table that we were sitting around in the control room, and said crossly: "Boris, you need to keep your mouth shut. You're going to jinx us again, with two days of flight still ahead." He hit the nail on the head—the next trouble occurred toward the end of the day on 9 June during the 12th communications session. Up until that session, all systems had been running precisely according to the program. All commands were executed and telemetry was good. The last trajectory correction needed to be performed for a precise approach to the landing area. And suddenly during correction, the engine failed to shut down at the moment specified from the ground and continued to operate! It ran until it had burned up the entire fuel supply, including what was intended for the braking session. The off-nominal burn carried the station far off to one side. After the first measurements, the ballistics specialists calculated that it would pass approximately 160,000 kilometers from the Moon!

What had happened? As usual, when the on-board telemetry information was compared with the recording of commands issued from the ground, operator errors were figured out very quickly. Before the correction maneuver, the command "mark T-2" failed to pass from the ground to the sequencer specifying the engine operation time. Control personnel who weren't paying very close attention to proper setting, or "phasing" of the timer, committed the error. The mark limiting the engine operation time was lost. "They didn't phase the marker," explained Boguslavskiy.

As Colonel Amos Bolshoy later told me, Korolev and all of us were a pitiful sight to behold. Colonel Bolshoy coordinated the work of the industry specialists and those of the Command and Measurement Complex who were subordinate to the military. He was a very knowledgeable radio engineer, a decent human being, and a good-hearted comrade. We were on friendly terms. I later heard from him that, in those hours of general despondency, I had been the one who came up with the idea of using the station's flight past the Moon to test its functionality to the maximum extent possible. I don't remember whether it was my idea, but it gave us the opportunity to take action again, overcoming our funereal mood. To everyone's surprise, Korolev did not excoriate Ryazanskiy, Boguslavskiy, and their coworkers for the flight control errors committed. He participated directly in developing the subsequent program of experiments and demanded that it be executed precisely. Thus, we tested the operation of the radio system to a range of 600,000 kilometers, once again verified the normal operation of the attitude control system, performed the separation of the ALS from the KTDU using a command from the ground, and even verified the normal pressurization of the rubber bladders of the shock absorbers.

This launch enabled us to test everything in real flight mode except establishing a vertical, performing a braking session, and transmitting images. But as always, initiative is a punishable offense. We accumulated additional glitches that required modifications to the circuitry and logic.

In June, after returning from NIP-10 center in the Crimea, Korolev assembled all the "lunatics" together with the factory directors. Georgiy Babakin was invited. The production stock needed to be checked to see if it would allow the soft landing program to continue. With his inherent thoroughness, Turkov showed that his factory would not be capable of producing, testing, and shipping a Block L or a new ALS to the firing range before September. As always, S.P. took a good look at the schedules, corrected them with his beloved blue pencil, and demanded that the next launch take place "without going beyond August."

Korolev introduced Babakin as the person who would take over the lunar automatic station program. "It's time for us to make a complete switch to the piloted program. We are sending a man to the Moon in about three years. During that time, Georgiy Nikolayevich is supposed to land so many automatic stations on the Moon that there won't be any room left there for the Americans," joked S.P.

There was a serious discussion as to when we would finish up and when Babakin would begin. Babakin announced that he did not want to duplicate our work by making reproductions of the Ye-6. His conceptual designers were already working on a lunar satellite and even on a self-propelled lunar vehicle. But if we needed help, he was ready at his production facility to manufacture a couple of Ye-6 type ALSs.

"Our mission," S.P. summarized, "is to conclude the soft landing program and prove that the Moon is solid. Turkov needs to assemble another two or three ALSs from production stock, and for insurance and practice," he said turning to Babakin, "you will start up their production at your facility."

And with that, it was decided. Korolev got what he wanted. The factory demonstrated wonders of efficiency. Every day Turkov walked through all the shops that determined the fate of the ensuing Ye-6 launches, and made quick decisions in a style he liked to call "everything for the front, everything for victory."

I will remind the reader that, for the first time, two stages of the 8K78 launch vehicle were manufactured at the Progress Factory in Kuybyshev and delivered directly from there to the firing range. Our factory, "where Comrade Turkov was director," manufactured the third and fourth stages, i.e., Blocks I and L, and the Ye-6 lunar rocket itself, which was supposed to "land" the automatic station on the Moon.

In August, the latest 8K78 launch vehicle with Block L and Ye-6 No. 11 was delivered to the engineering facility at the firing range. The next, 10th, Moon shot was scheduled for 4 September 1965.

Failed launches and mishaps in the programs of flights to the Moon, Mars, and Venus had made many participants superstitious. Before blastoff, they avoided uttering any prognoses aloud. Everyone thought to himself that this time "Lady Luck" would not betray us after all. But nobody could have predicted what happened.

The rocket simply failed to lift off! As it was approaching launch readiness, the apparent velocity regulation (RKS) system issued a malfunction notification. A sensor in the RKS system had been prematurely uncaged. It was not possible to replace and reset the system on the fueled rocket. There were smart alecks who allegedly recalled seeing a cat run across the railroad tracks before the rocket was hauled out of the MIK to the launch site. Others were sure that there was a woman in the bunker during preparation, and this was just as dangerous as a woman on a naval vessel. One way or the other, the rocket would have to be "drained" and returned to the engineering facility. Everyone who had been making a mess of the lunar program for three years could expect to be punished for this launch disruption.

However, once again—for the umpteenth time—one had to admire Korolev's foresight. At his initiative, with Keldysh's support, one of the automatic stations designated for a launch toward Mars had been modified and converted into a space probe. This *Zond-3*, which we launched on 18 July 1965, passed over the invisible, far side of the Moon. The photo-television unit developed to transmit images of Mars was used to photograph and transmit images of the far side of the Moon. The quality of the photographs obtained was significantly better than what we had obtained in 1959 [with *Luna-3*].

This gave our press reason once again to talk about a major victory of Soviet science. In 1964 and 1965, while approaching the Moon, the American Ranger spacecraft had provided very good quality photographs of areas of the visible side of the Moon.[31] But when it came to photographing the far side of the Moon, once again we were ahead. We obtained excellent images of regions of the Moon that had been invisible beforehand. The images were transmitted in the microwave band using the parabolic antenna installed on board. During communications sessions, with the help of the attitude control

---

31. NASA's *Ranger 7, 8,* and *9* spacecraft, launched during 1964–65, provided the first closeup images of the surface of the Moon. These photographs were taken on the approach to lunar impact.

system, this dish antenna aimed with great precision at the large ground-based antenna. Commands from the ground prompted the image transmission.

OKB-1 developed the entire control system and the antenna technology. This was the work of teams that Rauschenbach, Yurasov, Kalashnikov, and I managed, and also of our department chiefs: Karpov, Bashkin, Legostayev, Babkov, Krayushkin, Shustov, Kupriyanchik, Chizhikov, and Vilnitskiy. They did not conceal their jubilation in this connection. After their successes controlling piloted spacecraft, not without pride, they said that they had fared better controlling the Molniya-1 satellite and interplanetary automatic stations than Pilyugin's and Morachevskiy's specialized enterprises had done controlling the Ye-6 system.

This "major victory of Soviet science" [i.e., the *Zond-3* mission] muted our Ye-6 failures for a while. Nevertheless, after the September embarrassment—the launch cancellation—tension was growing "at the highest level" and needed to be relieved. Korolev warned me not to rush off to the firing range to prepare the next Ye-6 launch. He had been informed that Ustinov, who was then secretary of the Central Committee for all defense, missile, and space issues, had demanded that Smirnov hold a meeting of the VPK with a report about our lunar "crimes" and issue appropriate organizational conclusions.

At best, "organizational conclusions" meant a reprimand at the governmental level, and at worst—removal from one's job. It was impossible to remove Korolev from the job and improper to reprimand him. Consequently, one of his deputies needed to be dealt with. Everyone knew that Korolev had appointed me as his deputy for the Ye-6 program. There were many individuals who were guilty of disrupting the program. Among them were: Pilyugin—who was responsible for the control system; Ryazanskiy—responsible for the radio complex; Morachevskiy—responsible for astronavigation; Konopatov—responsible for the failures of the Block I engine, etc. But they couldn't punish them all because "misery loves company." In this situation, I turned out to be the scapegoat for the accumulated sins of all the program participants. Which meant I needed to be singled out for punishment at the Kremlin to teach everybody else not to spend so much time debugging systems from now on.

My former colleagues, who had been promoted to jobs in the VPK and had studied the inside facts of similar sessions, warned me in advance that, regardless of what the report contained, the decision would be prepared in advance, and the organizational conclusions deliberately coordinated with the majority of the VPK members, were inescapable. At the same time, a draft of sanctions was already being prepared, in which my name would be at the top with a proposal to the minister to examine my "adequacy [for the job]" and, as regards the systems' chief designers—Pilyugin, Ryazanskiy, and Morachevskiy—the ministers were instructed to take all necessary actions.

However, there was no VPK meeting in September. Our friends in the office of the Council of Ministers, who were keeping a close watch on discussions of space plans, explained that a lot of urgent issues had accumulated, and they never got around to Moon matters.

Taking advantage of the postponement of the Ye-6 discussion at the Kremlin, Korolev decided to continue the launches. He needed the support of Keldysh and Tyulin. If the next attempt at a soft landing was successful, and theoretically, there was nothing to prevent that, the threat of sanctions at the highest level would be removed. Those involved in this program understood perfectly well that the road they had traveled had been too long and difficult to give up. "Victors are not judged" is a saying that proved true on multiple occasions in the rocket-space field.

There was a lull in piloted flights. Consequently, Chairman of the State Commission for the Ye-6 and First Deputy Minister of the new Ministry of General Machine Building Tyulin needed arguments in favor of the Ye-6 and MV programs, both of which he actively supported. The American activity in similar programs was a convincing argument. Korolev and Keldysh needed success, not for the sake of honor and glory, but to be certain of work on the N1-L3 program. All three—Tyulin, Korolev, and Keldysh—joined forces and pressed the offices of the Central Committee and VPK with the request that we be allowed to continue launches.

In all, at the end of the year, three 8K78 launch vehicles in the Ye-6 configuration and, correspondingly, three ALSs had been prepared. Two ALSs were again manufactured within extremely short timeframes by the heroic efforts of our factory, while the production of the third was completed according to our documentation at the Lavochkin OKB.

Our two ALSs received serial numbers 11 and 12. At Babakin's suggestion, the third was assigned No. 202 and not No. 13. Babakin had assigned No. 201 to the ALS that he had managed to manufacture for ground testing. This was undoubtedly his greatest service. The ground testing process was legitimized for all of Babakin's subsequent new developments. ALS Ye-6 No. 11 was the first to be prepared according to the launch schedule.

The launch schedules were cut and recut so that they wouldn't interfere with the preparation of three more spacecraft headed for Venus. At the same time, we needed to have a month in reserve to launch ALS Ye-6 No. 12. As a result, they managed to select the historic date of 4 October for the launch of No. 11. This was the eighth anniversary of the launch of [*Sputnik*] the world's first artificial satellite.

Korolev was not in Tyura-Tam for this launch. Descending into the bunker early in the morning after a sleepless night spent at the launch site, Kirillov tried to cheer us up. "At the front," he said, "I had the feeling that in this battle none of my soldiers or I would be killed. And it came true. Now I have the feeling that we'll have a normal orbital insertion. But then, after that at the Moon, that's not my battle. You're on your own there."

It turned out he was right. All four stages of the launch vehicle operated without a glitch, and the fifth headed for the Moon. TASS issued a communiqué about the launch of automatic station *Luna-7* toward the Moon. Twenty-four hours later, all the program chiefs and essential specialists once again flew to the Crimea.

The mid-course correction maneuver was performed at the calculated time without any glitches. After processing, data was received about the movement of the fifth stage in the calculated landing area. Korolev, Keldysh, and Tyulin checked into Simferopol's central hotel. Early in the morning, they drove to NIP-10 and listened to my general report about that night's events and the remarks of the chief designers on the systems; they accepted proposals on the procedure for subsequent flight control operations. Boguslavskiy usually did the latter, explaining in detail when and to where the radio command would be issued and how its passage would be monitored. The I-100 thermal mode gave us a lot of trouble. Pilyugin required that this primary instrument be shut down frequently to cool, which tremendously complicated control and monitoring during the last flight phase.

But in general, everything went so well that Korolev stopped cursing the ballistics specialists for their imprecise knowledge of the Moon's albedo at various distances.[32] The most heated emotional moment came on the fourth day, before the lunar vertical setup and braking session. The calculated time from the moment the session began until landing was around 2 hours.

Nervous tension mounted with every passing minute. A command from the ground activated the astronavigation system at the calculated time and the dramatic and complex orientation process began. An automatic system executed a sequential search for the Sun, Moon, and Earth. After the entire fifth stage settled down, the disk of the Moon would be in the center of the field of view of the astronavigation system optics. At the same time, another optical system would keep Earth in its field of view. If all went well, the antennas of the radioaltimeter would be pointed perpendicular to the surface of the Moon and it would begin to measure the rapidly decreasing distance; and at the proper moment, the KTDU would automatically execute

---

32. Albedo describes a body's reflectivity. It is denoted by the ratio of a body's reflected electromagnetic radiation to incident electromagmetic radiation.

a braking burn. If, due to some random reason or an error in the settings, Earth drifted out of the field of view of the "Earth scope," KTDU startup would be inhibited.

At the calculated time a report rang out over the public address system: "We have lunar vertical setup!" A few minutes later: "Radioaltimeter measurements are in progress—5,000 kilometers to the Moon." The minutes of agonizing waiting dragged on: "Four thousand kilometers to the Moon."

And suddenly, a shock for our nerves: "Loss of Earth registered."

"What does this mean?" Korolev asked, crestfallen.

"Losing the Earth," I explained, "inhibits KTDU firing. There won't be a braking burn."

"We've got to do something! Why aren't you giving any commands?"

But what could we do?

Only after studying the telemetry and comparing its data with the parameters of the astronavigation system settings could one understand the causes of the latest control system failure 4,000 kilometers from the Moon. Radio control established a lunar impact of enviable precision. The latest reports elicited only bitter smiles. At a brief gathering right then and there in the control room, keeping their emotions in check, Korolev and Tyulin gave out instructions obliging all the chiefs to submit operational reports and for me to write up a summary of the events and issue general findings concerning the causes of the failure.

Keldysh, as usual remaining calm even in the most critical situations, proposed: "We should give orders for the urgent preparation of the next launch so that we can squeeze it in this year."

Keldysh and Tyulin edited the TASS communiqué, published on 9 October, about the completion of the flight of station *Luna-7*. It reported that *Luna-7* reached the Moon in the vicinity of the Ocean of Storms. It went on to say:

> . . . *during approach to the Moon, a majority of the operations necessary for the execution of a soft landing on the lunar surface were performed.*
>
> *Certain operations were not performed in accordance with the program and require additional optimization.*
>
> *During the flight of station* Luna-7 *a large amount of practical material was obtained for subsequent operations.*

What's true is true. We did obtain a large amount of practical material. The last phrase of the TASS communiqué gave rise to great debates during the process of approving the text. It left open the issue of what was meant by these "subsequent operations." But they simply couldn't come up with better wording.

All night we studied the telemetry recordings and tormented Morachevskiy's staff. By morning, we had put together a plausible scenario of the failure. The angles at which the optical Earth-seeking sensors had been set when the equipment was adjusted contained an error. After Earth appeared in the optical sensors' field of view, control was carried out for 30 minutes. In the process, Earth's image fell on the very edge of the sensor's field of view, giving a signal that Earth was present. Some random factor, which in a normal situation would have been completely acceptable, caused Earth to drift out of the sensor's field of view, and the engine's braking burn was inhibited. At the same time, errors were also discovered in the devices issuing the lunar vertical setup signal. In conclusion, our group drew up measures that Morachevskiy would have to implement on the next astronavigation system package that was already at the firing range for Ye-6 No. 12.

On our return flight, in the cabin of our Il-14, Tyulin predicted, "Now 'Uncle Mitya' won't pass up his opportunity—you can expect to be called out on the carpet at the Kremlin."[33] Sure enough, two days later Mozzhorin telephoned.[34] As director of the head institute, he had received personal instructions from Ustinov to prepare a report on the status of operations on the Ye-6, explaining the causes of all the failures and including the surnames of the specific guilty parties.

"Keep in mind," added Mozzhorin, "his patience has run out and he's ready to mete out punishment."

---

33. "Uncle Mitya" was the nickname for Central Committee Secretary for Defense Dmitry Fedorovich Ustinov.

34. From 1961 to 1990, Yuriy Aleksandrovich Mozzhorin (1920–98) was director of NII-88 (later known as TsNIIMash), the leading research institution of the Soviet space program.

Chapter 14
Last Launches Under Korolev

After *Luna-7* crashed into the Moon, they really did decide once and for all seriously to "sort us all out." This was Ustinov's initiative. We were faced with reporting about the Ye-6 program at a meeting of the USSR Council of Ministers' Commission on Military-Industrial Matters (VPK). The VPK administered and coordinated the work of all the defense branches of industry, science, and technology. In principle, it replaced the specialized interdepartmental committees created soon after the war.[1]

After Khrushchev's overthrow, the "reverse reconstruction" of industry began. The republics, *krays*, and *oblasts* lost their economic independence in the management of the domestic economy.[2] The regional Councils of National Economy (*Sovnarkhozy*) were dissolved and their authority was transferred once again to central ministries. The ministries of the military-industrial complex were created on the basis of the old State Committees for the appropriate branches of science and technology, since their chairmen already held ministerial rank.

During 1965, the VPK consolidated eight ministries. To the great disappointment of the leaders of the republics and *oblasts*, the *Sovnarkhozy* were abolished and the production capacity of the defense industry was returned to the ministries. The authority of the VPK was not limited to industry. The president of the Academy of Sciences, deputy chairman of the USSR *Gosplan*, commanders-in-chief of the Air Force, Strategic Rocket Forces, Navy, and Air Defense Troops, and deputy minister of defense for new types of armament were also VPK members. The VPK chairman was simultaneously a deputy chairman of the USSR Council of Ministers.

---

1. Chertok has in mind the top-level "Special Committees" that were created after World War II to manage the atomic, rocketry, and radar programs.

2. A *kray* (territory) is an administrative region in Russia, similar to the *oblast* but less common.

Such a representative body made it possible to make decisions that had defining significance for the nation's military might. Essentially, the bulk of the scientific and technical intelligentsia was involved in one way or another with the projects of the military-industrial complex. The ministries of the electrical industry, ferrous and non-ferrous metallurgy, transport and heavy machine building industries, instrument building industry, and many other industries were not officially subordinate to the VPK, but they were tasked with plum orders for the military industry considerably more than they were for civilian production. Institutions of higher technical learning and universities received secret assignments for scientific-research projects. Consequently, the number of secret dissertations, both at the candidate's and doctoral levels, increased.

In terms of the VPK makeup, it seemed that tremendous authority was concentrated in the hands of its chairman—the entire military-industrial complex and four branches of the armed forces. In fact, the authority and might of the VPK were strictly limited. The VPK did not have the right to independently issue government resolutions. However, the offices of the VPK prepared, coordinated, and sorted out the draft resolutions of the Central Committee and USSR Council of Ministers. It was only after the General Secretary of the Central Committee [Brezhnev] and Chairman of the USSR Council of Ministers [Kosygin] signed them that any resolution acquired the force of law, and the ministers comprising the VPK and the industries on which this honor had not been bestowed were obliged to execute them.

The chief of the Central Committee's Defense Department and his staff vigilantly monitored the actions of the VPK to head off any attempts to display excessive initiative that might pose a threat to the sovereignty of the Politburo and Party bureaucracy. The minister of finance and *Gosplan* chairman were not officially subject to VPK decisions. The VPK could not be in charge of the distribution of material supplies and budgetary appropriations. A Central Committee resolution or a direct order from the chairman of the Council of Ministers was needed to obtain budgetary assets, funding for materials and equipment, or hard currency. The Ministry of Finance, *Gosplan*, and *Gossnab* recognized no other authority.[3]

Personnel matters were also outside the authority of the VPK. The VPK was not authorized to remove or appoint a director of even a minor defense factory. The Central Committee Secretariat dealt with high-ranking personnel.

---

3. *Gossnab—Gosudarstvennyy komitet po materialno-tekhnicheskomu snabzheniyu* (State Committee for Material and Technical Supply) ensured that Soviet state enterprises were furnished with producer goods.

The ministers themselves were supposed to be in charge of "local" personnel, but with the approval of the Central Committees of the local national Communist parties, regional committees (*obkom*), territorial committees (*kraykom*), and municipal committees (*gorkom*).

Nevertheless, VPK meetings and their subsequent decisions were of great importance. Here, leaders of industry and scientific institutions were afforded the opportunity to present their ideas on new types of weaponry and prospective technologies in considerable detail. The ministers turned to each other for assistance, appealing to the chairman when necessary. It was possible to critique projects, the setup of experiments, and introduce proposals that went beyond the scope of the capabilities of a single ministry.

Cosmonautics was completely accountable to the VPK. After discussing the technology and approving the proposals for the next launch, which had not only military but also political significance, a meeting usually ended with a decision to "approve the text of a letter to the Central Committee on this matter."

The Central Committee, rather than the VPK, even appointed and replaced the chairmen of the State Commissions on missile and spacecraft flight-testing. But the chairmen of the State Commissions were accountable to the VPK. The VPK brought up the most crucial strategic problems for discussion before the Defense Council. The chairman of the Defense Council was the Central Committee General Secretary.[4]

The VPK also issued its own decisions that did not require the expenditure of new budgetary funds and did not affect the interests of branches of industry outside the VPK's scope of authority. The VPK discussed plans for the development of new types of armaments, programs, and dates for spacecraft launches, and they appointed accident review commissions. Lists of particularly outstanding defense industry workers singled out for government awards also passed through the VPK.

Weekly plenary sessions of the entire VPK were held at the Kremlin in the Oval Hall, where the USSR Council of Ministers usually convened. After undergoing a triple check of their credentials, those invited to a VPK session entered the former Senate building designed by Kazakov. The great architect harmoniously blended an austere layout with rich stucco molding and bas relief.

Even though I frequently visited the Council of Ministers building, these occasions always made me feel a sense of majesty and pride from being involved with achievements of national importance. Again and again, I realized that

---

4. For most of the early space era, the general (or first) secretary was either Khrushchev (1953–64) or Brezhnev (1964–82).

many of my friends also experienced similar feelings. I must admit that, in those days, just being in the inner rooms of the Kremlin evoked certain solemn patriotic feelings that were not kindled in me in the offices of the Central Committee on Old Square or in the offices of various ministries.

Korolev made me prepare very seriously for the upcoming report on the Ye-6 program at the Kremlin. Twice he scrapped posters that I had prepared with the input of lead designers, conceptual designers, and contractors. The third time around he finally approved the "visual advertising" for the VPK members. The posters depicted the technology and flight program with a list of all the launches and the causes of the failures. Separate posters for each failed launch listed the measures we had implemented in large print so that it wouldn't be necessary to peek at a crib sheet during the report. The general picture proved to be so convincing that Bushuyev, who was present during our meetings with Korolev, said, "After a report like that, they should give you an award rather than punish you."

Korolev pored over the thesis of my report, and then he corrected, rewrote, and rebuked me for not emphasizing the significance of the actual process of mastering a totally new system while I was stating the facts.

Criticizing my thesis S.P. said, "If one were to listen to you, they'd get the impression that after each failure we instantly understand everything. That begs the question: where were we beforehand? We waited for a failure so that we could quickly make modifications and conduct a new launch? And we conducted 10 launches like this? Do you want them to think that we're just bumbling along irresponsibly? But you know better than I do—that's not the way it is at all! You need to explain that we are going through a comprehensive learning process. Our ministers are hardly stupid—they will understand! We are gaining tremendous experience!"

After reworking the report thesis and posters many times, I came up with an illustrated history of the lunar launches beginning in 1958. Considering that the Ye-6 program used the four-stage 8K78 launch vehicle, which was also used for Mars, Venus, and Molniya launches, a large amount of time in the report was devoted to issues of its experimental development, its reliability, and explaining the differences between the versions of its third and fourth stages in the various programs.

During my last rehearsal, I spent 40 minutes on the report. Korolev was satisfied, but he warned, "They won't give you more than 10 minutes to report before they'll start interrupting, asking questions, and then they'll launch into a free-for-all discussion between Kalmykov, Dementyev, and

Tyulin. Afanasyev, our new minister, is not yet up to speed on these projects.[5] He'll keep quiet. It's a good report for Keldysh's interdepartmental council, but not for the VPK.[6] You hold on to it, but go through it again and cut it down to 10 to 12 minutes."

Unfortunately, I no longer have the text of the report, but I have reproduced the gist of it from notes jotted down on notepads and from memory.

*The **first** objective was to reach the Moon and deliver a USSR pendant. This was accomplished using the three-stage 8K72 launch vehicle. The first two stages were standard R-7 rockets, and the third stage was the Block Ye with Kosberg's engine. The capsule containing the pendant, which we proudly referred to as the 'automatic interplanetary station,' weighing 360 kilograms, crashed when it reached the Moon. The durable pendant survived. The instant loss of radio contact with the station served as proof that it had hit the Moon. It took six launches to accomplish the mission. Of these, two failures took place during the stage-one flight phase; two occurred during the stage-two phase; one launch missed the Moon by 5,000 kilometers due to a ground control station error; and finally, in September 1959, the mission was accomplished.*

Having expended our efforts on six launches, we solved the following basic problems:

- *optimized the three-stage launch vehicle and mastered startup of stage three for the first time*

- *increased the reliability of the first stages after studying the phenomenon of longitudinal vibrations, and*

- *conducted radio communication for the first time at lunar distances.*

These six launches proved to be very useful for the development of technology, but did not provide any new scientific data about the Moon itself.

*The **second** objective was to photograph the far side of the Moon and transmit the images to Earth. We accomplished this objective in two phases.*

During phase one, using the same 8K72 launch vehicle, we inserted a Ye-2 model automatic lunar station on a path toward the Moon. The first Ye-2 launch was successful. After that, two launches with an improved automatic lunar station called the Ye-3 failed due to the launch vehicle.

---

5. Valeriy Dmitriyevich Kalmykov, Petr Vasilyevich Dementyev, and Sergey Aleksandrovich Afanasyev were respectively the ministers of the radio industry, the aviation industry, and general machine building (i.e., space and missiles). Georgiy Aleksandrovich Tyulin was Afanasyev's first deputy.

6. This is a reference to the Interdepartmental Scientific-Technical Council for Space Research (MNTS-KI) headed by Keldysh which was an inter-branch forum to preside over technical issues related to "civilian" space research during the 1960s and 1970s.

High-quality photography of the far side of the Moon was not achieved until 1965. Launch vehicle model 8K78 with Blocks I and L, the same models used for the automatic interplanetary stations of the Mars and Venus programs, were used for this mission. A Mars automatic interplanetary station model 3MV-4, which was called Zond-3, *performed the photography.*

In all, it took four launches, of which two were successful, to obtain images of the far side of the Moon. The mission provided unique scientific results.

It must be admitted that the **third** *mission, which is being examined today—soft landing (the Ye-6 program)—has exceeded our expectations in terms of its difficulties.*

We began implementing this program in 1963. We conducted launches of the Ye-6 automatic lunar station concurrently with our Mars and Venus flight program. Launch vehicle 8K78 was used for these programs. However, we were forced to modify the Venus and Mars version of the launch vehicle for a soft landing, and in its lunar version it actually has five stages rather than four. A space rocket performs the role of the fifth stage. It has a triple-mode KTDU with a stabilization and braking system.

Ten launches of the 8K78 launch vehicle were performed before the Ye-6 program began. A wealth of experience was gained with stage four—the Block L. For that reason, one might rightfully ask, why was it necessary to develop another version?

For a soft landing, it was necessary to increase the payload weight, in this case, this meant 500 kilograms of propellant for braking. That is why we asked Pilyugin to create an integrated control system for Blocks I, L, and the space rocket carrying the automatic lunar station for the soft landing. This system was created along with the new astronavigation and radio control system, but it called for other cooperation and considerable time for optimization.

Ten launches were performed for the soft landing program. In the course of these launches, we did not have a single failure in the first and second stages [of the launch vehicle]. *Two failures resulted from problems with the power supply system of the engine of the Block I. Seven failures were the fault of the new control equipment of the Block L and the spacecraft, and one more failure was due to an error in the radio control procedure.*

Due to the insufficient accuracy of measurements in the astronavigation system, we also installed Mayak *radio equipment on board. Yevpatoriya, Ussuriysk, and Moscow receive the* Mayak *signals. The measurement results are processed at three ballistics centers—at Mstislav Vsevolodovich Keldysh's OPM, at NII-4, and at NII-88.*

All systems and all modes were tested for practical purposes under actual flight conditions. More than 100 measures were implemented for modifications to increase equipment reliability. A tremendous amount of flight control experience was gained. At this time, thanks to our programs, two spaceflight control centers are operating in the Crimea: IP-10 in Simferopol for lunar programs and a flight control center in Yevpatoriya for flights to the planets of Venus, Mars, and beyond. In 10 to

12 days, we will finish the assembly and factory tests on the next Ye-6 No. 12. Its launch is scheduled for early December. Comrade Babakin is collaborating with us to manufacture another Ye-6 unit as a backup. We are fully confident that we will solve the very difficult problem of a soft landing with the upcoming launch.

I feel it necessary to point out that we have not had a single serious glitch with Chief Designer Isayev's new KTDU system. As for the Ministry of the Aviation Industry's astronavigation system, Chief Designer Pilyugin's emergency control system, Chief Designer Ryazanskiy's radio complex, and the Ministry of the Electrical Industry's converters, action has been taken in response to all their glitches, providing confidence in their reliable operation for the next launch. On the instructions of the technical review team, I request that the commission grant permission to conduct a Ye-6 program launch during the period from 25 November through 5 December.

When deciding who would comprise our delegation to the meeting at the Kremlin, besides me, Korolev named Mishin, Bushuyev, Tsybin, and Tikhonravov. In the morning before our departure, S.P. unexpectedly announced: "Chertok is the only one going with me. The rest of you stay here and work." I was left without the moral support of my comrades.

VPK meetings were usually attended by the senior administrative staff of the Council of Ministers, Central Committee instructors, chief designers involved with the program in question, and chiefs of the main directorates of the ministries, depending on the matter under discussion. A long table was placed down the middle of the Oval Hall of the Council of Ministers building.

The VPK members were seated around it in previously assigned places. All other individuals who had been summoned and invited

A model of the Ye-6-type lunar probe on display at the Tsiolkovskiy State Museum on the History of Cosmonautics in Kaluga. The package on top is the lunar lander (the ALS). The three silver balls visible in the foreground are three of the four gas storage bottles for the attitude control system. The four black-and-white nozzles facing down are verniers. The main KTDU engine is at the bottom of the entire spacecraft.

Asif Siddiqi

421

seated themselves in two groups on either side of the central table in chairs with fold-out reading desks.

Minister of the Aviation Industry Petr Dementyev, Minister of the Ship Building Industry Boris Butoma, Minister of the Defense Industry Sergey Zverev, Minister of General Machine Building Sergey Afanasyev, Minister of the Electronics Industry Aleksandr Shokin (quite recently appointed in March), and Minister of Communications Nikolay Psurtsev, who infrequently appeared at the meetings, sat to the right of VPK Chairman Leonid Smirnov.

Three or four chairs in this row on the right-hand side of the table usually remained unoccupied in case new ministers showed up. Before long, that's what happened—the Ministry of Machine Building split off from the Ministry of the Defense Industry.[7] In essence, this was the ministry of munitions and all sorts of detonators. Minister Vyacheslav Bakhirev occupied the corresponding place.

On the left side of the table, also in a permanent seating order were: deputies of the VPK chairman, Academy of Sciences President Mstislav Keldysh, Minister of Medium Machine Building Yefim Slavskiy, Minister of the Radio Industry Valeriy Kalmykov, Chairman of the VPK Scientific-Technical Council Academician Aleksandr Shchukin, then the commanders-in-chief of the four branches of the armed forces, usually represented by their deputies.

Having arrived with Korolev about 20 minutes before the meeting began, I was busy setting up the posters on the special stands. The hall gradually filled. After taking their seats, Ministers Dementyev and Zverev opened bulky folders and began to study their daily correspondence, believing that the upcoming discussion wouldn't concern them. To my surprise, Korolev sat at the ministers' table and not in the "guest" rows. When I started to look for a place to sit, he suddenly called, "Boris, sit here next to me." He pointed to a chair to his right and added, "Don't worry, nobody's going to kick you out."

I objected: "It will be awkward to get over to the posters for my report from here."

"And you won't have to; you're not going anywhere."

I was crushed. So, it had already been decided—they weren't even going to hear my report.

At 10 o'clock sharp, VPK Chairman Smirnov and Central Committee Secretary for Defense Ustinov appeared. Smirnov opened the meeting and announced that the first issue under consideration would be the status of

---

7. This happened in 1968.

operations for a soft landing on the Moon. He reminded the gathering that the 1959 decree had not yet been fulfilled. OKB-1, the head organization on this project, along with its collaborating enterprises, must explain what is going on. The VPK would have to assess the situation and figure out whether this objective would be achieved or whether we had decided to relinquish precedence for a soft landing to the Americans.

"Comrade Chertok has the floor to report on this matter."

Korolev managed to stand up before me, and I felt his hand on my left shoulder pressing down, preventing me from getting up.

"Allow me to speak on this matter," he said.

"But Sergey Pavlovich, you're the one who named Chertok to deliver the report, and we're waiting to hear what he has to tell us."

"True, but doesn't the chief designer have the right to take the floor in place of his deputy? I request that I be given the floor."

Smirnov hesitated. Evidently, Korolev's move messed up the previously arranged order. During this brief pause, Keldysh entered the fray: "This is a very serious matter. I support Sergey Pavlovich's request."

Two ministers—Dementyev and Kalmykov—supported Keldysh. Smirnov delayed another moment and looked over at Ustinov sitting to his left. The latter said nothing, but nodded slightly as if supporting the request.

"Sergey Pavlovich Korolev has the floor," announced Smirnov.

"Everything that Chertok was about to tell you will take a lot of time. An explanation of what caused all the failures while solving the soft landing problem is presented in detail on these posters, focusing on each individual launch. But there is one general cause that explains everything—this is a learning process. In our plans and schedules, we did not make provisions for the expenditure of resources and time on the learning process. That is where we made our mistake; we have paid for it, and I dare say that in the very near future the mission will be accomplished. Our learning process has taken us down a rough road, but we have gained invaluable experience. I request that the commission permit us to conduct a launch and make the final decision, if you deem it necessary, based on the results of that launch."

Korolev fell silent, but remained standing, waiting for the decision in response to his démarche.

Smirnov also stood up: "Will there be any questions for Sergey Pavlovich?"

Apparently, with Central Committee Secretary Ustinov present, no one felt compelled to ask substantive questions. They needed to comment on Korolev's proposal right then and there. Being well attuned to the mood of such meetings, Ustinov relieved the suspense with the brief remark: "I support Sergey Pavlovich."

For appearance sake, Smirnov asked: "Will there be any other proposals?"

There were none. The entire discussion lasted 10 minutes instead of the scheduled hour-and-a-half. The individuals who were supposed to report on a second matter hadn't shown up yet, and Smirnov announced a recess.

While I took down the posters and rolled them up, Ryazanskiy, Pilyugin, Morschevskiy, and Tyulin came up to me, and smiling, congratulated me on my "brilliant" report. Korolev had rescued not just me, but them as well. After my "brilliant" report that never took place, and Korolev's minute-long speech, Tyulin gave an assessment of the VPK decision.

"We have been granted remission of our sins until the next launch. One more failure and there will be no mercy. Sergey really bailed all of us out. But he also put us on the spot, assuring them that the 'learning process' is over and that there will absolutely be a soft landing on the Moon this year."

We only had a month to prepare for this decisive launch. During this same month, three back-to-back Venus launches were scheduled. Preparation for this program was proceeding more confidently than for the Ye-6. This was explained by the fact that the control system for the Venus and Mars spacecraft was completely in our hands. We shared responsibility only with the radio specialists, Ryazanskiy and Boguslavskiy.

THE DAYS WERE NUMBERED UNTIL OUR CREW WOULD DEPART ONCE AGAIN FOR THE FIRING RANGE; and it was late in the game to undertake any serious modifications to increase equipment reliability. Morachevskiy's mechanics had implemented the measures that we had come up with right at the firing range the night after *Luna-7* crashed. However, for reassurance, our conceptual designers performed thorough calculations of the disturbances that the spacecraft would be exposed to during the lunar approach segment up until the braking burn began. They analyzed all the instrument temperature modes one more time. They found no more new crimes.

After the Kremlin meeting, S.P. had devoted a great deal of time to the N1-L3 issue and to a new variant of the piloted lunar fly-by using Chelomey's *Pyatisotka* (UR-500). Several times he slipped away to the "little Kremlin"— the hospital on Granovskiy Street where all the "big shots" were sent. They had already "sentenced" him to an operation and, evidently, they were in a hurry. At lunch in our cafeteria, he let down his guard: "We're headed for Venus; we're done with the Moon, and I'm leaving you for a week or so. Those doctors won't leave me alone." Only after his death did we find out that S.P.'s hospital absences had to do with rectal bleeding that he had also experienced during his trips to Tyura-Tam.

After making arrangements with Ryazanskiy, Pilyugin, Morachevskiy, and Khrustalev, I convened a technical review meeting on control systems reliability for the upcoming Venus and Ye-6 launches at OKB-1. The supervisors and specialists from the subcontracting firms took part.

The meeting was well-attended and raucous. During the general report on the MV and Ye-6 programs, I was forced to level serious charges against my friends. The main complaints had to do with the fact that equipment had been delivered to us at the pilot production plant in a form that had not been optimized to an appropriate degree of reliability. The instrumentation of the radio complex on the MV and Ye-6 spacecraft had been replaced several times each in the process of testing at the plant and at the firing range engineering facility. For the MV, we had created a mockup spacecraft through which every flight unit passed. This enabled us to screen out instrumentation that had been released with obvious defects. Nevertheless, new failures occurred during the testing process in the control and test station (KIS) both at the factory and at the firing range engineering facility. On the whole, the situation was better with the reliability of the MV instrumentation than for the Ye-6. On the last three Ye-6 spacecraft, the I-100 instrument had been removed for replacement and repairs five times! We had introduced a strict testing regime for MV equipment at the factory, which had helped us achieve a substantial reduction in the number of failures during tests at the firing range. Unfortunately, we could not allow ourselves this for the Ye-6. The systems from NII-885, NII-944, and from the NII-923 branch arrived so late that the flight units were loaded onto airplanes, having had no opportunity to be checked out at the factory. It wasn't until they reached the firing range that they had to go through inspection for the first time as a unit.

Addressing Pilyugin, I complained that his deputies had no sense of their leading role—responsibility for the Ye-6 control system as a whole. My long report was full of incriminating examples. I cited the last three Ye-6 failures as proof of the lack of a systemic approach and strict control over errors in the adjustment of equipment. "All three [failures] are on the conscience of control system developers," I declared. A very heated discussion broke out.

The meeting was attended by the military representatives of all the invited organizations.[8] My complaints regarding quality and reliability threatened suspension of the acceptance process and a disruption of delivery schedules.

---

8. During the Soviet era, each industrial organization in the defense industry hosted a military representative on staff to monitor the R&D process to ensure close compliance with military requirements. The representative also certified experimental models for testing.

On numerous occasions, the result was that after our claims, a military representative stopped acceptance at contracting organizations until he had obtained exhaustive proof that all the glitches had been corrected. Thus, we stepped on our own tail, so to speak. Just raise a fuss over reliability and then the delivery dates get disrupted, and consequently, our deadlines, which were approved at the "very top."

After my report, our testers and lead designers voiced their shrill accusations. The conversation proved to be very useful. Listening to the accusations directed at his new NII, Pilyugin first turned sulky and began to chew his tongue, which was usually a sign that he was angry or offended.[9] But toward the end, he expressed many useful thoughts about an operating procedure for the future.

"We have an integrated stand for each rocket, and we put the entire flight complex through it before we send it for installation," he said. "But for the Ye-6, after three years, such a stand hasn't been set up for Morachevskiy's equipment, or for Isayev's automatics, or for Boguslavskiy's radio system. This is our failing. We will correct it. I'm giving this assignment to Finogeyev and Kovrizhkin. Of course, if we're sitting on the Moon in a month, maybe the stand won't be necessary."

After all the contractors had left, our cohort, taking stock of the difficult conversation, decided that it was time that we not only take over supervision of testing at the firing range, but it was essential that we also take over interplanetary flight control. For Vostoks and Voskhods, Korolev himself fulfilled the role of flight director, having as his assistants Feoktistov, Gagarin, and all the chief designers, who did their duty in shifts in Kirillov's office and in the MIK conference room. For long-duration interplanetary flights, it would be necessary to switch to having a staff professional flight director. For the time being, Yevgeniy Boguslavskiy was the one at NIP-10, while Pavel Agadzhanov began to be in command of the MV flights at NIP-16 in Yevpatoriya. So far, we had not had our own flight directors. We would begin to argue how best to direct, but quickly cooled down when we remembered with annoyance that beginning next year the entire automatic interplanetary station program was moving from us to Babakin. Let him take up the problem of who would control these flights.

Korolev's proposal to transfer the Molniya communications satellite to new Chief Designer Mikhail Reshetnev in Krasnoyarsk, and to transfer all projects involving automatic interplanetary stations for the Moon, Mars,

---

9. Pilyugin's Complex No. 1 separated from NII-885 in April 1963 to become the new and independent Scientific-Research Institute of Automation and Instrument Construction (*Nauchno-issledovatelskiy institut avtomatiki i priborostroyeniya*, NIIAP).

and Venus to Chief Designer Babakin in Khimki, was accepted by the VPK without objection and made official by the appropriate Central Committee and Council of Ministers resolution.[10] Korolev defied the objections of his deputies and other close colleagues, who had invested perhaps the best years of their creative lives to these projects. Rauschenbach, his staff, and I were among the offended. So much work and now, when we're finally so close to victory, we give it all away! What other chief designer could be so generous! I can't cite comparable examples. Such generosity was also a Korolev phenomenon!

KOROLEV HIMSELF DIDN'T FLY OUT, BUT HE TRIED TO FORCE ALMOST ALL THE CHIEFS OUT TO THE FIRING RANGE. Shabarov and Ostashev had been toiling at the firing range for a month without budging from there. They complained that many contractors' representatives were not resolving issues until their chief designer arrived.

On 25 October, we flew out to the firing range with a large group of chief designers. Here, in the most literal sense, everything was at a standstill. The thing was that the MIK at Site No. 2 and the pad at Site No. 1 did not have the traffic-handling capacity to satisfy the plan for the number of launches by the end of the 1965. With the approval of the State Commission, the firing range command transferred part of the launch vehicles to Site No. 31 for preparation and launch.

For the first time, two four-stage launch vehicles were being prepared at a new MIK for launch from a new launch pad. New squads of military specialists had been brought in for the job. Doubling the firing range's launch capacity also required increased numbers of launch participants from industry. An ever-greater number of specialists and workers left their enterprises and were at the firing range for long-term business trips. The day after Ostashev, Shabarov, and I arrived, we drove out to Site No. 31. The work conditions here were incomparably worse than at our Site No. 2. In the new MIK, which was not yet habitable, the air was saturated with dust; the floors were filthy; the laboratory rooms had not been completed; there were no chairs, no desks, and people sat at the consoles on boxes.

More than 1,000 people had already been housed in the residential village 1 kilometer from the engineering facility. Up to five people had been squeezed into a single 15-square-meter room in the hotels. The bed linens

---

10. These transfers were part of a broader move on Korolev's part to focus only on piloted space programs. Korolev transferred robotic reconnaissance satellites (to Kozlov's Kuybyshev branch), communications satellites (to Reshetnev's OKB-10 in Krasnoyarsk), and automated lunar and deep space probes (to Babakin's Lavochkin OKB in Khimki).

were changed only when there was a change in lodger, and even that did not always happen. One had to stand in line at the dining hall for almost 2 hours to get something to eat. A group of young engineers approached me. The young fellows complained that "Military officers have separate tables with a pretty waitress in a starched apron, and the rest of us get tables cluttered with dirty dishes, and the only utensils left for us are aluminum spoons. Meanwhile, you walk 1 kilometer from the MIK, stand in line, eat lunch, walk another kilometer back, and it's already time for dinner."

Despite all of this disarray in living conditions, which was so normal for our new construction sites, our people worked heroically. A particularly difficult lot fell to the installers from our factory. The "working class," as we referred to them, knew how to adapt to inconveniences quicker and better than the technical intelligentsia.[11] They did not use the dining hall—a brigade arranged for food right in the work place. They kept their work stations and hotel rooms clean and orderly, unlike the engineering staff.

"We work as much as we have to," said the foreman. "You just do a bad job providing us with documentation for preparation. A lot of time is spent deliberating. There aren't any instructions and there's no one to ask. But don't worry—we're doing our job and we won't let you down." By and large, they really did learn to prepare the fourth stage—the Block L—and the launch vehicle calmly, without flaunting their heroic efforts.

The State Commission approved the following schedule: Venus launches would take place on 12, 16, and 25 November. Ye-6 vehicle No. 12 would be launched to the Moon from Site No. 1 on 3 December. We argued for a long time over which naval ships to send and where to send them to monitor the flight. During this period, Yangel, who was conducting maximum flight range tests on combat missiles, had also laid claim to the ships. This time, instead of the *Dolinsk*, they sent the *Krasnodar* to the Gulf of Guinea. Two motor vessels, *Chazhma* and *Suchan*, covered the Pacific Ocean basin.

The launch of 3MV-4 No. 6 to Venus on 12 November was the first launch of a four-stage launch vehicle from the new launch site. At 7 a.m. we were still at T-minus 60 minutes when from Yangel's launch site just

---

11. "Technical intelligentsia" was a Soviet-era term to describe a substantial class of Soviet citizens who had applied skills gained from both technical education and practical experience at R&D facilities or factories.

4 kilometers away, the latest R-16 lifted off like a shot and headed for *Kura*.[12] The cold pre-dawn steppe lit up brightly for several seconds and a rapidly fading roar rolled over it.

Having descended into the new spacious bunker, I ensconced myself on a box next to State Commission Chairman Tyulin. He was already on the high-frequency phone line reporting to Moscow. In distant Kaliningrad at NII-88, Korolev, Mozzhorin, Afanasyev, Pashkov, Serbin, and many "fans" from the Central Committee, VPK, and ministries had gathered. Launch chief Kirillov complained, "This bunker is spacious but blind. There are just two lousy periscopes. This is because they were special-ordered for rocket specialists, while the old bunker has excellent periscopes from old submarines."

Launch preparation proceeded without any mishaps. Our four-stage 8K78 carrying AMS 3MV-4 No. 4 lifted off right on time. Reports began to rapidly pour into the bunker: "Preliminary," "Main," "Separation."

Tyulin reported to Moscow: "From the new launch site all three stages have operated normally. We're going over to Site No. 2."

After a sleepless night, we slowly climbed the stairs and stepped out of the bunker into the cold wind. The driver raced the vehicle along a gray ribbon of concrete at top speed. We needed to cover the 30 kilometers to the communications center at Site No. 2 before Moscow asked, "What happened there with you guys this time?" We made it.

Settling in at Kirillov's warm office, we waited for information from the Far East tracking stations. NIP-15—Ussuriysk—reassured us: "After Block L separation all parameters normal." From the Pacific Ocean, the *Chazhma* pinpointed the estimated startup time of the fourth stage. For a brief period of time, all the tracking stations fell silent—these were the most suspenseful minutes. Hundreds of people were waiting to see whether the latest Venera had entered the design orbit or would remain an Earth satellite—the umpteenth new Kosmos.

Suddenly reports rolled in one after another: all the southern tracking stations had pinpointed a strong signal right above the horizon. First question: "What trajectory are you following?" Answer: "We're flying precisely according to target designation."

Hoorah! It meant the spacecraft was en route to Venus. Information from the Gulf of Guinea was no longer of interest. We hugged and congratulated

---

12. *Kura* (Hen) was the name of the RVSN proving ground in the Kamchatka peninsula on the eastern coast of the Soviet Union. The range was established near the Klyuchi settlement on 29 April 1955 as a location to monitor the reentry of test warheads flown on the R-7 ICBM. At the time, the proving ground was code-named *Kama*.

each other, passing our congratulations to Moscow and the Command and Measurement Complex stations.

"It's early for us to be celebrating," I warned. "It takes more than three months to reach Venus—a lot can happen."

And indeed, 3 hours later NIP-16 reported that telemetry data had disappeared. We began experimenting with sending commands to the spacecraft to switch to backup units. Here was another advantage of the 3MV over the Ye-6. When any radio complex instrument failed, a command from the ground could switch to the backup unit. After trying all sorts of combinations, telemetry came back and a quick report followed: "Antennas deployed, radiators deployed, solar arrays deployed, 'solar current' normal, signal strong."

There were 107 days of flight ahead. On 27 February 1966, *Venera-2*—as it was officially announced in the latest TASS communiqué—passed by the surface of Venus at a range of 24,000 kilometers at 0552 hours Moscow time, but Korolev did not live to see this moment.

On 16 November, almost all of the events described above were repeated. Spacecraft 3MV-3 vehicle No. 1 (*Venera-3*) lifted off for Venus. On 1 March 1966, 105 days later, *Venera-3* delivered a pendant with the emblem of the Soviet Union to the surface of the mysterious planet.

On 23 November 1965, as a backup to 3MV No. 4, we launched 3MV-4 No. 6. But our lucky streak was over. A failure of the stage-three propulsion system turned this next Venera into an Earth satellite. It was called *Kosmos-96*.

The Venus launch on 23 November 1965 was the last in the interplanetary program of Korolev's OKB-1. Beginning in 1966, all operations related to the automatic stations for Venus and Mars exploration were handed to Babakin at the Lavochkin OKB.

Over a period of five years, we had conducted 19 launches toward Earth's closest neighboring planets. Of those, only one ended in success—delivering a USSR pendant to the surface of Venus on 1 March 1966. En route to those planets, we conducted an extensive set of scientific studies, accumulated a wealth of experience for all the systems, and gained crucial experience in the design of interplanetary automatic stations. For the sake of this program, the nation's most powerful center for deep space communications and flight control was created in the Crimea. Hundreds of specialists took science courses that were not offered in a single university, and for which there was not a single textbook.

The very process of gaining this experience, the "learning process," as Korolev called it, came at a high price. Cosmonautics was going through its infancy, when everything was learned by touch, by trial and error, right there in space. Spacecraft were installed on launch vehicles that were not yet tried-

and-true. It seemed that if we performed many launches, we would quickly achieve our goals, while rigorous ground developmental testing would require time and resources that we would not be given. And indeed, we were not limited in resources to order dozens of launch vehicles or manufacture dozens of spacecraft. But just try to ask for much fewer millions of rubles to build a laboratory for thermal vacuum, vibration, electromagnetic ("anechoic") and all kinds of other tests, or to procure special stands and measuring instruments, and you would hit a wall of incomprehension.

Engine specialists Glushko, Isayev, and Konopatov had long understood that the reliable operation of propulsion systems in flight could only be assured after numerous experiments and hundreds of firing rig tests in various modes with thorough analysis of the results of each one.[13] In spite of this, in-flight failures caused by engines occurred from time to time. It was not that the designer overlooked something during testing; rather, these failures were primarily the result of production defects that occurred after the many months of experimental development.

Experimental engines, as a rule, were rig-tested long before the first flight units were manufactured. If test rigs for the testing and verification of spacecraft systems were even manufactured by then, it was so late in the game that the temptation arose to ship the spacecraft to the firing range, and perhaps, even launch it into space without waiting for the results of the ground testing. The time of working according to strict standards and multi-step reliability assurance programs was yet to come. It took five to seven years not only to develop laboratory and experimental facilities, but also to alter the mentality of the entire engineering corps.

On 28 November, without having caught our breath yet after the three Venus launches, the State Commission at the firing range was discussing the matter of the Ye-6 launch. We began with the earliest date—30 November. Everyone was so eager to leave the cold and windy Kazakhstan steppes as soon as possible and go to the still-warm Crimea or home. But, as packed as the schedules were, the launch was moved back to 3 December. Korolev and Keldysh flew to the firing range before this launch.

---

13. Aleksandr Dmitriyevich Konopatov (1922–2004) succeed Semyon Kosberg as chief designer of OKB-154, which produced several major upper stage engines for Soviet launch vehicles.

Shabarov, Kirillov, Samokhin, and I drove to the airfield for the traditional meeting. The reader deserves a special introduction to Samokhin. Mikhail Ivanovich Samokhin was a colonel-general of aviation, a Hero of the Soviet Union, and had been awarded many combat medals. Retiring from active military service due to age, he accepted Korolev's offer to become deputy chief of OKB-1. There had been a sharp increase in the scales of the organizational problems at the firing range in connection with the construction of the N1 assembly factory. A tremendous burden was placed on our air transport. We needed a strong organizer to manage the business of industry trips, vehicular transport, and our aviation detachment, which had grown so rapidly that Vnukovo airport was forced to allocate a separate area for it, calling it "Vnukovo-3."

Once again, one must note that Korolev had not erred in his selection of a manager for a very troublesome sphere of activity. A former "Don Cossack" and cavalryman, after the Civil War, Samokhin asked not to be discharged and entered a school for naval pilots.[14] He was more successful in aviation than in the cavalry. In addition to building the "world's first proletariat air force," Samokhin climbed up the ladder of military service. A rank-and-file pilot, then squadron commander, commander of an aviation regiment, at the beginning of World War II he already commanded Baltic Fleet aviation, participated in the defense of Leningrad, and completed his wartime service near Königsberg (Kaliningrad). Samokhin's last post was as deputy commander of the nation's Air Defense Troops. At that time, the *first* deputy commander of the Air Defense Troops was the legendary pilot Colonel-General Georgiy Baydukov.

When I told Samokhin that I had become acquainted with Baydukov back in 1936, when he was a test pilot at Factory No. 22, he took a shine to me. Soon thereafter, after drinking to "Bruderschaft," we started addressing one another as *ty*.[15]

Samokhin abounded with energy and the desire to render us all kinds of assistance using his high general's rank and connections with the Ministry of Defense. He quickly got acquainted with all the chiefs and lead specialists. With his convivial nature, extraordinary organizational skills, and an inexhaustible supply of breathtaking old war stories, Samokhin won everyone over. Soon his presence at the firing range became absolutely essential for solving organizational, business, transportation, and domestic problems.

---

14. Don Cossacks were Cossacks who settled along the lower Don river in the late 16th century.

15. *Ty* is the familiar form of "you" in the Russian language, in contrast to *vy*, the formal term.

The minister turned directly to him if an airplane was needed; chief designers called him regarding the transport of urgent cargoes and hotel reservations; newcomers, who didn't yet have transport, asked him to get them a car. Everyone praised the extensive menu in the dining halls and grumbled if there was no hot water. But Samokhin's primary concern was supervising new construction.

The military command at the firing range was forced to contend with the fact that a colonel-general represented industry in Tyura-Tam, while the director of the entire firing range back then was only a major-general. So as not to put the officers and generals at the firing range in an awkward position, Samokhin only wore his general's uniform on solemn occasions, or in this case, to meet high-ranking guests.

Before we departed, Samokhin informed us that Korolev wanted to go straight from the airfield, before any meetings, to the construction site of the big new MIK for the N1. Samokhin also warned the military builders about the impending visit and asked Barmin to go to the construction site.

Barmin usually arrived for the launches a few days before the other chiefs. He oversaw the readiness of the launch system and listened to Boris Khlebnikov, his almost permanent representative at our firing range launch sites. Now Barmin had acquired many new worries at even more launch sites, too. Construction began at once on two launch facilities for the N1 rocket. A government decree named Barmin the main chief designer for the development of launch facilities for this rocket of unprecedented dimensions. The new launch pads, created using Barmin's design for Chelomey's *Pyatisotka*—the Proton—were also extremely complex. But there was no comparison between them and the scales of the launch pads for the N1. Many dozens of organizations were already involved in creating the systems associated with the future launch of the N1. All this abundance of systems would have to be integrated into a single huge system for which he, Barmin, was the head and principal responsible party.

Usually the director of the firing range and his chief of staff came to meet the arrivals at the airfield. That's how it went—there were as many people in the welcoming party as there were arrivals when the "chief theoretician" [Keldysh] and chairman of the State Commission flew with the chief designers. Samokhin received a cable listing all the passengers on board. I noticed that Glushko and Kuznetsov were missing from the list, but Isayev and Pilyugin were on it.

The first day of December was unusually warm and calm. All those who had arrived on the plane were happy to stretch their legs after disembarking in Tyura-Tam. Right off the bat, Korolev and Tyulin announced that the State

Commission meeting would take place at 1700 hours at Site No. 2, and first they would drive out to the N1 construction site. It turned out that almost all the arrivals and those who had turned out to meet them also wanted to see the construction "project of the century" fortuitously located almost right on the way to Site No. 2.

At all of our big construction sites, good access routes appeared only after construction was over. The new construction for the N1 was no exception. It was not easy to drive right up to the construction site in light vehicles over a road that had been beaten up by hundreds of trucks.

Reinforced-concrete columns already marked the outlines of the new, enormous assembly shop. The total area of the main building was 30,000 square meters. They were simultaneously building the main building, the auxiliary service areas, and—1.5 kilometers away—a new town with hotels, barracks, dining halls, and various sundry services for six to seven thousand people.

The military project managers never failed to complain to visitors that just five thousand soldiers from construction battalions were working on the project, and in order to complete all the facilities in one year's time—as the Ministry of Defense had ordered—they needed to double the number of men on the job. However, the orders were not backed up with real assistance.

For safety considerations, two N1 launch sites were built 10 kilometers from the large assembly factory. The general scope of construction and assembly operations at the launch sites was just as great. Barmin, who had driven up from the direction of the launch sites, added that the scale of the construction was now many times greater than what it was when the firing range was being created, but the rate of construction was that much slower. He noted that the Ministry of Defense, which was responsible for the construction, took no interest in it.

"The Americans have not completed their launch site for the Saturn V either," said Barmin, "but, after all, it's there in Florida, and we have the Tyura-Tam winter right around the corner."

Korolev, Barmin, Tyulin, and Keldysh immediately agreed that when they returned to Moscow after "landing on the Moon" they would make the appropriate presentation to Smirnov and to the Central Committee.

THAT EVENING AT THE STATE COMMISSION MEETING WE DISCUSSED AND CONFIRMED THE DATE AND TIME OF THE LAUNCH of the next Ye-6 for 3 December at 1346 hours 14 seconds. In this case, the estimated time of the Moon landing was 6 December at 2357 hours. The ballistics specialists calculated down to the second the coverage areas for NIP-10, the time of powered flight for all the stages and rocket blocks, the distance to the Moon for establishing the lunar

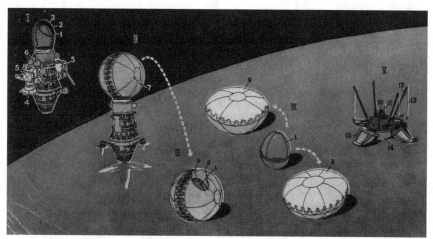

Nauka i zhizn

A model of the Ye–6–type lunar lander shows its ingenuous "petal" design. In scene I, we see the Luna spacecraft with (1) Automatic Lunar Station (ALS) which is covered by (3) a thermal covering. Ideally, at an altitude of 75 kilometers, the padding (2) around the ALS would be pressurized as we see in scene II. At a command from the radio–altimeter located at (4), two side packages (5) from the main bus are jettisoned and the main KTDU retro engine is turned on. At a given altitude from the lunar surface, the pressurized ball (7) would be separated from the main bus and would impact on the surface as shown in scene III. The pressurized covering surrounding the ALS would then separate into two parts (8 and 9) revealing the inner ALS. As shown in scene V, the petals of the lander would then unfurl, stabilizing the main body of the ALS (14), allowing a small suite of scientific instruments to operate. These included antennae (12 and 15) and a camera (10).

vertical, and the altitude for the beginning and end of the braking burn. All these time-tagged events were reported as if no one had any doubt that the flight would go without a hitch according to this schedule. The conceptual designers gave the weight report with the same scrupulous precision, beginning with the total weight of the fourth stage down to ALS Ye-6 No. 12. It turned out that the goal of our labor was to gently lower just 104.5 kilograms onto the surface of the Moon. It was this 100-plus kilograms that would secure the next dramatic success for Soviet cosmonautics.

Korolev demanded that data on the fueling of Isayev's KTDU with propellant and oxidizer, and the time of its operation, be reported. There were many other questions for which precise answers were provided on the spot. The great deal of experience gained from the preceding launches was evident. Next came the traditional reports of all the chief designers regarding the measures that had been implemented and the certification of the absolute

435

reliability of the system for which each of them was responsible. After that, everything went according to standard procedure.

In the bunker, I manned the station at the guest periscope. The most dramatic seconds of a launch usually pass without telemetry reports. An experienced eye can ascertain the situation quicker than the monitors can indicate it.

Ignition, the roiling fire and smoke of the preliminary stage—and the flame of the main stage! The launch suspension arms fell back, and the roaring rocket smoothly lifted away from the swirling flames and rapidly disappeared in the gray shroud of the low cloud cover. The actual liftoff differed from the calculated one by just 0.3 seconds. The slightly agitated reports from the tracking stations resounded over the loudspeakers in the bunker, enveloped in silence: "Ninety seconds . . . Flight normal."

All the stages functioned according to design. One hour later, after driving to the engineering facility, we received the report that the Block L had finished its burn and Ye-6 No. 12 was flying to the Moon. The previous failures had taught us all to be restrained, and we did not exhibit any particular emotion in this regard.

Little more than 80 hours remained until the Moon landing. The first scheduled correction burn was to take place on 4 December at around 2000 hours. Korolev and Tyulin suggested that everyone go to bed a little early; the next day at 7 a.m., everyone would be flying off to their different destinations: some to Moscow and the State Commission and all the "lunatics"—to the Crimea.

Korolev would depart from Tyura-Tam for the Crimea. He would never participate in another launch. None of us knew at that time that Korolev was spending the night for the last time in the "Korolevian cottage" that he had stayed in for eight years. Korolev had no way of knowing that this would be the last time that he would ride along the paved road from his *Dvoyka* (Site No. 2) to the airfield, that the steppe he was viewing would be called the "Baykonur Cosmodrome," and that the world's mass media would extol the name of the hitherto unknown Academician Korolev.

Despite my fatigue, I felt that the emotional stress of the last few days wouldn't let me fall asleep, and after dinner I stopped by to see Isayev. He was obviously pleased and right away handed me a pack of his beloved *Belomor* cigarettes. The days of our youth spent in the romantic aviation field and our adventures on German soil had long ago receded into the past. But as the years had passed, the two of us had an increasing need for visits that allowed us to share our innermost thoughts. Isayev said that he had asked Korolev and he would not be flying to the Crimea—he was having a lot of trouble with the naval missiles. He needed to fly to Krasnoyarsk immediately.

"I simply can't deal with these bandits," said Isayev, referring to the production engineers at the Krasnoyarsk factory where his engines were in series production. "But, after all, this is just a trifling issue. In the end, we'll finish them off. And then, you guys and the N1—in my opinion, you're really stuck in a mess. I don't want to be a prophet. I'm certain that Kuznetsov won't have an engine any time soon.[16] That is, they can produce lots of metal in Kuybyshev. They have colossal capacities there. But to optimize reliability for such a rocket—especially when you have delivered 30 "bottles," each weighing 150 metric tons, just for the first stage!"[17]

"The day before yesterday, I was looking at that gargantuan construction project and thinking that it wasn't in your best interests to speed it up. They'll round up another 10,000 soldiers and they'll build it. We know how to do this brilliantly. But as for the engines, with all the fittings, and particularly with the new closed-loop configuration . . . Vasya Mishin and Misha Melnikov described these engines as their own personal achievement. Supposedly, they convinced Kuznetsov to pick this configuration. But I trust Vanya Raykov more. He doesn't share their optimism. In '68 we won't have the engines."

Then Isayev told me about the scene that he happened to witness on the airplane the other day.

"You've noticed that when I fly with your managers, I don't like to sit in the forward cabin—I sit in the coach-class cabin where I can read or sleep without having anyone bother me. I had just dozed off when I was invited into the forward cabin, supposedly to drink tea. Well, there was tea and cookies, of course, and even about 15 drops of cognac apiece. But that's not the point. Keldysh got it into his head to ask me, in front of Korolev, my opinion of Kuznetsov's engines. At the same time, he also suggested that perhaps Glushko's high-boiling component engines would achieve greater thrust earlier. After all, he was getting ready to make large engines for Chelomey and Yangel."

"You understand, in this group of people—not to mention the gloomy Korolev—I didn't want to say what I was really thinking. And why did I need to talk about this so that later Keldysh could quote me somewhere to people in high places? They will drag me to expert commissions *ad nauseum*."

---

16. This is a reference to Nikolay Dmitriyevich Kuznetsov (1911–95), Chief Designer of OKB-276, which designed cryogenic engines for the heavy-lift N1 lunar booster. His design bureau was based in Kuybyshev (now Samara). When Glushko refused to produce cryogenic engines for the N1, Korolev turned to the inexperienced Kuznetsov to deliver the engines for the rocket. The rift between Korolev and Glushko seriously estranged the two giants of the Soviet missile and space programs. Chertok will describe the history of the N1 in detail in Volume IV.

17. The N1's first stage had 30 NK-15 engines, each with a thrust of 150 tons.

"I just started to give noncommittal responses and then Korolev blew up: 'Mstislav Vsevolodovich! Stop this game. As it is, you've already done quite a bit to the detriment of the N1. And your posturing with Glushko is only doing harm'."

"Well, I can honestly say that I had not expected this turn in our amicable teatime conversation. Keldysh's refined smile immediately vanished; it seemed to me that he even turned pale. I have noticed for some time that Korolev and Keldysh address one another as *ty*. But Korolev switched to the formal *vy* and Keldysh responded to him in kind: 'Sergey Pavlovich, please don't forget yourself! I am not about to meddle in your personal relations with Valentin Petrovich [Glushko]. But the N1 project is a matter of such a scale that I have a right to take an interest in the opinions of specialists without having to contend with your biases'."

"He really told him off! But then Tyulin intervened. He shouted: 'That's enough, fellows! Put an end to this talk! This is not the right place. Does anyone have another bottle?'"

"You know, it's amazing, but we couldn't find another bottle. Then Tyulin went to the crew and arranged to get coffee for everyone. And so I suffered through the flight in this cabin until we landed instead of getting a nap in a nice comfortable seat!"

Isayev and I continued talking on many other subjects long into the night. I hoped that I would be able to catch up on my sleep on the plane.

If Korolev or Keldysh were on board, squadron commander Khvastunov piloted the flight from Tyura-Tam to the Crimea without making a stopover in Astrakhan. The flight usually lasted around 4 hours. Each person tried to use these 4 hours in his own way. Napping, unwinding from the constant tension, discussing something important, or simply collecting one's thoughts, thumbing through notebooks, remembering something and jotting it down.

This time, Korolev, having grabbed a bulging folder, pretended to be absorbed in the daily mail that he had brought from Moscow. Keldysh was snoozing, having set aside some thick book on the seat. Tyulin had decided not to disturb either of them. Ryazanskiy was being tormented by thoughts that demanded discussion. Neither Tyulin nor Ryazanskiy would let me sleep and they whisked me off to the aft cabin, which was half-empty.

"Let's discuss the situation calmly," Tyulin proposed.

He pulled out a piece of paper covered with notes and enumerated: "This year alone for the soft landing we have conducted four launches, counting the one yesterday. Three failures. Counting the *Zond-3*, we've made four Venus shots. Two Veneras are in flight, so it's still unclear what will happen with them. Belyayev and Leonov, even though they landed in the taiga, they returned to Earth safe and sound, thank God. After all, just between us, it could have been a tragedy."

"We launched two Molniyas, and, to your credit, Boris, kudos, they are still working. I know that the Far East *kray* committee secretary thanked Brezhnev personally for the successful relay of the Moscow programs."

"Things have also been going pretty well with the Zenits. In any event, in the Defense Council they said that there is no need to be stingy with resources for this technology. Even though the results were not outstanding, Chelomey launched two Protons with a lot of hubbub. The *Pyatisotka* [UR-500] has begun to fly. That's beneficial for all of us. After long delays, the *Devyatka* [R-9] has finally been put into service.[18] Sergey did not quiet down, and they're launching solid-propellant missiles from Kapustin Yar."

"If you add to this all of our space reconnaissance projects, then that means we already have 64 space launches this year. And together with Yangel's and Chelomey's combat missiles, the number has far surpassed 100."

"This is in just one year's time! Imagine what a burden this is on industry and on all of us. We can absolutely compete with the Americans in terms of the effectiveness of our combat missiles. But when it comes to space, I am afraid they will soon be ahead of us. It's time for me to give serious thought to concentrating our efforts on our main projects. Construction on the N1, of course, is behind schedule, but that's not where the main danger is. Korolev, Glushko, Yangel, and Chelomey are not capable of working together as a team. My relations with the minister [Afanasyev] are complicated. I know that 'Uncle Mitya' also has few friends in the Politburo right now. The famous Council of Six Chiefs has turned into a council of 66, and within it Nikolay and Viktor are at each other's throats and as are you, Mikhail, and Lesha Bogomolov. Sergey's relationship with Valentin has gone bad permanently. Incidentally, Mishin's aggressiveness has contributed to this. The military men are debating whether we need piloted flights. Malinovskiy is obviously displeased with the Air Force's activity in this field, and he doesn't think lunar landings are necessary."

"I'm saying all this," continued Tyulin, "so that you will give some thought to how we can affect this headstrong foursome—Korolev, Glushko, Chelomey, and Yangel—not from above, but from below. You, control specialists, are sort of nonpartisan. Especially you, Mikhail—you work on everything. Boris should reconcile Nikolay and Viktor, and let's give some thought to how to patch up the relationship between Sergey and Valentin."

---

18. The R-9 ICBM had been officially declared operational on 21 July 1965.

Tyulin continued talking for a long time under the droning of the aircraft engines. I jotted down his monologue from memory many days later and now I am reproducing the gist of it without details. Mikhail and I interrupted, clarified, and complained about our own behind-the-scenes complications. The upshot was, we convinced one another that we needed to resist more actively the chiefs' dysfunctional tendencies.

I didn't pass up the opportunity to remind them of Korolev's attempts at rapprochement with Chelomey. I told them about my meeting with Khrushchev's son, who was quite close to Chelomey.[19]

"You bear in mind," said Tyulin, "that Chelomey doesn't have Sergey Khrushchev anymore. And your former contacts with him have little clout."

I was annoyed when I noticed that, because of this long conversation, I had missed my opportunity to admire the Main Caucasian Range one more time. The airplane was descending, and we returned to the "government" cabin.

KOROLEV, KELDYSH, RYAZANSKIY, AND TYULIN CHECKED INTO THE BEST HOTEL IN SIMFEROPOL. When he needed to travel from town to NIP-10 or return to the hotel, Korolev invited Ryazanskiy or Tyulin to ride in the car that had been secured for him. On his initiative, Colonel Bugayev wheedled an additional vehicle from the Crimean authorities for the president of the Academy of Sciences. Ryazanskiy had also been a witness to the scene on the plane that Isayev had recounted. He believed that it would be some time before Korolev and Keldysh would be able ride in the same car. Nevertheless, during all the communications sessions, they listened amicably, and then discussed detailed plans for the next coverage zone.

We all usually ate dinner together without leaving NIP-10. Those who were staying in Simferopol hotels said that the food was much tastier at NIP-10 than in a restaurant. Here, Babakin joined us. During dinner he told about his young design bureau's plans in the field of interplanetary automatic stations. During such conversations, Korolev and Keldysh acted very interested. There was no outward sign of their falling-out.

When we appeared at NIP-10, Boguslavskiy reported that on board everything was A-Okay, but someone had proposed conducting a "false correction" session that evening for greater confidence. The plan was to switch on all the attitude control systems, but not permit KTDU startup. Korolev and Keldysh supported the proposal.

---

19. Khrushchev's son, Sergey Nikitich Khrushchev, worked as deputy chief of the control systems department at Chelomey's design bureau.

Turning to me, Korolev said, "Boris! Don't go anywhere, let's sit down, and you'll explain everything to me."

On a sheet of red graph paper I prepared a schedule of the session containing a list of commands to be issued from the ground, on-board sequencer tags, and actions of the on-board systems.

On 4 December 1965 at 2030 hours in the Simferopol NIP-10 control room with all the participants gathered around, a tense silence set in.

"Yevgeniy Yakovlevich!" Korolev said loudly. "You take charge of all the sessions. Don't pay any attention to us. Just, please, give the commands loud and clear. I also want to understand what's going on."

At 2047 hours Boguslavskiy announced: "Transmitter on. We're starting to measure range and radial velocity."

I put the first check mark on my schedule.

Korolev moved the schedule over to himself and demanded: "Explain quietly!"

The commands came one after the other. Boguslavskiy announced: "Correction program on . . . SAN on . . . Searching for the Sun . . . Search for the Moon has begun . . . I-100 switched on for warm-up . . . Number check in progress . . . I-100 checkout in progress . . . ."

At 2219 hours: "On-board equipment off . . . Spin-up initiated . . . .

At 2242 hours: "End of session. No anomalies observed!"

"It seems like we vented," said Korolev. He pulled out his handkerchief and wiped his forehead.

The next, full-fledged correction with a KTDU burn on 5 December also proceeded "without a hitch"—if one overlooks the demonstrational reaming-out that Korolev gave to Bezverbyy and Lidov for a slight velocity underrun during correction. NII-88, NII-4, and OPM had performed the calculations for the correction settings. Actually, OPM, considered to be completely Keldysh's domain, had developed the whole calculation procedure. It was difficult to figure out which of the three organizations was more guilty of the slight error. But the tongue-lashing had such a demonstrative nature that for a while it defused the tension that had reigned among those working at NIP-10.

Tragedy fell on the night from the sixth to the seventh of December 1965. Everyone understood that, in any event, we wouldn't be sleeping. We all had an early supper together and at 1830 hours we were involved in the communications session, during which all the systems, except for the firing of the KTDU, were checked for the last time. I once again prepared a detailed schedule on graph paper of all the operations in Moscow time. A lot of people were milling about the control room. There were those directly involved, those who "rooted" for us even though not involved in the work, and those who just worried on our behalf. Even the Crimean regional committee (*obkom*) representative showed up.

Having assessed the situation, Korolev called over a bunch of us—station chief Colonel Bugayev, Ryazanskiy, Boguslavskiy, and me—and loudly announced, so that everyone heard: "Everyone who is not involved in this session, go into the adjoining rooms—you'll have all the information there. No one but Boguslavskiy is going to issue any control commands. I forbid anyone to bother him. State Commission member Ryazanskiy will ensure compliance with this order. Colonel Bugayev is responsible for seeing that the personnel execute all instructions precisely and for checking all the equipment. I ask all participants one more time to make sure the systems are in good working order. Keep your performance and telemetry data reports short and to the point. Speak briefly and succinctly. Is that clear?"

It was all clear. Having removed his jacket, S.P. was now wearing the woolen shirt that we sometimes referred to as the Korolevian uniform. After taking his seat, he turned to me and said: "Sit by me, set the schedule down. You're going to give me a blow-by-blow account in a low voice."

Keldysh and Tyulin settled in at an adjacent work station. Boguslavskiy, microphone in hand, walked about the control room and gave his last instructions.

On 6 December 1965 at 2230 hours, Boguslavskiy, clearly with trepidation, and almost solemnly in the midst of the silence that had fallen over the room, issued command No. 17: "Switch on on-board transmitter."

Next came the reports: "Receiving numbers . . . Monitoring settings . . . Stage five telemetry on . . . ."

At 2244 hours, the command "permission to start up on-board program—set up lunar vertical" was issued three times. The commands were issued continuously. And reports flowed continuously: "SAN activated . . . Precision telemetry activated . . . Spacecraft is stable . . . Receiving control numbers for final adjustment of settings . . . Search for the Sun complete . . . Searching for the Moon . . . ."

Korolev, flinching just a little with each report, placed a checkmark and checked the time. I reassured him: "So far we're right on schedule—the lunar search entails turning the entire spacecraft relative to the direction toward the Sun."

"I understand. Spare me the chatter," grumbled Korolev.

At 2330 hours, they began the search for Earth, and the SAN switched to precision mode.

At 2351 hours 6 seconds came the triumphant report: "We have a lunar vertical."

Korolev shuddered. This was the spot where we messed up the last time.

"We're out of the woods," I said.

"Shut up!" Korolev crossly cut me off and put down two checks one after the other, after hearing the reports: "I-100 on."

"Altimeter on for warm-up."

Next came the reports: "Altitude 8,272 kilometers!"

"Tags T7 and T8 have passed—shock absorbers are pressurizing, we're waiting for KTDU burn!"

According to the schedule, the KTDU was supposed to perform a braking burn when it reached the specified altitude of 74.8 kilometers at 0051 hours. Everyone froze in anticipation. The Ye-6, oriented about the lunar vertical, was falling toward the Moon. The last time, the failure took place before the lunar vertical was set up. Through my shoulder I felt Korolev stiffen. He listened attentively and placed checkmarks.

We've jumped over the area of the previous crash and gone farther.

Another report: "We see a loss of pressure in the spherical tank. To the Moon 1,200 . . . 1,000 . . . 800 kilometers."

And suddenly: "Altitude measurements have stopped."

Korolev reeled in my direction: "Maybe it's a mistake? A random failure?"

But Boguslavskiy had already come running up with the microphone: "Sergey Pavlovich! Something inexplicable has happened. As I see it, the spacecraft spun around and the altimeter stopped its measurements, then it once again 'got on course' for the Moon!"

Some hope still lingered. Soon the engine will fire; stabilization will be restored; braking will take place; and all of this spinning is just a bad dream. After all, this was the 11th launch! We were so close to our goal! I explained that if the altimeter lost the Moon, then the KTDU couldn't execute a braking burn.

Reports came in that extinguished all hope of a happy ending: "51 minutes, 30 seconds. End of reception!"

"No signal!"

"The command to fire the engine passed."

"The engine operated for 9 seconds instead of 42. It's got to be a crash landing on the Moon."

At 0051 hours 29.6 seconds, communication abruptly ended. I pushed my way into a huddle of flight planners, control specialists, and radio operators already whispering amongst themselves. Everyone rushed to share his impressions and hypotheses. I finally managed to pacify them and get them to report calmly.

Gradually the picture cleared up. Up until tags T-7 and T-8, all systems had operated normally. When these tags were issued, the valve of the high-pressure spherical tank opened and the rubber bladders of the shock absorbers started to pressurize. The altimeter began to lose its orientation on the Moon 13 seconds after this command. Loss of the Sun was detected, and then Earth as well. The control specialists managed to calculate the rate of spin at the moment the engine started up was 12 degrees per second. Beresnev and Kovrizhkin together with the SAN specialists determined that the destabilizing moment at such a rate of spin was three times greater than the control moment capabilities.

I returned to Korolev. He was giving no instructions and was sitting unusually passively, listening with obvious annoyance to Tyulin and Keldysh, who had already hashed out the form and content of the report to Moscow.

"Well, what?" S.P. asked at last.

"The first impression is that a hole formed in the shock absorbers and the leaking air created a destabilizing moment that the control system couldn't cope with."

Babakin came up and tried to reason: "We already have a new automatic lunar station ready. Let's do this again in two months. Over that period of time, we'll figure out what happened. It's clear it was something stupid."

"Yes, we'll do it again if they trust us after this," Korolev responded sadly.

Everyone shared a feeling of guilt. Rather than divide us, it rallied us. No one wanted to rest, despite it being the dead of the night. Five days of thorough investigation was needed to understand the true cause of the failure. It seems Pallo was the first to get to the bottom of things. He and the "rubber specialists"

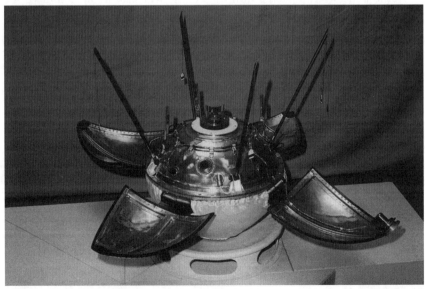

A model of the Ye-6 ALS showing its ingenious petal design.

had conducted a whole gamut of experiments to debug the inflatable rubber shock absorbers. When he was informed that the bladders had probably leaked, he categorically rejected the accusation and began a thorough investigation.

When they are pressurized, the shock absorbers rest in the fiberglass support bracket securing the lobe antennas. They checked the bracket. It broke easily, forming sharp edges. During pressurization, the broken piece of bracket had pierced the rubber bladder. This is what caused the destabilizing moment.

But why hadn't the previously manufactured brackets broken during testing? Pallo discovered that the fabric, which formed the basis of the material, had been placed in the mold improperly. This was a fundamental process error of the worker who had placed the workpiece in the mold before curing and pressing. The bracket's breakage and the piercing of the rubber bladders were entirely reproduced in full on an experimental unit.

Mozzhorin once again received Ustinov's order to prepare a reliability and quality report that would make it possible to deal not only with Chertok, but also with Korolev himself. Really, what kind of quality control system is this anyway? If the error of a worker who doesn't know where this fiberglass cloth is going, or what it will be used for, can cause consequences like these? And what kind of design is this if a bracket that broke inadvertently lacerates the rubber and generates destabilizing moment, causing such a spacecraft to spin?

The VPK administration began preparing for the next meeting to discuss what had caused all the Ye-6 launch failures. This time there would be a report from Korolev and Tyulin. For this VPK meeting, I prepared a chart showing all the spacecraft launches using the four-stage 8K78 launch vehicle (see Table 2). It provided some idea of the scale of our operations at that time. I am citing it in this book in a somewhat supplemented form. For the reader's convenience, I am also citing a table that explains the designations of the R-7 launch vehicle modifications (see Table 1).

SUBSEQUENT TRAGIC EVENTS PREVENTED THE VPK FROM HEARING KOROLEV'S REPORT. On 14 December, at the age of 51, Leonid Voskresenskiy suddenly died. In 1963, due to heart disease, he had withdrawn from active testing duties at the firing range. On his initiative, a flight vehicle testing department was set up at the Moscow Aviation Institute (MAI). In early 1964, a heart attack sent him to the hospital, but he did not succumb, and after recovering, he returned to work performing light duties. As Korolev's deputy for testing, Voskresenskiy had to participate in design work on the N1 and sign off on the relevant draft plan materials. In this regard, serious differences of opinion arose between him and Korolev and Mishin. Voskresenskiy was unable to convince them that they needed a firing rig to perform tests on the first stage.

445

Table 1—R-7 Launch Vehicle Modifications

Purpose	Public Identification Number	Name for Mass Media Use
1. Flight-ready weapon-type two-stage (range 8,000 km)	8K71	ICBM
2. Flight-ready weapon-type two-stage with reduced warhead	8K74	ICBM
3. Two-stage satellite modification	8K71PS 8A91	Sputnik
4. Launch vehicle of the first lunar craft and Vostok vehicles, stage three—Block Ye	8K72	Vostok
5. Four-stage launch vehicle for the program of launches to Venus and Mars and for the insertion of Molniya high-elliptical orbit communications satellites, stage three—Block I, stage four—Block L	8K78	Molniya
6. Three-stage launch vehicle for the Zenit-2 spacecraft, stage three—Block Ye	8A92	–
7. Three-stage launch vehicle with an enhanced-reliability control system for Voskhod and the Zenit-4 spacecraft, stage three—Block I	11A57	–
8. Five-stage launch vehicle for the soft lunar landing program (in contrast to item 5, stage five, the Ye-6, controls stages three and four)	8K78/Ye6	–
9. Three-stage launch vehicle of piloted Soyuz and Progress vehicles. Subsequent modifications of the control system, telemetry, and propulsion systems	11A511 11A511U 11A511U 2	Soyuz
10. Three-stage launch vehicle of piloted and automated spacecraft with a modified control system and propulsion system	–	Soyuz-2

Finding no support within the staff at OKB-1, he gradually drifted away from his operational testing work and limited himself to the development of procedures and scientific research projects; he immersed himself more and more into organizing his academic activity in his department at MAI.

During the summer of 1965, a couple of times we arranged family outings together. We drove out in our private cars to the woods along Moscow's first air defense missile ring and enjoyed picnics with campfires and mushroom gathering. These picnics were attended by all of Korolev's deputies who lived at 3 Ostankinskaya Street: Bushuyev, Voskresenskiy, Okhapkin, and I—and "chekist" Yakovenko—Korolev's deputy for security.[20] We were joined by MVTU Professor Feodosyev, who was a friend of Okhapkin. Our picnics were light-hearted and very laid-back. Our wives wouldn't let us get carried away with work-related tirades, and being in our own cars prevented us from "overindulging."

During these rare excursions into nature, Voskresenskiy livened up; he cracked jokes and told stories about stage, theater, and film premieres. In this regard, the Voskresenskiy family was unrivaled.

---

20. Chekist was a term commonly used for a KGB operative. It is derived from the acronym ChK—*Chrezvychaynaya kommissiya* (Extraordinary Commission), the first Soviet secret police organization headed by Feliks Dzerzhinskiy and established shortly after the October Revolution in 1917.

Table 2–List of Launches to the Moon, Venus, and Mars and of Molniya-1 Communications Satellites by the 8K78 Launch Vehicle in the period 1960-1966

Item No.	Launch Date	Spacecraft Name	Goal of Flight	Results	Results Obtained / Reasons for Unfulfilled Mission
1	10/10/60	1M No. 1	Flight to Mars	Mission not accomplished. Failure of launch vehicle stage three.	Failure of Block I gyro horizon 309.9 seconds into the flight. Emergency engine unit shutdown (AVDU) command issued at 324.2 seconds.
2	10/14/60	1M No. 2	Flight to Mars	Mission not accomplished. Failure of launch vehicle stage three.	Freeze-up of fuel in the ducted pipeline of Block I due to a leak in the oxidizer dividing valve.
3	02/04/61	1VA No. 1	Flight to Venus with a pennant	Mission not accomplished. Failure of launch vehicle stage four.	Breakdown of bearings or brushes of converter PT-200 in vacuum in the Block L control system.
4	02/12/61	1VA No. 2 *Venera-1*	Flight to Venus with a pennant ix;n the descent module	Mission not accomplished. Automatic interplanetary station (AMS) was inserted on an interplanetary trajectory to Venus. Failure of spacecraft attitude control system and sequencer (PVU).	Overheating of the sun position indicator. Failure of the event controller in the radio system.
5	08/25/62	2MV-1 No. 3	Flight to Venus with a pennant in the descent module	Mission not accomplished. Failure of launch vehicle stage four.	Failure of the stage four control system. Low degree of effectiveness of the Block L propulsion system effector system. Block operated for 45 seconds without orientation.
6	09/08/62	2MV-1 No. 4	Flight to Venus with a pennant in the descent module	Mission not accomplished. Failure of launch vehicle stage four.	Failure of the fuel valve in the Block L propulsion system heating system.
7	09/12/62	2MV-2 No. 1	Flight to Venus with a photo-television unit (FTU) for photography during fly-by	Mission not accomplished. Failure of launch vehicle stage three.	Abnormal separation of the nose cone and Block I. Launch vehicle stage three broke up 531 seconds into flight. Attitude control and stabilization system failed to activate. Block L propulsion system broke down 0.8 seconds after firing.
8	10/24/62	2MV-4 No. 3	Flight to Mars with a photo-television unit (FTU) for photography during fly-by	Mission not accomplished. Failure of launch vehicle stage four.	Breakdown of the turbopump assembly (TNA) during startup of the Block L propulsion system. Jamming of the turbopump assembly spring in vacuum.
9	11/01/62	2MV-4 No. 4 *Mars-1*	Flight to Mars with a photo-television unit for photography during fly-by	Mission not accomplished. Automatic interplanetary station was inserted on an interplanetary trajectory to Mars. Communication with the station lasted 140 days.	Failure in the effector system. Loss of working fluid. Valve leak in the attitude control system. Resin particle under the valve seat.
10	11/04/62	2MV-3 No. 1	Flight to Mars with a descent module	Mission not accomplished. Failure of launch vehicle stage four.	Failure of the Block L propulsion system. Propulsion system functioned at reduced power. Premature swing-out of ignition support during startup.
11	01/04/63	Ye-6 No. 2	Soft landing on the Moon	Mission not accomplished. Launch vehicle stage four propulsion system failed to start up.	Block L remained in intermediate Earth orbit. Failure caused by the control system due to failure of converter PT-500/1000.
12	02/03/63	Ye-6 No. 3	Soft landing on the Moon	Mission not accomplished. Abnormal operation of the launch vehicle control system.	Failure of Block L to be inserted into intermediate orbit. Disparity in pitch angle after 105.5 seconds in the trajectory monitoring system. Setting error in instrument I-100.

447

Item No.	Launch Date	Spacecraft Name	Goal of Flight	Results	Results Obtained / Reasons for Unfulfilled Mission
13	04/02/63	Ye-6 No. 4 Luna-4	Soft landing on the Moon	Mission not accomplished. Spacecraft was inserted on a trajectory to the Moon. Stellar trajectory correction failed to be executed.	Failure of the Yupiter astronavigation system. Unequivocal cause was not determined. Spacecraft flew past the lunar surface at a distance of 8,500 kilometers.
14	11/11/63	3MV-1A No. 2 Zond-1	Flight to Mars to optimize launch vehicle and spacecraft systems	Mission not accomplished. Abnormal separation of stages three and four. After 1330 seconds, there was no information from Block L	Detection of Block L in Earth orbit by Air Defense Troops (PVO) facilities. Failure of the Block L propulsion system. Cavitation at a pump inlet, possible turbopump assembly explosion.
15	02/19/64	3MV-1A No. 4A	Flight to Venus to optimize spacecraft systems	Mission not accomplished. Failure of launch vehicle stage three.	Failure during startup of the Block I propulsion system. Cutoff valve leak. Fuel freezeup while still on the launch pad.
16	03/21/64	Ye-6 No. 6	Soft landing on the Moon	Mission not accomplished. Launch vehicle stage three propulsion system failed to start up.	Failure of the oxidizer valve to open during the power stage.
17	03/27/64	3MV-1 No.5	Flight to Venus with descent module landing	Mission not accomplished. Launch vehicle stage four propulsion system failed to start up.	Power supply system failure. Block L control system power failure upon receiving an "air-to-air" command. Error in electrical circuitry.
18	04/02/64	3MV-1 No. 4 Zond-1	Flight to Venus with descent module landing	Mission not accomplished. Two stellar trajectory corrections were performed. Loss of radio communication after two months.	During the second stellar correction, there was a 20 m/s loss of speed. Loss of pressurization in the orbital compartment, breakdown of the thermal mode, and malfunction of main transmitters. Throughout the entire flight, radio communication was conducted using the descent module transmitters.
19	04/20/64	Ye-6 No. 5	Soft landing on the Moon	Mission not accomplished.	Failure of power feed from Block I to Block L. Failure of converter PT-500/1000 or I-100.
20	06/04/64	Molniya-1 No. 2	Test of communications satellite systems and method for inserting it in orbit	Mission not accomplished. Failure of launch vehicle stage two.	Failure of the Block A tank emptying system (SOB).
21	08/22/64	Molniya-1 No. 1 Kosmos-41	Repeat of the unsuccessful launch mission of 06/04/64	Mission not accomplished. Tested plan for the orbital insertion of a new class of vehicle into elliptical orbit and its correction, operation of all satellite systems except the relay system. Acquired nine months of experience in spacecraft control.	Failure of parabolic antennas to deploy. Defect in the deployment mechanism.
22	11/30/64	3MV-4 No. 2 Zond-2	Flight to Mars with a photo-television unit for photography during flyby	Mission not accomplished. Automatic interplanetary station was inserted on an interplanetary trajectory to Mars, but radio communications were lost after one month. Communications via parabolic antenna were tested for the first time.	Failure of solar arrays to fully deploy and the sequencer malfunctioned. Vehicle missed Mars by 650,000 kilometers due to an inability to perform orbital correction.
23	03/12/65	Ye-6 No. 9	Soft landing on the Moon	Mission not accomplished. Stage four control system did not function during intermediate orbit around the Earth.	After "air-to-air" command, converter PT-500/1000 failed—balancing ring grazed the housing closure.

Item No.	Launch Date	Spacecraft Name	Goal of Flight	Results	Results Obtained / Reasons for Unfulfilled Mission
24	04/10/65	Ye-6 No. 8	Soft landing on the Moon	Mission not accomplished. Failure of launch vehicle stage three.	Failure of the Block I engine to build up to design mode—leak in the pressurization system.
25	04/23/65	Molniya-1 No. 3	Test of the long-range satellite communications system	Mission accomplished. Operation of all satellite systems was tested, including the communications system. Execution of a Moscow-to-Vladivostok TV transmission. May 1 and 9 holiday programs were broadcast. Up to thirty channels of telephone and radio broadcasting were tested.	Operation of the satellite for only 7.5 months due to deterioration of solar arrays from being supercooled in the Earth's shadow.
26	05/09/65	Ye-6 No. 10 *Luna-5*	Soft landing on the Moon	Mission not accomplished. Spacecraft was inserted on a trajectory to the Moon. Stellar correction session was performed with large error.	Large errors in setting the stellar navigation and control system because device I-100 failed to warm up.
27	06/08/65	Ye-6 No. 7 *Luna-6*	Soft landing on the Moon	Mission not accomplished. Spacecraft was inserted on a trajectory to the Moon. During the stellar correction session, correcting braking engine unit (KTDU) failed to shut down. Spacecraft missed the Moon by 160,000 kilometers.	Error during issuance of commands from the ground for phasing of calculators in the command radio link system.
28	07/18/65	3MV-4 No. 3 *Zond-3*	Flight to Mars involving photography of the farside of the Moon during flyby and a large amount of scientific research	Mission accomplished. Photographs taken of the farside of the Moon during a flyby on the way to Mars. Stellar trajectory correction was performed. Spacecraft Earth-sensing orientation system and high-volume radio link were tested. Radio communication was conducted for 7.5 months.	Large amount of scientific research performed. Quality of lunar photos was good. There was no assignment to photograph Mars. During flyby, the 1960 mission of Ye-3 was accomplished and photo-television transmission was performed from a distance of greater than 3 million kilometers.
29	09/04/65	Ye-6 No. 11	Soft landing on the Moon	Mission not accomplished. Launch was cancelled. Launch vehicle was removed from the launch pad and returned to MIK for repairs.	Failure of the apparent velocity control system in the launch vehicle main control system.
30	10/04/65	Ye-6 No.11 *Luna-7*	Soft landing on the Moon	Mission not accomplished. Spacecraft was inserted on a trajectory to the Moon. Loss of spacecraft attitude control during braking burn. Hard landing.	Error in stellar navigation Earth-sensor settings.
31	10/14/65	Molniya-1 No. 4	Performance of experiments via color TV transmission	Mission accomplished. Experiments conducted entailing black-and-white and color TV transmission, including with France.	Communication sessions limited in duration because of a rapid drop in solar array current. Operations stopped on 02/18/66 due to solar array deterioration.
32	11/12/65	3MV-4 No. 4 *Venera-2*	Close flyby of Venus	Mission partially accomplished. Radio communication with the automatic interplanetary station ended 17 days before the planetary flyby. Spacecraft passed at a distance of 24,000 kilometers from Venus on 02/27/66 at 0552 hours without performing a correction maneuver.	Poor transmission of commands and final loss of communication caused by a high temperature in the compartment due to failure of temperature control coatings.
33	11/16/65	3MV-3 No. 1 *Venera-3*	Landing on Venus	Mission accomplished. A pennant with the emblem of the USSR was delivered to Venus after 3.5 months on 03/01/66 at 0950 hours.	

Item No.	Launch Date	Spacecraft Name	Goal of Flight	Results	Results Obtained / Reasons for Unfulfilled Mission
34	11/23/65	3 MV-4 No. 6 *Kosmos-96*	Close flyby of Venus	Mission not accomplished. Failure of stage three.	At the end of stage three operation, fuel was cut off before the main command to shut down Block I. The Block L engine failed to start up due to flight destabilization. Separation was accompanied by great disturbance.
35	12/03/65	Ye-6 No. 12 *Luna-8*	Soft landing on the Moon	Mission not accomplished. Flight along the path to the Moon and all corrections were executed normally. Crash at the lunar surface during touchdown. Hard landing.	Broken bracket punctured a rubber shock absorber and caused disturbance, disrupting orientation relative to the lunar vertical.
36	01/31/66	Ye-6M No. 202 (No.13) *Luna-9*	Soft landing on the Moon	Mission **accomplished**. World's first soft landing of a spacecraft on another heavenly body and transmission of landing site images to Earth.	

Notes: 1. Spacecraft Ye-6M No. 202 (Luna-9) was produced and modified by the S. A. Lavochkin Machine Building Factory.
2. This and all subsequent launches to the Moon were conducted under the leadership of Chief Designer G. N. Babakin.

Item No.	Launch Date	Spacecraft Name	Goal of Flight	Results	Results Obtained / Reasons for Unfulfilled Mission
37	03/01/66	Ye-6S No. 204 *Kosmos-111*	Lunar satellite	Mission not accomplished. Failure of Block L.	Failure of the Ye-6 spacecraft's I-100 control system gyroscope.
38	03/27/66	Molniya-1 No. 5	Establishment of communication with the Far East	Mission not accomplished. Failure of launch vehicle stage three.	
39	03/31/66	Ye-6S No. 206 *Luna-10*	Lunar satellite	Mission **accomplished**. World's first artificial lunar satellite created. Launch dedicated to the Twenty-third Communist Party Congress.	
40	08/24/66	Ye-6LF No.101 *Luna-11*	Photography of the lunar surface and its exploration from lunar satellite orbit	Mission partially **accomplished**. Photography not performed due to a disruption of the spacecraft's attitude control in orbit (control system malfunction)	
41	04/25/66	Molniya-1 No. 6	Establishment of communication with the Far East	Mission **accomplished**. Communication restored via Molniya-1 after Molniya-1 No. 4 stopped operating.	Communication ceased on 09/02/66 due to deterioration of solar arrays.
42	10/20/66	Molniya-1 No. 7	Communication restored after No. 6 stopped operating	Mission **accomplished**. Spacecraft functioned until 10/20/68.	
43	10/22/66	Ye-6LF No. 102 *Luna-12*	Photography of the lunar surface and its exploration from lunar satellite orbit	Mission **accomplished**. Photographs of the lunar surface obtained from low lunar satellite altitudes. Series of scientific space studies conducted.	
44	12/22/66	Ye-6M No. 205 *Luna-13*	Second soft lunar landing	Mission **accomplished**. New stereo panoramas of the lunar surface were obtained and lunar soil was studied.	

Note: Beginning in 1967, Chief Designer M. F. Reshetnev supervised the preparations for and launches of Molniya-1 communications satellites. Academician Reshetnev headed the NPO Prikladnoy mekhaniki (Scientific-Production Association of Applied Mechanics, former Post Box G-4805) in Atomgrad, also known as Krasnoyarsk-26. This NPO is the head enterprise for the development of Russian communications and navigational satellites.

On 14 December, the Voskresenskiys had been with friends at a concert in Tchaikovsky Hall. After the concert, they dropped in on friends. And here, Leonid himself asked them to call an ambulance. When the ambulance arrived, he was already dead.

The next morning, Korolev called Mishin, me, Shabarov, Turkov, Yakovenko, and Party and labor union leaders. He looked shaken and despondent. After he announced what had happened, he said: "Leonid was a trailblazer. We must repay him. Despite the fact that he was not a Party member, I propose that his memorial service be held in the Palace of Culture. I've already telephoned Ustinov. He remembers Leonid well. He expressed his condolences and promised to arrange for his burial in Novodevichye Cemetery. Now, as for the funeral commission, I've prepared a draft decree. Boris, you were his good friend; you will be chairman, and Shabarov—you're Leonid's successor; you'll be deputy." The makeup of the funeral commission had been specified, and Korolev signed the decree right then and there.

Thus, I became chairman of the commission responsible for organizing Voskresenskiy's funeral. Dozens of volunteers showed up to help Shabarov and me. For the first time, I felt that organizing a funeral called for a systemic approach. In terms of the number of interactions and things to keep track of, it was as daunting as preparing for a rocket launch. Not a single thing could be missed, and all the proceedings had to be executed strictly in real time. On the day of his memorial service and burial, thousands of people passed by his casket in the Palace of Culture. Delegations flew in from Leningrad, Kuybyshev, Dnepropetrovsk, the Northern Fleet, Kapustin Yar, and Tyura-Tam. Dozens of Moscow delegations continued to arrive, filling every nook and cranny of the Palace with wreaths.

Opening his memorial service, I said, "How awkward it is to say Leonid Voskresenskiy's name in the same breath as the word 'funeral.' Thousands of people knew the talented and daring engineer, our first tester, who became a Hero of Socialist Labor, professor, and doctor of science. We, his close comrades, loved our remarkable friend, who knew how to encourage us with his sharp wit and wise advice in a difficult moment, to lift our spirits, make us smile, to always love life, and place a high value on friendship. I don't know what official historians will write, but Leonid—he represents an entire epoch in our work, imbued with romanticism, courage, and the hard work of the conquest of space."

"It is painful, dear friend, that you departed so soon. Farewell."

Having reread the text of my speech after 40 years, I would not take back a single word.

At Novodevichye Cemetery, the first gravestone with the image of a flying rocket appeared.

I suppose Voskresenskiy's death was harder on Korolev than the deaths of other close comrades of his. Right after the funeral, Korolev checked back into the hospital and was, for once and for all, sentenced to an operation. Several days later, I dropped in on him to brief him on the progress of the Venera flight and to clarify our relationship with Babakin for the last Ye-6 program launch scheduled for January.

I was surprised to see him display indifference for the first time. But, that familiar Korolevian twinkle flashed in his eyes once again when he started to ask about the status of the development of the L3 lunar spacecraft control system. I told him that we calculated that we could gain another 200 to 300 kilograms for the L3 because of the launch vehicle, referring to the last studies on the installation on stage three (Block V) of a variable current turbo generator. Iosifyan and Sheremetyevskiy had made this suggestion one year before. They had developed an unusually light brushless AC generator. It was supposed to generate power up to 25 kilowatts and provide power for all three of the rocket's stages. Arkhip Lyulka had begun developing the high-speed turbine to drive the generator.[21] Korolev was already acquainted with Lyulka from the Academy of Sciences. He had several meetings with him, persuading him to work on the development of a liquid-hydrogen rocket engine.[22] I heard that Korolev was quite enthralled with Lyulka's distinctive Russian-Ukrainian diction. Right away, Korolev supported the turbogenerator project. Three years later, this idea was implemented and paid off completely.

---

21. Arkhip Mikhaylovich Lyulka (1908–84) was a pioneer of Soviet jet engine design. In his later years as chief designer of OKB-165, he contributed to the Soviet space program by developing components for power sources and cryogenic rocket engines.

22. Chertok is referring to the creation of the 11D57 cryogenic high-performance engine with a thrust of 40 tons.

Chapter 15
The Molniya-1 Communications Satellite

This chapter and the next contains recollections not only about the creation of the first satellite communications systems in the Soviet Union, but also about certain contemporaneous events of the first half of the 1960s.

The cooperation of the technocratic structures that coalesced around Korolev's staff accounts for most of the top achievements of Soviet cosmonautics during 1961–1965. Among their triumphs were the piloted flights of six Vostoks and two Voskhods, the secret Zenit photoreconnaissance satellites, a series of launches to test out a soft landing on the Moon, almost 20 launches of automatic spacecraft headed for Venus and Mars, and four Elektron satellite launches to study Earth's radiation belts. During this same period, development began on new piloted spacecraft including prototypes for the future Soyuz series and an enormous launch vehicle and spacecraft for the N1-L3 lunar landing project. At the same time that this highly sophisticated space technology was being developed, the R-7A (8K74) ICBM and its three- and four-stage launch vehicles were being modified; the R-9 (8K75) combat missile, the GR-1 (8K713) global missile, and solid-propellant missiles also were being developed. Central Committee and Council of Ministers decrees, VPK decisions, ministers' orders, and the schedules themselves packed all of the programs into a very tight timeframe.

It seemed impossible to find an opening for any new space project in this work schedule, which was incredibly tight in terms of production and human intellectual resources. Our attempts to find such an opening reminded me of preparing to set off on a mountain-climbing trek. We would stuff our knapsacks until they were about to explode and we could hardly lift them. And then after they were finally packed, we discovered we'd forgotten something. We would open up the knapsacks, pack things tighter, and squeeze in the forgotten item.

We found just such an opening for communications satellites. Design studies began in 1961, and on 23 April 1965, the third in number and first successful launch of the first Soviet Molniya-1 (Lightning-1) communications satellite took place. The very next day, the Soviet Union's first communication session via space between Moscow and Vladivostok was conducted. This event was not accompanied by the propagandistic hoopla that was customary for those times. Its anniversary is rarely celebrated in space history. My contemporaries did not appreciate the true importance of this event!

I will describe the history of the creation of Molniya-1 in greater detail below. Developed almost 40 years ago, it has undergone several modernizations but has not become obsolete.[1] The methods and work style on the Molniya project are quite instructive. Modern space enterprises have already accumulated so much experience that continuity is an essential condition in new developments. The greater the number of adopted ideas, developments, and mastered technologies, the sooner the authors will earn awards. The more innovative new proposals are—thus raising the prospect of future changes in design and production—the greater the resistance they provoke, frightening away conservative managers who are afraid of the personal costs of unpredictable results.

In the early 1960s at OKB-1, work was officially spurred using the banner "maximum use of ready equipment." But this was a slogan to calm upper management who, for the sake of decorum, pretended that that's the way it was, although they understood full well that spacecraft were literally created "from scratch." Molniya-1 is an example of a development that started from scratch.

Officially, the organization of operations at Korolev's OKB-1 at that time conformed to policies approved at the ministry, and the "selection, placement, and training" of personnel was in keeping with Party directives.[2] According to administrative division organizational charts, there were no "interest-based" brigades. Nevertheless, such groups of energetic, self-motivated, young engineers who had concentrated their efforts to achieve a common goal did join forces without regard for which of Korolev's deputies were their official bosses.

Above all, Molniya-1 owed its birth to Korolev's unbridled initiative. He "squeezed" it into his "jam-packed knapsack." False modesty aside, I will say that my personal contribution and the contributions of my comrades in the creative aspects of the development of Molniya-1 exceed much of what we did for the Vostoks and Voskhods. But in addition to initiative, Korolev's

---

1. The Russian government continues to launch uprated versions of the Molniya communications satellite well into the 21st century. A Molniya-1T-class satellite was launched into orbit on 18 February 2004. A more recent variant, a Molniya-3K, was launched on 21 June 2005 (although it failed to reach orbit).

2. In other words, personnel assignments were required to fulfill quotas from lists of the Communist Party members (or *nomenklatura*).

clout, and the creative enthusiasm of the immediate "authors," considerable material expenditures were required for a project whose potential was still quite vague.

The top managers, on whom the very existence of Molniya-1 depended, already had every government award imaginable. The industrial teams couldn't complain of a dearth of decorations and medals either. Every time a new missile was put into service, it brought with it, as a rule, yet another series of awards. The military saw no acute need to have a special communications satellite. Nor did the conservative bureaucracy of the Ministry of Communications, which had an absolute monopoly in radio and television broadcast technology, fully appreciate the prospects that cosmonautics would open up.

At the Ministry of Defense, orders for new communications facilities were under the jurisdiction of Marshal of the Communications Troops Aleksey Ivanovich Leonov.[3] The command subordinate to him was promoting work on tropospheric radio channels and radio relay links, but first and foremost, it was focused on secure wire and cable links for encrypted multi-channel high-frequency communications.[4]

Rather than the military communications operators, it was the military *missile* specialists subordinate to the Strategic Rocket Forces Commander-in-Chief Marshal of the Soviet Union Nikolay Ivanovich Krylov, who supported the idea of satellite communications. The military missile specialists were in dire need of reliable communications between the central staff and the missile divisions where dozens of new missiles carrying nuclear warheads had gone on duty. Subordinate to the missile chief command, Military Unit 32103—the actual proprietor of the Command and Measurement Complex (KIK)—also actively supported the creation of new long-range communications facilities.

The Main Directorate of Rocket Armaments (GURVO) was formed at the Strategic Rocket Forces headquarters in 1960.[5] The chief of the Third Directorate of GURVO was Kerim Aliyevich Kerimov—our comrade-in-arms from Bleicherode and Kapustin Yar.[6] He actively supported the

---

3. Marshal Aleksey Ivanovich Leonov was marshal of the communications troops from 1961 to 1973.

4. Chertok uses the abbreviation ZAS (*Zasekrechivayushchaya apparatura svyazi*) to denote "secure communications equipment."

5. GURVO—*Glavnoye upravleniye raketnogo vooruzheniya*—was actually formed much earlier in 1946 (under a different name) but it was formally subordinated to the RVSN in 1960. GURVO was the branch of the RVSN which supervised requirements, testing, and acceptance of missiles for the Rocket Forces.

6. The Third Directorate of GURVO, established in 1960, was charged with all issues related to the procurement and operation of Soviet space assets.

Molniya-1 project in the bureaucracy of the Ministry of Defense. Minister of Communications Nikolay Demyanovich Psurtsev, who until 1948 had been deputy chief of the Soviet Army communications troops, at first showed no particular interest in the ideas of satellite communications.[7] But there were several devotees in the Ministry of Communications system who supported our proposals from the very start. This helped us obtain the necessary ministry authorizations on the draft governmental decrees. When it came to the launch of the first Molniya-1, Psurtsev's attitude had changed and he consented to take on what was at that time the risky post of chairman of the State Commission responsible for testing it. This enhanced the prestige of the entire program. That was the broad context. Now let's return to the story.

The idea of using artificial satellites for super long-range radio communications and radio broadcasts coincided with the appearance of the first artificial satellites. Historian debates as to whose idea it was to use artificial satellites for these purposes seem pedantic to me. The story of the practical implementation of this idea is much more interesting.

Without exaggeration, it can be said that not a single achievement of cosmonautics is of such economic, political, and cultural significance in today's post-industrial society as satellite communication systems. According to futurologists' predictions, our civilization will soon be transformed from "post-industrial" to "informational." All inhabitants of the planet will receive unlimited opportunities for communication with one another. The concept of "universal" audio- and video-communication will be realized using space systems in the near future. Direct television for any program for any inhabitant of Earth in real time is no longer a technical problem, but an economic one.

The Americans began the first experiments in communication via special artificial satellites in 1960. On 12 August 1960, they inserted into space the inflatable *Echo-1* passive relay satellite with a diameter of 30 meters. The metal-coated surface served as a reflector of radio signals sent from one ground station to another. *Echo-1* had an orbital altitude of a little more than 1,500 kilometers. The research conducted using this passive satellite enabled scientists to gain greater knowledge in the field of radio wave propagation in near-Earth space.

The large metal-coated surface of *Echo-1* was a good reflector not only of radio waves but also sunlight. *Echo-1* was very clearly visible in the night sky over Kazakhstan. During the flights of our Korabl-Sputniks carrying

---

7. Nikolay Demyanovich Psurtsev (1900–80) served as USSR minister of communications from 1948 to 1975, i.e., for 27 consecutive years.

dogs, we used to joke that our dogs barked loudly whenever they passed beneath the orbit of *Echo-1*. Our dogs did not bark in vain—there were no subsequent communications systems developed on the basis of passive reflector satellites.

The first American *active* relay satellites were the *Telstar-1*, launched on 10 July 1962, and the *Relay-1*, launched on 13 December 1962.[8] Both satellites were inserted into low near-Earth orbits and were stabilized by rotating about their longitudinal axis. Their transmitters had a power of just 2 watts. Reception on the ground required a large-diameter antenna. Their low orbits prevented long-duration communication sessions, and their rapid movement over the horizon required continuous automatic tracking.

In 1963, for the first time, the Americans attempted to insert Syncom satellites into geostationary orbit using the three-stage Thor-Delta launch vehicle.[9] In near-Earth space, only one stationary orbit is possible. A satellite's angular velocity in such an orbit is equal to the angular velocity of Earth's rotation. The plane of the geostationary orbit coincides with the plane of Earth's equator. In this orbit a satellite is constantly over a specific point of the equator at an altitude of 35,880 kilometers. The orbital period of such a satellite is equal to a sidereal day—33 hours, 56 minutes, and 4 seconds. Such a geostationary satellite has continuous, round-the-clock line-of-sight coverage of one third of Earth's surface. The presence on the satellite of an active re-transmitter made it possible to conduct radio communications between any points located within this area. *Syncom-3* became the first satellite inserted into geostationary orbit on 19 August 1964 (after the first two unsuccessful attempts).

In the ensuing years, geostationary orbit began to be used intensively. Thirty years later, hundreds of communications satellites, both operating and expired, are located in that orbit. Strict international rules have been implemented due to overcrowding in space. According to these rules,

---

8. Active communications satellites have equipment on board (such as transponders) that receives a signal from Earth, amplifies it, and then retransmits it back to Earth. A passive satellite only reflects or scatters incident radiation, some of which may be in the direction of Earth.

9. *Syncom-1* and *Syncom-2* were launched into orbit on 14 February 1963 and 26 July 1963, respectively, using Thor Delta B launch vehicles. *Syncom-1* failed but *Syncom-2* successfully reached a geosynchronous orbit (but not a geostationary orbit). A geosynchronous orbit is an orbit around Earth where the orbital period is the same as Earth's rotation period. In other words, from a fixed location on Earth, a geosynchronous satellite is at the same point in the sky at the same time every day. A geostationary orbit is a subset of a geosynchronous orbit where the orbit is circular and directly above Earth's equator. There is only one type of geostationary orbit but several different types of geosynchronous orbits.

after a time-consuming, expensive, and arduous process, the International Telecommunication Union—permanently headquartered in Geneva—issues a "ticket" granting the right to be in geostationary orbit.[10]

Despite the obvious advantages of geostationary orbit, a different orbit was selected for the first Soviet Molniya-1 communications satellite. Before 1961, work on communications satellites in our organization was limited to such talk as, "We have the staff; we have a place; we have a partner, but we have no time." Really, none of us had any time! The timid hints from radio specialists that the Americans supposedly had already begun and we were still waiting for something, ran up against our backbreaking workload.

The irritating glow in the night sky of the U.S. Echo, and information about the American preparation of new satellites (active relay satellites), were the impetus for the first Central Committee and USSR Council of Ministers decree for the development of an experimental communications satellite on 30 October 1961. The idea was to conduct experiments to study radio wave propagation and gain experience transmitting in telephone and telegraph mode. It stipulated the involvement of the leading radio engineering organizations (including NII-885, NII-695, OKB MEI, and the USSR Academy of Sciences), State Committees, and of course, the Ministries of Defense and Communications.

We were authorized to manufacture two experimental satellites. The VPK was supposed to develop a detailed plan to support implementation of the decree. For those times, that was a weak decree. In the early 1960s, the majority of scientists in the space industry still underestimated the prospects of satellite communications systems.

Korolev was the first one who, openly in the press and "covertly" in the form of letters to the Central Committee and government (at that time under Khrushchev, these were almost one and the same), raised the issue about developing a satellite communications system. On 31 December 1961, *Pravda* published his article "Soviet Soil Has Become the Shore of the Universe." As usual, Korolev was using his pseudonym "Prof. K. Sergeyev." In the article, he reflected on 1961, a year rich in space-related accomplishments. After paying tribute to the flights of Gagarin and Titov, he wrote:

---

10. The International Telecommunication Union (ITU) is a UN agency, although its formation preceded the existence of the UN. The ITU was formed in 1906 by combining older international agencies involved in telegraph communications. The ITU name formally came into usage in 1934.

Outer space, as yet little explored, undoubtedly is of great practical interest for solving a whole series of applied problems of national, economical, and scientific importance.

In the near future one can expect the development of a system of satellite stations for communications and to relay radio- and television broadcasts, for ship and airplane navigation, for systematic weather observation, and in the future, perhaps, also to exercise some active influence on weather formation . . . .[11]

"Prof. K. Sergeyev" did not have the right to divulge information about the beginning work on communications satellites and about the fact that the Central Committee and Council of Ministers, at the initiative of the author of this same article, had already adopted their first decree.

At that time we had already begun design work. In early 1961, S.P. had already tasked me to meet "only" in that regard with Kotelnikov, Ryazanskiy, Bogomolov, Bykov, and with whomever else I deemed it beneficial to gather information to select our "approaches."

Meetings with the colleagues listed above took place quite often anyway, but they were filled with such a number of current, hot, and urgent issues that it was simply inappropriate to begin a conversation about satellite communications. It was all the more surprising that, having instructed me to initiate these special conversations, Korolev was already conducting them himself. And he personally had written the corresponding paragraphs in *Pravda* after these meetings.

I should take this opportunity to note that, despite his enormous workload, Korolev did not assign anyone to write the articles for "Prof. K. Sergeyev"— not even a first draft. The thought and writing that went into the *Pravda* articles was all his.

It didn't matter at whose initiative the work on the new spacecraft was being performed, it always originated in the conceptual design department. In 1960, spacecraft designs began to be developed in Department No. 9 where Mikhail Tikhonravov was in charge. The subject matter in the department was divided between two sectors. Piloted spacecraft designs were developed in Konstantin Feoktistov's sector and unpiloted spacecraft—"automats" (*avtomaty*) in our terminology—in Yevgeniy Ryazanov's sector.

---

11. The *Pravda* article from which Chertok quotes was later reprinted in M. V. Keldysh, ed., *Tvorcheskoye naslediye akademika Sergeya Pavlovicha Koroleva* [*The Creative Legacy of Academician Sergey Pavlovich Korolev*] (Moscow: Nauka, 1980), p. 433.

Groups conducting work on the Zenit photoreconnaissance satellites and the numerous interplanetary spacecraft were subordinate to Ryazanov. These projects required the department chief's particular attention and didn't mix well with the piloted projects.

Four of Korolev's 'high command' at OKB-1 shown here during the May Day parade in 1964 at Red Square: From left to right are Leonid Voskresenskiy, Igor Sadovskiy, Sergey Okhapkin, and Pavel Tsybin.

While Korolev farmed out projects dealing with ballistics, design, and control systems to Vasiliy Mishin, Sergey Okhapkin, and me, when it came to conceptual designers and testers, they were sort of his own personal guard, despite the fact that he had Bushuyev and Voskresenskiy as the respective deputies for these areas. He personally and very meticulously monitored the conceptual designers, the organization of the design work, the editing of the baseline data, and the draft plans.

After discovering that Tikhonravov's department was clearly overloaded, Korolev divided it into two departments: No. 9 and No. 29. Feoktistov was named chief of Department No. 9 and Ryazanov of No. 29. Tikhonravov received the job title of deputy chief designer, and Department No. 9 was made subordinate to him.[12] Pavel Tsybin returned to us at that time from the aviation industry. Korolev also appointed him as his deputy and made Department No. 29 subordinate to him. Despite their high positions, Konstantin Bushuyev was placed over Tikhonravov and Tsybin as deputy for space projects.[13] I never observed friction, antagonism, rivalry, or any rank pulling among the Bushuyev, Tikhonravov, and Tsybin design trio. The main disputes flared up while distributing the job assignments of good junior specialists who had already gained some recognition.[14] All three were distinguished by such refinement and decorum that other deputies agreed with Voskresenskiy, who once remarked: "In front of them, as in the company of ladies, you can't allow yourself to use strong language."

Despite the workload of Department No. 29 when Korolev demanded that Ryazanov start work on the communications satellite, the latter decided to pull several men away from work on the beloved Zenit for this and then shore up that field with new junior specialists as needed.

From the author's archives.

Operations on the communications satellite were entrusted to 27-year-old Vyacheslav Dudnikov. He already had two years of experience on Zenit layout designs and was considered to be a very sharp, inventive, and self-sufficient designer. Dudnikov put together a small group of even younger guys, who had not yet learned that "you can't do it that way because nobody has ever done it that way." To a great extent, the credit

Vyacheslav Nikolayevich Dudnikov was chief of the design group in charge of developing the Molniya-1 communications satellite.

---

12. Tikhonravov was officially made deputy chief designer on 25 August 1963.

13. In other words, at the time, Korolev's main deputies (deputy chief designers) for spacecraft development were Bushuyev, Tikhonravov, and Tsybin. However, the latter two were also subordinate to Bushuyev. All the deputies were also supervised by the "first deputy," Vasiliy Mishin.

14. In the Soviet Union, a university graduate was ordered by the state to work at a specific enterprise for a period of three years, during which time he/she was called a junior specialist. His/ her job assignment was initially determined by the personnel department. After a few months passed, managers would try to get the best and brightest reassigned to their own sections.

for the top new ideas in the early stages of design development goes to a team of young designers—Slava Dudnikov, Volodya Osipov, Tolya Buyanov, Boris Korolev, Viktor Stetsyura, and other members of the design group in which the average age was barely 26 years.

Vyacheslav Nikolayevich Dudnikov, the young chief of the design group, was presented the opportunity to demonstrate his capabilities, having started work with a "clean slate." He was the heart not only of the small design group, but quickly won recognition as unofficial leader among everyone brought in to work on the first domestic communications satellite. Dudnikov possessed a quality essential to a designer—that of having a thoughtful and even empathetic attitude toward ideas and suggestions that came not only from his own but also from "outside" specialists. Rather than stifling his subordinates, Dudnikov's inexhaustible energy, initiative, and exacting nature stimulated them to show their own initiative while maintaining a sense of responsibility.

FOR THE TIME BEING, WE COULD ONLY COUNT ON THE FOUR-STAGE 8K78 LAUNCH VEHICLE FOR THE LAUNCH INTO SYNCHRONOUS ORBIT. In 1961, it had seemed to us that after the two years that would be needed to produce the satellite, this launch vehicle for the Moon, Venus, and Mars shots would be well optimized.

Young ballistics specialists Mikhail Florianskiy, future cosmonaut Georgiy Grechko, Yevgeniy Makarov, and other ballistics specialists subordinate to battle-tested luminaries Svyatoslav Lavrov and Refat Appazov got to work on the calculations for orbit selection.

Since 1947, in NII-88's Department No. 3, Korolev had entrusted the management of computational and theoretical ballistics calculations to Vasiliy Mishin. Mishin had managed to create a very strong team that knew how to quickly calculate the optimal trajectories for missiles and orbits for spacecraft. It is hard to explain how, before the appearance of the computer, these young ballistics specialists managed in a day, and sometimes in hours, to calculate several alternative options. The leading lights of ballistics were up to their ears with other programs and had been threatened with being packed off on the first train from Tyura-Tam to Moscow if they missed a deadline.

Surprisingly, for all its multi-echelon structure, the work on Molniya-1 was taken up "from below" without strict master schedules and orders "from above."

Geostationary orbit was the starting point for research. The ballistics specialists' calculations showed that our four-stage 8K78 launch vehicle, which had sent almost 1 metric ton on its way to Venus, was not capable of inserting more than 100 kilograms of payload into geostationary orbit. And it was all because of the need to match the orbit with the equatorial plane! Even the boldest of the young enthusiasts could not imagine such a small satellite. Having given up on the requirement for geosynchronicity, the designers and the ballistics specialists began to calculate what would happen with the duration of communication sessions for our expansive territory if there were consecutive orbital periods of 4, 6, 8, and finally, 12 hours. During a brainstorming session, they came up with the idea of using a highly-elliptical orbit. At an inclination of 65 degrees—a value determined by the site of the launch from our firing range—the orbital apogee was 40,000 kilometers above the northern hemisphere and the perigee was 400 kilometers above the southern hemisphere. Such an orbit provided continuous coverage of Moscow and Vladivostok from the satellite for 8 to 9 hours. The most amazing thing was the fact that a satellite with a mass of 1,600 kilograms could be inserted into such an orbit.

However, a strict weight discipline was maintained even for such a "heavy" satellite. A battle was waged for each kilogram of mass, each milliampere of power consumption, and each control command. As a result of the design process, a unique situation was achieved—after development, there was a 10 percent mass margin.

Right off the bat, we obtained the biggest, and consequently, the most powerful communications satellite in the world! The American communications satellite designs at that time did not exceed 300 kilograms. We could have large-area solar arrays, not skimp on the mass of the buffer batteries, and have a good margin of propellant for multiple orbital corrections. Due to redundancy, it was possible to promise a satellite service life of many months and perhaps even a year, and finally, the radio engineers could be allowed to make a repeater that was so powerful that on the ground it was not necessary to build enormous antennas to receive television information.

As for the radio engineers, I went contrary to Korolev's expectations. Rather than pester Ryazanskiy and Bogomolov, I recommended that NII-695 be tasked with developing the on-board equipment for the communications link, namely the repeater or relay station. Korolev was very well acquainted with *Zarya* Chief Designer Yuriy Bykov and asked whether such an additional project would bother him.[15] I assured him that at NII-695 they were working on a different project, which Murad Rashidovich Kaplanov was managing. He had no idea of what we were going through with the piloted flights and agreed to work with us on communications satellite problems.

S.P. didn't know how to put such decisions on the "back burner." He immediately called up Bykov on the Kremlin direct line and said, "I've got

---

15. *Zarya* was the name of the voice communications system for the Vostok spacecraft.

Chertok sitting here in my office. He proposes assigning development of the repeater for the Molniya-1 communications satellite to Kaplanov. I don't know him. What's your opinion as institute director?"

Bykov's response put Korolev at ease. After that he instructed me to introduce him to Kaplanov.

"Just do this diplomatically. Don't go and pass on an order. I simply want him to tell me about his ideas."

Kaplanov and his close associate, Ivan Bogachev, had to meet often with our designers and antenna specialists who couldn't do their job without getting a handle on the main payload—the powerful repeater.

It was simple to arrange for Kaplanov to visit Korolev. When they met, I was already quite convinced of Korolev's skill at performing "strength checks" on people in person, without farming out the task to his deputies. He artistically came up with improvised psychological tests; and based on the results of his observations, he formed an opinion about the individual.

Contrary to my assurances that for the time being only an experiment had been conceived, Korolev began to talk about a crucial government assignment. He said that we would not be developing a satellite for experiments, but would immediately build a communications system for the entire territory of the Soviet Union! The satellite's service life was supposed to be at least six to nine months, and to ensure the beginning of the trial operation they needed to put at least five to seven satellites into production. I must confess that this improvisation was a surprise to me. Later Kaplanov accused me of hiding our real plans from him. What could I say?

Kaplanov belonged to the category of people that you trust the moment you meet them—akin to the feeling of "love at first sight." Right off the bat, I developed a feeling of sympathy and trust for him. He presented his ideas simply and convincingly, without trying to make himself the center of attention, and was very respectful in arguments with his listeners. During one of our many vigils at the firing range in the course of piloted launches, while waiting for the next communication session, I was conversing with Bykov about Molniya-1. He said that we were fortunate to have Kaplanov, and for the first time he mentioned Kaplanov's difficult life. Later in 1964, also at the firing range while preparing to launch the second Molniya-1, Kaplanov, Sheremetyevskiy, and I were sharing one of the four now historic cottages at Site No. 2. Nights spent at the firing range brought people closer together than years of normal on-the-job work. With his melancholy sense of humor, Murad Kaplanov told us a little of his unconventional biography.

Back then the personnel forms contained the infamous item No. 5, "ethnicity." Personnel department employees who had studied Kaplanov's personal information became confused when they found that item filled in with "Kumyk." In today's parlance, he was a "person of trans-Caucasian ethnicity."

Sheremetyevskiy and I confessed that although we had traveled through the Caucasus on a number of occasions, we had not heard of that ethnicity, and it was not right to accuse the personnel office employees of being ignorant.

It turns out that the Kumyks are a small ethnic group in Dagestan.[16] Kaplanov's grandfather had been a Kumyk prince and wealthy landowner. The wealthy Kumyks emulated the Russian aristocracy in their zeal to fit in with Western culture. Murad's father had been sent from "wild" Dagestan to Paris to the Sorbonne—the most prestigious institute of higher education at that time— and received a law degree. In Paris, Rashid Kaplanov fell in love with a young Jewish woman and married her. The prince cursed his son for marrying outside of his religion.[17] However, in 1914, as he was dying, he summoned his son back to Russia and left him his land. When filling out autobiographical information and personnel forms, Murad had been forced to respond to the "social origins" question with "Father—landowner (before the revolution), attorney."

During the Stalinist repressions of 1937–38, among all the social strata of the Caucasian peoples, the intelligentsia suffered most of all. Attorney Rashid Kaplanov, former landowner, son of a prince, a son who had received his education in France, automatically ended up on the list of enemies of the people and disappeared in 1937.[18]

Before his father's arrest, Murad had managed to finish his course of studies in the electrical physics department of the Moscow Energy Institute. In 1938, Murad was arrested and convicted for "failure to report a crime." After spending five years in a "correctional labor camp," he was "completely rehabilitated" in the Red Army, and after being discharged, he received the right to work in a "P.O. Box" under the vigilant supervision of security authorities.[19]

It wasn't until the Twentieth Party Congress [in 1956] that Kaplanov was freed from the constant feeling of social inferiority. "No matter what they say about Khrushchev, I am grateful to him." On that note, Kaplanov ended his nocturnal tale at the firing range. He had talent, enthusiasm, optimism, and kindness to spare.

---

16. The Kumyks are a Turkic people, most of whom live in the lowlands of northeastern Dagestan, a Russian republic bordering Chechnya and the (now independent) Republic of Georgia.

17. Kaplanov was a Muslim.

18. This is a reference to the peak of the Great Terror in 1937–39. Among the many hundreds of thousands arrested were specific populations targeted for their ethnic origins.

19. Soviet defense-related enterprises were openly known as "P.O. Boxes" to disguise their "true" names.

In early 1962, Kaplanov helped to prepare the first prospectus publicizing the Molniya-1 proposal, which was distributed to the Central Committee, VPK, and all interested ministries.

This prospectus announced Molniya-1 as an experimental communications satellite intended to test the design concepts and reveal the operational capabilities of long-distance radio communications systems. For the time being, the experimental system via satellite was proposed only between Moscow and Ussuriysk.[20] Connecting this system to the existing radio relay lines was supposed to link the Far East (Vladivostok) with the main cities of the European portion of the USSR and Western Europe. The prospectus said that, "satellite relay equipment makes it possible to transmit a television program (one channel) or to carry out multi-channel telegraph or telephone communications (40–60 channels)."

After reading this document 33 years later, I appreciated the clarity of the mission statement and the listing of stumbling blocks and main problems to be studied; I felt a new sense of respect for Korolev, who approved this prospectus, Kaplanov, myself, and everyone who took part in its development. In essence, the prospectus was a program document that justified the need for two or three experimental launches to obtain data on the operation of the entire communications complex and develop recommendations for the construction of an operational system. In the program, particular attention was given to the experimental development of systems and equipment to achieve reliability and prolonged operation. It was proposed that this work be conducted concurrently with the execution of the satellite's first launches.

Further along in the prospectus, these prophetic words appeared: "Managing the Molniya-1 satellite project this way will prepare a reliable base for implementing a rather long-term operational communication system using artificial satellites in 1964–1965." Unlike the promises we made for programs involving a soft landing on the Moon and reaching Venus and Mars, this promise was fulfilled without much ado.

From the first days of 1962, the very design of the satellite and the development of its structure and all its systems exceeded the capacities of Dudnikov's design group. Guidance and control specialists, electrical engineers, thermal engineers, and on-board antenna specialists were actively brought into the project.

The Molniya-1 project was fortunate in the sense that no one interfered with the young staff's tendency to let their imaginations run wild. At first, it seemed

---

20. Ussuriysk is located in Russia's Primorskiy Territory, which is in the absolute southeastern tip of the Russian landmass, on the Sea of Japan.

that instead of working, they were like children playing "make-believe." In actual fact, the "little kids" were growing up fast. They took into consideration the failures on other automatic spacecraft that had already flown.

When it came to Molniya-1, S.P. granted me considerably more rights and freedoms than he had on any other spacecraft. From among the deputy chief designers, I was teamed up with Tsybin, who oversaw Grigoriy Boldyrev's design department. Other deputies tried not to interfere in this project. I believed that my primary goal was to join the forces of the specialists of our departments and the subcontracting organizations for a systemic, integrated handling of the problems. Understanding people, maintaining a creative atmosphere, and establishing close, amicable relations for the solution of each specific problem was a lot more difficult than sorting out technical glitches.

Distracted by their own local problems, narrowly focused specialists sometimes forgot about the assembly process, usability, testing procedure, production capabilities, and ground preparation. Dudnikov's young team of designers didn't stand on the sidelines where these issues were concerned either.

Dmitriy Slesarev, whom Korolev had appointed as the "lead designer" of Molniya-1, also proved to be a good assistant. Korolev enjoyed saying that lead designers were "a chief's eyes and ears." That he was. But Slesarev did not abuse his right to report to the chief. His involvement facilitated the solution of a multitude of day-to-day conflicting issues.

In mid-1962, the engineering drawings started to come in. Our factory—perhaps the most essential performer—got involved in the Molniya-1 project. Roman Anisimovich Turkov, director of the pilot factory, was at the same time first deputy to the chief of OKB-1, i.e., to Korolev. At the factory, he was an absolute master in the best sense of the word. The two of us established not simply a business relationship but also trust and friendship. Having been through the grueling school of heavy gun production during World War II, he had not grown callous like many veteran directors.[21] Outwardly he was stern and demanding, but inwardly he was a kind, sensitive man who enjoyed and appreciated frankness and a sense of humor in others. Turkov had a brilliant memory. He knew how to tell apropos and instructive tales about wartime production events. All the while, he would look at the younger members of his audience with a barely perceptible genial smirk.

---

21. During part of World War II, Roman Anisimovich Turkov (1901–75) served as director of the famous Barrikady Factory (Factory No. 221) in Stalingrad, specializing in the production of guns, mortars, and bombs.

Isaak Khazanov was another production chief on whom the Molniya-1 project depended. Exuberant energy, initiative, and extraordinary organizational skills soon made him the number two man at the factory. One of his valuable qualities was his ability to quickly establish contacts with subcontracting production facilities and other factories. More often than not, we found ourselves in hopeless situations in terms of production capacity. In these cases, Khazanov had the rare ability to find outsiders who could save us.

I painted quite a rosy picture of the prospects of satellite communications for Turkov and Khazanov. I briefed them in depth on the project and warned them that it would be impossible for us to avoid a large number of changes in the production process. I implored them to speed up the manufacture of the first assemblies and mechanisms so that we could run pre-flight service life tests for six to eight months.

It was mid-1962 before we got the green light to put Molniya-1 into production at the factory. However, as soon as production got under way, a stream of design changes ensued.

Sometimes, after yet another design change in a drive, antenna, or control instrument during the course of production, Turkov would tell me over the phone that I was like a husband who was the last to find out about the unseemly behavior of his wife.

"But I for one," shouted Turkov, "am sick and tired of your beloved Molniya disrupting operations on all of our other articles."

We agreed that Khazanov and I would look into the situation and report all of our findings to Turkov. These investigations with Khazanov, right in the shops, always ended with solutions that increased reliability, regardless of the number of modifications they involved. Korolev couldn't stand it if we didn't inform him of any changes made in the drawings of Vostoks and Voskhods, no matter how minor. He was not disturbed by changes in the technical documentation of Molniya-1. The designers were more afraid of Turkov himself than his threat to "tell Korolev."

Usually Turkov began his workday by making a tour through the shops. During the process and without fail, he selected the biggest trouble spots. After returning to his office, he would call up the managers and ask, "In shop No. 5, after all the modifications, according to your leak integrity tolerances such-and-such a valve doesn't pass. What are we going to do?" The chief deputy or department chief didn't know anything about that yet that morning and urgently summoned his subordinates and reamed them out, "Why has the factory director found time to call me and you don't tell me anything?"

Trouble spots were identified and rectified quicker using that method than during a tedious "talking-to" at logistics meetings. The manufacture of

Molniya-1, its assembly, the installation of all its components, experimental operations, and finally, the last factory tests were concentrated at the "second production territory." That is what we started to call the forested area that we had inherited from Grabin.[22] German Semyonov was in charge of all of the production facilities in that area. He was an all-around production man. At NII-88 he had been the chief of Korolev's pilot shop. Then he went to Dnepropetrovsk where he learned series missile production.[23] His kinship with Korolev (German was married to the sister of Korolev's wife, Nina Ivanovna) did nothing to facilitate his situation. On the contrary, his bosses and subordinates thought that they could demand more and more from him, supposedly because Korolev would always help him.

More than once I realized that S.P. was more demanding of German than others. Sometimes during production briefings, German would parody Korolev in a good-natured way, ending the meetings by slamming his fist on the table and exclaiming: "Make it happen! Or you'll be walking home on the railroad tracks!"

After Bushuyev was transferred to the first territory—a bit closer to Korolev—I occupied his place in Grabin's former office; and German and I actually bore the responsibility for everything that went on at the second production territory.[24]

At any aircraft or rocket-space factory, the site most frequently visited by upper management is usually the assembly shop. Before our merger, we had just one assembly shop, No. 39 at the first territory. Here, missiles, the first *Semyorkas*, and the first spacecraft were assembled. The control and test station (KIS) was at the end of the shop's production line.

In 1962, spacecraft assembly and testing was transferred to the "second production territory." For this purpose, another assembly shop—No. 44—was set up. Grigoriy Markov was appointed chief of this facility. Molniya-1 was to be assembled in that shop. With the physique of a weightlifter and wearing a snow-white lab coat, Markov met guests with the hint of a bashful smile. He really was a kind-hearted and shy person, even though he established strict order, cleanliness, and exemplary discipline in the shop. Despite the stream of changes that my departments brought down to the assembly shop, there were never any conflicts between Markov and me. In late 1962, Markov completed the assembly of the first Molniya-1.

---

22. See Chertok, *Rockets and People, Vol. II*, Chapter 27.

23. This was at Factory No. 586, which later became the heart of Mikhail Yangel's OKB-586 or KB Yuzhnoye, the designer of many Soviet strategic ICBMs.

24. Korolev transferred Bushuyev out of his job as chief of the second territory and back to the first territory in May 1963.

Testing is the finishing touch for the engineering process in missile and spacecraft production. The KIS, formerly part of the assembly shop, became an independent subdivision. Administratively, the KIS was part of the factory. However, the abundance of engineering problems and the complexity of the test processes, which required the involvement of dozens of engineers (the systems developers), made the KIS a place where it wasn't so much the quality of the factory's production that was being tested but the intellect of the engineers who had developed the hardware. The most difficult thing was beginning and ending the tests. It was difficult to begin because there was always something missing: equipment, instructions, test consoles, and all sorts of other hardware. It was difficult to end because, during the testing process, a multitude of glitches would occur and a decision would have to be made on each of them.

Unlike any shop foreman, a KIS foreman must deal many times a day with dozens of the development engineers in the most varied specialized fields from his own and subcontracting organizations. A KIS foreman must be able to patiently listen to each one, intelligently articulate the claims, and without holding up the testing schedule, if possible, find a way out of an initially inextricable situation.

"Inextricable" situations usually occurred at the interfaces of various systems. Invariably something in the apparatus would not match up with the electrical circuits (either logically or physically) or did not correspond to the instructions. The worst were parasitic coupling and cross-coupling between systems that came as a total surprise to the systems' creators.

When there was an obvious failure of an instrument during the testing process, the decision was made to replace it. But removing and replacing an instrument is only half the battle. Testing cannot be considered complete until the developer of the failed instrument submits his analysis of the causes of the failure and issues a document guaranteeing the reliability of the newly installed unit.

The bureaucratic procedures for obtaining findings with many signatures took time. This prolonged the tests but instilled discipline in all the participants. Each began to understand that making it through KIS with the hope of performing follow-up modifications at the firing range was possible only if done legally with the consent of the chief designer and senior military representative.

Anatoliy Nikolayevich Andrikanis was a foreman at the KIS at OKB-1 during the development of the Molniya-1 communications satellite.

From the author's archives.

Decisions on complex, integrated matters were made at the KIS during management's regular meetings. Where Molniya-1 was concerned, I was usually in charge of them. Anatoliy Nikolayevich Andrikanis was the KIS foreman at the second production territory. Testing the entire Molniya-1 series was a difficult exam for the 28-year-old foreman. In the most complicated situations, one had to have the skill to prove that one more day or week was needed for testing. As a rule, deadlines were missed even before the object was handed over to KIS. Managers at all levels, including the deputy ministers responsible for the program, tried somehow to make up for lost time by shortening the work cycle at the KIS. This situation required fortitude and forbearance on the part of the KIS foreman to resist the temptation to push a spacecraft that had not been fully tested out to the firing range. Back then, Andrikanis was just learning to be firm and amicably obstinate. In the ensuing 30 years of managing factory tests, he released so many spacecraft [into flight] that it is quite possible that he can claim an entry in the *Guinness Book of World Records*.

Out of all the wealth of technical inventions from the first space communications system, I would single out the following: the on-board repeater, the control system, the Earth-tracking high gain antennas, and the ground stations.

The on-board repeater actually consisted of five transponder units. Kaplanov correctly considered that a disruption of communication caused by the repeater threatened to compromise the entire concept. He justified the selection of a single frequency of 800 MHz for the "ground-to-space" link and 1000 MHz for the "space-to-ground" link (I have rounded off the numbers). The transmitters of the three transponders had a radiated power of 40 watts each. The true service life of the transmitters was not yet known. We believed that if each one operated until its first failure, it might last up to one year. In the event of a power shortage, the repeater had two more transmitters with a power of 20 watts each. The transmitter's most critical element in terms of reliability was its traveling-wave tube (TWT).[25] That is where the power from the on-board power source was converted into high-frequency current. The TWTs had a very low coefficient of efficiency for such conversion. The bulk of the energy was lost as heat. Therefore, our thermal engineers Oleg Surguchev and Yevgeniy Belyavskiy proposed placing all of the TWTs in a separate assembly and devised liquid cooling for it. The temperature mode for the entire unit was maintained by keeping the satellite's longitudinal axis constantly pointed at the Sun.

A radiator-heater constantly illuminated by the Sun was installed in the same plane as the solar arrays on the pressurized compartment of the body. Its surface was covered with photovoltaic converters, thus increasing the overall area of the solar arrays. A radiator-cooling unit was installed behind the radiator-heating unit around the cylindrical shell of the pressurized compartment. By automatically reversing the flow of the liquid circulating in the radiators, it was possible to cool the TWT unit and maintain the thermal mode in the entire spacecraft.

The mode of the repeaters required particular attention when they were activated during ground testing. They could "burn up" while still on the ground not just from overheating, but also when the antenna was switched off. Because their energy could not be converted into radio waves, it was converted into heat. While the KIS personnel were accumulating operating experience, they managed to burn up one repeater. When Kaplanov found out about this, he put a pill under his tongue. I forbade activation of the repeater without Kaplanov's representatives present.

The development of the control system complex required the greatest number of inventions. We didn't use a single one of the many systems that had been created by that time, not even as a "starting point." A small team from Rauschenbach's group devised a new multi-mode control system for Molniya-1, which has been basically preserved to this day. It was possible to combine many of the functions that the system was to perform during its long service life thanks to the multipurpose use of a fundamentally new type of gyroscope—a free control moment gyroscope with controlled rotor speed.

The gyroscope played a leading role in virtually all attitude control system operating modes. Yevgeniy Tokar developed the rather complex theory for its use. Since this instrument was a specialized electrical device, we were unable to manufacture it ourselves. Specialists at Andronik Iosifyan's institute set to work on this project—without coercion and with enviable enthusiasm. Nikolay Sheremetyevskiy was in charge of the project. The Molniya-1 gyroscope was

---

25. In Russian, the term would be *Lampa begushchey volny* (LBV).

the source of a new scientific and technical field at [Iosifyan's] All-Union Scientific Research Institute of Electromechanics—control moment gyroscopic stabilization for spacecraft.

The attitude control system began to operate once the satellite's angular rate had dissipated after its separation from the launch vehicle. Next, a special Sun sensor searched for the Sun and the satellite's longitudinal axis, perpendicular to the surface of the solar arrays, was pointed at the Sun. By changing the angular rate of spin of the gyro-flywheel, the entire spacecraft was made to rotate about the direction to the Sun until one of the two parabolic antennas was in a position enabling it to track Earth. A special optical sensor monitored the necessary angular orientation.

To achieve the maximum influx of electrical energy from the Sun's illumination of the solar arrays, it was necessary to maintain continuous orientation toward the Sun for the entire "long" orbital segment before the satellite entered Earth's shadow over the southern hemisphere. During flight through the "sunny" segment, one of the repeater's two high gain antennas was supposed to orient itself continuously, tracking the direction to the center of Earth. An ingenious orbital correction maneuver was devised whereby, before reaching perigee, the satellite oriented itself so that at the point of perigee, the propulsion system directed a correction pulse along a tangent to the orbit. In those cases when the powered gyroscope did not have enough control moment or it needed to be "desaturated," the thrusters operated in the most basic "relay" mode.

Vladimir Branets, Vladimir Semyachkin, and Yuriy Zakharov developed the dynamics of the artificial satellite, controlled by a single electric motor flywheel and all the "relay" control modes. The average age of the dynamics specialists and developers of the control system hardware was 30. They were a whole four years older and more experienced than the conceptual designers. These four years proved sufficient to make real what had seemed to be a "make believe" system.

The engineers who designed Molniya-1 in the early 1960s became respected scholars with numerous disciples. But neither they nor their students would have solved such a problem today without dozens of personal computers and a pool of powerful machines at the computer center and without the experimental development of the system on analog and digital models! At that time, no one had dreamed of such possibilities.

Incidentally, James Watt, the inventor of the centrifugal governor for the steam engine, had only mastered four mathematical operations, while George Stephenson's first steam locomotives, which reached a speed of up to 50 kilometers per hour, were produced in the early 19th century without even the use of a slide rule.[26] In the 20th century, the theory of centrifugal governors and of locomotive construction was complicated by so much mathematical baggage that if those same inventors were alive, they would not quickly make sense of it.

For the control of modern satellites, it is simply impossible to develop a system without using an on-board computer that assumes the responsibility for several different processes: selecting dynamic operating modes according to the flight program, activating the on-board equipment at the proper time, performing diagnostics, monitoring propellant consumption, and performing many more calculations that "way back when" were entrusted to the on-board sequencer (PVU) and ground services.

Isaak Sosnovik and Nina Kvyatkovskaya invented a transistor-based sequencer that controlled on-board systems according to a program that could be uploaded from the ground via command radio link. The command radio link itself, which was combined with the orbital control and telemetry transmission hardware into an integrated complex, had been ordered for development by SKB-567. Khodarev, Malakhov, and the entire radio team that had been around for the bitter experience of the Mars-Venera (MV) automatic spacecraft developed the radio system for Molniya-1.

The selection of the chief developer of the electrical system and logics for the on-board control complex was difficult. Yuriy Karpov was usually in charge of this work. He had experience in the integrated developments of the principles and electrical circuitry for all the first satellites, the Vostoks, and the Venus-Mars spacecraft. His next assignment was the new piloted 7K-OK spacecraft, or article 11F615 (the Soyuz).

After some conflict and hesitation, we made the following decision for Molniya-1: to develop the on-board complex simultaneously with the ground testing equipment, so that it would be possible right away to test the entire spacecraft. In addition, instead of dozens of operators at consoles for each system, we would have just one or two at a central console, having supplemented it with a device to automatically record their actions, whether correct or erroneous. Anatoliy Shustov's radio department undertook this complex development. The department had grown significantly by dint of specialists who had joined it after our merger with Grabin's staff.

---

26. The Scottish inventor James Watt (1736–1819) made key contributions to the development of the steam engine that were crucial to the Industrial Revolution. Watt did not actually invent the centrifugal governor (which regulates the admission of steam into the cylinder) although he perfected its use on the steam engine. George Stephenson (1781–1848) was an English engineer who designed one of the most famous steam locomotives of the 19th century, the *Rocket*, which served the line between Liverpool and Manchester.

Viktor Popov's laboratory began to develop the entire complex under Shustov's management. However, engineers who had left us earlier to work in the VPK administration lured Popov away to the Kremlin office suites. After working at the Kremlin for a short while, he transferred to the administrative offices of the Party Central Committee and until 1991, he remained "our man" in the Central Committee Defense Department. Petr Kupriyanchuk took over Popov's laboratory. He correctly judged that a simpler "on-board" system would make it easier to automate its testing but would also make the "ground" system more complex.

IN THE LITERATURE ON COSMONAUTICS, I HAVE ENCOUNTERED NO REFERENCES TO THE PROBLEMS OF DEVELOPING GROUND-TESTING EQUIPMENT, despite the fact that such equipment is always featured in the background in numerous photographs and in movie and television footage on rocket and space subjects. Developing ground-testing complexes and all the ground-based equipment was implicitly considered less prestigious work than developing on-board equipment. This prejudice came to rocket-space technology from aviation.

In modern aviation, the "ground" (airfield, testing, navigation, and air traffic control (ATC)) is acquiring ever-greater significance.[27] Although each passenger on a modern airliner knows that he or she is flying on an Ilyushin, Tupolev, or Boeing airplane, not one knows about the company or the chief designer who developed the incredibly complex radio electronic equipment for ATC.

In rocket-space technology the "ground" plays a much greater and more vital role than in aviation, and therefore, for people in the rocket-space business not to appreciate the "ground" is a harmful mistake. Only a good "ground" makes it possible to rigorously test and launch a reliable "on-board" into space. A bad "ground" can result in a failure even before launch. The disaster of 24 October 1960 was a grim but instructive example of this.[28] We had hundreds of instances—fortunately, without human losses—of "on-board" failures while still on Earth due to false commands issued by the test "ground."

Petr Kupriyanchuk arranged for joint development of the Molniya-1 on-board layout and test "ground." The ground station that was designated 11N650 performed the bulk of the electrical tests. Anatoliy Maksimov, Boris Bugerya, and Artur Termosesov developed its operational principles. I came

---

27. In Russian, the term for air traffic control is UVD—*Upravleniye vozdushnym dvizheniyem*.

28. This is a reference to the infamous "Nedelin disaster" when scores of engineers and military personnel were killed during preparations to launch an R-16 ICBM. See Chertok, *Rockets and People, Vol. 2,* Chapter 32.

up with the idea of completely automating testing. This required a system for recording all the test operations. Demand gave rise to supply. Engineer Boris Barun proposed to Kupriyanchik a device that automatically issued commands to the "on-board" using perforated tape and a system that automatically printed all the results onto another tape. This automatic system was called *Volna* (Wave). The drawings of this test equipment, which was complex for those pre-computer times, were developed in Chizhikov's instrumentation design department under the supervision of Ivan Ivanovich Zverev.

"I have a bird's surname," said Chizhikov. "I don't get along well with wild beasts."[29] Accommodating "the wishes of the working classes," we split the design departments. Chizhikov kept the "on-board," while Zverev was assigned the entire test "ground." Thus, the "ground" obtained a full-fledged design department.

Our instrument production facility manufactured only two sets of 11N650 test stations along with the *Volna* system: one for the KIS and one for the firing range engineering facility. Even during testing of the first Molniya-1, we understood that the age of full automation had not yet arrived. We needed to develop a compromise semi-automatic system that was more general-purpose than the 11N650. As often happens, difficulties encountered in the process of developing something new lead to solutions that are better than those originally intended.

The idea of developing an integrated ground test station was "in the air." Searching for the optimal designs, Yuriy Karpov and his firing range-seasoned comrades-in-arms Vladimir Kuyantsev, Vladimir Shevelev, and Rem Nikolayev proposed a structure for the test station that was not fully automated, yet provided thorough diagnostics, flexibility, and the capability to quickly modify the test technology. The system involved a multiplex, multi-channel remote-signaling line.

This system was not going to make it in time for the first Molniyas. However, the need to develop such a system was so obvious that I proposed combining and accelerating the development of the entire "ground" in a single project team. Petr Kupriyanchik and Artur Termosesov were in charge of the project. The first model of the general-purpose test station appeared in 1964. The appropriate directorate within the Ministry of Defense validated this development as a universal facility for spacecraft ground tests and assigned the station the designation 11N6110.

---

29. Chizhikov's surname is derived from *chizh*, the Russian word for "finch," while Zverev's surname comes from the Russian word *zver*—wild animal.

We first used this station to test the first 7K spacecraft, the future Soyuz. The experience proved to be so successful that it created a demand for series production. This was still beyond the capacity of our production facility. Without external coercion, Nikolay Vasilyev, Director of the Azov Optical-Mechanical Factory, took on series production of the 11N6110. Eventually, the station was so widely used that, in all, more than 100 of them were produced. These successful technical designs proved to be long-lived despite their obsolescence. Even in the 1990s, during the triumphant march of digital computer technology with its boundless capabilities for automating the most varied types of tests, diagnostics, and data processing, the old 11N6110 stations (after 30 years!) remained in operation in factories and at firing ranges. During 1966, when we began to hand over Molniya-1 to Mikhail Reshetnev at the Krasnoyarsk branch, testing was geared only to the 11N6110.

In 2004, I visited the Mozhayskiy Military-Space Academy in Saint Petersburg. To my great surprise, among the operating training systems, I saw the 11N6110 station! It was a graphic and functional instructional aid. This was 40 years after it was developed!

ON MODERN COMMUNICATION SATELLITES, ONE OF THE PRINCIPAL PROBLEMS IS SELECTING THE TYPE OF ANTENNAS AND THEIR DESIGN. For Molniya-1, without any experience, we solved the problem "head-on" and proposed two parabolic antennas, each with a diameter of 1.4 meters and each one serving as backup for the other. They were installed on special booms and controlled by an electric drive. Kaplanov supported our proposal. This reassured our antenna specialists, who were responsible for converting the power of its repeater transmitters into the "end product," radio wave energy.

To transmit signals from "on board" to all the ground stations located simultaneously in the coverage zone, it was necessary to develop an on-board high gain antenna to receive and transmit simultaneously. By that time, Mikhail Krayushkin's antenna laboratory had grown and distinguished itself as an independent department. The department staff announced that they were going to take on the radio engineering aspects of antenna problems, including everything from designs and simulation to acceptance testing. Vladlen Estrovich, Ivan Dordus, Gennadiy Sosulin, Nadezhda Ofitserova, and the mechanical engineers of the mockup workshop performed the most difficult part of the task.

This and many subsequent developments clearly demonstrated how important it was to have the ingenuity and magic touch of skilled workers in the immediate vicinity of the laboratory, not just in factory shops. The emergence of computers made it possible to significantly shorten the amount of time required for the theoretical design work that preceded the release of the drawings. However, for all the power of modern computer technology, it has not been possible to create a single high gain antenna without preliminary experimental development on mockups. In this laboratory modeling process, one cannot overstate the role of the foreman and worker who understands the engineer before he barely says a word and who needs no detailed drawings. Only after repeated modifications did the factory receive official by-the-book drawings for the manufacture of flight models of the antennas. The individuals who finally fabricated the antennas out of metal would not have guessed that their conceptual design was rooted in the solution of a system of differential equations discovered the century before.

"Medvedev's tubes" were installed on the antenna feed. That was what we called the optical sensors, which, after capturing the edge of Earth's disc in their field of vision, sent signals controlling the antenna drive, thus turning the entire object so that the antenna would be pointed at the central portion of the visible disc throughout the entire communication session. At that time, Boris Viktorovich Medvedev, an optical electronics engineer from Geofizika, had just begun to develop what would later become his extensive inventory of all manner of sensors for space technology and then for submarine-launched missiles.[30]

The electromechanical antenna control drive proved to be a complex mechanism. It was supposed to operate under space vacuum conditions continuously during each communication session. Ensuring its reliability was one of the most difficult tasks. Vladimir Syromyatnikov and Lev Vilnitskiy, the chief of the department of control surface actuators, drives, and mechanisms, spent most of their time in the factory shops waiting to find out when they could get their hands on the first model of the drive for development tests. I implored Turkov and Khazanov to speed up the manufacture of the first mechanisms so that before the flight we could test their service life for six to eight months.

They also reinvented the on-board power supply system (SEP in our terminology).[31] Powering the primary consumer—the repeater—plus various and sundry service systems during a communication session (which lasted 8 to 9 hours) required receiving up to 1500 watts continuously from the solar

---

30. Geofizika was located at the Moscow-based Factory No. 589 and officially known as the Geofizika Central Design Bureau (TsKB Geofizika). The factory, which specialized in the production of optical, geodesic, and medical instruments, traced its roots back to the mid-19th century. Currently, the organization is known as NPP Geofizika-Kosmos.

31. SEP—*Sistema elektropitaniya.*

arrays. In 1961, such an output for a spacecraft seemed as enormous as the output in 1921 of the Volkhov hydroelectric power plant, which was the first project in the State Commission for the Electrification of Russia (GOELRO) plan.[32] At the time, its 60,000-kilowatt output also seemed fantastic.

Aleksandr Shuruy, who distinguished himself in the art of radio control for an anti-tank missile while working for Vasiliy Grabin, developed the power supply for Molniya-1. "I have every right to be proud of the SEP for Molniya-1," said Shuruy, recalling the heroic epic work of the early 1960s.

I decided to use the development of the Molniya-1 for a "revolution" in cosmonautics: introducing a new, uniform standard of 27 volts for everything in the "on-board" and in the "ground" instead of the "mishmash" that existed on spacecraft. The aviation sector also had been contemplating switching to a standard of 24–27 volts. We were in danger of falling behind.

After merging with Grabin's team, the size and professional skill level of our electrical engineering groups increased so much that we could assume the leading role in drawing up a new standard and proving its advantages on an actual spacecraft. Molniya-1 was a very suitable vehicle for this. At the same time, a similar revolution would also be enacted on the Zenit reconnaissance satellite. Shevelev and the Petrosyan brothers were developing the electrical equipment for it under Karpov's supervision. After I had accused them of hard-headed conventionalism, they became part of our alliance under the new 27-volt standard.

Having determined that I would have support "from below," I needed to find some allies among the subcontractors. Our primary customer, Kaplanov (of NII-695), supported me without any reservations. The switch to 27 volts made it possible to reduce the mass of the on-board cable network twofold. We understood that Molniya-1 was just the beginning. During 1959, decrees had been issued calling for the design of large launch vehicles and new heavy spacecraft. We argued all the advantages of 27 volts taking these prospects into account. After discussions during which a majority did not accept the new rating value with unanimous approval, I announced the decision in favor of 27 volts as an ultimatum from the head organization. I seldom resorted to that sort of tactic, trying to avoid conflicts that might result in Korolev's arbitration.

Ryazanskiy unexpectedly objected. He was supposed to take over from SKB-567 the manufacture of the control radio complex, which had been

---

32. GOELRO—*Gosudarstvennaya komissiya po elektrifikatsii rossii*—was the famous commission founded in 1920 and entrusted by Lenin to introduce electricity on a national level to all of Russia. The Volkhov hydroelectric power plant, one of the first Soviet regional hydroelectric dams, was opened on the Volkhov River in 1926 as part of the GOELRO plan.

derived from the 12-volt Venus-Mars spacecraft. This required modifications, and as usual, complications arose at the factories. Korolev supported me in the most decisive way. A standard of 27 volts ± 3 volts was authorized, and it is in effect to this day in all rocket-space technology.

The second problem with the SEP was selecting buffer batteries with a guaranteed service life in a "charge-discharge" cycling mode of at least one year. Silver-zinc batteries had indisputable weight advantages, but they could not compete with nickel-cadmium batteries in terms of the number of cycles. After consulting with us, Viktor Tenkovtsev of the All-Union Scientific-Research Institute of Batteries in Leningrad developed a new type of sealed nickel-cadmium battery with a built-in pressure sensor. This sensor enabled us to develop a master controller, which ensured voltage within a range of 24–31 volts by adjusting the number of individual storage batteries comprising the on-board battery that were connected to the on-board network at a given time.

A spacecraft's primary source of electrical energy is the Sun, and therefore without Nikolay Stepanovich Lidorenko it was impossible to prepare for even a single spaceflight. By this time, the All-Union Scientific-Research Institute of Current Sources (VNIIIT), where Lidorenko was both director and chief designer, had actually gained a monopoly in the production of solar arrays. We developed the structure of solar arrays and the mechanics of their deployment following separation from the launch vehicle after coordinating all the parameters of the silicon photovoltaic converters with Lidorenko. Arkadiy Landsman and Valeriy Kuznetsov were the primary developers of solar energy converters at VNIIIT. Skipping ahead, I will say that Molniya-1's "growing pains" were above all associated with the solar arrays.

During Molniya-1's third and subsequent spaceflights a rapid degradation of the photovoltaic converters (FEP) was detected.[33] It was the manifestation of a little known effect of irradiation when passing through near-Earth radiation belts. Another factor affecting the performance of the solar arrays was thermal cycling—a temperature differential from plus 120 degrees in the Sun to minus 180 degrees in the shade during each orbit.

In 1966, to reduce degradation and extend the life of the solar arrays, Lidorenko's institute introduced a silica glass coating to the working surfaces of the photovoltaic cells. We had to give in and increase the mass, but we had some leeway there, thanks to the efforts of Dudnikov's designers. Additional solar arrays shaped like special pleated curtains were installed, which made

---

33. FEP—*Fotoelektricheskiye preobrazovateli.*

them heavier. When needed, the curtains opened and fresh elements that hadn't been harmed by radiation or thermal cycling were put into operation.

One of the few systems appropriated from the Venus-Mars vehicles was the correcting engine unit (KDU).[34] Isayev gave us the "go-ahead" to use it. He had made minor changes to it, having been assured that we would fire it for correction maneuvers no more than three or four times. This was enough for a year of operation. Our dreams did not extend beyond a year. The KDU was placed on the hull so that the thrust vector coincided with the longitudinal axis constantly pointing at the Sun. Infrared sensors (IKV)—local vertical sensors sensitive to the infrared area of the spectrum along the edges of Earth's disc visible from space—were installed on both ends of the hull.[35]

Two-and-a-half hours before Molniya-1 approached perigee, when the satellite was still in the coverage area of the Shchelkovo tracking station, continuous orientation on the Sun was switched off. The antennas, equipped with their own infrared sensors, also "lost" Earth. After this, one of the two infrared sensors located on the hull was switched on and the entire satellite turned until Earth was in its field of vision. The longitudinal axis constantly pointed at the center of Earth until the satellite reached a point in which its axis was parallel to the velocity vector at perigee. At this moment, the infrared sensor switched off and the satellite continued its flight in a state of inertial attitude, having memorized the orientation of the longitudinal axis before the point of perigee. At that point, the KDU was activated and fired a correction burn to accelerate or brake, depending on which of the two infrared sensors had been selected from Earth. After KDU shutdown, the solar sensor switched on, the solar arrays were once again pointed at the Sun, and the Molniya-1 was again ready to conduct communications sessions. The sequence of the described operations could not be transmitted from the ground because the maneuver took place over the southern hemisphere outside the coverage zones of our ground tracking stations. In this case a sequencer—the primitive predecessor of the modern-day on-board computers—took the place of the ground operator and command radio link.

TsKB Geofizika on Stromynka Street developed new optical-electronic sensors for Sun and Earth orientation according to our specifications. Vladimir Khrustalev, the chief designer of the system, did not let us down: we received the complex instruments on time. Khrustalev said, "This is because, thank

---

34. KDU—*Korrektiruyushchaya dvigatelnaya ustanovka*. This unit contained the 11D414 engine with a thrust of 200 kilograms.

35. IKV—*Infrakrasnaya vertikal*—or "Infrared Vertical."

God, you don't require celestial orientation for Molniya." The truth was that the star orientation sensors for the MV spacecraft and Ye-6 gave them a lot of trouble. Khrustalev had more than enough headaches in that regard.

ANY PROBLEMS THAT CROPPED UP DURING THE DEVELOPMENT AND MANUFACTURE OF THE FIRST MOLNIYA-1S, to one degree or another, fell into my purview. I was directly involved in resolving the main issues; I gave advice on other issues; on a third type of issues, I gave instructions such as, "That's your business—you decide;" and on a fourth, I simply made note of them. Design documents, calculations—which we referred to as "RS" (from *raschety*, or calculations), electric schematics for the entire satellite, and descriptions of the main systems were all works of a collective nature. It was not possible for a single manager to study the thousands of working drawings for the entire spacecraft, instruments, and various and sundry internal components, even if he were a genius and worked around the clock. At best, his job was to familiarize himself with the general configuration and give the go-ahead to hand over the entire set of drawings to the production facility. According to our established procedures, detailed drawings required signatures no higher than department chief. Korolev personally approved RS that determined flight trajectory, launch vehicle parameters, and general layout with a specific spacecraft. He read and approved each volume of the draft plan without fail.

As the dates for the beginning of flight tests approached, there were more and more problems that required Korolev's attention. He had not brushed aside Molniya-1; but more and more one sensed his desire to expand the piloted flight program and his distraction with problems concerning the circumlunar mission and the N1-L3 program. Humankind had seized its first tiny foothold in space. It needed to be secured, and on the wave of our first successes, expanded. Korolev sensed this better than we did.

The draft plan for Molniya-1 was completed in 1962. By that time, it was not yet completely clear what the structure of the ground facilities for the first communications satellite system would be. The primary work on the "ground" was performed by NII-695, under Kaplanov's supervision, and a new subcontracting organization on satellite communications systems, NII Radio, headed by Aleksandr Dmitriyevich Fortushenko.

New people joined our work from the Ministry of Communications. Among them, Nikolay Vladimirovich Talyzin stood out as the most active, striving to coordinate within his ministry all the projects pertaining to the system. I first met him in Shchelkovo while discussing the preparation of NIP-14 for upcoming flight tests. Talyzin was a specialist who had a thorough understanding of the intricacies of communications problems, which I was encountering for the first time. He had an attitude of distinct respect for us "rocketmen" (*raketchiki*), as the ground communications specialists referred to rocket engineers, and did not begrudge the time to recount the specific problems of ground radio and telephone communications. In 1965, 36-year-old Talyzin was appointed deputy minister, and in 1975—minister of communications for the USSR.[36] It was fortunate that the technology of Soviet satellite communications systems had this future minister of communications as a pioneering lead engineer.

In late 1962, we were confident that the new satellite would be ready for launch in one year. It was impossible within that timeframe to produce a special "ground" for Molniya-1 by the beginning of flight development tests (LKI). Having studied the pool of ground-based equipment that was already in operation for all the other space projects, Fortushenko, Kaplanov, and Talyzin proposed using an array of *Saturn* system ground stations, which had been developed at NII-695 under Guskov's supervision.[37] In 1963, NII-695 merged with NII-885, and Ryazanskiy tasked Guskov and Khodarev with adapting *Saturn* to perform these new tasks. Almost all ground tracking stations were equipped with *Saturn* system stations. They were used in systems monitoring rocket trajectories, spacecraft orbits, and in command transmission systems. Parabolic antennas with 12-meter diameters provided the requisite communications radio link power margin for Molniya-1. That is how a simple solution was found for the first years for ground-based communications facilities.

Joining the Molniya-1 project were officers from Military Unit 32103, whom we knew quite well from previous projects—Agadzhanov, Fadeyev, Bolshoy, and the chiefs of four tracking stations: NIP-15 in Ussuriysk, NIP-4 in Yeniseysk, NIP-3 in Sary-Shagan, and NIP-14 in Shchelkovo. High-frequency cable and radio relay lines connected NIP-14 with the Moscow broadcasting center and long-distance automatic telephone exchange. Communication between NIP-15 in Ussuriysk and Vladivostok was conducted via radio relay line.

Fortushenko and engineers from NII-380 in Leningrad developed equipment to link up the ground tracking stations with the broadcasting centers in Moscow and Vladivostok. To link up satellite channels with the long-distance telephone trunk lines, Kaplanov developed the *Ruchey* (Stream)—a single-channel multiplexing and encryption system. Somewhat

---

36. Nikolay Vladimirovich Talyzin (1929–91) served as USSR minister of communications from 1975 to 1980. Before his retirement in 1989, he served in several high governmental posts including as chairman of *Gosplan*, the state economic planning commission.

37. The *Saturn* stations were originally placed at NIP-3, NIP-4, NIP-14, and NIP-15 to support communications with deep space robotic spacecraft.

later, when Molniya-1 was no longer an OKB-1 project, Kaplanov would develop the *Korund* (Corundum) satellite system for troop control. This was the first [Soviet] space link with digital data transmission.

In April 1964, a decree of the CPSU Central Committee and Council of Ministers appointed Psurtsev, despite his personal objections, to the post of chairman of the State Commission for testing Molniya-1. Korolev let slip that he had exerted quite a bit of effort for the minister of communications to become chairman. "This will make him delve more deeply into the substance of the matter and devote greater attention to satellite communications, and we will still be in charge of business that first year," he said. While preparing the decree, Korolev had sensed indifference and even antagonism toward satellite communications ideas in the office of the Ministry of Communications. In his opinion, appointing Psurtsev as chairman of the State Commission would be a turning point and we would obtain new allies.

So that someone would "be in charge of business," we hurried to staff operations groups to manage testing. I was slated to be the technical leader of the State Commission. The chief executive, i.e., a chief designer, had been appointed to that post in all previous State Commissions. In 1964, Korolev had already occupied that post on three different State Commissions: piloted launches, flights of automatic spacecraft to the Moon, Mars, and Venus, and the R-9 combat missile.

During flight tests, the chief of the main operations group was also appointed. The main operations group bore primary responsibility for management and decision making during flight control. This was an organ of the State Commission that was entrusted with operational decisions, leaving the Commission to deal only with strategy.

Through a process of heated debates involving the project chiefs, the leadership of the Central Directorate of Space Assets (TsUKOS), Military Unit 32103, and the many organizations involved in the new project, we came up with a list of working groups and their makeup.[38] The main operations group was in charge of specialized duties such as program development, flight control, communications complex testing, analysis and decryption of telemetry information, spacecraft control, groups responsible for the operation of NIP-14 and NIP-15, and even a separate group responsible for writing up

---

38. TsUKOS—*Tsentralnoye upravleniye kosmicheskikh sredstv*—was the precursor to the current-day Space Forces in Russia. TsUKOS was officially established in October 1964 as part of the Main Directorate of Rocket Armaments (GURVO) within the Strategic Rocket Forces.

TASS reports. There were as many as 100 specialists on the working staff of each group. This included the chief designers of all the systems, the chiefs of the ministries' main directorates, the Command and Measurement Complex leadership, and conceptual designers and developers of all ranks. This was both the intellectual and command-and-control nucleus upon which the system's future depended.

Modern flight control services that depend on powerful computers and automated data processing and display systems are staffed with professionals who have no other obligations beyond working at the Flight Control Center (TsUP). In the 1960s, only the ballistics centers had computers. Flight control operations groups had a simple telephone at their disposal as the primary means of receiving, transmitting, and processing information. However, the majority of operations group members had already passed through all the stages of spacecraft development: design, production, and experimental development at the factory KIS and the firing range. The regular staff of the operations groups viewed Molniya-1 not as someone else's child, but as their own flesh and blood, for which they must do their best to send it on its way.

Back then the control centers were free of hierarchical subordination and formal lines of control. A spirit of camaraderie, mutual trust, and solidarity, regardless of departmental affiliation, prevailed. I was certain that Korolev would recommend to Psurtsev that he approve my candidacy as chief of the main operations group. In that case, I was hoping that after a successful first launch, I would leave the firing range along with my friends and set up in the outskirts of Moscow in Shchelkovo. I dreamed of plunging into the study there of space applications for the promising spacecraft that our Molniya-1 could become. But Korolev had decided otherwise. And he had reasons.

One day in March 1964, I was headed down a corridor in building No. 65 to Korolev's office. Suddenly S.P. literally flew out of the reception area followed by Konstantin Feoktistov. Seeing me, S.P. said, "Oh, good, come with us!" It was no use asking questions on the way. We went through the passageway to the new building No. 67. Here, at S.P.'s initiative, exhibit items for our future museum had been collected in a half-empty hall. The main exhibits were the three descent modules (SA) that had returned Gagarin, Titov, and Tereshkova from space.[39] The other descent modules had been donated to other exhibitions. Along one entire wall of the hall lay a fully-assembled four-stage *Semyorka* stack. Korolev walked up to Tereshkova's descent module and began silently and thoughtfully to examine the interior

---

39. SA—*Spuskayemyy apparat.*

layout through the open hatch. Then he quickly returned to Feoktistov and me and said, "Here's an assignment for you. Instead of one person, we need to fit three in here."

This order was the beginning of transforming the Vostok into the Voskhod. Truly revolutionary decisions needed to be made. It was impossible to accommodate three cosmonauts in spacesuits. Without spacesuits, Feoktistov just barely managed to pack them in like sardines. Feoktistov had a very strong incentive for this reconfiguration—he saw an opportunity to make room for himself in the crew. Here was his chance not just to design for others, but also to fly to space in his own spacecraft.

Why did Korolev suddenly, without preliminary design studies and discussions, make a decision so quickly for a three-man flight that previously hadn't been part of our plans? I found out about this from Aleksandr Sergeyevich Kasho after the *Voskhod* carrying the crew of Komarov, Feoktistov, and Yegorov had successfully returned from space.

Kasho, the lead designer for the R-7 launch vehicle, had been in Korolev's small office giving a briefing on the latest glitches and operations. The Kremlin line rang and Korolev picked up the receiver and motioned him to "be quiet." As soon as Kasho heard: "Yes, Nikita Sergeyevich!," he simply cringed in his chair. Korolev listened in silence. Then he said, "That's impossible." Again he listened in silence. Once again he said, "That is difficult and will require a lot of time." But the conversation ended with a promise: "We'll size things up, and I'll give you a report."

Kasho recounted that the expression on Korolev's face had completely changed—it was somber and distant. He listened no further to Kasho's report. "He wants me to make room for three cosmonauts right away!" said Korolev. Feoktistov was called in immediately and Kasho was sent out of the office with instructions to keep his mouth shut. The meeting described above and the conversation next to Tereshkova's descent module took place after this conversation.

The reconfiguration of the single-seater Vostok into the three-seater vehicle that would be called Voskhod was the primary reason why I was not going to be afforded a little breather in Shchelkovo. Korolev already knew something that we, his comrades-in-arms, did not know—that regardless of the outcome for Molniya, we would be handing it over to someone else, just as we did with the Zenit reconnaissance satellites.

My duties as technical leader of the main operations group for the Ye-6 [lunar probes] was the second reason. A third reason to free me from the leadership of the operations group was Korolev's upcoming trip to Czechoslovakia. This was quite incredible! The highly-classified chief was being allowed to go on vacation abroad with his wife, Nina Ivanovna, for nearly a month. Of course, our embassy had been given all the necessary instructions, and the Czech government had given the corresponding guarantees.

All the more so, Korolev decided that during this "special period" all of his deputies should be in their places: at the firing range or in Podlipki. Under these hectic circumstances, Korolev appointed Pavel Tsybin chief of the main operations group for control of the first flight of a Molniya-1. Nikolay Fadeyev, a colonel from Military Unit 32103 and two high-ranking supervisors, Viktor Bogdanov from the Ministry of Defense Central Scientific-Research Institute of Communications (TsNIIS) and Petr Gobets from the Ministry of Communications, were approved as his deputies.[40]

I was left with the responsibility for preparing the first Molniya-1 at the firing range. The first Molniya-1 complex along with the first experimental automatic test "ground" underwent testing at the KIS for 14 months! This was a record-setting length of time for factory tests. At that time not a single spacecraft had been debugged for so long.

In that regard, Kupriyanchik and Andrikanis said, "Nobody bothers us; nobody rushes us. Any glitch is thoroughly analyzed. We continue to make changes to the "on board" and "ground.""

Preparation began on Molniya-1 vehicle No. 2 in the KIS before testing was completed on vehicle No. 1; No. 2 was clearly farther along. It turned out that they could be shipped to the firing range almost simultaneously. Notification that Molniya-1 No. 2 was ready for shipment to the firing range was issued on 19 April 1964, on the eve of the launch of Ye-6 vehicle No. 5. In a radiogram from Turkov, Korolev wrote simply: "Set up acceptance." There were no instructions for me.

The next Ye-6 launch took place on the dark Monday of 20 April. The wake-up call at the hotels came at 6 a.m. After a quick breakfast, we were seated in our vehicles at 7 a.m. and departed for the launch site. The launch took place at 1308 hours 30 seconds. The "flight normal" reports ended abruptly 295 seconds into the flight.[41]

We dragged ourselves up the steps out of the bunker. A feverish item-by-item search for the possible causes of the latest failure alternated with thoughts about the 3MV on its way to Venus.[42] That night a report arrived from Yevpatoriya saying that commands were not getting through. To make contact with the spacecraft, which was already millions of kilometers away, NIP-16 boosted the power of its ground transmitters to the maximum. Resorting to all kinds of

---

40. TsNIIS—*Tsentralnyy nauchno-issledovatelskiy institut svyazi.*
41. See Chapter 13 of this volume.
42. A 3MV spacecraft (vehicle No. 4) had been launched earlier on 2 April 1964 toward Venus. The space probe had been publicly named *Zond-1.*

tricks at a radiated power of 50 kilowatts, we managed to push through several commands to prepare for a correction session using a star tracker.

Meanwhile, for the failure of the Ye-6, I was once again appointed chairman of the failure investigation commission. By midnight, we believed we had identified the PT-500 converter as the culprit. Korolev insisted that I fly to Moscow at once. We needed to decide what further action to take with the PT-500 and, if possible, save the automatic interplanetary station headed for Venus. I was faced with an inglorious and joyless return to Moscow. "Now comrade Levitan won't get to talk himself hoarse," said Anatoliy Kirillov in parting. "You can leave and come back with your summer clothes. When you return for Molniya-1 it will be over 30°C [86°F]."

On 21 April, I was discussing the program of operations for Molniya-1 with its lead designer, Slesarev, who had just flown in. He was very well-versed in the special features of all the systems and did not attempt to pressure the developers needlessly. His composure combined with his genuine sense of responsibility favorably impressed me. In the work of a lead designer, it is very important that his large and motley team of specialists and subcontractors heed his instructions and feel that he is a wise assistant in the organization of operations, rather than a slave driver.

Upon my return to Moscow, after making arrangements with Bushuyev, I convened a council to discuss how to increase the reliability of Ye-6 and make maximum use of the MV experience. After listening to my proposals, Bushuyev, Tsybin, and Ryazanov doubted that Korolev would agree with them. The spacecraft needed to be radically modified. An interruption in launches of one to one-and-a-half years was inevitable. Rauschenbach came to the council meeting with Bashkin and Skotnikov. It was proposed that they undertake the development of the on-board control system, replacing Pilyugin's and Morachevskiy's system on the Ye-6. All three in unison said it would be two years until the next launch, and only as long as there weren't any other new projects. Boguslavskiy and Pikovskiy, whom I had also invited to the discussion, said that modifying the radio system for the sake of integrating it with the 2MV would mean losing the experience accumulated on the Moon shots. Both were also opposed to my idea because it was unrealistic to transfer control of a flight to the Moon from Simferopol to Yevpatoriya. I gave in and announced that I would not take this proposal to Korolev.

I am remembering this incident as an example of a mistake of what, at first glance, seemed to be reasonable demands for the standardization of spacecraft, but turned out to be not so.

Still to come was a series of scandalous sessions of the investigation commission, during which Pilyugin and Iosifyan crossed swords over the culpability of the PT-500 in the latest Ye-6 failure. Nor did Aleksey Fedorovich Bogomolov pass up the opportunity to rub salt in the open "lunar wounds." He proposed a complete replacement of Boguslavskiy's radio system with one developed by his OKB MEI, as well as an altimeter and a lunar vertical. I advised him to switch his energy to the piloted lunar landing vehicle for the L3 project. I voiced this recommendation, hinting at the recent tirade that Korolev had unleashed upon Bogomolov at the firing range before the launch.

Unaware of the situation, Aleksey Fedorovich had approached Korolev offering his services on the radio system for the lunar program, suggesting that he could guarantee the reliable passage of commands. At this time, Kirillov had been informed of a failure of Bogomolov's trajectory-monitoring instrument. Kirillov approached Korolev and Bogomolov recommending that a delay be called so they could replace the instrument. Korolev flew off the handle and shouted at Bogomolov: "Get out of my sight and don't ever come back!"[43]

After this launch, other things overshadowed this incident, but by all accounts, it was "like water off a duck's back" to Bogomolov. At the next session of my investigation commission, Colonel Yevgeniy Panchenko was harshly critical. He declared that the Ye-6 could not be launched with the control system in its current condition. Four out of the five launches had failed due to the control system! Of the five primary control instruments on the latest spacecraft, two had been replaced in the last few days at the engineering facility before the launch!

Meanwhile, on 12 May, there were plans to attempt to carry out a star correction session on the 3MV [*Zond-1*] spacecraft. The Venera continued to lose precious nitrogen from the instrument compartment. The fan no longer circulated air over the instruments and the temperature rose menacingly. To be "closer to the Venera," I flew to Yevpatoriya with a small group of developers.

The Crimea in early May is magnificent. Carpets of red poppies spread along the roadways from Saki all the way to NIP-16.[44] The air is lush with intoxicating scents making us all the more reluctant to enter the smoke-filled rooms of the service buildings that were grandiosely referred to as the "control center."

The work proved to be incredibly hard. Commands to load the settings for the star correction were repeated many times. Energy-conserving modes were selected to save the mortally wounded interplanetary station. The first correction got through but with errors. The AMS flew right past Venus. We began each day by developing a program to issue commands in a series consisting of several combinations.

---

43. See Chapter 9 of this volume for background to this incident.
44. Saki (or more commonly, Saky) is a town in eastern Crimea known for its "healing" mud baths and other health recreation facilities.

Finally, on 14 May, at a range of more than 13,000,000 kilometers, we managed to start up the KDU. Everyone thought that the star correction had at last been accomplished! But the telemetry provided no solace. The engine had fired for 20 seconds less than the calculated time. We did not lose hope. We decided to repeat the star correction session. This time it worked!

The NII-4 ballistics center reported that the [delta-V of the] last burn amounted to 53 meters per second.[45] This was clearly a shortfall. Preliminarily calculations indicated a miss by at least 100,000 kilometers! Once again we had gotten a Zond off the ground, but there had been "no pennant on Venus."

Khodarev and Malakhov tried to convince everyone that the transmitter's carrier frequency shift was a clear sign that the on-board instruments were overheating. Shuruy corroborated their claim: "Hot batteries don't take a charge. If we activate the transmitters, they'll be shut down immediately to protect them against 'U minimum.' Low voltage!"

Reporting the details of our latest attempts to revive the Venera to Korolev over the high-frequency line, I said, "We are like general practitioners at the bedside of a seriously injured patient. We need a surgeon to mend the hole. Our patient is dying from a high fever."

Korolev replied, "I already promised Turkov I'd weld his process engineers' holes shut because of the quality of their work. Let things cool down, we'll make one more attempt at correction on 30 May, but you get back here right away."

After returning from the Crimea, Korolev received me as if I had been lounging around and enjoying myself at a resort.

"Fly out and get to work on Molniya-1. I'm not going to be at the firing range. You and Shabarov will be in charge. Give me daily reports. But after the launch, I will not allow you to take a holiday in Shchelkovo! Psurtsev has asked me to head up the review team. So we agreed that you would be the chief of the operations group permanently based in Shchelkovo. But for the time being, Tsybin is going to replace you. You supervise all the operations groups. Give Tsybin the best people. But don't let yourself get sidetracked. Organize work on the L3 design. I cannot accept the weights that Bushuyev quoted to me. You understand that we cannot come out with a design that has a 2-ton deficit! I have to have a margin!"[46]

---

45. Delta-V represented the magnitude in the change in velocity.

46. This is a reference to the struggle in the mid-1960s to reduce extra mass from the L3 lunar landing payload so as to fit the launch capability of the N1 rocket.

Chapter 16
Molniya-1 in Space (and more)

On 24 May [1964] I flew out to Tyura-Tam with an intermediate stopover in Uralsk for refueling. The traditional Uralsk breakfast of calf's tongue and a glass of thick sour cream took the edge off the rest of the flight over the steppe, which had not yet been scorched under the hot sun.

On 25 May, upon my arrival at the engineering facility, we held a review team meeting. Compared with the stream of defects that we had experienced with the Ye-6 and MV, the state of preparations with Molniya-1 according to the report from Slesarev and those responsible for systems was, as Kirillov said, "amazingly, knock on wood, quite satisfactory." Naum Alper, whom I remembered from my student years at MEI, accounted for the reliability of the powered gyroscope. Kaplanov reported briefly that the *Alfa* (Alpha) repeater had been rechecked and wasn't needed any more in the tests, but they needed to straighten things out with the antenna feeders. The poor matching of the antennas was reducing the output power from 20 to 8 watts. Tenkovtsev and Shuruy mentioned that, due to an error in the design of the automatics of the power supply system, there had been a dangerous accumulation of hydrogen, but now everything was fine, and the problem had been solved by resoldering.

On 30 May, Kerimov held a meeting of the State Commission in the MIK. As Korolev had anticipated, the official chairman of the State Commission, Minister of Communications Psurtsev, assigned Kerim Kerimov—who was in charge of the Molniya project at TsUKOS—to conduct all ongoing work. I had been on excellent terms with Kerimov since the days of Bleicherode and Kapustin Yar. The State Commission meeting proceeded "in an atmosphere of complete mutual understanding."

Meanwhile, the last communications session with 3MV-1 No. 4 (*Zond-1*) spacecraft was on 30 May. Yevpatoriya reported that they had barely managed to find the carrier signal (without any useful information) among the noise on the frequency. They tried to communicate using the descent module transmitters.

The Molniya launch was scheduled for 2 June. On 31 May, the traditional rollout took place from the cool MIK onto the sun-baked, and now rust-

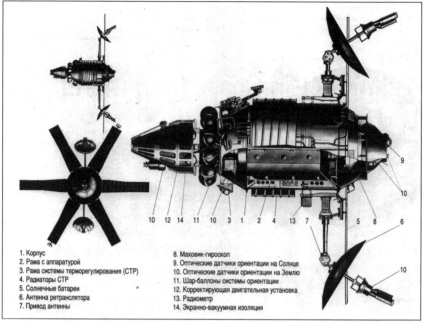

1. Корпус
2. Рама с аппаратурой
3. Рама системы терморегулирования (СТР)
4. Радиаторы СТР
5. Солнечные батареи
6. Антенна ретранслятора
7. Привод антенны

8. Маховик-гироскоп
9. Оптические датчики ориентации на Солнце
10. Оптические датчики ориентации на Землю
11. Шар-баллоны системы ориентации
12. Корректирующая двигательная установка
13. Радиометр
14. Экранно-вакуумная изоляция

RKK Energiya

A diagram of the Molniya-1 communications satellite shows its primary systems. The legend reads (1) body of the satellite (2) framework for instruments (3) thermal control system (4) radiator of the thermal system (5) solar panel (6) transponder antenna (7) antenna drive (8) flywheel gyroscope (9) solar orientation sensor (10) Earth orientation sensor (11) balls for pressurized gases for the attitude control system (12) correction engine unit (KDU) (13) radiometer and (14) vacuum screen insulation.

colored, steppe. Everything was going according to the pre-launch preparation schedules until T-minus 3 hours. Right before the fueling operation, during a standard check of the strap-on boosters' separation system, a pyrocartridge designed to jettison the nozzle lid to vent the Block G strap-on booster activated. A study of the circuit, which had flown dozens of times, showed an anomaly in both the circuit and in the instructions. The two overlapped. It was impossible to repair at the launch site. The entire launch vehicle was taken back to the MIK. At the MIK, they unmated the Block G and replaced the pyrolocks that had fired. On 3 June, the rollout to the launch site took place again. Since morning, the air had been a scorching 35°C [95°F].

New *bobiks* appeared on the launch vehicle during the second preparation process—the instructions didn't match up with the measurement results. During breaks, the launch crew uttered "Molniya's balking—this doesn't bode well," or words to that effect.

On 4 June 1964 at 8 a.m., the 8K78 launch vehicle lifted off from the launch pad carrying the first Soviet communications satellite, Molniya-1 vehicle No. 2. After 287 seconds of flight, the second stage deviated from the trajectory, without having "passed the baton" to the third stage. This Molniya-1 didn't even make it far enough to be called the latest Kosmos. In the local folklore, such launches were popularly referred to as "beyond-the-hill shots." As far as the eye could see, there were no hills on the steppe, but this phrase covered all sorts of launch vehicle failures.

During the day the heat was stifling in the MIK rooms where we were studying the telemetry tapes and arguing over what caused the failure of the second stage. After dreaming of a flight to Moscow to begin experiments with Molniya, the devastated Molniya team was released to go rest until evening.

When the heat let up, I assembled the managerial staff to lift their spirits and give them an assignment—to speed up preparation for the next launch. It turned out that it would be one month before a suitable launch vehicle would be at the firing range, and I granted permission to the main Molniya contingent to go enjoy a week in the cool climate of Moscow.

Meanwhile, the latest investigation commission determined the cause of the malfunction: a failure in the Block A tank emptying system. The kerosene fuel got used up prematurely. Without fuel, the Block A turbo pump assembly started racing, and the automatic engine unit shutdown (AVDU) command was issued. The chief designer of OKB-12 was the primary hardware developer for the tank emptying system (SOB).[1] But the tanks' fuel level sensors, which issue commands to regulate the throttle drives, and the throttle drives themselves, were produced in my departments. Technically, I was also at fault for the failure in the SOB. Konstantin Marks' department not only designed the sensors, but he also oversaw the work of OKB-12.

While the commission was working on the launch vehicle, changes took place in the leadership of the State Commission for Molniya-1. Psurtsev correctly judged that the State Commission for Molniya-1 needed a chairman who could devote considerably more time to it than the minister of communications was able to allow himself. The VPK made the proposal, the Central Committee approved it, and Kerim Aliyevich Kerimov was appointed the new State Commission chairman. A year later he was transferred from TsUKOS to the new Ministry of General Machine Building and became chief of its Third Main

---

1. SOB—*Sistema oporozheniya bakov.*

Directorate in charge of space technology.[2] He remained chairman of the State Commission for Molniya-1 for a long time, despite these career moves.

As chairman, Kerimov announced that the technical leadership of the main operations group must be made up of high-ranking specialists. Korolev yielded, and after the unsuccessful launch of Molniya-1 No. 2, I was confirmed as its chief. At that time, Korolev really did not have enough time for Molniya. A mockup dropped from an airplane to test new landing systems had crashed in Feodosiya. Titov's *Vostok-2* descent module had been used as the mockup.[3]

On 21 August, Korolev was supposed to give a status report to the VPK about the preparation of the future *Voskhod*. Disputes continued as to the makeup of the crews. There had been no experimental development of a soft landing system. It was clear that Molniya was far from being a priority for him.

Turkov managed to send Molniya-1 vehicle No. 1, the second in the series, to the firing range without prompting. Smart alecks at the firing range had a simple explanation for the first failure: "We shouldn't have launched No. 2 out of order." Now on 22 August 1964, preparations were under way to launch No. 1. Compared with the previous experience, testing at the engineering facility proceeded quietly, without sensational failures. Kirillov even complained to me, "This 11F67's behavior during testing has been suspiciously calm and smooth! Can any good come of this?"[4]

It turned out that it couldn't!

On 22 August 1964, an altogether trouble-free launch placed Molniya-1 vehicle No. 1 into an intermediate trajectory, then the Block L fourth stage operated normally, and according to the assurances of the NIP-14 ballistics center [at Shchelkovo], the spacecraft was placed in the planned trajectory. Slesarev checked and gave to me for my approval the list of those who would be flying to Moscow early in the morning. They would have to go directly from Vnukovo to NIP-14 in Shchelkovo to work in the operations groups.

That night the attendant on duty woke me up for an urgent conversation over the high-frequency communications line with NIP-14. And now our superstitious premonition was confirmed. Shustov, who was in charge of the data analysis group, and Popov, supervisor of the telemetry data analysis

---

2. Each industrial ministry in the Soviet Union supervised roughly a dozen "main directorates," each responsible for particular functional profiles. In the Ministry of General Machine Building, the Third Main Directorate was responsible for the development and production of spacecraft.

3. This was a test of the Voskhod soft landing system. To save time and resources, Korolev's engineers used Titov's *Vostok-2* descent module to test the new system.

4. Article 11F67 was the secret designation for Molniya-1 used in all production and acceptance documentation.

group, reported an incredible defect. According to preliminary data, both parabolic antennas, each one acting as the backup for the other, failed to deploy. They had moved but not set up in the nominal position. I asked to speak with Dudnikov. After confirming the report, he said that it seemed as if someone were holding both antenna booms and wouldn't let them move.

"We are developing an operating program," Dudnikov tried to reassure me.

Deputy for the main operations group Arkadiy Bachurin asked me what commands I would give in this unusual situation.

I responded, "Do a thorough check on all the other systems. But the most important thing is—try to understand the cause of the antenna failure."

Dudnikov picked up the receiver again and asked, "Should we wake up Sergey Pavlovich? After all, the first report he received was that Molniya was placed in the planned orbit."

"Wake him up," I replied. "There's no point in concealing this."

There was no way I was going to get any more sleep in the time remaining before our departure. I telephoned the State Commission chairman and woke him up. I told him about the report from Shchelkovo. I sensed that my message was a heavy blow for Kerimov. To replace Minister Psurtsev as chairman and get off to such a bad start. After all, this was the second launch, and practice was supposed to make perfect! We arranged to meet at the airport and come up with a further plan of action on the airplane.

The flight from the firing range to Vnukovo Airport gave us the opportunity to relax for several hours. On the airplane, no one gave anybody marching orders; there were no telephones; and each of us enjoyed his own time free of cares and emotional stress. Groups formed for card sessions; someone was immersed in an unfinished book; some caught up on lost sleep having tuned out those around them; and those who had missed breakfast pooled together the snacks they'd managed to grab and divvied up the last of the alcohol in glasses solicited from the crew. Repeatedly I found that on an airplane, even Korolev (who never passed up an opportunity to give instructions or obtain information) would organize his notebook and try not to talk about pending business. He also gave in to the general tendency to relax for a few hours.

Il-14 airplanes made it possible for us to spend 7 or 8 hours relaxing on board with a 1-hour stopover for refueling in Uralsk or Aktyubinsk. With the advent of the jet age, our flight unit received an Ilyushin Il-18, then a Tupolev Tu-134 and Antonov An-12 cargo planes. Our blissful flight time was cut in half, but the traditions continued. This time there was no bliss for us. For half the trip, we discussed with Kerimov possible causes for the antennas' failure to deploy. Deep down, each of us hoped that during the course of our flight "everything would straighten out," and the antennas would go into their nominal position.

There was mayhem at NIP-14. All the shifts of the necessary and not very necessary operations groups had assembled. They waited for us with the hope that we would bring some life-saving prescriptions from the firing range that were still unknown in Shchelkovo. In countless debates about the causes of the failure, we raised the usual questions that were eternal for such cases: "who is to blame?" and "what is to be done now?"

As early as the first decade of the rocket age, a system had developed to search for answers to these eternal questions. It consisted of a series of unwritten rules that are worthy of listing here.

First, call a halt to noisy, chaotic discussions.

Second, reject the legal principle of "presumption of innocence." When searching for the causes of a failure, a maximum number of potential guilty parties are accused. In the process of searching for the true causes of a failure, each accused party is obliged to prove his or her innocence without the assistance of an attorney.

Third, to do this, commissions and groups are formed that represent all the possible ways to investigate: studying schematics and designs; analysis of telemetry, factory test procedures, and glitches during preparation at the firing range; gathering documents and reports on developmental history and other "evidence."

Fourth, they are tasked with reproducing the flight situation on the next spacecraft available at the factory or engineering facility—a sort of reenactment of the crime.

Fifth, the members of operations groups who are not under suspicion, and specialists called in for the job, must develop and execute a checkout program for all the operating systems and provide flight control so that (God forbid) the patient will not be harmed. Molniya-1 vehicle No. 1 entered an orbit very close to the planned orbit and, according to the laws of celestial mechanics, was not about to leave it in the near future. We immediately decided to take advantage of that to check out all the systems. While commands were being received and executed, and while telemetry was being transmitted from the spacecraft, we needed to gain control experience and accumulate statistics on the performance of the systems.

Sixth and finally, without waiting for the results of the work of all the groups and discovering the true causes of the incident, a drafting group needed to be formed right away to begin writing the "Findings of the investigative commission . . . ." For the time being, they'd have enough work writing up what had already been determined.

As a rule, the leader of the "most important" working group "privately" already has some (one or two) of his own hypotheses (for some reason, we like referring to them as "hypotenuses"). He can seek ways to very quickly check them out, but in doing so he must not exert one-sided pressure on the minds of others who are also looking. Without drawing out the process, a report needs to be given to the higher-ranking officials about the broad plan for finding the answer to the question "who is to blame?" For a while, this provides the opportunity to analyze the groups' work in a relatively calm atmosphere.

Korolev instilled us with the habit that whenever problems occurred during testing—whether on the ground or in flight—that we had to report to him immediately and independently of reports to any individual in the government bureaucracy. The chief designer was always guilty in the eyes of the bureaucracy standing over him. But he needed to be shown mercy and forgiveness, while his deputies were to take the heaviest fire.

Appointed with the chief designer's approval, the technical leader was, in the eyes of his superior and by definition, a potential culprit responsible for an incident. There was a vast assortment of higher-ranking senior officials who demanded reports: the chief of the main directorate, the deputy minister, the minister himself, the chief of the VPK department, the VPK deputy chairman, the VPK chairman himself, the chief of the Central Committee Defense Department, and finally, the secretary of the Central Committee for defense issues. Each one was supposed to find out the cause and plan of action "first hand." Clerks who had been specially assigned to the review team to gather information and reports for superiors zealously tried to make sure that they were privy to what another representative might already know. It was very important for their future careers to be first! Military officers were in the best position. All the groups had "customer's representatives." In the line of duty each military representative was obliged to "delve" into the technology, and, if irregularities were to occur, to immediately report up the chain of command. Finally, there are a lot of friends and acquaintances on the job who are not directly involved with the incident, but who are interested in advancing the common good, and during every encounter they ask "who is to blame?" and "what is to be done?" It would be an insult to refuse to converse with them. It might be that one of them could give some good advice.

Generally speaking, those were the tactics involved in the investigation of various accidents. I adhered to this procedure in this particular case as well. I gave all the groups marching orders that had been deliberated during the flight. After barely managing to get away from the many people who were keeping track of my movements, I left for the factory with Kalashnikov and Dudnikov.

At shop No. 44, Khazanov and Markov were already showing Korolev the antenna deployment mechanism on the next Molniya-1, vehicle No. 3. After the release of the pyrolocks, springs actuated the antenna deployment booms and were supposed to be rigidly secured in the open position. A load-relieving device, simulating conditions of weightlessness, was used during the tests. The device did not do a very good job and the deployment process was sluggish. Korolev requested that they demonstrate the consecutive deployment of the solar array panels and then the antennas. He was getting edgy. If the solar arrays didn't open all the way, then they were hindering the deployment of the antennas.

"Where is your proof that the arrays have opened?"

He lost his temper when we appeared: "They intentionally made two antennas for backup! Can you explain why both of them failed at the same time? How did you test them?"

Using the aforementioned procedure, I tried to report about the organization of the commissions and groups, but S.P. did not let me speak.

"Surely you and your pal Kalashnikov already know something that you're not telling me. You've screwed up such a great spacecraft! There's no forgiving you! Call Tsybin and Boldyrev over here!"

In the meantime, shop handymen fixed the consecutive deployment process of the arrays and antennas. When Tsybin and Boldyrev showed up out of breath, S.P. asked them which one of them had been personally involved in checking out all the deployment mechanics. Without listening to their confused responses, Korolev said loudly so that everyone gathered around could hear: "Find the problem! Until we have absolute clarity and assurance, we will not deliver the next spacecraft! Boris! If you find out what happened up there, report to me at once, no matter where I am. In Feodosiya they smashed up Titov's descent module due to the electricians' error. Now it's history. Now all our work is for naught! Tsybin and Markov, you are personally responsible for all the work on the deployment mechanics. For the time being, on the ground, it is moving sluggishly. There's not enough margin of momentum on the springs."

Korolev walked off, removing his white shop coat on his way. Markov accompanied him to his car. We stood in pitiful silence. The telemetry would hardly inject a great deal of clarity. Judging by the current of the solar arrays, they had deployed normally. But something was still impeding both antenna booms. The force of the springs was clearly insufficient to overcome the unknown resistance.

Where should we look for the answer? I headed over to the KIS to see Andrikanis. He was supposed to keep track of all the details of the Molniya factory tests before their shipment to the firing range. Surely something would occur to him. After all, the two spacecraft were at the KIS for more than a year! But Andrikanis also claimed that at the KIS the antenna deployment tests had run smoothly. Andrikanis mentioned that after all the tests, they had discovered damage to the insulation of the cable running to the antenna boom. They called in the designer. He decided to additionally wrap the cable with vinyl chloride tape. After that, the deployment was checked only at the firing range. That was a loose end that shouldn't have been allowed.

Konstantin Fedorovich Gorbatenko was the permanent representative of the shop at the engineering facility. Kostya Gorbatenko's term of service in the rocket field began back in General Tveretskiy's Special Purpose Brigade. In 1947 and 1948, he was a rank-and-file soldier, and a member of a crew in the electrical firing section. During preparation of a missile at the launch site in Kapustin Yar, his position was on the upper service platform by the instrument compartment. Korolev remembered him well. Once, when for some reason another soldier turned up on the upper service platform, he demanded that the command put Gorbatenko back up there.

After Gorbatenko received his discharge from the service, he came to work for us. He became one of the foremen and then a deputy manager of the assembly shop. Most of the time Kostya Gorbatenko was at the firing range. He had come back with us this time, and I wanted to ask him one more time how he conducted the antenna deployment tests. Kostya Gorbatenko admitted that he also thought the deployment process seemed sluggish.

"A very thick cable is needed to straighten out during the deployment process," said Gorbatenko.

Tsybin and design department manager Boldyrev were both old hands. But anything's possible! Tsybin, Boldyrev, and I reviewed the testing process in reverse order. Of course! We checked out all the structures in a cold chamber at minus 50 degrees without the vinyl tape wrapping.

"Grigoriy Grigoryevich," I said to Boldyrev, "Quickly wrap the cable with the same tape and put it in the chamber! Freeze it!"

All three of us had gotten the very same hunch!

"There's no need to freeze it, I already know—the tubing becomes petrified!" Boldyrev predicted.

"Until it petrifies, not a word to anyone!"

I got on the phone and started calling around NIP-14 for Slesarev. I asked him to get together with the thermal specialists and Rauschenbach's team to figure out the spacecraft orientation in which the Sun will heat the points where the cable bends.

"I don't think the Sun will heat them up. These areas are shaded by the solar arrays."

After being in the cold chamber for 3 hours, the already inflexible cable wrapped with vinyl tape really was "petrified." Thus, the most probable cause was determined. Now we needed "guaranteed" measures. This was a job for Vilnitskiy and Syromyatnikov. They had a skeptical attitude toward designs of spring mechanisms developed in other departments without their involvement. I proposed that they develop an electromechanical drive for "final compression," so that after deployment there would be an absolute guarantee that the antenna booms would be driven into the nominal position.

"For the sake of the cause, we'll save Tsybin and Boldyrev," Vilnitskiy promised.

For Syromyatnikov—future designer of the docking assemblies for all Soyuz, Progress, and Salyut spacecraft, the *Mir* space station, and even the Space Shuttle—"final compression" was a simple task, but at that date it saved Molniya-1. This relatively simple measure made it possible to assert at all official levels that "everything will be okay; we have introduced radical improvements to the design."

When the work was finished, aside from strong language from S.P., there were no sanctions against any of the commissions and groups. The operations groups, and consequently, all the services involved with the development and future operation of Molniya, received the opportunity to test and experiment with the spacecraft without restrictions. With the exception of the repeater system, all of the systems worked failure-free.

In the TASS report, satellite Molniya-1 vehicle No. 1 was referred to as *Kosmos-41*. Except for its primary system—the communication and relay system—its systems worked failure-free for nine months. As was customary for us, TASS did not report about the failure to use Molniya-1 for its direct purpose. However, almost 40 years later, on 9 April 2004, the American space monitoring service reported that the *Molniya-1* spacecraft inserted into a high elliptical orbit on 22 August 1964 under the name *Kosmos-41*, during its 22,130th orbit at 0839 hours Moscow time, entered the atmosphere, and burned up over the Atlantic Ocean. The Americans were wrong only about the launch date. However, this could be attributed to the time zone difference between Baykonur and the U.S. Of course, it is vexing that with the approach of the fortieth anniversary of the flight of our first Molniya, we learned this from the Americans rather than our own services.

With *Kosmos-41*, we gained confidence in the orbital insertion program for a new class of satellites. We had managed to test the operation of systems under conditions of long-term flight and to gain experience in control and orbital correction—in other words, everything except for communications.

MODIFICATIONS OF SUBSEQUENT SATELLITES BASED ON THE RESULTS OF THIS FIRST FLIGHT TOOK A LOT OF TIME. It wasn't until March 1965 that Molniya-1 vehicle No. 3 was shipped to the firing range. The launch took place on 23 April 1965. The eight-month interval can be chalked up to mistakes made during the ground experimental development of the two preceding satellites.

The period from August 1964 through April 1965 was so packed with other events on Earth and in space that work on Molniya-1 faded into the background. We were preparing for the launch of a *Voskhod* carrying three cosmonauts—Komarov, Feoktistov, and Yegorov. The development of new landing concepts using a solid-fuel engine for a soft landing entailed its own "little tragedies." One of these was the aforementioned crash of the *Vostok-2* Descent Module, used at Korolev's direction as the experimental mockup. Museums of cosmonautics lost a rare exhibit—the descent module of cosmonaut No. 2 German Titov.

On 10 October 1964, the *Voskhod* was launched. Twenty-four hours later, despite the crew's request to prolong the flight, they made a successful landing. This flight coincided with the "palace revolution"—the ousting of Khrushchev and the beginning of the Brezhnev era. For the first few months after these events, the Party and governmental offices were not concerned with us.

Meanwhile, we also had problems that made us forget about Molniya-1 for the time being. Right after the successful flight of *Voskhod*, we began preparation of *Voskhod-2*, which included a spacewalk by one of the crewmembers. Once again this was Korolev's initiative. No one at the top obligated us to do this, and no one from below insisted on such an experiment. A Soviet citizen must be the first human being in open space! This mission required substantial modification of the spacecraft, the creation of a special spacesuit, and a large volume of experimental operations.

On 18 March 1965, Aleksey Leonov spent 12 minutes 9 seconds in open space. After his safe return to the spacecraft, and thus, his fulfillment of the main mission—"a world first"—a much-described incident occurred in which the automatic attitude control system failed and the crew had to use manual control to return to Earth. This was followed by an epic rescue mission lasting several days after the crew landed in a remote area of the taiga.

During 1969, Pavel Belyayev and I performed night duty in Yevpatoriya during the flights of *Soyuz-6*, *Soyuz-7*, and *Soyuz-8*. I had often heard stories about the events surrounding the flight of *Voskhod-2*. It wasn't until that night of calm conversation with Belyayev that I realized how close we had been to a space tragedy—the possibility of Belyayev and Leonov dying in orbit. To this day I still consider their return to Earth, given the technology of that time, a great stroke of luck. Less than a year after our nocturnal conversation a tragedy did befall Belyayev. He died during an operation in the same Burdenko Hospital where I had been treated for a mysterious disease in 1957.[5]

---

5. Pavel Ivanovich Belyayev (1925–70) died on 10 January 1970 after complications arising from a chronic ulcer of the duodenum, which had been diagnosed too late. Despite two operations in late December 1969, Belyayev's health deteriorated rapidly soon after. He died of the effects of purulent peritonitis.

A launch to Mars, which made a name for itself as *Zond-2*, and two successive unsuccessful attempts for a soft landing on the Moon preceded the flight of *Voskhod-2*.[6] During this same period, fierce debates raged over N1-L3.

It is all the more surprising that Molniya-1 vehicle No. 3, launched on 23 April 1965, was in fine shape. A storm of cheers erupted in the control room at NIP-14 at the report that the antenna had "locked on" to Earth during its first attempt at orientation. I congratulated Boris Medvedev, whose optical sensors had persistently tracked "the ends of Earth," and Vladimir Syromyatnikov, who had provided the "final push," and for the first time, using Medvedev's signals, had managed to turn the "parabolas"—as we referred to the on-board antennas.

However, once again we realized how dangerous it is in our business to celebrate prematurely. After receiving the reports from the telemetry specialists that all systems were functioning properly on board and that the drives were operating steadily—orienting one of the parabolic antennas over the expanse of our nation from Moscow to Vladivostok—I looked questioningly at Kaplanov. Now it was up to him.

"Shall we activate?"

Everyone impatiently waited for the most triumphant moment—the command to activate the repeater.

"Activate," Kaplanov agreed.

His voice breaking from tension, Bachurin reported to Ussuriysk that we were activating the repeater and requested an immediate signal occurrence report. Ussuriysk reassured us that they were ready. Their orbital monitoring system and telemetry were in steady contact with the spacecraft. The ground-based *Saturn* antennas were tracking the spacecraft using target acquisition. After issuing the command to activate the repeater, they waited for a report from the telemetry operators about a surge in the consumption of current, and most importantly confirmation based on data from the radiation detectors that the repeater had begun to operate.

These reports did not follow. Was it possible that the command had not gotten through?

"Repeat the command!"

What the hell was going on? Bachurin checked and received the report: "We have acknowledgement that the command got through to the spacecraft twice!"

---

6. *Zond-2* was launched on 30 November 1964 while a single Ye-6 lunar probe was launched on 12 March 1965 (publicly named *Kosmos-60*).

Therefore, radio engineering had nothing to do with it. Something had happened in our on-board circuitry. Kupriyanchik, Shustov, and I traced the route of the command on the well-worn circuit drawing from the on-board decoder to the power switch feeding power to the repeater. Along the way there was a single solitary relay. If its contacts had become oxidized or if a "foreign particle" had found its way under them, this type of failure would occur. In situations when jubilation threatens to turn into dismay, everyone awaits the chief's instructions. I had seniority, both in terms of position and degree of moral responsibility.

In such moments, long forgotten cases resurface from the depths of your memory. My old experience as an electrician (where, when, whence—I do not remember!) told me that if the contacts had oxidized or if something had gotten under them, we needed to try to clean them off by repeated impact. Now it seems to me that this was some internal voice.

"Keep repeating the commands!" was all I said.

The repetition of commands began at intervals of 20 seconds. A tense hush hung over the room. We had already made more than ten attempts.

Kaplanov looked at me inquiringly. He said nothing, but I understood him: isn't it about time to stop?

"Continue issuing commands!" I said now with stubborn rage.

I cannot recall whether it was the 15th or 17th try, but suddenly a shout of excitement rang out: "We have activation!" I didn't believe it. Shuruy confirmed that current consumption had suddenly increased to the calculated value. "Number 35 reports carrier frequency reception," shouted the duty officer for communications with Ussuriysk. Kaplanov walked up and hugged me. Only then did I realize that my back was wet (I don't think mine was the only one). Handshakes were exchanged all around with no regard for rank. I have absolutely no recollection what words were spoken during this first historic [Soviet] conversation via space with Ussuriysk and then with Vladivostok.

Wasting no time, we decided to switch to image transmission experiments. We gave test patterns to both sides. Talyzin, Fortushenko, and VNIIT representatives counted 350 to 400 lines and 7 to 8 tonal gradations. This was quite satisfactory for a start!

Throughout these first hours Korolev did not telephone a single time. I sat down by the telephone and began to call around looking for him. Glancing at the clock, I realized that he should be home by now. But I found S.P. in his office. In a tired voice he asked me whether I was certain that everything would be okay. There were no congratulations. Evidently his mind was no longer on Molniya. Suddenly he said, "Well, thank God, we're going to give Reshetnev an operating system rather than a half-finished product."

That would have been better left unsaid! It was a blow that he had already braced himself for mentally and in his actions, but we were still hoping that perhaps the chief would give in and change his mind.

We tested the operation of the entire communications complex on the Moscow-Vladivostok line with the passion of gamblers who had suddenly gotten lucky after a losing streak. We so wanted to show the Far East the May Day rally and parade celebrating the 20th anniversary of the "Great Victory," and to show Moscow the naval review of the Pacific Fleet in Vladivostok.[7] We managed to pull it all off! Finally we had recouped our losses! We received personal thanks from the secretary of the Primorskiy Kray Communist Party Committee. At first he thanked Psurtsev. Then the Central Committee hinted to him that he should call Korolev on the high-frequency line. Korolev passed along the Far Eastern congratulations to us at NIP-14.

After telephone communications had been checked out over 30 channels and the capability for radio broadcasts had been confirmed, we decided to take a risk and perform an orbital correction in order to increase the duration of communications sessions. Back then we still considered starting up a correction engine in space to be a risky undertaking. What if it jerked or the integrator wouldn't actuate when the specified burn was achieved? If there was any seal failure there might even be a fire! And even worse—a loss of stabilization, and then the orbit would be messed up. Even today, no matter how brave a face the producers of rockets may put on, during lift off, as long as the engine is firing, something resonates within them when the telemetry operators report: "Pressure in chamber—steady, flight—normal!" Our first orbital correction with a Molniya-1 worked brilliantly!

At that time, color television was still a "rare delicacy." Nevertheless, Igor Rosselevich—who had arrived from Leningrad—and Aleksandr Fortushenko, begged us to start conducting experiments on the exchange of color programs with Vladivostok. And it worked! I derived just as much satisfaction from the color pictures transmitted from Vladivostok to Shchelkovo as I did from the first images of the far side of the Moon obtained six years before in Simeiz.

Rehabilitation was complete. At the factory they stepped up the production process and finished the next Molniya-1, vehicle No. 4. The Central Committee insisted that we guarantee television broadcasts of the November festivities. Ustinov telephoned Korolev and said that the Far East no longer wants to live without Moscow television, and if we let them down, then the Territorial

---

7. The "Great Victory" (*Velikaya pobeda*) was the date marking the end of the Great Patriotic War (World War II) in the Soviet Union.

Committee (*kraykom*) Secretary will complain directly to Brezhnev. One cannot help but compare the reaction of the country's top political leaders in 1965 to the Far East's request for television, and the central authorities' blatant neglect of the infrastructure of Primorskiy Kray and Northern Russia 30 years later!

We sensed that our work was essential not only for defense, politics, governmental prestige, science, and posterity, but also for thousands of ordinary people—our contemporaries. They were happy for our successes, directly appreciating them in their homes. We, too, were jubilant. We found the color transmissions from Vladivostok gratifying, not by dint of their artistic content, but by the very fact that they were happening. Once an artist has finished a picture, he gazes at it with a sense of creative satisfaction. We experienced something similar looking at the screens of the first color kinescopes during the transmission of the test pattern from Vladivostok.[8]

But our joy was mixed with disappointing bitterness. There were three reasons for this. First—the signal strength on the television screens proved to be significantly lower than expected. Ordinary viewers who were looking at satellite pictures for the first time didn't notice this, but specialists noticed it right away. Power was being lost somewhere. Checking out all the communication channel interfaces, we realized that either ground-based or on-board systems were at fault. Our investigation showed that the on-board antennas were responsible for the amplification shortfall. Our antenna specialists, who had gained valuable experience developing the high gain parabolic antennas for the Venus and Mars interplanetary stations considered the Molniya antenna problems to be relatively simple. A gain of 60 was trifling compared with 1,000 for the interplanetary antennas! They paid dearly for their overconfidence. Ultimately, an interdepartmental commission was created, calling in highly respected antenna specialists from the outside. Long searches were conducted at the engineering facility. Losses were first detected during local tests of the antennas with the entire feeder circuit. Such tests were not part of the normal testing procedure. Modifications for the upcoming launch were conducted "in-house," right at the engineering facility. Subsequently, the feeders were completely replaced.

Second—a rapid drop in solar array power had been noticed back on Molniya-1 No. 1. This phenomenon had not particularly troubled us, because with the antennas still undeployed, there was nothing to waste the electric power on. But on No. 3, after three months of operation, the power taken from each square meter began to decrease faster than the most pessimistic calculations.

---

8. A kinescope is a recording of a TV program made by filming the video monitor.

Nikolay Lidorenko and I developed a broad program of investigations to determine the cause of this phenomenon. According to the prognoses, given deterioration of this intensity, the satellite's life would end sometime in November. If, by that time, the fourth Molniya-1 had not been launched, or if it went "beyond the hill," the Far East television audience would inevitably be outraged, leading to troubles with the Party. Therefore, together with the factory we rushed to complete the fourth unit without detaining it to modify the solar arrays. A lot of "solar" operations were already accumulating, and we decided to implement them on Molniya-1, vehicles No. 5 or No. 6.

The third reason for our bitterness was Korolev's firm resolve to get rid of Molniya projects altogether. "We are going to hand them over to Reshetnev in Krasnoyarsk." He stated that firmly in early 1965, but there had been talk of it back in 1964. After our first successes, we had hoped that our chief, having seen firsthand how grateful the people were, would change his mind and the project would still be ours. Dudnikov, Shustov, Kupriyanchik, Rauschenbach's team, and our subcontracting communications specialists had so many new and interesting proposals!

We had only just recently parted with the Zenits, to which we had contributed many new original concepts.[9] Now once again we had to give up a creation that had already become near and dear to us! Dudnikov took a difficult step and wrote Korolev a tearful report in which he argued the deleterious effects of Korolev's plan to transfer the Molniya project. Korolev perceived it as my scheming.

We had a difficult conversation during which I was once again driven into a corner over botched operations and the enormous weight of the hardware in the lunar landing project. It ended with S.P. declaring that he intended to personally fly to the Yenisey and decide everything right there once and for all.[10] There remained the small hope that there on the Yenisey, Reshetnev, who had alluded to being overloaded with missile-related projects, to the factory's bad work, and difficult relations with the local authorities, would not feel like taking on a new assignment.

HERE I'M GOING TO TAKE THE LIBERTY TO DIGRESS SO THE READER WILL UNDERSTAND THE LOCATION OF OUR BRANCH NO. 2, where Mikhail Reshetnev was the chief and deputy chief designer.[11] While we were developing rockets and had only just begun to astonish the world with our successes in space,

---

9. Korolev handed over further work on the Zenit reconnaissance satellite project to OKB-1's Kuybyshev branch (Branch No. 3). See Chapter 12 of this volume.

10. Krasnoyarsk is located on the Yenisey River.

11. In other words, Reshetnev was the chief of the branch but still a deputy chief designer in the overall OKB-1 hierarchy.

atomic scientists had built closed cities and underground factories that made the famous German *Mittelwerk* near Nordhausen look like an 18th-century textile mill in comparison. These cities were officially called "closed administrative-territorial formations" (ZATOs).[12] At first, ZATOs were distinguished by post office box numbers. Later, they were named after the closest large cities. Thus, Arzamas-16, Chelyabinsk-40, Penza-19, Krasnoyarsk-26, Krasnoyarsk-30, Tomsk-7, etc. appeared. Each had its own specialization in the highly complex technological process of developing and manufacturing nuclear weapons.

ZATOs were towns that could easily be detected in photographs taken from space, but could not be found on a single geographical map available to the public, despite the fact that the population of some closed cities was as high as 100,000. ZATOs began to be built soon after World War II, as early as the days of the First Main Directorate under the command of Lavrentiy Beriya. After 1953, the entire atomic industry was consolidated into the Ministry of Medium Machine Building (MSM). All Ministry of Medium Machine Building towns had a lot in common besides just their missions. Built in regions that differed in terms of natural environment, they nevertheless had a lot of similarities in architectural layout, living conditions, and the everyday life of the populace.

The town where Branch No. 2 of OKB-1 would be working was called Krasnoyarsk-26.[13] Military construction workers and prisoners began to build this town in 1949. Sixty kilometers north of Krasnoyarsk on the bank of the Yenisey River, there was an area of still untouched taiga on low hills. The main facility, for which the city was erected, was a so-called Mining and Chemical Combine—a grandiose underground factory for the production of atomic explosives. Nuclear reactors—the main equipment of the factory—were located at a depth of 250 meters. This earthen dome was a reliable shelter in the event of a nuclear missile war. At the same time, if something happened at that depth underground, there would be no danger for the surface environment (the experience of Chernobyl later underscored this principle).

The first reactor went into operation in 1958. Its product—the crucial strategic material plutonium-239—serves as the explosive for atomic bombs and the detonator for hydrogen bombs. With the startup of the first reactor, the Mining and Chemical Combine became an operational facility of strategic importance.[14] Underground construction continued for another 10 years! In 1961, a second reactor went into operation, and in 1964, a third, which began

---

12. ZATO—*Zakrytoye administrativno-territorialnoye obrazovaniye*.
13. Krasnoyarsk-26 is now named Zheleznogorsk.
14. The actual designation of this facility was Combine No. 815.

providing heat to the entire city. It wasn't until 1969 that the underground giant reached its full design capacity.

On 4 July 1959, the State Committee for Defense Technology—which oversaw OKB-1—and MSM (which managed Krasnoyarsk-26), jointly issued an order for the creation of OKB-1's Branch No. 2. So that it would have standing among the local authorities, Branch No. 2 was referred to as OKB-10.[15] The projects of our new branch had no relation to nuclear technology. But the leadership of two ministries had judged correctly: it makes no sense to establish a new closed area if a city is already being built on the surface near the underground Mining and Chemical Combine. That is where they proposed transferring Reshetnev and his group of enthusiasts.

The town was located in a closed area with a diameter of more than 18 kilometers. The entire perimeter of the area was securely guarded and protected with several rows of electrified barbed wire. A barbed-wire fence also stretched for many kilometers along the bank of the Yenisey.

In 1959, construction of the "socialist" city was under way at full speed. A movie theater, the first schools, a Pioneer camp, polyclinic, hospital, stores, music school, recreational park, and recreation center for Pioneers and school children were already up and running.[16] A dam had also been built, blocking off the Kantat River, a tributary of the Yenisey, thus forming a wonderful municipal lake. Intensive housing construction was a great service of the creators of Krasnoyarsk-26. In terms of social and household amenities, the personnel of Branch No. 2 were better off in the middle of Siberia than our three other newly formed branches in Dnepropetrovsk, Uralsk, and Kuybyshev—even better off than the personnel at the main OKB in Podlipki. At that time, we still had many dozens of barracks where workers and rocket design specialists resided without basic living conveniences. Thousands of people dreamed simply of having a small room in a communal apartment. Hundreds of individuals registered for a personal meeting with Korolev and Turkov; of these, 90 percent were coming to request housing. That was how it was 18 kilometers from Moscow. It was no better in Moscow itself. It was dramatically better in the taiga, 60 kilometers from Krasnoyarsk.

---

15. The sequence of events was slightly different than that enumerated by Chertok. OKB-1 established an experimental design department at Factory No. 1001 at Krasnoyarsk-26 on 4 June (not July) 1959. This department became OKB-1's Branch No. 2 in August 1960 and finally an independent entity known as OKB-10 in December 1961.

16. The Pioneers (or the Young Pioneer Organization of the Soviet Union) was a mass youth organization in the USSR for children ages 10 to 15 that was established in 1922. The organization, a kind of Communist version of the Boy or Girl Scouts, was dissolved at the end of the Communist era.

Reshetnev's first task was design accompaniment for series production of the intermediate range R-11M missiles that we had developed and put into service.[17] The Krasnoyarsk Machine Building Factory mastered their series production. Production of Isayev's engines was also set up there both for these missiles and for Makeyev's submarine-launched missiles.

I visited Krasnoyarsk-26 for the first time in the summer of 1969. By this time, Branch No. 2 no longer existed.[18] Reshetnev was chief designer and director of an independent "P.O. box." The second time I visited Krasnoyarsk-26 was in November 1994, in conjunction with the seventieth birthday of Academician, Hero of Socialist Labor, General Director and General Designer of the NPO Prikladnoy Mekhaniki (Scientific-Production Association of Applied Mechanics) Mikhail Fedorovich Reshetnev.

Both during my first visit and my second, 25 years later, the closed city pleasantly surprised me with its cleanliness, coziness, patriarchal discipline, and refinement. The geometric austerity of the street system, the blending of the housing blocks into the natural relief (unusual for provincial towns), and the absence of ramshackle wooden houses, sagging fences, and barrack dormitories leapt out at me. During the years of acute shortage, the stores of the closed city were supplied a lot better than the nearest major regional city.[19] The hospitals, polyclinics, kindergartens, and preventive medicine clinics differed from their "average Russian" counterparts in terms of cleanliness, lavish equipment, and lack of lines. The labor discipline and very strict order, which tied in with the specific nature of production, were higher than anywhere in the country. Even permission to leave a closed city—not only permission to enter—was by no means granted to everyone.

Korolev and his wife Nina Ivanovna had already gone to the Yenisey in the summer of 1960. The city did not yet look as finished as when I saw it eight years later, but even then it had surprised Korolev, who had seen a lot in his life. He spent several days in this genuine "atomic city." Given the opportunity to see with his own eyes the living conditions and work of Branch No. 2 personnel, Korolev was convinced once and for all that he had made the right decision in selecting the time and place to establish the new organization and realized that he had not erred in the appointment of its

---

17. "Design accompaniment" typically meant small design changes and improvements during series production of a missile or spacecraft.

18. OKB-1's Branch No. 2 became independent in December 1961 and was hence known as OKB-10.

19. The years of "acute shortage" refer to the so-called "Brezhnev stagnation" when the Soviet economy began to slow down, especially in the civilian sector.

chief. Young Reshetnev, of course, complained about trifling matters, but at the same time he did not whine and remained an optimist.

Sensing a great future, Reshetnev and his young team were not afraid of their space assignment, further development of Molniya-1. There could certainly be no talk of reconsidering the decision now. Moreover, a directive followed to accept Reshetnev's specialists, show them everything, and prepare for the transfer of documentation. Turkov also received a corresponding command: finish a batch of seven units and prepare to transfer production equipment and all inventory; if need be, help with the manufacture of labor-intensive parts.

The successful operation of Molniya-1 No. 3 (of course, "No. 3" was not mentioned in print) changed the attitude toward the prospects for satellite communications within the conservative circles of Ministry of Communications specialists. At the same time, there was an ongoing discussion about selecting a standard for color television. The SECAM system had been developed in France, but the U.S. and Great Britain did not want to adopt it.[20] Soviet advocates of the French system came up with the idea of setting up an exchange of color television programs between Moscow and Paris. To do this, they prepared Molniya-1 vehicle No. 4. This launch took place on 14 October 1965—10 days after the launch of Ye-6 No. 11 (*Luna-7*), which ended in a failure during a braking session near the Moon. Molniya-1 No. 4 began to successfully operate on the second day after its launch. In the press it was declared the second Molniya-1 satellite. We had needed this launch not just to set up a communications system. We also obtained the necessary proof of the increased reliability of the 8K78 launch vehicle, and to a certain extent, vindicated ourselves after the latest "hard" landing on the Moon.

From 6 to 8 November, holiday programs were broadcast via the new satellite. Experiments involving the broadcast of color images to France began on 27 November. French specialists gave high ratings to the image quality and proposed that we prepare to receive color programs from Paris. At this time, the current of the solar arrays on the preceding Molniya-1 No. 3 had begun to drop at a catastrophically rapid rate. According to prognoses, the satellite had no more than two weeks of life left!

But two weeks later, Molniya-1 No. 4 also showed signs that it had started to deteriorate. In a meeting of the State Commission, we made a decision to limit communications sessions to 4 hours per day to conserve electrical power so that the buffer batteries would fully charge.

---

20. SECAM—*Séquentiel couleur à mémoire* (Sequential Color with Memory) is an analog color TV system used in the areas of the former Soviet Union, parts of French West Africa, and France (where it was originally developed).

A French delegation headed by their deputy minister for post and telegraph arrived in Moscow for talks concerning a color television exchange via space between Moscow and Paris. Neither I nor other OKB-1 employees were allowed to attend a meeting with foreigners. An exception was made only for Dudnikov. We were secret personae. God forbid that we should blab something about the trouble with the solar arrays during a banquet. But my operations group deputy, Arkadiy Bachurin, told me that at the banquet officially thrown by the Ministry of Communications, after he changed out of his officer's uniform and into a civilian suit, he sat between the head of the French delegation and the latter's mistress.

A "masquerade" held in Shchelkovo at NIP-14 for the demonstration of the communications satellite control center for the French generated a lot of amusement. All the officers and soldiers "dressed up" as civilians. The officers had their own "civil" costumes, but there was quite a scramble with the soldiers. The majority of the soldiers had burr cuts. An outside observer could easily guess that something was amiss here.

The French were very happy with their excursions, receptions, and first television exchanges via space. They departed, and we continued our work now on the development of the system. There were already three satellites in production.

A lot of proposals requiring serious modifications had accumulated for subsequent Molniya spacecraft. We realized that the solar arrays determined their lifespan in space. Therefore, we decided, beginning with vehicle No. 6, to change the manufacturing process of the photovoltaic converters themselves, strengthening their protection; we installed additional backup solar arrays that would remain closed until, when the primary arrays deteriorated, they would be opened and connected to the power supply system.

Rosselevich and Bratslavets insisted on installing a special television camera, making it possible to observe Earth from different orbital points: "Molniya is not just for communications—it will enable us to forecast the weather and will direct the Zenits to sites for detailed photoreconnaissance!"

It would have been sinful to reject such a proposal. The launch dates were moved to the first quarter of 1966, and it wasn't possible to do everything by that time. Before New Year's, Kerimov and I urgently assembled the main operations group and listened to reports about the "on-board" system status with regard to our assignment—to broadcast the Central Committee General Secretary's New Year's address to the Far East. The buffer batteries needed to be charged so that the broadcast would not be disrupted during the communications session. This mission of "national importance" was accomplished.

Meanwhile, the solar arrays were continuing to deteriorate so rapidly that from one day to the next, one could expect the complete failure of the electric power system of satellite No. 4. However, it was destined to fulfill one more historic mission. Korolev's funeral was successfully broadcast via Molniya-1 to Vladivostok. One month after Korolev's funeral, the satellite ceased to operate.

Molniya-1 vehicle No. 5 arrived at the firing range in February 1966. Testing at the engineering facility coincided with the demise of No. 4. The Far East was left without Moscow television programming. Complaints from the Primorskiy Kray Committee to the Central Committee followed immediately. Kerimov, as chairman of the State Commission and simultaneously the chief of the Third Main Directorate of MOM, was supposed to endure the complaints of the Central Committee, the VPK, and his own minister.

Remarkably, the Ministry of Communications behaved calmly. It was a lot more tranquil there without Molniya-1. The thing was, at that time, bitter debates had flared up between the ministries as to who should be the customer and future proprietor of the satellite communications system. For the time being, the State Commission was fulfilling this role with the help of the operations groups and our OKB. We proposed transferring our leading role (and, consequently, also the headaches) to the Ministry of Communications. But that ministry nodded toward the Ministry of Defense. The debate continued. This created a foolish situation. It suited the Ministry of Communications officials just fine that the new system's developers bore the responsibility for its operation. Talyzin tried to help us, but the Ministry of Communications had made it known to him that as long as breakdowns were possible, the space system could not be put into service. We solemnly promised to restore communications during the month of March.

The launch of Molniya-1 No. 5 took place on 27 March. And of all things to happen! The launch vehicle headed "beyond the hill." Once again it was a third stage failure! Luckily, at that time the next Molniya-1, vehicle No. 6, had arrived at the firing range. Considering the internal political pressure, we instituted a round-the-clock work regime to prepare it; and on 25 April, it was inserted into its nominal orbit. The next day, television communication with Vladivostok was restored. However, No. 6 proved to be ill-fated. In September, communications were once again terminated by a power system failure. It wasn't until No. 7 that the Molniya-1 satellite communication system began to operate in a completely stable manner. All the measures that we and VNIIT had devised to prolong the performance of the solar arrays were implemented on No. 7. Launched on 20 October 1966, this last satellite in our series was in continuous operation until January 1968! Molniya-1 No. 7 was the baton that we handed off to Krasnoyarsk-26 in space.

The short life cycles of the first Molniya-1 series simply did not make it possible to develop a system made up of three simultaneously operating satellites that would provide communications 24 hours a day. The solution of this problem was passed on to the new team. We stopped grieving and were happy over their rapid successes. On 25 May 1967, the folks in Krasnoyarsk launched the first satellite manufactured according to our documentation with virtually no changes.[21] Before the 1960s ended, seven more satellites had been put into the Molniya orbit.

In 1965 Kaplanov and I began working on a new promising design of a satellite communications system. This system was called Molniya-2. Boris Suprun, a conceptual designer with God-given talent, as they used to say, worked on the design of the new satellite. The primary difference between Molniya-2 and Molniya-1 was the use of the super-high frequency band. This made it possible to increase the transmission capacity, combine television broadcasting with a multi-channel telephone, and use a high-gain antenna, which selectively covered the time zones rather than the whole territory visible from the satellite. The military was interested in the super-high frequency band for its secure communications channels with spatial division. We also gave Krasnoyarsk the draft plan for Molniya-2.

We secured successes in space by creating new ground-based stations developed especially to relay television programs and telephone communications. The initiative for creating these stations belonged to Fortushenko and Bogomolov. OKB MEI developed the antenna system design for these stations. Aleksey Bogomolov was very proud of this work. The stations were given the name *Orbita* (Orbit). Twenty such ground-based stations were built before 1967. Any populated area that built an *Orbita* nearby obtained the opportunity to receive Moscow television programming. By 1984, there were more than 100 of them! *Orbita* towers, crowned with parabolic antennas 12 meters in diameter, became the objects of particular pride for the local authorities who affirmed their involvement in the successes of cosmonautics through such a monument.

NPO PM, headed by Reshetnev, was constantly improving the subsequent Molniya-1 spacecraft. Space communications received full recognition; the number of satellites needed to be increased, but it was beyond the power of the pilot plant at Krasnoyarsk-26 to set up series production. The Ministry of General Machine Building set up series production at the Omsk Machine Building Factory. There, they also found devotees who established their own

---

21. This was the fifth satellite launched into orbit under the open Molniya-1 designation.

OKB. Now their organization is called PO Polet, in memory of the famous combat aircraft that were manufactured in Omsk during World War II.[22]

The total number of Molniyas launched have exceeded 150. New, successful modifications (Molniya-2 and Molniya-3) began to crowd out the Molniya-1. Since the day NPO PM was founded, it has developed more than 30 types of spacecraft for various communications, television, and navigation systems. However, if Molniya-1 was to die, it wouldn't be due to obsolescence; rather, it was the result of the catastrophic deterioration of the Russian economy.

In Krasnoyarsk-26, development began on the first domestic communications satellites for geostationary orbit. For that, once again our participation was required. To put the new satellites into geostationary orbit, Reshetnev needed a launch vehicle that was more powerful than the 8K78. Our old, distinguished four-stage 8K78 launch vehicle inherited the name Molniya from the Molniya-1 satellite.[23] Both of them, the launch vehicle and the satellite, had long since been cured of their growing pains. The new satellites that needed to end up in geosynchronous orbit had to switch over to Chelomey's Proton, for which we developed a special fourth stage, the Block D.[24] Nikolay Pilyugin was assigned to develop the control system for this stage.

There was a work overload at the main territory of NIIAP, and Pilyugin assigned the development of the Block D control system to his branch at Fili. Its manager was Vladimir Lavrentyevich Lapygin, who assumed the post of general designer and chief of NPO AP after Pilyugin's death.

When I visited Lapygin on Block D business for the first time in Fili, I discovered that I was on a site that had been referred to as "the dachas" in the early 1930s. Here, on the high bank of the Moscow River there really had been dachas where German specialists, representatives of the Junkers firm, had once lived. In 1930, I had somehow offended the foreman of the electrical shop at Factory No. 22, and to "teach me a lesson" he had sent me to these dachas to replace the wiring. I was paid by the job, and not very gainfully. But the place was marvelous. Two years later, beneath this steep bank, I passed the GTO norms for swimming with a grenade.[25] Now a solid

---

22. PO—*Proizvodstvennoye obyedineniye* (Production Association). *Polet* is the Russian word for "flight."

23. This has been the custom for several Soviet launch vehicles, including the Vostok, Soyuz, Molniya, and Proton, which took their "open" names from early or major satellite payloads.

24. The Block D upper stage was originally developed as part of the L3 lunar landing payload designed for launch on the N1 super booster.

25. GTO—*Gotov k trudu i oborone* (Ready for Labor and Defense). GTO was a program that promoted physical fitness in the USSR. Individuals could earn GTO badges by passing written and practical exams.

concrete wall topped with barbed wire extended along the bank. The murky water of the once pristine river was filled with construction debris.

I shared my memories of this once beautiful area with its current proprietors. They rationalized that, "Back when you were young, this river bank and these waters were clean. Now we are working here in clean laboratories so that we can provide you with clean instruments." The organizational culture of the laboratories and the cleanliness in the production shops of NIIAP and its branch were certainly to Pilyugin's personal credit. After the initial hardships that were normal for any new space development, the Block D control system became firmly entrenched as one of the enduring systems in modern cosmonautics.

Beginning in 1964, and lasting until the end of the 20th century, all models of domestic satellites put into high elliptical (or Molniya) orbits or geostationary orbit, used only two models of launch vehicles, the Molniyas and the Protons. Their structures and systems continue to be a synthesis of the creative work from the schools of Korolev, Chelomey, Glushko, Kosberg, Pilyugin, Kuznetsov, Ryazanskiy, Barmin, and Isayev. When the information society replaces the post-industrial one, these names should be preserved in its long-term history.

After taking the baton from OKB-1 in the Siberian taiga, on the grounds of an "atomic city," Mikhail Fedorovich Reshetnev created a powerful leading organization for the development of satellite communications systems for the Soviet Union. He proved to be a worthy manager of this mission, which was vitally important for defense, culture, and economics. A Hero of Socialist Labor, Academician Reshetnev departed this life early.[26] Albert Kozlov then headed NPO PM. A difficult task fell to him: to keep a constellation of communications satellites in space in the face of the complete collapse of the Russian economy. Satellite communications systems, like all of domestic cosmonautics, have been dealt a heavy blow. However, we have not lost hope that domestic satellite communications technology will emerge from its difficult crisis in the 21st century.

---

26. Mikhail Fedorovich Reshetnev (1924–96) died on 26 January 1996.

Chapter 17
Korolev's Last Days, Death, and Funeral

During the Khrushchev era, Korolev and his wife, Nina Ivanovna, would ring in the New Year at the Kremlin. The new Party leadership, however, felt that it needed to make a show of modesty, and for the time being, refrained from pomp-filled New Year's receptions.

Under Khrushchev, the Party and government elite, Army generals, legendary marshals, ministers, and the most "distinguished and deserving" personalities of art and literature were invited to the New Year's receptions at the Kremlin. The Presidium of the Academy of Sciences, general designers and most prominent chief designers of aviation, rocket, and atomic technology were included, as well as other particularly distinguished scientists. Since all of these illustrious guests were invited with their spouses, the receptions were well-attended.

The post-Khrushchev cancellation of the Kremlin's New Years' receptions enabled the former "New Year's cohort" to take their own initiative. In 1966, the Korolevs rang in the New Year at the dacha of Central Committee Secretary Boris Ponomarev.[1] In 1962, at a regular session of the USSR Academy of Sciences, Ponomarev was elected as a full member of the Academy in the history department. He probably felt that he was beholden to Keldysh and the Academy Presidium for his election.

Boris Ponomarev made a favorable impression on academic circles in many respects. His name was not associated with the repressive machine of the Stalin regime. Before Stalin's breakup of the Comintern, Ponomarev had served on its executive committee and later, in the Communist Party Central Committee, he had been in charge of contacts with the Communist parties of

---

1. Boris Nikolayevich Ponomarev (1905–95) was a leading Communist Party ideologue who headed the international department of the Central Committee for over three decades, from 1954 to 1986. As an official historian of the Communist Party, he authored the official *History of the Communist Party of the Soviet Union* in 1959.

other countries.[2] Boris Ponomarev's brother Aleksandr was the chief engineer of the Air Force. I had the occasion to meet with Aleksandr Ponomarev a number of times at State Commission meetings for piloted space flights and at technical meetings, where Korolev saw fit to send not just Kamanin, but other senior Air Force officials as well. The Ponomarev brothers were from Zaraysk.[3] This gave me the occasion to tease my wife Katya: "There's someone from your hometown who's almost in the Politburo." "Almost" meant that Ponomarev was not a Politburo member at that time.

My jokes were not unfounded. The ancient city of Zaraysk was proud of Anna Semyonovna Golubkina, the famous Russian sculptor. After Anna Semyonovna's death in 1927, the efforts of her nieces Vera and [my wife] Katya, to a great extent, determined the fate of her sculptural legacy. In the late 1920s, when Anna Semyonovna's brother was stripped of his voting rights, high-ranking Party official Ponomarev helped the Golubkin family emerge from disfavor. After the death of the childless Anna Semyonovna, a decree of the Presidium of the All-Russian Central Executive Committee granted personal pensions to all of her minor nieces and nephews. Sergey Gorbunov, Ponomarev's comrade from the Zaraysk Komsomol, became director of the major aviation Factory No. 22 in 1930. Anna Semyonovna's niece, Katya Golubkina, came to work at this factory after graduating from an aviation special services technical school.

All of Anna Golubkina's relatives signed a grant allowing all of her works to be bequeathed to the state without compensation. Her relatives did not lay claim to the works that remained in her Zaraysk home or Moscow studio. Cousins Vera and Katya Golubkina expended a great deal of effort to gather and preserve Anna Semyonovna's sculptures and convert her studio on Bolshoy Levshinskiy Lane into a memorial museum. Vera Golubkina became the director of this museum. During the ideological persecution of expressionists, which Anna Semyonovna was considered to be, the authorities tried to close the museum and disperse the most valuable sculptures around the vaults of the Tretyakov and Russian Museums.[4] Once again Vera Golubkina turned

---

2. The Comintern (Communist International) was an international Communist organization founded by the Bolshevik Party in 1919 to advance the cause of Communism worldwide. The organization was abolished in 1943.

3. Zaraysk is a town about 100 kilometers south of Moscow on the Osetr River whose origins date back to the Middle Ages.

4. Expressionism originated following World War I as a rejection of conventional notions of representation in the arts (painting, sculpture, literature, etc.). It emphasized subjective expression, exaggerated emotionalism, and distortions of reality. The expressionist movement in the Soviet Union was supressed in favor of "socialist realism," a Stalinist aesthetic that focused on glorifying the struggles of the proletariat against the bourgeoisie.

to Boris Ponomarev in the Central Committee, appealing to his Zarayskian pride. He actually intervened and helped to preserve the rich artistic legacy of Anna Golubkina.

Now, after renovation, the Golubkina Museum has been reopened in Moscow. Admirers and connoisseurs of her work have no idea that at one time, Central Committee Secretary Boris Ponomarev helped to preserve this very valuable art collection.

But now let's return to Korolev's last days. According to Nina Ivanovna, couples had gathered at the dacha: the Ponomarev brothers and their wives, the Keldyshes, the Korolevs, and the Posokhins. Mikhail Posokhin was Moscow's chief architect.[5] He had designed high-rise buildings, the Palace of Congresses in the Kremlin, New Arbat, and many other architectural masterpieces of that time. Nina Ivanovna Koroleva recounted: "When Sergey was getting ready for the Ponomarevs' party, he brought along so much cognac that there was no way everyone there could have drunk it all."

During the last few years, I had repeatedly witnessed that Korolev had virtually given up drinking alcoholic beverages. Keldysh drank only champagne. Nina Ivanovna recalled that at the party at Ponomarev's dacha in 1966, everyone felt at ease and the evening passed without excessive eating and drinking.

On 1 January, Korolev relaxed at home. On the second, despite the fact that it was Sunday, he called for a car and set out for Podlipki. He had not managed to look through all the mail that had piled up before the New Year.

On 3 January, Korolev worked in his office, continuing to get his correspondence in order. He didn't call anyone in, he simply telephoned when he needed a clarification on some document or other.

Months later, one Sunday in early autumn 1966, Bushuyev, Okhapkin,

From the author's archives.

Sergey Korolev giving a talk.

---

5. Mikhail Vasilyevich Posokhin (1910–89) served as chief architect of Moscow from 1960 to 1982 and contributed to a rennaissance in architecture in Moscow in the post-Stalinist era.

Yakovenko, and I were relaxing around a campfire after a mushroom hunt in the woods where the pickings had been slim. At such gatherings we always ended up reminiscing about S.P. We tried to recall which one of us Korolev had conversed with on 3 January, and what instructions he had given. Everything that I could remember had to do with days before this one. Bushuyev and Okhapkin said rather uncertainly that S.P. had not telephoned them that day either. Only Yakovenko said that S.P. had called him with a request regarding documents that he had not yet signed off on that were stored in a special archive for secret documentation. S.P. asked him to make sure that they weren't given to anyone until he returned. And if it was a long time before he got back, then Mishin should be allowed to look at them.

On the morning of 4 January, Vasiliy Mishin convened a follow-up meeting to discuss how the weight of the lunar landing modules might be reduced so that the already manufactured new super-heavy N1 launch vehicle could cope with the task of landing at least one man on the Moon.

The offices of Korolev and Mishin in new building No. 65 shared a reception area. As they gathered for the meeting, everyone felt compelled to ask, "Where's S.P.?" Korolev's Secretary Antonina Alekseyevna [Zlotnikova] had been working with him for 15 years. Ever calm and kind to anyone who entered the reception area, she endured the most unexpected Korolevian outbursts. After yet another incident during testing or the disruption of a production or project schedule, Korolev needed to blow off steam at somebody, and quickly. That is when it was Antonina Alekseyevna's duty to hunt down that somebody, even if it meant going underground.

It would happen that at that given moment, the individual in question could not be tracked down quickly enough. In such situations Antonina Alekseyevna was told she was fired on the spot. But if she had taken offense and actually walked off the job, there would have been a much more ferocious tongue-lashing.

Before we would drop in on Korolev, as a rule, we would look questioningly to Antonina Alekseyevna. If she smiled and gave a slight nod, that meant that all was calm and it was all right to go in. If she had a plaintive smile while shaking her head ever-so-slightly, that meant: "It might be a good idea to stop by another time." And if we had been summoned with a menacing phone call, then her sympathetic gaze followed us.

I don't know whether Antonina Alekseyevna believed in God, but every time one of us from Korolev's cohort passed by her to enter his office, it seemed that she saw us off saying, " . . . may God help you!"

That day, all one had to do was look at Antonina Alekseyevna to ascertain that S.P. would not be receiving anyone on business today. In any event, when we gathered in the waiting room Antonina Alekseyevna said, "He's leaving soon."

Around 10 of the main managers of the lunar landing project had gathered in Mishin's office. This was not their first meeting. Fierce arguments usually raged over the lunar module's power generation, "working fluid" reserves, and possible alternative layouts and dates. Kilograms of spare weight—the chief designer's "untouchable reserve"—had already been used up, and now everyone was trying to look in somebody else's pocket to see it there wasn't a little something "stashed" there.[6]

But on this day, the atmosphere was amicable. Even Mishin, who usually flew off the handle when weights were being distributed among the systems and stridently demanded that everyone cough up all their reserves, review, reshuffle, and sort it all out again, was not chastising anyone today. Something had mollified all of us, and our conversation was very calm.

Suddenly the door opened. S.P. stopped in the doorway without entering the office. He was wearing his coat and fur hat and holding some sort of package. He looked at us with a tender and wistful smile. Mishin stood up and the rest of us followed suit.

"Sergey Pavlovich, come in . . ."—but he stopped short.

We looked at Korolev and couldn't help smiling. It was unusual on a workday, during work time to see our dear S.P. run-down, melancholy, and lost in his thoughts rather than our masterful and strong-willed commander. Someone said something else. And Korolev said something in reply. He gave a nod and said, "Well, carry on!"

We vied with each other to wish him well. He did not come into the office. He slowly withdrew from the doorway, turned, and without closing the door behind him, left the reception area. It did not occur to any of us that we had said our last farewell to Korolev. Mishin quickly adjourned the meeting. We all hurried back to our own domains.

IN NOVEMBER AND DECEMBER 1965, I SPENT MOST OF THE TIME IN TYURA-TAM, SIMFEROPOL, AND YEVPATORIYA. During my brief periods in Podlipki, I encountered so many problems concerning new developments and programs that the many sleepless days and nights I'd spent at the firing range no longer seemed all that troublesome. In late December, and especially when Korolev left for the hospital, these problems got even worse.

---

6. This discussion is a reference to strenuous efforts made by Korolev's engineers in the mid-1960s to increase the payload margin of the N1 lunar rocket by decreasing the weight of any and all components on the rocket.

During the first days of January, in Korolev's absence, the Ministry of General Machine Building (MOM) decided to be particularly active in supervising our OKB-1. Avoiding the risk of being responsible for resolving pivotal, thematic issues in cosmonautics, the minister assigned his deputy, Viktor Litvinov, and chief of the ministry's Third Main Directorate, Kerim Kerimov, to be particularly vigilant in monitoring the spacecraft production process. It is true that the deadlines Korolev had promised for the factory release of the piloted spacecraft in late 1965 could not be met. The culprit wasn't so much the factory as it was the frequent changes in the piloted spaceflight program.

The Korolevian Council of Chiefs lacked its former unity. Glushko did not limit his criticisms of programs to the Council. He also spoke out in his typical calm and cultured manner at meetings in Keldysh's and Ustinov's offices. Barmin was also grumbling. He could not hide his resentment that before, when only six men comprised the Council of Chiefs, "everyone was a Chief," and now there was "only one Chief Designer and one Chief Theoretician."

We, Korolev's deputies, always supported his initiatives both out of a sense of duty and conviction. But at the same time, perceiving the many external conflicts, we sensed Korolev's isolation. Outside of our organization I could not name a single person who could be considered a true close friend of Korolev, with whom he could share his innermost thoughts, ideas, and plans. Incidentally, I could say the same thing about Glushko and Keldysh. Perhaps such is the fate of particularly distinguished individuals.

For us, the Ministry of Defense's inner conflicts between the Air Force and Rocket Forces were no secret. Minister of Defense Malinovskiy and Air Force Commander-in-Chief Marshal Vershinin disagreed on the use of cosmonauts in military programs. These conflicts also affected our activities.

The program that Korolev proposed in 1963 for the piloted circumlunar program using a stack made up of three types of spacecraft—what was referred to as the "7K, 9K, and 11K" program—was halted.[7] Neither the military nor Khrushchev supported it. We, too, could see its incredible complexity, labor-intensity, and unreliability.

Having taken on the role of head developer of spacecraft control systems, we deprived chief designer of control systems Pilyugin of this interesting work. Chelomey's company followed suit. He also had enterprising engineers

---

7. The 7K-9K-11K program involved the use of the 7K (Soyuz), the 9K (a trans-lunar injection stage), and the 11K (a propellant tanker). Three or four tanker spacecraft would successively dock and fuel the 9K stage after which a piloted 7K vehicle would dock with the upper stage in Earth orbit and then be boosted on a trans-lunar trajectory.

there who decided to develop spacecraft control systems on their own. One of them was Sergey Khrushchev, Nikita Khrushchev's son. These circumstances also served to further exacerbate relations within the Council of Chiefs.

In connection with the lunar program, the decision was made to include automatic rendezvous and docking in the upcoming flight programs. This forced us to speed up development of the first docking assemblies and the *Igla* (Needle) radio system measuring the parameters of relative motion.

This set of operations had a backup scenario, which Korolev had accepted largely due to Kamanin's initiative and under pressure from the Air Force. The reasoning was persuasive: if we were to miss the deadlines for the developmental testing and flights of the 7K (Soyuz) vehicles, then we would need to have tried-and-true Voskhod vehicles in reserve. Korolev made the factory put five Voskhods into production.

We had not lost our ambition to be the "world's first" at any cost. I don't remember whose idea it was to use one or two Voskhod vehicles for an experiment to generate artificial gravity; perhaps it was Raushchenbach's. The idea was enticing. We developed a special control system, automatic winches, and other mechanisms to start two spacecraft spinning about a common center of mass. It was very interesting for theoreticians and medical specialists, and difficult for designers and the factory, but things went along despite the military's declaration: "We don't need this yet."

We couldn't abandon the Molniya-1 communications satellites that had just gone into operation. We were obliged to prevent an interruption of communication with the Far East and in case the two Molniya-1 satellites already in orbit malfunctioned, we were rushing to prepare another two satellites like them for backup.[8]

However, no matter how successful these programs might be, they could not compensate for our loss of superiority if the Americans were to become the first to fly around the Moon. The GRU had acquired a film from the U.S. about the American lunar landing program. This film had been shown to Malinovskiy and high-ranking generals in the Ministry of Defense who were involved with rocket-space technology. Then, one more exclusive group of senior officials, including Korolev and Chelomey, viewed it. Such "obvious agitation" made us think about alternative scenarios for flying around the moon that could be implemented within shorter timeframes, even before the N1 rocket was produced.

---

8. See Chapter 16 of this volume on the flight-testing of the Molniya-1 communications satellite system.

The programs were chock full of new instrumentation developments that were continuously being modified and caused tremendous difficulties in production. My colleagues and I developed good working relations and comradeship with factory Director Turkov, Chief Engineer Klyucharev, supervisor of instrumentation production Khazanov, supervisor of "secondary" production German Semyonov, and the factory's main shop supervisors. They were all very decent, kind people, who were devoted to their work. Despite being on very good terms, the production managers had serious bones to pick with me and my colleagues involved with on-board systems development for the continuous stream of modifications that caused deadlines to be missed. The broad range of projects demanded that we create a vast collection of instruments that, for the sake of deadlines, were often put into production using crude documentation before all the systems had undergone the final functional and layout integration process. Even when the production workers covered for us, one way or another, Korolev found out when the main deadlines were missed from the lead designers, i.e., those who supplemented the chief designer's insufficient number of "eyes and ears." If we pointed out that the modifications to our already installed instruments or cables were the result of changes that our subcontractors had made, S.P. double-checked that information immediately. Depending on which subcontracting chief designer's organization had been called out per the lead designer's report, right in front of Raushenbach, Kalashnikov, Yurasov, or me, Korolev would telephone the relevant chief designer we had dared to name.

These telephone conversations always varied. He would shout angrily at Bogomolov over the phone: "I can't work with you anymore. You behave like a little boy!"

He advised Ryazanskiy to straighten things out: "Misha! You, of course, can't possibly keep track of everything that's going on in your facility. You're too overloaded. But you put me in an extremely difficult position."

He talked threateningly to Chief Designer of Optical Instruments Khrustalev: "Vladimir Aleksandrovich! We are forced once again to replace your instrument and modify our configuration. When will this end? I will have no choice but to report to your minister."

Korolev played out these scenes on the telephone or at meetings of the technical review team. He did not complain to any of the ministers regarding the latest changes. He knew full well that such instructional talks would prompt phone calls from subcontractors who were now miffed at me: "Why did you complain to S.P. himself?" Nobody held anything against S.P. himself. A couple of days later, Korolev would be talking with those same chiefs about new projects, plans, and deadlines as if nothing had happened.

At the next business meeting, after conversations regarding the main lessons of the most recent Ye-6 launch and the instructions that S.P. had given concerning the handover of projects to Babakin, he briefly mentioned that, for several days in his absence, I should "drop everything," and without stopping by my own office, I should be on the production floor from morning until night.

"I can't tell Turkov that he has to meet these already crushing deadlines, because at any moment he might say that your department chiefs are going into the shops and warning that this won't work anyway, or that it won't be long before there will be more changes. If you don't want me to kick them out for talk like this, then go to the production floor yourself and sign the schedules along with the factory. But remember, after that, you personally will come to me with any change."

THE FIRST OF JANUARY WAS A LEGAL HOLIDAY, WHICH WAS AN ABSOLUTE NECESSITY AFTER OUR ALL-NIGHT NEW YEAR'S PARTY. In 1966, officially we had two days off since 2 January fell on a Sunday. Some wisecracker among us came up with the "Murphy's Law of Space." According to this law, during ground preparation and in flight the worst *bobiks* occur on Saturday night and the night before a holiday.

I had been hankering to put on my skis for the first time this season and set off on the fresh snow. But alas! That morning the telephone calls started to come in about incidents during the night on the *Venera-3* trajectory and the steady decrease in current from a Molniya-1's solar arrays. Babakin, who had taken over management of ground preparation for Ye-6 vehicle No. 13, consulted with me as to whether it was worth it to disturb Korolev regarding the launch date. He wanted to personally report that testing was going well, that there was absolute certainty as to the launch date—31 January, and he wanted to receive assurance from S.P. that he would be arriving in Simferopol on 1 February. Babakin and I had enjoyed an amicable relationship for quite some time. We had never experienced any conflicts. But this time, I flew off the handle, snapping at him and telling him that he shouldn't tempt fate, even over the telephone, and he shouldn't count his chickens when there was a whole month left before they were supposed to hatch. Babakin wasn't offended, but nevertheless the next day (I found out about this later) he talked to Korolev via the Kremlin direct line.

An extremely difficult situation developed at the factory with the manufacture of the new 7K (Soyuz) vehicles. Supervisor of secondary production German Semyonov was directly responsible for the assembly and integration operations on these vehicles. This fair-haired, stocky, sturdily built individual had been accustomed to hard work since the war years. As a rule, when we met, he smiled broadly and, with a hearty handshake, affirmed that "everything is going fine." But now, after his return from a business trip, I saw that he had lost a lot of weight and had dark circles under his eyes from fatigue and insomnia.

Turkov, who was Semyonov's immediate supervisor, was jealous of Korolev's direct meddling in routine production matters. German complained to me, "I have three immediate bosses: Sergey Pavlovich, Roman Anisimovich, and Viktor Yakovlevich." The latter was Deputy Minister Litvinov. "Each one demands that I submit a schedule with a feasibility rationale for all the deadlines. But if I draw up a schedule with a feasibility rationale, then nobody is going to agree with these deadlines."

During the first days of January, fulfilling Korolev's directive, my comrades and I immersed ourselves in factory business. Sometimes we had to make very difficult decisions, denying our developers what seemed to be indispensable changes. If test results clearly showed that these changes were unavoidable, we wracked our brains for hours trying to figure out how to do this as painlessly as possible using existing production stock or "with a follow-up modification," or by "authorizing a replacement at the testing facility," and sometimes even at the processing facility. In Korolev's absence, the factory workers still had Litvinov to complain to about my actions. But they conducted themselves quite honorably.

On 12 January, the factory managers gathered in German Semyonov's office to finally reach an agreement on the production schedule for all the spacecraft for the next few months. The schedule called for the factory to be ready to deliver three Voskhod vehicles to the firing range with deadlines of January, February, and April; two Molniya-1 spacecraft—one in January and one in March; and four Soyuz (7K) vehicles—two in March and two in April.

The schedule was submitted to Litvinov for approval. Viktor Yakovlevich had been director of Factory No. 1 in Kuybyshev throughout all of World War II. This was the main factory for the production of the famous Il-2 fighter bombers. During "life's trying times," Litvinov loved to tell us that the most difficult days of the war were easier than one post-war winter when he had just barely started working on a new jet airplane and he was given the assignment to immediately produce a series to take part in an air show. "The blank production shop was still just getting started on the airplane parts and they wanted an air show in three months!"

He repeated this story once again as he studied our schedule. Turkov said, "All of us, including Chertok, are signing with blood!"

"No," answered Litvinov, "You're going to sign with ink so that your signatures will be legible on the copies of the blueprints. I am also signing my approval of the schedule in ink. There will be plenty of blood when the minister lets loose at us during the review."

"Incidentally," said German Semyonov, "If Sergey Pavlovich weren't in the hospital, then he would certainly move up some deadline, and then he would have signed across the whole schedule in blue pencil."

We all burst out laughing. Turkov noted, "There's no point in laughing. According to my information, Sergey Pavlovich will be getting out of the hospital in about 10 days and he will still have time to shift the April deadlines to March, based on his old adage "no later than May!"

Copies of these schedules, which I miraculously still have, are signed: "Chief Designer . . . /Korolev/"

On 12 January, after looking at the schedules with all the signatures, Vasiliy Mishin, who officially remained acting chief, placed a slanted line in front of "Chief," signed his name, and said: "If we disrupt the schedule, then we'll tell Sergey Pavlovich that we signed it after drinking to his health."[9]

None of us doubted that, in the next few weeks, Korolev himself would review this extremely demanding schedule.

From the author's archives.

Boris Chertok speaks with Mariya Balandina, the mother of Sergey Korolev.

---

9. When signing a document on behalf of another, Russians draw a slanted line before that person's name or title.

ON THE MORNING OF 5 JANUARY, KOROLEV WENT TO THE KREMLIN HOSPITAL ON GRANOVSKIY STREET. By coincidence, his mother—Mariya Nikolayevna Balanina—was also there. On 6 January, his wife Nina Ivanovna visited the hospital and met up with Konstantin Rudnev there. He was visiting his own wife in the hospital. Rudnev hadn't been officially involved with our line of work since 1965.[10] He had been demoted from the high rank of deputy chairman of the Council of Ministers to chairman of some very obscure State Committee for the Coordination of Scientific-Research Work, and then he had been appointed minister of "instrument building, automation, and control systems." Rudnev had not lost his optimism. He always had a supply of jokes to cheer you up. He told Nina Ivanovna that they would cope with S.P. here "in the best way possible."

After several days of examinations, an operation was scheduled for 14 January. On the eve of the operation, we learned that Soviet Minister of Health Boris Vasiliyevich Petrovskiy, a cardiac surgeon, would be attending. Then it turned out that Petrovskiy wasn't attending, but that he was actually performing the operation. In the midst of the daily hustle and bustle, the majority of those close to Korolev, even the members of the Council of Chiefs, who took an interest in each others' health, did not give much thought to this operation. "It's no big deal, some polyps somewhere. A lot of people have had this kind of surgery in both the duodenum and the rectum."

The consensus was that this type of surgery was less complicated than an appendectomy. "Korolev will be back at work in a week," explained Tyulin at the State Commission meeting held to discuss the launch readiness of Ye-6 vehicle No. 13. This time Babakin gave a detailed briefing on all the operations. I gave a quick briefing on the results of implementing the measures from all the latest accident review boards. A final launch date of 31 January was set.

"If he doesn't come to the firing range, then S.P. will at least come to Simferopol," assured Tyulin.

Tyulin had consulted with Deputy Minister of Health Burnazyan in that regard. Avetik Ignatyevich Burnazyan was responsible for the healthcare of atomic and rocket scientists. But the Fourth Main Directorate of the USSR Ministry of Health was outside his sphere of influence—he was not in charge of Kremlin hospitals.[11] Evidently, the selection of the operating surgeon was the result of a mutual agreement between Korolev and Petrovskiy. Korolev

---

10. Rudnev's greatest claim to fame was as the chairman of the State Commission for the launch of Yuriy Gagarin's Vostok mission in 1961. He served as a senior official in the Soviet missile and space industry (under Dmitriy Ustinov) from 1952 to 1961.

11. Burnazyan was in charge of the Third Main Directorate in the Ministry of Health.

was impressed that he was going to be operated on, not just by a famous surgeon, but by the USSR Minister of Health to boot!

AROUND NOON ON 14 JANUARY, I WAS ALONE IN MY OFFICE. I was studying a bulging folder of classified mail that had accumulated during the past few days. I asked Zoya Grigoryevna not to let anyone disturb me. Suddenly Tsybin ran in and shouted, "Sergey Pavlovich died!"

"Are you out of your mind?! Which Sergey Pavlovich?"

"Ours, our Sergey Pavlovich Korolev! Nina Ivanovna telephoned from the hospital!"

I stood absolutely dumbfounded, having no idea what to do next. This can't be! This really shouldn't be happening! A few seconds later it occurred to me: "Let's call Burnazyan!"

I found his Kremlin telephone number in the government phone system directory, and hoping to hear that this was some sort of mistake, dialed 2759.

When I asked Avetik Ignatyevich what he knew about Korolev, he was silent for a few seconds. Then he said, "Petrovskiy and Vishnevskiy performed the operation. In addition to polyps, they found a sarcoma. They did all that they could, but his heart couldn't take it. The operation lasted over 4 hours."

After calling for a car, Tsybin and I drove over to see Mishin at the first territory.[12] When we entered, the main managers were gathering in Mishin's office at his request. Leonid Smirnov had already telephoned Mishin from the Kremlin. Smirnov told him that a Central Committee resolution had appointed him chairman of a government commission on funeral arrangements. This time the office of the Council of Ministers and the Central Committee had acted both quickly and properly. The procedures for high-level funerals were well drilled. Brezhnev and Politburo members were informed of Korolev's death straight away. After consulting with one another over the telephone, without hesitation, they decided to "out" Korolev, to announce to the world who he was. The very fact of the death of such an individual needed to be used to promote the achievements of Soviet science. In the first hours after receiving the news of his death, what upset us more than anything was, "How could this happen?!" No one could come up with any sensible explanation.

---

12. The "first territory" was OKB-1's main headquarters facility. Chertok's office was at the "second territory," which had been inherited from TsNII-58 in 1959. Both were in Kaliningrad (Podlipki).

The ringing Kremlin telephone line did not leave us time for much thought.

"Serbin's calling," said Mishin, "He needs the text of the obituary that will be in all the newspapers tomorrow in 1 hour. Boris! You sit down and write the draft and take it over to Serbin at the Central Committee in a half hour. Your pass has already been ordered."

Less than 1 hour had passed since my conversation with Burnazyan. I had the hardest time gathering my thoughts to write an obituary. Finally, I snatched away from the typist what I had written about the great scientist, academician, and scientific and industrial organizer, *without mentioning his work in rocket and space technology.*

Instead of 1 hour, an hour-and-a-half later I entered the office of "Ivan the Terrible"—head of the Central Committee Department of the Defense Industry Ivan Dmitriyevich Serbin. I handed him the text of the obituary. He glanced over it and, with a barely perceptible smile, said, "It can't be this modest. We have to tell people the truth about Korolev. Look, read this. We started to write something ourselves while you were driving over from Podlipki."

I couldn't believe my eyes:

With the passing of S. P. Korolev, our nation and world science has lost a distinguished scientist in the field of rocket-space technology, the designer of the first artificial satellites and the spacecraft that ushered in the era of man's conquest of space. . . . Until the end of his life, all his efforts were devoted to the development of Soviet rocket-space technology . . .

S.P. Korolev was the foremost designer of rocket-space systems that achieved world firsts . . .

Under the leadership of S. P. Korolev, the Soviet Union produced the piloted spacecraft that carried the first human being into space and from which the first spacewalk was performed.

Serbin was clearly pleased with the effect that these phrases, which during Korolev's life had been considered more than top secret, had on me. He decided to use part of my text too: "What you said about his keen sense of intuition and his great creative daring when solving highly complex scientific and technical problems is very good. You need to continue the phrase and describe his brilliant organizational capabilities and moral virtues."

Further down in my text there was an insertion:

Korolev remained an ardent patriot and steadfastly pursued his goal—to fulfill the dream of spaceflight, despite years of unjust persecution.

In this mild form, I was attempting to inform the reader that S.P. had suffered repression. Serbin frowned, and without saying a word, firmly crossed out those lines.

We spent another 30 minutes or so composing the text. That is how the obituary that was published on Sunday, 16 January, in all the national newspapers came to be. The delay of an entire 24 hours was explained by the fact that the medical report needed to be published at the same time, a good photograph needed to be provided, and funeral arrangements announced.

That evening Bushuyev, Mishin, and I listened to Nina Ivanovna's faltering story. On the day of the operation, she had been at the hospital since early in the morning and managed to see "Serezha alive."

Chief of Surgery Blagovidov assured her that the operation would last no more than an hour. But 1 hour passed, then 2. The nurses shot out of the operating room with some instructions. She understood that they were adjusting the respirator. They rolled in some other equipment. Then Blagovidov, who was assisting Petrovskiy, came out and told her that there were complications. Suddenly Soviet Army Head Surgeon Vishnevskiy appeared from somewhere and quickly strode into the operating room. Again, agonizing uncertainty. More than 4 hours had passed since the operation began when they announced to Nina Ivanovna that they "had done all that they could." They had been unable to stop the hemorrhaging after they removed the polyps. They made the decision to open up his abdominal cavity. When they started to get closer to the site of the hemorrhage, they discovered a fist-sized tumor. It was a sarcoma—a malignant tumor. Petrovskiy made the decision to remove the sarcoma. In so doing, they removed part of the rectum. They would have to take out the remaining part through the abdomen. In the end, cardiac failure developed very rapidly.

It turned out that for a long time they had been unable to insert the respirator tube into Korolev's trachea. Such was the peculiar nature of the way his large head was positioned on his short neck.

They didn't tell Nina Ivanovna the whole truth. At that time, she was in no condition to grasp everything, much less pass it on to us. Vishnevskiy knew Korolev and Nina Ivanovna very well. But he did not go out to explain things to her.

I WOULD LIKE TO SUPPLEMENT MY OWN RECOLLECTIONS ABOUT THE LAST DAYS OF KOROLEV'S LIFE WITH OTHER HISTORIC MATERIALS. In January 1996, what have now become the traditional "Korolev lectures" were held. Director of the Korolev Memorial Museum Larisa Filina gave a speech about "The Last Days in the Life and Work of Academician S.P. Korolev" at this lecture series.[13] In my opinion, Filina's speech is a very precious historical document. It was prepared using pages from S.P. and N.I. Korolev's archives stored at the Academician S.P. Korolev Memorial House-Museum in Moscow. Larisa Aleksandrovna has kindly permitted me to cite the following excerpts from her speech:

> . . . *Soon after returning home* [from Simferopol after the launch of *Luna-8*—comment by B. Chertok], *Nina Ivanovna does not remember the precise date, Sergey Pavlovich started bleeding from his rectum in the middle of the night. This had happened before, but this was the first time the bleeding was so profuse. Extremely alarmed, Nina Ivanovna insisted on calling an ambulance, but Korolev categorically forbade it. He remained completely self-reliant . . . . They coped.*
>
> *But even in the morning Sergey Pavlovich refused to go to the hospital, saying that he was going to be extremely busy that day . . . . This day was a very trying one for Korolev. He would have to face the music for the* Luna-8 *failure . . . .*
>
> *Nor did Korolev go to the hospital on the following day . . . . On Tuesday, 14 December, Korolev went in for examination at the Kremlin hospital surgical department on Granovskiy Street. In those days, Dmitriy Fedorovich Blagovidov was in charge there . . . .*
>
> *Judging by the remarks in his notebook, he brought materials to the hospital for the "article for Azizyan"* (one of the editors of the newspaper Pravda *who helped Korolev during the 1960s when he was preparing articles published under the pseudonym "Professor K. Sergeyev"). He brought along organizational matters and materials, notes for drafting a memorandum by 1 January 1966 regarding Ye-6, and a lot of other things . . . .*
>
> *On Thursday, 16 December, Leonid Aleksandrovich Voskresenskiy suddenly died . . . . Sergey Pavlovich couldn't miss his old comrade's funeral . . . .*
>
> *The examinations that were performed over almost three full days were certainly not thorough, but a polyp discovered 9 to 11 centimeters from the sphincter enabled the doctors to reach a diagnosis of "rectal polyp." D. F. Blagovidov told S. P. in Nina Ivanovna's presence: "Go home, Sergey Pavlovich, have a nice New Year's celebration with Nina Ivanovna, and then come back and we'll remove the polyp. A simple little operation."*
>
> *On Friday, 17 December, Korolev was released from the hospital. The same day he was back at work. There are many documents that could tell what Sergey Pavlovich was working on that day and subsequent days, but we will focus on a petition from the*

---

13. Chertok's excerpts here are from the complete published text of Filin's paper published as "O poslednykh dnyakh zhiznii i tvorchestva S. P. Koroleva" in L.A. Filina, ed., " . . . *Byl veku nuzhen Korolev": po stranitsam arkhiva Memorialnogo doma-muzeya akademika S. P. Koroleva* [" . . . *It Was a Century That Needed Korolev": From the Pages of the Archive of the S. P. Korolev Memorial House-Museum*] (Moscow: Russkaya istoriya, 2002), pp. 145–154.

director of a boarding school for blind children, Hero of the Soviet Union M. I. Mochalov from Bolshevo. He turned to Korolev requesting that a truck be allocated to deliver food, clothing, and other goods to the school. Korolev's resolution reads as follows:

1. *To Comrades A. P. Ivanov, G. M. Azarov (dept. 34), and P. I. Lyubovin (head bookkeeper):*
   - *Permission is granted to transfer a truck from the motor pool in operation at P.O. Box 651 to the boarding school's inventory;[14]*
   - *The vehicle needs to be fixed up, perhaps, make a canopy body for it and perform preventive maintenance and repairs—may our patronage be in deed, rather than word.*

2. *What else can we do to help this school? I await your suggestions.*
   17.12.65
   Korolev.

Voskresenskiy's funeral took place on 19 December. The museum contains the memoirs of a member of the first firing team of 1947, Lieutenant Colonel Konstantin Andreyevich Obukhov (he first met Korolev on 3 May 1946, in Potsdam, at the artillery personnel department): "The last time I saw Sergey Pavlovich was on the day of his Deputy L. A. Voskresenskiy's funeral—19 December 1965. He said with such anguish: "How hard it is to attend the funeral of my closest and most dependable assistant and devoted friend who has died before his time . . . ."

And at the Korolevs' home, an elegant and succinct invitation awaited them: "Dear Comrade S. Korolev. Please join us with your wife at a casual dinner party on 20 December 1965.

Signed: Rauschenbach
Boris Viktorovich will be celebrating his 50th birthday."

I am interrupting my citation of Larisa Aleksandrovna's speech in order to add my own recollections about the birthday party at the Zvezdnyy restaurant on Tsander Street. I very often have the occasion to walk or drive by the site of that cozy, and for those times, quite accessible restaurant. The building has been torn down, and a high-rise office building has sprung up in its place. Now, in the neighborhood surrounding it, the "new Russians" park their expensive foreign cars. The business elite, preoccupied with their mega-incomes, are unaware of the fact that on this tiny patch of the planet called Tsander Street, Academician Korolev spent several happy hours before he died.

---

14. P.O. Box 651 was the "open" name of Korolev's design bureau.

Rauschenbach's friends and coworkers managed to fill the evening with a stream of good humor and good-natured ribbing about our work conditions. One of the well-wishers, conjuring up a caricature of the birthday boy immersed in the creative process, said, " . . . and to the left you see the elbow and forearm of the Chief Designer . . . ."

The text of a facetious speech parodied official congratulatory ceremonies. It was delivered with tremendous pathos by Yuriy Levitan himself, who had been talked into making a recording in advance. Yaroslav Golovanov brought about 15 special issues of a mock special edition of *Komsomolskaya pravda*.[15] Everyone really had a good time. Korolev and Nina Ivanovna laughed so hard they cried. It had been a long time since we had seen Korolev so jolly and roaring with laughter.

Now I will continue my citation from Larisa Filina's speech.

. . . on 23 December there was another birthday. Pavel Vladimirovich Tsybin turned 60. The page marked "Thurs. 12/23" in Korolev's notebook contained the following notations: "Regional Committee plen. Session. Mosc. Municipal Council, Marble Hall—09:00, Comrade Tyulin—proposal for Soyuz, see—Abramov, Chertok re ground. Barmin—L-1" and then: "15:00 Tsybin—birthd. mtg."

On Saturday, 25 December, Korolev was at the Kremlin for a meeting about the five-year plan. Several people gave speeches—M. V. Keldysh, V. D. Kalmykov, P. V. Dementyev. Sergey Pavlovich jotted down synopses of them in his notebook . . . .

The next day, when he met with cosmonauts, Korolev touched on the subject of the outlook for the development of cosmonautics . . . .

. . . It was Sunday. The cosmonauts hosted the Korolevs at the Zvezdnyy, and then they all visited the new apartments that Yu. A. Gagarin, V. V. Tereshkova, and A. G. Nikolayev would be moving into. They took a dip in the pool, dined, and even had their picture taken. Everyone was in a good mood . . . .

From the memoirs of N. I. Koroleva: "Before New Year's 1966, we stopped by the hospital on Granovskiy Street to visit Mariya Nikolayevna. In the corridor outside the staff lounge, the doctors surrounded Sergey Pavlovich and proposed that they drink some champagne to the New Year. And suddenly, one of the surgeons (Sergey Dmitriyevich Belov) came up to Sergey Pavlovich and said in a low voice: 'Don't worry, Sergey Pavlovich, the operation is no big deal. They're just concerned about your heart.' 'I'm not worried,' responded Serezha."

---

15. Yaroslav Kirillovich Golovanov (1932–2003) was one of the greatest Soviet/Russian space program journalists. The author of nearly 1,400 articles and 20 books, he worked for the major newspaper Komsomolskaya pravda. He is most well-known for his magnum opus, *Korolev: Fakty i mify* (*Korolev: Facts and Myths*), undoubtedly the most well-researched and highly regarded biography written about the Soviet chief designer.

But knowing her husband and being very much attuned to him, Nina Ivanovna instantly realized that Belov's words spoken in passing had jolted Sergey Pavlovich out of his carefree mood . . . . Once they had returned home, Sergey Pavlovich complained of pains in his heart, noting that they were unusual. "Well, then," reasoned Nina Ivanovna, "we'll give you an unusual medicine today too." And instead of the usual Validol, she gave him a dropper of Valocordin.[16] Soon the pain went away.

. . . From the memoirs of Ivan Matveyevich Ryabov (the manuscript is kept at the museum): "on 31 December 1965, relatives and close friends gathered at the home of Nina Ivanovna and Sergey Pavlovich to ring in the New Year of 1966. After everyone had arrived, the majority was ready to start sending off the old year. But at that moment the telephone rang. When the conversation was over, Sergey Pavlovich said: 'Please excuse Ninochka and me. Central Committee member Ponomarev has invited us to his New Year's party. We'll try to return a little before midnight so that we can continue our New Year's party together.' They returned home at 4:30 . . . ."

That same day, one of S. P. Korolev's most interesting articles, "Strides into the Future," was published in Pravda. Bright and passionate, it ended with the words: "The human mind knows no limits!" . . . .

4 January—Korolev's last work day . . . .

That day after work, as Sergey Pavlovich and Nina Ivanovna had arranged that morning, he stopped by to pick her up at the home of some relatives. They lived not far from the Korolevs' home in Ostankino.

From the memoirs of N. I. Koroleva: "Sergey Pavlovich arrived on time, very tired, and dismissed his car . . . . The time came to walk home . . . . Just before we reached our gate, Serezha suddenly said, 'I want to have a word with you.'

I detected alarm in his voice.

'Okay, Serezhenka, we're almost home; we'll talk there.'

'No, no, you listen to me. What I've got to tell you is difficult for me and for you, too. If something happens to me, please move out of this building.'

'Good God, Serezha, what are you talking about?!'

'That's all I've got to say,' he abruptly broke off further discussion.

Once we were home, we did not raise the subject again."

On Wednesday, 5 January, S. P. Korolev entered the hospital of the Fourth Main Directorate on Granovskiy Street. Every day Nina Ivanovna was by his side . . . .

---

16. Validol is the trade name for Methyl pentanoate, a sedative that was used for mild heart disorders in socialist countries. Valocordin is a barbiturate-based cardiac medication made from the sedative valerian root popular in the Soviet Union.

On Tuesday, 11 January, B. V. Petrovskiy performed a biopsy for a histological analysis. The excision of the polyp caused heavy bleeding. They were barely able to stop it. Ward doctor R. V. Reznikova told Nina Ivanovna about this. However, the case management team did not discuss it. An operation was scheduled for 14 January.

The 12th of January was Sergey Pavlovich's 59th birthday. Nina Ivanovna came to the hospital with Mariya Nikolayevna to wish him happy birthday . . . .

On Thursday, 13 January, shortly after dinner, anesthesiologist Yuriy Ilich Savinov came to the ward to see Korolev while Nina Ivanovna was visiting. He brought a sheet of paper containing the biopsy results. Nina Ivanovna remembered the crux of those findings: "Polyp not suspicious." This meant that the tumor was benign. But the doctor asked that we not tell anyone about his inexcusable candor with the patient.

During the conversation Sergey Pavlovich sorrowfully asked Savinov: "Doctor, you are our friend. Tell me, how long can I live with this? . . . ." And, now silent, he placed his hand on his heart . . . .

Our archives contain Nina Ivanovna Koroleva's narrative, which is penetrating in its detail. Throughout all the long and excruciatingly uncertain hours of the operation, she sat outside the operating room. She saw and heard a lot . . . .

From the documents directly related to the tragic event, in addition to her own memoir entries, Nina Ivanovna has handed over to the museum Sergey Pavlovich's hospital chart detailing his stay in the hospital from 5 through 14 January 1966, containing the preliminary diagnosis "rectal polyp," as well as his death certificate, with the same diagnosis as the final diagnosis on his hospital chart—"leiomyosarcoma." But why then did our medical luminaries allow their own great patient to die of heart failure? After all, according to the medical report that appeared on 16 January in all the national newspapers, " . . . death was due to acute myocardial ischemia"!!!

It wasn't until 16 January at 6 in the morning Moscow time that Levitan read over the radio the government announcement of Academician Korolev's death. All of the morning newspapers contained an obituary in which I found no changes after the final approvals the evening of 14 January in Serbin's office.

The first signature under the obituary was that of Brezhnev, followed by all the Politburo members, Central Committee secretaries, marshals, Keldysh, and ministers of the military-industrial complex. The last signatures belonged to Georgiy Tyulin, Moscow Regional Committee Secretary Vasiliy Konotop, and Academicians Mikhail Millionshchikov, Anatoliy Blagonravov, and Leonid Sedov.[17]

Following the obituary was a statement from the government commission on funeral arrangements. They announced that the casket containing the body of S. P. Korolev would lie in state in the Hall of Columns in the House of Unions. The building would be open for the public to pay its last respects on 17 January 1966 from noon until 8 p.m. The funeral would take place on 18 January at 1 p.m. on Red Square.

Despite the fine-tuned protocol established by the administrative machinery of the Council of Ministers and the Kremlin commandant's office, and the tried-and-true ritual procedures for high-level funerals that had been thought out to the smallest detail, assistance was needed from our enterprise. Our factory director Turkov organized a work brigade to decorate the Hall of Columns for the funeral. The photo lab made copies of portraits of Korolev for the press. Special portraits needed to be produced for the casket and the urn.

Using the same good photograph that was published in the newspapers, artists painted an enormous portrait in a day. On the night of 16 January, it was fastened to the façade of the House of Unions. An honor guard roster was prepared. All available automobiles were enlisted and additional vehicular transport was called in to bring people from Kaliningrad to the House of Unions. Each OKB facility, production facility, and organization in Kaliningrad ordered wreaths of fresh flowers, cleaning out the floral farms in the vicinity.

But for us, the main event of 16 January was the unexpected election of a new chief designer. We, all of his deputies, were crushed by the suddenness of what had happened. It seemed sacrilegious to even discuss the subject of a new chief before Korolev had even been buried. Moreover, everyone had long ago come to terms with the established procedure for an appointment to such a high post. Usually the appointment procedure was a multi-step process. After conversing over the telephone with Central Committee officials, receiving instructions from Ustinov, and speaking with the closest deputies, the minister of general machine building was supposed to draw up a formal proposal for a candidate. This proposal was supposed to be reviewed by the Central Committee Secretariat. The decision of the Secretariat was approved

---

17. Vasiliy Ivanovich Konotop (1916–95) was the first secretary of the Moscow regional Communist Party from 1964 to 1985. This was one of the most powerful positions in the Communist Party bureaucracy.

From the author's archives.

Vasiliy Mishin, Anatoliy Kirillov, and Boris Chertok. Kirillov was a senior official at Tyura–Tam who was responsible for launch operations.

in the Politburo. After the Politburo's decree was issued, the minister was supposed to issue an order appointing the new chief designer and chief of OKB-1. The entire process would take four to five days.

But Viktor Makeyev, who had flown in from Miass, took the initiative.[18] After assembling all of Korolev's deputies, he announced that we shouldn't waste a single day and wait until one of the brass foisted somebody on us. We should decide our own fate. Speaking of the need for immediate action, Makeyev radiated such stirring conviction that everyone assembled there agreed with him. For some reason, Mishin did not take part in the discussion.

I proposed that Makeyev himself agree to accept this post under the condition that we all immediately appeal to the Central Committee with a collective letter. But Makeyev categorically refused this high honor. He declared that he'd finally gotten things squared away in Miass, he was caught up in work producing submarine-based missile systems and was not going to abandon it. Construction had begun on nuclear submarines, which would carry not three, but 16 missiles. Could anyone really abandon such a project?

---

18. Makeyev, a young protégé of Korolev's, was chief designer of SKB-385, the main developer of strategic submarine-launched ballistic missiles.

After a brief pause, Bushuyev and I proposed Mishin's candidacy. Sergey Kryukov was against it.[19] He reasoned that Mishin was too impatient with people whose positions on technology differed from his own. A chief designer must listen patiently to everyone before making a decision on fundamental issues. Mishin would not want to listen to anyone and would always do things his way. In Kryukov's opinion, Mishin would not be able to lead the Council of Chiefs.

Having taken on the role of chairman, Makeyev put the issue to a show of hands. Mishin's candidacy was accepted. Kryukov was the only one to vote against him. Without delay, Makeyev himself wrote a letter addressed to the Central Committee and we all signed it. And that was the genesis of the collective letter that Serbin received on the morning of 17 January, with copies for the minister and Smirnov in the VPK.

Our letter was a surprise for the Central Committee, Ustinov, and the minister. Some time later, we learned that, before Ustinov received the letter, he had arranged with Tyulin for him to be appointed to this vital post. Our collective letter prevented this appointment from happening.

On 17 January at around 9 a.m., Bushuyev, Okhapkin, and I drove to the House of Unions. We needed to arrange for employees from OKB-1 and subcontracting organizations to be able to gain free access without having to stand in line, to verify the admission of all family members, to get ourselves into one of the first shifts of the honor guard, and to advise the colonel—the honor guard director—on whom to post on the honor guard and when. Each of us had all kinds of instructions. But the main thing was that we wanted to be with Korolev a little longer, even though he was no longer with us. We had been given a last opportunity to see him.

On the way to the House of Unions we were startled when we saw the crowd of people dressed in black. It had been announced that the doors would not open for the public to pay its last respects until noon. In spite of that, at 9 a.m. in the freezing cold, a line had formed stretching all the way to Soviet Square. It would be 3 hours before those standing at the head of the line could count on entering. Mostly these were people who had not known or heard anything about Korolev until his obituary had appeared the day before. What force, whose will made them stand about in the freezing cold for so many hours in order to quickly walk through the Hall of Columns past

---

19. Sergey Sergeyevich Kryukov (1918–2005) was one of the leading creative designers at Korolev's design bureau. He played a critical role in the development of the R-7 ICBM and its derivative launch vehicles. He had been serving as a deputy chief designer under Korolev since 1961.

the casket and glance at the dead face of a person who had been completely unknown to them before?

Forty-two years before, I had also stood in line with three young friends at the House of Unions. True, it was much colder during the days of mourning for Lenin. But Red Army soldiers built bonfires and we took turns running to warm ourselves by the fire. People went to say farewell and to take a look at the deceased, who while alive had been our leader, well-known to millions of his supporters and enemies. At that time in January 1924—and in March 1953 when Stalin died—no one had been specially selected or coerced to pass by the casket.

The people went to pay their final respects to Stalin sincerely, considering him the greatest of the great leaders, a father figure. No one forced the people to risk their lives in the tremendous throng. This was the uninhibited adoration of leaders. Many thousands of people also paid their final respects to Kurchatov because while alive he had been well-known and famous.[20] His work had protected the Soviet people from the cunning Americans. Everyone knew that he was the "father" of our atomic and hydrogen bombs. What's more, Kurchatov promised to give an unlimited amount of almost free energy to the whole world in the near future by controlling the thermonuclear reaction. He was a great Soviet scientist, and every patriot of Soviet science and technology considered it his or her personal duty to pay their final respects to the legendary "beard." Millions had known about Kurchatov long before his death.

But Korolev? After all, except for several thousand people who were allowed to enter the Hall of Columns without waiting in line, no one had ever heard of Sergey Pavlovich Korolev. I believe that hundreds of thousands of people came on Monday, 17 January, to file by Korolev's casket, and on Tuesday, 18 January, to view the urn that contained his ashes, because a particle of truth had finally been revealed to them. They had finally been told who deserved tribute for human civilization's greatest triumph. There was a general sense of being party to a partially divulged secret. He might be dead, but at least one could catch a glimpse of him. The viewing line illustrated the people's cosmic oneness. There was a shared grief and a shared pride. It was late in coming, but the people had been given the opportunity to pay tribute to the great Korolev for his great achievements. It was as if everyone who passed by his casket brushed up against these historic achievements. By no means will everyone agree with me, but that is how I have explained to myself the eagerness of those many thousands of people to file through the Hall of Columns.

---

20. Kurchatov died on 7 February 1960.

All the chief designers and Korolev's deputies each stood watch in the honor guard twice. If they could have, they would have stood three times. But there were too many others who wanted to. Politburo members, ministers, marshals, generals, admirals, academicians, workers from our factory, managers of Moscow's largest industrial enterprises, and delegations from Leningrad, Kapustin Yar, Tyura-Tam, Dnepropetrovsk, and the capitals of the Union Republics stood watch in the honor guard. It was amazing how the Moscow funeral services managed to meet the demand for wreaths. The House of Unions was overflowing with hundreds of wreaths and baskets of live flowers.

At 8 p.m. the Hall of Columns was supposed to close to the public, but the line still stretched along Pushkin Street. The funeral commission made the decision to extend the viewing period for another hour. For the last time we approached the casket surrounded by a sea of flowers. Korolev's face was peaceful and serene. S.P. had accomplished what he was supposed to in this world, and then some. That night Korolev's body was cremated.

The next day, in the Hall of Columns, soldiers stood at attention paying their last respects by the pedestal supporting the urn containing Korolev's ashes, and we changed guard every 3 minutes on our sorrowful watch. The funeral procession through the Hall of Columns lasted until 1 p.m. Smirnov,

From the author's archives.

Three of Korolev's leading associates carrying the urn with his ashes at his funeral in January 1966. From left, Sergey Kryukov (deputy chief designer), Kerim Kerimov (chairman of the State Commission for Soyuz), and Boris Chertok.

Keldysh, Tyulin, and Gagarin carried the urn out of the House of Unions. After the members of the funeral commission, Mishin, Shabarov, Abramov, and I were the first to take the bier holding the urn. We carried it along the façade of the USSR *Gosplan* building. After that, the bearers rotated out frequently. At the Museum of History, cosmonauts took the urn. On its snow-dusted bier, the urn was placed in front of the mausoleum. The Politburo members and those who were supposed to speak climbed onto the rostrum. The marshals and generals seated themselves on the right wing of the mausoleum.In his speech, Keldysh said, "Our nation and the entire scientific world have lost a scientist, whose name shall forever be associated with one of the greatest conquests of science and technology of all time—the dawning of the age of human space exploration . . . ."

After the ceremony, Brezhnev and the other Politburo members descended from the mausoleum, they lifted up the urn and carried it to the Kremlin wall. Leonid Smirnov removed the urn from the bier and placed it in its niche. The thunder of an artillery salute rang out. What other nation and what other people bury scientists with such honors? The niche was covered with a standard slab of stone with the inscription:

KOROLEV
SERGEY PAVLOVICH
30.12.1906–14.01.1966

Korolev's date of birth was given according to the Julian calendar (Russian Old Style).

After the funeral, a funeral reception was held at "Korolev's cottage." By the Moscow standards of that time, the square footage of Korolev's two-story cottage was more than sufficient living space. But the house had not been designed for a funeral feast on the scale required after an interment at Red Square. The only way all those invited to the reception could be accommodated was by waiting their turn in line. Our factory kitchen staff took charge of everything needed for the table, freeing the family of this hard work. However, as wife and hostess, Nina Ivanovna was supposed to endure every toast that has to be made on such occasions, all the sorrowful speeches, and vows to be eternally faithful to the memory of S.P.—all without a break. I drove from Red Square to the reception with Katya. She had walked from the House of Unions in the funeral procession with Korolev's family. According to Katya, Nina Ivanovna was in such a state that it seemed she might not withstand the lengthy formalities of the reception.

The next day, each of us would have work—all sorts of problems waiting for us that would first dull the pain of loss, and then remove it altogether. Only Nina Ivanovna would have no distractions capable of damping her inconsolable grief.

After the first wave of guests that had settled around the table gradually began to show signs of excessive energy, I arranged with Bushuyev and Okhapkin that we would attempt to take away some of the invitees in groups and distribute them among our apartments to continue the reception. I asked Gagarin to help me. The two of us assembled a group, which made its way across the street and continued the reception at my apartment. Thus, Katya and I ended up hosting Gagarin, Titov, Popovich, Komarov, Bykovskiy, and Feoktistov at our home.

In addition to other treats laid out for the guests with cognac and tea, Katya opened a box of assorted chocolates produced by the Rot Front Factory in honor of achievements in space.[21] The wrapper of each piece depicted some space-related event that was described in detail on the inside surface of the box top. Katya demanded that each cosmonaut eat a piece of chocolate with his own portrait, while it was recommended that I "eat the first *Sputnik*." Gagarin protested: "This is a rare souvenir. Guys, I beg you not to touch this box, but let the host and hostess keep it in remembrance of this day." After a lengthy debate, half of the chocolate was used for its intended purpose. After that, Bykovskiy drew three elaborate sine waves on the box top along with the inscription "That's how we flew," and signed his name. The other cosmonauts added their autographs, too.

Despite the hostess' insistent appeals to make themselves at home, Gagarin asked his comrades not to overindulge. But these words of warning were not necessary. Everyone had something personal in their memory about Korolev, and here one could freely talk about even one's innermost feelings.

Now, I shall not begin to attempt to reconstruct from memory everything that the first cosmonauts said. I remember the questions and the speculation that the most senior, experienced, and sober of the cosmonauts—Vladimir Komarov—expressed. He had reasonable doubts as to the true causes of Korolev's sudden death.

During the war, our surgeons worked miracles when confronted with the most severe wounds. Even when a person's entire abdominal cavity was torn open, they sewed it up. And how many pilots they saved! After the war, test pilot Mosolov had such a bad crash that the situation seemed hopeless, but the story goes that

---

21. The Rot Front Factory is one of the oldest Russian candy factories, established in the early 19th century. It was named "Rot Front" ("Red Front" in German) in the early 1930s in honor of the slogan of the German Communist Party.

[Korolev's first wife] Kseniya Maksimilianovna Vintsentini at the Botkinskaya Hospital put him back together bone by bone. Maybe he'll never fly again, but he's alive. And at the nation's best hospital, equipped with technology that doctors in our other hospitals, much less our field surgeons, have never dreamed of, they can't save this man's life! They didn't haul him in from the battlefield. He walked in under his own power, alive and unharmed until his last day.

Addressing his comment primarily to Gagarin and me, Komarov said that this needed to be looked into. At the time, it never occurred to any of us that Komarov himself only had a little more than a year to live.[22] A government commission enlisting the assistance of dozens of relevant organizations and conducting special experimental work to test various hypotheses would perform a lengthy and painstaking investigation into the causes of his death. There was no investigation of the cause of Korolev's death. Very many people sought answers to the questions that Komarov had posed.

A QUARTER OF A CENTURY AFTER KOROLEV'S DEATH, WHEN I ASKED NINA IVANOVNA TO FILL ME IN ON CERTAIN DETAILS OF HIS LAST DAYS, SHE RESPONDED, "I remember everything that happened back then perfectly. I forget what happened yesterday, but what happened long ago—I remember just fine. This is what your memory does when you get old."

Nina Ivanovna told us what she saw and heard on the day of the operation several days after the tragedy. Now she added something that she had not mentioned then.

"Aleksandr Aleksandrovich Vishnevskiy was called in to the operation quite late. You know, we were good friends, as they say, like family. Three or, I don't know, maybe four years after Serezha died, Vishnevskiy's wife died. She was out in the country. Somehow, a nanny goat that was tied to a tether got its rope wrapped around her leg, tearing through a vein. Blood poisoning set in and she couldn't be saved. The day of Vishnevskaya's funeral I was at the funeral reception. Aleksandr Aleksandrovich sat down next to me on the divan. He was shaken by his wife's death. But he started to reminisce about Serezha: 'When I arrived at the hospital on Granovskiy Street, I was told to continue the operation. I said: "I don't operate on corpses".'"

"'The first draft of the medical report read: "Petrovskiy operated, Vishnevskiy assisted." Then this page was ripped out of the medical report and replaced with: "Petrovskiy and Vishnevskiy operated." The text of the

---

22. Komarov was killed on 24 April 1967 during reentry of the *Soyuz-1* spacecraft. See Chapter 20 of this volume.

official medical report published in the newspapers contains five signatures and no mention is made of who operated'."

The official medical report was published on 16 January 1966.

Medical report about the illness and cause of death of Comrade Sergey Pavlovich Korolev:

Comrade S. P. Korolev suffered from rectal sarcoma.

In addition, he had atherosclerotic cardiosclerosis, sclerosis of the cerebral arteries, pulmonary emphysema, and a metabolic disorder.

An operation was performed on S. P. Korolev to remove a tumor, which entailed the extirpation of the rectum and part of the sigmoid colon.

Comrade S. P. Korolev's death resulted from cardiac failure (acute myocardial ischemia).

The USSR Minister of Health, Full Member of the USSR Academy of Medical Sciences, Professor B. V. Petrovskiy; Full Member of the USSR Academy of Medical Sciences A. A. Vishnevskiy; hospital Chief of Surgery, Docent, Candidate of Medical Sciences D. F. Blagovidov; Corresponding Member of the USSR Academy of Medical Sciences, Professor A. I. Strukov; Chief of the Fourth Main Directorate of the USSR Ministry of Health, honored scientist, Professor A. M. Markov.

15 January 1966.[23]

Of all the signatories of this report, I knew only Professor Vishnevskiy. The first time I met with him it was at Korolev's insistence, or rather, his orders. In 1961 or 1962, the Korolevs were vacationing in Sochi. It was either on the beach or on the tennis court that Nina Ivanovna made the acquaintance of a charming woman—the wife of the chief surgeon of the Soviet Army, Colonel-General Aleksandr Aleksandrovich Vishnevskiy. Korolev and Vishnevskiy did not play tennis and did not sun themselves on the beach, but once they got to know each other, they liked each other very much. Korolev's testimonials about Vishnevskiy were accompanied by the epithets—"a wonderful, brainy, jolly fellow, who has cut open hundreds of people."

Vishnevskiy had a great knack for telling fascinating stories about the outlook for surgery. He marveled at the power of the technology that Korolev created and lamented the egregious backwardness of our medical technology.

S.P. did not remain indifferent to the compliments paid to cosmonautics, and promised, once he had returned to work, to send his own instrumentation engineers to Vishnevskiy at the Institute of Surgery to see "what kind of real

---

23. *Pravda* no. 16 (17333), 1966.

help we could offer." He kept his word. After returning from vacation, without delay, he called Isaak Khazanov and me into his office. He told us about the remarkable surgeon Vishnevskiy and then requested that we, without putting it off or passing it on to subordinates, meet personally with the famous surgeon and do everything in our power to address his problems and needs.

"Keep in mind," said Korolev, "when you see and understand everything, I won't need to persuade you."

Right then and there, Korolev called up Vishnevskiy, and the next day Khazanov and I were at the Aleksandr Vasilyevich Vishnevskiy Institute (named after the younger Vishnevskiy's father). Short in stature, very agile, with his head shaved, and still wearing his scrubs after performing an operation, Vishnevskiy flew into the reception area, about 10 minutes later than the time we had agreed upon, telling off his secretary on the way, "You silly woman, why are you keeping them in the reception area? You haven't even offered them tea or cognac."

Turning to us, as if we were old friends who had just parted a little while ago, he said, "Let's go right into the ward; you'll have a look at everything there."

He tossed us a couple of lab coats and we quickly walked through a spacious ward that accommodated about 20 people.

"Just look how we live! Crowded! One toilet for the entire floor. But that is not the main problem."

He led us up to two pale young fellows who were lying on cots by a large window. Uninhibited by the inquisitive glances of the patients, Vishnevskiy said loudly, "These are sailors. Fine, healthy, young guys. They should be cruising at sea now, or at the very least, chasing skirts. But they can't do anything. Their spinal columns got damaged in some storm so that nothing works from the waist down. They can't even urinate without assistance, not to mention everything else. Neurosurgery might help, but for the time being that is a distant problem. What we really need is an electrical stimulator. I'll operate to install electrodes where they are needed and when the time comes to do something, the sailor will be able to press a button. An electrical impulse will go where it's needed, replacing the normal signals that are unable to get through the broken nerves."

The sailors' gaze fixed on us spoke louder than words. Khazanov looked at me and announced, "We'll do it, Aleksandr Aleksandrovich!"

And we really did do it. Our engineers became insiders in the laboratories of Vishnevskiy's institute. Among its small staff were individuals keen on introducing electronics and all kinds of mechanization into medical practice.

Two months later, the first experimental model of an electronic bladder stimulator was tested on one of the sailors with splendid results. We would need to prepare for production of a limited series. Thanks to the perseverance of Khazanov at the factory and of the deputy chief of our instrument design department, Yevgeniy Volchkov, at the KB and in the laboratories, our assistance to the Vishnevskiy Institute was quite tangible. And not only in terms of the stimulators. Cardiopulmonary bypass machines were perfected at the instrumentation factory and special devices were manufactured for hospital beds to control the position of a paralyzed patient. An entire medical technology laboratory was established. Vishnevskiy would admit any of our colleagues to an examination or operation at Khazanov's or my request.

From the very first day, Vishnevskiy and I started to address one another using the familiar *ty*. When I appeared in his office, after dropping what he was doing, he would make a show of chewing out his secretary (who was grinning with delight), if tea, coffee, or cognac had not appeared quickly enough. Despite a certain affected, even theatrical rudeness with personnel and patients, Vishnevskiy was loved and idolized. More than once he excoriated his staff— well-known surgeons—in front of me in a manner reminiscent of Korolev. In private, his colleagues intimated that this spectacle was for the benefit of outsiders. His book *Diary of a Surgeon* came out in 1967.[24] Vishnevskiy gave me a copy with the inscription:

"With hope for your help and with great admiration.
9.1.68 A. Vishnevskiy"

Marshal Zhukov wrote the foreword to *Diary of a Surgeon*.

From the first to the last days of the war, military surgeon Vishnevskiy performed hundreds of operations. His *Diary* is by no means just for medical specialists. His simple description of the everyday life of war, in and of itself, provided a wealth of material for thought. In field hospitals, our surgeons snatched people with the gravest internal injuries from the jaws of death.

So, why then did outstanding surgeons send Korolev from the operating table to the morgue? Perhaps the specifics are found in S.P. Korolev's case history, which should be on file in the archives of the Kremlin Hospital.

Korolev should not have died on 14 January 1966, on the operating table. Vishnevskiy confirmed this four years later. In the winter of 1970, during a business meeting, I let it slip to Vishnevskiy that I had recurring pain in my leg from an old gunshot wound. He responded instantly, "Come see me. I'll

---

24. A. A. Vishnevskiy, *Dnevnik khirurga: velikaya otechestv. voyna 1941–1945 gg.* [*Diary of a Surgeon: The Great Patriotic War, 1941–1945*] (Moscow: Meditsina, 1967). A subsequent volume was published in 1970.

perform a Novocain block that will rejuvenate your entire body and for your leg—the Vishnevskiy hot ointment procedures."

It wasn't until spring that I had the opportunity to take advantage of this offer. Vishnevskiy warned me, "You saw for yourself—we have no amenities here. But I'll put you in our only double room rather than in the common ward. Your roommate will be the secretary of our Party district committee. He has cirrhosis of the liver. There's nothing I can do to help him. He is a veteran of the front and has blind faith in surgeons. He thinks we can do anything. He's imploring me to operate on him. With the cirrhosis that he has, there is not a single surgeon who can save him. He'll feel more cheerful with you there."

My roommate turned out to be a man with a powerful build, affable, and with a good sense of humor. He went to war as a sailor in the Baltic Fleet and fought honorably for four years, finishing up his military service with the Battle of Königsberg. As third secretary of the Zamoskvoretskiy district committee, he was responsible for Party work in small district organizations, including dining halls, restaurants, and the whole social services sphere. There were several hundred such "spots" in his district.

"Here's how it happened," complained my roommate. "Wherever you'd show up for an inspection, things began and ended with drinks. It was always possible to find faults, for which managers should have been removed from the job or even kicked out of the Party. I wasn't always able to turn them down. That's how the disease became so acute, and it began back during wartime."

It was difficult to come to grips with the idea that this jolly individual, brimming with vitality, was doomed.

Without much delay, Vishnevskiy subjected me to his favorite Novocain block treatment. Despite its apparent simplicity, all the procedures were performed adhering to the strict rules of major operations, and I was afforded the opportunity to observe Vishnevskiy's artistry in the business to which he had devoted his life. The ailing leg was wrapped in multiple layers of bandages saturated with Vishnevskiy's hot ointment and tightly bound over its entire length.

When I found myself once again back in my room, I was surprised to see a bouquet of flowers and a bottle of cognac on my bedside table. It turned out that Yevgeniy Volchkov—Semyon Chizhikov's deputy and our leading specialist on medical technology for the Vishnevskiy Institute—had been instructed to present me a medal that I had been awarded in honor of the 100th anniversary of V. I. Lenin's birthday [in April 1970]. That day, on behalf of the district committee, my roommate was supposed to present the same medal to Vishnevskiy. Just about 4 hours later, Vishnevskiy stopped by straight from the operating room.

"It was a difficult job," he said, "an aortic aneurysm. You might have seen this very young Georgian fellow walking around here with blue lips. It was a miracle he lived to be 18. In two days I'll make him get up and walk; and in six or seven days, we'll release him. Now he can get married."

Volchkov poured a round of cognac and we drank to my latest decoration. Vishnevskiy even allowed my roommate to toss back a drink, assuring him that 100 grams wouldn't hurt him at all after those hundreds of liters that he had consumed as he developed cirrhosis of the liver. After offering our traditional toasts, I asked, "Why, after heart surgery for an aortic aneurysm, will the patient begin to walk in two days, and after having a rectal polyp removed Korolev never left the operating table?"

There was a long silence. After taking another sip of cognac, Vishnevskiy said, "It was late when they came for me. I had gone out for a stroll. It's a good thing I hadn't gone far. But, all the same, when I arrived at Granovskiy, I knew right away that there was nothing I could do. If you're looking for guilty parties now—there are many. I'm guilty for not advising and for not saying where and by whom the operation would have been best performed. Korolev himself is guilty—he insisted that a minister of the Soviet Union operate on him. A minister has quite enough to do without performing operations; an operation is by no means the only thing on his mind. I don't think they did a thorough pre-op examination. They could have discovered the sarcoma even earlier. It was the size of a fist and completely encysted. He could have lived a lot longer with it—there were no metastases—everything was clean. And the way they got to it was not at all according to the operation plan. First they removed the rectal polyps. They weren't able to stop the hemorrhaging. Then they decided on abdominal surgery—open him up, suture up everything there, and perform a colostomy. Naturally, it wouldn't be easy for him to have a stoma [an opening—ed.] in his side. It's a lot of trouble, but later they could perform reconstructive surgery. Then they saw the tumor and they decided to remove it."

"But the clock was ticking. Under general anesthesia, not everyone's heart and lungs function reliably. And the anesthesiologists had overlooked something. In addition to everything else, Korolev had an unusual neck structure. They needed to quickly set up artificial respiration but they couldn't manage to insert the tube into his throat. They hadn't thought of or prepared for that eventuality. The cardio-pulmonary bypass machine hadn't been prepared in advance. Whatever the operation, everything is supposed to be on hand and ready. A surgeon is supposed to think through all of his actions in advance, to the last detail. The entire surgical team must be close-knit and more highly trained than a hockey team. Heart surgery is still a recent development, yet fatalities are rare. Korolev's case is a tragedy; he was a casualty of our medicine. He shouldn't have died."

Now a believer in the rejuvenating effect of the Novocain block, I was released from the hospital a week later and got to work. A month later I went to the Zamoskvoretskiy district committee for the funeral ceremony of my hospital roommate. One of the witnesses of Vishnevskiy's confession had departed this life. I believe that Aleksandr Aleksandrovich expounded to his closest specialist friends, in medical language and in great detail, about what happened with Korolev. However, no doctors were to be found who wanted to reconstruct in detail the true picture of what killed Korolev. No commission was created to investigate the circumstances of Korolev's death.

Writer and journalist Yaroslav Golovanov undertook the most extensive private investigation. In 1994, Golovanov's book *Korolev: Facts and Myths* was published.[25] In the last chapter, the author cites the "testimonies" of a wide range of people who had witnessed this tragedy. Even before Golovanov's book came out, the newspaper *Komsomolskaya pravda* published his article "Korolev Did Not Die of Natural Causes!"[26] Yet, investigative journalism, the honesty and integrity of which I do not doubt, still cannot replace the investigation that should have been undertaken by qualified medical specialists with authorization to access hospital records.

To date, everything that is known about Korolev's death allows me to agree with Yaroslav Golovanov's assertion that, "Korolev did not die of natural causes!"

Gagarin's triumph and subsequent flights into space gave rise to natural pride in the nation's achievements among all strata of Soviet society in the 1960s. Successes in space were the real proof that our people "hadn't just fallen off the turnip truck." We might still lag behind America and Europe in terms of the number of automobiles per 1,000 inhabitants and square meters of decent living space per person, but "we were ahead of the whole planet" in the most cutting-edge science and technology. The cosmonaut heroes were the end product of this same cutting-edge technology. But who created it? The hundreds of thousands of people who stood in line to bid farewell to Korolev during those bitterly cold days in January 1966, passing silently through the Hall of Columns, were demonstrating that they knew "who was who." People wanted to know much more than the official lines of the obituary and the brief commemorative speeches. For the first time, the public had access to a deceased, completely unknown, great chief designer, whose identity had been kept a highly classified secret from his own people. This was the first breach in the multi-layer information

---

25. Ya. K. Golovanov, *Korolev: fakty i mify* [*Korolev: Facts and Myths*] (Moscow: Nauka, 1994).

26. Yaroslav Golovanov, "Korolev ne umer. Korolev pogib!," *Komsomolskaya pravda* no. 153, 6 August 1994.

insulation surrounding all the true creators of rocket-space technology. Now, at last, the Chief had been named. But society wanted to know how he had accomplished all of this. Was it really just because he was a remarkable scientist and engineer with great organizational skills?

EVERYONE UNDERSTOOD THAT NO MATTER HOW GREAT A GENIUS KOROLEV WAS, THERE MUST BE OTHERS AS WELL. Would the public really have to wait for the obituaries of these "others"? TASS reports, public press conferences, and enthusiastic magazine articles were simply not enough to satisfy the information famine. Bona fide information about Korolev's life and work and about his school was meted out in stingy portions. Each new publication raised more questions than it answered. Professional journalists made their first attempts to show the process that had formed Korolev's persona and to describe the people who created cosmonautics.

In 1969, Politicheskaya Literatura publishing house issued 200,000 copies of A. Romanov's book *Spacecraft Designer*.[27] Note that the price marked on the book is 24 kopecks![28] That same year, Mashinostroyeniye publishing house issued P. T. Astashenkov's book, *Academician S. P. Korolev*. Voyennyy publishing house supplemented and reprinted 200,000 copies of the book at a price of 75 kopecks each in 1975.[29]

A. Ivanov's short book *First Stages*, published by Molodaya Gvardiya publishing house, caused a sensation in 1970. It was actually written by Oleg Genrikhovich Ivanovskiy, the former lead designer of Vostoks. The real heroes of this work are not fictitious, but at the censor's request, the names were changed "to protect the innocent." S. P. Korolev was the only name left untouched! Oleg Ivanovskiy did not manage to correct and supplement his literary work until 2005. But 100,000 copies of the first book were printed, while the latter, truthful and uncut version had just 3,000 copies.[30]

---

27. A. P. Romanov, *Konstruktor kosmicheskikh korabley* [*Spacecraft Designer*] (Moscow: Politizdat, 1969). Subsequent editions were published in 1971, 1972, 1976, and 1981. Unexpurgated editions, entitled simply *Korolev* and published by Molodaya gvardiya publishing house, were issued after *glasnost* in 1990 and 1996.

28. At that time, according to the official going rate, 24 kopecks was approximately equivalent to 25 cents.

29. P. T. Astashenkov, *Akademik S. P. Korolev* [*Academician S. P. Korolev*] (Moscow: Mashinostroyeniye, 1969); P. T. Astashenkov, *Glavnyy konstruktor: o S. P. Koroleve* [*The Chief Designer: On S. P. Korolev*] (Moscow: Voyenizdat, 1975).

30. Aleskey Ivanov, *Pervyye stupeni* [*First Stages*] (Moscow: Molodaya gvardiya, 1970). Subsequent updated editions were published in 1975 and 1982. The full version of his memoirs appeared in 2005 as O. G. Ivanovskiy, *Rakety i kosmos v sssr: zapiski sekretnogo konstruktora* [*Rockets and Space in the USSR: Notes of a Secret Designer*] (Moscow: Molodaya gvardiya, 2005).

Mark Lazarevich Gallay, a former test pilot, talented writer, and Hero of the Soviet Union, expended a great deal of effort, first in periodicals, and then at Sovetskiy Pisatel publishing house, to publish his work *Man On Board*, a real page-turner. The book came out in 1985, but the first attempts to publish it began in 1975.[31]

I cite this partial list of the first literature about Korolev and, consequently, about Soviet cosmonautics "for reference." The prices for books on Korolev have grown on the average 300-fold over the course of 35 years, while the number of copies have decreased 100-fold. I don't remember where and when I read that during Napoleon's years of triumph, when he took over almost all of Europe, a sycophant compared his might with that of God. "At your will you can make a soldier a marshal and a marshal a king!" Napoleon replied that he really was capable of making a soldier a marshal and even a king. "But I cannot make a scientist. Only God is capable of that." Even if some historian made up Napoleon's response, it is a very clever story.

Just 30 years after Korolev's death, major works have begun to emerge that investigate Korolev's true biography, his creative work, unusual larger-than-life character, and amazing organizational talent more fully and credibly. Of these printed works, I shall single out three academic publications that together most comprehensively reflect the "Korolev phenomenon."

The **first** work is the 800-page tome *Korolev: Facts and Myths.* Its author is the talented journalist, writer, engineer, and science columnist for the newspaper *Komsomolskaya pravda* Yaroslav Kirillovich Golovanov. The Nauka academic publishing house issued the book in 1994, and not because the manuscript had been waiting in line for a long time. Golovanov had begun to collect material about Korolev, make sense of it, and double-check it in 1968. He finished in 1994. A quarter century for a book about a great man of the 20th century! I received a copy of this book on 1 February 1995 with this inscription from the author: "To my dear consultant—Boris Yevseyevich Chertok for your invaluable contribution to the creation of this book." Golovanov enjoyed the "invaluable" consultations of all of Korolev's deputies, members of the Council of Chief Designers, cosmonauts, and specialists from Korolev's school. Golovanov's book is a fascinating, artistic work of historical literature, but not strictly a scholarly study.

---

31. M. L. Gallay, *S chelovekom na bortu: dokumentalnaya povest* [*Man on Board: A Documentary Story*] (Moscow: Sovetskiy pisatel, 1985). Mark Lazarevich Gallay (1914–98) was one of the greatest test-pilots in Soviet aviation history. He was invited by Korolev to contribute to the training of the cosmonauts in the early 1960s.

A historian who spares no time or effort searching for the necessary documents in secret and public archives can produce such a study, free of the emotions of the author and his consultants. But a historian's desire and even his enthusiasm are not enough. He must be a specialist in the rocket-space field. And luckily for the history of our cosmonautics, such a man was found: Georgiy Stepanovich Vetrov.

In 1998, Nauka publishing house also released a **second** major scientific work, *S. P. Korolev and His Affairs: Light and Shadow in the History of Cosmonautics.*[32] My name also appears among the long list of names of the scholarly editorial staff. The general editor was Academician Boris Rauschenbach. Georgiy Vetrov wrote the opening article, "The Phenomenon of Sergey Pavlovich Korolev." A study of the historical materials assembled in the book confirms that Korolev was a phenomenon. Vetrov began working in Korolev's OKB-1 in 1947. In 1950, he was manager of a group, and then of a larger sub-division dealing with rocket motion dynamics. Before Korolev's death, Vetrov was involved in the design of all combat missiles and launch vehicles. He didn't need to sort out "what's what" and "who's who." Under Vasiliy Pavlovich Mishin, Vetrov was appointed academic secretary of the Academic Council of the Central Design Bureau of Experimental Machine Building (TsKBEM).[33] He defended his candidate's and doctoral dissertations. The history of rocket technology became his hobby. I worked rather closely with him both in our primary field of work and on history. He was a historical research fanatic. Unfortunately, the administration of NPO Energiya did not judge his work in the field of history according to its merits. Back in 1980, with the support of USSR Academy of Sciences President M. V. Keldysh himself, Vetrov managed to publish *The Creative Legacy of Academician Sergey Pavlovich Korolev: Selected Works and Documents.*[34] A Commission for the Development of the Scientific Legacy of the Pioneers of Space Exploration was denoted as the author. But Vetrov was actually the one who compiled the materials and authored the commentaries.

---

32. B. V. Raushchenbach, ed., *S. P. Korolev i yego delo: svet i teni v istorii kosmonavtiki: izbrannyye trudy i dokumenty* [*S. P. Korolev and His Affairs: Light and Shadow in the History of Cosmonautics: Selected Works and Documents*] (Moscow: Nauka, 1998).

33. TsKBEM was the official name of OKB-1 after Korolev's death in 1966.

34. M. V. Keldysh, ed., *Tvorcheskoye naslediye akademika Sergeya Pavlovicha Korolev: izbrannyye trudy i dokumenty* [*The Creative Legacy of Academician Sergey Pavlovich Korolev: Selected Works and Documents*] (Moscow: Nauka, 1980). Note that Keldysh was not alive in 1980, having passed away in 1978.

Academic publishing house Nauka released a **third** book, the two-volume work *Otets* (*Father*) in 2001–2002. Its author is Korolev's daughter, Nataliya Sergeyevna Koroleva.[35] In the foreword to the first volume I wrote:

This book proves to the reader that Professor N. S. Koroleva, Doctor of Medical Sciences, is both a gifted writer and historian . . . . She set out to study the life of a family and man very dear to her, from distant ancestors to her father's last hour. And she managed to fill in a lot of blanks in Korolev's biography. This primary source, in the truest sense, has provided completely new and at times surprising information about the inner world of the man who we thought we already knew very well. In order to be as close as possible to the truth when writing about her father's life, N. S. Koroleva did what no other biographer had been able to do. She walked, drove, and flew to all the places where he had lived, worked, loved, suffered, triumphed, celebrated, and once again tirelessly toiled. Zhitomir, Nezhin, Odessa, Kiev, Magadan, Omsk, Germany, Kapustin Yar, and Baykonur. And of course, Moscow and the city of Kaliningrad, which, not without the involvement of N. S. Koroleva, was renamed Korolev. N. S. Koroleva's work synthesizes the virtues of a literary novel, a nonfiction narrative, historical research, and a personal confession . . . .

Natasha was 12 when her family fell apart. Korolev fell in love with Nina Kotenkova and severed relations with his first wife Kseniya Vintsentini. Under her mother's influence, Natasha remained estranged from her father for a long time. Korolev wrote to his daughter, who turned 18 in 1953: "My dear Natasha, I don't think that you are being fair with me. Please give it some thought." Those who knew Korolev well, i.e., Vasiliy Mishin, Sergey Kryukov, Konstantin Bushuyev, Petr Meleshin, and Leonid Voskresenskiy knew how painfully he endured his daughter's defiantly cold attitude toward him. One might blame Natasha for causing her father undeserved grief during his "star-studded" years. I feel compelled to say that, in writing her two-volume work *Father*, Korolev's daughter truly has more than atoned for all the hurt that she willfully caused her father.

Her remarkable two-volume work on Korolev's life is one of the authoritative documents describing this epoch of the 20th century. However, this book is only part of the multi-faceted work that Natasha Koroleva is conducting to perpetuate her father's memory.

---

35. Natalya Koroleva, *Otets*, 2 t. [*Father*, in 2 vols.] (Moscow: Nauka, 2001–02). These books were completely updated and published as a three-volume work in 2007.

During the period from 1973 to 1976, breaches in the information isolation surrounding our space operations increased as a result of the Soyuz-Apollo mission. The Central Committee and VPK had long understood that it was better to reveal at least a glimpse into our space operations than to force people to tune into Voice of America and to discuss matters in smoking rooms and kitchens. For the foreign media, and in particular for NASA, all of our achievements were explained away as activities of the all-powerful Academy of Sciences. Therefore, the Academy was also authorized to lift the veil of secrecy ever so slightly. President Keldysh was too busy and shifted this difficult task onto Academicians B. N. Petrov and V. A. Kotelnikov.

The Academy Presidium tasked Boris Nikolayevich Petrov, Academician and Secretary of the Department of Mechanics and Control Processes, to establish order in the conduct of various conferences, symposia, and lectures.[36] The first academic resolution was issued calling for cooperative scientific lectures in memory of prominent Soviet scientists—pioneers of space exploration—beginning in 1979 in Moscow. The aim of these lectures was to examine individual issues of the development of rocket-space science and technology and their history, to familiarize the scientific-technical community with the most important achievements in the field of space exploration and exploitation, and to promote the achievements of Soviet cosmonautics. The first lectures took place in February 1979, almost 10 years after the American lunar landings that had been accompanied by a deafening worldwide media campaign.

In April 2001, I was confirmed as chairman of the Commission of the Russian Academy of Sciences for the Development of the Scientific Legacy of the Pioneers of Space Exploration. Before me, Academician B. N. Petrov and then Academician B. V. Rauschenbach had performed this honorable duty. The organizational work entailed in conducting the "Korolev lectures" after censorship restrictions were lifted has proved considerably more complex and labor-intensive than during the time of classified information. It turned out that there were hundreds of people who wanted to participate—to present scientific reports or reminiscences and historical research. More than 1,000 people took part in the last lecture series.

---

36. The Presidium was the Academy's highest deliberative body.

All these years, the Academy's Institute for the History of Natural Sciences and Technology (IIYeT) has been entrusted with the main organizational problems.[37] For successful work in this field and for the development of cosmonautics itself, driven professionals are needed. Ever since the first "Korolev lectures," IIYeT Department Chief Viktor Nikolayevich Sokolskiy and Academic Secretary for Lectures Alla Konstantinovna Medvedeva have performed this work. The recent death of Sokolskiy was a great loss to work in the field of the history of cosmonautics. The new bosses of our nation have launched unruly and destructive projects to reconstruct the center of Moscow, first and foremost, in the interests of the top dogs of the oligarchy and functionaries of the corrupt elite of Russian society. IIYeT's former location on Staropanskiy Lane, a sturdy, early 20th-century five-story building, which is quite well-suited as an office building, is going to be torn down. The institute's work has been virtually paralyzed. It is a good thing that the baton has been passed to the N. E. Bauman Moscow State Technical University to organize the work on the lectures. Its assistance over the past two years has been invaluable. The joint efforts of the USSR Academy of Sciences, the space community, the municipal authorities of Moscow and Kaliningrad, and the staff of the Korolev school, named after him, have yielded tangible results.

In 1968, Third Ostankinskaya Street in Moscow was renamed Academician S. P. Korolev Street. At one time, this street was the main street of the suburban Moscow dacha village Ostankino, well-known for its monument of 18th century architecture, the Sheremetyev family palace. Now, Academician Korolev Street is the main thoroughfare connecting the television broadcast center with the street Prospekt Mira, crowned with the famous Ostankino television tower. And at the top of the street on Avenue of Heroes, which leads to the space obelisk, a memorial bust of Korolev was unveiled in 1967. The sculptor Faydysh-Krandiyevskiy was commissioned by the USSR Council of Ministers to create this work.[38] Also in 1967, by decision of and at the expense of TsKBEM, a memorial bust by the same sculptor was unveiled on the memorial square on the grounds of TsKBEM. The fact that this monument was unveiled on 4 October 1967 is to the great credit of TsKBEM Chief and Chief Designer Vasiliy Pavlovich Mishin.

---

37. IIYeT—*Institut istorii yestestvoznaniya i tekhniki*. Since 1991, this institute has been known as the S. I. Vavilov Institute for the History of Natural Sciences and Technology.

38. Sculptor Andrey Petrovich Faydysh-Krandiyevskiy (1920–67) was responsible for a number of space-related monumnets in Moscow, including the giant Tsiolkovskiy statue next to the Kosmos Hotel in Moscow, which was unveiled in 1964.

On 7 October 1988, a monument to Korolev was unveiled in Kaliningrad on Korolev Prospekt. Sculptor V. M. Levin and architect S. A. Ryazin collaborated on the monument, which in my opinion is the best of all the monuments immortalizing scientists that I have ever seen. After the name of the city of Kaliningrad was changed to Korolev in 1996, this monument became one of its sightseeing attractions, left by the twentieth century to its distant progeny, who someday will leave their footprints "on the dusty pathways of distant planets."

Chapter 18
Birth of the Soyuzes

Today, Soyuz piloted spacecraft are world-famous. Until 1995, they were the only means for delivering the crews of the Salyut and *Mir* orbital stations into space and returning them to the ground. In 1995, experiments were successfully conducted for the delivery of international crews to the *Mir* station on board the reusable Space Shuttle. At the end of the 20th century, the world had only two types of vehicles on which people were delivered into near-Earth space: non-reusable Soyuz vehicles and reusable Space Shuttles. The Soyuz had come on the scene 10 years before the Space Shuttle.[1] The history of its development dates back to 1961, when the first Vostoks had just begun to fly. By contrast, the first Chinese piloted vehicles appeared in the 21st century.

The original mission of the Soyuz design was not to fly to orbital stations, but to perform a circumlunar flight with a person on board. The precursor of the modern Soyuz was the piloted 7K vehicle, which was a component of the Soyuz complex. In terms of design, the Soyuz rocket-space complex included a 9K booster unit, three 11K space refueling vehicles, and the piloted 7K vehicle.

Our conceptual designers came up with the idea of a piloted circumlunar flight using "materials on hand," without a super-heavy launch vehicle, after President John F. Kennedy announced to the entire world that the U.S. was accepting our challenge and beginning its fight for supremacy in space.[2] As we launched Vostoks into space, we had to be thinking about our future prospects.

---

1. The first Soyuz spacecraft was launched into orbit in 1966, 15 years before the first flight of the Space Shuttle in 1981. However, the first piloted Soyuz mission was in 1967 and the first Soyuz taxi mission to a space station was in 1971, 15 years before the Shuttle.

2. Yuriy Gagarin's Vostok mission on 12 April 1961 served as a catalyst for President Kennedy's historic speech of 25 May 1961 when before a joint session of Congress he set the goal of landing an American on the Moon before the end of the decade.

Mikhail Klavdiyevich Tikhonravov (1900–74) was one of the giants of the Soviet space program. He played a critical role in almost every major space project at Korolev's design bureau until his retirement in 1966.

O.V. Gurko

Korolev approved the initiative that came out of Tikhonravov's design department calling for a circumlunar flight using materials on hand, and requiring at least five *Semyorka* launches. The 9K booster unit would be inserted into near-Earth orbit first. Next, three 11K refueling vehicles would be launched in succession and transfer fuel to the 9K using pressurized gas. The three refueling vehicles brought the fuel margin in the 9K up to 25 metric tons. The 7K carrying two cosmonauts would be launched last. It would dock with the 9K, and after one or two orbits around Earth, the 9K boosted it toward the Moon. Before flying around the Moon, the unneeded 9K would be jettisoned. The 7K was supposed to return to Earth by itself and enter the atmosphere at reentry velocity.[3]

In order to execute a circumlunar flight according to such a complex plan, first and foremost it was necessary to solve the problem of reliable automatic rendezvous and docking in space. The second problem was the development of a new piloted vehicle with a new control and navigation system ensuring return to Earth within a very narrow spatial corridor using two atmospheric passes to reduce the speed.

Despite all the complexities of this proposal, Korolev took it to the "higher ups." The psychological effect of the flights of Gagarin and Titov on the nation's top political leadership was so great that, without a serious technical evaluation, the USSR Council of Ministers authorized the proposal for a circumlunar flight with

---

3. Russians typically use the phrases "first cosmic velocity" and "second cosmic velocity" to denote the velocities required to achieve Earth orbit and translunar injection, respectively. For bodies returning from lunar distance, the reentry velocity approximates translunar injection velocities, i.e., the second cosmic velocity.

a decree on 16 April 1962. The Soyuz program was mentioned by name for the first time in this decree. All three spacecraft had been developed in drawings, but the manufacture of the vehicles at the factory was proceeding far from splendidly. Turkov had had no trouble proving to Korolev that our factory was in no condition to fulfill the programs for Vostoks, Zenits, and interplanetary and lunar automatic stations if it was also going to be loaded down with an armada of Soyuzes. Korolev decided to broaden the front of operations and on 3 December 1963, another decree was issued, entrusting the manufacture of the unpiloted 9K and 11K vehicles to the Progress factory in Kuybyshev. "We aren't going to hand over the piloted 7K vehicle to anyone, we'll only make it at our factory," decided Korolev. The 7K was the most complicated part of the Soyuz complex. We needed to "jump the gun" and create it. Without it, the 9K and 11K lost their meaning.

All of the departments subordinate to me viewed the development of the control systems for the 7K as a natural continuation of the work that had been started on the Zenits and Vostoks. Here, wide-ranging opportunities for implementing new inventions for automatic flight control really were presenting themselves. Rauschenbach's departments were tasked with developing a system that would perform the entire spacecraft motion control cycle in near-Earth and lunar space, and in the atmosphere during return to the ground. The system would have to support the following: orbital, inertial (relative to fixed stars),

From the author's archives.

Korolev drinks a toast at Boris Rauschenbach's 50th birthday celebration in January 1965. Visible in the picture are, from left to right: N. I. Paramonova, Mikhail Tikhonravov, Korolev, M. G. Chinayev, A. I. Patsiora, Boris Chertok, and L. F. Ivin.

and solar (for illumination of the solar arrays) orientation; various programmed maneuvers; orbital corrections; precise stabilization of the spacecraft during operation of the correcting engine; rendezvous, final approach, and docking with another spacecraft; as well as guided descent to Earth both from near-Earth orbit at orbital velocity and during return from the Moon at reentry velocity.

The memory of the automatic part of the on-board complex control system would have to recognize setpoints: the required position of the spacecraft's axes in space, flight delta velocity, and selection of the requisite timeline for activating or deactivating equipment. These setpoints would be uploaded via command radio link from the ground or from the display and control panel for subsequent execution at the specified time of the in-orbit operations. In addition, it was still possible to make real-time program changes by commands from ground stations and from the crew display and control panel.

Nikolay Kamanin and a few other cosmonauts were quite pleased when we promised that the crew would have the capability for complete functional backup of the control of all operations, including docking and descent.[4] The design called for visual data display on electronic indicator screens and optical sights for the direct observation of external reference points. The control system being designed made it possible, at the discretion of the ground or the cosmonaut, to select automatic, "manual," or mixed control modes.

Isayev had developed a brand-new approach and correction engine unit (SKDU).[5] It had two engines: a main approach and correction engine (SKD) and a backup correction engine (DKD).[6] The systems of effectors for attitude control and final approach employed highly concentrated hydrogen peroxide. Under Knyazev's supervision, the staff at OKB-1 had developed two systems of effectors—approach and attitude control engines (DPO) and attitude control (only) engines (DO).[7]

---

4. Nikolay Petrovich Kamanin (1909–82) served as the deputy chief of the Soviet Air Force's General Staff from 1962 to 1966, responsible for space issues. From 1960 to 1971, he was responsible for cosmonaut training.

5. SKDU—*Sblizhayushche-korrektiruyushchaya dvigatelnaya ustanovka.*

6. SKD—*Sblizhayushche-korrektiruyushchiy dvigatel;* DKD—*Dubliruyushchiy korrektiruyushchiy dvigatel.*

7. DPO—*Dvigateli prichalivaniya i orientatsii;* DO—*Dvigateli orientatsii.* The Soyuz (7K-OK) spacecraft had several sets of engine systems. The DPO system was designed for attitude control during approach and docking and included 14 engines for docking and orientation (10 kilograms thrust) and 8 engines for orientation (1–1.5 kilograms thrust). Besides attitude control, the spacecraft also included the main orbital correction engines known as the SKDU, which comprised two engines, a primary (SKD) at 417 kilograms thrust and a backup (DKD) at 411 kilograms thrust. During the work of the primary engine, the crew could carry out attitude control using the thrusters of the DPO. The SKD would also be used for deorbit.

A fundamental innovation was the use of an electronic instrument that measured the intensity of the ion flux hitting the spacecraft and generated signals indicating the deviation of the spacecraft's axes from the direction of flight. The ionic orientation (IO) system had undergone preliminary testing on Zenits with encouraging results.[8]

Initially, the 7K spacecraft had been designed to carry two cosmonauts. For that reason, the Vostok landing method was rejected out of hand. No ejection seats! The Voskhods had confirmed that spacesuits were only needed for spacewalks. Anyone not performing a spacewalk would get by without a spacesuit!

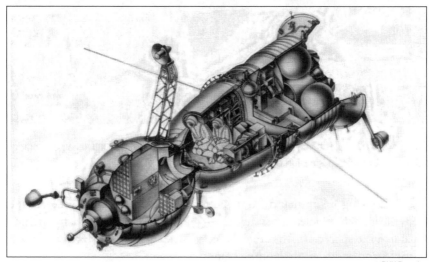

RKK Energiya

A cutaway of the original version of the Soyuz (the 7K–OK) which flew between 1966 and 1970. Visible are the three major segments of the spacecraft. From the left, the living compartment, the descent module, and the instrument compartment.

A descent module shaped like an automobile headlight was designed to ensure a guided descent into a specified area with great precision and acceptable g-loads. The American Mercury and Gemini capsules had the same sort of "headlight" shape.[9] For landing, a soft landing rocket-assisted parachute system needed to be developed. An emergency rescue system for an emergency during launch or during the initial flight segment was developed.

---

8. IO—*Ionnaya orientatsiya*.

9. In reality, the shape of the Mercury and Gemini reentry vehicles significantly differed from that of the Soyuz descent module. The former had shapes closer to cones with blunt surfaces than the "headlight"-shaped descent module of the Soyuz.

NASA

Soyuz orbital modules being assembled at the Factory for Experimental Machine Building (ZEM) next to Korolev's design bureau.

Upon receiving a command from the ground or an emergency command from the launch vehicle, this system, using powerful solid-propellant engines, separated the spacecraft from the launch vehicle, moved it as far away as was possible, and activated the nominal landing system.

A spacious habitation module was provided to create additional comfort during prolonged flight.[10] Its forward section contained the docking assembly and airlock hatch. We wanted to consolidate the large number of radio systems into one complex. Ryazanskiy proposed creating an integrated radio system. Boguslavskiy set about developing a long-range radio complex (DRK), which made it possible over a single radio link to transmit individual commands and files of setpoint commands, to measure orbital parameters and current coordinates, and to ensure the transmission of a television signal and voice communications at distances from the Moon to Earth.[11]

---

10. Most Westerners know the habitation module as the "orbital module."
11. DRK—*Dalnyy radiokompleks*.

For reliable long-range communications and the transmission of "pictures," a parabolic antenna was designed that opened after orbital insertion. However, they couldn't bring themselves to do away with Yuriy Bykov's now-familiar system—the *Zarya* radio telephone voice communications system. They had the same attitude toward Aleksey Bogomolov's orbital monitoring system, and the radio transponder remained on board despite the DRK. Under Mikhail Ryazanskiy's strong pressure, the new BR-9 telemetry system was installed on all Soyuz-type vehicles. Bogomolov's *Tral* systems remained only on the launch vehicles.

In December 1962, Korolev approved the draft plan for the circumlunar flight by a Soyuz, which called for five launches and four dockings. In 1963, the space design departments, completely free of Vostoks, and having started work on the Voskhods, spent their time developing the new descent module and—jointly with Fedor Tkachev—the rocket-assisted parachute landing system. LII was getting ready to conduct flight experiments with the prototype of the emergency escape system. Models of the descent module were subjected to wind tunnel tests at TsAGI. The designers haggled fiercely with the developers of various systems about their weights. Konstantin Feoktistov was the main authority on the 7K spacecraft.

The set of problems associated with the rendezvous and docking modes was the most complicated. There were two departments vying to develop the docking assemblies (at that time they were called "docking nodes"): Semyon Chizhikov's design department and Lev Vilnitskiy's department of control-surface actuators and mechanisms. By that time, Vilnitskiy had acquired a strong staff of "electromechanics," while Chizhikov was extremely busy with the development of instrumentation. After agonizing indecision and debates, Kalashnikov and I gave preference to Vilnitskiy's department.

In his own department Vilnitskiy had dared to entrust the design side to junior engineer Vladimir Syromyatnikov. Kalashnikov had complained to me about this: "Lev didn't take this assignment seriously. First he put that Syromyatnikov in charge of developing the winch for the artificial gravity experiment, and now he's also entrusted the docking assembly to him.[12] If the artificial gravity system doesn't pan out, that's no big deal, but a Soyuz with no docking—that's a program failure. That's a risk he shouldn't take."

Preoccupied with trying to find a developer for the rendezvous radio system, I set Kalashnikov's mind at ease reasoning that, by the time we finished developing the radio search and rendezvous system, Syromyatnikov

---

12. The artificial gravity experiment, originally planned for a later Voskhod mission in the 1966 timeframe, was never flown.

wouldn't be a junior engineer anymore. Our decision of 1962 decided Syromyatnikov's fate "for the rest of his life." Professor Syromyatnikov— Member of the International Academy of Astronautics, Doctor of Technical Sciences—earned worldwide recognition, having founded the school of the mechanics of spacecraft docking systems.[13]

My friend Semyon Chizhikov passed away some time ago, but his son Boris, a student of Syromyatnikov, has become one of the leading specialists on docking assemblies, and his granddaughter Marina is a specialist in the dynamics of this complicated mechanical complex. No American or European firms have been able to develop competitive docking assemblies. Three decades later, both the Russian Soyuzes and the American Space Shuttles dock using mechanisms that veterans of Korolev's OKB-1 began to develop.

The task of two spacecraft coming together in near-Earth space seemed more complicated to me than the problems of intercepting targets with surface-to-air missiles or even intercepting a missile with an anti-ballistic missile. Using powerful (for those times) ground-based computers, the "ground" was actively involved in the anti-aircraft and anti-ballistic missile systems. We were unable to use ground-based guidance radar or ground-based computers to directly control the rendezvous process in real time. Our "ground" was capable only of preparing for rendezvous. The actual rendezvous process, based on the ballistics specialists' first approximate calculations, could last two and even three orbits. It turned out that an on-board computer was needed for control. But where could we get one that was ready to go, and if we couldn't find one, then how much time would we have to spend developing a new one?

The docking mechanism was another issue. It was an engineering project requiring no computer. I was certain that through the efforts of its designers and of the factory we would master the task of docking before we mastered rendezvous. Docking was a matter of electromechanics and dynamics. We already had experience in that area. But rendezvous . . . .

In this chain of sequentially solved problems, the on-board digital computer (BTsVM) seemed to be the most complicated link.[14] Time showed that we had learned to make docking mechanisms and to solve the

---

13. Vladimir Sergeyevich Syromyatnikov (1933–2006) led the development of Soviet and Russian docking systems for more than three decades. His name first came to light in the West in connection with his design of the androgynous docking unit used on the Apollo-Soyuz Test Project (ASTP) in 1975. Syromyatnikov was elected a Corresponding Member of the Academy of Sciences in May 2006, just four months before his untimely death.

14. BTsVM—*Bortovaya tsifrovaya vychislitelnaya mashina*—or literally, "on-board digital computing machine."

problems of rendezvous and docking long before the BTsVM appeared on board of our vehicles.[15]

Operations on the Voskhods diverted everyone involved with the piloted flights away from the Soyuzes. And this was the biggest and best part of our staff and those cooperating with us. Using the "big names" of the cosmonauts who had already flown, the Air Force Command continued to insist on ordering 10 more Vostoks or Voskhods for new military purposes. The VPK and State Committee on Defense Technology wavered, having no clear idea which program to pursue. Minister of Defense Rodion Malinovskiy and his deputy, Andrey Grechko, were unequivocal in their contention that Vostoks were not necessary for near-Earth flights and certainly not for lunar flights. And they were right!

The weak points of our overly complicated circumlunar flight project were so obvious that Vladimir Chelomey, who had confidence in his own UR-500 launch vehicle, came out with a proposal to execute the circumlunar flight using a single-launch plan. He proposed using his own piloted LK-1 vehicle, boosted toward the Moon by its own booster unit. According to his proposal, the UR-500 rocket was converted from a two-stage into a three-stage rocket, and in that version it received the designation UR-500K, or for public use, the Proton-K.[16] Before Khrushchev's downfall, Chelomey managed to have him sign the decree calling for this version on 3 August 1964.

One aspect of this decree was sensational news to us: the announcement that this decree obligated Chelomey to build 12 spacecraft for circumlunar flights by the beginning of the second quarter of 1967! A second organization had been created that had dared to take on the responsibility of piloted spaceflight.

At our OKB-1 virtually everyone was in solidarity with our chief. To hell with rockets—there really were many different kinds of them, so let them come up with different chief designers for them. We would generously give away automatic spacecraft to Babakin and his affiliated enterprises. But the development of vehicles for piloted flights must remain our monopoly.

It became pointless to work on the circumlunar flight model of the Soyuz after this decree [of 3 August 1964] came out. When we analyzed the reliability of our version more thoroughly, we came to the conclusion that it wasn't worth

---

15. The first digital computers installed on Soviet spacecraft intended for piloted flight were the *Salyut-1* models used on the 7K-L1 (Zond) circumlunar spacecraft launched between 1967 and 1970.

16. For details of the UR-500/LK-1 project, see Asif A. Siddiqi, *Challenge to Apollo: The Soviet Union and the Space Race, 1945–1974* (Washington, DC: NASA SP-2000-4408, 2000), pp. 395–396, 443–446, 497–502.

the trouble. During 1964, a feverish search for ways out of this dead end led Korolev to the idea of joining forces with Chelomey. Our Block I (third stage) of the R-7 rocket could be converted into a booster unit for the UR-500K, and Chelomey would not have to develop his own upper stage. Similarly, the 7K could be modified so that Chelomey would not need to develop his LK-1. Chelomey rejected this proposal out of hand, but the political situation changed abruptly after October 1964.[17] At least from the standpoint of the highest governmental interests, it was now possible to argue that the ideological and technical merger of the operations of OKB-1 and OKB-52 was a sensible option in terms of solving the common high-priority task of a circumlunar flight.

On 25 October 1965, a new decree was issued stipulating that, for the first time, OKB-1 and OKB-52—Korolev and Chelomey—through their combined efforts must solve the problem of a piloted circumlunar flight using the 7K spacecraft in 1967. This same decree called for the landing of a man on the Moon in 1968. The Soyuz circumlunar flight program was terminated.[18]

Instead of the piloted 7K spacecraft for the Soyuz program, two different spacecraft had to be developed: one for flights in near-Earth space using the R-7 launch vehicle and one for circumlunar flight using the UR-500K launch vehicle. The former was assigned the designation 7K-OK and the latter, the 7K-L1, or simply L1. Both designations were secret. The "open" designation that appeared on the drawings for 7K-OK was 11F615, and for 7K-L1 it was 11F91.[19] At Korolev's suggestion, the 7K-OK orbital spacecraft was named "Soyuz" (Union) for future publications in honor of its origins.

Thus, Chelomey would have to prepare the UR-500K for the circumlunar flight and we would have to develop the new 7K-L1 vehicle and the booster upper stage. This stage, which was assigned the designation "D," later proved to be surprisingly enduring and became an indispensable part of the UR-500K rocket for launching all communications satellites into geosynchronous orbit as well as launching new automatic interplanetary stations.

By the time the decree was issued on 25 October 1965, the design work for the transformation of the 7K into 7K-OK had already been developed. The working drawings were revamped at full speed with a lot of overtime.

---

17. This is a reference to the fall of Nikita Khrushchev and the rise of the new leadership of Leonid Brezhnev and Aleksey Kosygin. Khrushchev had been a strong supporter of Chelomey's projects.

18. Technically, the Soyuz circumlunar program was actually terminated in August 1964 when Chelomey was tasked with the same goal using his own LK-1 spacecraft.

19. Of course, even the "open" designations were secret since these were known only to a small group of people and completely unknown in the West until the early 1990s.

Boris Chertok and Yevgeniy Bashkin.

Korolev assigned Feoktistov to supervise the design work on the new Soyuz. Discussions heated up over the motion control, rendezvous, docking, on-board complex control systems, and problems of electric power supply and distribution. My compatriots—Rauschenbach, Legostayev, Bashkin, Karpov, and Shustov, who had just enjoyed the fruits of victory with the Vostoks, Voskhods, Zenits, and Molniya—were sure that the cooperation that had already been established in our role as the head organization would make it possible to solve this problem.

Indeed, the development of a mutual understanding and good working relations with the engineering staffs of Iosifyan, Lidorenko, Bykov, Khrustalev, Kuznetsov, Bogomolov, Mnatsakanyan, Antipov, Tkachev, Zaychenkov, Voronin, Kartukov, Chachikyan, and many others was a much more complicated and protracted process than developing the protocols to make systems parameters compatible with one another.[20]

Behind each item on such protocols were living individuals with their own interests: personal, company-related, and agency-related. The formal relationships in such a complicated cooperation, regardless of the number

---

20. These were the names of the top chief designers of the Soviet piloted space program who acted as subcontractors to Korolev's design bureau. Each of them headed a large and independent design bureau or research institute within the Soviet defense industry.

of "proper" papers, could scuttle a large system. The papers are necessary in any case; they instill discipline. But having good friends in subcontracting enterprises is a lot more important. People involved in the production of a system must make it not only mechanically sound, but self-organizing, self-controlled, and structurally reliable. If the relationships between the individuals involved in the program are strictly formal, it is impossible to develop such a system.

In planning discussions, Korolev flatly condemned the tendency of a powerful organization, like Pilyugin's NIIAP, to completely remove itself from the development of spacecraft control systems. After countless arguments (in house, in the ministries, and in the VPK) and discussions during meetings at the firing range with Keldysh's intervention, they arrived at a compromise arrangement for the distribution of work.

OKB-1 would remain the head developer of all on-board systems for 7K-OK, i.e., the Earth-orbit Soyuz. This meant that I was actually entrusted with the work of the chief designer for the Soyuz control system. But at OKB-1, aside from Korolev, there could be no other "chiefs." Therefore, Korolev remained the chief responsible for the control system.

Pilyugin was named chief designer of the control system for the circumlunar flight complex. This was noted in a government decree. Nevertheless, we had not completely resigned ourselves; and over and above the decrees, we "amicably" negotiated with Pilyugin's deputies Finogeyev and Khitrik, and they persuaded their boss that those tasks of lunar navigation involving the cosmonaut and also problems of manual control and issues related to sensors for sun and star orientation be left for us.

A real all-hands rush job to develop and manufacture Soyuzes got underway in the second half of 1965. After *Voskhod-2*, an ideological vacuum, disorder, and vacillation cropped up in the piloted launch program. There was no clear-cut answer to the question, "what's and who's next?" In August, the Americans had launched the two-seater *Gemini V* into space. In it, astronauts Gordon Cooper and Pete Conrad set a new flight duration record: eight days.[21] America celebrated. Following our example, the Americans sent their astronauts on a triumphant tour around the planet. The Soviet monopoly on space records had been dealt a serious blow. The Americans' successes were very painful for Korolev to endure. The fact that he was unwell also exacerbated his reactions.

---

21. *Gemini V* broke the previous duration record set by Valeriy Bykovskiy on *Vostok-5* in 1963.

It wasn't until August 1965 that the wavering ended over setting the priorities between the new series of Voskhods, the artificial gravity experiments, and the construction of Soyuzes. Several times, Turkov and the director of operations at the second territory, German Semyonov, came out of the Chief's office carrying production schedules for the first Soyuzes covered with strike-throughs. Soyuz lead designer Aleksey Topol quickly drew up new ones. I explained to Semenov and Topol that they did not have the right to change dates that have been previously approved without consulting us control specialists. But they had already lost the ability to smile and advised me to talk it over with S.P. by myself.

I seriously prepared for a conversation with S.P. I armed myself with reference data, memoranda of agreement with subcontracting enterprises, and essentially with statements based on common sense and the clamoring of my comrades. On the evening of 28 August, Aleksey Topol dropped by my office and placed the "Schedule for the final handover of the experimental units and handover of the main assemblies to KIS for 11F615 vehicles." In the upper left hand corner, it said "Approved—Chief Designer," followed by the clear signature of Korolev. In the upper right-hand corner was Factory Director Turkov's signature of approval. Below it was printed: "Do not display. Dissemination per list." Only German Semyonov and Aleksey Topol signed the schedule.

"S.P. ordered us to remove all the approval signatures," said Topol. By some miracle, I have held onto this "disseminated per list" schedule with the signatures of Korolev and Turkov.

At the time, in August 1965, the deadlines set by the schedule really infuriated me. The schedule called for the production of mockups for 13 large-scale experimental operations, including thermal tests, airplane drops, sea tests, developmental testing of the emergency rescue system (SAS), static and vibration tests, developmental testing of the docking systems, and spacewalks under zero gravity conditions. Sergey Darevskiy demanded the manufacture of a full-scale mockup as a simulator. And all of the mockups, each of which constituted the entire vehicle or a significant part of it, had to be fabricated and fitted out before December of that year, 1965!

I was infuriated not just because of the unrealistic deadlines, but also because S.P. had disregarded my report about the need for an "engineering mockup" of the 7K-OK—an exact ground analogue of the flight vehicle. Back on 6 July, at the meeting I conducted, the conceptual designers, testers, and designers had spoken ardently about the need for such an analogue for ground developmental testing of the interaction and cross-coupling of all the systems. German Semenov was the only one to object, believing that it was impossible to produce it within a reasonable timeframe.

Topol said that he had briefed S.P. on this during their discussion of the schedule. Turkov persuaded Korolev that it was unrealistic to produce yet another vehicle in advance of the flight models, and the issue was "closed." Later, indignant testers Yurasov and Dorofeyev wrote a lamenting report to S.P. requesting that an engineering model of the vehicle be included in the factory plan. Over the telephone S.P. explained to both of them that this was unrealistic, then he brought me into the loop and asked that I stop these harmful conversations.

According to the schedule, three flight vehicles were supposed to be assembled and handed over for testing at the KIS: two in December and a third in January. Two days later Korolev summoned Nikolay Kamanin to his office and they had a serious discussion about cosmonaut candidates for the first two Soyuz flights. The program stipulated that the very first Soyuzes must rendezvous and dock. One vehicle would carry one cosmonaut and the other—three. In the event of a successful docking, two of the cosmonauts would transfer through open space (we referred to it as "outside") from one vehicle to the other and return to the ground with a different crew.

After his meeting with Kamanin, Korolev confided that Gagarin was also in training as part of the Soyuz flight crew. I had the imprudence to remark that the cosmonauts still needed a good year to train for their flight on the Soyuz.

"It will take at least a year before we get a reliable *Igla* from Mnatsakanyan," I reasoned.

"Me and my big mouth," I thought when S.P. pounced on me, incensed by my prognosis.

"If that's the case, we'll shut down our work on the record-breaking Voskhod, but I won't allow the Soyuz deadlines to be disrupted anymore."

At the Air Force's insistence, Turkov had been ordered to produce a batch of Voskhods. One of them was supposed to "rob" the Americans of their flight duration record. (After Korolev's death, this mission was completely "forgotten," and we never seriously returned to the order of new Voskhods.)[22]

Through some of his "military" channels, Tsybin found out that Gagarin and a group of cosmonauts who had already flown had sent a letter to Brezhnev criticizing the minister of defense's space policy and proposing that all space matters be transferred from the control of the Rocket Forces to the Air Force.

---

22. Chertok is referring to one of a number of Voskhod missions planned for the 1965–66 period. One of them, Voskhod-3, would have carried two cosmonauts on a long-duration flight of about 18 days. See Siddiqi, *Challenge to Apollo*, pp. 507–511, 522–527.

I was surprised that S.P. received this news calmly and was not miffed by the fact that Gagarin had not informed Korolev about his appeal to Leonid Brezhnev. According to his information, the letter supposedly also contained a request to speed up the manufacture of the new Voskhods in view of the delay in Soyuz production.[23]

Korolev called in my comrades and me to report on the state of affairs with the rendezvous and docking system. Kalashnikov, Vilnitskiy, and Khazanov reported on the progress of operations on the docking assembly. Beforehand, they had reached an agreement with the factory not to complain about each other. Vilnitskiy and Khazanov could serve as a model for the creative alliance of a talented designer and a talented process engineer—a production organizer. Both of them reported in such a way that there was no one to chew out and nothing to chew him out for. They really completed a colossal amount of work in a short period of time. In December, it would be possible to begin "docking" in assembly shop No. 39 on the dynamic mockup, but in the meantime "static" testing was underway. There were a lot of glitches, but that was natural since it was a complex assembly. Korolev promised to drive out to the second territory in the next few days to have a look at the first "live docking node," as he called it.

"So, why don't we get a whole group together and go have a look at how things are going with Mnatsakanyan? Who in your department is overseeing Mnatsakanyan directly?"

I said that Boris Nevzorov and Nina Sapozhnikova were working on hardware and Igor Shmyglevskiy and Anatoliy Shiryayev were working on theory. These names didn't ring a bell with him, so I explained: "Rauschenbach and Legostayev are supervising all of them."

"Well, these guys I know."

With his inherent engineering intuition, Korolev sensed that spacecraft rendezvous control constituted one of the most complicated problems of cosmonautics. Solving this problem would require a set of elements, including the integrated development of a theoretical foundation that would optimize control methods, the creation of fundamentally new on-board systems, and the use of a ground-based flight monitoring and control network. He had already discussed this with Rauschenbach and me. Independently of us, he had consulted with Keldysh and Ryazanskiy.

---

23. Chertok probably refers to a letter written by cosmonauts Gagarin, Leonov, Belyayev, Titov, Nikolayev, and Bykovskiy. Dated 22 October 1965, the letter is reproduced in N. P. Kamanin, *Skrytyy kosmos: kniga vtoraya, 1964–1966 gg.* [*Hidden Space: Book Two*] (Moscow: Infortekst, 1997), pp. 245–248.

Korolev knew, in general terms, about the job assignments to solve this problem, but he obviously wanted to test his knowledge. During the past few months, he had not had the opportunity to delve deeply into the arcane matters of rendezvous control. In due course, he had read and approved the draft plan and reports where all the proposals were written with the requisite number of equations and diagrams, plus a description of the hardware. These were documents aimed at specialists who, unlike Korolev, were untroubled by orders for a new batch of Voskhods; the latest emergency hard-landing of the Ye-6 on the Moon; a cutback in funding for the N1 program; the emergence of Chelomey's competing UR-700 design; disruption of the housing construction schedule; dozens of tragedies and requests made at private consultations; preparations to receive the latest in a series of Party activists; difficult tests on the solid-fuel RT-1 missile at Kapustin Yar; and hundreds—not just a handful but literally hundreds—of problems ranging from minor to very major.[24] How interesting it would be to spend a whole day or two with these guys who were true believers in automatic docking!

Korolev was forced to trust Mishin, Rauschenbach, Lavrov, Bushuyev, Legostayev, Bashkin, and me. We made mistakes! He chewed out each of us when he saw us. But, after all, we hadn't betrayed or deceived him. "We're going through a learning process!" That's what he told Smirnov at the VPK when they wanted to make short work of me for the [Ye-6] lunar soft landing failure.[25] At that time, he strategized perfectly—all the ministers understood and supported him.

Lately, in meetings, S.P. had not concealed the fact that he was tired, very tired of this incessant race, dozens of phone calls per day over the Kremlin hot line, reports from two firing ranges over the high-frequency phone lines, complaints from the "leads" (or lead designers) about schedule disruptions, and a never-ending stream of letters stamped "secret," "top secret," and "high importance." Yes, if each letter addressed to him and each report had been read "cover to cover," he would have had to sit in his office 20 hours a day, 7 days a week.

Before our predictably "devastating" field trip to NII-648, Mnatsakanyan's institute, S.P. assembled a small group and requested that they bring him up to speed on the fundamental difficulties in the tasks of rendezvous, our final proposals for job assignments, and the current state of affairs.

---

24. The UR-700 was an alternative "super-booster" proposed by Vladimir Chelomey in the mid-1960s for use in a piloted lunar landing. The proposal was formally approved for development in 1967, although the effort was cancelled in 1969 before any significant work was carried out.

25. See Chapter 13 of this volume.

IN MODERN AND FUTURE COSMONAUTICS, THE PROBLEMS OF RENDEZVOUS AND DOCKING ARE SO CRUCIAL that I shall allow myself to very briefly write about their beginnings, bearing in mind that new generations of readers don't have time to search for special literature and there have simply been no popular scientific works published on this subject.

Design developments of the various methods for spacecraft rendezvous began back with Tikhonravov's group, which worked in NII-4. When he transferred to work at Korolev's OKB-1, these developments were continued exclusively on a theoretical basis.[26] However, we didn't start serious work aimed at producing a rendezvous control system at OKB-1 until early 1961 when Rauschenbach's group transferred there. Around the beginning of 1963, when discussing the problem of controlling the multi-launch Soyuz complex for the circumlunar flight, we reached the conclusion that rendezvous control must be an integral part and one of the modes of the overall motion control system. Therefore, we must be the head developers of the entire rendezvous problem. In addition to the subcontracting organizations that were developing gyroscopic and electronic optical instruments for orientation and navigation per our orders, we needed to find people willing to develop something that we clearly were not capable of developing ourselves: radio systems or optical systems measuring parameters relative to the motion of two rendezvousing spacecraft.

From the very start it was clear that the on-board equipment could be simplified if we were to include an already existing ground complex and its ballistics centers in the rendezvous control process. In this case, the distance and relative velocity between the two spacecraft inserted in orbit at different times, and perhaps, from different launch sites, was to be reduced to the minimum value with the help of the ground Command and Measurement Complex (KIK), using the tried and true principles of orbital correction via commands from the "ground." This minimum value was determined by the KIK measurement errors and the on-board systems' errors during correction sessions. Three groups of ballistics experts—Lavrov and Appazov at OKB-1, Elyasberg and Yastrebov at NII-4, and Okhotsimskiy and Platonov at OPM— separately analyzed the problem of rendezvous based on commands from the "ground," and then they got together and came to a mutual conclusion. Using this method it was possible for the spacecraft to approach to a distance of 20 to 30 kilometers—depending on how lucky you were. The spacecraft would be at this distance for a short period of time and then once again they would

---

26. Tikhonravov transferred from NII-4 to OKB-1 on 1 November 1956 while a few of his group members transferred in the following months.

drift apart. To prevent this from happening, as soon as the far approach phase was completed, the on-board autonomous system, which was independent of ground-based measurements and commands, would come into play. "Radio lock-on" would occur when the autonomous systems detected one another at a range of at least 20 to 30 kilometers—the spacecraft would "catch sight" of each other and then execute rendezvous and final approach autonomously, without "ground" participation. Here, one of the vehicles is the active one: it moves toward the passive one, which must turn with the open base of the cone—the drogue of the passive part of the docking assembly—toward the probe of the active vehicle.

Two types of control were possible during the autonomous rendezvous and final approach segments. Purely automatic, without human intervention, and manual, during which the cosmonaut, using the system's measurement results and the cosmonaut's own observations in the optical sights, uses the controls to fire the attitude control engines and docking and attitude control engines and brings the active vehicle to the point of "mating." We made it our primary task for Soyuzes to execute rendezvous in the automatic, unpiloted mode.

It was obvious that the 11K active refueling vehicles would only dock with the 9K booster unit automatically. The 7K called for automatic mode, and for reliability, manual mode as well. Despite the suspension of operations on the Soyuz circumlunar flight, the absolute requirement for automatic rendezvous and final approach remained in effect. We already knew that the version the Americans had adopted only worked with human involvement. We did not copy the Americans, and this would prove to be a future strong point of our cosmonautics.

Heated arguments erupted over the methods for guiding the active vehicle to the passive one. The thing was that, back then, all sorts of air defense missile guidance systems used "dog leg" pursuit paths, when the pursuer follows the target the whole time and runs after it at a constant or increasing velocity.

We had the capability to use guidance methods relying on the laws of celestial mechanics, i.e., the natural motion of the vehicles in Earth's gravitational field. In this case, it is assumed that based on the results of the ground measurements, data about the orbital motion of both vehicles relative to ground coordinates is loaded into the "memory" of the active vehicle. Algorithms making it possible to predict the relative motion and to compute a solution resulting in rendezvous need to be loaded into the active vehicle's system. As I mentioned, this method results in large errors. Correction is needed based on the results of reciprocal measurements of the relative range and velocity provided by the on-board radio or optical system. To correct natural orbital motion, it is necessary to precisely calculate the magnitude, direction, and time to fire a correction burn. This sort of correcting effect based on the prognoses of theoreticians required no more than two firings of the active vehicle's approach and correction engine unit (SKDU) before it entered the final approach zone, which was estimated to be 50 to 200 meters. However, the very determination of the parameters of the control actions for this type of two-burn rendezvous maneuver required the solution of a complex problem, which assumed that each of the correction burns was calculated by the on-board digital computer such that the vehicle would end up in the final approach zone after a certain period of free flight. This sort of rendezvous required the very close collaboration of control experts with ballistics experts—they were rightfully considered the masters of free flight in near-Earth orbits. They also came up with the name for this, seemingly, natural rendezvous—the "free trajectory" method.

This self-evident method promised a very economic consumption of the on-board SKDU and attitude control system fuel margin, but required calculations that could not be executed on board without an on-board digital computer. In 1964 it was already clear to us that we would not have a computer that was reliable and suitable for similar tasks in the next two to three years. I still hope to write about the tragic history of the development of a space computer in the USSR.

We already knew that, on their Gemini and future Apollo spacecraft, the Americans were using a high-speed computer, making it possible to solve rendezvous tasks semi-automatically with the required involvement of an astronaut. Though they already had reliable on-board computers, they did not risk using just automatic rendezvous based on the "free trajectories" method. Having "artificial intelligence" at their disposal, the American rendezvous specialists combined it with the natural intelligence of a well-trained astronaut and thus solved the problem in tandem. We had to solve the problem without having artificial or natural intelligence!

We had to set aside this attractive method until the time was right and instead teach the active vehicle to act like a hunting dog, which, even after rigorous training, will never run to the predicted convergence point beyond the scampering rabbit. Now it wasn't the ballistics specialists, but Rauschenbach's dynamics theoreticians—Legostayev, Shmyglevskiy, Shiryayev, and Yermilov who, using the experience of homing missile systems, coupled it with ballistic space flight and proposed the "line-of-sight rendezvous" method, or simply the "parallel rendezvous" method.

With this method, the role and responsibility of the on-board relative motion parameters measurement system increased. This required the precise measurement of several parameters: the relative range between the vehicles, the rate of change of this distance, i.e., the rate of approach or back-off, and the angular rate of rotation of the line of sight, the latter of which coincides with the straight line connecting the two vehicles in space. This information is fed in the approved form to the analog electronic instruments, which do not have an on-board digital computer's ability to predict the relative motion and foresee the consequences of corrections during a prolonged period of time. Therefore, the on-board instruments controlling rendezvous will turn the spacecraft throughout the rendezvous process and issue commands for small, but frequent, correction burns—now one to accelerate, then one to brake—in order to bring the active vehicle inside the calculated corridor of relative rate values, which change depending on the range. During the rendezvous process, the active vehicle not only accelerates or brakes, but must also make lateral corrections to reduce the line of sight angular spin rate to zero.

As the range decreases, the relative rate of approach must also decrease. In final approach mode, to keep from breaking the docking assembly, rendezvous had to take place at a rate on the order of 5 to 7 centimeters per second and the relative angular rate had to be reduced to a minimum.

To convert the ideas of theoreticians—who were divorced from real possibilities—into instruments, we required experienced specialists who also possessed the common sense of a real equipment developer. We were fortunate. Boris Nevzorov, who had arrived along with Grabin's staff, was brought in to work on the rendezvous system. He possessed precisely those qualities that made it possible to find a compromise between the brash theoreticians, the requirements for the radio measurement systems' parameters, and the capabilities of the developers of our electronics instruments.

We had barely "revealed" (in "top secret" letters with references to the VPK decision) our need to develop on-board relative motion parameters measurement systems, when we received four mutually exclusive proposals almost simultaneously.

NII-158 in Leningrad was the first to come forward. That institute specialized in aircraft radar, in particular, for homing in on fighter jet targets flying under cover of darkness. NII Chief Engineer Veniamin Smirnov demonstrated to me that they were the only institute in the Soviet Union that had the experience to produce the system we needed. Alas! The 200 kilograms that they reckoned for the active vehicle were more than we could lift off. And then, they said they needed at least three years. We needed delivery in a year!

The second to respond to our appeal was Yevgeniy Kondaurov, who worked at NII-648 on the problem of missile homing systems. He and a group of comrades had received the Lenin Prize for this work. Kondaurov saw a real opportunity to implement many of his ideas in space. NII-648 Chief Armen Sergeyevich Mnatsakanyan was less than enthusiastic with the proposal to use these developments for spacecraft. But when Mnatsakanyan heard that we were ready to hand over this assignment to Aleksey Bogomolov at OKB MEI, he put away his misgivings and gave Kondaurov the opportunity to get started with the project per our requirements for development. Bogomolov had his own "rendezvous enthusiast"—Petr Kriss. He decided to use all the potential of pure radio engineering and began the original design of a "phase" system.

Finally, at his own initiative, David Khorol, the chief designer of heat-seeking and optical homing heads at TsKB Geofizika, came forward. He proposed a system based on laser optics.

Aside from the initiative they had shown, for any of the four organizations to work, a VPK decision was needed. We couldn't put off our final selection of the system very long. The design layouts, drawings, schematics, and instruments could be produced within an acceptable period of time only for one option. As the chief curator of these systems, Nevzorov rather convincingly argued that NII-158's system would not pass in terms of mass, overall dimensions, and power consumption, while Khorol's system could only work at short ranges; and even so, for the initial search, it would require scanning radio technology.

Mnatsakanyan and Bogomolov remained. We had backers of both men. For the Soyuz we decided in favor of Mnatsakanyan. We promised Bogomolov a place on the lunar spacecraft if he were to beat out Mnatsakanyan in terms of mass, overall dimensions, and accuracy of measurements.[27]

Thus, the *Igla* (Needle) radio measurement system—which during 12 years caused us countless hardships, joys, and a heap of trouble accompanied by reprimands from the minister and being called out "on the carpet" at the Kremlin—made its name in the history of cosmonautics. In this history, *Igla* is inextricably intertwined with the name of Mnatsakanyan. It helped establish an institute, but it was also one of the causes of Mnatsakanyan's downfall.[28] The first successes in rendezvous using *Igla* significantly raised the institute's prestige and the prestige of its director. Unfortunately, spacecraft rendezvous did not bring Kondaurov, the primary author of the development, and the institute director closer to one another.

---

27. Bogomolov's OKB MEI later developed the *Kontakt* (Contact) rendezvous and docking system slated for the Soviet piloted lunar landing project.

28. Armen Sergeyevich Mnatsakanyan (1918–92) was fired from his post in 1977 after a spate of docking failures involving the Soyuz spacecraft. From 1953 to his dismissal, he had served as chief designer of NII-648, which is now known as the Scientific-Research Institute of Precision Instruments (*Nauchno-issledovatelskiy insitut tochnykh priborov*, NII TP).

Igla-1 was developed for the active spacecraft, and *Igla-2*—for the passive. The first sets' guaranteed range at the beginning of relative measurements was 25 to 30 kilometers. The active *Igla* required the installation of three antenna arrays on the Soyuz: scanning antennas, which operated in the preliminary rendezvous partner search mode; a gyro-stabilized antenna, which, after detecting the partner, set up the automatic tracking mode using the passive *Igla*'s responder and tracked it as its own spacecraft executed attitude maneuvers; and special antennas for the final approach mode. After the vehicle's orbital insertion, five different antennas were deployed for the *Igla* alone. And in all, the first Soyuzes had so much varied radio technology on board that they required 20 antennas.

"Instead of a piloted spacecraft, you've made a veritable antenna carrier for me," seethed Korolev when we showed him a poster of a general view of the Soyuz, which, out of all the exterior accessories on the hull, deliberately depicted only the antennas. Aside from the number of antennas, S.P. was particularly unhappy with the complicated structure that we called the "roof." This "roof" protected the *Igla*'s receiving antennas against radiation reflected from the spacecraft hull.

Everything that the *Igla* antennas transmitted and received passed into the transmitters, receivers, amplifiers, signal processing units, and devices communicating with our own calculation units and telemetry units—instruments containing sophisticated electronics. After establishing radio contact between the active and passive vehicles, a closed autonomous loop controlling the two vehicles was formed. The ultimate objective of this loop was to carry out rendezvous until receipt of the "Contact" signal. Upon receiving this flag, *Igla* shut down.

S.P. listened very attentively and with great interest to our rather confusing reports about the system's operating principles. Questions tended to deal more with the status of specific issues, particularly the manufacture and developmental testing of equipment. "White TASS" reported that *Gemini VI*, carrying two astronauts, was set to launch on 25 October, further stirring up S.P.'s interest in the rendezvous problem.[29] It was announced that the primary objective of the flight was to attempt to dock with the Agena Target Vehicle. If the American experiment were successful, then it would be a heavy blow to our prestige in space. The next day Korolev telephoned me in a good mood.

---

29. White TASS—the equivalent of secret news—included very candid accounts of domestic and international events prepared only for government ministries and Communist Party offices.

"The Americans' docking attempt failed. Something happened up there with their Agena. It didn't go on orbit. Warn Armen that we are coming over to congratulate him in such a way that he'll never recover."

Rauschenbach, Legostayev, Bashkin, Nevzorov, and I drove out with Korolev to Otradnoye, where NII-648 was located. The ZiS automobile that had been passed from Grabin to Korolev was fully "loaded."[30] Indeed, this visit by Korolev and his entourage went down in NII-648 history as an event that sharply accelerated the development of *Igla*. Interrupting one another, Mnatsakanyan and Kondaurov gave their explanations. The latter repeatedly tried to say that the deadlines were being disrupted because the institute's management had started this project half-heartedly. Things were also bad in production. The factory was swamped with other orders.

We went from laboratory to laboratory. In each of them, individual panels that were ready for demonstrations were lying on tables, stands, and workbenches; mockups assembled for experiments were hooked up to oscilloscopes. I wasn't so much listening to the reports and explanations as I was admiring Korolev's artistry. In one of the laboratories, after picking up an open electronic unit from the table, he asked: "Is this just for show or is it already a finished unit?"

"It's a mockup that we use to optimize parameters, revise and correct drawings, and then we send them to production to produce the flight model," explained Kondaurov.

"So, when will you give me the finished instrument for vehicle assembly?"

"In about five or six months."

"So that's it! I'll be going now. I'm leaving you your comrades, who have already been deceiving me for six months, assuring me that *Igla* would be completely ready in just a little while. Supposedly, only these "evildoers" from military acceptance are finding fault, and that's why production is lagging. If they were deceiving me, I will take care of them myself, and if you really said that, I won't work with you anymore! I need a system no later than December. Either you work at our pace, or we will find others. The Americans are finding a simpler solution to the problem: an astronaut is controlling rendezvous and they don't need such a complex system. Maybe we can do without the *Igla* altogether? Aleksey Bogomolov is promising us a simple system for operation with a human being immediately."

---

30. ZiS stood for *Zavod imeni Stalina* (Stalin Factory) which produced this model of automobile.

I was about to open my mouth to utter a quip about "deception," but S.P. gave me such a look that none of us even dreamed of interfering. There were several similar excoriating performances in various laboratories. Korolev wanted to really heat up the situation in such a way that the people who really felt it were not the handful of managers, but rather the many dozens of engineers who, with undisguised interest and excitement, were seeing and hearing the legendary chief designer for the first time.

The last stop on our tour was the antenna laboratory. Here, laboratory chief Adolf Shpuntov stole the show. He spoke so persuasively about the technology for optimizing antennas in an anechoic chamber and the miraculous properties of the gyro-stabilized parabolic antenna, that all previous doubts were assuaged. S.P. lightened up and simply asked everyone to think things through one more time, to review the deadlines, and in one week's time show him the schedule (coordinated with me) for the delivery of the first three sets.

Attempting to clear the way for the Soyuz, Korolev gave Turkov the order to stop work on the new Voskhods and the experiments with artificial gravity. Around the same time, the doctors called him in again to the polyclinic on Granovskiy Street. Once again he had to fly out to Tyura-Tam, for the last Venus and Moon launches of his life.

IN LATE NOVEMBER, AT THE MINISTRY COLLEGIUM, AN ATTEMPT WAS MADE TO DEAL A CRUSHING BLOW TO OKB-1'S STYLE OF OPERATION as punishment for the disruption of programs, primarily, for the disruption of the Soyuz program. Not wishing to waste time delving into our shortcomings, Korolev very deftly beat off their attacks, after showing them that we failed to receive deliveries from our primary subcontractors. The vehicles' hulls were empty; assembly could not move forward; the schedules for experimental operations had also been derailed—there had been no deliveries by way of "cooperation," and the ministry and VPK were berating OKB-1 rather than the subcontractors for their poorly organized operations. Korolev managed to fend off the first attack on the Soyuz program from the "powers that be" while he was still alive. But the Americans were once again pouring salt into our open wounds.

On 4 December 1965, *Gemini VII* was launched with two astronauts on board (Frank Borman and James Lovell). On 15 December, *Gemini VI* was inserted into orbit, also carrying two astronauts (Walter Schirra and Thomas Stafford). The two American spacecraft performed an in-orbit rendezvous. They approached one another until 50 feet separated them. They did not risk approaching any closer. And there was no need. After all, they had no docking assemblies. We were once again blindsided by the duration of the flight—*Gemini VII* stayed in orbit just less than 14 days, and, in addition, the Americans demonstrated to the world their capability for rendezvous in space.

Back in 1965, I absolutely could not have imagined that *Gemini VI* astronaut and future Lieutenant General Thomas P. Stafford would write a very warm foreword to the American edition of my memoirs *Rockets and People*, the first English-language version of which came out in 2005. Not quite 40 years after his first space flight, in the office of the general designer of Rocket-Space Corporation Energiya, I thanked General Stafford for this foreword.

Our plans for piloted flights were shifted to the following year. The ministry and VPK, without having delved very deeply into the state of affairs at our production facility, believed that we would manage to execute a flight lasting 18 to 20 days in early March.[31] Only Deputy Minister Viktor Litvinov, who had been at our production facility everyday, had a good understanding of the real situation. But he didn't dare report the whole truth to the brass, either.

Tyulin, who was then the first deputy minister directly responsible for space, took advantage of Korolev's absence—he was in the hospital—and called Bushuyev and me into his office to discuss the schedule of piloted flights for the next two years. He intended to propose launching two Voskhods and two pairs of Soyuz vehicles, for a total of four, in 1966. I opposed the Voskhod launches. Regarding the Soyuzes, I proposed that the first pair be launched to perform an experiment for the unpiloted automatic docking program, and that the second pair be piloted. Bushuyev proposed adding one more pair of piloted Soyuzes to the plan. We argued passionately and failed to reach an agreement.

Korolev's unexpected death was a blow that knocked us flat for the entire month of January. Once we had recovered, we once again started up a round-the-clock work regime at the factory on Soyuz production. In January 1966, the minister personally approved the new schedule, replacing Korolev's long-since disrupted schedule from 28 August of the previous year. The new schedule was painstakingly prepared with Litvinov's participation, but it too, in keeping with established tradition, was loaded with deadlines that could not be met.

In early February, the OKB-1 Party committee decided to look into the status of operations on the Soyuzes. Party Committee Secretary Anatoliy Tishkin received the necessary instructions in the Central Committee office. He formed a commission to prepare a decision and proposed that Turkov and I give a progress report.

---

31. The mission that Chertok mentions would have been in the Voskhod program.

The Party committees of large organizations had real clout. They could "recommend" to an enterprise's administration that the management of one department or another or shop be "strengthened." This meant the end of one supervisor's career and the beginning of another's. The decisions of the Party committee were supposed to be respected. Lower Party organizations made sure they were carried out. Their independence manifested itself primarily in the organization of political propaganda, Party ideological circles, and the monitoring of the public activities of Party members. They had an impact on the production process itself in several ways: they kept track of the awarding of prizes and other material benefits, they helped the department supervisor maintain labor discipline, they organized "socialist competition," and they took part in the investigation of labor conflicts. If the secretary of the lower Party organization turned out to be a decent person, then his activity promoted the cohesion of the workforce and had a positive impact on the fulfillment of the missions.

At the Party committee meeting, first Turkov and then I gave the situation reports. Turkov stressed the disruptions of deliveries from the subcontractors, and I spoke primarily about our internal arrears and the lag in experimental operations. The primary fire of Party criticism was directed to the shortcomings of the factory's work.

The commission which prepared the meeting did a great job and neither the shop foremen nor the "individual managers," who had exhibited "irresponsibility and had underestimated the seriousness and complexity of the operations," escaped their attention. My staffers were no exception—the commission enumerated the personal transgressions of Rauschenbach, Bashkin, Chizhikov, Shustov, and Pogosyants, who had "failed to conduct a systematic analysis of the work of the subcontractors responsible for deliveries of instrumentation and assemblies, and had not been aggressive and persistent enough in making sure the deliveries were made . . . ."

The Party committee decreed:

. . . to consider the work on object 11F615 to be among the most crucial and critical, and to focus the attention of the Party organizations of the shops and departments on the need to mobilize the workforce to the fullest extent . . .

To warn comrades R. A. Turkov, V. M. Klyucharev, P. V. Tsybin, V. A. Kalashnikov, G. Ya. Semyonov, I. B. Khazanov, and V. D. Vachnadze that if they do not guarantee delivery of the back orders by 15 February for the production of the experimental objects and units according to the schedule approved by the minister, then the Party will hold them accountable . . . .

Comrades B. Ye. Chertok, P. V. Tsybin, Ya. I. Tregub, V. A. Kalashnikov, and K. D. Bushuyev must take steps to ensure that the necessary experimental work is conducted before they depart for testing on objects Nos. 1 and 2 and Nos. 3 and 4 . . . .

Comrades B. Ye. Chertok, Ya. I. Tregub, I. Ye. Yurasov, and Ye. V. Shabarov must take the necessary steps to ensure the timely preparation for operations with the object at the military installation.

I have intentionally cited such extensive excerpts from the Party committee's resolution so that the reader might understand that, in those days, the Party organizations in industry were not only involved with policy, ideology, and the "struggle against nonconformist thought," but tried to get involved in technology and production engineering. Wielding real authority over people who were Party members, they had the opportunity to affect the production process. With few exceptions, every chief designer was a Party member. It was far more dangerous to receive a Party reprimand than a reprimand ordered by the head of an enterprise or even a minister.

The Communist Party of the Soviet Union was a party of power. This was a party that actively meddled in the production process not only from the top—through the Central Committee or Politburo—but also from the bottom. Things did not always turn out as planned, but as a rule, they had the best of intentions. The Party attempted to encompass all aspects of a person's life with its ideological influence. Any job was supposed to be a "thing of virtue, honor, and heroism," not for the sake of personal prosperity, but to strengthen the power of the state. "So long as our motherland lives, there are no other cares"—these words succinctly and rather accurately reflected the spirit of a myriad of Party propaganda campaigns. Any deviation from the Party line was punished mercilessly. The Party allowed no liberalism within its ranks. In the course of history, there has not been and there is no other Party like it. A possible exception is the Communist Party of the People's Republic of China, but for a time it studied at the feet of its "big brother."

The Communist Party of China learned the bitter mistakes of the Communist Party of the Soviet Union, corrected its own mistakes, and during the past 15 years, has achieved impressive success in all fields of science and the national economy. Judging by its rates of development, Chinese rocket and space technology will overtake the Russian space program in 10 to 12 years; and perhaps it will overtake the American program as well. After a series of errors, the Chinese Communists exhibited greater wisdom than their Russian "big brothers."

Objective reality was stronger than a minister's appeals or a Party directive. Deadlines, even the minister's timetables, were disrupted at various points for one or two months, and for the first flight vehicle—for five months.

Testing of the first Soyuz flight model began at the monitoring and testing station (KIS) on 12 May 1966 and continued for four months . . . instead of the planned 13 days!

We made it a point not to release the first flight-ready Soyuz from the KIS until we had eliminated all the defects and glitches we had discovered. This time, we were completely convinced of the validity of Murphy's Law, published in one of the American manuals for guided missile specialists: "The actual time required to solve a problem always turns out to be twice the time calculated in a reasonable preliminary estimate."[32]

The factory tests ended with results that were not as clean as we had wanted. KIS chief Anatoliy Andrikanis wrote in the memorandum and logbook entry that "Object 11F615 No. 1 was removed from the KIS on 30 August 1966 to perform stowage operations in accordance with the review team's decision." The review team's patience ran out. Indeed, 2,000 glitches had been discovered during the testing process. Decisions needed to be made regarding each of these glitches: whether to replace instruments, make cable and structural modifications, introduce changes to the test procedures, update instructions, etc., etc. It is interesting that, of the 2,000 glitches, 45 percent were "closed" by amending the documentation; 35 percent required modifications of on-board instruments and cables; and 20 percent required modifications of ground testing equipment.

Object 11F615 was removed from the testing facility and sent for "stowage and assembly" operations before being shipped to the firing range with a dozen glitches that still hadn't been completely sorted out. They decided to examine them thoroughly at the engineering facility.

Testing lasted so long that many subcontractors, having discovered their own wrongdoings, used the time to modify the instruments before delivering them to us. The 7K-OK was the first spacecraft whose factory tests were notable for a substantially greater degree of depth, and the process of troubleshooting identified defects was made easier thanks to the use of universal testing complex 11N6110.

On the Soyuzes, for the first time we fully appreciated the advantages of this ground-testing complex that we had developed. The *Kardan* unit was another completely new ground-testing stand. It was a massive structure that accommodated the *Igla*, its gyro-stabilizing antenna, and the vehicle's main gyroscopic instruments. The platform holding the hardware was installed on the inner ring of the gimbal mount. The *Kardan*'s 3 degrees of freedom made it possible to simulate the behavior of *Igla* and the attitude control system and their phasing and operation while simulating the various positions of the active vehicle. We were still unable to spin and rock the entire assembled vehicle on the KIS.

---

32. The most common and basic formulation of Murphy's Law goes, "if anything can go wrong, it will." Obviously there are infinite variations on the basic form.

Only those sensitive elements that responded in flight to changes in angles and angular rates were placed on the *Kardan*. At the engineering facility at Site No. 31 and later at *Dvoyka* (Site No. 2), we built large anechoic chambers enabling us to roll in the entire vehicle in one unit. There, we checked out operation of the whole rendezvous and attitude control system considerably more thoroughly than when we had used the *Kardan*. But this was only at the firing range.

IN THE HISTORY OF OUR OKB-1, 1966 WAS A TIME OF TROUBLE.[33] IN MARCH, ALL THE DEFENSE ENTERPRISES UNDERWENT ORGANIZATIONAL CHANGES, the upshot of which was that their names and post office box numbers changed. OKB-1 was renamed the Central Design Bureau of Experimental Machine Building (TsKBEM). Our customary post office box No. 651 was replaced with P.O. Box No. V-2572. The factory "where comrade Turkov was director" had a classified name—Factory No. 88 (V-8711). Now it was publicly referred to as the Factory of Experimental Machine Building (ZEM). Our neighbors in Podlipki also got new names: NII-88 became the Central Scientific-Research Institute of Machine Building (TsNIIMash) and Isayev's OKB-2 received the name Design Bureau of Chemical Machine Building (KB Khimmash).[34]

In theory, these new names—without changing location—were supposed to confuse enemy intelligence services. A lot of jokes and funny stories cropped up in this regard. Earlier, when someone who worked at a classified enterprise was in contact with the outside world, he or she would say or write that they worked, for example, at P.O. Box 651—and no other names were mentioned. It was permissible to use the name OKB-1 only in secret correspondence. Now they had decided for once and for all to trip up the CIA station chiefs. It was forbidden to use the name P.O. Box V-2572 in documents and conversations classified any lower than secret, while TsKBEM, on the other hand, could be mentioned anywhere—even at medical clinics, housing offices, and police departments.

After the death of Korolev, the "powers that be" procrastinated with their decision to name Mishin chief designer and enterprise chief. It wasn't until 11 May [1966] that the minister's order came out naming Mishin our chief designer and TsKBEM chief. Mishin inherited the R-9A combat missile, which had been put into service, the RT-2, which had begun flight-testing, the colossal N1 launch vehicle, which had been in the throes of birth in the factories, and new piloted vehicles: 7K-OK, 7K-L1, and the L3 complex.

---

33. This is an indirect reference to the "Time of Troubles" (*Smutnoye vremya*), the period in Russian history from 1598 to 1613 characterized by war, famine, and political instability.

34. TsNIIMash—*Tsentralnyy nauchno-issledovatelskiy institut mashinostroyeniya*; KB Khimmash—*Konstruktorskoye byuro khimicheskogo mashinostroyeniya*.

It took three years to manufacture the first 7K-OK. True, they really weren't working on "lead designer Topol's article 11F615 (7K-OK)" until the second half of 1965.[35] They had hoped that vehicle 7K-L1 would proceed more rapidly, but it was simply impossible for it to appear before 1966 drew to a close.

With respect to the L3, this future lunar complex consisted in turn of the Lunar Orbital Craft (LOK), the Lunar (landing) Craft (LK), and rocket Blocks G and D.[36] There were still no working drawings for the L3.

Dmitriy Kozlov's and Mikhail Reshetnev's branches didn't want to remain under the double oppression of MOM and their own alma mater.[37] There was nowhere to hide from MOM, but it might be possible to get out from under Mishin without leaving completely; then at least they could gain maximum independence. This was a natural process and it made no sense to keep the Kuybyshev and Krasnoyarsk branches on a short leash. Kozlov's young and spirited staff, which had achieved great success developing spy satellites, came up with the idea of producing a piloted military spacecraft. The design was assigned the designation 7K-VI, i.e., a 7K from the same Soyuz complex produced for military activity.[38] The patrons of the Kuybyshev TsSKB in the Ministry of Defense hoped that it would be considerably easier to come to terms with Kozlov than with Mishin regarding the development of a piloted military spacecraft.[39]

This was already the fourth design of a Soviet piloted vehicle. But proposals for piloted vehicles in the Soviet Union did not end there. Despite the decision of the Communist Party Central Committee and the Council of Ministers that our 7K-L1 (assisted by launch vehicle UR-500K) would execute the circumlunar flight, Chelomey did not suspend operations at OKB-52 on his own version of a piloted complex. He came up with the idea for the Almaz (Diamond) military reconnaissance orbital station. Its own reentry vehicle was developed to return the crew from the station.

---

35. Full-scale work on the 7K-OK model of the Soyuz, i.e., Earth-orbital version, did not begin in earnest until 1965 because earlier work was focused on the circumlunar version.

36. LOK—*Lunnyy orbitalnyy korabl*; LK—*Lunnyy korabl.*

37. Chertok is referring to the fact that both Kozlov and Reshetnev had started work as branches of Korolev's OKB-1 design bureau, the former working on reconnaissance satellites and launch vehicles and the latter on communications satellites. They attained "full" independence from the center in 1974 and 1961 respectively.

38. The "VI" stood for *voyennyy issledovatel* (military researcher). For more on the 7K-VI, see Konstantin Lantratov, "Soyuz-Based Manned Reconnaissance Spacecraft," *Quest* 6 no. 1 (Spring 1998): 5–21.

39. After Kozlov's branch formally separated from TsKBEM in 1974, it was named TsSKB— *Tsentralnoye spetsializirovannoye konstruktorskoye byuro* (Central Specialized Design Bureau).

All of this was taking place at a time when the Americans had successfully completed their Gemini program, and a year later they were supposed to begin flights on the Apollo, the United States' only piloted vehicle at that time.

Under these conditions we, the managers of OKB-1/TsKBEM, resolved to concentrate maximum efforts on the 7K-OK vehicles with automatic docking, which were moving forward in terms of readiness. We wanted to prove their reliability during the first flights, and then refine them, having made them the sole multipurpose vehicle for any scientific and military objectives that required the presence of a human being in near-Earth space. Korolev assumed that the 7K-OK vehicle would be the basis for the Lunar Orbital Craft of the N1-L3 complex. Mishin continued this line, and we stood behind him on this. We regarded the second 7K-L1 piloted complex for the circumlunar flight as a minor thing—a concession to space ambitions.

The first months after he assumed his leadership role, Mishin often consulted with Bushuyev, Tregub, Feoktistov, me, and other engineering, policy, and personnel placement leaders. When Korolev was still alive, Mishin had no need to sort out all the details in the massive amount of technical problems of piloted cosmonautics. Bushuyev, Tsybin, Feoktistov, Rauschenbach, and I, as well as some "narrower" specialists, were in constant communication with all the chief designers. All the fundamental issues and, above all, reliability and safety problems were under the vigilant control of S.P. himself. One of the problems that Korolev devoted a great deal of time to was the tricky relationship with the Air Force when it came to cosmonaut selection, training, and crew assignments. He ardently supported the initiative to create, within the Ministry of Health system, a special Institute of Biomedical Problems that would counteract the monopoly held by the military's Institute of Aviation Medicine. This was the source of many conflicts between Korolev and the Air Force command, especially General Kamanin.

Mishin would have to shoulder the entire burden that Korolev had carried, or share it with his own deputies. I hoped that my old camaraderie with Mishin would enable me to give him advice on matters that did not lend themselves to being solved by engineering methods. In particular, I advised him to entrust "human" problems completely to Bushuyev, Tsybin, and Pravetskiy—the latter being the chief medical officer of the Third Main Directorate, who had transferred to work with us.[40] They would find common ground more

---

40. Vladimir Nikolayevich Pravetskiy (1921–80) served as chief of the Third Main Directorate in the Ministry of Health until his transfer to TsKBEM. Earlier, during the 1950s, Pravetskiy had worked in the Soviet nuclear program.

quickly with Yazdovskiy, Gazenko, Karpov, and other military medical officers. Furthermore, the famous test pilot Sergey Anokhin was now working with us to train the cosmonauts. It would be easier for him to come to terms with the other pilot hero—Cosmonaut Training Center Chief Nikolay Kuznetsov.[41] At first, Mishin agreed with me; but by the end of 1966, he had been so distracted by conflicts over training and crew assignment issues that his relationship with Kamanin and Rudenko became very strained. They had disagreed over all sorts of inconsequential issues, taking a lot of time away from technical matters that had required the chief designer's meticulous attention.

In many respects, the job was more difficult for Mishin than it had been for Korolev. Korolev's authority, his gruff nature, and notoriously predictable reaction to guidelines and all manner of compliance inspections acted as a deterrent against nitpicking by government agency bureaucrats. Now they had the opportunity to prove how many sins TsKBEM had committed. The origin of these sins dated back to Korolev's administration, but nobody would speak ill of the dead. Admonitions and criticism from the "brass" usually contained the implication that "if Korolev were around, it would have been different; but Vasiliy Pavlovich is acting up and doesn't listen to good advice."

Kerim Kerimov was named chairman of the State Commission on Soyuz flight tests. The two of us had been "broken in" together on the Molniya State Commissions. The Molniya launches were still going on, and he would chair both State Commissions. After Kerimov's commission, approval was granted to establish the State Commission for the 7K-L1 circumlunar flight. Tyulin was named its chairman. Thus, right from the beginning, Mishin found himself with two State Commission chairmen that we had started working with back in Bleicherode.

Nevertheless, during one of my meetings, Tyulin, who was now a first deputy minister [of the MOM], let me know that, although we were old friends of his and we knew each other inside out, he would cut us no slack, neither as deputy minister, nor as State Commission chairman. In particular, he would give no quarter to my friend and boss Vasya Mishin. Tyulin complained that for the time being, his dealings with the minister were complicated.

"He [Afanasyev] is an experienced production engineer and manufacturer, but he's not up to speed yet on what we do.[42] He understands your factory

---

41. Nikolay Fedorovich Kuznetsov (1916–2000) served as director of the Cosmonaut Training Center between 1963 and 1972.

42. During his early career, from 1946 to 1957, Afanasyev had served in a number of factory management positions in a major defense industry ministry where he had been responsible for production processes.

difficulties just fine, but he doesn't immediately grasp why a vehicle has to undergo months of testing at the KIS after it's been manufactured."

Actually, this was difficult to explain to a new person. Two or three months of testing at the KIS, a month for "packing and shipping," and another two months of round-the-clock rush jobs at the cosmodrome engineering facility—in all it was six months after the official end of the production cycle. In fact, the explanation for all of that was simple. Before the first flight model was manufactured, a precise engineering analogue would have had to be manufactured. It would be used to practice the test procedures; to find and correct all the design errors, circuit tie-ins, and cross-coupling; to modify test equipment and update the testing instructions; to make all the repairs to the flight complex equipment, cables, and instruments; and only then to set about testing it.

We accepted aviation technology without the proper analysis. Usually the first developmental prototype from the assembly shop rolled out onto the airfield, and after a few days of taxiing and takeoff approaches, it made its first flight. After each flight, some modifications and improvements were made based on the comments of the test pilots. But a spacecraft could not make a takeoff approach and return for modifications. The flight of each spacecraft was its first and last. We had to foresee, anticipate, and presume all possible nominal and off-nominal situations in a spaceflight and make sure that the complex combination of spacecraft systems would enable the crew to safely return to the ground.

The tests on 7K-OK No. 1 at the KIS lasted 112 days or 2,240 work hours. During testing, 2,123 defects were discovered in instruments, cables, and documentation, and 897 modifications were required. Overall, it took 600 hours to investigate the glitches and correct the defects. To correct errors caused by parameter discrepancy, 25 adapter cables connecting interfacing systems had to be remanufactured. We called them "pears." They really looked like this fruit. Smart alecks contended that Karpov's department was second to none when it came to "knocking down pears."[43]

Despite this titanic work conducted in the KIS, at the cosmodrome engineering facility, after two months of preparation for the first flight, another 300 new defects and 100 documentation glitches turned up. Ninety percent of the defects and glitches were inherent to the Soyuz model, and not just to this specific spacecraft. They should have been first discovered and eliminated on the analogue that we had not produced. Then, from the thousands of glitches

---

43. This is a Russian expression equivalent to "twiddling one's thumbs."

that occurred on the first 7K-OK flight model at the KIS, we would have had no more than one hundred that pertained specifically to that particular complex.

But . . . , but . . . , if we imposed a testing cycle like the one that I, and later Yurasov, Dorofeyev, and Shabarov had advocated back in early 1965 under Korolev, then the launch of the first flight model would be officially pushed back almost a year! This loss of a year was very obvious for management. Moreover, the factory and all the subcontractors had to manufacture one more "complete package that no one will need afterwards" and there wasn't enough manpower even for the flight models.

In any event, we lost a year in getting through the ground testing process. But at the same time, we also lost in reliability, using up the service life of the flight systems and limiting the volume of experiments out of a constant sense of caution. When ground testing was completed on the first two 7K-OK vehicles in November 1966, having conducted a detailed analysis of the glitches identified in the KIS, three deputies of the chief designer—Tregub, Shabarov, and I—wrote a memorandum to Mishin in which we argued that the main cause of the abundance of defects was the lack of preliminary developmental testing of the integrated circuitry on stands and the engineering analogue.

We warned that in the future, such an abundance of defects was unavoidable for any new object if we were to continue the practice whereby various systems come together for the very first time on the flight model during flight-ready tests. The spacecraft's complex electrical system required preliminary stand developmental testing, as did the complex propulsion systems. We argued that, on the 7K-L1 and especially the L3, which were even more complicated objects, the situation would not improve unless a system of preliminary developmental testing was introduced.

There's a saying: "penny wise and pound foolish." The history of our cosmonautics is overflowing with examples confirming this folk wisdom. Conserving on ground development testing, test units, and integrated stands leads to subsequent losses that far exceed the apparent gain in time and money.

We began preparing "Soyuz-1" for testing at the engineering facility in January. The *Dvoyka* launch site, which had seen a lot of use over the years, had outlived its usefulness and was undergoing a major overhaul. Launches were scheduled to take place at Site No. 31, and consequently, from the Assembly and Testing Building (MIK), which was poorly equipped for the preparation of spacecraft. But the amount of work required to install new testing equipment, stands, and gear was so great that in September, when the first vehicle had been unloaded at the MIK, the adjustment and acceptance process of the ground-based equipment was still under way.

After repeated discussions of the state of affairs with *Igla* and the docking assembly, Bushuyev and I proposed that we toss the first single developmental launch out of the flight development test (LKI) program. If we were taking the risk anyway, we should go all out! We also arranged with Mishin to do away with the single unpiloted launch. If we were going to fly, we should just fly! We already had an experience where *Vostok-3* and *Vostok-4* had been launched from the very same launch site just 24 hours apart. Now we proposed launching "Soyuz-2" two days after "Soyuz-1," and then immediately, during the first or second orbit; however it panned out, going for rendezvous and docking!

The State Commission accepted our proposal with an amendment: the launch of the first pair of Soyuzes should be considered tests. Its primary mission should be to checkout all the vehicles' systems, and most importantly—automatic docking. The second pair—"Soyuz-3" and "Soyuz-4"—would be prepared in the piloted version. We should immediately set about training the crews and preparing the flight program for the second pair of Soyuzes.

Thus, for the first launch we now needed to prepare not one, but two vehicles. In June at the KIS, we had a second workstation, and testing began on 7K-OK No. 2 just a month after No. 1. Staggering the dates like that was sufficient to reduce the total number of defects and glitches on No. 2 to nearly half that of No. 1. During the period from September through November, 7K-OK No. 3 and No. 4 passed through the KIS. Progress was obvious. The total number of glitches was 736 and 520, respectively. For 7K-OK No. 4, the testing time was almost one-third that of No. 1!

Rush jobs at the factory did their part. From September through December 1966, four 7K-OK vehicles were shipped to the cosmodrome. By today's standards, the rates and volumes of work would be impossible not only for Russian cosmonautics, but also for the American and European aerospace industries!

The military testers at the cosmodrome longed for real work. They greeted our specialists like long-lost friends. Here people gathered from more than 50 different organizations from many different cities. Over the course of 10 years of work at Tyura-Tam, not only the "rocket veterans," but also the neophytes learned to understand one another surprisingly quickly, diving right into the commotion. The experience of the Vostoks, Voskhods, Molniyas, the MV program, and the Ye-6 had not been in vain. And still, the authors who proposed something new and additional during testing were pressed unmercifully.

For the sake of the deadlines and schedules, the chiefs of testing and lead designers developed a dangerous self-confidence. If all the tests had been conducted in accordance with approved instructions and methods and had produced positive results, then any new initiatives among skeptics were stopped. Yet, life had shown: one should never disregard proposals for additional tests.

On the first 7K-OK, Boris Nevzorov had doubts, natural for an experienced radio engineer; he was concerned about the procedure for testing the command phasing in the attitude control system. Bashkin's deputy in charge of testing was conducting this labor-intensive test over two channels: pitch and heading. Everything had been okay, and he decided against performing the troublesome roll test, which required modification of the ground equipment. Later, again and again, he would find that the smallest bit of arrogance was punishable. Blind faith in sheer luck is unacceptable.

American rocket specialists had formulated yet another Murphy's Law for testers and designers: "*If anything can go wrong, it will.*"

In October 1966, spurred on by American successes in the Gemini program and the latest reports about progress in the Apollo program, senior management was uneasy about the lull on our space front. By autumn 1966, the break in our piloted launches had lasted a year-and-a-half and no one had dared name the precise date for the next sensational flight.[44] There had never been such a lull under Korolev and Khrushchev.

From March 1965 through September 1966, the Americans had launched their Gemini spacecraft into space nine times, each one carrying two astronauts.[45] In all, if you start counting from the Mercury flights, the Americans had 21 astronauts, and three of them had each flown twice. So, in terms of the number of astronauts (21 astronauts versus 11 cosmonauts) and the total number of piloted flights (15 versus 9), the Americans had shot far ahead. The U.S. announced that the last Gemini flight would take place in November 1966. It would last up to four days and carry astronauts who were preparing for an Apollo flight.[46]

Speaking before Congress, the President of the United States reported about the successful progress on the Apollo program. He assured Congress that the U.S. had already achieved supremacy in space, but in order to hold onto it, they needed to speed up the Apollo program.

Central Committee Secretary Ustinov made a promise to Communist Party Central Committee General Secretary Brezhnev to deliver new magnificent triumphs in space, counting solely on Soyuzes. The uneasiness of the Party and government leaders could also be attributed to the upcoming anniversary year. The fiftieth anniversary of Soviet rule in 1967 was supposed to be celebrated with great achievements in all areas of economics, science, and

---

44. The previous Soviet piloted mission had been *Voskhod-2* in March 1965.

45. These missions were *Gemini III* (March 1965), *Gemini IV* (June 1965), *Gemini V* (August 1965), *Gemini VI-A* and *VII* (December 1965), *Gemini VIII* (March 1966), *Gemini IX-A* (June 1966), *Gemini X* (July 1966), and *Gemini XI* (September 1966).

46. *Gemini XII* astronauts James A. Lovell, Jr. and Edwin E. "Buzz" Aldrin, Jr. were soon to be recycled for training for Apollo missions but this was no different from almost all other Gemini veterans preceding them.

culture. And suddenly, there was this lull, which was so incomprehensible for the Soviet people and for the nation's higher political leadership!

Ustinov (in the Central Committee itself) and Smirnov (in the Council of Ministers) answered directly to the Politburo. Minister Afanasyev was accountable to both of them. Tyulin and Kerimov were now immediately subordinate to Afanasyev, and both were chairmen of State Commissions, under the aegis of which we were supposed to undertake our new achievements.

Tyulin came to one of the informal meetings in Mishin's office. After reviewing our proposals for the scheduling of upcoming launches, he warned that in October this matter would be put before the collegium of the ministries "with all due seriousness." Afanasyev had told him that he intended to "excoriate" Mishin and everyone who was guilty of the complete breakdown in the piloted launch program. After the panel, a discussion of this matter in the VPK was inevitable, and then a report to the Communist Party Central Committee would follow.

Tyulin added: "They're expecting us to resume piloted launches this year. Instead of you, the Americans are getting a gift ready for the forty-ninth anniversary of the October Revolution. They have announced the flight of *Gemini XII* in early November. There was never such disgrace when S.P. was around. For the fiftieth anniversary of the October Revolution, we have been assigned a mission to execute a piloted circumlunar flight on the 7K-L1, and in 1968 a Moon landing." This was the plan that Tyulin brought in after receiving a pep talk from Ustinov and his own minister.

I hazarded to remark that the Americans had already passed us by in terms of the number of piloted flights and astronauts when S.P. was still alive. Mishin pledged that we would launch the first pair of Soyuzes to perform a docking in October. If we were to dock in space, then we would pass up the Americans in this area.

As Tyulin had promised, the collegium really did meet. Our work on the 7K-L1 and the Soyuzes was declared unsatisfactory. "Korolev guaranteed us that the 7K-OK would be launched in the spring of 1965. Soon it will be 1967, and we still have no spacecraft. OKB-1 and Mishin personally got cocky. For them, you know, directives from the Central Committee are not the law"—that is more or less how the minister summed up the charges against us expressed in the previously prepared speeches of his deputies Tyulin, Litvinov, and Tabakov.[47]

---

47. Georgiy Aleksandrovich Tyulin (1914–90), Valentin Yakovlevich Litvinov (1910–83), and Gleb Mikhaylovich Tabakov (1912–95) served as deputy ministers in the Ministry of General Machine Building (MOM). Tyulin was more senior to the other two, since he was a "first deputy."

Soon after the collegium, a VPK session took place in which a directive from the higher-ups designated "Luna" as "mission No. 1." At the meeting, Georgiy Babakin's latest achievements were held up to us as an example; in addition, Chelomey's proposals for circumlunar flight, which OKB-1 had allegedly obstructed, were mentioned, and once again the following mantra was repeated: "Don't give away the Moon to the Americans!"[48]

None of the managers standing over us in the administrative and Party hierarchy—particularly Central Committee Secretary Ustinov, Minister Afanasyev, and their deputies—were outsiders by any stretch of the imagination. Each had passed through the rigorous school of arms production during the war. Individually, each of them understood the difficulties we faced, and in their own minds realized how unrealistic it was to attempt a piloted circumlunar flight in 1967 and a Moon landing in 1968. But as soon as they gathered for official meetings with the actual executives in attendance, they gave in to a sort of collective hypnosis and demanded the impossible.

At one point, I couldn't stand it any longer and during the next one-on-one meeting with Deputy Minister Tyulin, I hazarded to remark that, "No one would propose a Central Committee resolution to shorten a woman's pregnancy from nine months to six in order to increase the population. They could send you to a psychiatric hospital for such an initiative. And here they are essentially demanding that we develop a highly sophisticated system with unrealistic deadlines. And according to the old Ukrainian saying: 'Whatever is made in haste is stillborn'."

"Don't get any ideas about spouting off anywhere with old sayings like these," said Tyulin." "It was Sergey who proposed these deadlines in his day, not the Central Committee."

About 10 days later, a terrible decision circulated, once again committing the participants of the 7K-L1 and N1-L3 lunar programs to complete all the work on the lunar spacecraft and launch vehicles "as top priority and as especially crucial state assignments." Alas! We were not able to begin the Soyuz launches by the forty-ninth anniversary celebration of the October Revolution [in 1966].

---

48. The reference to Babakin's recent achievements is a nod to the lunar successes of the first soft landing of a probe on the surface (*Luna-9* and *Luna-13*) and also of orbiting a probe around the Moon (*Luna-10, -11,* and *-12*). All of these missions were accomplished in 1966.

Chapter 19
Flying the Soyuz

Before he left Moscow for the upcoming Soyuz launch, State Commission Chairman Kerimov telephoned all the members of the State Commission requesting that all the "bigwigs" go to Tyura-Tam. Among the "bigwigs" who arrived at the firing range were Keldysh, Mishin, Rudenko, Karas, Kamanin, and Pravetskiy. Ryazanskiy was the only one from the old Council of Chiefs who came. Kerimov was miffed, but it was easier for those of us who were directly involved with preparation to deal with the deputies.

On the morning of 18 November at Site No. 31, Mishin held a meeting of the technical leaders. They decided to propose launching the unpiloted pair of Soyuzes 24 hours apart: the active "Soyuz No. 2" on 26 November and the passive "Soyuz No. 1" on 27 November. According to the factory documentation, the active Soyuz vehicles were assigned even numbers and the passive ones, odd numbers. In TASS reports the Soyuzes were numbered according to the order in which they were inserted into space. The unpiloted vehicles were referred to as Kosmos vehicles with corresponding numbers in TASS reports. If, after orbital insertion, the passive vehicle turned out to be no more than 20 kilometers away from the active one, which was quite likely, then the command would be given right then and there to activate *Igla* to begin search and rendezvous.

If the distance between the vehicles turned out to be more than 20 kilometers, then a special maneuver would be necessary and rendezvous would take place 24 hours later. After rendezvous and docking there would be a test of the controllability of the link-up established between the two vehicles in space. The plan was to undock on the third day, and then 24 hours later, execute the controlled landing of the vehicles.

At the State Commission meeting, Ryazanskiy, Mnatsakanyan, Tkachev, Khrustalev, and Isayev presented reports on the results of systems' preparation. I reported on the general results and readiness of the two vehicles. Kirillov gave a detailed account of the testing process and main glitches, and he assured us that they had all been thoroughly investigated and the findings of the chief designers had been received. It was proposed that the launch dates be approved. Despite the extra exertion of the testers, factory installers, and military and civilian personnel, the State Commission still moved the launch from the 26 November to 28 November so that there would be no "loose ends."

The launches were to be unpiloted, but while we were occupied with the technology, Mishin and Pravetskiy were in a heated debate with Kamanin and Rudenko concerning the composition of the crews for the subsequent piloted Soyuz launches. The planned program called for two cosmonauts to transfer from vehicle to vehicle through open space. Kamanin insisted on "absolutely healthy military cosmonauts." Mishin demanded that two cosmonauts from OKB-1—Yeliseyev and Kubasov—be included in the main crew. Komarov was proposed as the commander of the main crew with Gagarin as the backup. Why Gagarin needed to be included in this game, especially as the "backup," was unclear. It seemed to us "civilians" that Mishin had good reason to be miffed. But he clearly crossed the line in his skirmish with Kamanin. When it became apparent in Gagarin's backup crew that Gorbatko had been approved instead of Volkov, whom Mishin had proposed, Mishin got angry. His conversation with Kamanin took place at the MIK, where there are always throngs not only of workers, but also curious onlookers waiting for the rocket to be rolled out to the launch site.

"You push your own people through and think that yours are better trained. Our engineers are more competent. We are not going to work with you like this!" Mishin declared. These declarations, which were uttered in a very irascible tone with a rather raised voice, were tactless under those circumstances. Everyone present understood that. That evening Ryazanskiy said to me: "You should impress it on Vasiliy that when he is out among a public hungry for spectacles, he should conduct himself more discreetly. Such flashes of emotions do not increase his authority, and it will be difficult for us to support such a quick-tempered leader of the Council of Chiefs."

While the Soyuz program was in development, the Ministry of Defense Central Directorate of Space Assets (TsUKOS), which Andrey Karas headed, came out with a proposal to control piloted flights from NIP-16 near Yevpatoriya. At that time NIP-16 was called the Center for Deep Space Communications. This proposal was met with enthusiasm. No matter what was happening in orbit, one could always find time, even at the expense of sleep, to take advantage of the benefits of the warm sea and sandy beaches of the western Crimea.

In any event, the information that we received at our primitive command center in the cramped room of the MIK at Site No. 2—from the ground tracking stations (NIPs) located throughout the entire country, from the spacecraft, and from the landing areas—was relayed through the General Staff communications center in Moscow. In this sense, Yevpatoriya had no particular advantages. The ballistics centers equipped with the computers that made flight control possible were located in Moscow and at NII-4 on the outskirts of Moscow. Moscow had decisive advantages. Subsequently, control centers for both piloted and unpiloted vehicles were created near Moscow. In addition to advantages of climate, NIP-16 near Yevpatoriya had advantages over the cosmodrome command center and Moscow in terms of its wealth of radio engineering equipment. However, the rapid processing of telemetry information—the primary source for near real-time decision making—still required the participation of a brigade of our telemetry operators.

The full State Commissions were duty-bound to gather at the cosmodrome to make the launch decision. After determining that the spacecraft had been inserted into orbit, the members of the State Commission could board their airplanes and scatter, some to Yevpatoriya and some to Moscow. The State Commission could not be in session all the time and make decisions, much less real-time decisions. For this, operations control groups were formed:

Group T   Tyura-Tam, i.e., the cosmodrome

Group Ye   Yevpatoriya or NIP-16

Group M   Moscow at OKB-1 or NII-4.

Initially, I was appointed chief of Group Ye. Pavel Agadzhanov was my deputy and simultaneously commanded "all military forces" involved with control. Endowed with talent as a good organizer and enjoying the rights of deputy commander of Military Unit 32103 (which included all NIPs, communications facilities, and the computer center), Agadzhanov was the first to master the new profession of flight control.[1] Before the end of the 1960s NIP-16 had no means for automatically processing and visually displaying information in real time. The main means of receiving and transmitting commands were secure telephone and telegraph communications. Equipped with dozens of electronic screens, the mission control centers at Cape Canaveral and in Houston seemed like a fantasy to us.

---

1. In effect, Agadzhanov was the first to hold duties analogous to that of "flight director" in the American context. Military Unit 32103 was the staff of the Soviet ground tracking network known as the Command and Measurement Complex (KIK).

The launch of an 11A511 launch vehicle with a Soyuz spacecraft in the late 1960s.

For the first two Soyuzes, Mishin and I arranged that before the first launch, all the main forces would be concentrated at the cosmodrome. Immediately after the launch of the first (active) vehicle, I was supposed to fly out with a group to Yevpatoriya, to join Agadzhanov to monitor the rendezvous and docking process from there. After the launch of the second (passive) vehicle, Soyuz No. 1—the Chairman of the State Commission, Mishin, and other State Commission members and "chiefs" who wanted to would also fly to Yevpatoriya. Group T included Shabarov, Yurasov, Topol, and, of course, Kirillov. He did not miss the opportunity to remark that the more detail the General Staff went into when describing the troop dispositions and movements, the quicker everything got into a mess. At the launch of 7K-OK on 28 November 1966, that's just what happened.

Sitting in the bunker at Site No. 31, we waited until we received reports that 7K-OK No. 2 had successfully entered into orbit and that the solar arrays and all the *Igla* antennas had deployed. Mishin, Kerimov, Mnatsakanyan, Ryazanskiy, Bushuyev, Feoktistov, Shabarov, Ostashev, Yurasov, and I, as well as all the other "chiefs and those in equivalent positions" had run to our cars and were racing to Site No. 2. I rode with Bushuyev. The two of us were in an excellent mood. We were finally about to see the opening of the Soyuz era. Bushuyev and Feoktistov had expended a great deal of effort to slow down manufacturing work on a series of unpromising Voskhods in favor of the Soyuzes. "Now 'Uncle Mitya' [Ustinov] will assign a mission to catch up to and surpass Gemini based on the number of cosmonauts, dockings, and flight durations," we assumed, swapping predictions.

Having arrived at the command post (KP) ahead of us, Mishin and Kerimov were already questioning the telemetry operators. While we were driving from Site No. 31 to *Dvoyka*, they found out that, after separation and before leaving the radio coverage zone of our Far East tracking stations, the vehicle had not settled down. An inexplicable spin perturbation continued.

About 10 minutes after our arrival at the control center, having flown around the globe, the new spacecraft entered the NIP-16 radio coverage zone and then the coverage zone of local tracking station IP-1. Both reported that all facilities were experiencing stable reception and good signal level.

An agonizing period of silence set in, while Golunskiy and all the telemetry operators at IP-1 were trying to reconcile their visual impressions. They had a lot of experience with the launch vehicles, and there was unusual information coming down from the Soyuz—from the new BR-9 system rather than the customary *Tral*. They were the first to suspect a problem, but were afraid to report without a thorough checkout.

Kirillov demanded permission from Kerimov and Mishin to roll out and prepare the second launch vehicle with the passive Soyuz. Less than 24 hours remained before the second launch. Should they roll it out or not? The launch crews and all the firing range services, inspired by the successful launch, were prepared to go without sleep for one more day for the sake of the long-awaited resumption of our space achievements. Samokhin, the commander of our auxilliary aviation, was keeping airplanes at the ready to depart for the Crimea and Moscow and needed answers to the questions: "Are we flying or not? And if yes, then who and where to?"

Agadzhanov was requesting further instructions from Yevpatoriya in order to prepare all the NIPs for the next session. Tension was at its height when at last the paper rolls containing the direct recordings of the on-board systems' performance began to arrive from the receiving stations to the "screening" room on the first floor.

Dozens of eyes were glued to the lines that were continuous, tangled, criss-crossing, and smoothly changing yet convulsively quivering, comprehensible only to the initiated. Despite the general hubbub, telemetry service officers were trying to interpret the results. Golunskiy whispered something to Feoktistov, and then he said to me: "It seems that all the propellant has escaped from the DPO tanks."

Feoktistov asked them to check one more time. Again and again scaled rulers were applied to the paper tapes. Doubts dissolved and with them hope as well. The spacecraft was spinning at a rate of two revolutions per minute. "Loss of roll stability,"—that is how I described the situation for the State Commission report.

When a doctor gives a diagnosis that says someone close to you has an incurable disease, there is still hope. Perhaps another doctor will say there's hope or you'll manage to find a folk healer. In a nutshell, in such cases there is always hope enabling one to fight for life. After deciphering telemetry that has informed you that the entire supply of propulsion fluid for the DPO system has been lost during a single orbit, it is pointless to search for another doctor. The spacecraft has already lost all capability to execute the primary mission assigned to it. Now, since it was already in space, we needed to quickly change the previously developed programs and check out the other still-living systems of its complex body.

The report to the State Commission was short: "There's no propellant in the DPO tanks." Replaying the data on the memory unit showed that right after separation, the approach and attitude control engines had performed a very intense roll burn. Telemetry was corroborated by calculations showing that, during an orbit in such a mode, it was possible to lose all the propellant from the DPO tanks.

It was now 3 a.m. Kerimov asked for the review team findings. It was clear that rendezvous and docking were impossible. Kerimov and Mishin made the obvious decision: "Stop the launch preparation of the second vehicle. Instead of leaving for Yevpatoriya with the operations group, Chertok is to make recommendations right here for a new flight program. Bushuyev, Feoktistov, and Rauschenbach will figure out where the DPO propellant went and why. Shabarov and Kirillov are to take steps to maintain the second launch vehicle and spacecraft. The ballistics experts should immediately ascertain the orbit and give a prognosis: how many orbits will the spacecraft survive without correction?"

In addition, Kerimov and Mishin were supposed to immediately prepare a report for the brass in Moscow and a draft communiqué for the morning newspapers and radio about the flight of *Kosmos-133* instead of the anticipated Soyuz.

All at once everyone was up to their eyebrows in work. I caught a glimpse of Gagarin's sad face. The next pair of Soyuzes were to have been piloted, and Gagarin had not lost hope of being in space one more time. Now everything was messed up. This was the first time he was involved in such a complex "kitchen." Not wanting to distract anyone, he went around from one arguing group to another, trying to understand what was going on and what the prognoses were. I must say that for a man who did not know the structure and special features of the spacecraft's systems, it really wasn't easy to figure out. But Gagarin conscientiously did without sleep, along with the rest of us.

The ballistics experts were the first to report: "The vehicle will begin to plunge into the atmosphere after the 39th orbit." I proposed performing tests of all systems. First and foremost, we should check the ion attitude control system and the backup correction engine (DKD) to make sure that it would be possible to return to Earth with its help. The thing was that the depletion of DPO propellant had deprived us of the capability for a retrograde burn to return to Earth using Isayev's main approach and correction engine unit (SKDU). When the SKDU was operating, the DPO provided vehicle control and stabilization. Now it was out of propulsion fluid. We began feverishly drawing up the programs for the upcoming tests. During subsequent orbits we found out that all the systems were working completely reliably. The primary task now was to test what was for us a fundamentally new mode of guided descent and soft landing. If it were to succeed, then it would show that the 7K-OK was safe for piloted flight.

Despite a second sleepless night, after each communications session, the indefatigable telemetry service spread out rolls of telemetry data and dozens of heads bent over them. However, the first 7K-OK continued to buck. In the correction mode, the controls on the DKD were the rocket nozzles, which used exhaust gas from the turbopump assembly. When the test was conducted on the DKD, the vehicle pitched and yawed in directions opposite the ones given in the commands. This might happen if the phasing of the commands to the controlling vernier nozzles had been reversed.

"The gyros are issuing the right command, but the vehicle is turning the other way," reported Zvorykin.

After a heated argument with Yurasov, Zvorykin, and Dubov, Isayev's engine specialists agreed that everyone understood the concepts of "clockwise and counterclockwise" as exactly opposite. Isayev's people agreed and plead guilty. But how could the vehicle return to the ground now?

We had delivered into space a vehicle that had the fatal coincidences of two of the most stupid, but most common, errors of polarity or sign: "plus" for "minus" and "clockwise" for "counterclockwise." Isayev's reliable orbital correction, maneuvering, and braking engine system for return to Earth was redundant on the Soyuzes. But similar mistakes were made in each of these two reliable systems [i.e., the DPO and DKD] during assembly, making stabilization and control during engine operation impossible. The error in the DPO system was the obvious fault of our TsKBEM designers. The error in the DKD backup system "slid" to the side of Isayev's designers.

When designing the Soyuz, we proved its reliability by fulfilling a postulate: "Any one failure in any system must not lead to the failure of a program; any second failure must not lead to a life-threatening situation for the crew." Redundancy of the systems enabled this condition to be fulfilled. And suddenly in the redundant flight control system with backup engines, we

Boris Rauschenbach, Boris Chertok, and Nikolay Sheremetyevskiy.

found two similar failures that wouldn't allow the vehicle to return to Earth. The most aggravating thing was that these errors were not discovered during the process of multi-stage ground testing.

What had happened was particularly painful for Yurasov and me. It was very difficult to come up with a reasonable explanation for the causes of such technical "slovenliness." The explanations of Rauschenbach and his coworkers in this regard were too tactful. I expressed everything in more scathing terms, using stronger language than usual, and promised upon returning to organize an action group for an on-the-record study.

At the factory monitoring and testing station for Soyuzes, we used the special *Kardan* stand, and at the engineering facility in the anechoic chamber, we used a special rocking platform making it possible to test all sorts of "phasing" and "polarities." And, as a matter of fact, we found them! Each finding of a similar type of error showed that on paper "everything was okay" but the designer, the person developing the circuitry, and the factory installer each had his own understanding of that paper. There was no guarantee that no errors would turn up on board during the installation process, even if—after the calculations, simulation, and testing of the circuit—the theoretician and circuitry man had agreed that everything was in order.

Bashkin and his deputies were supposed to find such errors. Yurasov and I grilled them and found out that it was precisely these two errors for which no clear instructions were provided in the testing documentation.

And nevertheless, we needed to find a way out of this hopeless predicament. I don't remember who along with me came up with the idea to fire the deorbit burn using the SKDU system with short pulses of 10 to 15 seconds instead of continuous operation for around 100 seconds. If we were to set up the vehicle for a deorbit burn using the attitude control system's low thrust engines and fire the SKDU, 10 to 15 seconds would not be enough time for it to veer sharply off course. Thus, with the sum total of many of these braking pulses, we would build up a burn sufficient to enter the atmosphere and land on Soviet territory. It was too early to speak about the accuracy of the landing. But at least the descent control system (SUS), landing system, and soft landing engines would be tested.[2]

Just short of two days, before the 34th orbit, we were busy with the agonizing operations to maintain attitude control and to fire brief braking burns. We were waiting for reports from the search services. None of the services or ground tracking stations in the areas where the vehicle was supposed to appear, not even the Air Defense Forces search facilities, detected anything. It was neither in space nor on the ground. In a sleep-deprived stupor I reported to Mishin and Kerimov: "This was supposed to happen. We didn't take into consideration that the spacecraft has an APO system. All those braking burns did not guarantee that the vehicle would land on Soviet territory, and the APO blew our spacecraft into so many bits that even the air defense radars were incapable of detecting them." We returned to Moscow in disgrace.

The next day, in keeping with the rules at that time, I assembled the "triangles" and main supervisory staff of all the departments.[3] In my speech— rather impassioned according to those who heard it—I said that we should not lay the primary blame for the loss of the first Soyuz on the one who committed the error. Errors always have been and always will be committed. In our complex system, we commit them everyday. The errors must be discovered during ground testing. Our testing equipment at the monitoring and testing station (KIS) and the engineering facility was perfectly suited for detecting those two errors. Therefore, the primary blame for what happened should be laid on the managers who did not display vigilance and who did not insist on high standards when developing the procedures and conducting the actual tests. During the turbulent discussions that followed, Aleksandr

---

2. SUS—*Sistema upravleniya spuskom.*

3. A "triangle" (*treugolnik*) typically consisted of an enterprise chief, a Communist Party secretary, and a trade union representative.

Pronin—one of the developers of the APO system—spoke. He proposed that no one be punished, but rather that a commendation be issued for the fact that, through their fault, we managed to check out and confirm the reliability and safety of the emergency spacecraft destruction system, especially since this system had never been subjected to full-scale ground testing.

Searching for the guilty parties was a more complex matter. It turned out that back on 14 May, six months before the launch, laboratory chief Nevzorov was reprimanded with an administrative order "for failing to quickly resolve simple technical matters, which resulted in a delay in the release of baseline data for updating technical documentation." These baseline data contained requirements for changing the order in which spin commands were sent to the DPO. This requirement arose, as sometimes happens, for reasons that had nothing to do with control principles.

Someone from among the thermal mode specialists came to the conclusion that the hot jets of gas from the DPO nozzles would blow on the solar array panels. They reported to Feoktistov. Without giving it much thought, he proposed that they turn them on their support bracket 180 degrees about their axis, so as not to undertake a complex modification of the spacecraft and look for other sites to install the engines. Rotating them this way changed the sign of the moment around the axis of rotation. To maintain the order in which the spin commands are given, it would be necessary to change the polarity or the phasing of the command issuance by the control system instruments.

Feoktistov sent an office memorandum to Legostayev on this subject. Legostayev referred the conceptual designers' request to Shmyglevskiy. Here, the problem's theoretical study began, and then it was recommended that Nevzorov give the baseline data to the developers to change the instrumentation circuitry of the docking and orientation engines control assembly (BVDPO).[4] They, in turn, were supposed to give written instructions to Chizhikov's design department. Finally, this department prepared the notification for the instrumentation plant to modify the BVDPO instrumentation. By this time the instrumentation had already been installed in the vehicle. Permission was required to remove it and return it to the manufacturer's shop for modification. This could only be done with the approval of lead designer Topol, who was outraged and gave instructions to his deputy Yuriy Semyonov to find out "where they'd all been before." He found out that "before" had been two months ago. At a hectic time, when every hour counts, someone needed to be punished for such a delay. It turned out that a large portion of the "failure to

4. BVDPO—*Blok vklyucheniya dvigateley prichalivaniya i orientatsii.*

act quickly" was attributed to Nevzorov, and he was immediately reprimanded to set an example for his associates.

But the reprimand did not free the factory from modifying the instrument. They checked out the circuitry of the instruments still in production. There everything was being done in accordance with the "dressing-down" notification. They examined the installation of the engines on vehicles No. 3 and No. 4, which had not yet been shipped. Everything was as Feoktistov had recommended.

"But, look, it wasn't that way on the first vehicle," one of the assembly shop installers suddenly announced. But how could they check it out now? Vehicle No. 2 had not only flown away, the APO system had blown it to smithereens. There was still time to check out vehicle No. 1, which was still at Site No. 31, and so we sent a high-frequency telegram ordering that the spacecraft be removed from the launch vehicle and thoroughly rechecked, focusing on the polarity of command execution.

No one was interested in internal reprisals or further aggravation of the external situation, which was already red hot. Bushuyev and Feoktistov made a reasonable proposal to speed up the launch of the passive vehicle for a solo program to thoroughly check out all the systems. Get the *next* pair, No. 3 and No. 4, ready for docking. After consultations via high-frequency communications with Kirillov, Shabarov, and Ostashev, who had remained at the cosmodrome, a 14 December launch date was set for a solo version of the 7K-OK.

On 10 December, Mishin held a meeting of the Council of Chiefs, during which they examined the amended Soyuz flight development tests and the draft schedule for the manufacture of 7K-L1 vehicles for a lunar flyby. The first piloted circumlunar flight was supposed to take place in June 1967. Before this, in the first half of the year, two vehicles were supposed to perform a circumlunar flight without returning to Earth followed by two that would return to Earth. The piloted circumlunar flight using the new UR-500K complex—7K-L1—was just six months away, and we were not even ready for the first unpiloted launch. Moreover, in addition to the two vehicles already shipped to the cosmodrome, three remained that had yet to be fitted out with dozens of instruments and for this reason could not undergo factory checkout tests (ZKI).[5] The primary, fifth, piloted vehicle was completely "naked." Pilyugin and Ryazanskiy declared in unison that these deadlines were absolutely unrealistic. Mishin wasn't about to argue, and referred to Ustinov and Smirnov, who had dictated these dates to him as a directive. Taking his leave after the Council session, Pilyugin expressed his

---

5. ZKI—*Zavodskiye kontrolnyye ispytaniya.*

displeasure: "Sergey would not have tolerated such a ridiculous schedule for the Council, and Vasiliy is reluctant to argue with Ustinov and Smirnov. You and Bushuyev should have explained it to him."

"You are a member of the old Korolevian Council—why can't you just explain this to 'Uncle Mitya' [Ustinov] yourself?" I asked in defense.

But Nikolay just waved his hand and headed to his car.

At the State Commission, Kerimov insisted that, to avoid redundancy and confusion, the main operations control group (GOGU) based in Yevpatoriya should take on mission control right after the vehicle goes into orbit.[6] The group was once again approved with the following personnel: Agadzhanov, Ryazanskiy, Tregub, Feoktistov, Rauschenbach, Levin, Pavlov, Anokhin, and me. We would have to be at NIP-16 within 24 hours for a readiness check and then report to State Commission before its last pre-launch session. During launch preparations at the cosmodrome, Grigoriy Levin was assigned the role of communications officer with GOGU. Mishin, Kerimov, Kamanin, and the future Soyuz crews were to fly to Yevpatoriya from the cosmodrome immediately after the launch.

On the morning of 13 December, I flew along with Ryazanskiy, Rauschenbach, Feoktistov, and a dozen or so members of various groups who lingered in Moscow against regulations, to the Crimean Naval Airfield in Saki and met up with Agadzhanov and Tregub for lunch at the dining hall of the "chief designers and generals." They voiced their dissatisfaction with the control center's state of preparation. They let us know that there would be no recreation at this resort. The stormy December sea was hardly inviting.

To someone new to the scene, the situation in this control center would have been reminiscent of a stirred-up anthill. During the flight control of the first Soyuz, organizational factors led to many mistakes. In view of the off-nominal emergency situation, Group T, at Tyura-Tam, taking advantage of the authority of Kerimov and Mishin, actually took over control. But the communications facilities and all the groups needed for operational control were located at NIP-16. The redundancy of commands, the inconsistency of decisions, and reciprocal charges of failing to make quick decisions unnerved people separated by a distance of 3,000 kilometers.

During the flight of *Kosmos-133* very many flaws were discovered in the ground control complex. Agadzhanov and Tregub organized individual and general training sessions simulating the flight mode in order to achieve mutual understanding between various groups consisting of military personnel who listened to their commanders, and industry specialists who believed that the

---

6. GOGU—*Glavnaya operativnaya gruppa upravleniya.*

"drill sergeants" only interfered with work; the goal was to get them to learn to work with one another.[7]

There remained a day before the launch, and Ryazanskiy and I insisted on conducting a "demonstrative meeting," during which each specialist, regardless of military rank or departmental affiliation could lay out his or her complaints and recommendations for putting things in order. Above all, control group chief Colonel Yevgeniy Rabotyagov was on the receiving end. He was the last link in a long decision-making chain. One of the duties of his service was to transmit the contents of commands to all the NIPs on Soviet territory and to ships on the seven seas. Within at least 20 minutes, each ground and sea station had to receive the exact time that coverage would begin, target designations for antenna set up, the sequence and codes of the commands transmitted on board, and a list of the top-priority telemetry parameters that needed to be processed. In turn, his service had to manage to keep track of the receipt of confirmations from the stations that directives had been properly executed and to give new instructions, etc.

The main sources of information for preparing GOGU decisions were Vladimir Yastrebov's ballistics group, Vadim Kravets's analysis group, and Colonel Rodin's telemetry service. At the meeting, Rabotyagov complained that, on the whole, there was no order; during a session, when there should have been silence and everyone should have been busy, the hectic scurrying about from room to room for information continued, a result of "responsible and irresponsible specialists searching for each other in order to be the first to report their ideas to the management." He also said that communications throughout the country on the whole could not cope with the telemetry traffic that needed to be transmitted to Yevpatoriya. After extracting the essentials, this traffic needed to be reduced. Rodin asserted that only the first real flight "opened our eyes to how we needed to work."

"We found out," said Rodin, "that we can't rapidly process such a stream of information."

He proposed using the Molniya satellite for communications. The telemetry service was one of those in which the military and civilian specialists worked in complete harmony. They had already started a tradition of resolving their internal conflicts on their own rather than "airing their dirty laundry." To a great extent, this was to the credit of the chiefs of our telemetry service: Golunskiy, Popov, and Vorshev. Incidentally, the ballistics specialists also formed a social

---

7. This is a reference to the fact that the flight direction under the GOGU was typically performed jointly by military and civilian controllers. The former represented the Strategic Rocket Forces and the latter, the various design bureaus and scientific research institutes.

class in which you could tell the military ballistics specialists from both our ballistics specialists (as well as academic ones) only by their military uniform.[8] The analysis group caught the most hell. This group included about two dozen leading specialists, each responsible for his or her own on-board system. Each demanded that the parameters they were interested in be processed first, and each hurried to be first to report to the supervisor that "everything is okay" or "it's a real mess, we need to do such-and-such and thus-and-so and not do this-and-that." Sometimes the information transmitted from the stations via telephone depended on what military brass was "on the horn."

Yastrebov admitted that they were late giving out the coverage zones because the ballistics calculations still hadn't been set up here in the local computer. Everything went through the NII-4 ballistics center in Bolshevo. They were ready to provide real-time computation here on site for the upcoming launch. Summing things up, Agadzhanov and I mainly gave out "dos" and "don'ts": don't shout during communications sessions, so-and-so should sit over there, communicate only such-and-such by telephone, etc. In the remainder of the meeting, it became clear that in order to control two vehicles simultaneously, a division had to be established in the groups for active and passive vehicles.

One way or another, on the morning of 14 December, we reported to the State Commission that all flight control services were ready for the launch of Soyuz No. 1. But this time GOGU and the entire ground Command and Measurement Complex (KIK) did not have a chance to verify their capability for real-time flight control.

We gathered in the large room that was called the central control room. Agadzhanov, Tregub, and I formed a triumvirate that would take over control of the Soyuz immediately after its separation from the launch vehicle. We sat shoulder to shoulder around a single table covered with dozens of telephones, among which a microphone stood out in front of "No. 12"— that was Agadzhanov's special call sign. A stand in front of us held placards with call signs, command designations, and a color-coded program of the first 24 hours of the flight. At that time we still didn't have any screens or electronic monitors displaying information; we received 90 percent of all our information by ear. We needed our eyes to look at reference materials contained in bulging files. All the documentation printed on pinkish-brown blueprint paper was secret.

---

8. In the early days of the Soviet space program, ballistics support for piloted space missions was typically provided by a group of military officers and civilians. The latter represented both the design bureaus and academic research institutes from the Academy of Sciences.

Behind us were the main representatives of the large analysis group. They too were supposed to receive all telemetry reports by ear and ballistics prognosis data over the public address system, monitor our conversations with the NIPs, receive reports about the execution of commands, rapidly grasp what was going on, and give us advice as quietly as possible.

There were several dozen specialists for all the systems. They were located in other rooms. If an individual responsible for one system or another was in our room and wanted to consult with his subordinates, he darted out of the main hall and ran to look for the people he needed. If he didn't get a clear answer, he usually brought several people back with him and then a raucous discussion started, interfering with the hearing of the operational information.

Reprimands from "No. 12" were of little help, and there was the danger that if we experienced several off-nominal situations on board, there might be chaos and unpredictable consequences on the "ground." Such apprehensions were expressed at the meeting the day before. But the next day, on 14 December, after announcing our "complete readiness" to the State Commission at T-minus 4 hours, and having calmly eaten lunch, we sent everyone off to their own places. After receiving Levin's announcement of T-minus 5 minutes from the cosmodrome, the public address system—as well as external "circular"—demanded total silence from us.[9]

The launch was scheduled for 1400 hours Moscow time. We received a broadcast about pre-launch preparation right up until T-minus 1 minute. The announcement of T-minus 1 minute passed through all the ground and ship-based tracking stations. Dozens of encouraging reports of "T-minus 1 minute accepted!" were received via circular. From Moscow and Yevpatoriya "all the way out to Kamchatka" at dozens of NIPs, hundreds of people in equipment rooms stood stock-still. The ships *Chazhma* and *Chumikan* awaited signals in the Pacific Ocean. In the Gulf of Guinea, the *Dolinsk* was rocking about on stormy waves and the telemetry operators were waiting for the first orbit.

T-minus 60 seconds dragged out. Levin was silent. A total lack of information has a worse effect on the psyche than bad information. Something had happened at the liftoff! After only 30 minutes, we received this command from the State Commission: "Switch off all equipment and systems!"

What had happened? On 14 December, the pre-launch tests, the fueling of tanks and all the final operations ran their course without a glitch. The pre-launch commission meeting went smoothly. All the chiefs once more concluded that there were no glitches and there was complete readiness.

---

9. "Circular" (*tsirkulyar*) was the term for the on-ground voice communication system at and between the flight control center and the NIPs.

From the author's archives.

The team of 11N6110 ground testing station developers. Seated, from left to right are: A. M. Termosesov, V. F. Ovchinnikov, B. V. Barun, K. M. Belov, and G. P. Sokov. Standing are Yu. A. Rumyantsev, V. I. Antonov, Yu. D. Beltyukov, and A. S. Derdenkov.

Upon reaching T-minus 15 minutes, in keeping with Korolevian tradition, a group left the launch site and went down into the bunker; they included Kirillov, Kerimov, Mishin, Shabarov, Dorofeyev (who had come to this launch from Kuybyshev), Dmitriy Kozlov, and his deputy for testing, Mikhail Shum. Ostashev and the vehicle testers were at the console of the 11N6110 station. Kirillov and Shabarov were standing by the periscopes. Site No. 31 now had combat status. The vehicle consoles were manned by officers from a combat missile crew. In terms of their level of discipline and knowledge of the technology, they were already on par with the fire commanders of the launch pad at Site No. 1.[10]

At T-minus 60 seconds, Kirillov began to monitor the rocket through the periscope, as if it might break away from the launch pad early. He dictated the traditional set of commands: "Feed one," "Key in launch position!," "Vent," "Feed two," "Launch!"[11] After that, the automatic control system was supposed to work according to the timeline for the ignition of all the engines. One after the other, the

---

10. Site No. 1 was the location of the original Sputnik and Gagarin launch pad where all Soviet piloted spacecraft had taken off, from 1961 to 1966.

11. "Feed" was a command to start recording telemetry data, at the time done on rolls of paper. "Key" enabled the passage of the "launch" command to fire the engines.

flickering display lights on the console indicated that operations had been performed and the ignition squibs in all the engine chambers had fired . . . except for one on a strap-on booster. It turned out that the igniter of one of the chambers was not ready for ignition and the automatic equipment "tripped" the circuit. It would be possible to reset and make another attempt at ignition only after inspecting all the chambers, replacing the igniters, and determining and correcting the cause of the failure.

Usually in such situations, the fire commander accepts responsibility. If Korolev had been here, Kirillov would have asked for his consent. Now Kirillov took the entire responsibility on himself. The launch crew was ordered to go up to the rocket, inspect the engines, and find the cause of the failure. The missile service platform was extended out toward the rocket, providing access to the engine nozzle. Dorofeyev and Shum joined the military crew. Kirillov, Kerimov, Mishin, Kozlov, and Ostashev climbed out of the bunker to the "zero" marker. From a clear sky, hanging low over the horizon, the winter Sun provided good illumination of everything happening at the launch site.

Suddenly somewhere above the rocket there was a blinding flash of light accompanied by a violent popping sound. The engines of the emergency rescue system (SAS), mounted on the exterior of the fairing, had fired up. Those who found themselves on the launch pad watched in amazement as the vehicle's descent module swung beneath its parachute a half kilometer from the launch site over the steppe. The nose fairing halves crashed down next to the launch pad. Kirillov managed to turn his attention in time to see the sparks that were dancing merrily above the ruined top of the rocket. It was not difficult to imagine what might come after the still innocuous fiery plumes that were flowing down.

He gave sharp commands over the public address system: "Everyone clear the launch pad and go immediately to the bunker! Leave the service platform and go down the postern toward the underground oxygen factory! We need water on the launch site!"

The tragedy of 24 October 1960 was still fresh in our memory. No one needed to be prodded. Everyone fled as fast as physically possible. The solid propellant engines of the emergency rescue system carefully carried the descent module away to an altitude of 700 meters and placed it under the parachute's care. As later determined, landing went absolutely normally—the soft landing system even activated.

By the way, I will note that I was very interested in the testing of the soft landing system. On my recommendation, Professor Yevgeniy Yurevich from the Leningrad Polytechnic Institute (LPI) was recruited for this development.[12] Korolev, in his time, didn't think much of this proposal, but he did not dismiss

---

12. LPI—*Leningradskiy politekhnicheskiy institut.*

it, either. Thus, the activity of a new, young OKB LPI began. For a long time Yurevich was its chief designer. Soon, its activity in the space field went way beyond its narrow niche of altimeters for soft landing.[13]

But let's return to the troublesome issue. Why did the SAS set the rocket on fire?

Later we realized that when the SAS engines lifted off the Descent Module, the pipelines of the liquid thermal control system ruptured. A special fluid, which had unique properties for a coolant fluid, was developed for this system. However, this fluid was more combustible than gasoline. It caught fire from the flame of the SAS solid-propellant engines. After the Descent Module was jettisoned, in the vehicle's Instrument-Aggregate Compartment, which had been left on the rocket, the peroxide system in the DPO-DO lost its pressure integrity.[14] The fire spread to the rocket's main boosters and was soon accompanied by explosions that showered buildings standing 1 kilometer from the launch site with glass and plaster.

The process developed in such a way that by the time of the most violent explosion, which destroyed the launch facility structure, people had managed to take cover in the bunker or in the postern. One officer perished. He took cover near the rocket behind a concrete structure, which withstood the explosion, and died from smoke inhalation.[15]

The obvious cause of the fire was the activation of the SAS engines. They dutifully executed their mission. If a living cosmonaut had been in the vehicle rather than a mannequin, he would have come to no harm after landing a half kilometer from the launch site. But who gave the command to start up the SAS? The fire suppression systems were still unable to cope with the fire; the launch site was still burning; and the electrical engineers in the bunker were frantically thumbing through thick albums of electrical diagrams and refreshing their memories on the SAS operating logic. Who was the culprit? The answer proved to be unbelievably vexing in terms of its simplicity and also because it wasn't one of Pilyugin's clever staffers or one of our circuitry experts who suggested it, but Nikolay Khlybov—a gyroscope

---

13. OKB LPI (now known as NPO Impuls) was the leading Soviet (and later Russian) organization responsible for the design of strategic command-and-control systems for all ICBMs, including the first generation system known as *Signal* developed in the late 1960s.

14. The instrument-aggregate compartment is more commonly known as the "service module" in the West.

15. Although Chertok does not name the officer in question, according to other sources, the individual who was killed was a Major Korostylev. See N. P. Kamanin, *Skrytyy kosmos: kniga vtoraya, 1964–1966 gg.* [*Hidden Space: Book Two, 1964–1966*] (Moscow: Infortekst, 1997), p. 443.

specialist from Kuznetsov's organization. The *Gorizont* and *Vertikant*—command gyroscopes of the rocket's central core booster—had been provided with emergency contacts for the rocket's emergency destruction system. The gyroscope rotors by their nature had their axes "tied in" to stationary stars or, as the theoreticians say, to the inertial coordinate system. If the rocket's angular deviation during flight relative to the direction of the gyroscopes' axes is at an angle many times greater than the design values, the contacts close. Such a generalized emergency signal is used to start up the automatic equipment of the rocket's emergency engine unit shutdown (AVDU) and the spacecraft SAS in flight.

In our case, the rocket did not fly, did not vacillate, and did not deviate. Then why did the emergency contacts of the gyroscopes that had already shut down after the circuit was tripped, close?

After power is removed from the gyroscope rotors, they have a prolonged run-out. It takes about 40 minutes for them to stop. All this time their axes "slip away" relative to the fixed housing containing the emergency contacts because the rocket spins along with Earth. When designing emergency systems, Earth was assumed to be stationary. You couldn't dream up what happened if you tried! Without delving into theory, in such cases, out of safety considerations, it was routine to provide an SAS power inhibit in the automatic equipment or instructions in case of an emergency. We provided such an inhibit for the APO system, on account of which these very emergency contacts had been introduced. No matter what mistakes we committed on the ground, it was impossible to feed power to the APO system to initiate destruction. But unlike the APO, the SAS had to operate from the moment of launch to save the cosmonaut in the event of a launch vehicle emergency at the launch site.

The very first time the system proved its reliability brilliantly. However, no one had figured that, while saving a cosmonaut, it would be capable of setting on fire and destroying a perfectly good rocket that hadn't even failed.

The launch pad at Site No. 31 was knocked out of action for a long time. The State Commission, which convened at the cosmodrome on 16 December, made the decision to immediately prepare launch Site No. 1 for Soyuz launches. Firing range chief General Aleksandr Kurushin approved a period of one month to perform everything necessary for this project.[16]

The next, third, launch of a single unpiloted Soyuz was preliminarily scheduled for 15 January 1967. Consequently, they decided to schedule

---

16. Aleksandr Aleksandrovich Kurushin (1922–) was a veteran of the Kapustin Yar test range who later served as chief of the Tyura-Tam range (NIIP-5) from 1965 to 1973.

the subsequent launches of two vehicles for docking in March. Levin, who was involved in the work of the State Commission, called from the firing range to inform those of us in Yevpatoriya. We were busy with organizational interfacing for two days and then flew back to Moscow.

The first space decade and the first year of our work without Korolev ended joylessly. After two accidents in a row of the new piloted vehicles, people started to say that under Korolev, this wouldn't have happened; Mishin hasn't displayed the necessary stringency in matters of reliability.

I believed then, and believe now, that Mishin was not guilty for the crashes of the first unpiloted Soyuzes. The primary causes of both crashes were put in place while Korolev was still alive. Neither Korolev, nor Mishin, nor any other chief designer could have foreseen all the errors that their deputies and many dozens of specialists standing behind them could have committed. Both crashes should be chalked up to errors committed at the interfaces of various systems.

In both cases, the following should be considered the guilty parties: I, who was responsible for the control system as a whole; my deputies Yurasov and Rauschenbach; and the department heads subordinate to them. Both crashes were the result, not of failures, but of errors that should have been discovered during testing. Also at fault were the testers, who—when checking the work of the developers—were not exhaustive enough in their test procedures.

On 14 December, each of the chiefs confirmed the reliability and readiness of their system in writing. The failure of a single system to ignite is a random failure that can occur in any system. But the catastrophe that followed this failure—which was by no means caused by an emergency—showed that the high reliability of each link in a complex system still does not guarantee the reliability of the system as a whole. Officially, I was the number one person responsible for the control system as a whole; and then after me, my deputies and the testers—certainly not Mishin. We were supposed to have thought up how to inhibit the SAS before the beginning of the actual flight and how to inhibit the damping system after separation so that all the DPO propellant would not be lost. After what happened, simple modifications and operations were conducted to keep similar events from happening again.

Speculation about what other errors remained undetected at the interfaces of other systems was agonizing. The events of the following year, an "anniversary" year, unfortunately confirmed these fears.[17]

---

17. The anniversary year was the 50th anniversary of the Bolshevik Revolution in 1967.

IN THE LAST DAYS OF THE DEPARTING YEAR, HOPING IN SOME WAY TO COMPENSATE FOR THE 7K-OK FAILURES, Tyulin, who had been appointed chairman of the State Commission for the 7K-L1, became particularly active. For the first time, he succeeded in bringing together Chelomey and Mishin and the chief designers of the leading organizations involved in the piloted circumlunar flight program "in honor of the 50th anniversary of the October Revolution."

At the State Commission meeting on 24 December, Chelomey reported that launch vehicle UR-500 had already flown four times.[18] The UR-500K differed from the UR-500 in that it had a third stage, which increased its payload in near-Earth orbit from 12 to 20 metric tons.

After Mishin's speech, in which he proposed a two-launch "stopover" (*podsadochnyy*) scenario for the piloted flight, we had a discussion. The gist of this plan was that, to begin with, the UR-500K would insert the L1 into orbit with no crew. Then a *Semyorka* would launch the 7K-OK carrying two cosmonauts. If everything was okay on the two vehicles, they would dock, and the cosmonauts would transfer from the 7K-OK to the 7K-L1 via spacewalk. They would set out for the Moon, and, after flying around it, they would return to Earth. This was a complicated scenario. We still hadn't had a single successful flight of the unpiloted 7K-OK vehicles or of the UR-500K launch vehicle, we had never executed a rendezvous or a docking, we still had no docking version of the vehicle, yet we had already decided that two cosmonauts would fly around the Moon no later than 10 months from now!

I told Bushuyev and Feoktistov, who supported Mishin's scenario, that to promise a flight like that before the anniversary holidays was pure folly. Barmin spoke out emphatically in favor of a *direct* flight on the UR-500K without the "stopover" scenario but under the condition that there would be at least four preliminary unpiloted flights. Pilyugin, who had a stake in the flight development of the stage-three control system and the L1 itself, supported Barmin. Tyulin gave instructions to work through both options.

Just one week later, he convened a second meeting of the State Commission simply to verify readiness for the first unpiloted launch of the 7K-L1 on the UR-500K. Chelomey, Mishin, Barmin, and Pilyugin reported on the readiness of the UR-500K, the first 7K-L1, the launch pad, and control system for a launch on 15–20 January. Ryazanskiy and Spitsa announced that 100 million

---

18. The first four launches of the (two-stage) UR-500 launch vehicle were on 16 July 1965, 2 November 1965, 24 March 1966, and 6 July 1966. On the third launch attempt, the rocket failed to deliver its payload, a scientific satellite known as Object N-4, into Earth orbit. Payloads for the other three launches were openly announced as *Proton*, *Proton-2*, and *Proton-3*.

rubles needed to be allocated to re-equip the KIK.[19] According to Kamanin's report, as of the New Year they were supposed to start training crews for the L1 vehicles regardless of the readiness for flights on Soyuzes. Everyone left the State Commission wishing one another a happy New Year.

Mishin, Bushuyev, and I dropped into Tyulin's office to have a private conversation with him and get his support for our long-range plans for 7K-OK launches. He said that recently Chelomey had been more actively pushing for his own version of a lunar landing expedition. The draft plan of the super-heavy UR-700 rocket—with new engines designed by Glushko—would allow the possibility to haul up to 140 metric tons into near-Earth orbit. This was at least as good, and perhaps better than the American Saturn V. An expert commission had perceived nothing illegitimate in the design, except that construction was already under way at the firing range for the N1, hundreds of millions of rubles had been spent, and everything needed to be started all over again for the sake of the UR-700. "Uncle Mitya" and Smirnov were not supporting Chelomey. Our minister [Afanasyev] was wavering, as was Keldysh. Much would depend on how things went with us.

"Before the October holidays piloted 7K-OKs need to dock. This will somewhat alleviate the discontent at the highest levels, even if we don't manage to pull off a circumlunar flight," Tyulin concluded.

ON 18 JANUARY 1967, WE WISHED MISHIN A HAPPY 50TH BIRTHDAY. Circumstances did not lend themselves to organizing a large-scale celebration. Mishin understood this and the event was limited to the arrival of guests with testimonials and a modest number of toasts in the office that Korolev had left just a year ago. The mood had been somewhat tainted the day before. At the State Commission meeting Keldysh had deemed it necessary to fling a reproach at Mishin: "The technical chief needs to take on a directing role. The Council of Chiefs meets extremely rarely."

And nevertheless, at the beginning of the year they managed to modify 7K-OK No. 3, implement all the well-thought-out measures, and prepare it for a single unpiloted launch in early February.

I got together a team to fly out to the Crimea, having arranged with Yurasov and Ostashev that the preparation of the vehicles at the engineering facility was entirely on their conscience. Preparation had been simplified by the fact that rendezvous had been eliminated from the program and we wouldn't have to test such a troublesome system as the *Igla*.

---

19. Ivan Ivanovich Spitsa (1919–92) served as chief of the Soviet Command and Measurement Complex (KIK) from 1965 to 1973.

We were all shaken by the news of the death of the three American astronauts Virgil "Gus" Grissom, Ed White, and Roger Chaffee on 27 January. Rather than having died during spaceflight, they had burned alive on the ground during a training session inside the Apollo spacecraft, which was sealed up according to procedure. The spacecraft had been installed on the Saturn 1B launch vehicle. The American media did not skimp in describing the details of the tragedy. Pure oxygen was used for breathing in the Apollo life support system. A spark generated by one of the instruments caused the flammable plastic material to ignite in the atmosphere of pure oxygen. The astronauts burned and suffocated. Attempts to quickly open the hatch from the inside failed, and for some reason help from the outside was delayed. NASA leadership was subjected to extremely harsh criticism. We decided we needed to immediately prepare a finding stating that this could not happen on our spacecraft. In fact, the atmospheric composition in our life support systems corresponded to ordinary air, but fires also happen in an ordinary atmosphere. For the upcoming launch we didn't have time, but for subsequent launches material engineers received a joint assignment with fire-fighting specialists to prepare a report about all the materials used inside the vehicles from the standpoint of their fire safety. Of the three astronauts who burned to death, two had already been in space.[20] Judging by the press reports, America was in shock, and the launch dates for the next Apollo were postponed for an unspecified period of time.

The 14 December disaster at Site No. 31 and the American tragedy prompted the development of additional safety measures. We swore to implement many of the sensible measures proposed before the next piloted launch. A great deal of work was done so that the SAS would be "rendered harmless." Experimental operations performed by Bushuyev's and my departments together with the parachute specialists, gyroscope specialists, and solid-propellant engine specialists enabled Bushuyev and me to issue a summary finding; we argued that, in the event of an emergency at the launch site, activation of the SAS was guaranteed upon receiving a command from the bunker. Meanwhile, the parachute system would activate at an altitude of at least 800 meters, and the landing range would be from 100 to 170 meters from the launch site. On 14 December, the descent module touched down 300 meters away.

As part of the entire primary GOGU contingent, we flew out to the Crimea on 2 February. After everyone had been given their group assignments and had performed the first training session, we realized that there was significantly

---

20. Grissom had flown on *Mercury Redstone 4* (1961) and *Gemini III* (1965) while White flew on *Gemini IV* (1965).

more order and no commotion at all during the training session. On 6 February at T-minus 4 hours, the firing range told us to stand down for 24 hours. They had found a "plus" on the vehicle hull. A "plus" or "minus" short-circuiting of the vehicle's electric power grid to the hull was determined by the special indicator at the 11N6110 station located in the bunker. The testers hated this indicator because the illumination of the red "Hull" indicator light was not a rare event, and searching for the specific site of the short-circuit took a lot of time and messed up the preparation schedule.

On 7 February at 0620 hours Moscow time, 7K-OK No. 3 lifted off. After it was determined that it had successfully entered into orbit and it could no longer be kept a secret, it was given the name *Kosmos-140*. Right off the bat the situation at our command center could not have been more intense. The first assignment was to verify the passage of commands to the spacecraft, run tests of the attitude control system using the 45K star tracker, check out the power supply system, and then check out the SKDU and DKD.

A group of cosmonauts arrived from the firing range along with Kamanin. They were candidates for the future piloted flights. Gagarin was part of our control group as a member of GOGU.

Up until the third orbit it seemed that everything was proceeding normally. After receiving the report "so far everything is okay," Mishin and Kerimov flew from the firing range to join us in Yevpatoriya. After that, Murphy's Law started to take effect. One principle of that Law says: *"If everything seems to be going well, you have obviously overlooked something."* This was a guiding principle for American engineers during rocket testing. It proved to be even more valid when applied to spacecraft.

The troubles started during the fourth orbit. According to the program, before it went into its "silent" orbit, i.e., out of communication range—the vehicle was supposed to orient the solar arrays toward the Sun and spin about the "solar axis," maintaining its attitude like a free gyroscope. This would charge the chemical batteries.

The command to spin failed to go through. We didn't have time to figure out whether the new long-range radio complex or our on-board automatic systems were at fault. Before the vehicle entered its silent orbits, the analysis group "cheered us up" with a report that the attitude control system was consuming propulsion fluid very rapidly. The cause of this was unknown. Around 50 percent of the propulsion fluid had been lost just during the period of time it took to run the stellar orientation tests. Overall, inexplicable wonders were happening with our stellar orientation system. Our optics specialists could not determine whether we had found the requisite star to set up orientation or whether the 45K star tracker had locked onto some point of light. Our early discipline was destroyed by many things: frantic demands for information, on-the-spot discussions, dozens of freelance suggestions as to how the program should proceed, summons to explain what was going on to Kerimov and Mishin, and reports to Moscow.

After ferocious discussions, a decision was taken to at least have an altitude margin to avoid plunging into the atmosphere before the supply of electrical power was exhausted. We would have to boost the orbit, and this would require firing the SKDU. If stellar orientation wasn't working, then we still had ionic orientation as a backup. During the 20 second orbit we managed to fire the correcting engine, which boosted the orbit so that the ballistics experts said, "That's enough for a month, now figure out how to get it back to the ground."

We decided to try to get it to spin on the Sun again using the 45K solar-star tracker. Again it failed to spin! What the hell was going on? Any flare light can interfere with stellar orientation, but how do you block out the Sun?

It was a small consolation that the automatic equipment that would be backed up by a cosmonaut during piloted flight, had failed to operate. Someone had told Gagarin that I was the main guru for fully automatic control and that I opposed cosmonaut intervention. Having found a moment in the flow of the discussion about "what to do during the upcoming orbit," he did not miss the opportunity to say with an innocent smile: "If I were up there, then wouldn't I be able to spin on the Sun?"

I had to agree with this. For a cosmonaut, orienting on the Sun and spinning was not a problem. It remained to decide what method of attitude control to use to fire the engine for a braking burn. *Kosmos-140* had already flown for more than two days without recharging the storage batteries. Where had such electrical power reserves come from? I broke away from the general throng and turned to Irina Yablokova—the "mistress of silver batteries"— with this question. She had not gotten involved in the general arguments, the whole time calculating the remaining ampere-hours.

"Irina Yevgenyevna! According to my rough estimates, you should be hysterical. Why are you so calm?"

"Now you're using up the NZ [emergency supply] that only I and a couple of colleagues know about.[21] It's enough for a maximum of one more day. Today it's quite enough for a landing."

This was an example of when playing it overly safe came in handy. Each of our subcontractors ought to have a resource in reserve that he or she does not divulge to anyone, including the general contractor.

---

21. NZ—*Neprikosnovennyy zapas.*

Here's an example of the opposite: at the regularly scheduled meeting, after listening to his coworkers, Rauschenbach announced that he could not guarantee the reliability of orientation in the ionic orientation (IO) mode. The engines' exhaust gases might prove to be fatal interference for the ionic tubes.

Meanwhile, loading the settings for the braking burn proceeded normally. Now all they had to do was wait: would the *ionka* go astray? Hoorah! During the descent session, telemetry from the spacecraft reported that the integrator had shut down the engine! The engine had fired for the calculated length of time! Separation took place and the descent module switched over to power from its own autonomous battery.

"Now I hope you won't ream me out for intentionally concealing the ampere-hours," said Yablokova with such a disarming smile that all I could do was ask for her consent to introduce these margins into our future calculations.

"But I can't give my consent for that. Ask Lidorenko. It's always a good thing for someone to help you out in a pinch."

On the descent trajectory, the ground didn't receive any signals. After the calculated touchdown time, weak signals began to come in, not from the nominal touchdown area, but from the vicinity of the Aral Sea. At first we didn't believe it, but after 4 hours the spacecraft really was detected on the ice of the Aral Sea. While we were analyzing the results of all the pre-launch recordings in Yevpatoriya, we got word that the spacecraft had sunk. Only the parachute remained on the ice.

"That's because it's ashamed that it missed the calculated point by 500 kilometers," joked those who were not guilty of the latest troubles.

It took four days to finally raise *Kosmos-140* from the bottom of the Aral Sea. On 16 February, the State Commission convened to review all the peripeteia of the flight. It turned out that a hole formed in the bottom of the spacecraft during its descent into the atmosphere. The burn-through occurred because the heat shield had been damaged during the installation of a temporary cap.

This was a good lesson! If this happened on a piloted spacecraft, the crew would have died without spacesuits. Even with a spacesuit, death would still be possible if incandescent gases burst into the descent module through the hole in the front shield. The State Commission formed a working group to investigate all the troubles that occurred with regard to the control system.

At this same meeting, Tyulin decided to check on the state of affairs with the 7K-L1. He announced that on 4 February 1967, the Communist Party Central Committee and Council of Ministers had issued a resolution noting the unsatisfactory state of operations for the fulfillment of the preceeding resolution of 3 August 1964. The new resolution resolved "to consider the circumlunar flight of a piloted spacecraft and Moon landing as projects of critical national importance."

"Soon you will receive the text of the resolution in the ministers' orders. But I can tell you that the resolution has set the deadline for the first piloted circumlunar flight at June-July 1967, and for the first lunar expedition at September 1968."

This report was not received with enthusiasm. It was clear that the deadlines had been set by the Central Committee and Council of Ministers as a "call to arms," and they were far from the realities of our lives.

Before June 1967, four unpiloted L1 spacecraft were supposed to be launched. I availed myself of the opportunity and gave a "scorching" speech directed at our subcontractor friends who had disrupted every possible delivery deadline and thereby made these plans unrealistic. Ryazanskiy took offense and after the meeting said to me: "You could have called me and warned me over the telephone instead of rearing your head at the State Commission with these complaints. You and Mishin raise a big stink over the deliveries to distract attention from your own sins!"

Nevertheless, we managed to fully equip, prepare, and on 10 March send off toward the direction of the Moon [sic] the first simplified vehicle No. 2P from the 7K-L1 series. TASS announced the launch of the latest Kosmos vehicle, *Kosmos-146*. The flight program of 7K-L1 No. 2P did *not* call for a lunar flyby and return to Earth at reentry velocity. We assigned a mission of testing acceleration to lunar reentry velocity using Chelomey's UR-500K launch vehicle and our Block D upper stage. The experiment was a success. Out of five launches, the UR-500K now already had four successful ones.

On the other hand, the status of the Soyuz program was not encouraging. The first Soyuz was destroyed in flight by the APO system. The second set the rocket on fire on the launch pad, but then proved the reliability of the SAS. The third racked up such a number of inflight failures across various systems that, after their thorough investigation and modifications, one ought to have executed another launch to have a clean unpiloted flight.

However, common sense was smothered by the tendency to privilege ideology—to obtain, come hell or high water, outstanding results and demonstrate the reliability of our technology in time for the anniversary of the October Revolution. At the very same time in the U.S., astronauts were burning alive while still on the ground.

Now, after three unpiloted failures, it was difficult to say who had the initiative to make the leap and accept a program calling for the launch and docking of two piloted Soyuzes. Active vehicle No. 4 was supposed to carry one cosmonaut. Twenty-four hours later, according to the program, vehicle No. 5 carrying three cosmonauts would be inserted in orbit. After docking, two cosmonauts from vehicle No. 5 would transfer through open space to vehicle No. 4. The process involved a double airlock operation that had never been tested involving egress from No. 5 and entry into No. 4. This had been conceived as a rehearsal for the lunar program (that program called for the transfer of one cosmonaut from the Lunar Orbital Craft (LOK) to the Lunar Craft (LK), and then, after walking on the Moon, liftoff and docking in lunar orbit, and the cosmonaut's return to the LOK also by external transfer).

The risk of having an overweight design and the existence of rigid deadlines at that time precluded making a radical decision—to develop a docking assembly with a hatch for internal transfer without performing acrobatic tricks in open space. This assembly was not developed until 1970, based on an idea proposed by designers Syromyatnikov and Utkin. The idea was snatched up by Vilnitskiy and found an enthusiastic supporter in the person of Feoktistov. The first such assembly was tested in flight in April 1971 on *Soyuz-10* and on the first *Salyut* long-duration orbital station. On the first attempt, transfer through the internal hatch didn't take place due to a malfunction in the retracting mechanism. It wasn't until June 1971 that the *Soyuz-11* crew successfully transferred through the internal hatches from their spacecraft to the orbital station.

The general front of operations to increase reliability was, by the standards at that time, very broad. We were improving our ground testing procedures at the control and testing station (KIS) and the engineering facility and making them more rigorous. We required that the subcontractors provide detailed conclusions with three signatures—the chief designer, the factory director, and the military representative—affirming that the delivered items were in compliance with regulations for 3KA vehicles.[22]

Mishin and Bushuyev were damaging our relationship with the Air Force in arguments about the makeup of crews. Rauschenbach's departments were setting up experiments with the 45K star tracker, trying to understand why stellar orientation and spinning on the Sun had failed. The idea of orientation using an infrared vertical came up. Technical assignments (TZ) were quickly approved and the Geofizika Factory began manufacturing the instrument, an analogue of those that were already tried-and-true on Zenits

---

22. Chertok is referring to the "Regulations for the 3KA" issued in 1960, which enumerated quality control procedures and assignment of personal responsibility to particular chief designers to ensure that each system would operate without failure in flight. The 3KA was the operational version of the Vostok piloted spacecraft.

and Molniyas.[23] The descent control system specialists modified the circuitry of the programmer, which had switched vehicle No. 3 out of a flat guided descent mode into a steep ballistic one and driven it into the Aral Sea!

In the quest to raise the numerical reliability indicators, the developers of the electric circuits and instruments provided backup for elements and circuits. The numbers really did improve, but during testing they failed to find many false connections and to prove that each of the parallel lines was in good working order. The hard-fought campaign to achieve reliability spread to the instrument-building factory in Ufa and then to Kievpribor and other factories. My comrades and I had to fly out to Ufa and Kiev to work with factory specialists to develop ways to make the instrument testing procedures more rigorous.

At the Air Force's experimental airfield near Feodosiya, we continued to accumulate statistics proving the reliability of the parachute system. Mockups of high-explosive bombs (FAB) and Descent Module mockups were dropped.[24] The primary parachute system (OSP) and backup parachute system (ZSP) were tested in many dozens of drops from an airplane.[25] And nevertheless, the cosmonaut who would fly on 7K-OK No. 4 was already condemned. No ground experiments or the most thorough pre-flight tests could have saved him.

On 25 March at the Kremlin, Smirnov held a meeting of the VPK at which they examined the progress being made in preparation for the piloted Soyuz launches. According to the program that Mishin laid out in his report, the active vehicle would be launched on 21–22 April ("on readiness"), and the passive vehicle, the next day. The active vehicle would carry one cosmonaut, and the passive one—three. After a successful docking, two cosmonauts were to execute a transfer "through open space" from the passive to the active vehicle. A day later, after undocking, both vehicles would return to Earth. Karas reported on the readiness of the Command and Measurement Complex (KIK). Kutasin reported on the readiness of the search and rescue facilities. And Kerimov, summing things up as chairman of the State Commission, confirmed that operations were going according to schedule and there were no doubts as to the reliability of the vehicles. Kamanin introduced the Soyuz crews—12 men in all. The primary crews included Komarov (for the active vehicle) and Bykovskiy,

---

23. The TZ—*Tekhnicheskoye zadaniye* (Technical Assignment)—was an important stage of the R&D process in Soviet industry which articulated the basic requirements of a particular system.

24. FAB—*Fugasnaya aviatsionnaya bomba.*

25. OSP—*Osnovnaya sistema parashutnaya*; ZSP—*Zapasnaya sistema parashutnaya.*

Khrunov, and Yeliseyev (for the passive vehicle). Kamanin's announcement that Gagarin would be Komarov's backup came as a surprise.

The meeting ended with a decision to approve the program and prepare a report for the Central Committee. Bushuyev and I remained in the reception room to wait for Mishin, whom Smirnov had detained along with Vershinin and Kamanin. Seeing his opportunity, Mrykin approached us. A guilty smile had replaced his usual stern, preoccupied expression.

"Everyone is satisfied with the decision to resume piloted flights. Have you thought twice about it? After all the reshuffling, we should do one more unpiloted test launch. Everyone is in such a hurry. I don't want to interfere in your business."

Bushuyev and I reassured him. Everything had been examined. All the systems would be certified. But, oh, how right Mrykin was! Mishin and Kamanin emerged excitedly from their talk with Smirnov. On the way to Podlipki, Mishin said that Smirnov had detained them to discuss a matter about Gagarin.

"We have no right to risk Gagarin," Smirnov announced on behalf of the Central Committee and government. Mishin supported him. But Vershinin and Kamanin strongly objected. In their opinion, Gagarin shouldn't be deprived of the opportunity for spaceflights.

Smirnov said that assigning crews was the business of the Ministry of Defense and the State Commission, and the Gagarin issue would be decided not by him, but by the Politburo. When the Politburo forbade risking Gagarin's life on Soyuzes or during the circumlunar flight on the L1, it hadn't dawned on them to prohibit him from flying in conventional training fighter planes. This simply hadn't occurred to anyone.[26]

Yurasov and Ostashev were in charge of preparing the Soyuzes at Site No. 31. According to their reports, aside from minor glitches that they dealt with on the spot, everything proceeded normally. Tregub, Agadzhanov, and I were tasked with verifying the readiness of personnel, documentation, and the work of all the ground services so that from the very start control could be conducted from the Yevpatoriya center. I was supposed to fly back to the Crimea before the launch and act as technical chief in Yevpatoriya until Mishin arrived.

But the second experimental launch of the L1 was wedged into this schedule. This time, the program called for it to fly around the Moon and train the "ground" in how to control the return program. Mishin flew out to the firing range on 6 April to participate in the launch, having first sent Tregub and me to Yevpatoriya.

---

26. Chertok is foreshadowing Gagarin's death in a regular air training accident one year later.

According to information from the firing range, the launch of the UR-500K carrying vehicle L1 No. 3P on 8 April at seconds past 1200 hours went very well. The spacecraft entered into an intermediate orbit and now, 24 hours later, from Yevpatoriya we were supposed to give the command for the second firing of the Block D to boost it toward the Moon. Mishin flew out from Tyura-Tam to join us in Yevpatoriya to supervise the experiment.

The second firing of the Block D failed, and the guilty party here wasn't the Block D, control system, or engine. The culprits were the people who had been tasked with changing the circuitry of the instrument's automatic equipment, which controlled the second firing. Someone was to blame for failing to do this. I felt miserable seeing Mishin get mercilessly berated by Tyulin over the high-frequency telephone line from the firing range. We could only imagine that Tyulin himself, as chairman of the State Commission had received all the requisite "usual compliments" from Ustinov for the latest Kosmos, no. 154. Our supply of simplified vehicles was used up. Now we would have to launch flight-ready L1 spacecraft fitted out with all the systems.

On 10 April we returned to Moscow feeling thoroughly depressed. On 12 April we celebrated Cosmonautics Day for the second time since Korolev's passing. Finally, on 14 April, Mishin and I departed for Tyura-Tam early in the morning. After a two-year hiatus, the piloted spaceflight program was about to start up again.

Chapter 20
The Death of Komarov

On the evening of 14 April 1967, in a packed hall at Site No. 2, Kerim Kerimov chaired a meeting of the State Commission, during which the test results for 7K-OK vehicles No. 4 and No. 5 were presented and the decision was made to fill the engine units with propellant and the attitude control and docking engine systems with propulsion fluid.

Igor Yurasov and Arkadiy Ostashev had supervised the testing and preparation of the vehicles at the engineering facility. In the State Commission meeting, Yurasov gave his report, providing a detailed account of all the test cycles of both vehicles. Colonel Anatoliy Kirillov gave a report as a co-presenter. He took the liberty of saying that the hundreds of glitches that occurred during testing were an indication that the vehicles were still "green." This enraged Vasiliy Mishin, and he sharply upbraided Kirillov, saying that he'd "teach him to work."

After the meeting, the insulted Kirillov turned to Yurasov and me and said, "Would you explain to your boss, if he doesn't understand this, that I am not a little boy and shouldn't have to listen to such rantings. I am just as interested in success as he. If something goes wrong, as an academician, there's nothing they can do to him, but for me, in the best-case scenario, I'll be charged with professional incompetence."

From the author's archives.

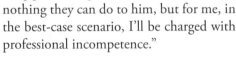

Anatoliy Semyonovich Kirillov (1924–87) played a key role in launch operations in the early days of the Soviet space program. As head of the First Directorate at Tyura-Tam from 1960 to 1967, he supervised launch operations for many of Korolev's key programs, including the Vostok and Voskhod launches. After a brief stint as deputy commander of Tyura-Tam, he later transferred to a civilian post in the Ministry of General Machine Building.

629

Alas! It would seem that neither a "tactless" academician, nor a very experienced tester, nor dozens of other specialists who had gone through "fire, water, and brass trumpets" could foresee what was to happen 10 days later.[1]

The next day Boris Rauschenbach and I were studying Nikolay Kamanin's claims regarding the cosmonaut training program. Actually, we had already coordinated the program with the Air Force. A conflict arose about the fact that only 4 hours were set aside in the training schedule for the crews to train inside the vehicles. We agreed that Rauschenbach would administer additional lessons with an assortment of all possible situations that might occur during the processes of approach, manual orientation, and manual solar inertial spin mode, paying particular attention to the constant monitoring of the consumption of propulsion fluid in the attitude control system.[2]

Our meeting with the cosmonauts was not without controversy. Gagarin and Komarov requested that automatic approach up to 200 meters be approved in the flight program and that final approach be performed manually. The previously approved program did not call for manual final approach.

In view of the disagreements, Mishin decided to bring this matter before the Council of Chief Designers. Actually, rather than a meeting of the Council of Chiefs, a wide-ranging meeting took place, attended by all the members of the State Commission, cosmonauts, Cosmonauts Training Center instructors, and firing range testers. Armen Mnatsakanyan, who spoke first, argued that rendezvous and final approach must be fully automatic. Konstantin Feoktistov supported the cosmonauts' proposal. Mishin viewed this as a betrayal of the chief designer's position. I advocated a compromise version, under the condition that, in the automatic mode, we would go up to the final approach zone—200 meters. If the Main Operations Control Group (GOGU) had no objections, based on the results of a preliminary analysis of the systems' inflight operation, the cosmonauts would be cleared for manual final approach.[3] And that's what was decided.

On the evening of 20 April, Kerimov once again convened the State Commission. Mstislav Keldysh, Valentin Glushko, Nikolay Pilyugin, and Vladimir Barmin, who had arrived that same day, appeared at the State Commission meeting. They were all in a very combative mood. Later it turned out that, the day before, Kerimov and Mishin had complained to Dmitriy Ustinov that the most

---

1. "Fire, water, and brass trumpets" is a reference to a Russian folktale in which the hero is tested by adventures in which he has to contend with fire, water, and being celebrated as a hero (brass trumpets).

2. The solar spin mode was to keep the Sun's radiation evenly distributed over the spacecraft's outer hull, a procedure U.S. astronauts refer to as "barbecue mode."

3. GOGU—*Glavnaya operativnaya gruppa upravleniya.*

crucial launches since Gagarin's flight were coming up and the State Commission was forced to make a decision without having a quorum. Ustinov responded and "recommended" that all "full-fledged members" depart immediately.

It was a great honor for Kerimov to hold a meeting of the State Commission with almost the same makeup as the one that Konstantin Rudnev had convened in April 1961. Six years had passed. Once again, it was April, and a light breeze brought the inimitable aromas from the endless steppe. And once again, the order of business was to approve launch dates and the crew compositions.

According to the ballistics specialists' calculations, the launch times fell within a window from 3 to 4 a.m. After a brief discussion, a launch date of 23 April at 0335 hours Moscow time was approved for 7K-OK vehicle No. 4 (the name used in open publications would be *Soyuz-1*). If there were no counter-indications within 24 hours, the launch of 7K-OK No. 5 (*Soyuz-2*) would take place on 24 April at 0310 hours.

Something was missing to raise our spirits to the level of April 1961. We didn't have the celebratory mood we'd had back then.

"I guess we're all missing something," said Ryazanskiy, with whom I had shared my doubts during some difficult moments. "We're missing Sergey and Leonid."

Asif Siddiqi

The original *Soyuz-1* and *Soyuz-2* crews having arrived at Baykonur in April 1967. From the left are Vladimir Komarov, Valeriy Bykovskiy, Yevgeniy Khrunov, and Aleksey Yeliseyev. In the background on the extreme right (in glasses) is Chief Designer Vasiliy Mishin. Academy of Sciences President Mstislav Keldysh is between Khrunov and Yeliseyev in the background.

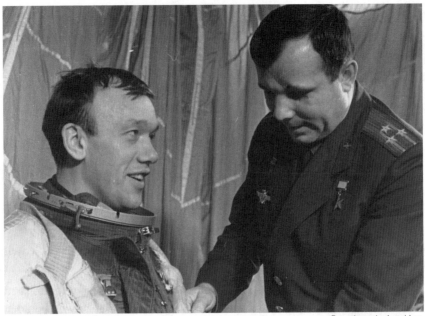

Cosmonaut Aleksey Yeliseyev and Yuri Gagarin during training for the abandoned *Soyuz-1/2* mission in the spring of 1967. By the time the mission was finally accomplished by *Soyuz-4/5* in 1969, Gagarin was no longer alive.

I agreed with him.

All the chiefs on the State Commission confirmed the readiness of the two launch vehicles and spacecraft. Kirillov reported the test results one more time, but this time he refrained from criticism.

On behalf of the Air Force command, Kamanin reported on the readiness of the crews and made proposals regarding personnel assignments. Vladimir Komarov was proposed as commander of the active spacecraft and Valeriy Bykovskiy as commander of the passive one. Gagarin and Nikolayev were named as backups for the crews. Aleksey Yeliseyev and Yevgeniy Khrunov were nominated to be the passive vehicle crewmembers who would perform the spacewalk and transfer, and their backups were Gorbatko and Kubasov.

The State Commission approved Kamanin's proposals without discussion. Keldysh, Mishin, Rudenko, Kerimov, and Karas congratulated the cosmonauts and did not forget to wish them a happy landing.[4] Both commanders—Komarov and Bykovskiy—gave speeches thanking everyone for their confidence and promising to fulfill the tasks assigned them.

After the State Commission meeting, I asked Mishin why Gagarin was named as backup. After all, Leonid Smirnov had told him quite recently that Gagarin could fly only with the Politburo's approval.

"That's all Air Force monkey business," responded Mishin crossly. "They convinced Ustinov that Gagarin couldn't be chief of cosmonaut training unless he flew again."

"I was supposed to leave for Yevpatoriya with Gagarin—a GOGU member—a day before the launch of the first spacecraft. What am I supposed to do now?"

"Leave with Rauschenbach, without Gagarin. He'll be here until the launch, and then we'll all fly out to join you," said Kerimov.

After the State Commission meeting, Keldysh, Mishin, and Pilyugin traded opinions about Gagarin's designation as backup on *Soyuz-1*. This clearly rubbed them the wrong way; nevertheless, no one in the State Commission voted against this proposal. Who knows, maybe Gagarin himself arranged this with Ustinov or even Brezhnev. Realizing that it was too late to change anything now, I put the question to Kamanin again anyway: "As members of the GOGU leadership, Rauschenbach, Gagarin, and I were supposed to be in Yevpatoriya before the launch. What should we do now?"

"Now you and Rauschenbach will fly out there, and we will send you Gagarin right after the launch. The following day, we'll send off the *Soyuz-2* and everyone will head over to you."

On 21 April, we departed for the Crimea without Gagarin. Once again I had the opportunity to delight in the panoramas of the Main Caucasian Ridge from an altitude of 7,000 meters. Very likely, alpinists who had grown weary from their heavy backpacks were now gazing up with envy at the airplane passing over them. But for some reason I envied them. Admiring the mountains from the airplane for the umpteenth time, I missed the challenge of hiking through the mountains; I also missed the friends and comrades at camp with whom I would share a can of condensed milk and then wash down this delicacy with water that was so ice-cold it made your teeth hurt. Yevpatoriya greeted us with beautiful spring weather. The Kazakh steppe in the bloom of springtime was good, but the Black Sea was better still, even when it was still cold.

---

4. These men represented the Academy of Sciences (Keldysh), TsKBEM (Mishin), the Soviet Air Force (Rudenko), the State Commission (Kerimov), and the Central Directorate of Space Assets (Karas).

THE ENTIRE DAY OF 22 APRIL WAS SPENT IN TRAINING SESSIONS, readiness checks of the various services and all the tracking stations, and assigning people to cover the two spacecraft in analysis and telemetry groups. We understood that for the first 24 hours, regardless of one's official assignment, everyone would be involved with the first spacecraft and our duty chart was just a formality. On paper, the breakdown into groups, vehicles, and shifts looked quite proper.

The main control room of the Flight Control Center (TsUP) in Yevpatoriya was located on the second floor of a building in the immediate vicinity of the formidable ADU-1000 facility. We would not need the eight 16-meter dishes of this beautiful engineering design to track the Soyuzes. Dozens of other individual antennas of various calibers covered the vast area of NIP-16. Through them we would receive telemetry information, monitor orbits, transmit commands, conduct telephone communications, and observe the cosmonauts over a television channel. The dozens of antennas were just the visible tip of the radio-technology iceberg. Low buildings near each antenna housed the support equipment serviced by hundreds of soldiers and officers.

The arrival of generals, State Commission members, and members of "the brass" communing with cosmonautics was a great trial for the local military authorities. Military service is considerably more peaceful when there is nothing of interest going on in space and there are no State Commissions at the tracking station. Of course, it was bliss here compared with Tyura-Tam. During the most intense minutes, you can walk right up to an open window, refresh your smoky lungs with pure sea air, and rest your eyes by gazing into the sapphire sea sparkling with flecks of sunlight in the distance.

The launch days at the firing range in Tyura-Tam began on 22 April. We were in constant contact with the firing range and kept fully informed about the preparation process. On the morning of 22 April, at the launch pad next to the rocket, a meeting was held of those involved in launch preparation and the launch itself. Feoktistov, Kirillov, officers, and sergeants gave speeches reassuring the crews of the spacecraft that everything had been prepared reliably and the cosmonauts could rely on the technology. Spacecraft commanders Komarov and Bykovskiy gave speeches in response. No one, not a single person on the launch pad, in Yevpatoriya, at the factory, or anywhere else had any way of knowing what was going to happen. And no pre-flight tests could have detected the danger that was hidden in each of the two flight-ready spacecraft still at the factory. These first piloted Soyuzes contained a production error that hadn't been present in the previous launches or during all the tests had been conducted earlier. No one could yell, "Stop! You mustn't launch these spacecraft!"

To be refreshed when it came time to report to the State Commission, all GOGU members headed to their quarters to sleep once they had gotten off duty and eaten dinner.

At 2330 hours, the State Commission pre-launch meeting began at the firing range. Grigoriy Levin relayed to us that all the chiefs and all the services had given their readiness reports. They had read the telegram signed by Pavel Agadzhanov, Yakov Tregub, and me concerning GOGU readiness and the readiness of all the Command and Measurement Complex services.

At the launch site, the process of fueling the rocket began and was completed by 3:00 a. m. on 23 April. A bus brought Komarov and Gagarin to the launch site. Later Gagarin recalled how he had ridden in the elevator with Komarov to the upper deck of the service tower and stayed by the vehicle until the hatch was closed. "I was the last one to see him alive and said, 'See you soon!'"

After Gagarin came down to the bunker, he and Nikolayev conversed with Komarov and exchanged information on the preparation process. Everything went without a hitch, according to schedule. All the events were also relayed to us clearly and without a hitch. The rocket lifted off precisely at the calculated time, at 0355 hours. Information from the NIPs monitoring the powered flight segment raised no doubts. After 540 seconds of flight came the report that the vehicle had separated and had entered orbit. The first Soyuz spacecraft with a man on board! We applauded. But right then we remembered. Now flight control authority has officially been transferred to us.

Agadzhanov, Tregub, Rauschenbach, me, and two dozen people who had fallen silent behind our backs, were waiting for the first telemetry and for Komarov's first reports. The first report from the telemetry experts struck our strained nerves: "Per data from NIP-4 and NIP-15, all antennas are deployed. The left panel of the solar array has not opened yet . . . we're rechecking for solar current."

There was the hope that the solar panel had deployed, but that the sensor wasn't functioning. The spacecraft slipped beyond the radar horizon, which stabilized after the disturbances of separation. All we could do was wait almost 1 hour until it appeared in our coverage zone. Agadzhanov reported to the State Commission, which was waiting for information at Site No. 2 in Kirillov's office:

"This is number 12 (*dvenadtsatyy*)! Telemetry data indicates the deployment of the left solar array as not detected. All other parameters are normal. Cabin pressure and temperature are normal."

"This is number 20 (*dvadtsatyy*)!" Mishin responded. "Recheck thoroughly one more time and report! You understand that we have to make a decision about the next operation."

We knew this perfectly well without having to be reminded.

Then came the report from the analysis group. They had discovered that the backup telemetry system antenna had failed to deploy, and that the shield protecting the 45K Sun-star sensor from engine exhaust contamination had failed to retract. The solar array panel that had failed to open was in their way. The backup antenna, we could deal with that; we'd get along, but the 45K! If it didn't find the Sun and stars, there would be no spinning of the spacecraft and no solar or astral orientation for correction maneuvers.

While we debated how to report to the State Commission, they gave the 5 minute warning before the beginning of the communications session during the second orbit. The ballistics experts managed to break through and announce: "Perigee 196.2 kilometers, apogee 225 kilometers, inclination 51 degrees 43 minutes, period 88.6 minutes."

These parameters were very important if there was to be a rendezvous. But now, although we still hadn't said anything to each other, each of us understood in his own mind that there would be no rendezvous.

Finally, there was a report from Komarov. His voice was clear and calm. (The *Zarya* system worked well.)

"This is *Rubin* (Ruby). I feel fine. Cabin parameters normal. Left solar array has not opened. Spinning on the Sun did not happen. Solar current—14 amperes. KV-communications not operating. I tried to spin manually. Spinning did not happen, but pressure in the DO tanks has dropped to 180."[5]

We understood that, with the asymmetry caused by the failure of the solar array to deploy, spinning on the Sun would not be possible in automatic or in manual mode. We reported this to the State Commission. We had no time to waste: cancel the launch of the second Soyuz and decide on a landing for Komarov. It was dangerous to drag out the decision. We risked running down the buffer batteries and then . . . simply unthinkable! But the State Commission made the decision itself and transmitted the command to the spacecraft to make another attempt to spin.

"Well, this is Vasiliy Pavlovich's stubbornness," I assumed.

Flight control had split into two centers of power. Evidently, the chiefs there at Site No. 2 could not decide on the spot to cancel the second launch and the rendezvous program they'd promised Moscow. A message arrived saying that Gagarin was flying out to join us to participate in control.

That night and morning were agonizing. Only after the fifth orbit, at around 10 a.m., did we finally receive the State Commission's decision to

---

5. KV—*Korotkovolnovyy* (Short-wave); DO—*Dvigatel orientatsii* (Orientation Engine).

cancel the second launch and the command to develop a landing program for Komarov during the 17th orbit with the 18th and 19th orbits as backup. At noon Gagarin showed up, ashen from worry and lack of sleep. True, he boasted that he tried to get 3 hours of sleep on the airplane.

We had to go without sleep, and there would be no reprieve until the landing. The State Commission dictated that, "Agadzhanov, Chertok, Gagarin, Yastrebov, Raushenbach, and Tregub are responsible for the landing."

Our main difficulty was deciding which method of orientation to use before firing the engine for the braking burn. According to Komarov's report, the first orientation attempt using the ionic system had failed. During the 13th orbit, the cosmonaut made one more attempt to spin the spacecraft. But the "Solar current" did not rise higher than 12 to 14 amperes. Twenty-three to 25 amperes were needed to charge the buffer battery. After calculating the balance until the 19th orbit, the electric power group warned that after the 17th orbit, a switch to the backup battery was possible. They advised against putting off landing beyond the 19th orbit. We already understood that this was out of the question! In order to select correctly the orientation method before braking, it was necessary to analyze critically the results of all the tests and listen to the conflicting reports of specialists from various groups. It wasn't until 11 a.m., after the spacecraft had entered "silent" orbits, when there was a lull in the communications sessions, that we finally had the opportunity to more calmly mull over what had happened on the spacecraft.

Everyone agreed that three clearly pronounced failures had taken place. **First,** the left solar array failed to deploy. This not only kept the vehicle from replenishing its supplies of electrical power and limiting its orbit life, but also, the deployed half of the solar array was operating inefficiently. The mechanical asymmetry that had formed prevented the deployed half of the solar array panel from being oriented on the Sun. The mechanical imbalance caused the disruption of the spinning mode. That is why Komarov's repeated attempts to spin manually resulted in an increased consumption of propellant in the orientation engine (DO) system. It was pointless and dangerous to continue further attempts at spinning. When the approach and correction engine (SKD) is fired in the braking mode for landing, there would be a danger of a loss of stability due to the inability of the docking and orientation engines (DPO) to cope with the moment generated from the displacement of the center of mass.[6]

The **second** failure, or random glitch, was in the operation of the ionic system. Its use with the docking and orientation engines was, evidently,

---

6. See Chapter 18 for details of the 7K-OK Soyuz spacecraft engines.

incompatible. The engines' exhaust generated interference with the ionic tubes, and we risked wasting propellant and not being able to land the spacecraft at all.

The **third** failure—the 45K Sun-star sensor—could not be explained by the failed retraction of the shield. Something more serious was going on with the tracker itself.

I had no doubt that Komarov had long since grasped the complexity of the situation. He was not a young fighter pilot, but an experienced engineer and test pilot. More than once, he had risked his life during aircraft tests. Now his own composure and faultless actions, rather than automatic controls, would determine his return from space.

Had Komarov rested during the "silent" orbits? He had made attempts at spinning and manual orientation, the whole time pondering the situation like a true tester, trying to record and pinpoint in his memory everything that had happened. During the 13th orbit, only the Far Eastern tracking stations managed to hear Komarov. He reported that he had made repeated attempts to spin on the Sun. The attempts had failed. He activated the ionic sensor orientation system and once again observed malfunctions.

"On the dark side, it was difficult to orient along Earth's course manually," Komarov managed to transmit. We understood that he hadn't felt like sleeping. We had very little time left to argue among ourselves and clear the program for Komarov's return with the State Commission. The 15th orbit was already under way, and we were still arguing. The optimum orbit for landing was the 17th. During the 16th orbit, we had to transmit detailed instructions to Komarov about what actions he was to take.

Yastrebov's ballistics experts recalculated the various scenarios, trying to keep track of our arguments. During the 14th orbit, Yastrebov delivered an ultimatum: "If you don't make a decision in the next half hour, we are not going to have time to prepare all the input data for transmission to the spacecraft and a radiogram for Komarov."

I got in touch with the State Commission, and based on Mishin's fragmentary phrases I surmised that it was a madhouse there, too. Which was no surprise. Most of the members of the State Commission at the firing range and of GOGU in Yevpatoriya had gone two days without sleep. During negotiations we abandoned restraint and violated regulations concerning the use of secret communications.

After our latest showdown with the State Commission, we finally received imperative instructions to land during the 17th orbit using the ionic orientation system. Gagarin was supposed to understand everything down to the last detail in order to calmly read it to Komarov during the communications session on the 16th orbit.

The three of us—Agadzhanov, Rauschenbach, and I—checked the text that was prepared for transmission. Gagarin explained everything very well and calmly. He asked Komarov to keep talking during the 17th orbit, when the pre-startup operations would be underway, and not break contact under any circumstances. During the last seconds of contact, Mishin and Kamanin from their control station managed to wish Komarov a safe return to Earth. Now came the suspenseful wait for contact with Komarov and reports from the NIPs about events during the landing orbit.

We had a signal! We had contact!

The ballistics experts of our NIP reported that the orbital parameters had not changed. The spacecraft was not flying according to the predictions for landing. Komarov made contact and reported that, at first, ionic orientation proceeded normally, but near the equator the spacecraft pitched out of orientation and the system issued an inhibit preventing SKD firing. Landing during the 17th orbital fell through. We feverishly cleared the landing scenario for the 18th orbit with the State Commission. We felt like we were running out of time.

There, at the other end of the secure phone line, they were arguing once again. The communications session in the 17th orbit ended and we hadn't managed to transmit any new instructions for Komarov.

Finally, the only obvious and possible scenario was worked out. It had been a backup scenario, but now it had become the primary option. We proposed manual orientation on the daylight side "airplane-style," then, before entering the shadow, transferring control to the KI-38 gyroscopes. This device produced by Viktor Kuznetsov's enterprise had never let us down. After emerging from the shadow, [Komarov would] check the orientation and, if necessary, adjust it manually and issue all the proper commands at the calculated times for landing during the 19th orbit.

Meanwhile, we once again wrote a radiogram. I remembered the reserves of electricity. It occurred to Irina Yablokova, too.

"We have one or two more orbits, that's all! Then we'll automatically switch to the backup battery. That's a maximum of three more orbits . . . ."

From the author's archives.

Irina Yevgenyevna Yablokova worked at NII IT and was responsible for space-based batteries for various spacecraft.

I assured her that we would not allow the switch to the backup battery.

Once again the communications session began. Gagarin relayed the message: "Perform manual orientation on Earth's course at 5 hours during the daylight orbital portion; turn 180 degrees for landing orientation. Before entering the shadow, activate stabilization using the KI-38 gyroscopes. When you emerge from the shadow, adjust orientation manually. Hold it there! At 5 hours, 57 minutes, 15 seconds fire the SKD. The calculated time of engine operation is 150 seconds. After 150 seconds, if the integrator has not shut it down, shut down the engine manually."

Komarov understood everything. The cosmonauts had not trained for this landing scenario. We devised it out of desperation after the 16th orbit. But Komarov not only understood everything, he also executed it exactly.

Komarov's last report during the landing pass was barely audible—separation had occurred. The report was transmitted via the slot antenna of the descent module.

"The engine operated for 146 seconds. It shut down at 5 hours, 59 minutes, 38.5 seconds. At 6 hours, 14 minutes, 9 seconds the command 'Emergency-2' was issued."

After that the report was drowned out by noise. Rauschenbach was the first to come to his senses: "I got it! The DPO couldn't cope with the destabilizing moment caused by the asymmetry, and the gyroscope issued the 'Emergency-2' command after the 8-degree drift. But this isn't so bad—the braking burn was good enough. It's just that now after the 'Emergency' command, we're going to abort the guided descent and go into ballistic descent. The orientation system is shut down."

"The heat sensors will initiate separation," I relayed to Mishin.

Just then a report came over the loudspeaker: "We have separation initiated by heat sensors!"

The time was 0615 hours, 14 seconds. The analysis group managed to figure out and report that gyroscope KI-38 went into 8-degree contact at 0614:09. The SKD functioned normally. Separation occurred. At 0622, Air Defense Troops facilities detected the descent module and confirmed the ballistics experts' prediction. The descent module was headed for a landing 65 kilometers east of Orsk. The calculated time of landing was 0624 hours.

We did not expect anyone to report from the landing site. The State Commission no longer needed us. Even Gagarin was unable to find out over the complex Air Force communication system how the landing had gone.

"You'll never find out anything from General Kutasin's search and rescue service," grumbled Gagarin. "No one will get a clear answer from him until he reports to the Commander-in-Chief."

We managed to find out from our representative at the firing range that, "according to General Kutasin's report, the search and rescue service located the descent module on its parachute east of Orsk. The State Commission is departing: some for the landing site and some for Moscow."

On behalf of the entire GOGU management, Agadzhanov congratulated and thanked all those involved in the continuous round-the-clock watch and warned that, after a brief rest, by the end of the day each group must submit a report.

"Comrades! I request that you all report to the dining hall at 8 a.m. You deserve a good breakfast," announced the station chief. We accepted his offer with great enthusiasm. Leaving the duty officer to receive communications, we went our separate ways to freshen up before our festive breakfast. Breakfast really was excellent, especially since bottles of Georgian wine from the military brass' special stash—reserved for the arrival of the entire State Commission at the station—appeared on the table.

After satiating the first pangs of hunger and thirst, we finally felt that we could relax. Vying with one another for a chance to speak, each of us spoke about our experiences. We had to "pick apart" the designers of the systems that had put us in this critical situation. If only we had known that morning that we should have been thanking, rather than cursing, the ones whose fault it was that the solar panel had not deployed and the 45K sensor had failed! Gagarin did not pass up the opportunity. Turning to Rauschenbach and me with a cagey grin, he said: "What would we have done without a man on board? Your ionic system proved unreliable, the 45K sensor failed, and you still don't trust cosmonauts."

We were rather dazed; and admitting our mistakes, we promised to set up control so that a cosmonaut had access to all operations on an equal footing with "the ground." With our light-hearted debates in full swing, an officer entered and told Gagarin he had an urgent call.

"It's probably Moscow," someone guessed. "Now we will find out about the arrangements for the reception in Moscow."

About 10 minutes later, Gagarin returned. His usual genial smile was gone.

"I've been ordered to leave immediately for Orsk. The landing was off-nominal. That's all I know."

It wasn't until the end of the day, before we departed for Moscow, that we learned of Komarov's death. Late in the evening on 24 April, when I returned home, Katya met me with the instructions—call Mishin immediately! From Mishin I learned that a government commission had been formed to investigate the causes of Komarov's death. Ustinov was its chairman. Tregub, Agadzhanov, and I were supposed to quickly prepare a brief report concerning all the actions GOGU had taken, all the commands issued, and an analysis of the systems' operation. For the time being it was clear that the main cause had been the failure of the parachute system. Either the reentry control system (SUS) had functioned abnormally, or there had been a failure in the circuits issuing the commands to open the hatches—this needed to be studied.[7] Mishin was already at the landing site.

"It's a terrible scene. Komarov burned up. All the instruments burned. We must quickly find out what prevented the main parachute from unlatching."

While we were en route, all the specialists and designers were assembled at the OKB, particularly, the parachute specialists and electricians. A post-mortem analysis of possible scenarios was under way. At the firing range, commands were issued to prepare a detailed report about all glitches that had occurred during testing. That day a TASS report would be broadcast over the radio, and the next day it would be in the newspapers. The urn containing Komarov's ashes would be placed on display at the M. V. Frunze Central House of the Soviet Army (TsDSA).[8] We had a sleepless night ahead of us. This time it would be spent at the OKB analyzing diagrams and embroiled in heated arguments over different hypotheses.

The arguments quieted down when the latest news broadcast came over the radio. After a brief listing of the test mission events, the TASS communiqué said:

. . . *However, during the opening of the main parachute at an altitude of 7 kilometers, preliminary data indicate that the twisting of the parachute strap caused the spacecraft to descend at a high rate of speed, resulting in the death of V. M. Komarov. The untimely death of distinguished cosmonaut and spacecraft testing engineer Vladimir Mikhaylovich Komarov is a heavy loss for all Soviet people . . . .*

It was also reported that a decree of the Presidium of the Supreme Soviet of the USSR had posthumously awarded Komarov a second Gold Star medal and a bronze bust of the hero would be erected in his hometown.[9]

On the morning of 25 April, a large delegation of our co-workers headed by Mishin was at TsDSA. We brought a wreath and stood in the honor guard. The second Gold Star had already been pinned on red velvet next to the first in front of the urn awash with flowers. The stream of Muscovites who had come to say farewell did not dry up until late in the evening. On 26 April, the public was still being admitted to the TsDSA Red Banner Hall.

---

7. SUS—*Sistema upravleniya spuskom.*
8. TsDSA—*Tsentralnyy Dom Sovetskoy Armii imeni M. V. Frunze.*
9. Komarov was born in Moscow.

Cosmonaut Vladimir Komarov, the first Soviet cosmonaut to die during a space mission.

Asif Siddiqi

Our delegation drove in the long funeral procession from TsDSA to the House of Unions. The streets and squares were filled with crowds of people. Suslov, Keldysh, and Gagarin spoke from the rostrum of the mausoleum at the funeral. An artillery salute thundered as the urn containing Komarov's ashes was placed in its niche in the Kremlin Wall.

We returned to TsDSA for the commemoration. Of all the speeches, it is the words of Komarov's father that are preserved in my memory. He said that Vladimir's death was a heavy loss for all Soviet people. But the pain that a father experiences with the loss of a son is especially great. He understood that in exploring a new field of human endeavor, casualties are inevitable among the trailblazers. How many brave souls perished before aviation became a safe means of transportation. Volodya loved his parents, loved his country, but he had to fly. He died for a worthy cause, sparing the lives of others following after him. Neither Komarov's father speaking at that time, nor we listening to him, knew yet that these words would be fully, concretely confirmed. His ashes would haunt us until we unraveled the true causes of that disaster.

For the reader to understand what really caused Komarov's death, I must give a brief description of the parachute system of the first Soyuzes. The body of the descent module housed two containers for the parachute system, which were in the form of elliptical cylinders. The larger of the two was intended for the main parachute system and the smaller—for the backup system.

The parachute bundles are pushed with great force into the cramped containers after the entire descent module has undergone heat treatment in a special autoclave at a temperature of several hundred degrees to polymerize the thermal protective coating. Before this, the openings of the empty containers must be covered with flight lids since, being part of the exterior surface of the descent module, they have the same thermal protective coating. During descent, upon reaching an external pressure that corresponds to an altitude of

9.5 kilometers, a special pressure unit issues the commands to jettison the lid of the main parachute system (OSP).[10] After the parachutes are packed and the lids closed, the containers are airtight and normal atmospheric pressure is maintained inside them. When the container lids are jettisoned, the internal pressure [of the container] drops abruptly to the value corresponding to an altitude of 9.5 kilometers. The internal pressure of the descent module, which is close to 1 atmosphere, acts on the body of the container. Because of the pressure differential, compressive force acts over the entire surface of the container. The jettisoned lid sweeps the drogue parachutes into the main stream; these, in turn, pull out the braking parachute. The timing mechanism counts down a delay of 17 seconds needed for the braking parachute to balloon open and decelerate the descent module to the designated descent speed. Responding to the command issued after 17 seconds, the braking parachute begins to pull the bundle containing the main parachute out of the container. After the canopy of the main parachute enters the stream, the braking parachute flies away with the bag in which the main parachute was stowed.

During descent with the OSP, the impact velocity is around 7 meters per second. To reduce the g-loads upon impact, an independent soft landing system is used. One meter from the ground, a gamma-ray altimeter installed on the bottom of the descent module issues a signal to fire four solid-fuel braking engines. This reduces the landing speed from 7 to 2.5 meters per second. During soft landing, the bottom of the descent module is deformed only slightly. The shock absorbers of the seats act as a backup means to reduce the g-loads on the cosmonaut in the event the gamma-ray altimeter or solid-fuel engines fail.

In order for the gamma-ray altimeter and soft landing engines to operate, at an altitude of around 3 kilometers the main pressure unit issues a signal causing the massive heat shield enclosing the entire outer bottom of the descent module to be jettisoned.

Independently of the main pressure unit, a second pressure unit monitors the external pressure and at an altitude of 5.5 kilometers, it activates a barometric instrument that measures pressure change during a fixed period of time. If the rate of pressure change exceeds the normal rate for the descent mode using the main parachute, then a command is issued to jettison the lid of the reserve parachute system (ZSP) container.[11] During a landing using the ZSP, the soft landing system also reduces the impact velocity to 2.5 meters per second.

---

10. OSP—*Osnovnaya sistema parashutov.*
11. ZSP—*Zapasnoye parashyutnoye sisteme.*

Pyrocartridges are the final actuating elements of all the commands. They jettison the hatches and braking parachute and, on the ground, the parachute cords, etc. And where one pyrocartridge was sufficient in principle, we put in at least two of them for reliability. In the electrical circuitry, all command-issuing instruments, relays, and the cable network had redundancy. A single failure of any element of the electrical circuitry could not cause the failure of the OSP or ZSP. We developed the operating logic of the electro-automatics in the landing system in close cooperation with the Scientific-Research and Experimental Institute of the Airborne Assault Service (NIEI PDS), which was headed by Fedor Tkachev.[12] We gave Ruben Chachikyan, who developed the barometric instruments, a lot of headaches with our stringent reliability requirements.

The Soyuz landing system underwent thorough testing at the Air Force firing range near Feodosiya. Descent module mockups with standard parachute systems and our standard automatics were dropped from airplanes five times. We studied all the glitches that occurred during these drops and modified the systems repeatedly. Finally, we had the experience of landing the two previous 7K-OKs. Vehicle No. 1 (*Kosmos-133*) landed successfully using the ZSP when the emergency recovery system was actuated, and vehicle No. 3 (*Kosmos-140*) came down onto the ice of the Aral Sea, and there were no complaints about its parachute system.

Ustinov held the first meeting of the accident investigation commission in his office. After a long discussion, seven sub-commissions were formed. As one who potentially bore responsibility, I was included in the landing system sub-commission headed by Flight-Research Institute (LII) Director Viktor Utkin and the flight control sub-commission headed by Georgiy Babakin.[13] Gagarin was also a member of the latter sub-commission.

We meticulously developed electrical hypotheses for entire days. More than anything, we were alarmed by the possibility of two simultaneous failures when issuing commands to the pyrocartridges. It was crucial to understand what had actuated the pyrocartridges—was it the electrical commands or did they explode on the ground due to the high temperature of the fire? The developers of the landing system automatics were completely demoralized by the abundance of unsubstantiated electrical scenarios. I asked that Karpov and Petrosyan be brought in to assist them with their work in the sub-commissions.

Tsybin, whom Mishin had summoned to Orsk right away to help our group of descent module specialists and serve as the official TsKBEM representative, described

---

12. NIEI PDS—*Nauchno-issledovatelskiy eksperimentalnyy institut parashyutno-desantnoy sluzhby.*

13. LII—*Letno-issledovatelskiy institut imeni M. M. Gromova*—is one of the leading Russian establishments for testing high-performance aerospace systems.

in detail what they had found at the crash site. Tsybin, Sergey Anokhin (who had accompanied him), and all of their fellow travelers were devastated not only by the very fact of Komarov's death, but also by what they saw at the crash site.[14]

"I can't tell you how many burned up airplanes I saw during the war," said Anokhin, "but there was no comparison with what we saw there. Hydrogen peroxide proved to be much more horrible than gasoline."

Asif Siddiqi

A still from footage of the *Soyuz-1* wreckage soon after the crash that killed cosmonaut Vladimir Komarov. His remains had still not been recovered.

On impact with the ground, there was an explosion and a fire started. The descent module tanks still contained around 30 kilograms of concentrated hydrogen peroxide, which had served as the gas for the guided reentry system engines. It doesn't just burn; it expedites the combustion of anything that isn't burning by giving off free oxygen as it breaks down. Due to the off-nominal high rate of descent, rather than being jettisoned at an altitude of 3 kilometers, the front shield was jettisoned right at ground level. The command to power up the gamma-ray altimeter was not executed, and consequently, the

---

14. Sergey Nikolayevich Anokhin (1910–86) was a Soviet test pilot who gained fame during both World War II and the early postwar era while testing dozens of new aircraft at the M. M. Gromov Flight-Research Institute (LII). In May 1964, he transferred to Korolev's OKB-1 as chief of the design bureau's new flight-testing department (Department 90) in charge of training its own civilian cosmonauts. Anokhin briefly trained as a cosmonaut, from 1966 to 1968.

command to start up the soft landing engines was not issued either. Impact with the ground was so violent that the depression that was formed was more than a half-meter deep. Local residents were the first to arrive at the crash site. They tried to smother the fire by throwing dirt on it. When the helicopters of the search and rescue service landed, fire extinguishers were used. When Kamanin arrived, he demanded before all else that they search for what remained of Komarov. His burnt remains were immediately sent to Orsk.

The main parachute was melted inside the container. The drogue, braking, and backup parachutes were intact.

After all the scraps of structural parts and instruments had been removed, including the capsule containing cesium—the source of the gamma radiation—a small mound was raised at the crash site with the State Commission members present. Anokhin removed his service cap and placed it atop this memorial landmark.

Yevgeniy Utkin, who had been in charge of our group in the search and rescue service, delivered the remains of *Soyuz-1* from the crash site to Podlipki. They were laid out in the control and testing station (KIS) room. The spectacle was horrifying. The melted and charred instruments were so deformed and mixed with Earth that it was even difficult for their own designers to figure out what was what.

The magnetic tape recordings stored in the armored cassette of the *Mir-3* telemetry system were of greatest interest for developing hypotheses. However, Sulimov and Komissarov, whom we all urged to restore this "black box" recording by any means possible, said that the cassette was melted and the recording on the remains of the tape was indecipherable. This was a heavy blow for us electricians. Only the *Mir-3* telemetry could prove that all the automatics' commands had been issued and arrived at their destinations.

The work of the sub-commissions continued for more than a month. During the process of the commissions' work, I got to know Air Force Deputy Commander-in-Chief Mikhail Mishuk quite well. He proved to be an uncommonly erudite individual and, unlike many of his colleagues, capable of unconventional thinking. It was interesting and enjoyable to talk with him both as an individual and as a specialist. His objectivity also impressed me. He did not classify people as "us," i.e., aviation specialists, and "them"—missile specialists and "spaceniks." Specialists from the Scientific-Research Institute for the Operation and Repair of Aviation Technology (NIIERAT) very meticulously and objectively studied all the spacecraft's bits and pieces.[15] Investigations of almost all aviation disasters passed through the hands of the specialists of this NII. They had a

---

15. NIIERAT—*Nauchno-issledovatelskiy institut ekspluatatsii i remonta aviatsionnoy tekhniki.*

wealth of experience and were the first to answer the question tormenting the electricians, having confirmed that all the pyrocartridges had been actuated by electrical commands and not the high temperature in the fire.

The sub-commissions studied thousands of pages of documentation and conducted a myriad of experiments and calculations. The descent module and parachutes were tested in the wind tunnels at the Central Aerohydrodynamics Institute (TsAGI). At Feodosiya, mockups with deployed braking and backup parachutes were dropped from airplanes. It was unequivocally determined that the main parachute had not come out of its container, and that the backup had been pulled out but its canopy failed to open up.

According to the commission's findings, the most probable cause of the main parachute's failure to deploy was the insufficient force of the braking parachute. Because the main parachute system's braking parachute had not been jettisoned, it created an aerodynamic shielding effect preventing the backup parachute's canopy from opening up. The possibility of their simultaneous operation had not been looked at before. The commission explained that the reason for the insufficient force of the braking parachute was that, due to the pressure differential, the OSP container had deformed and constricted the parachute bundle so that the amount of force needed to pull it out of the container was significantly greater than the pull generated by the braking parachute. When asked why this hadn't been noticed during all the developmental testing drops, the responses were not very convincing. Regarding 7K-OK vehicle No. 3 (*Kosmos-140*), there had been no pressure differential, since after the bottom of the descent module burned through, the spacecraft had depressurized. During the work of the commissions, they were unable to provide a convincing explanation for the normal operation of the OSP during test drops from airplanes.

Independently of all the sub-commissions, a brigade of our factory specialists that had remained at the firing range decided to conduct its own experiment. They had reason to be skeptical. They opened the OSP hatch, pulled out the braking parachute, attached its cords to a hoisting crane through a dynamometer, and began to lift it to measure the force at which the main parachute bundle would begin to come out. Imagine the surprise when it turned out that the 2,800 kilogram mass of the descent module was not sufficient. And after all, in this case the container had not been subjected to any differential pressure, and consequently, the parachute bundle had not been exposed to any constricting deformation. They did not report this experiment to the commission.

All told, the commission made so many recommendations to increase reliability that it would have taken more than a year to implement them.

The main action was to alter the shape, increase the volume, and increase the stiffness of the OSP container.

Eduard Korzhenevskiy suspected one additional reason: the roughness of the container's interior surface. In addition to everything else, the friction force could have been very great. At his suggestion, they decided to polish the interior surface. The wise and experienced Designer Korzhenevskiy may have suspected something else. Photographing the entire packing process of the parachute systems operation-by-operation was psychologically a very effective measure. As happens in such cases, the logic of the control automatics was changed, the timeline was changed, and many recommendations were made regarding the organization of flight control.

It was recommended that dual authority during control be eliminated. From now on, only GOGU should be in control from Yevpatoriya. The most important result of the commission's work was the decision to conduct a preliminary checkout of all its recommendations during unpiloted launches.

Before the fiftieth anniversary of the October Revolution, we got the opportunity to launch at least two unpiloted 7K-OK vehicles to verify automatic rendezvous and the reliability of the landing process.[16] Subsequent launches proved that the measures that we had developed were correct. There has not been a single failure of the parachute systems since that time and to this day.

And nevertheless . . . Many years later, when the "statute of limitations" expired and thus it made no sense to resort to reprisals, yet another, perhaps, more credible cause for the disaster was brought up that not one single sub-commission had zeroed in on. A narrow circle of people at our factory had surmised but considered it best to keep silent, especially since a production violation had been the cause and it would not be at all difficult to eliminate it for the future.

According to the standard production process, after the thermal protective coating was applied to the descent module, it was placed in an autoclave, where the synthetic resins that make up the thermal protection were polymerized at high temperature. Deviating from the approved process, all descent modules up to No. 4 and No. 5 went into the autoclave without the parachute containers. As often happens, the manufacture of the containers lagged behind the deadlines for the entire body. This was, it would seem, an innocent violation of the manufacturing process. For unpiloted launches, such a deviation was permitted. For airplane tests of the descent module mockup, they simply covered it with plastic foam without any thermal protection. Therefore, no operation in the autoclave was required.

---

16. This was the joint *Kosmos-186/188* mission. See Chapter 22 of this volume.

Beginning with No. 4 and No. 5, which were designated for piloted launches, any deviation from the standard manufacturing process was categorically forbidden. The descent modules for No. 4 and No. 5 were placed in autoclaves along with the containers. But now it turned out that the flight lids for the parachute containers had missed the deadlines. If anyone remembered what was used in place of the lids and how the containers were sealed, he wasn't telling. When I got interested in the details for the sake of this memoir, it turned out that there were no longer any living witnesses. It has been suggested that, in all probability, the containers were sealed with something, but not tightly.

RKK Energiya

The original *Soyuz-1* and *Soyuz-2* crews during training in early 1967. From left to right are Valeriy Bykovskiy, Aleksey Yeliseyev, Vladimir Komarov, and Yevgeniy Khrunov.

In other words, it hadn't occurred to the process engineers of shop No. 1 that volatile fractions formed during the polymerization of the coating might precipitate on the interior surface of the containers causing the surface to become grainy, bumpy, and sticky. It actually turned out to be beyond the power of the braking parachute to pull the tightly packed main parachute out of such a container.

Now the successful tests of parachute systems during airplane drops were easily explained. The descent module mockups for these tests did not have thermal protection, they had not been put through the autoclave, the containers remained clean and the force of the braking parachutes had been sufficient to pull out the main parachute.

After the first edition of this book was published, I received criticism from a former shop No. 1 process engineer regarding the scenario alleging that volatile fractions of the polymerized coating precipitated on the interior surface of the parachute container.

"We carefully covered the containers or used flight lids. The scenario about the parachute getting 'glued down' should be thrown out."

Eduard Ivanovich Korzhenevskiy was still alive, and I thought I needed to consult with him once again.

"Yes, there was such a scenario," he said. "But my opinion is that the parachute was so severely jammed into the cramped, rough container that no additional adhesion was required."

Flight vehicles No. 4 and No. 5 were assembled using the very same process. If both solar array panels on *Soyuz-1* had deployed after it entered orbit and if the 45K sensor hadn't failed, then we certainly would have launched *Soyuz-2* carrying cosmonauts Bykovskiy, Khrunov, and Yeliseyev on 24 April. After docking, Khrunov and Yeliseyev were supposed to transfer to Komarov's vehicle. In that case, the three of them would have perished, and shortly thereafter, it is quite likely that Bykovskiy might have perished, too.

Experiments showed that the solar panel failed to deploy because it had gotten snagged on the vacuum-shield "blanket." The star visor of the 45K sensor simply fogged up.

The persons responsible for these defects were not very severely punished administratively. According to the most current unofficial version, rather than be punished, they needed to be thanked for saving three lives and inadvertently defending the prestige of Soviet cosmonautics. Fedor Tkachev was severely punished more than anyone. Two more accidents related to the work of NIEI PDS occurred soon after Komarov's death. Deputy Minister of the Aviation Industry Vasiliy Kazakov removed Tkachev from his position as chief designer and director of NIEI PDS. Nikolay Lobanov was appointed in his place.

The shakeup that our entire space industry experienced had a decisive influence on increasing the reliability of all systems and on the subsequent program for perfecting the Soyuzes. Everyone who has safely flown, is flying; and those who will fly into space on a Soyuz should remember that they owe their reliable and safe return to the ground not only to the creators of that spacecraft, but also to Vladimir Komarov.

Chapter 21
"On the Distant Star Venus . . ."

In 1967, fate presented us with the gift of *Venera-4*, which eased the grip of gloom that had settled over the "space" atmosphere after Komarov's death.

On the morning of 17 October 1967, Mishin, who was attending a session of the Supreme Soviet of the Russian Federation, gave me the following instructions over the telephone: "Tomorrow *Venera-4* is landing. Today Keldysh is flying out to the Crimea with the minister. You must fly with them."

"Excuse me? *Venera-4* isn't ours anymore. That whole project belongs to the Babakinites!"

"Have you forgotten how much effort we, and you yourself, put into the Venus project? If it hadn't been for S.P., Babakin wouldn't have this Venera. Keldysh himself demanded that we both fly with him, but it's inconvenient for me to leave the session. I gave all the orders to Khvastunov—get going!"

At 1600 hours, I was at Vnukovo, and at 1615 our An-24 was already taxiing to the take-off runway. We would fly to the Saki military airfield. From there it was 20 kilometers to the center in Yevpatoriya. The airplane was carrying a distinguished group: Minister Afanasyev, Keldysh, Karas, Kerimov, Mrykin, and Tsarev.

Babakin, Ryazanskiy, Boguslavskiy, and "Venusian" scientists from various academic institutes had gathered in Yevpatoriya one week before the anticipated landing on the planet. *Venera-4* had been manufactured at the S. A. Lavochkin Factory, drawing on the experience that we had gained from our interplanetary designs dating back to 1960. Our last Venus project was *Venera-3*. On 1 March 1966, it delivered a Soviet Union pennant to Venus.[1] We can say this with complete confidence, despite the fact that we lost radio contact with the spacecraft before its approach to the planet. For that reason, we received no information from the descent module.

---

1. *Venera-3* was part of a flotilla of three Venus-bound spacecraft launched in November 1965 by OKB-1. The first (*Venera-2*) was launched on 12 November and intended for a Venus flyby; the second (*Venera-3*) was launched on 16 November for a Venus impact; and the third (*Kosmos-96*), launched on 23 November was intended for a flyby but never left Earth orbit.

Georgiy Nikolayevich Babakin (1914–1971), one of the most famous chief designers of the Soviet space program. As chief of the Lavochkin design bureau from 1965 to 1971, he oversaw the major expansion of Soviet lunar and interplanetary programs. He also diversified Lavochkin's portfolio into military space operations by beginning production of buses for Soviet early warning satellites.

From the author's archives.

For more than a year Babakin updated, and sometimes redesigned, our existing design and documentation. You have to give credit to his staff, who found many flaws in our design and systems. The ground tests performed on the instruments, parts, and systems were more stringent and comprehensive. The entire spacecraft was thoroughly examined, with special attention given to the radio complex.

The enthusiasm with which Babakin's staff began working on the interplanetary automatic stations did not prevent them from handling reliability problems very responsibly. They thoroughly tested an analog of the station flying to Venus. On the analog, they were able to simulate situations occurring on the interplanetary route.

On the way to Venus, all the communications sessions and astral corrections went smoothly on this Babakinite interplanetary automatic station. For that reason, those of us flying on the An-24 were confident that at least the separation of the descent module would be recorded. I was apprehensive that the minister [Afanasyev] would ask me in mid-flight to tell him about the descent program to the planet. But Andrey Khvastunov saved me when he entered the passenger cabin and, smiling broadly, offered us coffee and cookies.

When the Minister, Keldysh, or the chief designer flew, it was Khvastunov, the commander of our aircraft squadron and a Hero of the Soviet Union, who sat in the captain's seat for takeoffs and landings. This procedure had been established while Korolev was still around. For his last two years, Korolev flew only with Khvastunov. It wasn't that he had less faith in the other pilots, it was simply the way it was supposed to be—the squadron commander would stake his life on the safety of the flights and reserved the right to be commander during flights carrying distinguished passengers. Khvastunov usually met the "chief" passengers by the service building at Vnukovo-3. His sturdy physique, cordial smile, and calm report that everything was ready and permission for takeoff had been granted, exuded composure and complete confidence that it could be no other way.

For the first hour we were completely engulfed in gray clouds. Somewhere over Kharkov, at an altitude of 7,000 meters, the aircraft broke through the multiple layers of clouds and the orange light of the setting sun burst through the windows. Below us fantastic cloud peaks towered above an endless, multicolored carpet. Keldysh tore himself away from the window and, addressing everyone, he said, "Isn't that just like the landscape on Mars or Venus?"

We looked down on the clouds. The setting sun tinted them in red tones. The sky above the clouds was pure and transparent. It changed color from deep crimson at the horizon to ultramarine closer to the zenith. Everyone was captivated by the landscape. Conversations about impending events trailed off.

To get from Moscow or Tyura-Tam to NIP-16, located 10 kilometers from Yevpatoriya, our airplanes used a naval airfield near the resort of Saki. The naval aviation command always gave us a warm welcome when we landed and a cordial farewell ceremony when we departed. The road from the airfield to Yevpatoriya passed along undeveloped beaches, and after leaving Yevpatoriya, it passed through a vineyard. Thus, on the car trips from the airfield and back, we observed life at the resort during all the seasons of the year.

That year there was an abundant grape harvest. Clearly they weren't coping very well with it. Masses of grapes were being loaded onto dump trucks and the road was strewn with ripe clusters of grapes for lack of people to gather them up.

We got to Yevpatoriya by dinnertime. The main contingent of "Venusians" had already gone to bed. Communications sessions with *Venera-4* began at 4 a.m. The flight was in its 128th day.[2] All along the flight route, communications had been quite satisfactory. The parabolic antenna developed by our antenna experts from Krayushkin's department and refined under Babakin was sending information back to Earth that usually resulted in the

From the author's archives.

A group of flight controllers with the ADU–1000 dish system in the background at the Center for Deep Space Communications near Yevpatoriya.

2. *Venera-4* had been launched on 12 June 1967.

report: "Signal good, reception stable." Eight rigidly connected 16-meter ADU-1000 parabolic antennas tracking Earth's rotation were reliably tracking the station's approach to the planet.

It seemed that the nearby and mysterious planet was resorting to tricks to hide the secrets kept beneath its cover of clouds. As the station drew nearer, Venus' gravitational field increased its speed. The Doppler effect altered the wavelength of the signals received on Earth. The radio operators needed to be particularly vigilant so that the information sent from the station consistently landed in the narrow "throat" of the ground-based receivers.

"Venus isn't just a bashful planet, it's also a crafty one," complained Boguslavskiy.

After breakfast, at 0530 hours we were already in the main hall. The planetary approach communications session via the on-board omni-directional antenna began at 0536. Commands had been issued to orient the on-board parabolic antenna toward Earth. Boguslavskiy was beaming, "The signal-to-noise ratio is 1,000! 40,000 kilometers to Venus!"

As the spacecraft approached the planet, its speed increased. We perceived this by sound—by the increased frequency of the clicks in the speaker, to which the audio signal was fed from the space "speedometer."

In honor of the occasion we also set up a concise running commentary.

"7 hours, 50 minutes Moscow Time, 20,000 kilometers to Venus! Reception—stable, ratio—1,400. On-board voltage—14.5 volts."

The calculated time for the separation of the bus and descent module was announced—0738 hours.[3] When separation occurred, the signal was supposed to disappear—the parabolic antenna would turn away from Earth. After comparing the ballistics experts' prognosis with the change in Doppler frequency, Boguslavskiy announced, "Arrival on Venus is going to be 10 minutes later than the ballistics experts' prediction. We made a correction."

"One can forgive 10 minutes over a period of 128 days," said Afanasyev.

At 0738 hours came the report: "Separation command!

"Loss of signal!"

"We have separation!"

Now the bus would disintegrate and burn up in the atmosphere of Venus, and we would be on tenterhooks awaiting the appearance of a weak signal from the small sphere of the descent module covered with a thin layer of

---

3. Each Venus probe in this class (internally known as the "4V-1" type) weighed about 1,100 kilograms and split into two sections, a bus and a descent module. The bus burned up in the atmosphere after releasing the lander.

durable thermal shielding. It would enter Venus' cloud cover at a velocity in excess of 11 kilometers per second. According to the calculations, the descent module would decelerate to a velocity of 300 meters per second. If the little sphere remained intact, then the jettisoning of the hatch that pulls the parachute behind it would be crucial. If the parachute held up even somewhat, and did not burn up in the hellish atmosphere, we would obtain invaluable information.

Babakin called for silence and patience. He advised us that the temperature, pressure, and altitude would be reported virtually in real time by a well-trained group of telemetry operators.

At 7 hours, 44 minutes, shouts of elation were heard: "We have a signal!"

"We're receiving from the descent module!"

"We have telemetry!"

Reports come in one after the other: "7 hours, 46 minutes: altimeter reading—28 kilometers; pressure—960 millimeters; temperature—78 degrees. Pressure is increasing rapidly!"

"8 hours: pressure—1400 millimeters; temperature 114 degrees."

"8 hours, 12 minutes: pressure 4.7 atmospheres; temperature 146 degrees."

"Temperature inside descent module—14 degrees."

"The radio altimeter has stopped giving readings."

Someone noted over the public address system that, "It's hard to believe the altimeter now. It's out of kilter."

"Quiet! We'll sort it out later!"

Because of the loud running commentary that was interpreting the information being sent from Venus' atmosphere in real time, all of us there in the NIP-16 operations rooms felt a rare feeling of involvement in this great discovery. In my mind, I thanked Mishin for sending me here. Tomorrow, no, even today, the whole world would learn about the discovery of this mystery! But we, the ones who had made this discovery, would remain to the world a classified state secret!

"8 hours, 18 minutes: temperature—167 degrees; pressure—5.6 atmospheres!"

"8 hours, 32 minutes: pressure—8 atmospheres; temperature—201 degrees. Temperature in the descent module is rapidly rising; it's now more than 20 degrees."

"8 hours, 53 minutes: one pressure sensor went off the scale at 9.3 atmospheres; temperature—250 degrees.

Keldysh said in surprise: "Our planetologists certainly didn't anticipate that the pressure could be that high. Now we won't find out the actual reading at the surface. We should have had another sensor with a greater range."

"What's Keldysh grumbling about?" I thought.

I felt that not only I, but all of us, had been caught up in the ineffable exhilaration of this discovery! And how! We were all enjoying the fresh sea air, safe and sound, while at the same time the descent module was plunging into a veritable hellish mass of boiling goop and would soon meet its doom.

"9 hours, 14 minutes, 9 seconds: loss of signal! Now the temperature is more than 280 degrees!"

In the breathless silence of anticipation Babakin could stand it no longer, and violating established procedure, he declared, "Now it is clear that these flights have not been in vain! No one is going to start planning piloted landing expeditions to Venus!"

At 0930 hours, once again a shout rang out: "We have a signal! It's clean, no telemetry, no noise! . . . We lost the signal!"

With that, history's first transmission from Venus ended. What the actual atmospheric pressure reading was at the surface, we simply didn't find out at that time. According to calculations, the descent module was supposed to have been crushed by the external pressure, the value of which it managed to report. One of the scientists ran over to us from the telemetry operators, and gasping with joy, reported: "The primary atmospheric component is CO_2! Oxygen makes up 1 percent. No traces of water! No nitrogen!"

Our tension was transformed into elation. We began embracing one another, exchanging congratulations; there were orders to accurately process the telemetry, prepare a good meal, and report to Moscow.

Now each of us was faced with so much work and so many headaches. But for those few minutes everyone was smiling and walking on air. The restraint ingrained in us by our previous failures prevented us from expressing our feelings of personal and collective happiness any other way. We felt like jumping for joy and roaring with laughter, but in our sober condition this was impossible. For each of the participants directly involved with the creation of the Venus interplanetary vehicles or their flight control, this was perhaps one of the happiest days since the flight of the first *Sputnik*. I witnessed similar feelings of collective happiness when the photograph of the far side of the Moon first appeared, and when, after 11 failures, we finally executed a soft landing and broadcasted a panoramic shot from the Moon's surface. But that had been the nearby and generally quite familiar Moon.

And on this day, during the descent module's 1 hour of descent on its parachute, we learned more about the secrets hidden under Venus' clouds than science had discovered during preceding centuries.[4] Fate had rewarded all of us who had gathered that day at NIP-16 with a generous gift for many years of hard work. The happiness of a scientific discovery is never complete. There and then, having just regained our composure, we began to bemoan the fact that, evidently, the descent module had not reached the surface and had been crushed by external pressure. None of the planetologists had expected that it would be greater than 5 atmospheres! It turned out to be greater than 9, but how many atmospheres had it really been? It was maddening that the pressure sensors had reached their limit, that the radio altimeter had failed, and that the chemical makeup of the clouds was still unknown.

We had blazed just the first trail. The descent module's 1.5 hour stay in Venus' atmosphere had shown that a piloted landing expedition on the surface of Earth's closest neighbor in the solar system was a thing of the very distant future—significantly more distant than a landing on Mars.

After a thorough processing of all the materials, it was proven that the *Venera-4* descent module made it to around 25 kilometers from the surface. During 1968, in light of its own experience, Babakin's group would produce interplanetary stations *Venera-5* and *Venera-6*, which would be launched on 5 and 10 January 1969, respectively. The equipment on both twin stations would operate flawlessly in flight.

On 16 and 17 May 1969, the stations approached Venus one after the other and their descent modules made the first corrections to the data from *Venera-4*. They confirmed the hypothesis that the *Venera-4* descent module had been crushed at an altitude of 25 to 27 kilometers. But the descent modules of these two "sisters" would not reach much deeper into Venus' atmosphere. Only *Venera-7*, which lifted off on 17 August 1970, and reached the planet on 15 December 1970, would finally actually execute a landing on Venus' surface. The temperature at the very surface was 470°C [878°F] and the pressure was almost 100 atmospheres!

At the meeting of the interdepartmental council in Keldysh's office, Babakin enthusiastically recounted the results of the flight of *Venera-7* and was even more passionate when he talked about the *Venera-8* project, which was supposed to

---

4. The first successful space mission to Venus, NASA's *Mariner-2*, had returned a significant amount of data on Venus during its own flyby in December 1962. This was the first successful planetary mission performed by any country. Another NASA spacecraft, *Mariner-5*, had also returned important information during its Venus flyby in October 1967 (almost simultaneously with the *Venera-4* encounter).

land on the illuminated side of Venus.[5] All the preceding Venus modules had landed only on the dark side of the planet. That way it had been more convenient and easier for the ballistics specialists to calculate and control the flight. Babakin's brainchild *Venera-8* lifted off on 27 March 1972, after his death.[6]

The day after the "victory over Venus," the military brass of the Yevpatoriya center invited a select few on a fishing expedition for Black Sea beluga followed by a celebration dinner featuring a magnificent fish soup. It was rare, but nevertheless there were hours when we managed to relax and enjoy good-natured jokes about our comrades, the brass, and ourselves. During dinner I was coaxed into recounting the events surrounding *Venera-1*. It was launched on 12 February 1961, and on 20 May that same year it silently passed 100,000 kilometers away from Venus, becoming one of the many vagabonds of the solar system.[7]

If someday our future generations find themselves doing a cosmic garbage cleanup of the solar system, then, after recapturing *Venera-1*, they will find in it a pennant of the Soviet Union and will place it in a museum not far from the Scythian gold ornaments.[8] Who will know hundreds of years from now that people named Korolev and Chertok signed the document calling for *Venera-1* to carry this pennant?

IN 1992, ON THE OCCASION OF MY EIGHTIETH BIRTHDAY, I RECEIVED ALL SORTS OF CONGRATULATIONS IN THE FORM OF TESTIMONIALS AND ALBUMS. Among them was a unique *Atlas of the Surface of Venus* published in a very limited edition.[9] The cartography of the surface of Venus, impenetrable by optical telescopes, had been produced using the radar equipment of artificial satellites *Venera-15* and *Venera-16* in 1983 and 1984.[10]

---

5. The "interdepartmental council" was probably the Interdepartmental Scientific-Technical Council on Space Exploration, an interagency advisory body composed of scientists, engineers, and military officers who were responsible for long-term strategy in Soviet space science research.

6. Georgiy Nikolayevich Babakin (1914–71) passed away on 3 August 1971. *Venera-8* landed on the Venusian surface on 22 July 1972 and transmitted data from the surface for about 50 minutes.

7. See Chertok, *Rockets and People, Vol. II*, Chapter 31.

8. The largest collection of Scythian gold is stored in the famous and massive State Hermitage Museum in St. Petersburg.

9. V. A. Kotelnikov, V. R. Yashchenko, and A. F. Zolotov, eds., *Atlas poverkhnosti venery* [*Atlas of the Surface of Venus*] (Moscow: GUGiK, 1989).

10. *Venera-14* and *Venera-15* were launched on 2 June and 7 June 1983, respectively. They conducted their primary scientific missions from October 1983 to July 1984.

The mastermind behind this titanic project was Academician Vladimir Kotelnikov.[11] The G. N. Babakin Scientific-Research Center had developed the spacecraft on the basis of the stations that had once delivered the descent modules to Venus.[12] The synthetic aperture radar systems had been developed at OKB MEI under the supervision of the irrepressible Aleksey Bogomolov. Two of the largest antennas in the Soviet Union were used on the ground to receive and record the information. One—with a 70-meter diameter reflector—was located near Yevpatoriya, having replaced the distinguished ADU-1000, and the other—with a 64-meter diameter reflector—was in Medvezhiye Ozera near Moscow. The creation of the unique 70-meter antennas near Yevpatoriya and Ussuriysk was the achievement of NII-885 and Ryazanskiy. The 64-meter antenna in Medvezhiye Ozera had been constructed thanks to the tenacity of Bogomolov. The USSR Academy of Sciences Institute of Radio Engineering and Electronics, under Kotelnikov's leadership, processed the radar information transmitted from the radar stations circling Venus. Mathematical equipment developed at the M. V. Keldysh Institute of Applied Mathematics (IPM) was used for this purpose.[13] The Main Directorate of Geodesics and Cartography under the USSR Council of Ministers had published the Atlas of Venus. This directorate's publications also included atlases for road trips that I have used successfully.

Incidentally, on Venus there can be no trees or leaves—not blue, not red, and certainly not green. The images of the planet's surface that the descent modules of *Venus-13* and *Venus-14* managed to transmit confirmed this.

I mentioned that in May 1961, we lost contact with the *Venera-1* due to a failure in the on-board radio complex. After its launch on 12 February 1961, however, we were elated by our first success and convinced that we would fly to Venus, as were millions of people who had heard Levitan's voice. On that occasion, world-renowned astrophysicist Iosif Shklovskiy received the urgent order to write an article in *Izvestiya*, the mouthpiece of the Supreme Soviet of the

---

11. Academician Vladimir Aleksandrovich Kotelnikov (1908–) was one of the leading public spokespersons for the Soviet space program. He served as director of the Academy of Sciences' Institute of Radio Engineering and Electronics from 1954 to 1987. During part of this time, from 1980 to 1991, he was chairman of Interkosmos, the international space cooperation agency established by the Soviets.

12. The Babakin Scientific-Research Center was basically the public name of the Lavochkin OKB in the late 1980s and 1990s.

13. IPM—*Institut prikladnoy matematiki*. This institute had originally been the Department of Applied Mathematics (OPM) of the V. A. Steklov Mathematics Institute of the Academy of Sciences (MIAN). It separated from its parent institution in 1966.

USSR.[14] In 1991, after the scientist's death, his book *Echelon* was published.[15] It comprises skillfully crafted [non-fiction] novellas. In one of these, entitled "On the Distant Star Venus . . .," Shklovksiy recounts how his article for *Izvestiya* appeared under the headline "On the Distant Planet Venus . . . ."

Shklovskiy's article was unorthodox in the way it began:

Many years ago the excellent Russian poet Nikolay Gumilev wrote:
"On the distant star Venus
The sun shines with gold and fire;
On Venus, ah, on Venus
The trees have leaves of sapphire . . . ."

Thus, thanks to our Venus automatic interplanetary station, a scientist mustered up the courage to pay tribute to the memory of a poet who was executed by firing squad in 1921. Even during the years of the "Khrushchev thaw," Gumilev's poetry was not printed.[16] Shklovskiy sent the article published by *Izvestiya* to the poet's former wife, Anna Akhmatova, herself a renowned poet by that time. He received no thanks. It turns out that Gumilev had dedicated the cycle of poems *To the Blue Star* (*K siney zvezde*) to another woman.

Studying the atlas of Venus, I discovered a crater named after a famous Russian woman—Sculptress Anna Golubkina—my wife's dear aunt.[17] I don't know who was responsible, but not far from the crater "Golubkina" on the map of Venus was the crater "Akhmatova."

By decision of the International Astronomical Union, only female names are bestowed on all geographic formations on the surface of Venus. On the occasion of Golubkina's 120th birthday and Akhmatova's ninetieth, they each "received" a crater in 1984.

On 5 May 1989, the Americans launched the automatic spacecraft *Magellan*, which became a satellite of Venus. For two years, *Magellan* conducted radar mapping of Venus, encompassing 98 percent of its surface. Thus, the Americans earned the right to adorn the map of Venus with the names of their own illustrious

---

14. Iosif Samuilovich Shklovskiy (1916–85) was one of the most famous Soviet astronomers and astrophysicists of the twentieth century. He specialized in theoretical physics and radioastronomy. His memoirs were published in English posthumously as *Five Billion Vodka Bottles on the Moon: Tales of a Soviet Scientist* (New York: W. W. Norton, 1991).

15. I. S. Shklovskiy, *Eshelon (nevydumannyye rasskazy)* [*Echelon (Non-fiction Stories)*] (Moscow: Novosti, 1991).

16. Nikolay Stepanovich Gumilev (1886–1921) was an influential Russian poet who founded a poetic movement known as acmeism.

17. For Anna Golubkina, see Chapter 17 of this volume.

women. Through their joint efforts, Soviet and American radar astronomers have discovered 543 craters, 42 mountains, 20 plains, 27 canyons, and 16 valleys on Venus. They ran out of names, and the astronomers had to resort to mythology and use the names of ancient Greek and Roman goddesses.

Chapter 22
First Rendezvous and Docking

In modern piloted cosmonautics, three flight segments are critical. The **first** is delivery into Earth orbit. The reliability and safety of this segment depends entirely on the launch vehicle. In the history of our [piloted] cosmonautics, there have been three instances of launch vehicle failure. In each instance of failure during the powered flight segment, the cosmonauts' lives were saved by the emergency rescue system (SAS). The absence of such a system during the liftoff segment on American Space Shuttles resulted in the death of seven astronauts in the *Challenger* disaster on 28 January 1986. The seven U.S. astronauts were the only victims of the first critical segment in the 20th century.

During the **second** critical segment—return to Earth—we experienced two catastrophes—the death of Komarov on *Soyuz-1* on 24 April 1967 and the death of the *Soyuz-11* crew (Dobrovolskiy, Volkov, and Patsayev) on 30 June 1971. On 1 February 2003, the U. S. lost the seven astronauts on the Shuttle *Columbia* during the descent flight segment.

The **third** critical flight segment—rendezvous and docking—has yielded a great deal of experience in terms of off-nominal situations during the 40 years since the first attempt.[1] However, mishaps during rendezvous and docking have never resulted in disasters, even when a spacecraft ended up colliding with an orbital station.[2]

After 1957, we became accustomed to achieving "world firsts" in cosmonautics. Nevertheless, 10 years later we were not confident that we could pull off the world's first fully automatic rendezvous and docking of two unpiloted Soyuz spacecraft. A sober reliability assessment showed that the chances for success were 50/50. It was a gamble. The Americans, who had begun rendezvous experiments with their Gemini spacecraft in connection

---

1. The first successful piloted rendezvous in orbit was between NASA's *Gemini VI-A* (Schirra and Stafford) and *Gemini VII* (Borman and Lovell) in December 1965. The first successful docking in orbit was carried out by *Gemini VIII* (Armstrong and Scott) in March 1966.

2. Chertok is probably referring to the collision between *Progress M-23* and *Mir* in June 1997.

with the lunar program, were breathing down our necks. According to "open" publications, astronaut participation was an essential requirement of the American rendezvous and docking process. Traditionally, we began our design process on the premise that the process would be fully automated. It was a principle that had been established beginning with the first *Vostok*.

In his distant romantic youth, Korolev himself had flown gliders and obtained a pilot's license. Nevertheless, under the influence of a purely rocket-centric ideology, he demanded the creation of fully automatic systems for spacecraft motion control. In aviation, the "human factor" was the primary cause of accidents. The automatic systems must be failsafe—such was the doctrine of the "Korolevian school." During cosmonautics' pre-computer phase of development, it was very difficult to perform rendezvous and docking in a purely automatic mode. But, here too, we managed to be "ahead of the entire planet."

On New Year's Eve, 31 December 1961, under his pseudonym "Prof. K. Sergeyev," Korolev published an article in *Pravda* entitled "Soviet Soil Has Become the Shore of the Universe," in which he pointed out the need to develop spacecraft docking systems.[3] But as early as three years before this article, Korolev and Tikhonravov had prepared a report for the government about prospective space exploration projects, in which their objective was to work out the process of rendezvous between two spacecraft moving in contiguous orbits.[4]

During 1962, under Korolev's supervision, design studies were conducted to develop orbital assembly facilities. The development of a semi-automatic complex using the Vostok spacecraft was proposed as the first phase of mastering orbital assembly.[5] In proposals sent to the government, Korolev wrote that one of the most crucial tasks facing technology at that time was solving the problem of rendezvous and the assembly of spacecraft in orbit.

---

3. The same article that was republished as "Sovetskaya zemlya stala beregom vselennoy" ["Soviet Soil Has Become the Shore of the Universe"] in M. V. Keldysh, ed., *Tvorcheskoye naslediye akademika Sergeya Pavlovicha Koroleva* [*The Creative Legacy of Sergey Pavlovich Korolev*] (Moscow: Nauka, 1980), pp. 432–436.

4. See S. P. Korolev and M. K. Tikhonravov, "O perspektivnykh rabotakh po osvoyeniyu kosmicheskogo prostranstva" ["On the Prospective Work on the Mastery of Cosmic Space"] in *Tvorcheskoye naslediye akademika Sergeya Pavlovicha Koroleva*, pp. 405–408. The letter is dated 5 July 1958.

5. This is a reference to the Vostok-Zh study of 1962–63. A prospectus from this study has been published as S. P. Korolev, "Predlozheniya po sozdaniyu sredstv dlya orbitalnoy sborki" ["Proposal on the Creation of Means for Orbital Assembly"] in *Tvorcheskoye naslediye akademika Sergeya Pavlovicha Koroleva*, pp. 445–449.

At first glance, the docking ideas that Korolev expressed in 1961 sounded like science fiction. Five years went into implementing an idea that initially engendered little faith among experienced engineers.

The rocketeers were proud of the fact that a missile's speed made it the ultimate weapon. Missiles would transfer their speed to spacecraft. Two spacecraft with a mass of 6 metric tons each fly around Earth, each on its own orbit, at speeds in excess of 8 kilometers per second. They needed to see one another, to establish reliable reciprocal radio or optical contact, to begin rendezvous using the power of their on-board propulsion systems, and to approach one another at a relative velocity of centimeters rather than kilometers per second. At the same time, their mutual alignment must be precise enough to ensure reliable mechanical engagement, subsequent latching, and the mating of multi-contact electrical and hydraulic connectors.

After the Americans had proclaimed their national mission to be a lunar landing by the end of the 1960s, our conceptual designers, headed by M. K. Tikhonravov and spurred on by Korolev, began to study intensively various circumlunar flight scenarios using "resources at hand." I say "at hand" because the heavy N1 launch vehicle would not be available soon for a flight to the Moon. We needed to hurry! For the time being, the only thing we had at our disposal was the reliable *Semyorka*. How could we use it to fly two cosmonauts around the Moon as a "world's first" and then safely return them to Earth?

In order to impart practical direction to these design concepts, Korolev handed over the problem of the rocket-space complex for the circumlunar flight to rocket designers Sergey Kryukov and Yakov Kolyako. Thus, the design of the rocket-space complex named Soyuz came to be. It required the development of three types of new spacecraft: 7K, 9K, and 11K. At the same time, four successive launches needed to be conducted using a *Semyorka*-based launch vehicle to accomplish a circumlunar flight.[6]

The 9K rocket stage would be the first to be inserted on orbit in an unfueled state. The 9K was intended to boost the piloted spacecraft toward the Moon. After the 9K, the 11K refueling spacecraft would be inserted in orbit in succession. One by one, they would rendezvous with the 9K and fill it with fuel and oxidizer; and with their job done, they would undock. When the ground was confident that everything was fine on board the 9K in orbit, the two-seater piloted 7K spacecraft would be inserted into orbit. After docking with the 9K,

---

6. These spacecraft were to be launched by two new three-stage versions of the *Semyorka* named the 11A55 and 11A56. See Timothy Varfolomeyev, "Soviet Rocketry that Conquered Space: Part 9: Launchers for An Early Circumlunar Programme," *Spaceflight* 41 (1999): 207–210.

using the engines and propellant of the 9K, this stack would be accelerated toward the Moon. After boost to circumlunar flight, the 9K would be discarded from the stack. Following the flight around the Moon, the piloted 7K would return to Earth at reentry velocity equal to the "second cosmic velocity."

In all, the design called for a minimum of three automatic rendezvous and dockings. Korolev approved the draft plan on 24 December 1962. The State Committee of Defense Technology supported the proposal; and in March 1963, a governmental resolution was issued committing us to carry out a circumlunar flight using a Soyuz spacecraft in 1966!

I must confess that this exotic multi-step circumlunar flight project didn't garner enthusiasm at OKB-1. Mishin was among the skeptics. He thought, and rightly so, that we needed a powerful launch vehicle for such a flight and that we should focus all of our efforts on the N1 rather than spread out in other directions. The aero- and gas dynamics specialists were scaring others by citing the exorbitantly high temperatures that would occur when the 7K entered Earth's atmosphere at escape velocity.

Korolev assigned the design process of the piloted 7K vehicle to Konstantin Feoktistov. Being a bold but level-headed designer, Feoktistov decided to take advantage of the hoopla over the Moon and develop a vehicle that would replace the Vostoks and Voskhods. Even if it was not going to fly to the Moon, at least we would have a new reliable vehicle for orbital flights around Earth.

Korolev was very sensitive to the moods of his deputies and lead designers. When Rauschenbach and I, after much searching, told him that, in the best case scenario, radio rendezvous systems would be available in four to five years, Korolev was furious and demanded to know whom he should talk to in person. During this period, Korolev was preoccupied with serious organizational restructuring. He had heard rumors to the effect that Bushuyev—his most obedient deputy for space programs, now ensconced in the historic office of General Grabin (the famous, but now former, chief designer of artillery systems) on the territory of the former NII-58—was planning to gain his independence and leave OKB-1. The reaction was swift and decisive.

Bushuyev was brought back to the first territory and given an office one floor above Korolev. In addition to my post as deputy chief designer, I was given the authority of chief of OKB-1 Branch No. 2 (the territory of the former NII-58) and deputy chief of the organization. I moved into Bushuyev's place in Grabin's office and was given special authority for the 7K spacecraft of the Soyuz complex. Soon, the developers of the Soyuz design for circumlunar flight lost their initial euphoria. For Korolev, there was a feverish rush to prepare the three-seat *Voskhod* and then the *Voskhod-2* plus a spacewalk.

I joined a group of "partisans" who convinced Korolev that, from the entire Soyuz, only the new piloted 7K spacecraft needed to be retained. Korolev agreed, but with an amendment that proved to be pivotal: the spacecraft must be a three-seater and must have rendezvous and docking systems for two spacecraft in automatic mode. He expressed his reasoning this way: "If we perform a structural linkup of two piloted spacecraft, it will be something like a space station in orbit."

In March 1965, the USSR Ministry of General Machine Building (MOM) was formed, and S. A. Afanasyev was named as minister. On 2 April 1965, Korolev sent the design of the Soyuz-K rocket-space complex to L. V. Smirnov, and on 3 April, to the new minister S. A. Afanasyev. The design proposed using two launch vehicles and two spacecraft for a single in-orbit assembly. The spacecraft was assigned the designation 7K-OK.

In the design departments and at the factory, work began on the manufacture of 7K-OK spacecraft, in "active" and "passive" models. Grigoriy Markov, the foreman of the space assembly shop simplified the nomenclature and said that he was making the spacecraft in "male" and "female" versions.

When distributing duties among his deputies and leading specialists, Korolev firmly emphasized a sense of accountability. Moreover, he strove for a cooperative style so that the most critical developments would have backups in addition to the main individual responsible, and would be under the supervision of specially appointed "lead designers." Korolev split the

Asif Siddiqi

Chief Designer Vasiliy Mishin, shown here at Baykonur some time in 1967. Behind him on the left is Baykonur Commander Aleksandr Kurushin and on the right is Kurushin's deputy Anatoliy Kirillov.

responsibility for the research and development of the rendezvous process between his two deputies: Vasiliy Mishin and me. In this case, we supervised each other. Mishin watched over the ballistics service, headed by Svyatoslav Lavrov at OKB-1. The ballistics specialists were responsible for "far approach." The ballistics specialists determined the date and precise time of the launches for this segment, processed the ground telemetry stations' orbital parameter measurements as rapidly as possible, and drew up projections for each orbit. They determined the radio communication zones and developed the orbital correction program for each of the two approaching spacecraft.

In the "far approach" segment, the rendezvous of the spacecraft was controlled by instructions from Command and Measurement Complex (KIK) ground stations. The ballistics specialists calculated the magnitude, place, time, and direction of the correction burns. The services subordinate to Mishin performed this work jointly with the NII-4 and NII-88 ballistics centers and the Academy of Sciences OPM.

Historically, a friendly cooperation had developed between the ballistics experts of OKB-1 headed by Ilya Lavrov, the Ministry of Defense NII-4, where Colonel Pavel Elyasberg was in charge, and the Academy of Sciences OPM, where Dmitriy Okhotsimskiy's young team worked enthusiastically under Keldysh's supervision. After checking each others' calculations, and time permitting, sending reports to the chief designer, the ballistics experts issued the baseline data for the execution of the next correction to the services subordinate to me.

When Korolev was no longer with us, the decision to carry out the ballistics experts' recommendations was made by the operations group, located at the control post in Simferopol at NIP-10 for the lunar programs or the control post in Yevpatoriya at NIP-16 for the flight programs of the first Soyuzes and their rendezvous.

The control systems departments subordinate to me were responsible for implementing the ballistics calculations for orbital correction. Before a stipulated correction burn, it was necessary to do several things: orient the spacecraft for a "boost" or "brake," sometimes at an angle to the orbital plane; load the set point determining the operating time of the correction engine unit (KDU) or the value of the apparent velocity into the control system memory; issue a complex sequence of commands at the right time to prepare and fire the KDU; ensure the reliable stabilization of the spacecraft during KDU operation; and shut down the engine when the stipulated apparent velocity is reached and immediately transfer the spacecraft into orbital orientation or into solar inertial spin mode to charge the on-board batteries from the solar array.

During the spacecraft's next orbit in the coverage zone, the ground telemetry stations took measurements and transmitted the results and express telemetry data to the ballistics centers. Everything had to be rapidly processed and reported to the flight control operations group.

The human factor played a decisive role in the reliability of the far approach phase. It wasn't just the ballistics specialists who could commit an error using complex ground-based computers. The myriad of auxiliary services, including the machine operators at the diesel electric power plants at the ground stations, could break up a correction session. Even during training sessions on the command radiolink control panels, ground radio operators sometimes entered an erroneous setting, or due to a magnetic storm or unexpected interference, communicators were in a position to delay the transmission of information to the center, and so on. One could come up with dozens of possible failures because hundreds of people supported this phase.

Korolev understood full well that it wasn't just Mishin and I who were responsible for the far approach phase. He demanded that everyone have an understanding of the situation and instilled a sense of responsibility from the top down.

Once the vehicles were less than 25 kilometers from one another, the "close approach" phase began. The first sign of this was the presence of the "lock-on" flag. This meant that the active vehicle (male) radio system had detected the passive (female) vehicle in space, having received its radio response. A completely automatic rendezvous and final approach began.

In 1967, our inexperience made us very bold. The automatic rendezvous segment in the flight program passed through all the coverage zones of the Command and Measurement Complex tracking stations. Intervention from the ground was impossible. The human factor was eliminated. Reliability was determined, first and foremost, by the *Igla* (Needle) radio system developed at NII-648 and by the relay-analog motion control system, developed by the control complex departments entrusted to me and manufactured at our factory in our own instrument production facility.

After the first successful automatic rendezvous and docking attempts, in keeping with probability theory, we had rendezvous system failures. The mishaps taught us to be more cautious and prudent. Many years later, when there was a reliable control computer complex on board, rendezvous and docking sessions were scheduled in the flight programs under the mandatory condition that they take place in the TsUP coverage zone with the possibility of control and intervention from the ground. Forty years later, after hundreds of rendezvous and docking events, the active vehicle receives clearance for final approach and docking only with permission from the ground in a telemetry and television coverage zone.

In October 1967, we could not have imagined what sort of actual monitoring facilities the marble palace on the outskirts of Moscow—the modern-day Flight Control Center—would possess. Millions of television viewers all over the planet can observe the spacecraft rendezvous process and marvel at the control rooms where dozens of operators anxiously watch the status of all the on-board systems displayed on their monitors in real time. In Yevpatoriya at the NIP-16 Control Center, the GOGU sat at a table covered with dozens of field telephones. And just think—it controlled both the flight to Venus and the docking of the first 7K-OK vehicles, named Soyuzes.

The "voyage" to Venus, which was unusual in that it offered us new experiences, lasted just 4 hours. The applause, congratulations, toasts, and touching good-byes with the departing "Babakinites" died down. We needed to get back to the Soyuzes. The next day, right there at the Yevpatoriya tracking station, we would have to immerse ourselves in the workaday business of preparing for the flight control of two unpiloted 7K-OKs, vehicles No. 5 and No. 6. Both of these vehicles would be called Kosmos, but for training purposes, in the hope of making the next two vehicles piloted, we assigned them the code names *Baykal* and *Amur*.[7]

Despite the experience we had already gained, preparing these vehicles was difficult. After Komarov's death, extensive modifications had been made on all the systems. Now, more test procedures were performed at the control and measurement station, especially at the engineering facility, and they had been made more stringent. Many glitches occurred with the modified parachute system during the airplane drops in Feodosiya. The military representatives and the developers themselves were particularly attentive to any glitch involving the descent and landing systems. On 16 October, the State Commission made the decision for the first launch during the period from 25 to 27 October.

Reporting at the State Commission, without any prior discussion with anyone, Mishin suddenly announced that the primary objectives of the joint flight of the two unpiloted 7K-OK spacecraft would be to test the reliability of all the systems, including the modified parachute system, and to execute maneuvers for rendezvous, but there would be no automatic docking mission. To the great satisfaction of Marshal Rudenko, General Kamanin, and the cosmonauts, Mishin said that we would practice final approach and docking during subsequent piloted launches. This made Armen Mnatsakanyan want

---

7. *Baykal* was named after the lake, and *Amur* after the river.

Armen Sergeyevich Mnatsakanyan (1918–1992) was chief designer of NII-648 (later NII TP) from 1953 to 1977. He was the main architect behind the *Igla* and *Kurs* rendezvous radar systems installed on Soyuz spacecraft.

From the author's archives.

to attack Mishin, but I held him back, explaining my position.[8] It wasn't appropriate for me to argue with the chief designer on the State Commission.

If there were no serious glitches on the active vehicle after checking out its systems on orbit, and if the second, passive vehicle was inserted into orbit, then, being part of the Main Operations Control Group (GOGU) as the technical chief, I would be able to persuade everyone necessary to take the risk. For this, all we would need to do would be to prepare a program at the two ground telemetry stations to supply power to the *Igla* radio system and switch the control system to rendezvous mode. Upon issuing our command from Yevpatoriya, without seeking permission, we would start up these programs on both vehicles. And after that, it would be up to the *Igla* system. If docking were to fall through, we would be able to claim that no such mission had been assigned. And if a miracle were to happen, then the victors would not be judged. Thus, "the wolves would have their bellies full and the sheep would be unharmed."

From the very beginning of the training sessions in Yevpatoriya, Agadzhanov and Tregub supported me, and accordingly, we tasked the ballistics, control, and analysis groups with ensuring the program provided for the express analysis of information and the actions of all services to activate rendezvous mode.

Having been taught by the bitter experience of the previous Soyuzes, we drew out the programs so that we could thoroughly check out everything necessary for the rendezvous system, and only after this would we risk going for rendezvous. The active *Amur* was supposed to fly solo for almost three days. After this, we were to send our findings about the status of its systems

---

8. Mnatsakanyan was the chief designer of the *Igla* automatic rendezvour radar system.

from Yevpatoriya to the State Commission in Tyura-Tam. If, in accordance with the ballistics experts' forecast, *Amur*'s orbital correction guaranteed that it would fly over the firing range launch site during its 49th orbit, then the *Baykal* launch would take place.

The end of the third day and the beginning of the fourth would be the most stressful. During the minutes after *Baykal*'s entry into orbit, the ballistics experts would have to determine its parameters, and GOGU would have to make the decision whether to go for rendezvous or execute an additional maneuver. Throughout the entire flight, it would be necessary to keep a particularly close watch on the consumption of propulsion fluid and the recharging of the batteries from the solar arrays.

We conducted the first training session on controlling the unpiloted Soyuz pair on 19 December. Besides me, the "managerial octet" included Agadzhanov, Tregub, Rauschenbach, Bolshoy, Rodin, Starinets, and our first cosmonaut, Gagarin.

For the upcoming launches it was not necessary to see off the cosmonauts, and, as of the first day of the training sessions, Gagarin decided to take part in the operations at the Control Center. With a guilty smile, he confessed that when it came to Soyuz flight control, especially during the rendezvous and docking segments, he felt like he was out of his element. We understood quite well: the deputy chief of the Cosmonaut Training Center and GOGU member could not be a passive observer—it did not fit with his active nature.[9] The training sessions, finalizing the documentation, talks with the test range, identifying new glitches, and developing the appropriate recommendations took each of us 10 to 12 hours a day. To regain our focus after heated arguments and analysis of various situations, we would take strolls to the seaside. We referred to this practice as a "sea-air brain-purging operation."

Gagarin accompanied me on one of these "purging sessions." After Komarov's death, he worked for two months on the accident investigation commission and still hadn't been able to come to terms with what had happened. Like me and our other comrades, the same unanswered questions tormented Gagarin. Why had we all—his commanders and the chief designer, the minister and the VPK, and Gagarin himself—decided on a piloted launch after having three failed unpiloted launches in a row before that?

Now we were going to act sensibly; we were preparing to fly two unpiloted vehicles at once, in which the errors of the preceding ones had been taken

---

9. Gagarin had been appointed a deputy director of the Cosmonaut Training Center on 20 December 1963.

into account. Even if we might be unable to dock them, at least we might be able to test controlled descent on one of them and the modified parachute system on both of them.

"The program that we are preparing now should have been implemented before Komarov. Then, Komarov would be alive and now we'd be preparing for piloted launches," was how Gagarin saw it.

I agreed with only half of what he'd said. True, Komarov would be alive, but vehicle No. 4 certainly was doomed in any case. If it had been unpiloted, the APO system would have blown it up because there was no way to return it to our territory. It had been able to land on our territory only because Komarov was on board and he had oriented it before executing the braking burn. If the APO system had been triggered on an unpiloted 7K-OK No. 4, not knowing there was a fundamental defect in the parachute system, we would have launched a cosmonaut on vehicle No. 5, and most likely, the end would have been just as tragic. Now we had wised up. If just one of the two unpiloted vehicles were to fulfill the entire program that we had prepared, then we could confidently prepare piloted launches.

Meanwhile, gloomy news had come to us from the firing range about a streak of bad luck during tests at the engineering facility. First the cables in the power supply system burned up after the polarity in the test equipment got mixed up, and then, while fueling the SKD tanks on vehicle No. 6 with propellant at the fueling station, the membrane in the propellant valve burst when a false command was issued to the pyrocartridges.

On 24 October, the State Commission and all the requisite chiefs finally flew out to the firing range. Mishin telephoned from Moscow and instructed us to be ready for launch no later than 27 October. On 25 October, Gagarin wistfully announced that he had to leave us. He had been summoned to the firing range to take part in a celebration for the builders.

"You see," explained Gagarin, "I can't say no. The builders have to be honored, but my real work suffers."

One can say without exaggeration that Gagarin was the most famous individual on the planet. Every schoolboy knew who Gagarin was. At the same time, he had absolutely no say over his own time and his own actions. He was given instructions to participate in a myriad of domestic and foreign social and political events. Where he should fly, where he should speak, whom he should receive—all of this he was ordered to do. And now he wasn't even given the opportunity to spend another two days with us before the launch to gain experience working at the flight control center.

On the morning of 27 October, we received the order to report to the State Commission about the complete readiness of GOGU and the entire

Command and Measurement Complex. At 0800 hours, T-minus 4 hours was announced. At 1230 hours, with an error of two hundredths of 1 second (!)7K-OK vehicle No. 6 (*Amur*) was launched from Site No. 31. TASS was notified of its name—yet another Kosmos, *Kosmos-186*.

Before the launch, it seemed that there were a lot of lounging around among us who had flown in to have a stroll along the seaside. Now the area seemed deserted. Everyone scattered to their workstations as happens on ships when the crew is given a call to battle stations. On the second orbit we confirmed that the solar arrays and all the systems' antennas were deployed. Rauschenbach reported that the spacecraft had stabilized normally, and that it was in solar inertial spin mode. Galin and Sorokin were able to verify the integrated modes of the long-range radio communications (DRS) system: orbital parameter measurement, telephone, and television.[10] Everything was working. Solar current was normal, the buffer batteries were charging, and even the usually fretful thermal engineers announced that there had been "no glitches." During the fourth orbit settings were entered to test the main approach and correction engine (SKD) and stabilization using the docking and attitude control engines (DPO)—everything worked out. The first day ended with optimistic predictions.

For purposes of "breaking in," to gain real experience in the flight control of a spacecraft as complex as the 7K-OK, the first day of actual flight proved more effective than the training sessions of the eight previous days. Trouble started on day two. Orbital correction during the 17th orbit failed. Once again, the Sun-star sensor let us down. Then we began to have malfunctions while loading settings into the DRS system. Another attempt at orbital correction during the 31st orbit failed due to a delay in issuing baseline data to the NIP. On day two, the ionic orientation failed, falling into "ion pockets." But one way or another, after three almost sleepless days, we brought *Amur* into a mode that was suitable for a rendezvous with *Baykal*. The third day ended in suspense awaiting the launch of *Baykal*.

Baykal lifted off successfully from Site No. 1 several 10ths of 1 second past 1112 hours 46 seconds on 30 October. TASS was informed that *Kosmos-188* had been launched. During the 49th orbit, the ballistics center reported from Bolshevo that the distance between the two spacecraft was just 24 kilometers, and for the time being, they predicted no clear tendency toward rapid separation. As arranged, when the spacecraft pair departed from our field of vision in pursuit, via circular order, Agadzhanov gave the command for rendezvous, which was transmitted

---

10. DRS—*Dalney radiosvyazi.*

to the spacecraft from Ussuriysk. I managed to report our initiative to Mishin and Kerimov at the firing range, but instead of the anticipated tongue-lashing, I was applauded.

Both vehicles departed from the coverage zone for commands and observations, and somewhere over the ocean, without our assistance and monitoring, they would attempt to rendezvous. All we could do was wait and speculate. And the speculation began. No one was patently optimistic. A significant majority were pessimists: "It's impossible to dock on the first try without monitoring and assistance from the ground!"

Tempered by research and development on air defense missile systems back in the 1950s, Yakov Isayevich Tregub reassured his colleagues: "Why are you all so worried? These complex systems aren't *supposed* to work the first time around. And nobody is going to hold it against us."

There was the hope that we would receive the first message about the ongoing process via what we referred to as low-capacity shortwave telemetry. But reception in the shortwave range was erratic. Stammering with excitement, Viktor Raspletin, the shortwave system developer, whispered to me that there was a sign of docking, but communication was so unsteady that it was better to wait until the beginning of the session.

The developers of *Igla* and the docking assembly were sweating it out more than anyone. For them, this was their first real exam. For 1 hour, as the two spacecraft maneuvered over the Pacific and Atlantic oceans and Africa, worries, debates, and predictions gripped literally everyone. We could not intervene in the process at all. And when Agadzhanov announced over the public address system the 5-minute warning until the beginning of the communications session, despite the authorized personnel list, an unusually large number of people crammed into the large GOGU room. Understanding our comrades' frame of mind, we did not turn anyone away.

The telemetry service and television experts understood that, first and foremost, they must give a rapid response to the question: "Has a historic event taken place or not?" No one needed a special pep talk about the need to be extremely vigilant. Nevertheless, Agadzhanov, attempting to ease the tension, made the following announcement to the telemetry service over the public address system.

"Colonel Rodin, report immediately!"

I had arranged this earlier with Golunskiy and Popov, warning them: "But only if it's a sure thing. If you say 'yes,' and then it turns out that you were mistaken, somebody's going to have a heart attack! And in the first report—to hell with details! We'll record everything, we'll replay the memory, and then we'll take our time analyzing it."

Petr Bratslavets placed the television monitor so that the GOGU directors could view the screen while still manning their telephones.[11] We were on a high point for space television. Our NIP-16 was the first tracking station on Soviet territory to encounter the vehicles flying from the west. The antennas of all the facilities were pointing toward the southwest, inclined to the horizon, and standing by! The report "All facilities have reception" usually evoked no emotions. This time, it jolted our nerves like a shot from a starting pistol would for runners crouched at the starting line for a world-championship race.

There's no way to know now which one of the telemetry operators was the first to discover the sought after characters on the paper tape, but they were great guys—in seconds they double-checked and shouted: "We have the capture and docking flag!"

Realizing not only the technical, but also the political significance of the information, Agadzhanov quickly replied over the microphone, "Carefully recheck and report once again."

Bratslavets, glued to his system's television receiver, literally howled: "They're docked!"

Sure enough, from the grainy image jostling about on the screen we made out the stationary contours of the 7K-OK structure. The television camera of the active spacecraft was transmitting an image of the passive spacecraft, which was stationary in relation to it. Now there was no doubt—docking had occurred! The hush exploded into applause! Someone even shouted "Hoorah!" We turned to one another with bear hugs, handshakes, and rebukes: "And you didn't believe it would happen!"

When the first wave of elation had finally subsided, and the analysis group had managed to get to work thoroughly processing the information obtained during the replay of the on-board memory recordings, we learned that rendezvous had gone awry, and the latching process had not been completed. A gap remained between the two vehicles, and the power connectors had not been mated. That evening, Mishin, Kerimov, Feoktistov, Kamanin, and Gagarin flew in from the firing range. At a joint session of the State Commission and the Technical-Operational Management (OTP) team, we listened to the dramatic reports of how the whole thing actually happened.[12] The rendezvous process began from a range of 24 kilometers between the two vehicles! Mutual orientation took 127 seconds. The vehicles were separating

---

11. Petr Fedorovich Bratslavets (1925–99) was a leading scientist at NII-380 (or NII Televideniya), the institute responsible for developing TV systems for Soviet spacecraft.

12. OTP—*Operativno-tekhnicheskoye rukovodstvo.*

at a rate of 90 kilometers per hour. *Amur* needed to stop separation and begin maneuvering using the SKD without releasing *Baykal* from radio lock-on.

Amur made more than 30 attitude maneuvers and fired the SKD 28 times. At a range of 350 meters, the rendezvous process automatically switched to the final approach mode, during which the DPO fired 17 times!

The entire rendezvous process up to the point of mechanical capture lasted 54 minutes. SKD and DPO propellant was consumed in excess of every calculation. As soon as the probe of the active docking assembly came into contact with the passive docking port, the *Igla* shut down. Mechanical capture superseded radio lock-on and the latching process began. Telemetry clearly pinpointed that the latching process was not completed; the connectors providing an electrical interface between the two vehicles failed to mate. Something clearly prevented the drives from completely joining *Amur* with *Baykal*. According to the analysis of the docking experts, a gap of 85 millimeters remained between them. But the spacecraft were rigidly mated. The television image convinced us of that. Docking on the first attempt, albeit not fully complete was our private concern; all the same, in time for the 50th anniversary of the October Revolution that was a victory!

The dynamic process of rendezvous was clearly abnormal. Because of this, working groups were formed for a detailed analysis of the operation of *Igla* and our rendezvous control unit (BUS).[13] An investigation would be conducted at the factory or at the engineering facility: what had prevented latching? It was risky to fly for a long period of time in a "not fully docked" condition.

After two orbits, the undock command was issued and we began to observe the television image of the process of the spacecrafts' slow separation. Now our concerns turned to preparation for the spacecrafts' return to the ground in the guided descent mode. However, once again the 45K optical sensor was displaying erratic operation. Lighting conditions were clearly jamming the star signal and we couldn't set up the spacecraft to begin a guided descent. After the triumphant docking, we had once again hit a streak of bad luck. On 31 October, during the 65th orbit, *Amur* made a soft landing in ballistic descent mode. We needed to verify the reliability of controlled descent. We tinkered with *Baykal* for another 24 hours trying to understand what was going on with the star sensor. Realizing that we couldn't rely on it on this spacecraft either, we made the decision to perform orientation for descent using the ionic system.

---

13. BUS—*Blok upravleniya sblizheniyem.*

We still had enough propellant for all sorts of experiments on *Baykal*, but Kerimov and Mishin, inspired by the congratulations from the Central Committee and Council of Ministers, insisted that we finish up our production in space and bring the spacecraft back to Earth by any means. On 2 November, after somehow setting up the spacecraft using the ionic system, the commands were issued to start up the descent cycle programs. The ionic system slipped up somewhere in the Brazilian Magnetic Anomaly, and the braking burn sent the spacecraft toward Earth on a long, flat trajectory that emerged beyond the limits of the authorized corridor. The APO system destroyed 7K-OK vehicle No. 5. This time our tracking system and the anti-ballistic missile system closely monitored the descent module's descent trajectory. The vehicle was blown up after it passed over Irkutsk. Had there been no APO system, it might have landed 400 kilometers east of Ulan-Ude. On 3 November, we boarded our planes and flew from the Crimea to Moscow.

On 1 November, *Pravda* published the following greeting: "We scientists, designers, engineers, technicians, and workers who took part in the development and launch of the two satellites, *Kosmos-186* and *Kosmos-188*, report the successful execution of the world's first automatic docking and undocking of two spacecraft on orbit. We dedicate this new achievement of Soviet science and technology to the 50th anniversary of the Soviet rule."

We read this in the newspapers that some Navy sailors had passed on to us at the airport. We brought many ideas and new instructions for control system modifications back to Moscow.

Ionic orientation was unreliable. The system needed to be supplemented with an infrared sensor. This would provide reliable pitch and roll orientation. Sensor 45K needed to be thoroughly studied and made to reliably track stars. The most difficult thing was to figure out how to modify the *Igla* and BUS in order to refine rendezvous dynamics. Finally, we needed to find the "demon" that was preventing the spacecraft from closing the gap completely.

Despite the festive mood after our farewell breakfast at NIP-16 and the additional regaling that we received from Navy pilots at the airport, we managed to hold an improvised OTR meeting, during which, Tregub, Agadzhanov, and I insisted that we absolutely must launch one more pair of unpiloted vehicles. Our arguments won out. The proposals were accepted, and it was decided that 7K-OK vehicles Nos. 7 and 8 should be prepared for unpiloted launch for a flight entailing a mandatory refinement of the rendezvous process and verification of the guided descent system. However, before the successful docking of 7K-OK Nos. 7 and 8, tragic events occurred.

Chapter 23

Heart–to–Heart Conversation

Upon returning from the Crimea, our good intentions to concentrate all available manpower on modifying the 7K-OK so as to eliminate control system flaws discovered in flight ran up against a major problem: the simultaneous mobilizations of specialists for both the 7K-L1 circumlunar flight program and the N1-L3 plan. Right after the festivities surrounding the 50th anniversary of the October Revolution, we felt that our pre-holiday successes, the landing on Venus, and the automatic docking had not relieved the pressure generated by the "powers that be." Government and Party bureaucrats had ample reasons for their dissatisfaction with the state of our space affairs.

In November, the Americans launched a Saturn V carrying a mockup of the Apollo spacecraft.[1] They inserted 140 metric tons into near-Earth orbit. After receiving this information, Central Committee Secretary Ustinov immediately asked Afanasyev what had changed in our operations on the N1-L3 recently and how we might respond to the Americans.[2]

From the conversations that Tyulin, colleagues from the Central Committee Defense Department, and VPK bureaucracy had started up with us TsKBEM deputy chief designers, we understood that Ustinov, Smirnov, and Afanasyev thought that Mishin was doing a bad job organizing the projects, especially N1-L3. Glushko and Barmin had the most candid conversations with Afanasyev and Ustinov about Mishin's character and his managerial methods. While Korolev was still alive, Glushko had not concealed his attitude toward Mishin. After Korolev's death, relations between Mishin and

---

1. The Apollo spacecraft launched on the first Saturn V, *Apollo 4*, was not a mockup but a Block I production model of the Apollo Command and Service Module (CSM). The Block II was the version designed for lunar flight, but the *Apollo 4* spacecraft included a number of key systems from the Block II.

2. *Author's note*: In addition to the successful test of the Saturn V carrying the Apollo, the Americans had executed a soft landing of the Surveyor automatic station on the Moon. *Surveyor-6*, launched on 7 November 1967, successfully landed on the Moon three days later.

Two of the men responsible for managing the Soviet space program in the late 1960s: Georgiy Tyulin (a senior official at the Ministry of General Machine Building) and Vasiliy Mishin (Korolev's successor at OKB-1), shown here relaxing at a dacha. Tyulin and Mishin had met each other in 1945 in Germany.

From the author's archives.

Glushko really fell apart. Pilyugin, who kept me apprised of the attitudes of his colleagues on the old Council of Chiefs, told me that even Barmin, who used to maintain a neutral position in private conversations, said that Mishin behaved tactlessly and had no knack for reaching compromises when this was required.

On 14 November, Tyulin flew out to the firing range to direct the latest unpiloted 7K-L1 launch. The launch was scheduled for 22 November. Before his departure, Tyulin asked that Chelomey and Mishin personally attend the State Commission meeting that would make the decision about the launch. Mishin told me that I did not have to leave for the firing range or for Yevpatoriya but could remain in Podlipki to speed up operations on the 7K-OK. He and Chelomey did not fly to the firing range on the same airplane.

We did not observe the nighttime launch of the 7K-L1, but we "listened" to it at TsNIIMash.[3] Mozzhorin finally had begun to set up his control center.

We'd had bad luck with the Moon while trying to achieve a soft landing. Nor were we so fortunate with the circumlunar mission. Four seconds after the second stage engines went into operation, the launch vehicle safety system (SBN) shut down the engines.[4] Based on the preliminary data, Chief Designer Aleksandr Konopatov, who headed the Voronezh OKB after Kosberg's death, pled guilty. The launch vehicle filled with its corrosive propellant components fell 300 kilometers from the launch site. However, every cloud has a silver lining!

---

3. The leading Soviet space research institute, NII-88, was renamed TsNIIMash (Central Scientific-Research Institute of Machine Building) in January 1967. It was located in the Moscow suburb of Kaliningrad (now Korolev).

4. SBN—*Sistema bezopasnosti nositelya.*

Our emergency rescue system (SAS) worked flawlessly. The spacecraft landed safely, having flown several kilometers away from the toxic vapors. This was now the fourth failure in our joint circumlunar flight tests with Chelomey.[5]

On 1 December, Afanasyev, whom the Council of Ministers had appointed as chairman of the "Lunar Council," held its first real business meeting. In essence, as head minister, the VPK had saddled Afanasyev with a significant share of the responsibility for problems with the lunar landing program. The primary task of the "Lunar Council" was the N1-L3 program. In behind-the-scenes office conversations, Ustinov and Smirnov were given credit for creating the "Lunar Council." As head minister, it was Afanasyev's duty to coordinate the operations concerning N1-L3. But now, in the event the program failed, the VPK had abdicated responsibility in advance—the Lunar Council would be the culprit. Another reason for creating the council was the desire to replace the Council of Chiefs, which Mishin headed after Korolev. The makeup of the Lunar Council, approved by the Council of Ministers, was quite representative. Its members included Marshal Nikolay Krylov; Marshal Sergey Rudenko; ministers of the aviation, defense, and radio industries; all the main chief designers; Academy of Sciences President Mstislav Keldysh; and VPK Deputy Chairman Georgiy Pashkov.

At the meeting held on 1 December 1967, Barmin reported that the first [N1] launch site was completely ready and could be put into service in the next few weeks. Deputy Chief of the firing range General Aleksandr Voytenko reported on the readiness of the services to begin flight tests. Based on these reports, the upshot was that it was now a matter of having the rocket ready. Mishin was forced to promise that all the work on the first flight-ready launch vehicle would be completed with a view to a launch in the first half of 1968. At the end of the meeting an argument flared up with the Air Force regarding the manufacture of simulators. A squabble started up between Kamanin and Mishin about who needed the simulators more: TsKBEM or the Cosmonaut Training Center. Mishin decided that his own cosmonauts needed to make up the crew for the lunar landing expedition. Therefore, the Air Force did not need the simulators that Darevskiy at LII had made according to our specifications. Afanasyev told Tyulin to get to the bottom of the conflict. With Tregub's input, Tyulin found a compromise despite Mishin's willfulness.

---

5. There were four launch attempts of the L1 spacecraft up this point: on 10 March (*Kosmos-146*), 8 April (*Kosmos-154*), 28 September, and 22 November 1967. Only the first one successfully achieved its mission objectives.

On 7 December, on the minister's instructions, Kerimov drove over to pay us a visit and look into the discord surrounding the new modification of the piloted 7K-VI spacecraft. Dmitriy Kozlov had proposed the development of such a spacecraft. He was chief designer of our Kuybyshev branch, which was striving to become an independent design bureau. The Air Force and Rocket Forces put aside their internal conflicts and unanimously supported Kozlov so as to facilitate the creation of a piloted spaceship that could be used exclusively for military purposes.

The 7K-OK design was used as the basis for the 7K-VI project, but the "stuffing" and control system were extremely different. It was assumed that the 7K-VI would be able to perform visual intelligence gathering, photographic reconnaissance, execute maneuvers for rendezvous, and in the future it would be able to be used to destroy a potential enemy's spacecraft.

Two years after work had begun on piloted vehicles for military use—Chelomey's Almaz and Kozlov's 7K-VI—Mishin proposed halting these projects.[6] I must say that Mishin's deputies and our leading specialists supported him in this local war. We did not want to lose our monopoly in piloted space flight.

The arguments against the 7K-VI were rather convincing: why duplicate the 7K-OK? We were capable of modifying it and performing all the requisite military tasks on the very same vehicle. If the generals so desired, we were even prepared to place guns on the spacecraft. As regards the Almaz orbital station, this was Chelomey's expensive and futile undertaking. A piloted orbital station for intelligence gathering was not necessary. The automatic systems that Kozlov was producing were capable of performing all the tasks. Mishin depended on Kerimov's support in the arguments about the 7K-VI. But when it came to the Almaz, I was trying to persuade Mishin not to fight and criticize Chelomey because he was our ally for the circumlunar flight. Moreover, coming out against the Almaz threatened to strain our relations with the minister of defense.[7]

"We need to seek unity with Chelomey," I convinced Mishin.

Discussing these problems with Okhapkin, Bushuyev, and Tregub, we reached the conclusion that clouds were gathering over us, not only because of the disruption of

---

6. The Almaz was a major piloted military space project proposed by Vladimir Chelomey in 1964. The project comprised a 20-ton space station (the Orbital Piloted Station or OPS) and a similarly sized transport ship (the Transport Supply Ship or TKS) to form a large reconnaissance station in Earth orbit piloted by two or three military cosmonauts. The original primary goal of Almaz was naval reconnaissance. The program was formally approved for full-scale development in 1967. For an English-language summary, see Asif A. Siddiqi, "The Almaz Space Station Complex: A History, 1964–1992," *Journal of the British Interplanetary Society* 54 (2001): 389–416 and 55 (2002): 35–67.

7. Then-Minister of Defense Andrey Grechko was reportedly a big supporter of Chelomey.

the circumlunar flight and N1-L3 programs, but also because of Mishin's belligerent behavior. We had received warnings from good friends in the offices of the ministry, the VPK, and Central Committee that an execution was looming.

In the process of signing off on the changes that we had introduced to the 7K-OK for the upcoming launches, I met with Aleksandr Mrykin, who had finally attained the long-deserved rank of general. After talking business, he switched to a confidential tone and requested permission to ask a question, warning me in advance that if I didn't want to, I didn't have to answer, that he would not be offended. The question had to do with my relationship with Mishin. I replied that I had known Mishin since before the war. He and I had always had a good rapport. We rarely had conflicts and only on technical problems—for example, a conflict about the development of the on-board digital computer. We had differing points of view on that problem. Mishin had taken Korolev's place because all of his deputies and the OKB-1 Party leadership, myself included, had written a letter to the Communist Party Central Committee.

Mrykin said that he had always valued my opinion highly, but in this matter it seemed to him that an old friendship was preventing me from making an objective assessment of the situation developing around Mishin. Mishin was uncompromising. He ruined relations with other chiefs for no reason. He allowed himself to say things that were demeaning to others. Mishin considered himself an absolute authority not only in technology, but also in missile strategy. Without having delved into the military's demands, he proposed putting a halt to operations on the 7K-VI, despite the fact that a government resolution was issued calling for its development. And he didn't want to listen to people with different points of view.

I had to agree with Mrykin that after Korolev's death, after Mishin's confirmation as chief, and his election to the Academy, his shortcomings in interpersonal relations had become more obvious.

"My relationship with Korolev was very complicated," said Mrykin, "but no matter how much we disagreed about something, ultimately we were able to compromise. Mishin sometimes takes absolutely irreconcilable positions for no good reason. This is not just harming him, but TsKBEM as a whole."

Regarding 7K-VI, I noted that this wasn't just Mishin's personal position, but also that of other specialists of ours, including me. In my opinion, we didn't need such a variety of piloted vehicles. Now that the Americans were so obviously outpacing us in the lunar program, it was all the more important for us to concentrate all of our efforts on developing a reliable 7K-OK orbital vehicle, further perfecting it, and testing systems for lunar vehicles on it.

I told Pilyugin and Ryazanskiy about my conversation with Mrykin. Out of the old sextet of chiefs, only they had remained on normal, friendly terms with Mishin. They both warned me that there were signs of reprisals in the offing against us. In Ryazanskiy's opinion, we would be flogged along with Mishin for any transgressions, but it was unlikely anyone would be removed from the job. Afanasyev wouldn't mind replacing Mishin, but without Ustinov's blessing, and then that of the Politburo as well, he couldn't do that. Ustinov won't agree to replace Mishin now because he was a "hostage" to the N1-L3 program. If the program failed, they'd be able to pin all the blame on him and on our firm. If there were a setback with the N1, Mishin would be responsible. If they removed him, then who would take the rap? One thing is clear—no one is going to turn to the Politburo with personnel issues now.

Pilyugin was more optimistic. In his view, the ministry, the Central Committee, and the VPK were so wrapped up with us by obligations, promises, and decrees, that "as long as there is no regime change, they will thrash us, but no one will venture to replace the chiefs."

In January 1968, a meeting was held in the Central Committee during which a "harsh" conversation took place with our minister. Afanasyev decided to convey us his own interpretation of part of what had been said to the ministry while also providing us with his organizational findings. On the morning of 23 January 1968, Mishin called together his main deputies and warned us not to run off. The minister would be coming by to have a serious conversation with us. We had been expecting this encounter, but had hoped that we would be allowed another two or three months to work calmly in order to finally execute a successful circumlunar flight, if only in an unpiloted mode, and to nail down our success with the 7K-OK and confidently begin a series of piloted Soyuz launches. The N1-L3 hung over us like the sword of Damocles. Here, it would be very difficult for us to make excuses. It was useless to ask for an extension of the deadlines—they were automatically slipping to the right.

Afanasyev arrived accompanied by Tyulin, Litvinov, and Kerimov. He began his speech saying that he had decided to make clear to us Ustinov's opinion and the mood that dominated the Politburo.

"After our rise in space, we have been experiencing a continuous fall. We are in a very serious situation. The Politburo no longer believes our promises. Very many organizations are just spinning their wheels. Your Factory No. 88 isn't busy. Everyone is trying to invent and remake his system without factoring in production stock and outside experience. Nobody has any respect for his or her own words and promises. There isn't a single decree that hasn't missed its deadline by a year, two years, or even more."

"Things are going very badly in the space field, and especially where comrade Mishin is concerned. We are like a rabbit facing a boa constrictor. It is the duty of all of us to correct Chief Designer comrade Mishin. We respect him as a scientist, but he must also consider other people. Many of them understand as well as he does, but he doesn't want to listen to them."

"In the U.S. they are working day and night on the Apollo program. They complain that the Lunar Module schedule has been disrupted by 80 hours! We find that amusing. Our schedules are disrupted not by hours, but by hundreds of days! You at OKB-1, the head organization, need to make a turnaround, and right away.[8] MOM views you as an organization that is operating very poorly. As of now, there is no serious analysis of the prospects for the UR-500K complex and the 7K-L1. And we have no confidence in the circumlunar flight. We don't even know why it is necessary."

"Comrade Mishin doesn't take into account the experience of Babakin and Reshetnev. Out of the 24 spacecraft you've planned, only 16 have been launched. Nuclear submarines are more complex, but Makeyev doesn't fail to meet deadlines like OKB-1 does. Mishin is bad when it comes to reliability processes. We need to change the work method. We need to review the managerial structure both at the Ministry and at OKB-1. Your people need to develop a special solid-propellant rocket complex. We can't continue to behave like amateurs. The 7K-OK, L1, and L3 programs are constantly changing. You're being given tremendous resources, and you're driving the entire country into a dead end! The Politburo has instructed me to review the space program yet again and make things happen faster."

"In the U.S., Saturn V is moving along without a hitch.[9] Over there, company directors are meeting every month. The capitalists have managed to dovetail their interests very well. And we can't make our chief designers—Communists—work together in harmony. Mishin's cockiness will lead to no good. We must correct the structure."

"We need a system that has:

- an OKB chief or director
- a chief designer, and
- a scientific leader and chief designers for each of the following programs: 7K, L1, N1, L3, RT-2, etc."

---

8. At the time, OKB-1 was actually known as TsKBEM.
9. NASA successfully carried out the first Saturn V launch on 9 November 1967.

"It's been pointed out to me that Comrade Mishin has many personal shortcomings and this is hurting our business. Everyone still remembers Korolev's style of operating."

"You need to urgently review the entire N1 program and finally tell the truth. You have committed design errors and now you're afraid to admit it. The N1 was built for a 75 metric ton payload, then it was adapted for 95 metric tons. The Saturn V gives 120 and promises 140! Who among you is responsible for such blunders? And the Saturn V launch weight is less than that of the N1.[10] I am not going to abdicate responsibility. These are the ministry's blunders. Where were the expert commissions and our scientists at the head institute looking?[11] Glushko requested that he be assigned the hydrogen-oxygen engine, but comrade Mishin is being capricious. He doesn't try to compromise with Glushko. On the contrary, he exacerbates relations with him. I was told that Mishin was the one who started the rift between Korolev and Glushko over the N1."

"We have a terrible fear of good cooperation between the chief designers. The N1 is way behind schedule and they're asking me in the Central Committee why I'm not punishing anyone. I can hand out reprimands; my colleagues would support me, but this will hardly help. How can Mishin and his deputies be made to answer for everything they do?"

Despite the small audience, Afanasyev spoke about Mishin in the third person. By this he emphasized that he wasn't expressing his own personal thoughts, but was passing on part of what had been said somewhere among "the powers that be." Who among "those powers" had participated in the discussion of these problems besides Ustinov, Afanasyev did not say.

Then Afanasyev switched from paraphrasing to the directive portion of his more than hour-long speech: "Projects need to be strictly distributed among Mishin's deputies. For example: Okhapkin—N1; Shabarov—L1; Sadovskiy—solid-propellant engines; Bushuyev—spacecraft; Chertok—all control systems, and so on! In the collegium [of the ministry] we will now regularly do a thorough examination, project by project. I will demand that all matters be brought to a head! No more studying schedules that we don't stick to! State discipline has fallen! All of you, and I along with you, are soon going to be called to account by the Party for idle chatter. At the lowest levels, among your rank-and-file coworkers, life goes on quietly and smoothly. There is no emotion, no stress.

---

10. In reality, the launch mass of the Saturn (about 3,000 tons) actually exceeded that of the N1 (about 2,700 tons).

11. The "head institute" is a reference to TsNIIMash (formerly NII-88), the leading space research institute of the Soviet space program.

An intense tongue-lashing is needed. You need to rally your people up the way you knew how to do under Korolev. After all, Korolev didn't achieve brilliant success all by himself. He did it with your help."

"We're going to develop a new complex for the RT-2.[12] I am issuing the order myself. Don't hold your breath waiting for comrade Mishin's proposals. We know that even when Korolev was alive, Mishin was against solid-propellant rockets. Sergey Pavlovich began this work, and we won't allow Mishin to lay this combat missile to rest. Space is a political issue and not just nuts and bolts! This is the nation's prestige! I am attaching a great deal of significance to today's meeting. We need to turn things around."

The minister wanted to stop; but after glancing at his notes, he continued: "Why is there no standardization of design, equipment, and developmental testing for the 7K-OK, L1, and L3? You promised that everything would be

From the author's archives.

Presentation of awards to leading engineers at Korolev's design bureau. From left to right they are I. B. Khazanov, G. Ya. Semyonov, Boris Chertok, Mikhail Khomyakov, Aleksandr Kasho, and Yevgeniy Shabarov.

---

12. The RT-2 was the first operational solid propellant ICBM in the Soviet arsenal. Its development was initiated while Korolev was still alive (in April 1961). The rocket was first launched in February 1966, just after Korolev's death. The Strategic Rocket Forces accepted it as operational armament in December 1968. In the West, the missile was known as the SS-13 Savage. See Chapter 6 of this volume.

identical and standardized! Does this mean this is all a sham? You proposed the 7K for the Moon, and virtually nothing remained of the 7K in the lunar designs. You promised to take the main system on all three vehicles for developmental testing. You yourselves assumed a sensible engineering policy, and then you abandoned it. You're pulling the entire ministry into a whirlpool!" After ending on that high note, the minister waited.

Other presentations started up, in which Afanasyev once again got involved, not allowing us to lapse into general political blather and excuses.

Mishin began his presentation with the words: "The causes for the disruptions of our operations and plans are considerably more profound that just the poor work of OKB-1. After the liquidation of the *Sovnarkhozy* . . . ."[13]

At the mention of the *Sovnarkhozy*, the minister interrupted Mishin: "Don't even think of reminding any of the 'brass' about the *Sovnarkhozy*. I really caught it for that before. Keep in mind, all of you, that everything we tell you here, and what you say is passed on to the 'brass' in a very filtered form. Don't remind anyone of the *Sovnarkhozy*! That's my advice for you, although I know that life was easier for you with them."

Mishin continued: "The difficulties involving weight on the N1-L3 and the difficulties involving deadlines are intertwined. Each of our managers, in terms of the level of problems he solves, outranks the chief designers of our subcontractors. I'm in favor of my deputies or managers of various complexes being called chief designers. But the most important thing now is that I ask you not torment us with organizational matters. The solid-propellant rocket project needs to be taken away from OKB-1! It's difficult to produce a vehicle when you don't believe in its destiny!"

The minister interrupted once again: "I'm not the one who foisted solid-propellant rockets off onto you. You began this project yourselves at Korolev's initiative. Why were you silent then?"

"I was against this project then and still am."

"Keep in mind, I will not give my support for the closure of these projects. For some reason, all the Americans' combat missiles are solid-propellant, and we just hang on to liquid-propellant missiles. But this is a whole different issue. And I advise you to be careful! You are still going to have a separate demand for combat missiles."

---

13. As the experiment with the abandoned *Sovnarkhoz* system was associated with Khrushchev, it was considered inappropriate in the Brezhnev years to suggest that it might have had some positive attributes.

At this juncture, Yakov Tregub could stand it no longer. Officially, as Mishin's deputy for flight testing, he bore no direct responsibility for the design and development of solid-propellant missiles. But, having been drawn into the process of inflight development testing of the RT-2, he had become steeped in faith in the prospects for this rocket. Along with Sadovskiy, often without asking for Mishin's blessing, he appealed to the "brass" and received assistance both from our MOM and from the Ministry of Defense.

Tregub said: "I don't agree with the confessions that have become so fashionable now at our OKB-1, that we have rashly taken on many projects, and nobody knows why. S.P. did not just take the L1 out of technical necessity. This was the result of the policy of consolidating all the lunar projects into the same hands. We need help, but not everywhere. We don't need help on the 7K-OK. If Chertok and Khazanov organize the production of instrumentation and the work among the subcontractors, if the deadlines for fitting out the vehicles aren't missed, *and* if they develop new testing stations at the KIS and at the engineering facility, we will catch up. As for the RT-2, this is a separate long conversation. But I must say that your deputies Udarov and Tabakov are a tremendous help to us.[14] I believe that we will put this system into service."

Bushuyev interrupted Tregub: "We have done an enormous amount of work on all the projects. The problem is that many issues were solved for the sake of deadlines and to the detriment of technical matters. We all understand and sense this perfectly. In many cases, S.P. took responsibility himself and others went along with him. Now it's easier to 'let us have it,' but this won't help matters. No one objected to Komarov's launch. But now everyone is so much smarter and they're telling us that before Komarov's launch, for the sake of 'clean operations,' we should have done an unpiloted launch. But they didn't tell us that [the reason for this omission] was again deadlines and deadlines, that we were lagging behind the Americans!"

Bushuyev's speech alluded to the most recent Central Committee and Council of Ministers decree and the subsequent orders for the ministries involved with the lunar program. Tyulin also felt compelled to put a word in: "We all need to take a serious look around. If we really give unrealistic deadlines, then who needs that? We need colossal efforts to support a piloted circumlunar flight. Where are we going to get this kind of manpower? Only from other projects. The L1, if we take this project seriously, will take on

---

14. Gleb Mikhaylovich Tabakov (1912–95) and Grigoriy Rafailovich Udarov (1904–91) were deputy ministers at the Ministry of General Machine Building.

different, very important political overtones. But you all know that this project also has opponents. Because of it, we buried Chelomey's lunar dreams. So, might this project be taken away from you?"

Mishin disagreed: "We still need to perform automatic launches. Regardless of whom they assign this work to. It's still early to be putting a human being on the UR-500K launch vehicle. A program needs to be put together, scheduling unpiloted launches for developmental testing not only of the spacecraft, but also the launch vehicle. After all, it's been a long time since we've needed unpiloted launches for the *Semyorka*. Now we only need to work out the spaceflight part."

Long conversations ensued about the number of unpiloted launches and the criteria that would allow us to make the transition to piloted flights on the new spacecraft and launch vehicles. During the debate, Tyulin pounced on Mishin for disregarding the American experience with the Saturn.

"If you're so smart, then why are we all sitting here? Is our conversation perhaps just a lot of hot air? Bushuyev says that now he wouldn't take the L1, so are we going to give up on the circumlunar flight? Are you going to give permission to prepare a report for the Politburo?"

The minister rarely supported Tyulin, his first deputy. But in this case he felt compelled to take his side: "We went to the government with such a proposal because you assured us that the L1 vehicle was the same as the L3. You assured me when Korolev was still alive that we would automatically get an orbital vehicle for a lunar expedition. This was the technical line and organizational policy espoused by Korolev. It turned out that the L1 and L3 have nothing in common. Now it is obvious that you are constantly revising your own decisions and don't want to consider the opinions of others. Both Keldysh and Glushko have justly criticized you for many N1-L3 matters. I have taken great pains to normalize relations between you and Glushko. After all, he announced that he was ready to develop a hydrogen-oxygen engine with 200-250 metric tons of thrust for the N1. But you don't want to sit down at the same table with him."

"How would you have me act when Glushko came to me and placed the layouts of the American Saturn and the N1 on my desk and argued that you are producing a rocket that will 'lift up air.'[15] I am a minister, but he is an academician, and unlike comrade Mishin, I must listen to him."

"I was told when I was not yet your minister that Vasiliy Pavlovich [Mishin] was Glushko's primary opponent and had turned S.P. against him.[16]

---

15. This is most likely a reference to the specific design of the N1, which used spherical propellant tanks, thus creating large "free" spaces within the internal structure of the vehicle.

16. Afanasyev became minister at the Ministry of General Machine Building in March 1965.

Now I am convinced that where there's smoke; there's fire. Glushko told me outright that he was ready to work with OKB-1, but just not with Mishin. What would you have me do?"

Sergey Okhapkin, who had long remained silent, intervened in the nick of time: "One must admit, Sergey Aleksandrovich, that we really don't study the design aspects of many issues in enough depth. The result of this is that we are now burdened with a heavy load of unresolved issues. We have attempted to correct the problem for N1-L3 and develop an integrated stand. But the ministry still isn't helping us, and it should play no small role in this matter. I am in favor of piloted launches of the L1 in 1968. This is a problem that will be difficult to solve, but it's possible. I am firmly convinced that our organization will measure up to the tasks assigned it. Our personnel assignments have been made based on background and for the most part correctly. We have rough spots, but what organization doesn't?"

"Back when Sergey Pavlovich was in charge, design errors were committed on the N1, which unfortunately were validated by decrees. We got a rocket, which, in terms of payload, is way behind the Saturn. We're looking for solutions. Don't accuse us of sitting idly."

Okhapkin's speech contained a clear allusion to the shared responsibility of our organization and the upper echelons of authority, who validated the N1 parameters with a Central Committee and Council of Ministers decree. It was dangerous to speak about this openly. The decrees were adopted before Afanasyev became minister. Now he was forced to bear responsibility for the mistakes of others.

I decided that it was also time for me to get involved in the debate, seeing that, in the final analysis, our conversations would result in the issuing of the minister's orders. Contrite conversations would be of no use. We needed to go on the offensive, drawing on Okhapkin's allusion.

"You can criticize us. We deserved it. But our work involves objective difficulties that in the next few years will be the decisive factor in our race with the Americans. The first of the problems is that our country lags behind in electronics in general and in computer technology in particular. We are not to blame for this, nor is our ministry."

The minister interrupted me: "I've heard this a million times. Do you want me to just pull new computers out of a hat for you or something?"

"No," I replied. "You, Sergey Aleksandrovich, need to arrange with Kalmykov and Shokin so that they be at least so kind as to give us access to the developments in their organizations.[17] There is the hope that in two years, no sooner, we will have on-board computers. Meanwhile, the circumlunar flight on the L1 will be done with a primitive computer. And then, for the L3, especially for the landing and liftoff from the Moon, we'll need one that is quite different, both in terms of speed and number of commands. For this reason alone the L1 system is different from the L3."

"The L3 has rendezvous and docking hardware that the L1 doesn't need. This is the second principle that makes the L1 fundamentally different from the L3. The third is the time in space. The L3 will be there twice as long as the L1. That means that we must have solar arrays, which are not essential on the L1. Unfortunately, our industry cannot provide the electrochemical generators [i.e., fuel cells] that the Americans have been putting on the Gemini spacecraft for a long time. Finally, there is no comparison between our electronics and the Americans' in terms of reliability and dimensions. That's where our weight, overall dimensions, and layout issues are coming from. And you can hardly help us there."

"Who made what mistakes and when is a matter that can be studied over a long period of time for history. We, of course, are the first individuals responsible, but we are not capable by ourselves of making the N1-L3 the national mission that Americans have declared their Apollo program to be. This is really their one and only and primary space mission for today. And we are plugging away with the 7K-OK, L1, and N1-L3; we're responsible for combat missiles; Chelomey is rolling out his Almaz program, which is also piloted. What are our subcontractors supposed to do under these conditions? For each of these programs they receive a decree that says that this is a crucial mission and it needs to be fulfilled 'as an absolute priority.' So they send us into a queue of 'absolute priorities'."

The entire supply of sandwiches that had been prepared in advance was devoured during the 5 hours of the continuous meeting. Kosyakov saw to it that for our subsequent conversations we were provided with tea and cookies as refreshments. But this too was quickly devoured.

Afanasyev realized that it was time to move on to the directives.

"Like it or not, I am going to issue the order to develop a special complex for the RT-2. And Comrade Mishin must fulfill it. Submit proposals for the appointment of managers and chief designers to me within three days. You'll

---

17. Valeriy Dmitriyevich Kalmykov (1908–74) and Aleksandr Ivanovich Shokin (1909–88) were ministers of the radio industry and electronics industry, respectively. As such, they were responsible for the development of electronic components and computers for the Soviet space program.

have another two days for the rest of the proposals concerning structure. Prepare for me an order of everything that you want from our ministry organizations. We will help you in every way we can. But don't think it's going to be easy. Don't be greedy! Tell us what to give away to whom to lighten your load. As for the 7K-OK and L1, there's nothing we can do. Just keep in mind, you will be held responsible and no one will cut you any slack for your problems!"

When we said our farewells, Valentin Litvinov, who had fallen behind the minister, gave us a firm handshake and, with a cunning smile, said: "How about this! During the war and right after it, when Stalin was in power, amicable conversations like these didn't happen."[18]

---

18. Valentin Yakovlevich Litvinov (1910–83) was a deputy minister at the Ministry of General Machine Building from 1965 to 1983. The "amicable conversations" is a reference to the fact that during the Stalin era, disagreements over political and/or industrial issues often resulted in arrest of the offending party.

Chapter 24
Zond-4

The "educational program" that Afanasyev, Tyulin, and Kerimov conducted on 28 January galvanized TsKBEM managers to a certain extent. We understood full well that this "heart to heart talk" had been at Ustinov's initiative. Neither Mishin nor his deputies could pretend in front of their staff that nothing had happened. We convened several times and attempted to formulate a program of priority measures.

By the 50th anniversary of the October Revolution we had, after all, achieved success—we had gained a foothold with automatic docking. But at the same time we were finding so many trouble spots in the Soyuzes! During rendezvous, the approach and attitude control engine (DPO) systems used up almost all of the fuel from the active vehicle. Instead of a executing a guided descent, one of the vehicles was blown up; the ionic system operated erratically. Despite the whole gamut of hardships, we kept our chins up.

Beginning with 7K-OK No. 7, we started to install the infrared vertical (IKV) sensor. Now it was possible at any time to control the vehicle's attitude around two axes: roll and pitch. All that was left for the *ionka* (ionic system), or any other system, was yaw attitude control. Experimental development of the parachute system continued. After all of the improvements, there was complete confidence in the primary parachute system (OSP), but the backup parachute system (ZSP) proved to be unstable for the increased weight of the descent module. If Tkachev could not increase the parachute's strength or if we could not reduce the mass of the decent module by 150 to 200 kilograms, then the three-man flight would have to be eliminated from the program. "Even winos only drink as a threesome. To depart from this tradition in space is unacceptable." We heard jokes like these during the debates about the future program of piloted flights.[1]

---

1. In Russian culture, it is often considered improper to drink alone or in pairs.

There were all sorts of other problems that ran up against each other in the course of discussions. The priority undertaking was the launch of the L1, which, based on its readiness schedule, was slated for early March.

Mishin was pleased with the new lead designer for the L1, Yuriy Pavlovich Semyonov.[2] Actually, after Semyonov and Topol (the lead for the 7K-OK) were separated and Semyonov was assigned his own independent sector, the L1, things at the production facility and during preparation at the firing range were a lot more pleasant. In any event, the deadlines for releasing the L1 vehicles throughout 1968 were not cause for particular misgivings. However, there were still more than enough technical problems, especially related to control and navigation, which we were handling together with Pilyugin's people. The "most crucial and most cursed" question for the L1 remained: "When will a human being fly around the Moon?"

This was the primary political mission. Now, as far as I can recall, and the more I talk to veterans, it seems that we didn't have a great deal of faith in a piloted L1 flight. Despite Komarov's death, each of us felt and was convinced that there would be two more unpiloted flights, and in a pinch, two more; and then after that, there would only be piloted 7K-OK flights. But as for the L1 . . . If we fly around the Moon in unpiloted mode and return the spacecraft to the ground at reentry velocity before the Americans, then why send a human being on this risky journey? It was another matter for the Americans. They needed this as a training run before landing on the Moon. For them it was crucial to perform a circumlunar flight using the Saturn V launch vehicle and Apollo spacecraft designed for the lunar landing expedition.

Our L1 spacecraft was a hybrid, the result of an opportunistic attempt to reconcile Korolev and Chelomey. Even in our own circle, we did not dare express such seditious thoughts. But we each knew what the other was thinking. The L1 program had been blessed by resolutions of the Central Committee and Council of Ministers. It was considered to be a project of critical national importance. In such cases Sergey Okhapkin used to say: "God forbid and have mercy if you should have any doubts."

Essentially giving me an order, Mishin told me that I should devote more attention to debugging the Soyuzes and get seriously involved with the L3.[3]

---

2. This is the same Yuriy Pavlovich Semyonov who served as general designer and chief of RKK Energiya from 1989 to 2005.

3. In other words, the implication was that Chertok should not get involved in work on the L1.

"For the Soyuzes, you need to be at the factory, at the engineering facility, and in Yevpatoriya. And the L1 can get along without you. Yurasov and Shabarov can cope with it, and Semyonov is a strong lead there."

Thus, I received permission to stay away from the firing range where the L1 launches were being prepared. And as for flight control, bearing in mind the decisive role of the control system, he proposed that I participate another "couple of times," and then: "Agadzhanov and Tregub will need to get along without you and Rauschenbach."

Often such conversations switched to the subject of how woefully far behind the Americans we had fallen in terms of computer technology.

"That's what we need to be concentrating on instead of all of us flying to the firing range and Yevpatoriya. The testers should get by without us."

These were dreams that were not destined to come true.

Three launches were scheduled for March and April: L1 No. 6 and 7K-OK No. 7 and No. 8. The technical review team and State Commission needed to hold meetings to approve the final flight programs, to verify and account for the implementation of all the commissions' recommendations regarding all the failures throughout 1967.

IT TURNED OUT THAT DURING THIS ENTIRE TIME, FROM THE END OF JANUARY TO THE END OF FEBRUARY, the cosmonauts who usually participated in these activities were occupied with completely different matters. They were graduating from the Professor N. Ye. Zhukovskiy Air Force Academy, or *Zhukovka* for short. Gagarin and all the cosmonauts who had already flown, with the exception of Tereshkova, had to defend their graduation projects by February.[4]

We had known for some time that Gagarin, Titov, Nikolayev, Popovich, Bykovskiy, and Leonov were students at the renowned *Zhukovka*, but we had thought that in any case they wouldn't have time to seriously pursue science, and for that reason during all sorts of meetings and conversations we never inquired how their studies were going. We engineers thought that

---

4. More precisely, all active members of the original group of 20 cosmonauts selected in 1960 *who did not have higher degrees* graduated from the N. Ye. Zhukovskiy Air Force Academy in 1968. In January, V. F. Bykovskiy, V. V. Gorbatko, Ye. V. Khrunov, A. A. Leonov, A. G. Nikolayev, P. R. Popovich, G. S. Shonin, B. V. Volynov, and D. A. Zaykin defended their diploma projects. Later, on 17 February, Yu. A. Gagarin and G. S. Titov followed. Of the remainder, I. N. Anikeyev, V. S. Varlamov, A. Ya. Kartashov, M. Z. Rafikov, and V. I. Filatyev were no longer active, while V. V. Bondarenko, V. M. Komarov, and G. G. Nelyubov were deceased. P. I. Belyayev was the only active cosmonaut who had a prior diploma before joining the team in 1960. Of the women cosmonauts selected in 1962, V. L. Ponomareva and I. B. Solovyeva graduated in July 1967, while the remainder, Zh. D. Yerkina, T. D. Kuznetsova, and V. V. Tereshkova, graduated in May 1969.

the bulk of the cosmonauts' time was spent studying spacecraft hardware or devoted to our space sciences, physical exercise, training sessions, parachute jumps, meetings with workers, trips to foreign countries, and the obligatory attendance of innumerable ceremonies.

We also took advantage of the cosmonauts' popularity; when we set off on difficult business trips to meet with subcontractors, we talked one of them into going with us on such expeditions. The presence of a cosmonaut immediately guaranteed a cordial reception at the Party regional committee, a meeting at the factory, and the subcontractors' commitment to do everything to strengthen the nation's power in space.

The cosmonauts themselves, either out of modesty or out of superstition, usually did not confide in us regarding their academic success. It wasn't that we weren't interested in that side of the cosmonauts' lives, but we simply thought that they were guaranteed diplomas, if only because you couldn't just fail a world-famous Hero of the Soviet Union just because he might forget some standard integral or get bogged down with calculations from a course on materials science.

Academy Professor Sergey Mikhaylovich Belotserkovskiy described very well how the cosmonauts were instructed at the *Zhukovka* on Leningradskiy Prospekt. He put his heart into his book, *Gagarin's Diploma.*[5] It reveals one more hitherto little-known page from the life of Yuriy Gagarin and his comrades. I received a copy of this book from the author with the inscription " . . . in fond memory of Yuriy Gagarin." I recommend that all who cherish the memory of the world's first cosmonaut and the history of cosmonautics read this book.

Gagarin and Titov defended their graduation projects at Star City on 17 February 1968. Here is what Belotserkovskiy writes:

Gagarin's defense was filmed, and his report was tape-recorded. For a long time we had thought that they were lost, but very recently we managed to find it all . . . .

When Yu. A. Gagarin received his higher engineering education and very successfully defended his graduation project, he was exceptionally emotional, enthusiastic, and animated. He displayed remarkably unadulterated, childlike, unconcealed joy. He wanted to hug everyone and to thank everyone who had helped him, to share his joy with and to do something nice for everyone.

---

5. S. M. Belotserkovskiy, *Diplom Gagarina* [*Gagarin's Diploma*] (Moscow: Molodaya gvardiya, 1986).

Asif Siddiqi

This still from a movie shows the transport of a 7K-L1 circumlunar spacecraft on its Proton booster on the way from the assembly building to the launch pad at Tyura-Tam. Note the cluster of solid-propellant rocket engines at the top of the launch escape tower. The hatch on the external fairing for cosmonaut entry into the actual spacecraft can be seen in the foreground as a dark oblong shape.

There were just 11 days between the joy of his Academy graduation and his next work assignment at the firing range, taking part in the work of GOGU in Yevpatoriya. He still needed to have time to train. Notwithstanding his *Zhukovka* diploma, Nikolay Kamanin would not give him a pass [to fly]. He was a tough and demanding taskmaster.

"Sergey Mikhaylovich, I am deputy chief of the [Cosmonaut Training] Center, I need to fly," Gagarin told Belotserkovskiy in parting on the evening after his brilliant defense. Gagarin had just one month and 10 days left to live!

ON 1 MARCH, I MANAGED TO CELEBRATE MY BIRTHDAY AT HOME.[6] On the morning of 1 March, Mishin, Chelomey, Barmin, Karas, Kazakov, Kamanin, and the cosmonauts flew out to the firing range. My comrades wished me a happy birthday the following day at Mozzhorin's TsUP.[7] Now at the Center it had become possible to have information simultaneously from the firing range, Yevpatoriya, and the ballistics centers.

---

6. Chertok turned 56.

7. This was the new flight control center at the premises of TsNIIMash in the Moscow suburb of Kaliningrad (now Korolev).

The launch of L1 No. 6 took place on 2 March 1968 at 2129 hours, 23 seconds. The most exciting segment according to our previous experience was the second firing of the Block D upper stage—the boost toward the Moon. It was successful. The ballistics experts figured that perhaps no correction would be required all the way to the Moon.

On 3 March at 12 p.m., I flew out to Yevpatoriya on an An-12 accompanied by Ryazanskiy, Boguslavskiy, and Khitrik, with a stopover in Saki, as usual. When we landed, there was already pandemonium at the airfield. An Il-18 had arrived from the firing range carrying more than 70 people. Mishin, Tyulin, Shabarov, and dozens of testers and guests, whom Mishin, for unknown reasons, had collared and who had nothing directly to do with flight control, had landed. As Shabarov managed to whisper to me, in celebration Vasiliy Pavlovich had "had a few," gotten mellow, and decided to reward people with a flight to Yevpatoriya. However, during the course of the flight, his benevolence had ended. When Mishin saw the landed An-12 and us, its passengers, he flared up: "And why are you here? Who summoned you? There's nothing for you to do here. Go home!"

Ryazanskiy began the negotiations. I was choked with resentment and made a show of walking to the airplane from which we had just disembarked. One of the local officers caught up with me, grabbed my suitcase, assured me that the "ceremonial part of the reception was over, and a *gazik* and your comrades are waiting for us."

Boguslavskiy and Khitrik were sitting in the *gazik*. Both of them had enjoyed the spectacles of the reception of the two airplanes and now were blowing off steam with jokes at my expense.

"You shouldn't complain," said Boguslavskiy. "After all, your organization has made progress: Korolev sent his people back from the Crimea 'walking along the railroad tracks,' but Mishin is providing airplanes!"[8]

"And Gagarin has unexpectedly stuck up for us and for you, in particular," said Boguslavskiy en route. "It turns out, he defended his graduation project and talked with Vasiliy Pavlovich so sternly that he [Mishin] gave up."

The cavalcade of cars rolled into NIP-16 for the beginning of the latest communications session with *Zond-4*, which Levitan had already announced.[9] This L1 flight really was just a probe. The spacecraft was supposed to fly not around the Moon but to a precalculated position 330,000 kilometers from

8. "Walking along the railroad tracks" was a reference to one of Korolev's favorite and most feared admonitions toward his engineers.

9. In the open press, the mission was announced as *Zond-4*. Earlier Zonds were unrelated probes launched as part of a test series of interplanetary probes. Six were actually launched between 1963 and 1965, but only *Zond-1*, *-2*, and *-3* successfully escaped Earth's gravity.

Earth. In this case, the Moon was not necessary. We had set as our main mission to test out the technology of control, stellar correction, return to Earth, entry into a precalculated corridor, deceleration with two atmospheric passes, and landing.

Having ascertained that all on-board systems were in order, we nevertheless decided to follow the recommendation of the ballistics experts and perform stellar correction on 4 March and fire the correcting engine (KD) for a small burn.[10] The first correction failed. What began next was an epic episode of a series of stellar corrections involving inexplicable failures of the 100K star tracker that later entered the annals of space folklore.

Excellent engineers, led by Vladimir Khrustalev, worked at the Geofizika Factory, which developed all sorts of optical sensors. None of us doubted their integrity and qualifications. Stanislav Savchenko oversaw all the developments from our side. Yevgeniy Bashkin had this to say about him: "We were really fortunate to have Savchenko. He is so well-versed in the developments at Geofizika that the developers are sick of him."

Khrustalev complained to both Rauschenbach and me about Savchenko's nit-picking ways and exacting nature. Despite this active participation from our side in the development of optical sensors, each stellar correction session, which began with search, lock-on, and then an attempt to hold the star in the field of vision, was reminiscent of a hockey face-off. We tried to lock onto Sirius or Canopus—the two brightest stars, and like hockey players, we tried to drive them into a small "goal"—the sensor's field of vision. But some unknown opponent appeared at the sensor's inlet, and the star, like a hockey puck, would not go into the goal.

This time the 100K star tracker worked "stupidly" at the very beginning. Next, it began to react to the streams of peroxide illuminated by the Sun that were being emitted from the attitude control engines. While there was time, we took a series of measurements to understand what was hampering us and learn which star could best be used for attitude.

Watching our ordeals, Mishin asked: "So, where are Legostayev and Bashkin?"

"You told them they couldn't come."

"Summon them immediately!"

On 5 March, Viktor Legostayev, Yevgeniy Bashkin, Oleg Babkov, Stanislav Savchenko, Vladlen Rastorguyev, and Aleksandr Sverchkov flew out to help us. They grabbed Rachitelnyy and Medvedev from Geofizika to help Anatoliy

---

10. KD—*Korrektiruyushchiy dvigatel.*

Azarov. The new arrivals gave this recommendation: if the 100K has lost so much sensitivity that it doesn't sense Sirius and Canopus, you need to ask the ballistics center to recalculate using Venus—its brightness will suppress any interference. On the night of 6 March we experimented with Venus! After prolonged tribulations Venus was found, but it turned out that the sensor produced "reverse polarity."

With a straight face Legostayev explained: "This is because Venus' shine is from reflected light rather than its own."

The observers from the VPK standing behind us took his joke seriously and began to grill me: "You *did* know that Venus is a planet, and not a star."

I got aggravated and blurted out, "We forgot!"

There was also no sense continuing sessions using Venus because the angular distance between Venus and the Sun had decreased to one that was risky for the sensor.

After using the command radio link to first set the 100K star tracker to a sensitivity corresponding to the luminosity of Sirius, we once again went back to the option of using Sirius. Troubleshooting the malfunctions and adjusting the sensitive electronic optical instrument at a range of 300,000 kilometers using the command radio link was hardly a simple operation back then. And it wasn't just three or four specialists dealing with this, but a raucous, argumentative staff of people who were skeptical of every new explanatory hypothesis. Faced with the shared danger of losing the vehicle, Mishin and Tyulin managed to establish good relations. From time to time, the minister walked in and, trying not to tear us away from our deliberations and formulation of commands, he asked us to explain what was really going on. The sessions were long and tedious. Once again we found that stabilization had been lost completely.

On 6 March the correction finally went through! The L1 passed through apogee and was returning to Earth. Now, to end up in the "goal" of the predetermined corridor, the next correction needed to be conducted at a range of 160,000 kilometers from Earth early in the morning on 9 March, and the landing was predicted for a few minutes past 2100 hours.

On 7 March, the temperature of the hydrogen peroxide was cause for alarm. It fell to minus 1–2 degrees. If it continued to drop and reached minus 4–5 degrees, we would no longer have fuel for attitude control! Meanwhile, three ballistics centers were calculating and arguing around the clock whether one more correction was needed to enter the "corridor." What if we botched it?

On 8 March at 3 a.m., successive measurement sessions began to refine the trajectory of the vehicle returning at escape velocity. Over the course of the flight we confirmed that the high gain antenna was operating, but for some reason the signal-to-noise ratio was significantly lower than the design value. Our antenna experts along with Boguslavskiy's radio specialists—Pikovskiy and Galin—were trying to get to the bottom of it, but they were debating and were unable to give an unequivocal report. The main ballistics group made the official announcement that for the time being, it was not possible to give a clear answer as to whether further correction was necessary. As the vehicle drew closer to Earth, a hectic and tense situation set in at the tracking station. It was as if cosmonauts were coming back to Earth and we needed to arrange a red carpet reception for them. As chairman of the State Commission, Tyulin demanded that the ballistics experts officially substantiate that the second and last correction was undesirable.

I spoke with Elyasberg, who was preparing the response at NII-4. He patiently explained: "We are solving a boundary problem, all the while updating baseline data as data comes in from you based on orbital measurements. We don't sleep here, but our computers do need routine maintenance and time for cooling down." Yes, the large ground-based computers of that era had a propensity for breaking down almost hourly. The ballistics experts patiently waited until the on-duty electronics engineers restored operating capability, and put us off with references to the complexity of the "boundary problem."

It was the seventh day of the flight of L1 No. 6. Now all attention was concentrated on loading the settings into the sequencer to start up the program for the descent cycle. In this cycle, everything started from stellar orientation; then the on-board computer was activated, and the gyroscopes of the dual-axis platform spun to provide stabilization during the first atmospheric pass. Having separated from all the parts of the spacecraft that would burn up in the atmosphere—the instrument-aggregate compartment [i.e., service module] containing the SKDU and high-gain antenna—the descent module shaped like an automobile headlight was supposed to graze Earth's atmosphere. After reaching an acceleration load of four Gs, the vehicle would begin to change its aerodynamic quality by spinning about its longitudinal axis, thereby altering the position of the center of mass relative to the center of pressure, and tear away from the atmosphere and into space. Having lost speed during the first atmospheric pass and having flown part of its way in an Earth orbit, 20 minutes later the descent module was supposed to finally dip into the atmosphere and land in the predetermined area.

By midday the cramped hall of the command post was packed to overflowing. People were debating; the latest telemetry data were being discussed; arguments continued concerning the strange behavior of distant Sirius and the reflected light of unruly Venus. Lists of glitches were being signed so that modifications could be made for No. 7, and a lot of various and sundry other things.

This photo dates from the late 1960s, probably during the flight of a Soyuz or Zond spacecraft. From left to right are Chief Designer Vasiliy Mishin, GOGU Chief (Flight Director) Pavel Agadzhanov, and Deputy Chief Designer Boris Chertok, fishing near the Center for Deep Space Communications on the outskirts of Yevpatoriya.

According to the updated calculation, the first atmospheric pass would take place on 9 March at 2119 hours, 18 seconds. Eight minutes before this, the descent module and the instrument-aggregate compartment were supposed to separate.

Beginning at 1800 hours, Agadzhanov started to take a poll of the readiness of all the ground facilities. The ship *Ristna* responded from the Atlantic Ocean. It was supposed to pinpoint the L1 meteor streaking over it at 2103 hours—before separation. The ship *Bezhitsa* would pick up surveillance after separation at 2120 hours in the Gulf of Guinea. Our NIP-16 would establish communication with the descent module at 2133 hours.

The times were estimated and could shift forward or backward by 1 or 2 minutes due to inaccuracy in the ballistics experts' calculations. Descent was above all a trial run for Pilyugin's control system. Turning to Khitrik, the minister warned: "Now we're going to make you squirm!"

By 2000 hours everyone was calming down. Mozzhorin reported from Moscow that his TsUP was also full of people.

The GOGU services started to deliver reports: "The object is nearing *Ristna*'s coverage zone."

Despite the presence of the brass, out of the semi-darkness someone added: "In Nigeria it's the rainy season."

"And they're eating shish kebab!"

"Silence! Separation should be taking place now!"

"*Ristna* has ended reception."

"*Ristna*'s report was received via Odessa."

"Attention! Separation occurred! All parameters and times are normal!"

"No word yet from *Bezhitsa* . . . . According to the time, the vehicle has already left *Bezhitsa*'s coverage zone, and *Bezhitsa* is silent! Airplanes have headed for the calculated landing area! The station in Tbilisi is receiving in *Zarya*'s frequency!"

"The object is flying per target indication!"

"NIP-10 has begun to receive! Axial accelerations 20 Gs!"

"From where?"

"That's from *Ristna*."

"Impossible!"

The National Air Defense Troops facilities calmly report that the "object" had not been identified in the nation's airspace.

Completely contradictory data. We attempt to make sense of it. Tbilisi and Simferopol had supposedly heard *Zarya*. *Bezhitsa* and we had not picked it up. That meant it was an obvious phony. They simply wanted very much to pick up something. Evidently, they picked up some local television stations' VHF signals reflected off the mountains.

After comparing various sets of data and throwing out the improbable hypotheses, we concluded that it wasn't necessary to search for the vehicle on Soviet territory. If the *Ristna* had confirmed accelerations of 20 Gs, it meant that the APO had activated! The system had been armed to go off in the event of a landing on the territory of Africa or Turkey. Twenty Gs—that's quite a lot, even for ballistic descent.

"We'll survive that!" Leonov said when he heard this number. "Only if you don't blow us up."

In order to make things add up, I constructed a diagram of radio communications allowing for the communications blackout period during the first passage through the plasma. My diagram matched up with the easy-to-understand picture that the ballistics experts had constructed in the event of a ballistic descent. Working together, we tied the events to the times when the *Ristna*'s and *Bezhitsa*'s radio contacts began and ended. The *Ristna* had contact with the vehicle from 2111 hours, 39 seconds until 2119 hours, 58 seconds. Reception ended when the descent module plunged into the atmosphere. The vehicle did not ricochet out of the atmosphere, but continued its descent on a steep trajectory, and the *Bezhitsa* intercepted it at 2121 hours as it left the plasma. At that point communication finally broke off. The APO system activated over the coast of Africa.

"I've no need for the beaches of Turkey, and Africa's nothing to me."[11] That's how I tried to sum up what had caused the APO system to destroy the vehicle. I was personally responsible for this system's reliability. "But why didn't it resurface after the first atmospheric passage? Khitrik has to explain that."

While we argued, the *Ristna* provided more details. During the orientation session before the atmospheric pass, a number of things happened: a strong light bias was detected; then the activation of the gyro-stabilized platform was inhibited; there was no star available; and finally, the autonomous control system went dead. What was going on? Perhaps the 100K had once again cranked out some random number, confused Sirius with Canopus, and issued an inhibit? We needed the films from the *Ristna* for thorough analysis. It was already heading for the coast of Africa at full speed so that an Il-18 could pick up the films in Somalia and deliver them to Moscow. But this would take six days! And meanwhile the minister really "made us squirm." Both Khitrik and I confessed that the 100K had not lived up to our expectations, but for the time being we had not come up with another attitude control method for guided descent.

Around midnight we decided to break up so that we could depart for Moscow at 9 a.m. the next morning. On the way from the command post to the hotels Leonov stopped us: "Today is Gagarin's birthday! Yuriy Alekseyevich really wants you to stop by the dining hall."[12]

Honestly, in this commotion we had forgotten to wish Gagarin a happy birthday! We stopped in. All the tables were covered with tablecloths and set in expectation of the guests. A normal landing had been anticipated at 2200 hours, and we had not only blown up the spacecraft, but then we had debated until midnight.

All the cosmonauts on the circumlunar flight team had gathered.[13] Yuriy himself was a bit embarrassed by the whole affair; he smiled and asked us to forgive him for robbing us of precious hours of sleep, and invited us before the cognac "got cold" to celebrate the first return to Earth at reentry velocity.

---

11. This is an allusion to a line from a popular patriotic song from World War II, whose lyrics were written by Mikhail Vasilyevich Isakovskiy (1900–73). Isakovskiy was most famous for the lyrics to "Katyusha," a love song which gave the name to the famed rocket launchers of the war.

12. Gagarin was born on 9 March 1934.

13. In January 1968, there were a total of 14 cosmonauts in the L1 training group, although it is not clear, as Chertok notes here, if all participated in the *Zond-4* mission. Candidates for crew commanders included V. F. Bykovskiy, G. T. Dobrovolskiy, P. I. Klimuk, A. P. Kuklin, A. A. Leonov, P. R. Popovich, and V. A. Voloshin while engineers included Yu. P. Artyukhin, G. M. Grechko, O. G. Makarov, N. N. Rukavishnikov, V. I. Sevastyanov, A. F. Voronov, and V. G. Yershov.

He didn't have to twist our arms, especially since the richly laid out table demanded action. We drank to the health of the first cosmonaut, to his brilliant defense of his graduation project, to his graduation from *Zhukovka*, to our harmonious joint work, to the health of all of us, and finally, to the future victories: launches of Soyuzes and L1 No. 7 were to take place no later than April!

Finally, after saying our goodbyes to Gagarin, completely exhausted but in an excellent mood, Agadzhanov, Ryazanskiy, Tregub, Boguslavskiy, and I strolled down to the seaside. During our walk everyone argued: "But, if there had been a cosmonaut, he would have taken control and your APO system would have been unnecessary, Boris!" I agreed and responded: "The Americans put more trust in people."

Far away on the dark horizon, a multidecker ship twinkling with lights plied the waters. There, people who had absolutely no idea about our lunar worries were enjoying themselves and relaxing.

At 9 the next morning, we were already at the airfield in Saki. The cosmonauts flew out ahead of us on their own airplane. The minister and Tyulin left with them. An Il-18 aircraft was waiting for us. Mishin, who had left the center before us, didn't show up at the airplane. With a guilty expression, Khvastunov told me that some sailors had taken Vasiliy Pavlovich to their place "for breakfast." Everyone looked hopefully to me as the senior member of the group. I proposed to Khvastunov: "You are the aircraft commander. Take this car and dash over to those sailors and tell Mishin that we need to depart immediately. We've been informed that Moscow might be shut down after 12 p.m., either due to weather or because of the arrival of dignitaries."

Khvastunov handled his mission successfully, but he received a reaming-out from Mishin for having created an awkward moment in front of our gracious airfield hosts. We flew out at 11 a.m. instead of 9. At 2 p.m. I was home. I took a hot bath and didn't feel very well. By evening my temperature had risen to 39°C [102.2°F].

The next morning I had to meet with Mishin in Tyulin's office to report about everything that had happened. Tregub reassured me over the phone that Tyulin understood our situation and would put off the meeting another day.

Mishin telephoned, inquired about my health, and cautioned me: "When we meet with Tyulin, let's insist that the next [L1] launch be no later than 25 April. And before that we will have to perform an automatic docking [using the Soyuz]. We need to understand what is going on with the stars. And we've got to keep our chins up. After all, the general result wasn't bad. Except for the star tracker, everything worked accurately the first time, without any glitches. The radio operators and ballistics experts provided terrific accuracy! They just informed me that they entered the corridor with an error of just 2 kilometers, when they allowed for 10! Get well. I've got to assemble a review team for the 7K-OK and then a State Commission."

I sensed that Mishin was in a good mood.

I don't remember who first uncovered the secret of the tricks that the star tracker played. The reason for its failures proved to be so instructive that later I included this example in a course of lectures that I teach to students when we get to the subject of reliability. The sensor's lens shade, in the form of a tube with light-occluding screens to protect against the Sun's glare, was painted black. In the vacuum of space under the effect of the Sun's direct rays, the shade warmed up; the unstable paint sublimated; and its particles nebulized in the zero gravity and settled on the lens. Khrustalev's team at Geofizika reproduced and experimentally confirmed this hypothesis.

"This is a case of 'you couldn't have come up with this if you'd tried'," I told Tyulin to preempt his wrath toward Geofizika, and I asked him not to call his colleague, the deputy minister of the defense industry, who oversaw Geofizika.

On 12 March, without our participation, Gagarin and Leonov gave an account of their impressions based on the results of the L1 flight. The first candidate for the circumlunar flight and leader of the lunar cosmonaut team, Leonov, the optimist, tried to argue that "it wasn't a catastrophe yet."

Kamanin was outraged that we did not disable the APO system: "Why blow up the descent module even if it is headed for Africa? We could test the parachute system. Ultimately, we would have found the module and learned a lot."

But the ideologues in the Central Committee believed that not a single tiny rivet should end up outside the cordon. Not because some technical secrets might be revealed from the fragments, but because the very fact that a failure had occurred must be concealed. We must not and cannot have failures! Kamanin telephoned Tyulin and Mishin and threatened that for the next launch he would officially demand that the APO system be disabled regardless of the landing area.

I immersed myself in preparations for the report on the 7K-OK at the State Commission. Having taken an interest in how things were going, Mishin said with a certain irony that Air Force Commander-in-Chief Marshal Konstantin Vershinin had already approved the cosmonaut training program for the N1-L3 lunar landing expedition, and yet we didn't even have a mockup of the LOK yet.[14]

Tyulin called us on the Kremlin direct line to check once again how things were going with the 100K star tracker. Everyone who had a Kremlin direct

---

14. Vershinin approved a two-and-a-half year training program for the lunar landing cosmonauts on 13 March 1968.

line assumed that their conversations might be recorded, but nevertheless sometimes allowed themselves liberties. Tyulin, in passing, said: "'Uncle Mitya' [Ustinov] is very upset by the pace the Americans are keeping. They told him that during 5–6 April there would be an experimental flight of the Saturn V with unpiloted lunar vehicles. He asked how we were going to respond. I promised that we would have one more automatic docking before 15 April and a circumlunar flight at the end of the month. But he demands assurances for a report to Leonid Ilyich [Brezhnev] that there will be a piloted flight. He's really upset about how we've fallen behind with the N1. He promised to pay you a visit at TsKBEM and conduct a thorough investigation. I warned Vasiliy about this, but you be ready too."

What could I promise Tyulin? For the time being we needed to finish up with the 7K-OK. This was assignment number one.

On 26 March two State Commission meetings were held one after the other. At 1000 hours Kerimov conducted a meeting of the State Commission on the 7K-OK. Mishin and Tkachev reported that the parachute systems had been tested and raised no doubts. I reported that, beginning with No. 7, we had improved the attitude control system, having installed an infrared vertical sensor on each vehicle. There were no doubts as to the reliability of the systems. Preparation at the engineering facility was proceeding normally; launches could be scheduled for 14–15 April. Kerimov proposed that all State Commission members depart for the firing range no later than 12 April.

At 1500 hours, Tyulin began the State Commission on the L1. Khitrik and I reported on the operation of the control systems. I said that there had been one serious glitch over the entire flight. The 100K star tracker locked onto, but then let go of Sirius. Without this, reliable correction could not be provided. It continued like that for the first three days of the flight. On the fourth day, correction proceeded successfully and the vehicle returned to Earth on the design trajectory. But before actual entry into the atmosphere, the star tracker failed again. Therefore, guided descent failed. According to the logic, we switched over to ballistic descent, and the APO system activated. Everyone was tired (so there were no more serious discussions at this point) and, without arguments, we accepted Mishin's proposal to launch L1 No. 7 on 23 April, under the condition that there would be a report on the complete clarity and reliability of the behavior of the 100K star tracker.

Gagarin was present at both State Commission meetings. In addition to his other medals and badges, he now had an academic rhombus on his jacket. He could not help but beam when anyone congratulated him on this new "merit badge."

Chapter 25
Gagarin's Birthday and Death

The 27th of March was always a difficult day for me. This was the date of Bakhchivandzhi's death.[1] Twenty-five years had passed since that day. It was also the date of my mother's death in Bilimbay. That had been 26 years ago. I arrived home late. Despite Katya's protest, to relieve the tension and commemorate the departed, I silently drank a shot of vodka.

At 10 p.m. the telephone rang. Katya answered it.

"That was Mishin," she said. "He said that I should first give you some valerian and then have you call him.[2] He has unpleasant news that he didn't want to tell me."

Instead of the valerian I drank another 150 grams of vodka and called Mishin at home.

"Listen! But keep your cool and have both feet on the floor, or better yet— sit down . . . Today at 11 a.m. during a training flight, Gagarin crashed."

I didn't say a word. I listened to Mishin's irregular breathing. The words he had uttered still hadn't registered with me.

Mishin continued: "This is all Kamanin's nonsense! Training! Gagarin and Seregin died in a MiG-15 UTI trainer fighter."

Finally it hit me. I was shaken. I told Katya. Then I telephoned Bushuyev and told him to meet me out on the street. We both paced back and forth in front of our apartment building, trying to comprehend the senselessness of what had happened.

The next morning at the KB practically no work was done; everyone was discussing the news of Gagarin's death. Gagarin's name was inextricably connected with our organization. Regardless of one's departmental and territorial segregation, everyone who was involved in the production of Vostoks, designers and workers, considered Gagarin "their very own," the most familiar

---

1. See Chertok, *Rockets and People, Vol. I*, pp. 197–199.
2. Valerian, often used as a sedative, is both a herbal treatment and a nutritional supplement.

This official picture of Yuriy Gagarin was probably taken around 1965 during a period when he was no longer actively training for space missions, having been busy with a desk job at the Cosmonaut Training Center. In late 1966, he resumed crew training as the backup pilot for Soyuz-1 although his training was suspended after that mission ended in disaster. At the time of his death, Gagarin had just resumed air training.

NASA

of all the cosmonauts. First of all, he was the first; second, you would always see him with us at gatherings with the staff and at meetings after the flights of each subsequent cosmonaut. He met with Korolev more often than the others. He did not refuse to come to Kaliningrad when Korolev invited him to the opening of the Palace of Culture.

Factory Director Viktor Klyucharev accurately described the mood of the workers on 28 March: "Today it's a lot more quiet in the shops than on a normal workday." There were more portraits of Gagarin in the shops and departments than there were portraits of the "leaders." The photograph of Korolev and Gagarin sitting side-by-side, smiling so happily, was particularly good. For the personnel who had produced the spacecraft, both of these men fit the concept of "near and dear."

At noon Bushuyev, Tsybin, Anokhin, and I drove to Star City.[3] There was nothing we could do to help anyone. We needed to find out at least something, to say some words of sympathy to Valya Gagarina, to the Star City administration, and to the cosmonauts, to share their grief. That's the way people are. When sudden disaster has fallen on everyone, taking action is somehow comforting.

Kamanin met us at the entry to Star City. We learned the first details from him. (Subsequently, the accident investigation commission precisely ascertained all the data). He told us:

---

3. Star City (*Zvezdnyy gorodok*) was the name given in 1968 to the suburban Moscow location of the Cosmonaut Training Center.

Gagarin took off on a training flight in a MiG-15 UTI trainer fighter. This is a two-seater fighter. These aircraft are produced in Czechoslovakia. This model of the MiG-15 jet fighter had been stripped of armament and the second seat was for an instructor. They flew into the zone in the area of Kirzhach at 1019 hours. At 1027 hours, Gagarin reported that their mission had been accomplished; the aircraft was in the zone near Kirzhach; and he requested clearance to return. His mission had taken just 8 minutes to fulfill. He had time left over.

At 1029 hours, he received clearance to return. He did not respond. At 1032 hours, he was asked repeatedly why he did not respond. They called for both Gagarin and Seregin to make radio contact. There was no contact. Before this, radar had indicated that they were heading toward the airfield and were descending. Leonov, Nikolayev, and Bykovskiy were there in the zone at that time. They said that they heard two explosions or bangs.

Helicopters and one Il-14 took off on alert status. The entire area was divided into squares, and the helicopters searched everywhere from an altitude of 50 to 100 meters. It wasn't until 1300 hours that one of the Mi-4 helicopter pilots spotted the signs of a crash and a huge hole in the forest 3 kilometers from the village of Novoselovo. It was very difficult to discern anything in the thick forest. Two more helicopters approached and the pilots finally managed to find the site that the first helicopter had indicated. I flew out there. It wasn't possible for a helicopter to land near this site. We made our way there from the road over very deep snow. We caught sight of a huge hole about 4 meters deep. Judging by the trees that had been cut down, one got the impression that the airplane had cut in at a 60-degree angle. Based on the debris field and the size of the crater, experts estimated that the impact velocity was around 700 kilometers per hour. They were unable to find anything intact. Everything was pulverized into tiny particles, mixed with Earth and snow, and scattered over a 200 by 100 meter area. The engine had the sturdiest parts. They were presumably deep in the ground.

We made no announcements and no reports until we found proof that the pilots had died. Maybe they had ejected. There was that glimmer of hope. By around 3 p.m., we had determined that one of them had died. That was clear. We found a piece of skin, someone's scalp. They were able to identify Seregin's jaw by his teeth.

Then, on a tree, they found part of a flight jacket with a pocket. It contained a breakfast voucher with the name 'Gagarin Yuriy Alekseyevich.' Only after this were Brezhnev and Kosygin informed. All the pieces of the bodies were gathered and sent to Moscow for blood tests. We quickly received an answer. They confirmed that it was Gagarin's blood.

The snow is very deep. Instructions came from Moscow that everything that could be should be quickly gathered for cremation. It is a very difficult assignment. In order to cremate them separately, a blood test needs to be performed on each little piece. Dozens of people are working there now. But for now there's actually nothing to cremate. Before it gets dark maybe we'll gather up something. The impact was very intense. The aircraft was 30 kilometers from the airfield at the moment we lost contact, and it followed a heading southwest of the airport. Why didn't they eject?

Kamanin gave his account in fragments with long pauses. It was difficult for him to speak, but he couldn't stay silent, either. Gagarin had been very close to him. He felt that he was partly to blame for the catastrophe. Kamanin hadn't slept all night, and now he would not be able to rest until the accident investigation commission appointed by the Ministry of Defense began to work. Air Force Commander-in-Chief Vershinin would be in charge of that commission. Suslov would be in charge of the commission for funeral arrangements.

During all of the spaceflights, no matter how much the descent module deviates from the predetermined landing point, the search and rescue service finds it within a matter of minutes. In the case of Gagarin's airplane, which was supposed to be continuously tracked by the radar service, it took more than 2 hours for them to find it! And this was near Moscow, 3 kilometers from a village!

The Moscow Air Defense Troops' two rings of missile complexes are capable of tracking the flight trajectory of any airplane at a range of hundreds of kilometers. In this case, they "lost" him before he crashed. Incidentally, you can't blame Air Defense. Unless it has special instructions, this service does not track such aircraft. We were able to ascertain this many years later from other instances.

We weren't about to torment Kamanin with questions, and we set off for the residential area. We met Leonov at the entrance to the apartment building where Gagarin had lived. We went up to the sixth floor with him. Valya was ill and had been in the hospital just yesterday. They brought her home yesterday evening. Tereshkova had been with her almost constantly. We mumbled some comforting words, adhering to tradition and ritual. Why?

Valya Gagarina, impassive, sat on the sofa wrapped up in a plaid blanket. There was no weeping or wailing. There was inconsolable grief. She was scarcely aware who had dropped in and what they were saying. By now the entire world knew about Gagarin's death.

The cremation of what remained of the pilots took place on the evening of 28 March. The urns for the funeral were delivered to the Order of the Red Banner Hall at the M. V. Frunze Central Building of the Soviet Army (TsDSA).[4] Long before 9 a.m. on 29 March, the time announced for the building to open

---

4. TsDSA—*Tsentralnyy dom sovyetskoy armii imeni M. V. Frunze.*

to the public, a line had already formed extending from TsDSA to the Sadovoye Ring. Our delegation drove up to Commune Square (*Ploshchad kommuny*) at 10 a.m. with wreaths and flowers—both sides of the boulevard were jammed with people and cars. We each took two turns standing in the honor guard.

We were at the TsDSA almost the entire day. Kamanin told us that after the cremation, during excavations at the crash site, by some miracle they found Gagarin's intact wallet. It contained his ID, driver's license, 74 rubles, and a photograph of Korolev—not his beloved wife, daughters, or mother, but Korolev!

The first day, more than 40,000 people filed through the TsDSA building. They stopped admitting people late in the evening. On the morning of 30 March, an endless line formed once again—the public would be admitted until 1 p.m. In the room where the members of the funeral commission and the "elite" representatives of the honor guard usually gathered, they started to talk about extending public admission another couple of hours. But the colonel from the Council of Ministers Directorate of Affairs who was in charge of the entire ceremony said that Red Square was already full of people and we couldn't take liberties with the procedure that Suslov had approved.

The funeral procession moved from TsDSA to the House of Unions along a corridor formed literally by two living walls of people held back by policemen and soldiers. The crowd was larger than the one for Korolev's funeral. For the majority of people, Korolev had been an abstraction. Before his official obituary appeared, most of our citizenry knew nothing about him.

Hundreds of millions of people all over the planet had seen Gagarin smiling joyfully in person or on television. He was theirs, familiar to everyone, and at the same time a "Citizen of the Universe." "If only it had been his destiny to die far away in the cosmos, in space, rather than in this senseless accident"—these were the thoughts expressed by those who were outraged that Gagarin had been allowed to fly in a fighter. It is more difficult to control the fate of a human being—at least without strong coercion—than it is to control a spacecraft. Gagarin himself was raring to go on this fateful flight; no one forced him. No State Commission heard readiness reports or made the decision for this flight.

Aide to the Air Force Commander-in-Chief Kamanin and Chief of the Cosmonauts Training Center Kuznetsov gave Gagarin no order, no command, and no directive—only their consent for the flight.[5] They decided that Seregin, an experienced pilot and the group commander assigned to the Cosmonaut

---

5. Kamanin's official position from 1966 to 1971 was Aide to the Air Force Commander-in-Chief in charge of cosmonautics.

Training Center, would check out and monitor Gagarin's actions. For Seregin this was a routine, quite uncomplicated flight. He had completed hundreds of flights that were far more complex and difficult.[6]

At the House of Unions the funeral urns were transferred to gun carriages and the procession made its way to Red Square.

At 2:30 p.m. an artillery salute thundered. The urns of Gagarin and Seregin occupied niches in the Kremlin wall right after Malinovskiy, Komarov, and Voronov.[7] I passed along the Kremlin wall with the crowd of mourners, unconsciously wondering about whose ashes would knock out the next bricks. At that time, it couldn't have occurred to anyone that in three years three more "space" niches would appear in this wall.[8]

A DECISION OF THE CENTRAL COMMITTEE AND COUNCIL OF MINISTERS CREATED A GOVERNMENT ACCIDENT INVESTIGATION COMMISSION. Ustinov was appointed commission chairman. To keep abreast of the activity of the numerous subcommissions, Mishin delegated Anokhin to visit LII and meet with friends and acquaintances working in these subcommissions. Sergey Anokhin, a very experienced test pilot, could assess the veracity of the different scenarios better than any of us. All the Air Force, civil aviation, and industry services needed for the investigation had been called in for the commission's work.

The people in our circles were happy with Ustinov's appointment as chairman of the government commission. We did not doubt his objectivity in this case. On Mishin's recommendation, Anokhin, Tsybin, and Feoktistov were recruited to work on the subcommission concerning cosmonaut flight training methods. The subcommission, which was made up primarily of generals and officers—Air Force pilots—was clearly dissatisfied as it listened to Feoktistov, who argued that flight and parachute training for cosmonauts was generally dispensable. In its findings, the subcommission had stated that flight training was an absolute necessity.

Now, after the passage of many years, I would say that it is wrong to categorically deny the usefulness of aviation flight training for the commanders of spacecraft. With the emergence of such winged vehicles as the Space Shuttle or *Buran*, the situation has changed. An experienced professional pilot will certainly master and execute control of the landing better than a cosmonautics

---

6. Colonel-Engineer Vladimir Sergeyevich Seregin (1922–68) was a top test pilot who was on assignment to perform standard air training with cosmonauts in the mid-1960s.

7. Rodion Yakovlevich Malinovskiy (1898–1967) and Nikolay Nikolayevich Voronov (1899–1968) were both famous Soviet war heroes. The former served as minister of defense from 1957 to 1967 while the latter was commander of Soviet artillery forces from 1941 to 1950.

8. Chertok is referring to the death of three cosmonauts on the *Soyuz-11* mission in 1971.

enthusiast who has little flight experience. However, for cosmonauts who spend hundreds of days on an orbital station, and in the future, a year or two on an interplanetary flight to Mars or working at a lunar base, flight and parachute training is hardly necessary.

The biggest battles took place between the subcommissions of the Chief Engineer of the Air Force M. N. Mishuk and another deputy Commander-in-Chief, B. N. Yeremin, who was responsible for the study of flight training, flight support, and its organization.

The first two weeks of the most intense work of hundreds of highly qualified specialists recruited for the work of the government commission shed no light on what caused the disaster. For the time being, everyone agreed that the airplane had been flying at a speed sufficient for horizontal flight before the crash. Neither pilot had attempted to eject. According to Gagarin's and Seregin's watches, which had been found, the crash occurred at 1031 hours, 1 minute after Gagarin's calm report. Had both pilots been able to perform seconds before the crash? That was the crux of the main arguments.

Virtually all the parts of the airplane and its equipment had been gathered. The Scientific Technical Institute for the Operation and Maintenance of Aviation Technology (NIIERAT) reported that, as a result of diligent excavations and sifting of Earth, everything that was gathered amounted to 95 percent of the total mass of the empty aircraft at the moment of the crash.[9]

They weren't able to find all the glass fragments from cockpit canopy. This gave rise to hypotheses that the canopy might have been destroyed before impact with the ground. But how? To destroy the canopy, there would have to be an explosion on the airplane. The possibility of sabotage was categorically rejected. Nevertheless, Kamanin and his supporters maintained that the pilots were for some reason unconscious.

MEANWHILE, AFTER SEVERAL DAYS IN A STATE OF SHOCK, PREPARATION CONTINUED FOR THE FLIGHTS OF THE SOYUZES AND THE L1. On 12 April, Cosmonautics Day was celebrated for the first time without Gagarin. I had to greet this day in the air with a group of my comrades, rather than at

---

9. NIIERAT—*Nauchno-issledovatelskiy institut ekspluatatsii i remonta aviatsionnoy tekhniki*— was the primary research organization dedicated to repair, quality control, and accident investigation in the Soviet Air Force. Its roots go back to the wartime Scientific-Experimental Base (NEB) formed in July 1942 to repair battle-damaged aircraft. In August 1957, it was converted to a scientific-research instititute, i.e., NIIERAT. In post-Soviet times, it is known as the 13th State Scientific-Research Institute of the Ministry of Defense. Soviet (and Russian) civil aviation has its own version of NIIERAT.

the Kremlin's Palace of Congresses. We took off from Vnukovo-3 at 0930 hours on an An-24 bound for Saki and then on to Yevpatoriya. We would be participating in the flight control of 7K-OK No. 7 and No. 8 with a mission to execute an automatic docking on 15 April and one more attempt at a circumlunar flight on the L1 before 1 May. In the gap between the launches of the 7K-OKs and the L1, one more Molniya-1 was supposed to be launched in order to provide, without fail, a television broadcast to Siberia and the Far East via the *Orbita* system for the May Day holiday.

Flying on the airplane were Ryazanskiy, Tregub, Bolshoy, Golunskiy, Popov, Rauschenbach, and star tracker specialists taken as "hostages:" Rachitelnyy from Geofizika and our Savchenko, and Viktor Raspletin, who had still not recovered after the funeral of his father, Academician Raspletin.[10]

After takeoff, as soon as we got above the clouds, we spontaneously pulled ourselves away from the windows and started to argue over what had caused the death of Gagarin and Seregin. The airplane was flying on autopilot and we drew the aircraft commander, an experienced pilot, into our arguments. Everyone was inclined to believe that if there was no evidence of a catastrophic hardware failure—for example, a jamming of the controls or an explosion on board—then the crew was to blame. By the time we reached the Crimea, we had managed to come up with a rationale for this scenario.

The day before [his death], Gagarin had sat on two State Commissions—one for the 7K-OK in the morning and for the L1 in the evening. His head was crammed full of problems concerning the management of the Cosmonauts Training Center; he was, after all, deputy chief. On 27 April he and Kamanin were supposed to argue once again as to who would be included in the crews for the future piloted 7K-OK and circumlunar flights. He was in a hurry.

The commission found no hard and fast instruction regarding what he specifically was supposed to do in the zone. This was on the conscience of Seregin and Gagarin himself. He had spent so much time trying to get permission for this flight. Finally, he gets to fly! The sensation of flight after a long break makes you want to have a little fun. There was still some time. Perhaps, by mutual consent, the pilots flew out of the radar coverage zone and held off responding to the ground's request to communicate. They performed some maneuver and lost altitude. After popping out of the clouds, they were momentarily disoriented, and held off responding to the ground.

Seregin was a very disciplined pilot and a group commander to boot. But in this case, he didn't manage to seize the initiative. Still, one must understand

10. Raspletin had died a year before, on 8 March 1967.

him psychologically: not everyone is assigned to fly as an instructor with the "first man in space." Perhaps he had realized that, given the maneuver that Gagarin had undertaken, they needed to eject. Per the instructions, the person in the second, or rear seat (i.e., Seregin), ejects first. But, could Seregin really bail out of the airplane if Yuriy Gagarin was sitting in front of him? And what about Yuriy? Maybe he still wanted to straighten out the airplane. After all, he'd been so lucky in life! He simply didn't have time to even think about ejecting. Now no commissions could reveal the truth: one dare not even think about the possibility of Gagarin being at fault—that would be the mindset of the highest commission.

On spacecraft, which had appeared just seven years before Gagarin's death, we provide continuous telemetry monitoring of all vitally important parameters. Even on the "distant planet Venus" we can tell, with accuracy down to the second, when and why the descent module stopped working! But on airplanes, despite the fact that millions fly on them, including the most distinguished and famous people in the world, to this day there are no reliable black boxes and no provision has been made for the same sort of continuous radio and telemetry tracking that we have on spacecraft.

After arriving in Yevpatoriya, that evening we heard a news report about the Cosmonautics Day celebration. Keldysh had made a rather standard speech. Almost all the Politburo members led by Brezhnev and Kosygin had been in the Presidium. This was a first for Cosmonautics Day. Unexpectedly, a recording of a speech by the chief designer was played. Was it really possible to forget the voice of Korolev? But his surname was simply not mentioned—again, this idiotic excessive caution.

I recalled that when we listened for the first time at home to the recording of the speech of the still unknown and secret chief designer, the entire family came to the conclusion that it is impossible to conceal and keep secret an intellect of such proportions. No one had ever spoken about the responsibilities of a chief designer better than Korolev had.

Arguments about the various scenarios of the catastrophe continued in the government commission, in all of its subcommissions, and among independent groups of specialists until the end of the year. They argued not so much about their substance as about their wording. It even went so far as a group of cosmonauts sending a letter to Ustinov defending the absolute innocence of Gagarin. Nikolayev, Popovich, Bykovskiy, Titov, and Belyayev wrote that it was impossible that during a relatively simple flight, Gagarin and Seregin, out of a fear of the clouds, would execute a breakaway maneuver and fall into a spin. They were incapacitated due to cockpit depressurization. In the opinion of this group of cosmonauts, the effect of a drastic change in cockpit pressure on the crew was what caused the accident. Mishuk's subcommission categorically opposed this scenario. It wasn't until December that Ustinov made up his mind at last to sign the final wording of the commission's decision. The upshot of the decision was that the cause of the accident had not been reliably determined. The most probable cause of the death of Gagarin and Seregin was a breakaway maneuver executed by the aircraft in order to avoid collision with a sounding balloon. A less likely cause was a breakaway maneuver executed from the top edge of the clouds. As a result of the abrupt breakaway, the aircraft reached the stalling angle; complex meteorological conditions made it difficult to control the aircraft; and the crew perished. A Central Committee decree called for a severe reprimand for Kamanin, a reprimand for Marshal Rudenko, and censure for some other guiltless generals.

To THIS DAY THERE HAS BEEN NO CLEAR ANSWER TO WHAT TOOK PLACE ON 27 MARCH 1968. After the work of the government commission was completed, scientists, experienced pilots, and even psychics have developed hypotheses as to what caused the disaster. Professor Sergey Belotserkovskiy— Gagarin's diploma project advisor—conducted a large research project, which included dynamic simulations of the possible conditions surrounding the accident.[11] In 1992, the Mashinostroyeniye publishing house issued S. M. Belotserkovskiy's book *The Death of Gagarin: Facts and Conjecture.*[12] By my reckoning, this is the most serious research published 24 years after the government commission completed its work.[13]

In his work, Belotserkovskiy cites the collective findings signed by a group of specialists in 1988, 20 years after Gagarin's death. The signatories included:

S. A. Mikoyan—Hero of the Soviet Union, distinguished test pilot of the USSR, Candidate of Technical Sciences, and retired lieutenant general of aviation[14]

---

11. Sergey Mikhaylovich Belotserkovskiy (1920–2000) was, from 1960, a senior faculty member at the Zhukovskiy Air Force Engineering Academy. He supervised the Academy diploma projects of all the early cosmonauts in the late 1960s.

12. S. M. Belotserkovskii, *Gibel Gagarina: fakty i domysly* [*The Death of Gagarin: Facts and Conjecture*] (Moscow: Mashinostroyeniye, 1992).

13. *Author's note*: In 1997, Mashinostroyeniye published S. M. Belotserkovskiy's book *Pervoprokhodtsy vselennoy: zemlya–kosmos–zemlya* (*Space Pioneers: Earth—Space—Earth*), in which the author expands on his hypothesis of the death of Yu. A. Gagarin and V. S. Seregin, and attempts to prove what caused it.

14. In the Soviet Union and its successor states, the Candidate of Sciences (*Kandidat nauk*) degree is the first graduate academic degree awarded, followed by Doctor of Sciences (*Doktor nauk*). The former degree roughly corresponds to a Ph.D. in the American system, while the latter higher degree has no equivalent in the U.S.

A. I. Pushkin—Hero of the Soviet Union, distinguished military pilot of the USSR, retired lieutenant general of aviation

S. V. Petrov—distinguished test pilot of the USSR, Lenin Prize laureate, Candidate of Technical Sciences, retired colonel

G. S. Titov—Hero of the Soviet Union, pilot-cosmonaut of the USSR, military pilot first class, Candidate of Military Sciences, colonel-general of aviation

A. A. Leonov—two time Hero of the Soviet Union, USSR State Prize laureate, pilot-cosmonaut of the USSR, Candidate of Technical Sciences, major general of aviation

S. M. Belotserkovskiy—USSR State Prize laureate, Doctor of Technical Sciences, professor, retired lieutenant general of aviation

A. V. Mayorov—USSR State Prize laureate, Doctor of Technical Sciences, professor, retired engineer colonel

P. G. Sigov—USSR State Prize laureate, retired engineer major general, and

A. M. Sosunov—USSR State Prize laureate, Candidate of Technical Sciences, retired engineer colonel.

There are no doubts as to the objectivity and expertise of the signatories of the "Findings" 20 years after the aviation disaster. I cite a brief excerpt from this document below.

. . . As was the case twenty years ago, we express our firm conviction that a scenario entailing improper actions or a lack of discipline on the part of the pilots must be rejected . . . Seregin could not have committed an outright and senseless violation of aviation procedure by permitting Gagarin to undertake an advanced aerobatics maneuver instead of fulfilling a command issued by the flight director to return . . . We have all arrived at the firm conviction that only some unexpected circumstance could have led to such an adverse turn of events . . . The pluralism of opinions in this matter still holds . . . .

After the first publication of the text of this statement in the journal *Grazhdanskaya aviatsiya* [*Civil Aviation*] (No. 7, 1989), S. M. Belotserkovskiy cites the results of additional research on a hypothesis entailing the near collision in the clouds of a fighter plane taking off from Ramenskoye airfield and Gagarin's airplane. Under Belotserkovskiy's supervision, a group of scientists developed methods for the computer simulation of the behavior of the MiG-15 UTI trainer fighter that Gagarin was flying in, focusing on a spin-stall caused by an evasive maneuver or by passing through the trailing vortex of another aircraft, or a combination of the two factors. The researchers succeeded with a high degree of reliability in reconstructing the final flight stage—from the last radio exchange until impact with the ground. This stage was only 1 minute long.

The results of this very conscientious, scientifically organized research in combination with other materials and evidence, which unfortunately the State Commission did not check out thoroughly enough in 1968, show that this is the most credible of all the possible scenarios.

The crew did everything possible to recover after the aircraft's sudden spin-stall. But they needed 1 or 2 seconds and 150 to 200 meters of altitude more. Back in 1996, Aleksey Leonov laid out this scenario for me again and again with great conviction when I met with him. I have to admit that he spoke very emotionally and demonstratively about facts that had each seemed insignificant in 1968, and that, for various reasons, had been rejected. However, taken all together, they could have had tragic consequences.

The thoughts we entertained earlier during our flight to Yevpatoriya in 1968, which I mentioned above, now seem naïve to me. However, after the publication of the collective findings of 1988 and after the release of the book *The Death of Gagarin* in 1992, scenarios devised by military aviation specialists continue to appear. One cannot consider the authors of these publications to be naïve or lacking expertise.

[Below, Chertok summarized six journalistic accounts on Gagarin's death published in recent years.—ed.]

March 1995 marked the 27th anniversary of Gagarin's death. Quite unexpectedly, while leafing through a copy of the newspaper *Argumenty i fakty* [*Arguments and Facts*] that I'd just pulled out of my mailbox, I discovered a brief article entitled "Gagarin Took a Risk for the Last Time."[15] According to the text of the article, its author, Anatoliy Grigoryevich Kolosov, had been a pilot instructor in the late 1950s and, among others, had taught the young Gagarin to fly. At that time Gagarin was a student at the Chkalov Military Aviation Institute for Pilots. In *Argumenty i fakty*, Kolosov writes that on that fateful day of 27 March 1968, he had met with Gagarin at the Chkalovskiy airfield.

Kolosov's version, published 27 years after the fact, is so close to the thoughts that we expressed in our airplane on 12 April 1968 that I am citing an excerpt from the publication here so that the reader will not have to set off searching for the old newspaper.

---

15. A. G. Kolosov, "Ob etom polete prodolzhayutsya spory. Gagarin risknul v posledniy raz," *Argumenty i fakty* no. 12 (22 March 1995): 8.

. . . The large commission created to investigate the accident confirmed that the aircraft and its equipment were in good working order; to the very end the pilots maintained their ability to perform; there were no collisions with other aircraft, with a sounding balloon, or with birds. So what happened?

After reporting that they had fulfilled their mission and after being cleared to enter the flight area corridor to return, Gagarin and Seregin went under the low cloud base and deviated to the side—the crash site confirms this.

Thus, they departed from the radar screen and communication stopped. They wanted to do stall turns and there was cloud cover in the aerobatic zone. Therefore, Gagarin and Seregin flew a little further and found, in their opinion, a safe place and began to "turn some somersaults."

As a rule, aerobatics figures—roll, loop-the-loop, and half-loop—are performed as a set. In the process, the MiG-15 UTI fighter trainer loses approximately 1200 meters of altitude. The altitude of the cloud cover was 1100–1300 meters. It was extremely risky to execute a loop-the-loop under such conditions. But the pilots took the risk and lost.

When the aircraft popped out of the clouds, its angle with the ground almost certainly was greater than 90 degrees. It became clear: the altitude was low and both pilots had to fight for their lives and for the airplane. It was already too late to eject. The pilots would have needed just a little more altitude to pull the aircraft out of the dive and resume horizontal flight . . . .

As sad as it is, dare-devilry and undue risk caused the death of Yuriy Gagarin.

ALMOST ONE YEAR LATER, in February 1996, *Moskovskiye novosti* [*Moscow News*] (No. 4) published an interview with the former chief of the Air Force flight safety service, Major General of Aviation Yuriy Kulikov. Kulikov categorically disagreed with the version that entailed the possible near-collision with another aircraft and Gagarin's aircraft ending up in the stream of its trailing vortex. This version casts a shadow on the ground services of flight control and monitoring of what goes on in air space. It is natural that the chief of the flight safety service would dismiss any thought about mix-ups that occurred in the air through the fault of the ground services.

A *Moskovskiye novosti* correspondent asks the straightforward question:

"So the victims of the crash are completely at fault?"

"I don't think so," answers Kulikov. "It's more that they fell victim to the procedures in place at that time. Everyone knew that Yuriy Gagarin was loaded down with all sorts of extracurricular activities, various and sundry official events, etc. But none of the chiefs dared pose the question either about suspending him from flying or improving his flight skills. The same could be said about Vladimir Seregin, of whom the first cosmonaut was very protective, as everyone knew."

Just one month after this interview, *Izvestiya* published an article by Sergey Leskov—a columnist for that newspaper—under the bold headline "Secret of Gagarin's Death Buried When the Trail was Still Hot."[16] Leskov cites the opinions of specialists who are also reputable and qualified, one of whom advised against meddling in investigations "where you won't dig out the truth." Nevertheless, at the end of the article Leskov cites the comment of one of our reputable test pilots and cosmonauts, Hero of the Soviet Union Igor Volk:

"It is unlikely that anyone will manage to reconstruct the circumstances surrounding the death of the world's first cosmonaut. And today, essentially, only one scenario, involving Gagarin's airplane getting caught in the jet stream of a supersonic fighter, is discussed seriously. This scenario is disputable, but there is no other."

On 27 March 2003, in connection with the 35th anniversary of Gagarin's death, the newspaper *Komsomolskaya pravda* published a hypothesis under the bold red headline "Russian Slovenliness and Disorderliness Killed Gagarin."[17] Despite references to documents from the KGB archives, which had become available, this shrill publication revealed nothing inherently and sensationally new. In this same newspaper, military pilot Colonel Spravtsev, who flew with Gagarin more than once, asserts that there was no spin and he declares: "I err on the side of a collision with a weather balloon." In this same newspaper, two-time Hero of the Soviet Union, USSR Pilot-Cosmonaut Boris Volynov, who had also flown more than once in the MiG-15 with hull No. 18 in which Gagarin crashed, writes: "There was something that we most likely will never know . . . ."

On Cosmonautics Day, 12 April 2006, *Parlamentskaya gazeta* printed an article by the General Director of the Center for Scientific-Technical Research and Expertise, Yevgeniy Kuzminov, entitled "Second Flight into Immortality." Arguing against the findings of the government commission, the author of the article asserts that, more likely than not, there was no "spin." He raises a scenario in which Vladimir Seregin suddenly falls ill. A brain hemorrhage caused such reflexive muscle tension that it bent the leg in the hip and knee joints, and it was not possible to straighten it at the knee completely. Yuriy Gagarin resorted to exerting tremendous muscular force to regain control of the aircraft. However, according to Kuzminov's scenario, his efforts were in vain. The final conclusion of this scenario: "Vladimir Seregin's falling ill

---

16. Sergey Leskov, "Taynu gibeli Gagarina skhoronili po goryachim sledam," *Izvestiya*, 28 March 1996, pp. 1, 5.

17. Andrey Pavlov, "Gagarina podstavili rossiyskoye razgildyaystvo i bezalabernost," *Komsomolskaya pravda*, 27 March 2003, p. 8.

resulted in the role of commander being transferred to Yuriy Gagarin, and after reporting to flight control that an off-nominal situation had occurred on board (to do this all he had to do was press the radio communication button on the control handle), right away he would have received the order to immediately eject. However, down to the last second of his short life, Yuriy Alekseyevich fought for the life of his comrade in distress."

During my 50-year career developing rocket-space technology, I have been a chairman, member, or an investigating specialist for many dozens of various sorts of accident investigation commissions. As a rule, we managed to determine the cause of the accident unequivocally—even when the spacecraft was lost millions of kilometers from Earth. It happened that there might be several probable causes of one failure or another in large, complex systems. Combinations of several events leading to a single tragic outcome are also possible. In this case, commissions use the following wording in their findings: "the most probable cause should be considered to be . . . ."

As a rule, after several launches or as a result of ground-based experiments, "the most probable" scenario turns out to be singular or another, unambiguous and indisputable one will appear. Almost 40 years have passed since the death of Gagarin! This is an entire generation. It is surprising that back in 1968, we unequivocally determined what caused the loss of the L1 spacecraft over Africa after it returned from its rehearsal of the circumlunar flight. But special commissions and many fully competent aviation specialists have been unable to determine what caused the crash of a training aircraft returning to a suburban Moscow airfield after a routine training flight. Thirty-eight years after the events described above, while preparing this book for republication, I am venturing to add one more scenario to all the previously published theories about Gagarin's death. I am doing this based on the "twin incident" principle that Professor of Medicine Vladimir Pravetskiy advised me to use. "If you can't form a diagnosis for a patient, take your time. Wait for another patient or the same one with similar symptoms. You'll have a breakthrough that will help you form a diagnosis."

The journal *Polet* [*Flight*] (No. 3, 2006) published an article entitled "Invisible Danger." The author of the article is Professor Aleksandr Ivanovich Zhelakninov, Head of the Department of Aerodynamics of the N. Ye. Zhukovskiy Air Force Engineering Academy (VVIA).[18] The article describes what causes a trailing vortex to occur behind an aircraft and cites examples of aviation accidents that they caused. It describes the most tragic incident, which occurred on 12 November 2001. An Airbus A300-600 began its

---

18. VVIA—*Voyenno-vozdushnaya inzhenernaya akademiya.*

takeoff run on the runway at Kennedy Airport in New York. Two minutes, 6 seconds after beginning the takeoff run and lifting off from the runway, the commander's alarmed voice was heard reporting that the aircraft had lost controllability. Nineteen seconds later, communication was lost.

Witnesses of the Airbus' takeoff watched with horror as the multi-ton vehicle began to fall apart in mid-air and spiraled into the ground. Two hundred and fifty-one passengers, nine crewmembers, and five persons on the ground perished. An investigation of the causes of the disaster showed that it all began when the Airbus A300-600 entered the trail line of a Japanese airline's Boeing 747, which had taken off less than 2 minutes earlier. The lateral forces caused by entering the trail line resulted in overloads sufficient to rip off the tail fin, and then the two engines as well. The process developed so rapidly that the pilots didn't have time to understand what was happening, much less report to the ground.

The article has more examples. A DC-9 weighing 80 metric tons, during a landing approach at an altitude of 40 meters enters the trail line of a 185-metric ton Tristar airplane 11 kilometers away. As a result, the DC-9 banks sharply. Despite the immediate actions of the pilots, the banking angle reached 35 degrees. The aircraft reached an altitude of 20 meters before the pilot managed to level off the airplane and avoid colliding with the ground.

On 17 January 1987, in the city of Tashkent, a Yak-40 airplane weighing 16 metric tons took off behind an Il-76 airplane weighing 160 metric tons. The time interval between the two airplanes was 65 seconds. At an altitude of 20 meters, the Yak-40 banked sharply to the left, lost altitude, and crashed into the ground.

The article cites several more examples of aviation incidents caused by an airplane entering the trail line of an airplane flying ahead of it. The trail line behind an airplane is a complicated vortex formation, the behavior of which depends to a great extent on the aerodynamic makeup, the flight speed, the atmospheric conditions, proximity to the ground, and other factors. The author of the article cites statistical data from the International Civil Aviation Organization. Twenty percent of all aviation incidents take place due to external influences, and according to the data of the U.S. National Transportation Safety Board (NTSB), of these, 13 to 17 incidents per year occur due to an airplane entering the trail line of airplanes flying in front of them.

ANOKHIN, WHOM MISHIN DELEGATED TO "MONITOR" THE WORK OF USTINOV'S COMMISSION, said that conversations started up again and again about the airplane that supposedly took off from the airfield in Zhukovskiy and could have turned up in the area where Gagarin's airplane was located. According to Anokhin, Air Force representatives stifled such conversations.

Anokhin did not find any documentary information about the possibility of the Gagarin-Seregin airplane suddenly entering into the trail line of an airplane flying in front of them. There were too many people interested in making sure that no scenario similar to this came out.

The probability of Gagarin dying on 12 April 1961, during the world's first flight into space was thousands of times greater than during the flight on 27 March 1968. Gagarin was also an official candidate for the L1 circumlunar flight program.[19]

Soon after Gagarin's death, while visiting Pashkov's office in the Kremlin on routine business, I spoke to the effect that if Gagarin had died while returning to Earth after flying around the Moon, it would have been a tragedy, but fitting and understandable. With a smile, Pashkov said, "Now I can say that the Central Committee never would have given its consent for Gagarin to fly on the L1. And the fact that he died in an ordinary fighter plane on a pointless flight, that's our fault. It was our duty to prohibit the world's first cosmonaut from such amusements. We are to blame."

I was not about to ask whom Pashkov had in mind when he said, "we."

---

19. Official listings of L1 trainees do not include Gagarin's name, and it is highly unlikely that he would have been allowed to fly if he had lived and the program had gone ahead.

Chapter 26
Academic Digression

In addition to all the other great debts of gratitude that I owe to the memory of my parents, I am obliged to them for the sense of reverence for science that they instilled in me from childhood. Our family certainly did not belong to the "upper crust" of the scientific intelligentsia. Nevertheless, the names of great scientists were held in much higher esteem than those of great political figures. Almost all of the teachers at the school that I had attended since the fifth grade still retained traits of the old Russian intelligentsia; and to a great extent, they helped shape their pupils' sense of what constituted a real scientist.[1]

In 1933, in my first year at the institute, after looking through the syllabus I was disappointed when I discovered that there was not a single professor among the lecturers, let alone an academician. Students, who were closer to the caste of the scientific intelligentsia than I, reassured me that this was how it worked at institutions of higher education. Nobody is exposed to any great scientists until their third year.

I saw my first "real, live" professors and academicians in the laboratories of the V. I. Lenin All-Union Electrical Engineering Institute (VEI) rather than in lecture halls.[2] With genuine respect and even awe, the staff of the laboratory, where I was sent for a consultation regarding the latest inventions, showed me the offices where Academy of Sciences corresponding member Karl Adolfovich Krug and Academician Klavdiy Ippolitovich Shenfer worked. Soon thereafter I saw them with my own eyes.

---

1. Chertok is using the term "intelligentsia" with a nod to its particularly Russian context. During the imperial era, the term had a specific connotation, representing a distinct social category (as opposed to class) of society comprising the intellectual and educated elite. Later, during the Soviet era, the Bolsheviks tried to eliminate traces of the old Russian intelligentsia, although they also acknowledged that post-Revolution rebuilding would require the services of the group.

2. VEI—*Vsesoyuznyy elektrotekhnicheskiy institut.*

Reverence for "pure" science, which supposedly originated in "ivory towers," was considered somewhat politically sinful among those who worked in production.[3] However, back then, under the influence of my factory experience, I began to understand that the slogan "During reconstruction, technology determines everything" meant becoming considerably more familiar with that form of science that the newspapers semi-ironically referred to as "pure." Now it is easy to explain this attitude without any politics. Since the beginning of the 20th century, the work of the overwhelming majority of scientists has been converted into a variety of collective industrial work projects. Historical literature has helped to develop a certain notion in our country about the style, way of life, and research methods of a scientist from the second half of the 19th century. The tumultuous industrialization of our country and the scientific-technical revolution that started in the world even before World War II have broken down the caste-like insularity.[4] The cognitive process has lost its purely academic interest in "science for the sake of knowledge." During the war, the line between pure science and the process of its industrial implementation was erased with particular urgency.

During the early post-war years, it was not just politicians and scientists, but also the broad masses who grasped the great significance of science as both a "productive and terrible destructive force."

Despite the internal and global upheaval in the upper crust of the old scientific intelligentsia and among the young scientists who had aligned themselves with recognized masters of science, the idea of selfless devotion to truth and scientific-technical progress remained. There were no regulations enumerating society's moral and ethical standards for a real scientist. Nevertheless, it was considered self-evident that ideals, a sense of responsibility, decency, and independent thinking should be qualities inherent to any representative of the upper tier of the scientific elite.

When I began to lecture in front of a student audience as a "major scientist" in the field of large control systems, I realized, not without personal satisfaction, that the reverence for science and its top representatives that had been common among young people of the 1930s still exists today. True, in recent years with

---

3. In the early Stalinist era, especially beginning the First Five Year Plan (1928–32), the Soviet government encouraged a deep suspicion of "pure" science on the grounds that such science was divorced from the realities of industrial reconstruction.

4. The term "scientific-technical revolution" (*nauchno-tekhnicheskaya revolyutsiya*, NTR) was a specific term usually applied by Soviet commentators to changes in science and technology in the 1960s when it was believed that the technical application of science would quicken the development and creation of fully socialist societies.

the excesses of democratic "freedoms" and the disappearance of the restraining force of ideological administrative control not only in the student body, but also in the realm of science in general, there is now a tendency for each individual to act according to his or her own thoughts to pursue their own personal interests. The concepts of conscience and decency have proved to be hindrances. Push your competitors out of the way, defeat and sink them; pillage, insult, or better yet—if you're lucky—simply leave this country.

In November 1968, I was elected to corresponding membership in the USSR Academy of Sciences. For me, this event was more significant that any distinguished awards I had ever received. The awarding of the Lenin Prize or conferment of the title Hero of Socialist Labor was a Party and government concern. If the Central Committee decided that such-and-such a project deserved distinguished awards, then the appropriate ministries, and then the enterprises and organizations were allotted "Hero" titles, orders, medals, and Lenin and State Prizes.

Lists of individuals singled out for awards underwent thorough scrutiny through the hierarchy from bottom to top, from the enterprise, through the ministry, VPK, Central Committee, and finally reached the Presidium of the Supreme Soviet of the USSR, which then issued the decree.

The situation was a little more complicated with the Lenin and State (formerly Stalin) Prizes.[5] Formality required a report in the appropriate section, proving the importance and high-priority nature of the project. Sometimes competing lists coincided. In this case, the committee members for the Lenin and State Prizes resolved the issue of conferment by a vote. In the overwhelming majority of the cases, the issue of who would be awarded a prize, and for what, had been predetermined; but for the sake of decorum, something akin to a competitive selection was still held.

As a rule, the actual decrees of the Central Committee and Council of Ministers on conducting the most vital projects in the defense industries already called for the participants to be given awards for producing new types of armament within the deadlines stipulated in the decrees. The bureaucrats of the Central Committee, VPK, and Council of Ministers considered themselves involved with the majority of these projects. Thus, some bureaucrats who were not overly modest accumulated more Orders of Lenin than any general designer.

---

5. The Lenin Prize was a high national award given to important Soviet figures in science, technology, architecture, literature, and the arts. The prize was bestowed annually from 1925 to 1935 and then from 1957 to 1991. The Stalin Prize was a similar award given annually between 1940 and 1954, later resurrected as the State Prize from 1966 to 1991.

It was quite a different matter when it came to conferring the distinguished academic title of corresponding member or active member, i.e., an academician. There was no chance any bureaucrat could become a member of the Academy of Sciences in this way.

I remember how Korolev could not conceal his joy and pride when he was elected to be a corresponding member in 1953. How could he help but be proud! Each individual elected to the Academy of Sciences of the Soviet Union could consider him or herself among the worldwide scientific elite, dating from the time of Peter the Great and Lomonosov.

The Academy of Sciences was Russia's oldest scientific institution, created by a Senate decree under Peter the Great on 28 January 1724.[6] The members of the Academy were the leading domestic and foreign scholars. Each academy member was justifiably proud to be part of an elite community, whose members had included Lomonosov, Euler, Bernoulli, Voltaire, D'Alembert, Franklin, Kant, Ostrogradskiy, Einstein, Rutherford, Bohr, Planck, Krylov, Tolstoy, Chekhov, Lyapunov, Vernadskiy, Pavlov, Kapitsa, and dozens of others whose names have gone down forever in human history.

In 1925, the USSR Central Executive Committee (TsIK) and Council of People's Commissars (*Sovnarkom*) adopted a joint resolution "On Recognizing the Russian Academy of Sciences as the Supreme Scientific Institution of the USSR."[7] From then on, the Academy has been called the Academy of Sciences of the Union of Soviet Socialist Republics. Until its last days, the USSR Academy of Sciences really was, both in terms of form and of substance, the nation's highest and most authoritative scientific institution and one of the leading centers of world science.

It is worth describing the importance of the Academy in our work and about our relationship with it. Intensive scientific research over a broad range in the fields of nuclear physics, radio electronics, rocket technology, and jet aviation during the war and early post-war years, produced obvious practical results. This contributed to the Academy's very strong rise in standing. The top leaders of the nations that had been at war, after switching their energy

---

6. When established in 1724, the Academy was originally known as the Petersburg Academy of Sciences. It has had many name changes during its nearly 300-year history: the Petersburg Academy of Sciences (1724–1747), Imperial Academy of Sciences and Arts (1747–1803), Imperial Academy of Sciences (1803–1836), Imperial St. Petersburg Academy of Sciences (1836–1917), Russian Academy of Sciences (1917–25), USSR Academy of Sciences (1917–91), and Russian Academy of Sciences (1991–present).

7. TsIK—*Tsentralnyy ispolnitelnyy komitet*—was the highest Party organization, while *Sovnarkom—Sovet narodnykh komissarov*—was the highest governmental body. This decree was issued on 27 July 1925.

from the "hot" war to the "cold" war, understood that supremacy in this war could be achieved with the help of scientists capable of offering more advanced technology than a potential enemy was producing.

The wartime people's commissars, the post-war ministers who replaced them, and leaders of industry were put in a situation in which the fates of their industries, and even their personal fates, were determined by the scientific achievements in the fields they were overseeing. Directives that came down from the Central Committee to the Party organizations demanded particular attention to scientific institutions and scientists. Under the slogan "Science has become a productive force," Party propaganda encouraged building an atmosphere of veneration for the omnipotence of science. The main scientific community—the Academy of Sciences and prominent scientists—were besieged with attention and honors.

The Academy was completely financed from the government budget. It had its own housing stock, hospitals, pharmacies, clinics, sanitaria, childcare facilities, dacha villages, motor pools, hotels, dining halls, bookstores, and even tailors' and dressmakers' shops where scientists could dress themselves and their family members well for a reasonable price.

From his early years as an administrator, while still at NII-88 and with no academic degrees or titles, Korolev established personal contacts with Academy scholars. His deferential attitude toward the so-called representatives of "pure science" set an example for his staff and members of the Council of Chiefs, who during the early "rocket" years didn't have any academic degrees either. I cited examples of the broad engagement of the Academy and its scientific institutions in the initial period of our activity in my first two volumes of *Rockets and People*.[8]

They assisted in the development of rocket technology, not only through their personal creative contribution, but also by promoting the actual creators of this technology up the hierarchical ladder of degrees and titles. The highest Party echelons encouraged this process. Even before satellites appeared on the scene, the scientific instruments of academic institutes were payloads for high-altitude launches of combat missiles. The scientists who developed experiments used the results for publications, dissertations, and subsequent scientific careers. Engineers, rocket specialists, and participants in experiments also received State Prizes and letters of commendation from prestigious academic institutes, which enabled them to receive academic degrees without the complex procedure of defending dissertations.

---

8. See Chertok, *Rockets and People, Vol. II*, Chapter 5.

Leading designers and military officials on the terrace at building "zero" in the town of Leninsk at the Baykonur Cosmodrome. Sitting (left to right) in the first row are: Georgiy Tyulin, Marshal Nikolay Krylov, Sergey Korolev, Boris Stroganov; second row: Vladimir Barmin, Mikhail Ryazanskiy, Nikolay Pilyugin, Aleksandr Mrykin, and Viktor Kuznetsov (all chief designers). Standing are: unknown, Aleksandr Zakharov (commander of Baykonur), and Chief Designer Andronik Iosifyan. Many of these chief designers became Corresponding Members or full Academicians of the Academy of Sciences. The photo probably dates from 1964.

Even during the war, when all the scientific institutions were working under the slogan "Everything for the front, everything for victory!" academicians found time at their general assembly in 1943 to elect Andrey Kostikov as a corresponding member. After his death in 1950, there was not a single rocket specialist in the Academy.[9] This void was not filled until October 1953, with the election of Korolev and Glushko as corresponding members. After that,

---

9. Andrey Grigoryevich Kostikov (1899–1950) was the infamous liquid oxygen rocket engineer who worked at NII-3; he was one of the was one of the primary individuals alleged to have been responsible for the denunciations that led to the arrest of Korolev and Glushko in 1938. Later, Kostikov was given an author's certificate for the famous Katyusha rocket launchers, although it is a claim that continues to be contested in the current day Russian space history community. For details of Korolev and Glushko's arrest, see Asif A. Siddiqi, "The Rockets' Red Glare: Technology, Conflict, and Terror in the Soviet Union," *Technology and Culture* 44 no. 3 (2003): 470–501.

there was a lull until the first artificial satellite. This achievement of Soviet science combined with the previous successes in combat missile technology served as impetus for Korolev and Glushko to be elected as active members (academicians) in June 1958. During this same Academy session, Barmin, Pilygin, Kuznetsov, Ryazanskiy, Mishin, and Chelomey were elected as corresponding members.

In 1960, taking advantage of Keldysh's support, Korolev exerted considerable effort to have one more of his deputies—Konstantin Bushuyev— elected as a corresponding member.

Korolev and Keldysh paved the way for the third large "breakthrough" of "applied" rocket scientists into the Academy in June 1966. By this time Korolev had passed away, but the general assembly session felt compelled to compensate for the heavy loss. Barmin, Pilygin, and Mishin were elected academicians. At the same time, "as an exception," Yangel and anti-ballistic missile Chief Designer Grushin were elected directly into the ranks of academicians.[10] Rauschenbach and Svyatoslav Lavrov were elected corresponding members. Thus, not counting Korolev himself, his staff now had four members of the Academy: Mishin, Bushuyev, Rauschenbach, and Lavrov. At this same meeting, Nikolay Lidorenko—the chief designer of rocket-space current sources—was elected as a corresponding member. Our rocket-space community was disappointed that Andronik Iosifyan's candidacy had been declined. The totality of his seminal works in the field of electrical machines, synchronous communication technology, electric drives, and finally, the development of a new class of spacecraft was more than sufficient for Iosifyan to be elected to the Academy.

This would be a good time to address the distinction that prevailed in society at that time between "fundamental" and

From the author's archives.

Andronik Gevondovich Iosifyan (1905–1993) was chief designer of NII-627 (later VNII Elektromekhaniki) from 1941 to 1974. He supervised the development of electrical equipment for Soviet missiles before moving on to develop various Soviet meteorological satellites.

---

10. In other words, they skipped over being "junior" corresponding members.

"applied" scientists, and in that regard, the role of Mstislav Vsevolodovich Keldysh. His name has deservedly (and forever, I hope) gone down in the history of the USSR Academy of Sciences, Russian and world science, and rocket technology and cosmonautics. And nevertheless, while writing these memoirs, as I reflected on Keldysh's career, it occurred to me that the actual historical role of Keldysh's work has not yet been properly evaluated.[11]

Officially, according to the rules, to this day all Academy members fall into two categories: "associate" corresponding members and "full-fledged" active members, or academicians. Unofficially, according to unwritten laws, or more likely according to taste, the members are divided into "fundamental" or "applied" scientists. It was understood that in the Academy priority must be given to scientists who create fundamental science, i.e., basic knowledge.

However, the boundaries between fundamental and applied science disappeared with the advancement of science and technology. The diffusion of fundamental research into the field of practical production, and vice versa, i.e., the influence of applied research on fundamental principles, started to snowball during World War II.

It is curious that the terms "fundamental knowledge," "fundamental research," and their respective counterparts "applied science" and "applied research" began to be mentioned in academic circles relatively recently. Perhaps it's been in the past 30 or 40 years that the opinion has emerged in the scientific community that such terminology increases the Academy's prestige, because Academy members deal with fundamental science, while the "ordinary" other scientists deal with applied science. "Fundamental" scientists work in scientific institutes, while applied scientists work in the NIIs of the various industrial sectors and in factories. But sometimes the boundaries disappeared in the most unexpected way. When Academician Mstislav Vsyevolodovich Keldysh was elected president of the Academy in 1961, he was described as a scientist of the highest caliber, whose work glorified our science.

However, even before his election to such an honorable post, Keldysh's name was well known in the scientific world because he had the talents of a theoretician in the field of mechanics and mathematics that brilliantly

---

11. There have been two major collected works about Keldysh's life published in Russian. They are: V. S. Avduyevskiy and T. M. Eneyev, eds. *M. V. Keldysh: izbrannyye trudy: raketnaya tekhnika i kosmonavtika* [*M. V. Keldysh: Selected Works: Rocket Technology and Cosmonautics*] (Moscow: Nauka, 1988); A. V. Zabrodin, ed., *M. V. Keldysh: tvorcheskiy portret po vospominaniyam sovremennikov* [*M. V. Keldysh: A Creative Portrait from the Recollections of Contemporaries*] (Moscow: Nauka, 2001).

combined with a purely engineering mentality. The results of his research on "flutter" and "shimmy" enabled airplane developers to rid themselves of the catastrophic consequences of these phenomena. In his works, Keldysh developed many purely mathematical problems, and he succeeded in finding a solution to problems that "pure" mathematicians had struggled with. Keldysh's main scientific works, which can now be classified as "fundamental," were conducted at TsAGI, which was not at all part of the Academy of Sciences system. However, in 1943, during the war, Keldysh was elected as a corresponding member, and just three years later, in 1946, he became an academician. When I returned from Germany in 1947 and transferred from NII-1 to NII-88, I was surprised to find out that "theoretician" Keldysh had quite recently been appointed scientific chief of NII-1.[12] In fact, the future "Chief Theoretician of Cosmonautics" became a chief because in his practical work he was not bent on artificially dividing research into prestigious fundamental and second-class applied research.[13] He strove to achieve their interpenetration and interrelation. Applied science enriched fundamental science. In 1953 Keldysh became a member of the Presidium, in 1960—vice president of the USSR Academy of Sciences, and in 1961—the Academy's president.[14]

In the process of my work over a period of almost 30 years, I often had the occasion to observe Keldysh, and then also to interact with him in all sorts of situations: at numerous routine meetings, at councils of the chief designers, at the firing range, during airplane flights, at mission control, in personal private conversations regarding off-nominal situations, and even at "stag" birthday parties. I never heard Keldysh say anything disparaging or demeaning about engineering work or so-called applied research. In 1963, speaking at the Academy's general assembly, he said: "It is the duty of scientists not simply to move science forward, but also to promote the quickest practical realization of scientific achievements."

Thanks in great part to Keldysh's active organizational efforts, members of the Academy of Sciences were in charge of the main NIIs and KBs that determined the development of rocket technology and cosmonautics. In 1975, when Keldysh gave up his duties as president, Academy members headed 28

---

12. Keldysh served as scientific chief of NII-1 from 1946 to 1961. He also served as director of the institute (an administrative position) from 1946 to 1955.

13. In the early 1960s, the Soviet press frequently referred to the "Chief Theoretician of Cosmonautics" who had played a key role in the birth of the Soviet space program. Keldysh's name was never explicitly linked with this anonymous title until the early 1970s.

14. The Presidium was the highest deliberative body in the Academy.

out of the 31 such NIIs and KBs. By comparison, in 1997, of those same 31 organizations, Russian Academy of Sciences members headed only 11.

In 2007, as I edit this chapter, Russian Academy of Sciences members head only four of the NIIs and KBs that survived the devastating reforms—TsNIIMash, the Scientific-Research Institute of Thermal Processes (formerly NII-1), the Scientific-Research Institute of Thermal Engineering, and the Institute of Biomedical Problems.

The main rocket-space organizations, which made historic breakthroughs in technology and science, made the Soviet Union the world's second superpower. Leaders who do not currently belong to the academic elite are in charge of the scientific schools of Academy members Korolev, Glushko, Pilyugin, Barmin, Ryazanskiy, Makeyev, Reshetnev, Semikhatov, Grushin, Konopatov, V. I. Kuznetsov and N. D. Kuznetsov, Iosifyan-Sheremetyevskiy, Guskov, and Lidorenko. Under the conditions of the market economy, the chief of any KB or NII must be a manager rather than a scientist. At least that's what Russian bureaucrats have explained.

In each specific case, authority can be shared—the manager is appointed as the "general" director, while the scientist is the "general" designer. Academician Glushko went through a similar experience back during Soviet rule. He was general designer, and the manager—NPO Energiya Director Vakhtang Vachnadze—was under his command.[15] The experiment proved to be a success. However, to this day, the problem of who should be the one in charge of a KB or NII is not decided by the staffs of these organizations (as a rule, the "people's" opinion is not sought), but by a multi-echelon bureaucratic hierarchy devised by the Russian government.

In the Russian Academy of Sciences, the director of academic institutes is preliminarily elected during a general assembly. The appropriate academic department makes the final decision on the director's appointment via secret ballot. This two-step democratic method, which was established in the Academy long ago, should also be extended to the non-Academy NIIs and KBs that play crucial roles in the nation's defense and economy.

The Soviet Union's Communist leaders made a conspicuous effort to bring outstanding scientists closer to power. The president of the Academy of Sciences was a member of the Communist Party Central Committee and a deputy of the Supreme Soviet. All communist rulers were concerned with the Academy's maintenance and development, and during the difficult war years,

---

15. Vakhtang Dmitriyevich Vachnadze (1929–) served as director of NPO Energiya from 1977 to 1991.

with its survival. The Soviet state sought to demonstrate to its people, and to the international community, the very high degree of esteem in which it held its scientists and science.

In the early 1990s, there was a real threat that the Academy would be disbanded; and the applied science institutes surrounding the Academy that were part of the industrial ministries were destroyed everywhere. The Academy fought it out, sustaining great losses. Rocket-space science also fought it out and sustained losses. At the beginning of the 21st century the flagrantly cynical attacks on the Russian Academy of Sciences abated. The President of Russia and the President of the Academy tried to find common ground.

Academicians Kurchatov, Yangel, Makeyev, and Kotelnikov were Central Committee members and deputies of the Supreme Soviet. Even under Stalin, Academy Member Nikolay Alekseyevich Vosnesenskiy was elected to the Politburo in 1943, and in 1958, Boris Nikolayevich Ponomarev was a candidate for membership in the Presidium of the Communist Party Central Committee. Historian Petr Nikolayevich Pospelov was elected academician in 1953, and soon thereafter, he was elected Central Committee secretary. One can cite many other examples of how the top political leadership of the Soviet Union sought to bring the regime and the scientific elite closer together. Sometimes, when it came to resolving critical issues concerning the selection of defense technology, decision-making authority was shifted from the minister of defense to the president of the Academy of Sciences. That is why the Academy of Sciences had exceptionally high standing in the Soviet Union and a person who received the title "academician" gained a certain new aura.

Our TsKBEM and the Collegium of the Ministry of General Machine Building nominated me as a candidate for election into the Academy in 1968. At that time, the total number of the scientific elite—academicians and corresponding members—was 600. Elections to the Academy were held once every two years. Unlike all other elections in our country at that time, this really was a democratic election and not an appointment under the pretense of an election.

Two months before the election, *Izvestiya* published a list of the candidates. Those who desired had the opportunity to approach the Academy Presidium to support or criticize any of the candidates. Next came the complicated procedure of preparing for the secret ballot. The first stage in this process was the allocation of open positions by departments and specialties. The Academy Presidium performed this task, having first opened up the maximum possible number of vacancies by a special decree from the Council of Ministers.

Expert commissions were created in each of the 14 departments.[16] During strictly closed sessions, these commissions prepared proposals for an assembly of the departments. From the plethora of candidates, these commissions were supposed to single out one or two for each vacancy and then report their recommendations to the department assembly. Several days before the assembly, each department was invited to a traditional "tea party with the president." The academicians in the given department took their seats around tables in the small palace hall of the Academy Presidium. The president and two vice presidents, who were familiar with the scientific discipline of the department invited to tea, were seated at the head of a Π-shaped table. At the tables set with bowls of cookies, there was only room for academicians. Corresponding members, who already knew that they would soon be "kicked out," humbly seated themselves in the rows of chairs that had been set out. At the "tea party," as a rule, there were no discussions. The members of the expert commissions only praised the recommended candidates. The unwanted ones were simply not mentioned. This was the first indication for whom one should vote. Corresponding members had no right to vote in the election of academicians. They could even have tea and cookies with the president while corresponding member candidates were under discussion. When it came to the discussion of academician candidates, they left the hall, being aware of their inferior status according to the Academy charter.

The first years that I participated in elections, I found this procedure offensive. Apropos of this, one of my colleagues who left the meeting hall with me remarked, "Don't be offended! Remember that Albert Einstein, Max Planck, Lev Tolstoy, Ernest Rutherford, and Thomas Edison were also corresponding members of our Academy in their time. And if these tea parties had been held back then, they would have had to leave the hall with their heads hanging, too."

The general assembly of a department was the defining stage of the election process. As a rule, academicians who were members of expert commissions spoke first. They did not reject anyone outright, but simply praised the more worthy candidates. Everyone followed the tradition: support the desired candidate and don't challenge even a clearly unworthy candidate. Each person who spoke "for" a candidate usually was a member of some particular faction or was affiliated with it.

My candidacy was in the Department of Mechanics and Control Processes. Almost all the chief and general designers of aviation and rocket technology were members of this division. Germogen Pospelov, who was elected to

---

16. Departments in the Academy were created according to particular scientific and applied scientific sub-fields.

this department two years before me, drove home the point to me that two "mafias" had formed in the division: aviation and rockets. Each faction was trying to ram through its own candidates for vacant positions. At that time, the rocket-space euphoria had not yet died down, and for that reason, Germogen thought my chances were rather good.

After a prolonged discussion extolling the scientific merits of each candidate, the actual process of the secret ballot began. Each department member received a ballot and sought out a secluded place, and rather than draw a line through the candidate's name, they crossed out the word "elect" or "reject." Each candidate who garnered at least two thirds of the department members' votes was considered elected. The work of the tabulation commission was also made difficult by the fact that the members were supposed to visit academicians who were unable to attend the meeting due to illness or old age.

During the first round, they rarely managed to fill all the vacancies. Once again discussion commenced with calls to focus on the most worthy candidates. They held a second, and sometimes third, secret ballot. There were instances when a department lost its open positions if the votes were distributed among candidates in such a way that not a single one could amass the requisite two-thirds of the votes in three rounds.

Academician secretaries reported the results of the department ballots to a general assembly of the entire Academy.[17] The general assembly then held a secret ballot to confirm the election results department by department.

Just as was the case in the departments, during the election of corresponding members everyone votes, but during the election of active academy members, only academicians vote. Instances in which a candidate failed to be elected in the general assembly after he or she had been elected in a division were very rare. But they did occur. For a candidate to be voted down in the general assembly, all it took was one person speaking negatively, no matter how much effort the division had put in for him. As I recall, there was a case when Academician Glushko went to the podium of the general assembly and declared that one of the candidates—a doctor of jurisprudence, elected by the Philosophy and Law Division, did not deserve to be in the Academy because in his works he referred to the legal theories of Vyshinskiy.[18] Glushko's statement was followed immediately by several passionate speeches in defense and support of the famous lawyer. But the deed was done. He was unable to amass the requisite two-thirds of the votes.

---

17. Each department in the Academy was overseen by a "secretary" who was usually a very senior academician in that particular field.

18. Andrey Yanyaryevich Vyshinskiy (1883–1954) was the infamous jurist who presided over many of the major show trials during the Stalinist Great Terror in the late 1930s.

Boris Chertok in his office at OKB-1. A painting of the "patriarch" of Soviet cosmonautics, Konstantin Tsiolkovskiy, hangs behind him.

In 1968, the academician secretary of the Department of Mechanics and Control Processes was Boris Nikolayevich Petrov, who knew me very well.[19] In fact, almost all the members of the department knew me. Nevertheless, I was not elected during the first ballot. Barmin was in charge of the tabulation commission at the department assembly. The three-ballot process lasted into the night. I wasn't worried, because I was certain that I wouldn't get the two-thirds of the votes needed for election. There were already too many big names on the ballot. At 1 a.m., Barmin called me at home and congratulated me on my election.[20]

The next day the general assembly at the Moscow House of Scientists confirmed the department's decision. And one day later, there was a big reception in the Academy Presidium where the old academicians congratulated the new members. On that occasion, the champagne, as they used to write in days of yore, "flowed like a river."

---

19. Academician Boris Nikolayevich Petrov (1913–80) was a major Soviet scientist specializing in automatic control systems. He made some important theoretical contributions to the Soviet space program. Although he was not deeply involved in the space program, he was a very visible public figure, representing the program during international cooperative programs such as Interkosmos and the Apollo-Soyuz Test Project (ASTP).

20. Chertok was officially elected a corresponding member of the Academy in November 1968.

With respect to the description of the election procedure of the 1960s, I will jump ahead 30 years to 1997. At the time, the Scientific and Technical Council of the Rocket-Space Corporation Energiya and a group of academicians nominated me to be elected as an active member of the Russian Academy of Sciences. I was a candidate for the one open vacancy for the control processes specialty in the Department of Problems of Machine Building, Mechanics, and Control Processes. During the three ballots in the department assembly, not one of the candidates was able to amass the requisite two-thirds of the votes. As a result, the department lost its vacancy—a place for an academician specializing in control processes. At this same assembly Viktor Legostayev was elected corresponding member for this same specialty. When I congratulated him on his election, he said: "But Chertok was invincible, no one got his place." This was small consolation for me.

In all, since 1953, 12 individuals from Korolev's OKB-1 have been elected to the Academy of Sciences of the Soviet Union.[21] In terms of this measure, only the nuclear physicists from the Kurchatov Institute have surpassed us.

From the author's archives.

Aleksey Isayev, Boris Chertok, and German Titov in the early 1960s. Isayev was the legendary figure who designed the first in-orbit rocket engines for the Soviet space program.

---

21. Chertok doesn't name the 12 individuals, but the Academy lists suggest that at least 15 former Korolev veterans have entered the ranks of corresponding members or academicians. They include, as academicians: S. P. Korolev (1958), V. P. Mishin (1966), V. P. Makeyev (1976), B. V. Rauschenbach (1984), M. F. Reshetnev (1984), and B. Ye. Chertok (2000); and as corresponding members: K. D. Bushuyev (1960), S. S. Lavrov (1966), D. I. Kozlov (1984), Yu. P. Semyonov (1987), V. P. Legostayev (1997), V. V. Lebedev (2000), V. S. Syromyatnikov (2006), Ye. A. Mikrin (2006), and V. P. Savinykh (2006).

During the 1960s and 1970s, it was improper to be the director of a large KB or NII in our government without having a distinguished academic title. Ministers, the Central Committee Defense Department, and the Academy Presidium exerted considerable effort to ensure that an entire industry was "equipped" with scholarly managerial staffs. Sometimes they met with completely unforeseen difficulties.

Due to his moral conviction and inherent old-fashioned decency, Aleksey Isayev flatly refused to accept the academic degree of doctor of technical sciences and to be nominated for election into the Academy. Korolev and Tyulin had me use "friendly" persuasion on Isayev to get him to at least fill out the necessary questionnaires and prepare a "list of scientific works" so that the Supreme Certification Commission (VAK) would have the official grounds for making a decision.[22] It wasn't easy, but I managed to overcome Isayev's stubbornness and he received his doctorate degree. But then, neither Korolev, nor Bushuyev, nor even "chief engine specialist" Glushko and the president of the Academy himself, Keldysh, could persuade Isayev to consent to be nominated as a corresponding member of the Academy.

He explained his stubbornness to me: "I am an engineer, and I want to remain an engineer. You talked me into the doctorate degree—to hell with you. Deadbeats need academic degrees so they can write scholarly works. I am not about to do that and I won't have time. The Academy is the only institution in the country where something similar to democracy exists. It has a secret ballot. Even if the minister, Keldysh himself, and the Central Committee support me, the academicians can simply black-ball me in a secret ballot. How will I look the guys on my staff in the eyes after that? Non-Party members in our country can live very well, but to become non-Party because you were excluded from the Party, that's quite a different story. And it's the same with the Academy. You don't have to be an Academy member, but if you fail to be elected, if you're black-balled during a secret ballot, that's shameful—it means you're not worthy."

If he had consented, Isayev certainly would have been elected. But we didn't manage to overcome his old-fashioned stubbornness. During my 38 years in the Academy, I have never encountered such behavior. On the contrary, dozens of candidates aggressively campaigned for written and oral support, fully aware that many of them had little chance of being elected.

The pursuit of high academic degrees and titles devalued the title of engineer. Regardless of talents, to be a "simple engineer" at an NII eventually

---

22. The VAK—*Vysshaya attestatsionnaya komissiya*—was responsible for certifying academic degrees and titles in the Soviet Union.

became less and less attractive both in terms of morale and materials. According to the legislation in effect at that time, a candidate of science's salary, let alone that of a doctor of science, was certainly higher than an engineer with no advanced degrees or a research associate, regardless of their talents. By his action, Isayev was striving to preserve the once-dignified title of engineer.

I can be proud of the fact that I was elected corresponding member in 1968 at the very same session during which the renowned aviation designers Sergey Ilyushin, Artem Mikoyan, Arkhip Lyulka, and yet another member of the Council of Chiefs—Viktor Kuznetsov—were elected academicians. Corresponding members elected along with me included developer of solid-propellant rocket engines Boris Zhukov, chief designer of aviation engines Nikolay Kuznetsov, and one more star from the Korolevian constellation—Viktor Makeyev.

In 1974, the celebration of the Academy's 250th anniversary was approaching. In honor of this event, a two-volume anniversary edition containing all the Academy members from 1724 through 1974 was being prepared for publication. Every active, corresponding, honorary, and foreign member was supposed to be represented by a portrait and a brief biographical sketch.

The Academy's history included dark days that for many years no one had wanted to remember. During Stalin's regime, a large number of prominent scientists became the victims of mass repressions, iniquity, and desecration of human dignity.[23] Some scientists were forced to emigrate. Under pressure from the system, Academy members who had been declared "enemies of the people" or who had failed to return from abroad were excluded from the Academy's ranks. There were also "enemies of the people" whom the regime had killed, but the Academy didn't remove their names from the roll.

A general assembly before the 1974 anniversary decreed that those who were unjustly removed from the Academy be posthumously reinstated, that repressed scholars who had been previously reinstated by specific orders of the Academy Presidium have their reinstatement as members of the Academy confirmed, and that foreign and honorary members be reinstated. Justice was restored for 48 prominent scholars. Their names found their way into the anniversary edition of all the Academy's members.

With the collapse of the Soviet Union, once again the Academy became the Russian Academy of Sciences. In the new Academy, scholars were not

---

23. See particularly, V. A. Kumanev, ed., *Tragicheskiye sudby: repressirovannyye uchenye akademii nauk sssr: sbornik statey* [*Tragic Fates: Repressed Scientists of the USSR Academy of Sciences: A Collection of Articles*] (Moscow: Nauka, 1995).

subjected to repressions or persecutions for "unorthodox thinking." The new bureaucracy practically separated science from the government and stripped it of its economic foundation, which had enabled it to be a powerful productive force for society. Hard times had come again. Many scientists left the country for economic reasons rather than political ones.

At the very end of the 20th century the Russian Academy of Sciences prepared to celebrate its 275th anniversary. During its almost three centuries of history, the Academy had often fought to establish new scientific fields of endeavor. It had come to the aid of and protected individual scientists. Now, for the first time, it was faced with the problem of survival and saving all of Russian science from degradation.

AT THE INITIATIVE OF ACADEMICIAN BORIS PETROV, a Scientific Council on Motion Control and Navigation Problems was created as part of the Presidium of the Academy of Sciences in the 1960s. At plenary sessions, section meetings, symposia, seminars, and issue-related schools organized by this council, specialists divided into their various departments routinely met. Here they had the opportunity to exchange experience and receive interesting information at a highly professional level. After becoming an Academy member, I was called upon to be actively involved in the Council's work. Academician Aleksandr Ishlinskiy, who was in charge of the section dealing with navigation systems and their sensitive parts in that council, soon thereafter transferred the management of this section to me. After thorough consideration, we made decisions in the form of recommendations, which we sent to the VPK and those ministries involved in the development of motion control systems.

We were one of a few councils that brought together broad-profile academic scholars and professionals from industrial organizations and institutions of higher education. The scientific and technical standing of our academic section was highly regarded in the machinery of the state.

After the collapse of the Soviet Union a "black hole" formed. Not only did the VPK disappear, but also the ministers that had showed the most interest in our work. However, rather than disunite the true enthusiasts of motion control for rockets, spacecraft, aircraft, and naval vessels, the economic crisis that erupted drew them even closer together. Now under the flag of the Russian, rather than Soviet, Academy of Sciences, we have all persistently been getting together at the Institute of Problems of Mechanics in Moscow and at TsNII

Elektropribor in Saint Petersburg.[24] The discussion of contemporary scientific and technical problems is interlaced with the survival stories of once flourishing organizations. As one of our colleagues said, to this day our meetings continue to be academically fulfilling and inspire optimism among all the participants.

WORKING ON MY MEMOIRS INEVITABLY LEADS ME TO ANALYZE THE PAST AND TO CONTEMPLATE THE UNFORESEEABLE FUTURE OF OUR ROCKET TECHNOLOGY, COSMONAUTICS, AND SCIENCE to which I devoted more than a half-century of my life.

The fundamental research conducted under the auspices of the USSR Academy of Sciences was organically intertwined with the applied works of hundreds of scientific-research institutes, design bureaus, and factory laboratories. The creative scientific interests and the very basis of the vital activity of most academic, industry, and higher-education scientific centers depended on the material resources of the most powerful military-industrial complex in the world. At the beginning of the 1980s, the Soviet Union possessed unique scientific and strategic potential. Science was not only a productive force of society, but also the very foundation of the nation's military might.

Intercontinental missiles and the shortest range missiles; the nuclear submarine fleet; all forms of nuclear weapons; supersonic and hypersonic aviation; and anti-aircraft and anti-missile systems are literally the brainchildren of science. It was scientists and military technocrats, not Party bureaucrats, who made the Soviet Union a real superpower. However, among the thousands of talented scientists, organizers of industry and the Armed Forces, there were no leaders who could convert this power into a political organization capable of staving off social catastrophe.

The paradox is that scientists, in alliance with the army, possessed power capable of destroying all of humanity many times over in a matter of hours. And at the same time we—scientists and other technocrats—did not take advantage of truly fantastic capabilities to preserve the superpower that our work had created. But nevertheless, the hope remained that the Russian Academy of Sciences would learn from these bitter lessons and ensure a future for Russia so that Russians would not feel "hurt and ashamed for their nation," so that each of us in the future might be proud not only of the past, but also of the present.[25]

---

24. TsNII Elektropribor is a naval scientific-research institute specializing in navigation and gyroscope systems. It was originally known as NII-303.

25. The phrase "offended on behalf of [the] nation" is from the classic Soviet movie *White Sun of the Desert* [*Beloye solntse pustyni*] (1969).

On 26 May 2000, I was elected an active member of the Russian Academy of Sciences, an academician. Vladimir Grigoryevich Peshekhonov was elected during that same session. We both failed to be elected in 1997, and at that time our department lost its opening for the specialty of "control processes." Now there were two openings, and we took advantage of them.

In addition to being a Russian Academy of Sciences academician, Vladimir Peshekhonov is president of the Academy of Navigation and Motion Control on a voluntary basis. Unlike other "honorary academies," the one that Peshekhonov heads is a truly active academy bringing together the best specialists on ship, rocket, airplane, and spacecraft control systems.

At this same session in May 2000, Gay Ilyich Severin and Yuriy Pavlovich Semyonov were elected as academicians. General Designer Boris Ivanovich Katorgin and, as a "world's first," Cosmonaut Valentin Vitalyevich Lebedev were elected as corresponding members.[26] I write "world's first" because it was seven years later that one more Cosmonaut, Viktor Petrovich Savinykh, was elected to the Academy.[27]

ELECTIONS ARE ELECTIONS, BUT THE FATE OF THE ENTIRE ACADEMY OF SCIENCES, of which I have had the honor to be a member for just less than 40 years, troubles you like the illness of a dear friend.

The Russian Academy of Sciences is Russia's great national possession, handed down from the disintegrated Soviet Union. It is the heir of the Academy of Sciences of the Soviet Union. This community has a right to be proud of the fact that the scientists who created the great superpower—the Soviet Union—were its members. According to my estimated total, more than 150 academicians and corresponding members of the Academy of Sciences, and now of the Russian Academy of Sciences, were directly involved in the creation of the nuclear missile shield, the space technology used in support of the powerful nuclear submarine fleet, and missile and aviation technology of all types.

And suddenly . . . Suddenly after the triumph of liberal democratic reformers and absolute dilettantes in governing our great state, the very existence of the Academy of Sciences was in question. The fact is, if we have an enormous amount of petrodollars, then we can buy everything abroad—even science.

---

26. Valentin Vitalyevich Lebedev (1942–) served as a civilian cosmonaut representing NPO Energiya from 1972 to 1989, during which time he flew two space missions, *Soyuz-13* (in 1973) and *Soyuz T-5* (in 1982), the latter a 211-day mission to the *Salyut-7* space station.

27. Viktor Petrovich Savinykh (1940–) served as a civilian cosmonaut representing NPO Energiya from 1978 to 1989. During this period, he flew three missions, *Soyuz T-4* (in 1981), *Soyuz T-13* (in 1985), and *Soyuz TM-5* (in 1988), amassing more than 251 days total time in space.

After the first attacks of various high-ranking crooks were beaten back, a purposeful campaign began to discredit the achievements of Soviet, and then, according to the logic, Russian science. The nation's vast wealth, previously considered as belonging to the people, transferred to the hands of a narrow group of oligarchs and the corrupt bureaucracy that served them. To the question "Does Russia need fundamental science?" the unequivocal answer was "No." We would be able to buy any developments and technology abroad. But alas, the Russian people, who are the true owners, are powerless to take charge of the petrodollars. The activity of the national Academy of Sciences did not at all fit into the ideology of giving the market economy free rein. There was a tremendous desire to shift at least part of the blame to the scientists for the nationwide crisis that had economically ruined the government's power. Even the most highly paid politicians haven't yet managed to pull that off. The oligarchs and the criminal element creeping around them have set their sights on the considerable assets of the Academy of Sciences.

The Russian Academy of Sciences has a staff of 115,000 persons. And if you add to the purely academic "fundamental" staff, the numerical strength of the design bureaus, NIIs, and "science cities" that are headed by Academy members . . . And then again there are Academy members in charge of universities. If all you add them all together, then we have a science army that surpasses that of the Ministry of Defense in terms of the number of "soldiers" (1,200,000). The Academy itself possesses 4,000 square kilometers of land and almost 15,000,000 square meters of buildings. And if, by this gauge, you add together everything that Academy members are in charge of in applied science, officially non-Academy organizations, then we have a powerful super state on a par with Holland or Belgium in terms of area and population. It is a shame that this strength, on which the might of the world's second superpower relied, did not stand up against the plundering of Russia during the 1990s.

On a sunny May morning in 2006, I was driving to a meeting of the Russian Academy of Sciences. Now they hold the meetings in an enormous new building at 34 Leninskiy Prospekt rather than in the cramped old House of Scientists on Prechistenka (formerly Kropotkin Street). I had prepared myself in advance and it seemed I was firmly convinced of whom I would support in the upcoming election—not simply with a secret ballot, but with a persuasive speech before the balloting began. The clarity of my convictions before the elections should have set my mind at ease.

Boris Chertok (left) and Vladimir Syromyatnikov (right) in June 2006 with *Rockets and People* series editor Asif Siddiqi. Syromyatnikov passed away only months later. The photo was taken at Korolev's old house in Moscow, which is now a museum.

At the May 2006 session of the Russian Academy of Sciences, an optimistic mood prevailed. Supposedly, the Academy president had convinced the president of Russia to curtail all talk of disbanding the Academy of Sciences. However, there remained the real threat of amending the charter as a concession to the bureaucracy.

I have digressed terribly in my narrative from the Ostankino—Sadovoye Ring—Profsoyuznaya Street route. The Academician Trapeznikov Institute of Control Problems is located on Profsoyuznaya Street. Our Department of Problems of Machine Building, Mechanics, Control Processes, and Power Engineering is located in this institute. (We recently merged with the Department of Power Engineering). The Department has become so large that the first stage of the secret ballot is conducted section-by-section—Mechanics, Control Processes and Machine Building, and Power Engineering.

After three rounds of balloting, two staffers from the Korolevian school of RKK Energiya—Vladimir Sergeyevich Syromyatnikov and Yevgeniy Anatolyevich Mikrin—were elected as corresponding members.[28] Syromyatnikov was well known to the international space community of scientists and designers. In 1956, for a reason I no longer recall, Korolev delegated me to meet the junior specialists who had been sent to OKB-1 and give them their work assignments. While talking with Syromyatnikov, I offered him a job as a designer of electrical mechanisms in Lev Borisovich Vilnitskiy's section. He unenthusiastically agreed. He had dreamed of beginning to design spacecraft straight away. Syromyatnikov became world-famous in the field of electromechanics. For a while, his work gave Russia a monopoly in the design and production of complex electromechanical assemblies allowing spacecraft to automatically dock with one another and with orbital stations. Syromyatnikov's international standing began to take off in 1975 with the Apollo-Soyuz program and reached its apogee during the International Space Station (ISS) program.[29]

During the last few years, at his own initiative, he developed a design for a reusable cargo transport vehicle with retractable wings (a *Syrolet*), for a controlled landing on an airfield like an airplane.[30] He also had other interesting ideas that were always completely practicable from an engineering standpoint. For example, he proposed "mirror satellites" that would illuminate northern lands during the dark winter months. During the elections, Syromyatnikov was in the hospital. We hoped that a joyful announcement of his victory in the very stiff academic competition would give him strength in his difficult battle with cancer. Alas, the extremely costly medications did not save him. Vladimir Sergeyevich was a corresponding member of the Russian Academy of Sciences for just three months.[31] An honorary certificate was given to Svetlana Ilinichna—his widow, by then.

Professor Yevgeniy Mikrin and I have offices off of the same corridor in building No. 5 at RKK Energiya's "second territory." During the past few years, he has been primarily involved in software for DOS, Mir, and ISS on-board systems. Like Syromyatnikov, he was standing for election to the Russian Academy of Sciences for the second time. His advantage over his

---

28. RKK Energiya—*Raketno-kosmicheskaya korporatsikya Energiya* (Rocket-Space Corporation Energiya).

29. In Russian, the ISS is more commonly known as MKS—*Mezhdunarodnaya kosmiches-kaya stantsiya*.

30. The nickname *Syrolet* combines the prefix of the designer's name with the suffix of the Russian word for airplane, *samolet*.

31. Syromyatnikov died on 20 September 2006.

competitors turned out to be his youth.[32] There was a special opening for "young people." I did not have much trouble making a case to the "old guys" that, among all the "young people" (up to 51 years old), he was the "best and the brightest." He won the secret ballot by a wide margin.

Cosmonaut Viktor Savinykh, two-time Hero of the Soviet Union, was elected—without any intervention from me—as a corresponding member in the Department of Ocean Science, Atmospheric Physics, and Geography. He became the second Academy of Sciences corresponding member among the cosmonauts, after Valentin Lebedev. He was not elected so much for his heroic achievements as for his actual scientific merits. He is a doctor of technical science, the president of the Moscow State University of Geodesy and Cartography, the chief editor of the journal *Rosskiyskiy kosmos* [*Russian Space*], and one of the active champions fighting to preserve the high status of Russian higher education.

In the 1970s Viktor Savinykh was an engineer in the department of sensitive elements and optical instruments that was part of the TsKBEM control systems complex that I headed. In my pre-election speech, I asked for the support of Valentin Lebedev's candidacy to become an active member, an academician. He was director of the Academy's Geoinformation Center. However, Director of the Irkutsk Computer Center Stanislav Vasilyev was elected to the only vacancy for an academician. Before the elections, I was at the Irkutsk Center that Vasilyev headed and at the Geoinformation Center headed by Lebedev. Both were completely deserving of being elected as active members. But there was only one opening, and in such cases one or two votes decide the election.

As an inheritance from academicians Boris Petrov and Boris Rauschenbach, and as an academic assignment by special decree, I was tasked to head an academic commission for the development of the scientific legacy of Academician Korolev and other pioneers of cosmonautics.[33] In the past few years, not only hundreds of space veterans, but also active enthusiasts of cosmonautics who are still quite young have been given the opportunity to mingle. A fresh breeze at the Korolev lecture series has given rise to optimism. But I will have more to say about this in the next volume.

---

32. Yevgeniy Anatolyevich Mikrin was born on 15 October 1955.

33. This commission that Chertok heads is officially known as the Commission of the Russian Academy of Sciences for the Development of the Scientific Legacy of the Pioneers of Cosmic Space.

Index

NASA History Series

Reference Works, NASA SP-4000:

Grimwood, James M. *Project Mercury: A Chronology.* NASA SP-4001, 1963.

Grimwood, James M., and Barton C. Hacker, with Peter J. Vorzimmer. *Project Gemini Technology and Operations: A Chronology.* NASA SP-4002, 1969.

Link, Mae Mills. *Space Medicine in Project Mercury.* NASA SP-4003, 1965.

Astronautics and Aeronautics, 1963: Chronology of Science, Technology, and Policy. NASA SP-4004, 1964.

Astronautics and Aeronautics, 1964: Chronology of Science, Technology, and Policy. NASA SP-4005, 1965.

Astronautics and Aeronautics, 1965: Chronology of Science, Technology, and Policy. NASA SP-4006, 1966.

Astronautics and Aeronautics, 1966: Chronology of Science, Technology, and Policy. NASA SP-4007, 1967.

Astronautics and Aeronautics, 1967: Chronology of Science, Technology, and Policy. NASA SP-4008, 1968.

Ertel, Ivan D., and Mary Louise Morse. *The Apollo Spacecraft: A Chronology, Volume I, Through November 7, 1962.* NASA SP-4009, 1969.

Morse, Mary Louise, and Jean Kernahan Bays. *The Apollo Spacecraft: A Chronology, Volume II, November 8, 1962–September 30, 1964.* NASA SP-4009, 1973.

Brooks, Courtney G., and Ivan D. Ertel. *The Apollo Spacecraft: A Chronology, Volume III, October 1, 1964–January 20, 1966.* NASA SP-4009, 1973.

Ertel, Ivan D., and Roland W. Newkirk, with Courtney G. Brooks. *The Apollo Spacecraft: A Chronology, Volume IV, January 21, 1966–July 13, 1974.* NASA SP-4009, 1978.

Astronautics and Aeronautics, 1968: Chronology of Science, Technology, and Policy. NASA SP-4010, 1969.

Newkirk, Roland W., and Ivan D. Ertel, with Courtney G. Brooks. *Skylab: A Chronology.* NASA SP-4011, 1977.

Van Nimmen, Jane, and Leonard C. Bruno, with Robert L. Rosholt. *NASA Historical Data Book, Vol. I: NASA Resources, 1958–1968.* NASA SP-4012, 1976, rep. ed. 1988.

Ezell, Linda Neuman. *NASA Historical Data Book, Vol. II: Programs and Projects, 1958–1968.* NASA SP-4012, 1988.

Ezell, Linda Neuman. *NASA Historical Data Book, Vol. III: Programs and Projects, 1969–1978.* NASA SP-4012, 1988.

Gawdiak, Ihor, with Helen Fedor. *NASA Historical Data Book, Vol. IV: NASA Resources, 1969–1978.* NASA SP-4012, 1994.

Rumerman, Judy A. *NASA Historical Data Book, Vol. V: NASA Launch Systems, Space Transportation, Human Spaceflight, and Space Science, 1979–1988.* NASA SP-4012, 1999.

Rumerman, Judy A. *NASA Historical Data Book, Vol. VI: NASA Space Applications, Aeronautics and Space Research and Technology, Tracking and Data Acquisition/Support Operations, Commercial Programs, and Resources, 1979–1988.* NASA SP-4012, 1999.

Astronautics and Aeronautics, 1969: Chronology of Science, Technology, and Policy. NASA SP-4014, 1970.

Astronautics and Aeronautics, 1970: Chronology of Science, Technology, and Policy. NASA SP-4015, 1972.

Astronautics and Aeronautics, 1971: Chronology of Science, Technology, and Policy. NASA SP-4016, 1972.

Astronautics and Aeronautics, 1972: Chronology of Science, Technology, and Policy. NASA SP-4017, 1974.

Astronautics and Aeronautics, 1973: Chronology of Science, Technology, and Policy. NASA SP-4018, 1975.

Astronautics and Aeronautics, 1974: Chronology of Science, Technology, and Policy. NASA SP-4019, 1977.

Astronautics and Aeronautics, 1975: Chronology of Science, Technology, and Policy. NASA SP-4020, 1979.

Astronautics and Aeronautics, 1976: Chronology of Science, Technology, and Policy. NASA SP-4021, 1984.

Astronautics and Aeronautics, 1977: Chronology of Science, Technology, and Policy. NASA SP-4022, 1986.

Astronautics and Aeronautics, 1978: Chronology of Science, Technology, and Policy. NASA SP-4023, 1986.

Astronautics and Aeronautics, 1979–1984: Chronology of Science, Technology, and Policy. NASA SP-4024, 1988.

Astronautics and Aeronautics, 1985: Chronology of Science, Technology, and Policy. NASA SP-4025, 1990.

Noordung, Hermann. *The Problem of Space Travel: The Rocket Motor. Edited by Ernst Stuhlinger and J.D. Hunley, with Jennifer Garland.* NASA SP-4026, 1995.

Astronautics and Aeronautics, 1986–1990: A Chronology. NASA SP-4027, 1997.

Astronautics and Aeronautics, 1991–1995: A Chronology. NASA SP-2000-4028, 2000.

Orloff, Richard W. *Apollo by the Numbers: A Statistical Reference.* NASA SP-2000-4029, 2000.

Lewis, Marieke and Ryan Swanson. *Aeronautics and Astronautics: A Chronology,* 1996–2000. NASA SP-2009-4030, 2009.

Management Histories, NASA SP-4100:

Rosholt, Robert L. *An Administrative History of NASA, 1958–1963.* NASA SP-4101, 1966.

Levine, Arnold S. *Managing NASA in the Apollo Era.* NASA SP-4102, 1982.

Roland, Alex. *Model Research: The National Advisory Committee for Aeronautics, 1915–1958.* NASA SP-4103, 1985.

Fries, Sylvia D. *NASA Engineers and the Age of Apollo.* NASA SP-4104, 1992.

Glennan, T. Keith. *The Birth of NASA: The Diary of T. Keith Glennan.* Edited by J.D. Hunley. NASA SP-4105, 1993.

Seamans, Robert C. *Aiming at Targets: The Autobiography of Robert C. Seamans.* NASA SP-4106, 1996.

Garber, Stephen J., editor. *Looking Backward, Looking Forward: Forty Years of Human Spaceflight Symposium.* NASA SP-2002-4107, 2002.

Mallick, Donald L. with Peter W. Merlin. *The Smell of Kerosene: A Test Pilot's Odyssey.* NASA SP-4108, 2003.

Iliff, Kenneth W. and Curtis L. Peebles. *From Runway to Orbit: Reflections of a NASA Engineer.* NASA SP-2004-4109, 2004.

Chertok, Boris. *Rockets and People, Volume 1.* NASA SP-2005-4110, 2005.

Chertok, Boris. *Rockets and People: Creating a Rocket Industry, Volume II.* NASA SP-2006-4110, 2006.

Laufer, Alexander, Todd Post, and Edward Hoffman. *Shared Voyage: Learning and Unlearning from Remarkable Projects.* NASA SP-2005-4111, 2005.

Dawson, Virginia P., and Mark D. Bowles. *Realizing the Dream of Flight: Biographical Essays in Honor of the Centennial of Flight, 1903–2003.* NASA SP-2005-4112, 2005.

Mudgway, Douglas J. *William H. Pickering: America's Deep Space Pioneer.* NASA SP-2008-4113, 2008.

Project Histories, NASA SP-4200:

Swenson, Loyd S., Jr., James M. Grimwood, and Charles C. Alexander. *This New Ocean: A History of Project Mercury.* NASA SP-4201, 1966, reprinted 1999.

Green, Constance McLaughlin, and Milton Lomask. *Vanguard: A History.* NASA SP-4202, 1970; rep. ed. Smithsonian Institution Press, 1971.

Hacker, Barton C., and James M. Grimwood. *On Shoulders of Titans: A History of Project Gemini.* NASA SP-4203, 1977, reprinted 2002.

Benson, Charles D., and William Barnaby Faherty. *Moonport: A History of Apollo Launch Facilities and Operations.* NASA SP-4204, 1978.

Brooks, Courtney G., James M. Grimwood, and Loyd S. Swenson, Jr. *Chariots for Apollo: A History of Manned Lunar Spacecraft.* NASA SP-4205, 1979.

Bilstein, Roger E. *Stages to Saturn: A Technological History of the Apollo/Saturn Launch Vehicles.* NASA SP-4206, 1980 and 1996.

Compton, W. David, and Charles D. Benson. *Living and Working in Space: A History of Skylab.* NASA SP-4208, 1983.

Ezell, Edward Clinton, and Linda Neuman Ezell. *The Partnership: A History of the Apollo-Soyuz Test Project.* NASA SP-4209, 1978.

Hall, R. Cargill. *Lunar Impact: A History of Project Ranger.* NASA SP-4210, 1977.

Newell, Homer E. *Beyond the Atmosphere: Early Years of Space Science.* NASA SP-4211, 1980.

Ezell, Edward Clinton, and Linda Neuman Ezell. *On Mars: Exploration of the Red Planet, 1958–1978.* NASA SP-4212, 1984.

Pitts, John A. *The Human Factor: Biomedicine in the Manned Space Program to 1980.* NASA SP-4213, 1985.

Compton, W. David. *Where No Man Has Gone Before: A History of Apollo Lunar Exploration Missions.* NASA SP-4214, 1989.

Naugle, John E. *First Among Equals: The Selection of NASA Space Science Experiments*. NASA SP-4215, 1991.

Wallace, Lane E. *Airborne Trailblazer: Two Decades with NASA Langley's 737 Flying Laboratory*. NASA SP-4216, 1994.

Butrica, Andrew J., ed. *Beyond the Ionosphere: Fifty Years of Satellite Communications*. NASA SP-4217, 1997.

Butrica, Andrew J. *To See the Unseen: A History of Planetary Radar Astronomy*. NASA SP-4218, 1996.

Mack, Pamela E., ed. *From Engineering Science to Big Science: The NACA and NASA Collier Trophy Research Project Winners*. NASA SP-4219, 1998.

Reed, R. Dale. *Wingless Flight: The Lifting Body Story*. NASA SP-4220, 1998.

Heppenheimer, T. A. *The Space Shuttle Decision: NASA's Search for a Reusable Space Vehicle*. NASA SP-4221, 1999.

Hunley, J. D., ed. *Toward Mach 2: The Douglas D-558 Program*. NASA SP-4222, 1999.

Swanson, Glen E., ed. *"Before This Decade is Out . . ." Personal Reflections on the Apollo Program*. NASA SP-4223, 1999.

Tomayko, James E. *Computers Take Flight: A History of NASA's Pioneering Digital Fly-By-Wire Project*. NASA SP-4224, 2000.

Morgan, Clay. *Shuttle-Mir: The United States and Russia Share History's Highest Stage*. NASA SP-2001-4225.

Leary, William M. *We Freeze to Please: A History of NASA's Icing Research Tunnel and the Quest for Safety*. NASA SP-2002-4226, 2002.

Mudgway, Douglas J. *Uplink-Downlink: A History of the Deep Space Network, 1957–1997*. NASA SP-2001-4227.

Dawson, Virginia P., and Mark D. Bowles. *Taming Liquid Hydrogen: The Centaur Upper Stage Rocket, 1958–2002*. NASA SP-2004-4230.

Meltzer, Michael. *Mission to Jupiter: A History of the Galileo Project*. NASA SP-2007-4231.

Heppenheimer, T. A. *Facing the Heat Barrier: A History of Hypersonics*. NASA SP-2007-4232.

Tsiao, Sunny. *"Read You Loud and Clear!" The Story of NASA's Spaceflight Tracking and Data Network*. NASA SP-2007-4233, 2008.

Center Histories, NASA SP-4300:

Rosenthal, Alfred. *Venture into Space: Early Years of Goddard Space Flight Center*. NASA SP-4301, 1985.

Hartman, Edwin, P. *Adventures in Research: A History of Ames Research Center, 1940–1965*. NASA SP-4302, 1970.

Hallion, Richard P. *On the Frontier: Flight Research at Dryden, 1946–1981*. NASA SP-4303, 1984.

Muenger, Elizabeth A. *Searching the Horizon: A History of Ames Research Center, 1940–1976*. NASA SP-4304, 1985.

Hansen, James R. *Engineer in Charge: A History of the Langley Aeronautical Laboratory, 1917–1958*. NASA SP-4305, 1987.

Dawson, Virginia P. *Engines and Innovation: Lewis Laboratory and American Propulsion Technology*. NASA SP-4306, 1991.

Dethloff, Henry C. *"Suddenly Tomorrow Came . . .": A History of the Johnson Space Center, 1957–1990*. NASA SP-4307, 1993.

Hansen, James R. *Spaceflight Revolution: NASA Langley Research Center from Sputnik to Apollo*. NASA SP-4308, 1995.

Wallace, Lane E. *Flights of Discovery: An Illustrated History of the Dryden Flight Research Center*. NASA SP-4309, 1996.

Herring, Mack R. *Way Station to Space: A History of the John C. Stennis Space Center*. NASA SP-4310, 1997.

Wallace, Harold D., Jr. *Wallops Station and the Creation of an American Space Program*. NASA SP-4311, 1997.

Wallace, Lane E. *Dreams, Hopes, Realities. NASA's Goddard Space Flight Center: The First Forty Years*. NASA SP-4312, 1999.

Dunar, Andrew J., and Stephen P. Waring. *Power to Explore: A History of Marshall Space Flight Center, 1960–1990*. NASA SP-4313, 1999.

Bugos, Glenn E. *Atmosphere of Freedom: Sixty Years at the NASA Ames Research Center*. NASA SP-2000-4314, 2000.

Schultz, James. *Crafting Flight: Aircraft Pioneers and the Contributions of the Men and Women of NASA Langley Research Center*. NASA SP-2003-4316, 2003.

Bowles, Mark D. *Science in Flux: NASA's Nuclear Program at Plum Brook Station, 1955–2005*. NASA SP-2006-4317, 2006.

Wallace, Lane E. *Flights of Discovery: An Illustrated History of the Dryden Flight Research Center*. NASA SP-4318, 2007. Revised version of SP-4309.

General Histories, NASA SP-4400:

Corliss, William R. *NASA Sounding Rockets, 1958–1968: A Historical Summary.* NASA SP-4401, 1971.

Wells, Helen T., Susan H. Whiteley, and Carrie Karegeannes. *Origins of NASA Names.* NASA SP-4402, 1976.

Anderson, Frank W., Jr. *Orders of Magnitude: A History of NACA and NASA, 1915–1980.* NASA SP-4403, 1981.

Sloop, John L. *Liquid Hydrogen as a Propulsion Fuel, 1945–1959.* NASA SP-4404, 1978.

Roland, Alex. *A Spacefaring People: Perspectives on Early Spaceflight.* NASA SP-4405, 1985.

Bilstein, Roger E. *Orders of Magnitude: A History of the NACA and NASA, 1915–1990.* NASA SP-4406, 1989.

Logsdon, John M., ed., with Linda J. Lear, Jannelle Warren Findley, Ray A. Williamson, and Dwayne A. Day. *Exploring the Unknown: Selected Documents in the History of the U.S. Civil Space Program, Volume I, Organizing for Exploration.* NASA SP-4407, 1995.

Logsdon, John M., ed., with Dwayne A. Day, and Roger D. Launius. *Exploring the Unknown: Selected Documents in the History of the U.S. Civil Space Program, Volume II, External Relationships.* NASA SP-4407, 1996.

Logsdon, John M., ed., with Roger D. Launius, David H. Onkst, and Stephen J. Garber. *Exploring the Unknown: Selected Documents in the History of the U.S. Civil Space Program, Volume III, Using Space.* NASA SP-4407,1998.

Logsdon, John M., ed., with Ray A. Williamson, Roger D. Launius, Russell J. Acker, Stephen J. Garber, and Jonathan L. Friedman. *Exploring the Unknown: Selected Documents in the History of the U.S. Civil Space Program, Volume IV, Accessing Space.* NASA SP-4407, 1999.

Logsdon, John M., ed., with Amy Paige Snyder, Roger D. Launius, Stephen J. Garber, and Regan Anne Newport. *Exploring the Unknown: Selected Documents in the History of the U.S. Civil Space Program, Volume V, Exploring the Cosmos.* NASA SP-4407, 2001.

Logsdon, John M., ed., with Stephen J. Garber, Roger D. Launius, and Ray A. Williamson. *Exploring the Unknown: Selected Documents in the History of the U.S. Civil Space Program, Volume VI, Space and Earth Science.* NASA SP-2004-4407, 2004.

Logsdon, John M., ed., with Roger D. Launius. *Exploring the Unknown: Selected Documents in the History of the U.S. Civil Space Program, Volume VII, Human Spaceflight: Projects Mercury, Gemini, and Apollo*. NASA SP-2008-4407, 2008.

Siddiqi, Asif A., *Challenge to Apollo: The Soviet Union and the Space Race, 1945–1974*. NASA SP-2000-4408, 2000.

Hansen, James R., ed. *The Wind and Beyond: Journey into the History of Aerodynamics in America, Volume 1, The Ascent of the Airplane*. NASA SP-2003-4409, 2003.

Hansen, James R., ed. *The Wind and Beyond: Journey into the History of Aerodynamics in America, Volume 2, Reinventing the Airplane*. NASA SP-2007-4409, 2007.

Hogan, Thor. *Mars Wars: The Rise and Fall of the Space Exploration Initiative*. NASA SP-2007-4410, 2007.

Monographs in Aerospace History (SP-4500 Series):

Launius, Roger D., and Aaron K. Gillette, comps. *Toward a History of the Space Shuttle: An Annotated Bibliography*. Monograph in Aerospace History, No. 1, 1992.

Launius, Roger D., and J. D. Hunley, comps. *An Annotated Bibliography of the Apollo Program*. Monograph in Aerospace History, No. 2, 1994.

Launius, Roger D. *Apollo: A Retrospective Analysis*. Monograph in Aerospace History, No. 3, 1994.

Hansen, James R. *Enchanted Rendezvous: John C. Houbolt and the Genesis of the Lunar-Orbit Rendezvous Concept*. Monograph in Aerospace History, No. 4, 1995.

Gorn, Michael H. *Hugh L. Dryden's Career in Aviation and Space*. Monograph in Aerospace History, No. 5, 1996.

Powers, Sheryll Goecke. *Women in Flight Research at NASA Dryden Flight Research Center from 1946 to 1995*. Monograph in Aerospace History, No. 6, 1997.

Portree, David S. F., and Robert C. Trevino. *Walking to Olympus: An EVA Chronology*. Monograph in Aerospace History, No. 7, 1997.

Logsdon, John M., moderator. *Legislative Origins of the National Aeronautics and Space Act of 1958: Proceedings of an Oral History Workshop*. Monograph in Aerospace History, No. 8, 1998.

Rumerman, Judy A., comp. *U.S. Human Spaceflight, A Record of Achievement 1961–1998*. Monograph in Aerospace History, No. 9, 1998.

Portree, David S. F. *NASA's Origins and the Dawn of the Space Age.* Monograph in Aerospace History, No. 10, 1998.

Logsdon, John M. *Together in Orbit: The Origins of International Cooperation in the Space Station.* Monograph in Aerospace History, No. 11, 1998.

Phillips, W. Hewitt. *Journey in Aeronautical Research: A Career at NASA Langley Research Center.* Monograph in Aerospace History, No. 12, 1998.

Braslow, Albert L. *A History of Suction-Type Laminar-Flow Control with Emphasis on Flight Research.* Monograph in Aerospace History, No. 13, 1999.

Logsdon, John M., moderator. *Managing the Moon Program: Lessons Learned From Apollo.* Monograph in Aerospace History, No. 14, 1999.

Perminov, V. G. *The Difficult Road to Mars: A Brief History of Mars Exploration in the Soviet Union.* Monograph in Aerospace History, No. 15, 1999.

Tucker, Tom. *Touchdown: The Development of Propulsion Controlled Aircraft at NASA Dryden.* Monograph in Aerospace History, No. 16, 1999.

Maisel, Martin, Demo J. Giulanetti, and Daniel C. Dugan. *The History of the XV-15 Tilt Rotor Research Aircraft: From Concept to Flight.* Monograph in Aerospace History, No. 17, 2000. NASA SP-2000-4517.

Jenkins, Dennis R. *Hypersonics Before the Shuttle: A Concise History of the X-15 Research Airplane.* Monograph in Aerospace History, No. 18, 2000. NASA SP-2000-4518.

Chambers, Joseph R. *Partners in Freedom: Contributions of the Langley Research Center to U.S. Military Aircraft of the 1990s.* Monograph in Aerospace History, No. 19, 2000. NASA SP-2000-4519.

Waltman, Gene L. *Black Magic and Gremlins: Analog Flight Simulations at NASA's Flight Research Center.* Monograph in Aerospace History, No. 20, 2000. NASA SP-2000-4520.

Portree, David S. F. *Humans to Mars: Fifty Years of Mission Planning, 1950–2000.* Monograph in Aerospace History, No. 21, 2001. NASA SP-2001-4521.

Thompson, Milton O., with J. D. Hunley. *Flight Research: Problems Encountered and What they Should Teach Us.* Monograph in Aerospace History, No. 22, 2001. NASA SP-2001-4522.

Tucker, Tom. *The Eclipse Project.* Monograph in Aerospace History, No. 23, 2001. NASA SP-2001-4523.

Siddiqi, Asif A. *Deep Space Chronicle: A Chronology of Deep Space and Planetary Probes 1958–2000.* Monograph in Aerospace History, No. 24, 2002. NASA SP-2002-4524.

Merlin, Peter W. *Mach 3+: NASA/USAF YF-12 Flight Research, 1969–1979.* Monograph in Aerospace History, No. 25, 2001. NASA SP-2001-4525.

Anderson, Seth B. *Memoirs of an Aeronautical Engineer: Flight Tests at Ames Research Center: 1940–1970.* Monograph in Aerospace History, No. 26, 2002. NASA SP-2002-4526.

Renstrom, Arthur G. *Wilbur and Orville Wright: A Bibliography Commemorating the One-Hundredth Anniversary of the First Powered Flight on December 17, 1903.* Monograph in Aerospace History, No. 27, 2002. NASA SP-2002-4527.

No monograph 28.

Chambers, Joseph R. *Concept to Reality: Contributions of the NASA Langley Research Center to U.S. Civil Aircraft of the 1990s.* Monograph in Aerospace History, No. 29, 2003. SP-2003-4529.

Peebles, Curtis, editor. *The Spoken Word: Recollections of Dryden History, The Early Years.* Monograph in Aerospace History, No. 30, 2003. SP-2003-4530.

Jenkins, Dennis R., Tony Landis, and Jay Miller. *American X-Vehicles: An Inventory- X-1 to X-50.* Monograph in Aerospace History, No. 31, 2003. SP-2003-4531.

Renstrom, Arthur G. *Wilbur and Orville Wright: A Chronology Commemorating the One-Hundredth Anniversary of the First Powered Flight on December 17, 1903.* Monograph in Aerospace History, No. 32, 2003. NASA SP-2003-4532.

Bowles, Mark D., and Robert S. Arrighi. *NASA's Nuclear Frontier: The Plum Brook Research Reactor.* Monograph in Aerospace History, No. 33, 2004. (SP-2004-4533).

Matranga, Gene J., C. Wayne Ottinger, Calvin R. Jarvis, and D. Christian Gelzer. *Unconventional, Contrary, and Ugly: The Lunar Landing Research Vehicle.* Monograph in Aerospace History, No. 35, 2006. NASA SP-2004-4535.

McCurdy, Howard E. *Low Cost Innovation in Spaceflight: The History of the Near Earth Asteroid Rendezvous (NEAR) Mission.* Monograph in Aerospace History, No. 36, 2005. NASA SP-2005-4536.

Seamans, Robert C., Jr. *Project Apollo: The Tough Decisions.* Monograph in Aerospace History, No. 37, 2005. NASA SP-2005-4537.

Lambright, W. Henry. *NASA and the Environment: The Case of Ozone Depletion.* Monograph in Aerospace History, No. 38, 2005. NASA SP-2005-4538.

Chambers, Joseph R. *Innovation in Flight: Research of the NASA Langley Research Center on Revolutionary Advanced Concepts for Aeronautics.* Monograph in Aerospace History, No. 39, 2005. NASA SP-2005-4539.

Phillips, W. Hewitt. *Journey Into Space Research: Continuation of a Career at NASA Langley Research Center.* Monograph in Aerospace History, No. 40, 2005. NASA SP-2005-4540.

Rumerman, Judy A., Chris Gamble, and Gabriel Okolski, compilers. *U.S. Human Spaceflight: A Record of Achievement, 1961–2006.* Monograph in Aerospace History, No. 41, 2007. NASA SP-2007-4541.

Dick, Steven J., Stephen J. Garber, and Jane E. Odom. *Research in NASA History.* Monograph in Aerospace History, No. 43, 2009. NASA SP-2009-4543.

Dryden Historical Studies

Tomayko, James E., author, and Christian Gelzer, editor. *The Story of Self-Repairing Flight Control Systems.* Dryden Historical Study #1.

Electronic Media (SP-4600 Series)

Remembering Apollo 11: The 30th Anniversary Data Archive CD-ROM. NASA SP-4601, 1999.

Remembering Apollo 11: The 35th Anniversary Data Archive CD-ROM. NASA SP-2004-4601, 2004. This is an update of the 1999 edition.

The Mission Transcript Collection: U.S. Human Spaceflight Missions from Mercury Redstone 3 to Apollo 17. SP-2000-4602, 2001. Now available commercially from CG Publishing.

Shuttle-Mir: the United States and Russia Share History's Highest Stage. NASA SP-2001-4603, 2002. This CD-ROM is available from NASA CORE.

U.S. Centennial of Flight Commission presents Born of Dreams ~ Inspired by Freedom. NASA SP-2004-4604, 2004.

Of Ashes and Atoms: A Documentary on the NASA Plum Brook Reactor Facility. NASA SP-2005-4605.

Taming Liquid Hydrogen: The Centaur Upper Stage Rocket Interactive CD-ROM. NASA SP-2004-4606, 2004.

Fueling Space Exploration: The History of NASA's Rocket Engine Test Facility DVD. NASA SP-2005-4607.

Altitude Wind Tunnel at NASA Glenn Research Center: An Interactive History CD-ROM. NASA SP-2008-4608.

Conference Proceedings (SP-4700 Series)

Dick, Steven J., and Keith Cowing, ed. *Risk and Exploration: Earth, Sea and the Stars.* NASA SP-2005-4701.

Dick, Steven J., and Roger D. Launius. *Critical Issues in the History of Spaceflight.* NASA SP-2006-4702.

Dick, Steven J., ed. *Remembering the Space Age: Proceedings of the 50th Anniversary Conference.* NASA SP-2008-4703.

Societal Impact (SP-4800 Series)

Dick, Steven J., and Roger D. Launius. *Societal Impact of Spaceflight.* NASA SP-2007-4801.

GPO U.S. GOVERNMENT PRINTING OFFICE: 2009—349-774